3rd International Conference on Recent Innovations in Engineering (ICRIE 2020)

IOP Conference Series: Materials Science and Engineering
Volume 978

Duhok, Iraq
9 – 10 September 2020

Editors:

James H. Haido
B. K. Mohammad Al-Sulaifanie
Lokman Hadi Hassan
Shaker Abdulatif Jalil
Salih Mustafa Saleem Atroshey
Arkan Fawzi Saeed
Dilsad Abdul Jabbar Haleem
Sarhat Mustafa Adam
Yaman Sami Shareef Al-Kamaki
Mohammed Salih Hussain
Nahla Hussein Aswad
Nashwan Shawkat Mahmood
Najeh Mohammed Mohammed
Ismail Amin Ali

ISBN: 978-1-7138-4793-9
ISSN: 1757-8981

Printed from e-media with permission by:

Curran Associates, Inc.
57 Morehouse Lane
Red Hook, NY 12571

Some format issues inherent in the e-media version may also appear in this print version.

Institute of Physics
Dirac House, Temple Back
Bristol BS1 6BE UK

Phone: 44 1 17 929 7481
Fax: 44 1 17 920 0979

techtracking@iop.org

Additional copies of this publication are available from:

Curran Associates, Inc.
57 Morehouse Lane
Red Hook, NY 12571 USA
Phone: 845-758-0400
Fax: 845-758-2633
Email: curran@proceedings.com
Web: www.proceedings.com

TABLE OF CONTENTS

IOP Conf. Series: Materials Science and Engineering **978** (2020) 011001 doi:10.1088/1757-899X/978/1/011001

Proceedings of the 3rd International Conference on Recent Innovations in Engineering (ICRIE) 2020

Preface

We are pleased to welcome you to the 3rd International Conference on Recent Innovations in Engineering (ICRIE 2020) in Duhok City, Kurdistan Region-Iraq. The main targets of this event are to bring together researchers, research scholars and leading academic scientists to share and exchange their research results and experiences on engineering aspects of research, innovation and designing. It will also offer a premier interdisciplinary platform for educators, researchers, and engineering practitioners to present and discuss the most recent trends, concerns and innovations as well as practical challenges faced and solutions implemented in the fields of engineering research. This will help Kurdistan Reign (KRG) of Iraq in following the current developments in the world and benefits from its effect on the economic, construction and industrial fields.

ICRIE 2020 provides an informative platform with pioneer keynote speakers from all over the world. Sessions will relate to a wide range of engineering fields such as civil engineering, electrical and computer engineering, architectural engineering, biomedical engineering and mechanical engineering. This will enrich the program of the conference in the direction of providing opportunities for the online attendees to interact with each other. We are looking for sustainable and fruitful experience of the online attendees with ICRIE 2020. Their participation will support this event to be continued its sustainable success.

I would like to thank the conference committees, keynote speakers and organizing staff led by Dr. Yaman Al-Kamaki who have made this conference possible. I would like to express my thanks to the reviewers for their cooperation and the author for contributing their paper outcomes to this event. My gratitude goes to Kavin Group in supporting this scientific occasion.

I encourage all participants to contribute actively in the interesting discussions over the conference days. I wish everyone a successful and productive conference.

Dr. James H. Haido, Assistant Professor

Conference Chairman

Proceedings Editors:

Assistant Prof. Dr. James H. Haido (University of Duhok, Kurdistan Region, Iraq), james.haido@uod.ac

Prof. Dr. Bayez Khorsheed Mohammed Al-Sulaifanie (University of Duhok, Kurdistan Region, Iraq) bayez@uod.ac

Assistant Prof. Dr. Lokman Hadi Hassan (University of Duhok, Kurdistan Region, Iraq), lokman.hadi@uod.ac

Assistant Prof. Dr. Shaker Abdulatif Jalil (University of Duhok, KRG, Iraq), shaker.abdulatif@uod.ac

Assistant Prof. Dr. Salih Mustafa Saleem Atroshey (University of Duhok, Kurdistan Region, Iraq), salihatroshey@uod.ac

Assistant Prof. Dr. Arkan Fawzi Saeed (University of Duhok, Kurdistan Region, Iraq), arkan.alhazeen@uod.ac

Assistant Prof. Dr. Dilshad Abdul Jabbar Haleem (University of Duhok, Kurdistan Region, Iraq), dilshad@uod.ac

Assistant Prof. Dr. Sarhat Mustafa Adam (University of Duhok, Kurdistan Region, Iraq), sarhat@uod.ac

Dr. Yaman Sami Shareef Al-Kamaki (University of Duhok, Kurdistan Region, Iraq), yaman.alkamaki@uod.ac

Dr. Mohammed Salih Hussain (University of Duhok, Kurdistan Region, Iraq), mohammed@uod.ac

Dr. Nahla Hussein Aswad (University of Duhok, Kurdistan Region, Iraq), mohammed@uod.ac

Dr. Nashwan Shawkat Mahmood (University of Duhok, Kurdistan Region, Iraq), nashwan.shawkat@uod.ac

Dr. Najeh Mohammed Mohammed (University of Duhok, Kurdistan Region, Iraq), najeh.mohammed@uod.ac

Dr. Ismail Amin Ali (University of Duhok, Kurdistan Region, Iraq), ismail@uod.ac

IOP Conf. Series: Materials Science and Engineering **978** (2020) 011002 doi:10.1088/1757-899X/978/1/011002

Peer review declaration

All papers published in this volume of *IOP Conference Series: Materials Science and Engineering* have been peer reviewed through processes administered by the Editors. Reviews were conducted by expert referees to the professional and scientific standards expected of a proceedings journal published by IOP Publishing.

- **Type of peer review: Single-blind**
- **Conference submission management system:** Open Conference Systems **OCS**
- **Number of submissions received: 218**
- **Number of submissions sent for review: 218**
- **Number of submissions accepted: 131**
- **Acceptance Rate (Number of Submissions Accepted / Number of Submissions Received X 100): 60 %**
- **Average number of reviews per paper: 2**
- **Total number of reviewers involved: 347**
- **Any additional info on review process: See below:**

Review Assessment Guidelines

Review assessment process:

1. Submission is submitted via online tool OCS:
http://conferences.uod.ac/index.php/icrie/ICRIE2020/login/signIn

a) Author to choose Topics of Interest/ theme (s) of relevance, following:
http://conferences.uod.ac/index.php/icrie/ICRIE2020

2. Track-editor will allocate Submission submitted online to 2 potential reviewers based on topic relevance.

3. Submission will be peer reviewed by 2 members registered with ICRIE 2020 via OCS:

a) OCS offer reviewers the following option:

• Score each Submission using the criteria established by the ICRIE 2020

• Provide/Leave a comment to the authors

IOP Conf. Series: Materials Science and Engineering **978** (2020) 011002 doi:10.1088/1757-899X/978/1/011002

• Provide/Leave a comment for the Track-editor

• Advise/Propose a Submission for a recommendation (accept, reject etc….)

b) Submission Revisions: an invitation to author/authors to revise would be sent if agreed and requested by the 2 potential reviewers. In that case the authors need to resubmit within a week. The Track-editor to inform the potential reviewers when resubmission has been done via OCS.

c) Disagreements: In case of disagreement from the potential reviewers, the Track-editor in liaison with the Director (Coordinator) and the ICRIE 2020 conference Chair will decide.

4. Once all submitted Submissions have been finally scored by potential reviewers, the coordinator(director) will develop draft full program with accepted Submissions in liaison with the ICRIE 2020 conference chairs.

5. The Organizing Committee and the Scientific Committee will revise and comment on proposed draft full program.

6. Once full program has been agreed the authors are informed of the decision. A call for full papers will start.

Criteria for Submissions assessment:

1. Thematic Relevance: Is the Submission relevant to 1 or more of the Conference topics? (Appendix A)
2. Presentation: Does the Submission track/follow the agreed guidelines for each type of abstract (Appendix B)?
3. Quality of Content: Is the Submission well written and understandable for people not familiar with the field?
4. Significance: Is the abstract relevant for its context?
5. Originality: Does the Submission brings evidence, approaches or new perspectives on a particular subject/area?

Appendix A: ICRIE 2020, Topics of Interest

1. Structural Engineering.
2. Manufacturing & Composite Materials Engineering.
3. Construction Safety & Fire Engineering.
4. Transportation Engineering & Traffic Planning.
5. Geotechnical Engineering.
6. Environmental Engineering.
7. Engineering Management.
8. Water Resources Engineering & Hydrology.
9. Architecture Engineering and Urban Planning & Design.
10. Electronic & Control Systems Engineering.
11. Communication & Computer Systems Engineering.
12. Electrical Power Engineering.
13. Surveying & Geomatics Engineering.
14. GIS & Remote Sensing.
15. Mechanical & Mechatronic Engineering.
16. Power & Applied Mechanics.

17. Biomedical Engineering.

APPENDIX B: submitted Submission

1. General guidelines

1) Each submitted Submissions should be appropriate/relevant to 1 or more of the conference topics

2) Every Submissions should contain/include full contact details of all authors

3) Each Submission should include a summary that is understandable to the readers who have not read the rest of the paper or do not know the practice and its context.

4) The Submission summary should be limited to 200 words.

5) The Submission summary should be organized and content key information requested in Section 2 below should be provided.

6) Provide/Include 4-6 keywords

7) All Submissions should classify/identify a lead presenter for inclusion in the programme

2. Structured Summary for Submissions to be submitted

1) An introduction (Background), covering background, problem statement, aims and objectives.

2) Methods/ Theory

3) Results & Discussions

4) Conclusions (comprising key outcomes/findings, discussion and lessons learned)

5) Highlights (innovation, Impact, outcomes and limitations)

- **Contact person for queries: Dr. Yaman Al-Kamaki yaman.alkamaki@uod.ac**

IOP Conf. Series: Materials Science and Engineering **978** (2020) 012001 doi:10.1088/1757-899X/978/1/012001

Qualitative Assessment for Walkability: Duhok University Campus as a Case Study

Layla M. Raswol

Architectural Engineering Department, Nawroz University, Duhok, Kurdistan Region, Iraq.

layla.raswol@uod.ac layla.raswol@nawroz.edu.krd Tel: +9647504964535.

Abstract. The University Campus as a learning environment needs to be a safe comfortable and accessible place, where students enjoy walking and social interaction. This study evaluates the quality of walking environment by analyzing both physical features of streets sidewalks and individual perceptions of pedestrians. The research approaches to answer the question: to what extent is the University Campus safe and attractive for the students to walk? The research used mix methods and case study approach to achieve study objectives. The mix methods include onsite observation, visual survey and questionnaire with students using walkability checklist to determine the walking quality on streets sidewalks. In this regard four variables are considered based on previous literature review as; safety and pedestrians' facilities, comfortable walking environment, walking connectivity and lastly walking attractiveness. The result shows that the lack of safety and pedestrian facilities, poor designed sidewalks and connectivity leads to low level of walkability and increase car dependency in the study area. The findings provide guidelines for increasing the efficiency of walking, in order to encourage students to choose walking over car driving.

Keywords: *walkability; sidewalks; pedestrian facilities; physical features; Duhok University Campus*

1. Introduction

1.1 Walkability concept

The term walkability refers to the ability to easy safe, convince and interest walk in an area (Speck 2012). Park (2008) defined walkability as the easier form of movement and an important physical activity to improve human health. The quality of the built environment encourages people to walk by providing safe, comfortable paths. However, the lack of adequate public transport as well as the inefficient sidewalk and pathway infrastructure leads to increase the using of the car as the main mode of movement on Duhok University campus. The study evaluates the quality of walking environment by analyzing both physical features and individual perceptions of the students on University campus. The research is aiming to enhance walking efficiency by introducing alternative physical design features for streets sidewalks to encourage walking over driving among students, who would prefer a convenient walking route between educational and recreational destinations.

1.2 The benefits of walkability concept

Walking is an enjoyable, convenient and healthy mode of transportation. It has many benefits for the human such as improving mental physical health, reducing travel cost, protecting environment, reducing pollution, as well as creating livable communities. It is also an efficient way of completing a relatively short trip, especially where parking is challenging. Many researchers and city planners have focused on the importance of creating walkable communities. Speck, J. (2012) in his book walkable city defined walkability as a useful way to move in an area. He identified four main attributes for increasing walkability as; useful, safety, comfort and interest, while Gehl, J. (2010) in his book Cities for people studied walkability in human dimension. Spoon (2005) identified the main advantages of a walkable community as an accessible, livable, attractive, safe, diverse and healthy living environment. Other benefits are defined by (Litman 2007) such as land use efficiency, cost reductions, equity and enhancing public health. Mc Nally (2010) identified the economic benefits of walkability as it is more cost effective and could provide business and economic opportunities for the user while walking. The social benefits of promoting walkability are building relationships by encouraging people to connect directly and increase social interaction. Additionally, walking as a green transport contributes to reduce noise, pollution and increase safety.

1.3 Design Factors influencing walking quality

The design quality of the built environment encourages people to walk by creating safe, comfortable path and providing walking friendly facilities. Researchers identify many influencing factors for the pedestrian friendly design such as accessibility, safety, connectivity and land use efficiency, which affect the walking quality of the built environment (Spoon 2005, Muhbach 2012, Choi 2012). There are other encouraging variables such as attractive nodes and parks. To this regards urban planners and architects have the significant role in creating walking friendly design to make walking safe, efficient and enjoyable by designing well safe streets sidewalks, well-maintained pavement, streets lighting, pleasant environment of sidewalks. Shay et al. (2003) identified five infrastructural factors of walkability; the first one is pedestrians facility, including sidewalks amenities; secondly are accessibility and convenience; the third factor is connectivity, measured by short distance and efficiency to destinations; then, the aesthetic aspect of walkability involves pleasant environment, attractive architecture, trees and landscaping of the sidewalks; the last one is street safety such as street lightings, curb extensions, street narrowing and sidewalk amenities. Ewing et al. (2006), on the other hand, studied the qualities of physical features and individual perceptions including safety and attractiveness as the major elements in the concepts of walkability. There are other elements influences the walking quality such as shading, trees and landscape, lighting and other pedestrian furniture (Petella 2009, Yeang 2000). The Austrians road guide to Road Design part 6A: Pedestrian and Cyclists Paths (Austroads 2009a) also identified five required key attributes of an environment to encourage walking as the 5Cs: Connectivity, Comfortable, Convenient, Convivial, and Conspicuous.

From the previous literature review the research concludes that there is a positive relation between the walking friendly design and the level of walkability. (Fig 1) shows some design features that increase walking quality. Using this framework this research evaluates the walking quality on university campus by measuring the quality of the physical elements of streets sidewalk and the individual perception of the students.

Figure 1. TU-Berlin University Campus pedestrian friendly design by using greenery, shading, seating furniture, planting strip between traffic and sidewalks, Source, Researcher (2019)

2. Case study

The University of Duhok (UoD) is a fast-growing institution in Duhok, Iraqi Kurdistan region. It is about 10 kilometers far from the city center of Duhok. The campus has an area of 3036 km2. The total number of Students that are currently studying in the campus is about (21,000) students. It has (1607) academic staff members and 2353 employees (Registration Directorate 2020). The area of the University Campus is equal to ((1450) Acres. The main mode of movement on Duhok university campus is private car as a result of inadequate public transport as well as the inefficient sidewalk and pathway. Promoting walkability on university campus is an urgent need to create livability and enhance students social interaction on campus. (Fig2) shows the study area and the main gates with the high traffic at the pick work hours.

Figure 2. Right: Duhok University Campus. Left the main two gates of the University. Source: Author based on university map, www.uod.ac

3. Research Methodology

The study approaches to answer the question, to what extent is the university campus safe and attractive for students to walk? The main objectives of the study are to assess the current situation of pedestrianisation on streets sidewalks in Duhok University Campus by identifying the main problems and provide guidelines and recommendation for improving the quality of walking. The research used mix methods and case study approach to achieve study objectives. The mix methods include onsite observation, visual survey and questionnaire. Furthermore, the survey used maps and photographic analysis to investigate the physical quality of streets side walk. The research focused on the most crowded streets sidewalk routes on the campus from college of education to the college of spatial planning and student center ended with the main entrance of the university campus as a case study

IOP Conf. Series: Materials Science and Engineering **978** (2020) 012001 doi:10.1088/1757-899X/978/1/012001

(Figure 3).

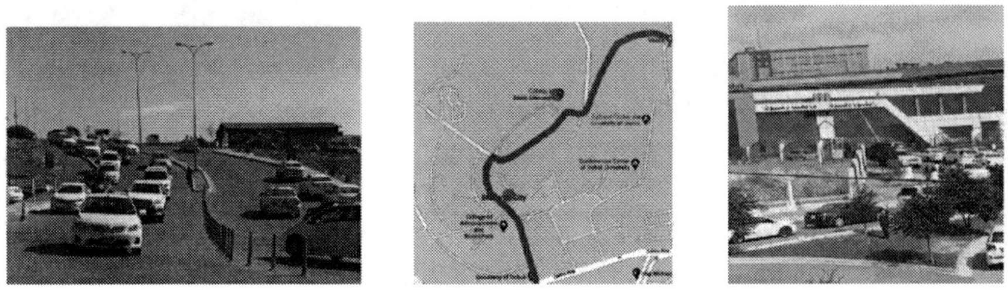

Figure 3. The case study street sidewalk connecting the main two gates of Duhok University campus Source: Photo Author 2020

Both sidewalks physical features and individual perceptions of pedestrians are studied. The direct observations focus on assessing the available physical features on the defined sidewalk route using the walking checklist and photographs. The research used qualitative assessment to evaluate sidewalks physical features for walking. The evaluation criteria based on four principles: pedestrians' facilities and safety, comfortable walking environment, walking connectivity and lastly the walking attractiveness. Each criterion has a set of sub criteria related to walkability concept (table 1). The questionnaire was conducted with 50 students, who were chosen randomly on the case study area, to find out their individual perception about the quality of walking on the campus. The questionnaire measures students satisfaction about the walking quality based on four basic elements: safety feeling, comfort feeling, connectivity and walking attractiveness.

Table 1: Research variables check list for evaluating walking quality (Author)

Variables		Attributes related to the Variables
1.	Pedestrian facilities & Safety	• Sidewalk condition (pavement material and maintenance) • Sidewalk amenities (public garbage, cans, benches, • Sidewalk width • Sidewalk obstructions (Poles, Signs, Trees, Garbage cans) • Lighting • Crossing walk lines are clear and available as needed • Safety feeling
2.	Comfortable walking environment	• Interesting buildings facades • Greenery, shading, vegetation, • Cleanness and maintenance • Comfort and enjoying feeling
3.	Walking Connectivity/ Accessibility	• Sidewalks completeness • Sidewalks continuity • Wayfinding signage • The accessibility of the sidewalk

4.	Walking Attractiveness	• Visual variety of physical environment richness • Landmarks • Public space for social activities • Attractive sidewalk furniture (Fountains, • Useful walking (short trip or including useful and enjoyable activities)

4. Results and Discussion

4.1 Visual survey

The collected data from the visual survey and the walkability checklist are focusing on evaluating the physical features to measure the quality of walking on the defined sidewalks. The following part shows the finding discussion of research variables.

4.1.1 Pedestrian facilities and Safety

The results of the observation and checklist for the first variable (pedestrian facilities and safety) focus on evaluating the sidewalk condition, sidewalk amenities (public garbage, cans and benches), sidewalk width, sidewalk obstructions (poles, signs, trees, and garbage cans), lighting and crossing walk line availability. The findings are shown in (fig 4). The analysis of the data shows poor pavement quality and lack of efficient amenities. The sidewalk material conditions of 60 % are completed with poor quality of pavement. There are many physical obstructions on the sidewalks such as signs and illumination structure. The crossing walk lines are not available as needed. As a result, the students cross the street irregularly and use street instead of the sidewalks. The sidewalks width is 2.40 m, but it is less in some places. The garbage cans are available, but the garbage is mostly located on the sidewalks and not in garbage canes. The lighting is a on the mid island of streets that disturbs the students from walking. There is a need to increase the safety level by improving physical features, sidewalks furniture such as the buffer space between street traffic and sidewalk. The buffer can be used as a planting strip to increase pedestrian perception of safety and provide nice landscape to the area. Furthermore, the walking safety can be improved through implementing better cross streets connection, decreasing traffic speed as well as clear signs such streets name, warning signs, which make students feel safer to walk between academic departments.

Indicator	Attribute	
Pedestrian facilities & Safety	Sidewalk condition: Photos right show poor pavement quality, many physical obstructions on the sidewalks	 
	Sidewalk amenities: Photos right show inefficient design and lack of sidewalk amenities such as benches and seating places	 

	Sidewalk obstructions: Photos right show many walking obstructions such as signs, trees and garbage cans	  
	Lighting: Photos right show inefficient design for lighting space Lighting structure prevents movement and access to walking	  
	Crossing walk line: Photos right show lack of crossing line. As a result, the students cross the streets irregularly	  

Figure 4: The assessment of the first research variable: pedestrian facilities and safety. Photos (researcher 2020)

4.1.2 Comfortable walking environment

The results of the observation and checklist for the second variable; comfortable walking environment focus on evaluating the availability of: interesting buildings facades, greenery, vegetation and shading, cleanness and maintenances, comfort and enjoyment feeling. The findings are shown in (fig 5). The built environment of the sidewalks near the college of education includes few interesting public spaces and greenery for the students to enjoy their walks, while there is lack of shading element and side walk furniture. On the other side there are many vacant lands, which can be designed as landscape area and create comfortable environment for pedestrians. The nodes are so crowded with cars and there is no separation between pedestrians and cars lanes. The comfort and enjoying feeling can be increased by implementing well-designed sidewalks furniture, promoting greenery and nice landscape.

Indicator	Attribute	
Comfortable walking environment	Interesting building facade: Photos right lack of interesting building facades	
	Greenery, Shading and Vegetation: Photos right show inefficient Inadequate shading and more car oriented places, which cause more pollution	
	Clearness and maintenance: Photos right show garbage on sidewalk and lack of maintenances	
	Comfort and enjoying feeling: Photos right show lack of comfort as result of the busy street and crowded nodes	

Figure 5: The assessment of research second variable: comfortable walking environment. Photos (researcher 2020)

4.1.3 Walking Connectivity/ Accessibility

The results of the observation and checklist for the third research variable; walking connectivity focus on the evaluation of the sub-variables: sidewalk, completeness, sidewalks continuity, way-finding signage and the accessibility to the sidewalk. The findings are shown in (fig 6). The sidewalk completeness is available near the college of education, but there are unpaved sidewalks near the college of planning. The sidewalks pavement is not completed as a result of construction or other physical barriers. The signs are not well placed, they cause visual and walking barrier for the students. Good signage along the walking routes should be available to guide the pedestrians to the desired destination.

IOP Conf. Series: Materials Science and Engineering **978** (2020) 012001 doi:10.1088/1757-899X/978/1/012001

Indicator	Attribute	
Walking Connectivity/ Accessibility	Sidewalk completeness: Photos right show uncompleted sidewalks	
Walking Connectivity/ Accessibility	Sidewalks continuity: : Photos right show lack of sidewalk continuity	
	Way-finding signage: Photos right show Insufficient design for way finding signage	
	Access to sidewalk: Photos right show lack of accessibility to sidewalks as a result of car oriented design	

Figure 6: The assessment of research third variable: walking connectivity/ accessibility. Photos (researcher 2020)

4.1.4 Walking Attractiveness

The results of the observation and checklist for the fourth research variable; walking attractiveness focus on the evaluation of visual richness of physical environment, landmarks, public space for social activities, attractive sidewalk furniture (Fountains, landscaping), useful walking. The findings are shown in (fig 7). The finding shows lack of walking attractiveness such as landmark, interesting landscape and sidewalks physical features. The attractiveness of a place depends on well-designed streetscape and the existence of sidewalk furniture, provision of shaded area and pedestrian facilities such as good walking routes, adequate seating and benches, greenery and well-designed landscaping.

Indicator	Attribute	
Walking Attractiveness	Visual richness of the physical environment: Photos right show lack of architectural visual richness	
	Landmarks: Photos right show lack of landmarks and interesting views. Mostly busy streets insufficient sidewalks	
	Public plaza: Photos right show lack of well-designed social activities	
	Attractive sidewalks furniture: Photos right show lack attractive sidewalk furniture (Fountains, greenery)	
	Useful activity: Photos right show crowding, busy streets rather than useful walking paths including mixed used activities	

Figure 7: Assessment of the fourth research variable; walking attractiveness. Photos (Researcher 2020)

4.2 Students Satisfaction

The findings of the student's perception and their satisfaction about safety feeling, comfort and enjoying feeling, connectivity and walking attractiveness, show lack of satisfaction (Figure 8).

The sidewalks are not safe for pedestrians, since only 34% of the user feel safe by walking. The results find out that the sidewalks are very uncomfortable. 70% of the students do not enjoy walking due to the lack of pedestrian facilities, lack of pedestrian amnesties as well as crowded street traffic and lack of separation between busy roads and sidewalks. Most of the pedestrians (66%) agreed that the sidewalks connectivity and accessibility is not as desired. The walking attractiveness is not sufficient, since more than half (62%) of the respondents are not satisfied. The finding of the questionnaire Figure out that there is an urgent need to increase pedestrian satisfaction by promoting the sidewalks design quality, providing well-designed sidewalks furniture, improving walking connectivity and increasing walking attractiveness.

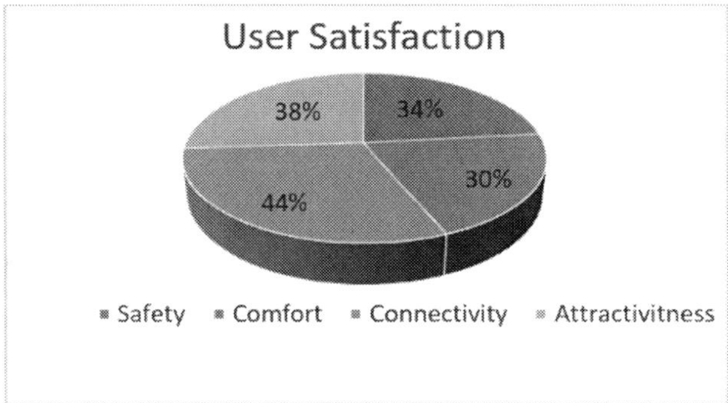

Figure 8: Findings of user satisfaction (Researcher 2020)

5. Conclusion

The research studied the quality of walking on Duhok University Campus at the aim of creating attractive and healthy urban environment for students to study in. The car free learning environment with well-designed and safe comfortable sidewalk will help students to interact in various social activities and enhanced health benefits. The research used case study approach and mix-method of direct observation, visual survey and questionnaire to evaluate the walkability concept on Duhok University Campus. The research defined four variables that can influence the walking quality as: pedestrian facilities and safety, comfortable walking environment, walking connectivity and lastly the walking attractiveness. The research findings conclude that the walking quality on the campus is not satisfying for the students as a result of poor pedestrians' facilities such as sidewalk conditions, inefficient design of lighting and pedestrian's furniture, inadequate sidewalk amenities, lack of landscape and shading elements, lack of connectivity, inadequate way finding signage and lack of attraction vibrancy. The sidewalks pavement is not adequate as a result of construction or other physical barriers. The signs are not well placed, they cause visual and walking disturbs for the students. The built environment of the campus lacks of vegetation and trees, in spite of the large vacant land, which can be designed as attractive landscape area for the student to interact. The nodes in the case study area are so crowded with cars without any separation between sidewalks and car lanes, which decrease safety, comfort and enjoyment feeling of walking. The finding shows lack of walking attractiveness such as landmark, interesting landscape and sidewalks physical features. The study conclude that the University has the opportunity to create a vibrant learning environment for the students by adopting walkability concept and create a car free area by designing safe, comfortable and interesting sidewalks to encourage students to enjoy walking on campus.

6. Recommendation

The walking quality on university campus can be enhanced based on creating comfortable sidewalks with useful and enjoyable attractive physical features and landscaping. The following recommendation can improve the walkability rate in the Campus with regards to research variables:

1. Increasing pedestrian Safety:

- Provision of safe walking surfaces and various pavement materials and colors
- Creating safe sidewalk design including space for signs, walking and buffer separating sidewalks from adjacent traffic such as providing planting strip
- Providing crosswalk facilities as needed
- Provision of pedestrian signages, speed humps and speed limits on the roads to control vehicles speeding on campus
- Improving sidewalk lighting taking solar power and LED lighting in consideration

2. Increasing Comfortability

- Improving the quality of the landscaping features, such as benches, greenery, to increase pedestrianization among the academic community.
- Increasing walking enjoyment feeling by designing comfortable pavement, nice vegetation and shading features
- Raising awareness to keep cleanness and building maintenance
- Encouraging free cars environment by designing comfortable and useful walking paths

3. Connectivity

- Providing ease of access for mobility such as ramps and curbs to enable people using wheelchairs and crutches to access the sidewalk
- Providing wayfinding signages according to sidewalks design standards
- Improving streets sidewalks continuity between academic departments
- Organizing useful public transport access to encourage students to use it instead of private cars

4. Increasing walking attractiveness

- Increasing visual attractiveness of Building facades and maintenance
- Provision of nice and enjoyable public spaces physical features such as plaza, fountains, greenery
- Provision of interesting landmarks and mix-use activities to encourage students to enjoy a useful walking through academic community.

References

[1] Austroads 2009a, Guide to Road Design Part 6A: Pedestrian and Cyclists Paths, by G. Veith, AGRD06A/09, Austroads, Sydney, NSW, Government of Western Australia Department of Transport WALKABILITY AUDIT TOOL July 2011 http://www.beactive.wa.gov.au/assets/files/Guidelines/Walkability%20Audit%20Tool%20July%202011.pdf

[2] Choi E., 2012. Walkability as an Urban Design Problem: Understanding the activity of walking in the urban.

[3] Ewing, R., S. Handy, Cervero, R. C. Brownson, O. Clemnete, E. Winston (2006): Identifying and Measuring Urban Design Qualities Related to Walkability. Journal of Physical Activities and Health

[4] Gehl, J. (2010). Cities for people: Island Press.

[5] Grava, S., 2004. Urban transportation systems, choices for communities. McGraw-Hill (www.digitalengineeringlibrary.com

[6] Litman, T., 2007. The Economic Value of Walkability. Victoria, BC: Victoria Transport Policy Institute.

[7] Marshal, W. and Garrick N., 2009. The Effect of Street Network Design on Walking and Biking. The 89th Annual Meeting of Transportation Research Board January 2010, Washington D.C

[8] McNally, K., 2010. Design Guidelines for Walkable Communities, Niehoff Studio https://www.uc.edu/cdc/niehoff_studio/programs/great_streets/w10/reports/design_guidelines.pd, (Retrieved 14.01.2020)

[9] Park, S., 2008. Defining, Measuring, and Evaluating Path Walkability, and Testing Its Impacts on Transit Users' Mode Choice and Walking Ditance to the Station. PhD. University of California. Berkeley.

[10] Pentella, R., 2009. Walkability and the Built Environment: A Neighborhood- and Street-Scale. Assessment of Diverse San Francisco Neighborhoods

[11] Registration Directorate (2020): University of Duhok, Registration Direction, www.uod.edu.krd

[12] Shay, E, Spoon, S. C. and Khattak, A. J. (2003). "Walkable Environments and Walking Activity". Final Report for Seed Grant Submitted to: Southeastern Transportation Center University of Tennessee Knoxville, Tennessee. December 2003.

[13] Spoon, S. Ch., 2005. What Defines Walkability: Walking Behavior Correlates, MSc, University of North Carolina at Chapel Hill.

[14] Speck, J. (2012). Walkable City: How Downtown Can Save America, One Step at a Time.USA: Farrar, Straus and Giroux.1 Approaches and lessons for enhancing walkability

[15] Yeang L. D., 2000. The Urban Design Compendium. 3rd ed. published by English Partnerships.

Influence of Long-Term Aging on HMA Made with Gel and Sol Asphalt Cements

Ziyad N. Aldoski[1,*], **Abdul-Rahim Ibrahim Al-Hadidy**[2], **Abdullah M. Rashed**[3]

[1]Road Const. Dept., Duhok Tech. Inst. Duhok Polytech Univ., Kurdistan Region, Iraq
E-mail: ziyad.nayef@dpu.edu.krd

[2]College of Eng., Univ. Mosul, Iraq - E-mail: abdet76@gmail.com

[3]Road Const. Dept., Duhok Tech.Inst. Duhok Polytech. Univ., Kurdistan Region, Iraq
E-mail: abdullah.mahmoud@dpu.edu.krd

[*] Corresponding author: Department of Road Construction, Duhok Technical Institute, Duhok Polytechnic University, 61 Zakho Road, 1006 Mazi Qr Duhok, Kurdistan Region, Iraq. Tel: +9647504577222.

Abstract. This study evaluated the effect of long term ageing (LTA) on the performance traits of dense graded asphalt mixes (DGAMs) made from gel and sol asphalt cement binders. The LTA was adopted at 85 °C and between four and eight days using forced draft oven to reproduce the in-service aging that occurs over many years in the field. The performance assays contains: Marshall traits, tensile and compressive strength under wetting and drying conditions, bending strength at zero and -10°C, cohesion at 60°C, tensile strength ratio, and recovered strength index were adopted on DGAMs. A mechanistically-design tool was explored for estimating the enhancement in paving life or thickness reduction of DGAMs and base course for similar service life caused from LTA of DGAMs. Assays results notified that the LTA increases DGAMs strength towards: (1) rutting; (2) moisture sensitivity; (3) cracking; and (4) cohesion failure. In addition, it was found that sol asphalt cement performs slightly better than gel asphalt cement, especially in the case of moisture susceptibility and flexural strength properties.

Keywords: *Aging; Asphalt; Performance traits; Service age; Pavement responses*

1. Introduction

Asphalt has long been classified as gel or sol type. Gel-type asphalts usually exhibit pronounced non-Newtonian behavior, whereas sol-type asphalts are more Newtonian. Gel-type asphalts generally contain large amounts of asphaltenes, and sol-type asphalts are characterized by low asphaltene content. Using the classical asphalt science terminology, sol-type asphalts are more compatible, while gel-type asphalts are less compatible. Less compatible (gel-type) asphalt is known to be highly susceptible to oxidative age hardening. While highly compatible (sol-type) asphalt is not susceptible to ageing (Shin-Che, 2008).

Aging of asphalt mixes and water in base course and roadbed are the two essential environment factors that affect performance of flexible pavements. Asphalt stiffness is affected by aging, which causes the stress in the pavement. In turn, the stress variation can affect other underlying paving courses, which depends on asphalt layer stress (Al-Hadidy and Tan, 2007).

Several studies reported aging influences on binders' traits using ASTM and SuperPave assaying methods. These methods are generally simple and consume little time (Soon-J. Lee et.al.,2009; Bell, 1994; Mansour and Amin, 2017; Shifeng et al , 2017; MinBai, 2017; Dongmei Zhangn et. al, 2017; Soon-J. Lee et.al., 2009). In contrast, researches on asphalt mixes ageing consume higher time and efforts. The evaluation of the mixes traits under ageing influences may be more benefits than for binders.

For asphalt mixes, the recommended lab method for short-term oven aging (STOA) is to heat the loose mixture in a forced draft oven (FDO) for 4 hr at 135 °C while for the hardener mixes, STOA of 2 hr at 154 °C is commonly utilized (Soon-J. Lee et.al., 2009). Whereas, the lab method for long-term aging (LTA) is to put the specimens in FDO for 4d to 8d at 85°C or 2d to 4d at 100°C. The LTA-conditions equivalent of older than 3yr mixes in the field as documented by Bell (1994) and Al-Hadidy (2018).

Mansour and Amin (2017) examined ageing and environment influences on recycled pavement asphalt (RPA) and steel slag aggregates (SSA) fracture resistance subjected to freezing and thawing (FT) cycles and LTA. It was notified that ageing improves fracture RPA/SSA-mixtures resistance.

Shifeng et al (2017) documented that ageing varied the physiochemical traits of rubber-bitumen blends (RBB). The RBB stiffness was found higher under influences of ageing.

MinBai (2017) examined ageing influences on the low temperature traits of styrene-butadiene- styrene (SBS)/bitumen containing rejuvenators utilizing: penetration, ductility at 5 °C, force ductility, Fraass point, dynamic shear rehometer, and bending beam rehometer assaying method. It was depicted SBS/ bitumen including rejuvenators give higher elasticity.

Dongmei Zhangn et. al, (2017) evaluated ageing influences on the rheological trait, physical traits and chemical constitutes of SBS-70 and SBS-90/ asphalts utilizing thin film oven assay, pressure aging vessel, and ultraviolet radiation methods. It was depicted SBS-70 is less damage to ageing than SBS-90.

Soon-J. Lee et.al. (2009) examined the influences of STOA on asphalt mixes utilizing the gel-permeation chromatography (GPC) method. Nine-asphalt mixes, adopting three various asphalt sources, were introduced and five-STOA procedures were utilized to examine the mixes. For comparison, the rolling thin film oven (RTFO) ageing was also performed for nine-asphalt binders. The asphalt mixes ageing, containing polymer-modified mixes, could be complied at different STOA factors. Statistical analyses of the GPC assay results depicted that ageing at 154°C for 2 hr and a 135°C for 4 hr are not significant. The RTFO ageing procedure was depicted to have less influence on binder ageing than the STOA procedures of asphalt mixes.

Al-Hadidy and Tan (2007) reported effects of aging on the performance traits of asphalt mixes containing polypropylene. The aging was adopted under 85 °C and between 2 and 4 days. Assays results notified that aging affects the performance traits of the mixes.

2. Study objectives

The research objectives are to document the influence of long-term ageing (LTA) of sol and gel asphalt cement binders on: (1) Marshall traits of dense-graded asphalt mixes (DGAMs); (2) Tensile and compressive strength of DGAMs at 25 and 60oC; (3) Moisture damage of DGAMs; (4) Bending strength traits of DGAMs at lower temperatures; (5) Cohesion of DGAMs at 60oC; and (6) Index of aging of DGAMs. Besides, explore regression formulas among the performance assays (Marshall stability, tensile strength, and compressive strength): a) for their utilized in the absence of the tensile and compressive strength assays, and /or b) the engineer does not select to utilize those assays, and reported the advantages of LTA-DGAMs course in pavements utilizing a mechanistically design tool. Gathering the above objectives could supply well understanding of LTA effects on the performance traits of DGAMs, so that their performance in the site can be exactly estimated.

3. Materials and Methods

3.1. Materials

Two types of 40-50 penetration grade asphalt cement binders, namely; gel type taken form Baiji refinery (200 Km. North Baghdad the capital) and sol type brought from Qayarah refinery (350 Km North Baghdad the capital) were adopted in this research. Traits of sol and gel asphalts are depicted in Table (1).

Aggregates were brought from Ashor co. in Mosul city located in the Iraqi northern region. Figure (1) depicts the used gradation tolerances by the ASTM (D3515) (2015) for DGAMs and the utilized gradation in this study at the middle of the tolerances.

The aggregates traits, such as crushing, wear, soundness, water absorption, and specific gravities values were calculated and the assay results are depicted in Table (2).

Table 1. Physiochemical traits of Baiji and Qayarah Asphalt Binders

Trait	Value		ASTM Spec.	SCRB Spec.
	Gel	Sol		
Penetration (25°C, 100g, dmm)	**42**	**41**	40-50	40-50
Softening point, °C	**58**	**52**	50-58	51-62
Ductility (25°C, cm)	>100	>100	≥100	≥100
Specific gravity	1.051	1.049	1.01-1.06	---
Flash point, °C	275	263	≥240	≥240
Loss due to heat and air, %	0.375	0.386	≤0.2	≤0.75
Kinematic viscosity (135 °C, Centi stock)	500.0	500	≥400	----
Sulphur (S3) (X-Ray), %	6.3	7.6		6.5 % Masson et.al.[16]
Asphaltene, %	11.2	6.70	---	---
Solubility in CCl4, %	99.4	99.7	≥99	≥99
Recovered ductility (25°C, cm)	80	97	---	≥50
Recovered penetration, %	66	68	---	≥55
Recovered softening deg., °C	5	4	---	≤10

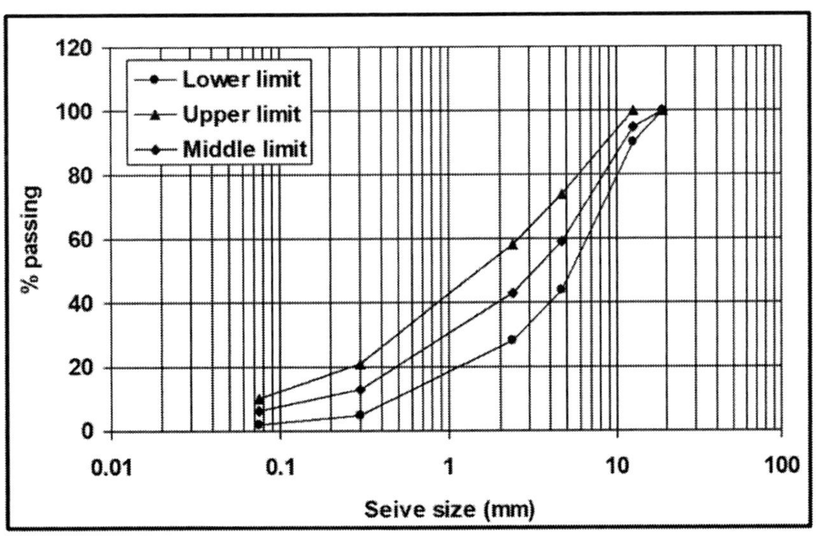

Figure 1. Gradation boundaries

Table 2. Aggregate traits

Trait	Coarse Agg.	Fine Agg.
Bulk spec.gr.	2.6520	2.5410
Apparent spec.gr.	2.6930	2.5920
Crushing,%	99.0	47.0
Wear,%	19.0	-----
Soundness,%,Na2SO4	0.968	0.664

Calcium carbonate ($CaCo_3$) was utilized as a filler. It was finer than 0.075mm sieve (ASTM D854, 2015) and 2.731 sp. gr.

3.2. Mixture Design

DGAMs optimal binder dosage (OAD) is selected to justify 4±1% air voids. Five binder dosages from 4% to 6% with increment of 0.5% for each type of asphalt were utilized in DGAMs design. It was noticed that at 4.0% air voids, the OAD was 5.0% for the two types of asphalts. This OAD was adopted in producing all other LTA mixes to obtain consistency throughout the research.

Three groups of specimens are blended at OAD (155±5 °C, 2 min) (Asphalt Institute, 1984). The 1st group was coded as unaged (DGAM0h). The 2nd group included placing of the consolidated specimens at (85°C, 4 days) into a Forced Draft Oven (FDO) and coded as (DGAM4d). The 3rd group included placing of the consolidated specimens at (85°C, 8 days) into FDO and coded as (DGAM8d). These LTA conditions used methods illustrated in (Bell, 1994).

Then, the loose specimens were compacted with 75 blow/face (This number of blow equivalent to 1379 kPa tire pressure) by Marshall hammer at viscosity of (280 ± 30 Cst)) (Asphalt Institute, 1984). After that, the samples were extruded from the mold (ASTM, 2015), and put in air for 1 day. 26 samples for each DGAMs group were assay with three for stability, cohesion at 60 °C, indirect tensile strength (ITS) at 25 °C, ITS at 60 °C, compressive strength (CS) at 25 °C, and CS at 60 °C, four for flexural strength (FS) at 0°C, and four for FS at -10°C. Thus, for each type of asphalt, seventy-eight specimens were assay for DGAMs groups.

3.3. Laboratory Tests Used

Four assays in lab were depended. The assays adopted were Marshall stability, ITS at 25 and 60oC, CS at 25 and 60oC, bending strength at 0 and -10oC, and cohesion at 60oC. Besides, the moisture influence assay was explored utilizing the ratio of tensile strength (TSR= ITS60oC/ ITS25oC) (ASTM D-4123, 2015), and index of retained strength index (IRS= CS60oC/CS25oC) (ASTM D-1074, 2015). Higher value of TSR or IRS notifies that DGAMs are more resistant to water effects. SCRB (2003) requires IRS ≥70%, whereas, SCDOT Bradley et al. (2004) requires a TSR≥85% and ITS60oC≥448kPa, respectively. Index of aging (A.I) was calculated by divided the LTA cohesion of DGAMs to unaged one.

Rutting strength was obtained from Marshall Quotient (MQ). Bending assay was performed on (300 × 50 × 48 mm) beams extruding from a consolidated (300 × 300 × 50 mm) sample. Four beam samples were tested at zero and -10oC utilizing a universal bending machine.

Resilient modulus (MR) of DGAMs is an important mechanical trait for pavement structural design. Thus, it is preferable to predict MR of DGAMs Dynamic ITS assay at 25°C was adopted (ASTM D-4123, 2015) for MR evaluation.

4. Mechanistic Empirical (M-E) Tool

M-E tool was utilized in this research to examine LTA-DGAMs advantages in forms of Layer Thickness Reduction (LTR) and Traffic Benefit Ratio (TBR). The adopted methodology has good capability of characterizing various material traits and loading options, and has the ability to examine various design options on an economic basis.

Two design options were adopted in this research are as follows:

1. Similar service life for LTA and unaged pavement models. It would lead to decreasing in base, or DGAMs thickness and has been notified in forms of LTR; and
2. Similar pavement models for unaged and LTA-DGAMs. It would result in more pavement service life due to LTA and has been notified in forms of TBR.

As input values, DGAMs modulus was examined, and typical moduli of asphalt, base, subbase and subgrade in Iraq were depended as shown in Figure (2).

The surface deflection (w_o), the horizontal strain in tension (εt) at the bottom of binder course, and the vertical compressive strains (εc) developed at the top of the subgrade in LTA pavement model were determined under the left tire utilizing BISAR program with (40×10^3 N and set of dual tires of 21cm dia.). The Wo, εt and εc responses were adopted for estimating the enhancement in paving service life or decreasing in DGAMs and base course thickness for similar LTA-DGAMs service life DGAM.

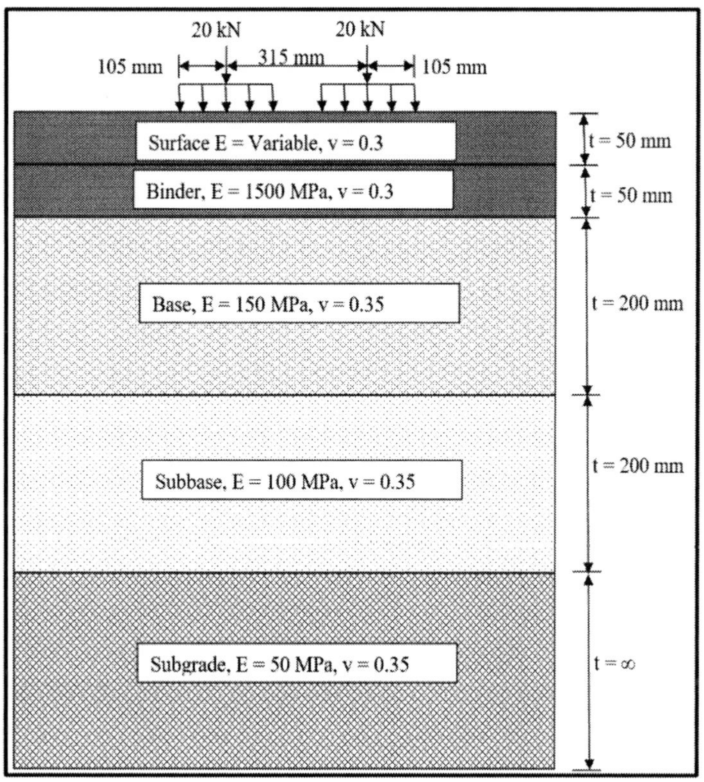

Figure 2. Pavement Design and Considered Loading: Dual Tires

5. Assays Results and Discussion

5.1. Statistical Analyses

Values of the Marshall, ITS, CS, TSR, IRS, FS, and cohesion assays were analysed statistically with a 0.05 α. For distinguishes, it should be notified that the samples were prepared at OAD.

IOP Conf. Series: Materials Science and Engineering **978** (2020) 012002 doi:10.1088/1757-899X/978/1/012002

5.2. Regression Analyses

The variations in the performance traits with LTA-DGAMs were reported at highest and ambient paving temperature. Statistical analyses (Pearson relation (r) and 2-tailed, probability) was performed to justified that a linear relation has adequate precision for calculation Marshall stability, ITS and CS traits of DGAMs for LTA.

Figure 3 depict the linear relations between the engineering assays and documents the LTA linear regression of DGAMs adopted at 5% significant level for gel and sol asphalt cement. The values are average of all lab mixes examined at 25oC and 60oC. The analyses satisfied that linear relations could be utilized for obtaining of ITS and CS traits of DGAMs at 25oC and 60oC. These relations were explored: a) for their utilized in the absence of the tensile and compressive strength assays, and /or b) the engineer does not select to utilize those assays.

5.3. DGAMs Optimization

Statistical analyses program (SPSS) was utilized to introduce an ideal table based from the DGAMs assays. Duncan internship was adopted to calculate the difference among groups. This difference was notified for each DGAMs group by addition of letter(s). If the same letter(s) among two or more groups, this means that no significant can be obtained among sets. Table 3 tabulated the relations from the analyses.

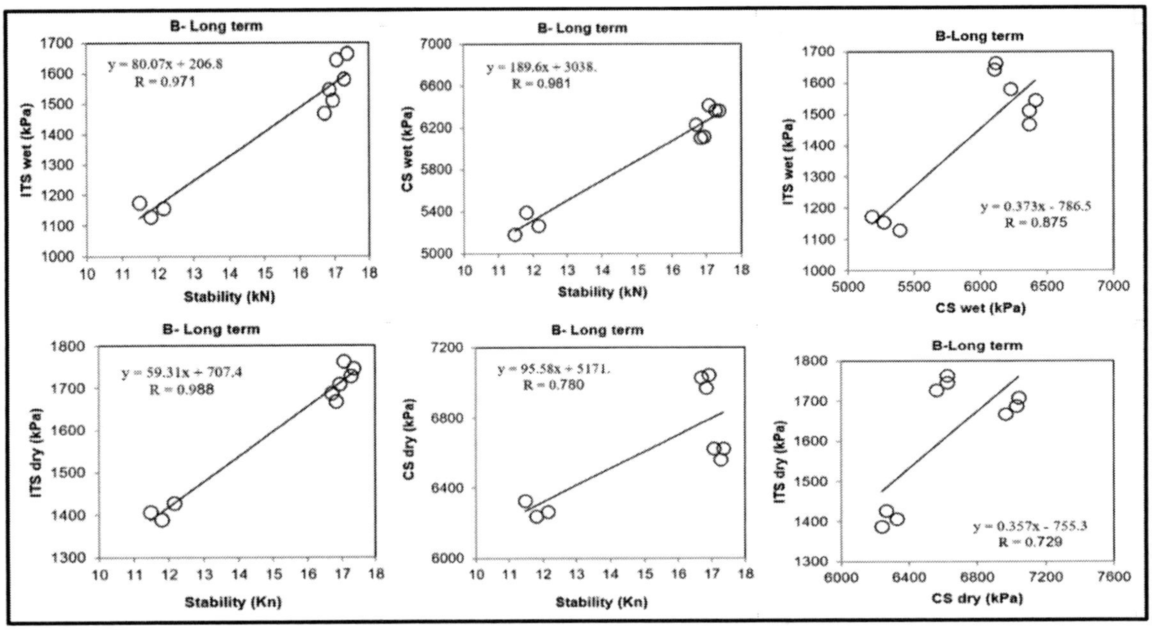

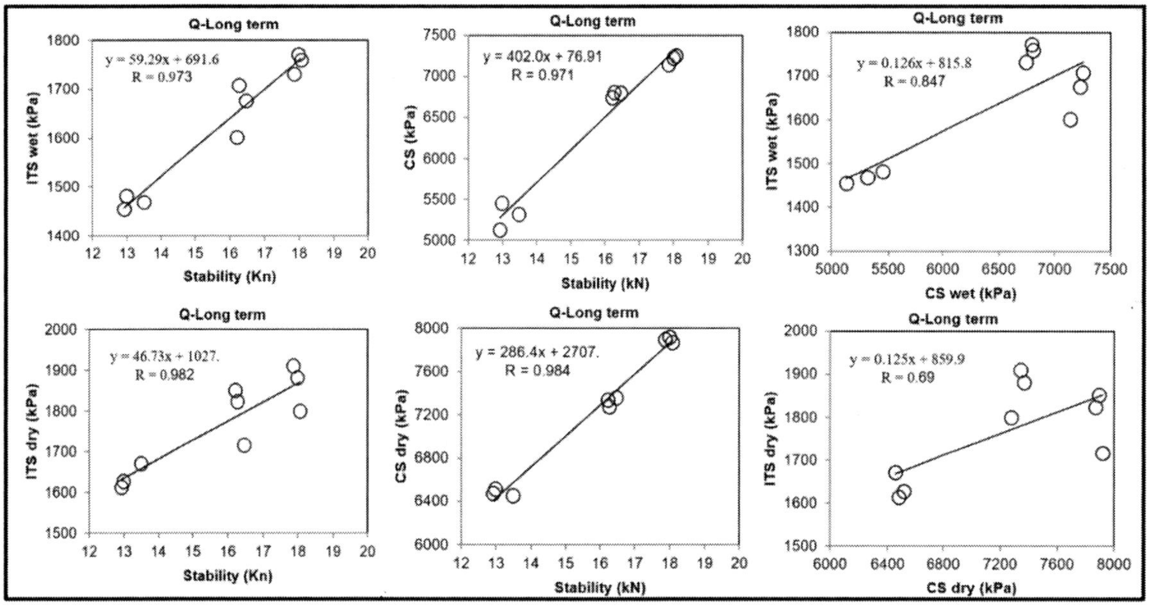

a- Baiji (gel) Asphalt Cement

b- Qayarah (sol) Asphalt Cement

Figure 3. Relations among DGAMs engineering assays with R value

Table 3. DGAMs Optimization Summary

Gel asphalt cement			
Mixture type	$DGAM_{0h}$	$DGAM_{4d}$	$DGAM_{8d}$
Stability, kN	11.80± 0.37	17.23± 0.17(s)	16.82± 0.14(s)
CS25°C, kPa	6278 ±50	1746± 19(s)	1688±27(s)
CS60°C, kPa	5279 ±118	1630± 44(s)	1508± 34(s)
ITS25°C, kPa	1408 ±22	6603±41(s)	7014± 45(s)
ITS25°C, kPa	1154 ±25	6151± 77(s)	6381± 32(s)
Flexural modulus, kPa, 0 °C	5012 ±132	6345± 158(s)	5974± 280(s)
Flexural modulus, kPa, -10 °C	5490 ±142	6846± 156(s)	6499± 243(s)
Sol asphalt cement			
Mixture type	$DGAM_{0h}$	$DGAM_{4d}$	$DGAM_{8d}$
Stability, kN	13.13 ± 0.35	17.79±0.11(s)	16.31±0.23(s)
ITS25°C, kPa	1637±34	1864±65(s)	1798±81(s)
ITS60°C, kPa	1648±16	1755±23(s)	1663±62(s)
CS25°C, kPa	6480±35	7327±51(s)	7894±28(s)
CS60°C, kPa	5300±182	6748±37(s)	7208±66(s)
Flexural modulus, kPa, 0 °C	5160±99	6523±114(s)	6393±121(s)
Flexural modulus, kPa, -10 °C	5642±175	6997±158(s)	6804±137(s)

5.4. Marshall Prospects

The Marshall prospects of LTA-DGAMs were examined and the values are found in Figure 4. From this Figure, it can be notified that 4d and 8d-LTA increases stability of gel and sol virgin DGAMs by 46% and 43%, and 37% and 24%, respectively, whereas, 4d and 8d-LTA flow values reduces by 37%, and 24%, and 35% and 30%, respectively.

Figure 4 also noticed that 4d and 8d-LTA increases the voids in mineral aggregates (VMA) for gel and sol by 0% and 1.5%, and 2% and 4%, respectively. This noticed that DGAMs durability increases with LTA. Besides, air voids（Av）was obtained between 3-5%.

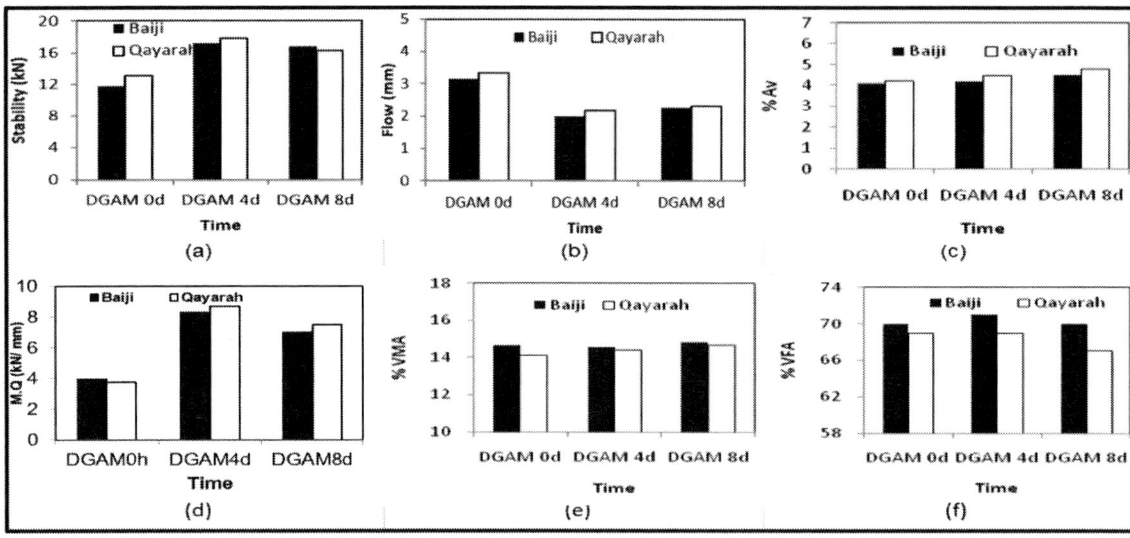

Figure 4. Marshall Prospects LTA-DGAMs (a- Stability, b- Flow, c- Av, d- MQ., e- VMA, & f- VFB)

MQ values of gel and sol virgin DGMAs were noticed higher by 132% and 99%, and 111% and 78% 4d and 8d-LTA, respectively. LTA-DGAMs notify well rutting strength and could be adopted for higher traffic pavements. Higher values of VMA and stability may be caused by the interlocking occurred between thiophen = oxygen, carbon = oxygen and carbon-hydrogen chain in bitumen.

5.5. Moisture Susceptibility

ITS for LTA-DGAMs are found in Figure 5. The results depict that ITS, and TSR raised for LTA. For DGAM0d, DGAM4d, and DGAM8d, average ITS values of wet-DGAM samples containing gel and sol asphalt cement are 1.154, 1.630, and 1.508 MPa, and 1.468, 1.755, and 1.663 MPa, respectively. For these, coefficients of variation (COV) of results were 1.96, 2.65, and 2.52%, and 0.95, 1.17, and 3.3%, respectively. Means values of ITS for dry-DGAM samples containing gel and sol asphalt cement are 1.408, 1.746, and 1.688 MPa, and 1.637, 1.864, and 1.798 MPa respectively. For these results, COV was 1.38, 1.0, and 1.18%, and 1.84, 3.09, and 3.96%, respectively. From these, % TSR values determined for DGAM0d, DGAM4d, and DGAM8d containing gel and sol asphalt cement 82, 93, and 89%, and 90, 94, and 93%, respectively. The higher ITS and TSR values after LTA may be related to that the asphaltenes /maltenes ratio.

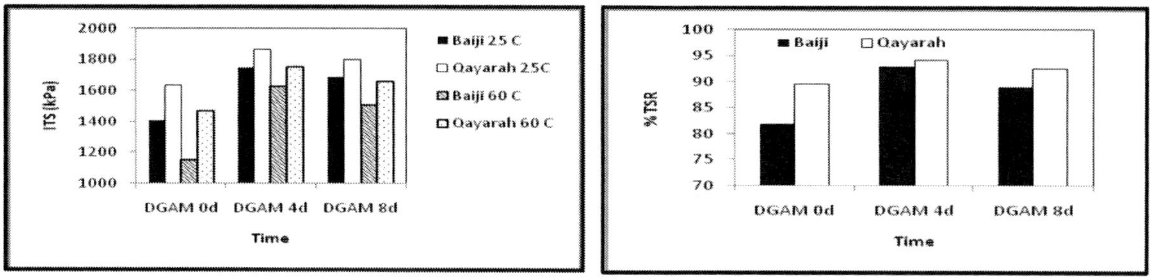

(a) (b)

Figure 5. ITS prospects of LTA-DGAMs (a- ITS, & b- TSR)

The CS variation with LTA is found in Figure 6. From this Figure, it can be depicted that CS for DGAM0d, DGAM4d, and DGAM8d samples containing gel and sol asphalt cement ranges from 6.278 to 5.279MPa, 6.603 to 6.151, and 7.014 to 6.381 MPa, and 6.48 to 5.3MPa, 7.327 to 6.784, and 7.894 to 7.208MPa at 25°C and 60°C, respectively. For these results, COV was 0.7 to 1.98, 0.54 to 1.1, and 0.56 to 0.44%, and 0.48 to 3.0, 0.61 to 0.48, and 0.32 to 0.8%, respectively. The research notified that CS of virgin DGAMs containing gel and sol asphalt cement at 4d and 8d-LTA was increased by 5 and 17%, and 12 and 21%, and 13 and 17% and 22 and 24% at 25°C and 60°C, respectively. From Figure 6, it was depicted that the IRS of virgin DGAMs containing gel and sol asphalt cement was increased by 11% and 8%, and 13% and 11% for DGAM4d, and DGAM8d, respectively. Besides, the notifications and statistical analyses depict that the ITS was the best assay for stripping examination of paving mixes.

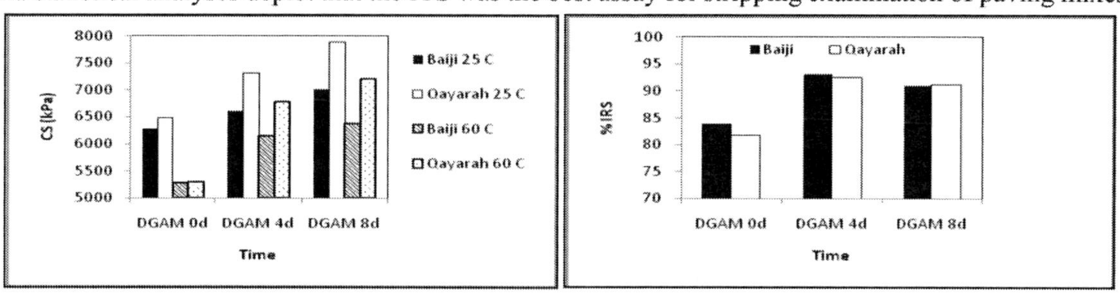

(a) (b)

Figure 6. Compressive Strength Properties of LTADGAMs (a- CS, & b- IRS)

5.6. Dynamic ITS Assay

Dynamic ITS assay was conducted at 25°C (ASTM D-4123, 2015) to assess the resilient moduli (MR). LTA-DGAMs depicted higher MR than virgin DGAMs as found in Figure 7. Assay results depict that MR for DGAM0d, DGAM4d, and DGAM8d samples containing gel and sol asphalt cement are 4010MPa, 6100MPa, and 5540MPa, and 4460MPa, 6100N/mm^2, and 5540N/mm^2 at 25°C, respectively, the MR raises by 52% and 38%, and 37% and 24% for 4d, and 8d-LTA, respectively. Thus, LTA has enhanced the MR of DGAMs. This is related to the LTA-DGAMs higher viscosity.

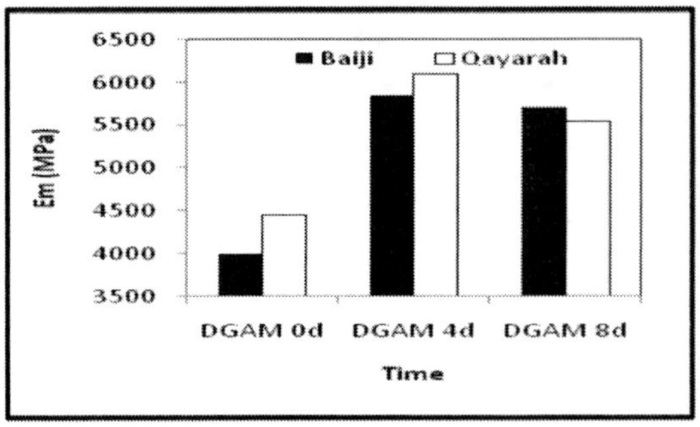

Figure 7. Resilient Modulus of LTADGAMs

5.7. Low Temperature Prospects

Paving materials at lower temperatures show elastic behavior that is defined by stress-strain curve (i.e. stiffness moduli) (Busby and Rader, 1972) which, in flexural examination, is named the flexural stiffness (S_t) and calculated from equ.1.

$$S_t = [(P \times L)/(b \times d^2)] \quad (1)$$

Besides S_t, rupture modulus (S_r) is an additional factor may be calculated from equ.2.

$$S_r = [(P \times L^3)/(6 \times b^2 \times d^2 \times \Delta)] \quad (2)$$

Where: (P= max. load, L= length of beam, b= width of beam, d= depth of beam, Δ = deflection at center of the beam).

The flexural assays result of LTA-DGAMs at 0 and -10oC are found in Figure 8. From Fig. 8, the S_r ranges for DGAM0d, DGAM4d, and DGAM8d specimens containing gel and sol asphalt cement were 5.012 to 5.49MPa, 6.345 to 6.846MPa, and 5.974 to 6.499MPa, and 5.160 to 5.642MPa, 6.523 to 6.997MPa, and 6.393 to 6.804MPa at zero and -10°C, respectively. For these results, COV was 2.7 to 2.63, 2.53 to 2.47, and 4.78 to 3.82%, and 1.95 to 3.12, 1.78 to 2.3, and 1.92 to 2.05%, respectively. Similarly, values of S_t for DGAM0d, DGAM4d, and DGAM8d containing gel and sol asphalt cement are 0.633 to 1.204MPa, 1.209 to 1.676MPa, and 1.117 to 1.557MPa, and 0.722 to 1.303MPa, 1.45 to 1.994 MPa, and 1.32 to 1.737MPa at zero and -10°C, respectively. From these results, LTA has positive effects on DGAMs.

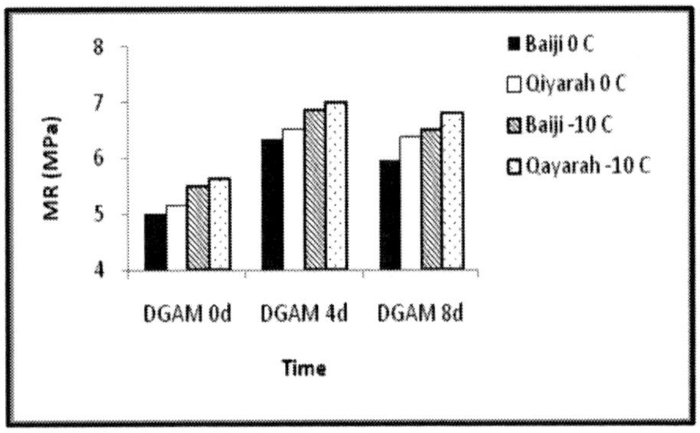

(a)

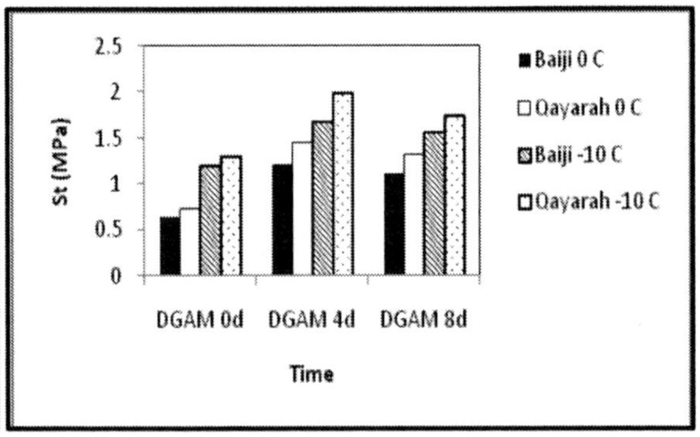

(b)

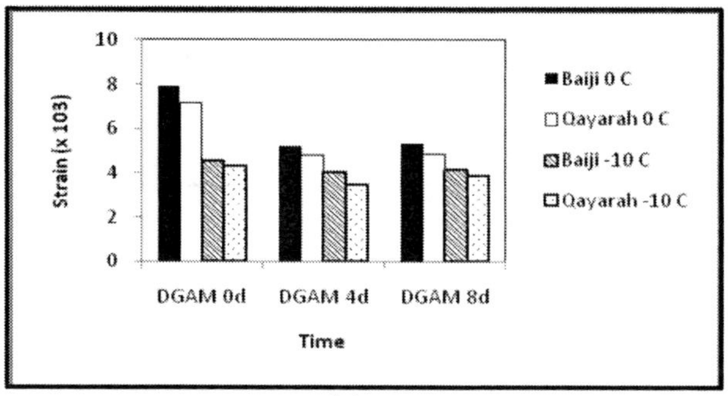

(c)

Figure 8. Flexural Strength Properties of LTADGAMs (a- MR, b- St, & c- Strain)

5.8. Cohesion Assay

Cohesion is a common assay to examine the binder viscosity. Cohesion of LTA-DGAMs was examined at 60°C (ASTM D-1560, 2015). LTA-DGAMs exhibited more cohesion than virgin DGAMs as found in Figure 9. Assay values depict that cohesion for DGAM0d, DGAM4d, and DGAM8d samples containing gel and sol asphalt cement are 210, 364 and 331, and 218, 339 and 314 at 60°C, respectively, the cohesion of virgin DGAMs containing gel and sol asphalt cement increases by 73 and 57%, and 56, and 44% for 4d, and 8d-LTA, respectively. Thus, LTA has positive effects on the DGAMs-cohesion. The higher cohesion values at 4d-LTA may be related to that the asphaltenes percentage at this ageing period remains within the specification limits (≤25%).

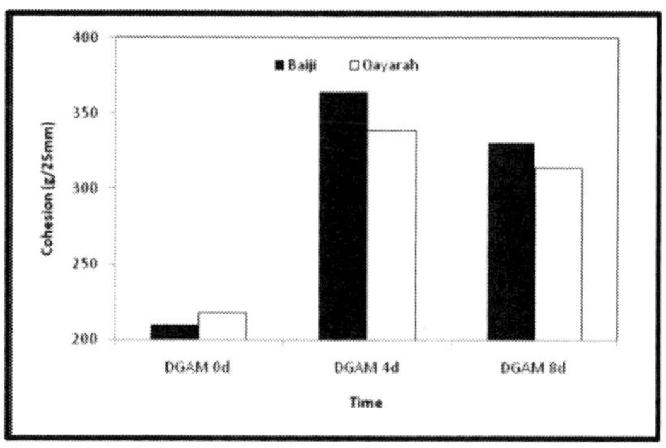

Figure 9. Cohesion of LTADGAMs

5.9. Aging Sensitivity

Aging sensitivity was examined by A.I. which equals to (cohesion of LTA-DGAMs /cohesion of virgin DGAMs). Fig. 10 depicted that LTA raises A.I. due to the higher bonds between binder and aggregates. Assay results depict that A.I. for DGAM4d, and DGAM8d samples containing gel and sol asphalt cement are 1.74, and 1.58, and 1.56, and 1.44 at 60°C, respectively. The findings agree with those reported by John and David (2003).

IOP Conf. Series: Materials Science and Engineering **978** (2020) 012002 doi:10.1088/1757-899X/978/1/012002

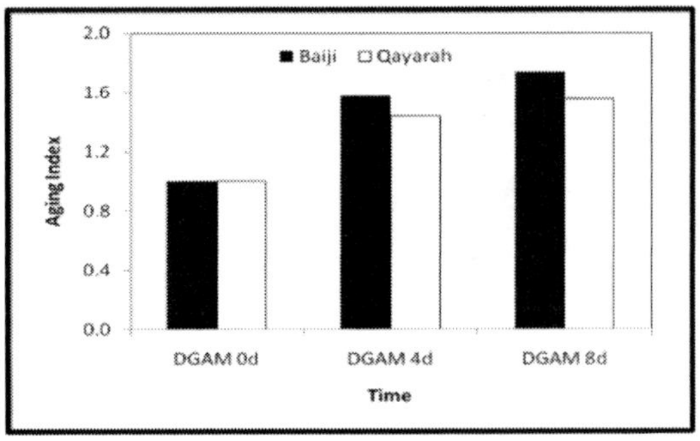

Figure 10. Aging Index of LTA-DGAMs

5.10. Aging Benefits

Commonly flexible pavements structural distresses are the surface cracking and rutting. Cracking in paving layers is caused from fatigue which is generated by the traffic repetitions, whereas, rutting is from the deformations occurs at other layers. εt and εc in pavement have been took as indices of fatigue and rutting. Yang (2012) takes 1.27 cm rut depth to be a failure limit for flexible pavement and the rutting in Eq. (3) is used:

$$N_d = 1.365/[10^9 * (\varepsilon_c)^{4.477}] \quad (3)$$

Where Nd = number of cumulative standard axles to produce a rutting of 12.7mm.

The w_o and εt in LTA pavement model were determined for different DGAMs and base course thicknesses utilizing BISAR program with (40×10^3 N and set of dual tires of 21cm dia.). For LTA model, thickness of the DGAMs of 5.0cm, binder course of 5.0cm, and subbase layer of 20cm were remained fixed and the thickness of base layer was changed. Similarly, thickness of DGAMs was changed for fixed binder (5.0cm), base (20cm), and subbase (20cm).

Figs. 11 and 12 depicts the w_o and the εt variation. The results of w_o and εt were utilized to examine the advantages of LTA-DGAMs containing gel and sol asphalt cement in forms of LTR and TBR. The TBR (equ. 4) depicts the pavement service life extension caused by LTA.

$$TBR = N_{dA}/N_{dU} \quad (4)$$

Where Nd = No. of traffic passes to produce w_o; and U and A notify unaged and LTA pavement models.

Al-Hadidy (2009) reported the advantages of polypropylene-asphalt pavement in forms of LTR (equ. 5) for similar service life.

$$LTR = [(D_U - D_A)/D_U]\times100 \quad (5)$$

DU and DA = DGAMs and base course thicknesses of unaged and LTA models.

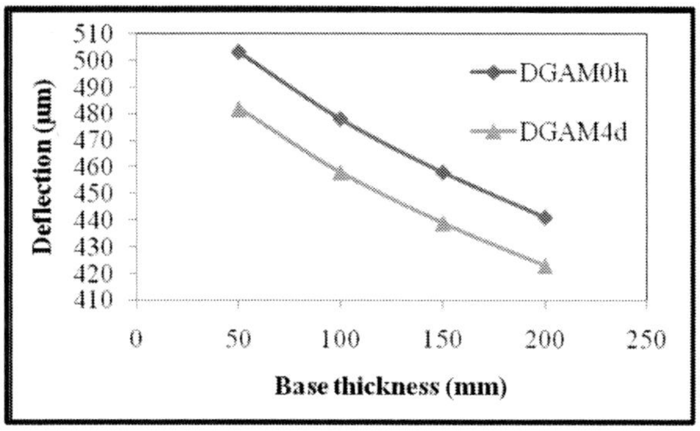

(a)

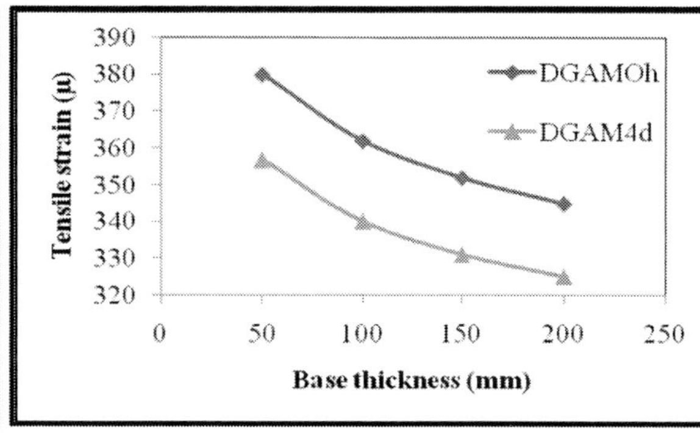

(b)

Figure 11. Variation of (a- W_o, and b- εt) with Base course thickness for Baiji Asphalt Cement

The results of BISAR analyses in Figure 11 depicted that the εt and the εc in the designed standard pavement model with gel DGAMs of 4010 N/mm^2 is 345μm and 251μm. Similarly, (Fig. 12), these values for gel DGAM with 4460 N/mm^2 is 340μm and 248μm, respectively. For fixed binder thickness, base course and subbase course, these levels of strain were gained for pavement model containing gel DGAM4d thicknesses of 43mm. Whereas, these values were found to be 46mm and 43mm, respectively.

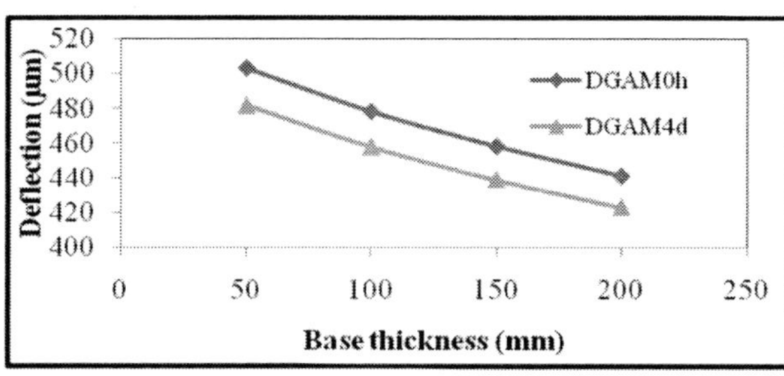

(a)

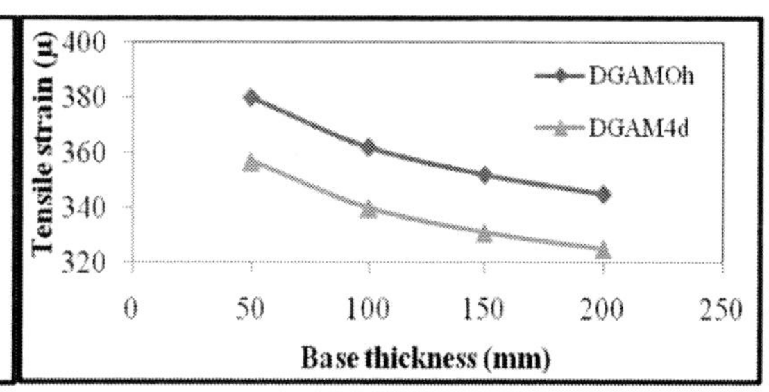

(b)

Figure 12. Variation of (a- W_o, and b- εt) with Base Thickness for Qayarah Asphalt Cement

If the pavement model is remained the same for unaged and LTA-DGAMs containing gel asphalt cement, the εc decreases from 251μ for unaged model to 240μ for DGAM4d; giving the TBR of 1.2. Similarly, for unaged and aged DGAM containing sol asphalt cement (Fig. 12), the εc decreases from 248μ for unaged model to 239μ for DGAM4d; giving the TBR of 1.2. It means that LTA pavement will have a life 1.20 times that of unaged pavement. These results depict that for a fixed thicknesses of gel DGAMs, binder layer, and subbase these levels of strain were gained for a base course thickness of 85mm, in DGAM4d pavement model, respectively. From these results, it was depicted that DGAMs and base course of unaged pavement model decreases by 14% and 57% for DGAM4d, for similar service life of unaged and LTA pavements. Similarly, (Fig. 13), for a fixed thickness of sol DGAMs, binder course, and subbase, these levels of strain were gained for a base course thickness of 90mm, for DGAM4d pavement model. From these results, it was depicted that DGAMs and base course thickness of unaged pavement model decreases by 14% and 55% for DGAM4d. Thus, the paving model can be designed using any of the discussed alternatives.

The pavement can also be designed for any intermediate thickness to decrease DGAMs or base course thickness, besides obtaining additional TBR. For e.g., for gel DGAM4d, the base thickness can be decreased theoretically below 85mm (57%LTR) for εc of 251μ. Besides, for all analyzed options, the W_o was reduced by 3 to 4%. It depicts that the overall pavement strength be higher due to utilizing LTA-DGAMs course.

6. Conclusions

Among the examination and analyses the following conclusions can be notified:

A- A review of Marshall, ITS, CS, FS, cohesion, IRS, and TSR results depicted the following:

1. Marshall prospects depicted that the MQ of unaged DGMAs containing gel and sol asphalt cement increased by 132% and 99%, and 111%, and 78% at 4d, and 8d-LTA, respectively. It can be depicted that the LTA-DGAMs give well rutting strength and could be adopted for higher traffic pavements.
2. Dry ITS containing gel and sol asphalt cement increases 24 and 20, and 14, and 10% at 4d and 8d-LTA, respectively.
3. The increase in CS for unaged DGAMs containing gel and sol asphalt cement at 4d and 8d-LTA was depicted to be 5 and 17%, and 12 and 21%, and 13 and 28%, and 22 and 36% at 25°C and 60°C, respectively.
4. The tensile strength ratio for DGAM0d, DGAM4d, and DGAM8d containing gel and sol asphalt cement was found to be 82, 93 and 89%, and 90, 94 and 93%, respectively. This depicts that LTA gives a positive advantage to moisture resistance.
5. It was depicted that the IRS of virgin DGAMs containing gel and sol asphalt cement increases by 11% and 8%, and 13% and 11% for DGAM4d, and DGAM8d, respectively
6. The Sr for unaged DGAMs containing gel and sol asphalt cement at 4d and 8d-LTA was notified to be 27 and 25%, and 19 and 18%, and 26 and 24%, and 24 and 21% at zero and -10°C, respectively. From these results, it is revealed that LTA-DGAMs exhibit well than virgin DGAMs at low temperatures, which is a preferable trait of LTA-DGAMs.
7. The resilient moduli for DGAMs containing gel and sol asphalt cement at 25°C temperatures increases by 52% and 38%, and 37% and 24% for 4d, and 8d-LTA, respectively.
8. The cohesion of virgin DGAMs containing gel and sol asphalt cement increases by 73 and 57%, and 56 and 44% for 4d, and 8d-LTA, respectively. Thus, LTA has enhanced DGAMs cohesion, which depicts that LTA make the asphalt viscosity higher.
9. It was depicted that A.I for DGAMs containing gel and sol asphalt cement increases by 74, and 58%, and 56, and 44% for 4d, and 8d-LTA, respectively. These findings agree with those gained by John and David (2003).
10. Statistical relations among the DGAMs assays are explored: a) for their utilized in the absence of ITS and CS assays, and /or b) the engineer does not select to utilize those assays.; and
11. Statistical analyses program (SPSS) was utilized to introduce an ideal table based from the DGAMs assays. Results gained from such analyses can be utilized to get the LTA options for any field conditions. Perfect choosing based on the option examined by the designer.

B- Mechanistically-empirical design tool results depicted the following:

1. If the pavement model is remained the same for unaged and aged DGAM containing gel and sol asphalt cement, the pavement gives TBR value of 1.2 for 4d-LTA.
2. It was depicted that gel DGAMs and base layer thickness of unaged pavement model decreases by 14% and 57% for DGAM4d, respectively, for similar service life of LTA and unaged pavements.
3. It was depicted that sol DGAMs and base layer thickness of unaged pavement model decreases by 14% and 55% for DGAM4d for similar service life of LTA and virgin pavement.
4. The LTA-DGAMs pavement as a wearing layer has advantages in decreasing the construction paving material. Perfect choosing based on the option examined by the designer for decreasing the individual layer thickness; and
5. It was found that sol asphalt cement performs slightly better than gel asphalt cement, especially in the case of moisture susceptibility and flexural strength properties.

References

[1] Al-Hadidy A.I., (2009), "Characterization of Polymer Modified Road Building Materials: Laboratory and Field Evaluation". Ph.D. dissertation, School of Transportation and communication, Harbin Institute of Technology, China.

[2] Al-Hadidy and Tan Yi-Qiu., (2007) "Long-Term Aging of Polypropylene Asphalt Paving Mixtures", Proceeding of the 26th Southern African Transport Conference (SATC 2007), 9-12 July 2007. P761-767. ISNB Number: 1-920-01702-X Pretoria, Accepted as a Word Cup 2010 Traffic Simulation.

[3] American Society for Testing and Materials (ASTM), (2015), Standard Specification, Section 4, Vol. 04-03.

[4] Asphalt Institute (1984). "Mix design method for asphalt concrete and other hot-mix types" (MS-2).

[5] Bell A. Chris, Alan J. Wieder, & Marco J. Fellin., (1994) " Laboratory Aging of Asphalt-Aggregate Mixtures: Field Validation", SHRP-A-390, Strategic Highway Research Program, Oregon State University, National Research Council, Washington, D.C.

[6] Bradley, J. Putman, and Serji, N. Amirkhanian. (2004) "Utilization of waste fibers in stone matrix asphalt mixtures". J. Resources Conservation and Recycling. (42):265-274.

[7] Busby EQ. and Rader LF. (1972) "Flexural stiffness properties of asphalt concrete at Low temperatures." Proceedings, AAPT, Vol.41:163-187.

[8] Dongmei Zhang, Henglong Zhang, and Caijun Shi. Investigation of aging performance of SBS modified asphalt with various aging methods. J. Construction and Building Materials, 2017, (145):445-451.

[9] John, R., David, W., (2003) "The Shell Bitumen Hand Book".5th Ed., Thomas Telford Ltd., pp.157-163,213,214.

[10] Mansour Fakhri and Amin Ahmadi (2017). Evaluation of fracture resistance of asphalt mixes involving steel slag and RAP: Susceptibility to aging level and freeze and thaw cycles. J Construction and Building Materials, 2017, (157):748-756.

[11] Min Bai. (2017). Investigation of low-temperature properties of recycling of aged SBS modified asphalt binder. J. Construction and Building Materials, (150):766-773.

[12] Shifeng Wang; Xinyu Zhao; and Qiang Wang. (2017). Rheological and Structural Evolution of Rubberized Asphalts under Weathering. J. Mater. Civ. Eng., 2017, 29(10): 500–507.

[13] Shin-Che Huang. Rubber Concentrations on Rheology of Aged Asphalt Binders. J. Mater. Civ. Eng.2008 20(3): 221–229.

[14] State cooperation of road and bridges, (2003). "Hot mix asphaltic concrete pavement", Iraqi standard specification, Ministry of Housing and Construction. Department of Design and Study, Section R-9.

[15] Soon-Jae Lee, Serji N. Amirkhanian, & Kwang W. Kim (2009), "Laboratory Evaluation of the Effect of Short-Term Oven Aging on Asphalt Binders in Asphalt Mixtures Using HP-GPC", Construction and Building Material.

[16] Soon-Jae Lee A, Serji N. Amirkhanian , Khaldoun Shatanawi , Kwang W. Kim , (2007) "Short-Term Aging Characterization of Asphalt Binders Using Gel Permeation Chromatography and Selected Superpave Binder Tests", Science Direct, Construction and Building Materials.

[17] Yang H.H, (2012), "Pavement Analysis and Design", Prentice-Hall, Inc., A Paramount Communications Company, Englewood Cliffs, New Jersey 07632, USA: 57.

A Review on Structural Behavior, Analysis and Design of RC Dapped End Beams

Qasim M. Shakir

College of Engineering, University of Kufa, Iraq
qasimm.alabbasi@uokufa.edu.iq Tel: +9647832460562.

Abstract. This paper represents a historical review of the theoretical and experimental studies carried out previously to study the behavior of reinforced concrete dapped end beams. Prestressing is one of the techniques that have been adopted to enhance the performance of such members. Furthermore, a dapped end needs to be strengthened due to various reasons to improve its performance. Consequently, the review has been extended to include studied such aspects. It was concluded that the ultimate capacity enhanced by reducing the shear-slenderness ratio and horizontal tension reaction and increasing the strength of concrete and fiber content. In addition, nib main reinforcement affects significantly the mode of failure, cracking and ultimate load. However, it was observed that the inclined hanger stirrups or bent performed better than that of vertical alignment and that the prestressing level and strand profile affect the shear capacity and performance significantly. On the other hand, it was found that the external bonding CFRP is the most adopted strengthening material and that the STM (Strut-and-Tie Model) approach results in better predictions of the ultimate load than the shear-friction method. Moreover, directions for future work are proposed, several conclusions have been drawn and a comprehensive list of references is provided.

Keywords: *Dapped end beams; Hanger zone; Re-entrant corner; Prestressed dapped end; Strengthening*

1. Introduction

Precast concrete (PC) was widely used beyond the 1950s, and its application spread dramatically around the world. PC structures have several features including rapid construction, more strength, and durability, energy- efficient, excellent appearance, noise minimizing, high durability. In general, such structures are segmented and joined using different types of linking elements. One connection detailing that is widely used in precast buildings and parking garages is the dapped end (DE). A dapped end is created when the web or stem of a beam is notched at the bottom corner; the bearing location is moved higher in the cross-section. The notch is termed as the "dap" and the portion of concrete being above the dap is called as the "nib". The dapped end allowed reducing the overall depth of a precast floor or roof structure by recessing the supporting corbel or L-edge into the supported beam. Thus, reducing substantially the overall height of a building, providing better lateral stability of the beams and providing an adequate and economical construction system (Mattock and Chan,1979).

The flow of stress at the dapped ends is interrupted by the sudden change in geometry, Figure.1. Thus, zones of disturbance in stress distribution are created in the vicinity of the re-entrant corner and within the nib. Such regions of discontinuity are referred to as disturbed regions (D-regions). Due to this complicated problem, traditional methods are not appropriate to design DEBs. Two methods have been used; shear-friction theory (Mattock, 1976) and STM models (ACI 318-19). The designers frequently ignore the aspects of proper details which may result in poor serviceability, severe cracking in the D-region, spalling of concrete cover, or even cause premature brittle failure. Figure.2 shows the most common failure modes of dapped ends.

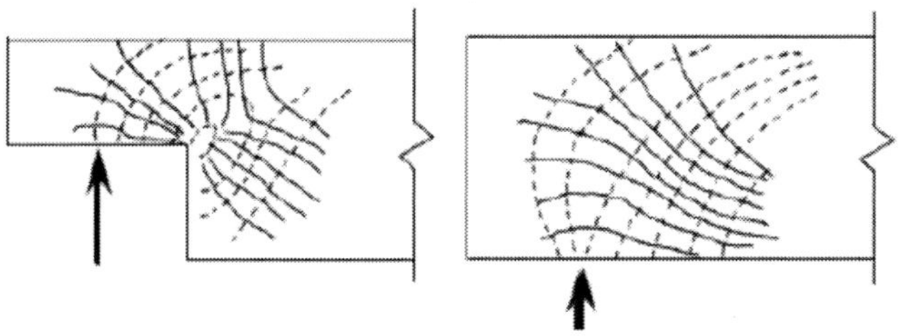

Figure 1.stress flow in DE and beams (Torris, 1998)

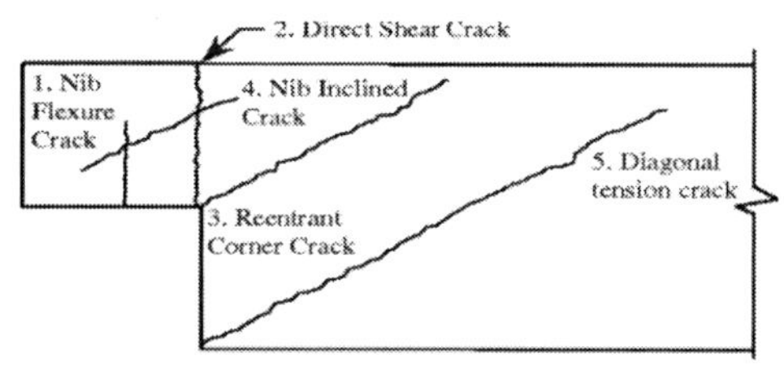

Figure 2. Modes of failure for DE (Precast /prestressed Institute (PCI) Handbook, 2004)

2. Shear friction Approach versus Strut and Tie Models

Several approaches and techniques have been proposed to study the behavior of RC dapped end beams (DEBs). One of the earliest studies was by Mattock and Chan (1979). Eight DEs with a shear span/depth (a/d) ratio of less than 1.0 were studied. The design was based on the procedure of Mattock (1976) with considering the DE as an inverted corbel and substituting the compressive force in a column that joined with corbel by a hanger reinforcement (A_{vh}) as shown in (Fig. 3). The DE support of some specimens was subjected to a horizontal force in addition to the vertical reaction. The first version of "shear friction" design procedure for DEs was proposed which include firstly checking the ultimate shear stress in nib not to be more than:

$$v_u = Min(\frac{V_u}{\emptyset bd}; 0.2 * f\acute{c}) \qquad ....(1)$$

Then, hanger (A_{vh}), reinforcement to be calculated from:

$$A_{vh} = \frac{Vu}{\Phi fy} \qquad ....(2)$$

The main nib (As) can be obtained from the condensed form based on Mattock and Chan (1979):

$$A_s = Max[Max\{\frac{\frac{Vu}{\emptyset*0.8} - kbd}{fy}; \frac{0.2bd}{fy}\} + \frac{Nu}{\o fy}; \frac{2}{3}A_{vh} + \frac{Nu}{\o fy}] \qquad ....(3)$$

The horizontal nib (A_h) can be obtained from:

$$Ah = 0.5(A_s - \frac{Nu}{\o fy}) \qquad ....(4)$$

In which Vu and Vn: ultimate and nominal shear capacities; b and d are the cross-section dimensions; fy and f'c: yield stress and compressive strength of steel and concrete respectively;

ø=0.85 and K= 0.5, 0.25 and 0.31 for normal weight, all-lightweight and sanded lightweight concrete respectively

The main findings were:

1- The DE may be considered as an inverted corbel providing "the shear span (a) is equivalent to the distance from the dap-vertical reaction to the center of the hanger stirrups (A_{vh})".
2- Cracking initiated firstly at the internal-corner of the nib (re-entrant corner) within the load level of 20% of the maximum capacity. However, cracks propagated at nearly 45° from the horizontal and penetrate a depth of about two-thirds the depth of the DE. However, in the final stage of loading; (70%-95% of the maximum load); the main nib reinforcement yielded accompanied with some compression spalling at the top face close to concentrated point load.

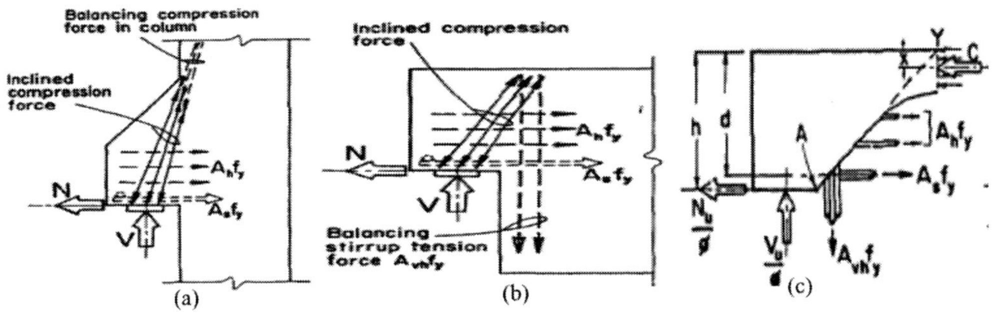

Figure.3 Internal forces in (a) a corbel (b) dapped end (c) Design model [Mattock and Chan,

In 1983, Liem adopted Mattock and Chan approach (1979) to predict the capacity of DEBs with inclined hanger reinforcement (HR). Eight full-scale DEBs tested with inclined bars instead of the horizontal bars. Various amounts of HR and several geometries of DEs were considered. it was reported that inclined reinforcement is ($\sqrt{2}$) times more than the horizontal reinforcement in terms of the amount of steel required and that the geometry of the dapped end has no effect on the ultimate shear strength.
A design procedure was proposed for DEs with a/d <1 based on equilibrium conditions and considering the inclined HR. The theoretical results yielded that the failure load of a DE with reinforcement of (45°) inclination is twice the strength of that with horizontal or vertical reinforcement.

In 1991, Barton et. al. verified the validity of the STM model to simulate DEBs with different reinforcement schemes. Four details of DEBs beam investigated. Two were based on STM models. Whereas, the other two DEs have detailing based on the PCI method and the Menon/Furlong (1977) design methods. The major finding reached by:
1- STM model seemed to be satisfactory to qualify the detailing of the DEB. However, all methods underestimated capacity of the tested specimens.
2- Failure of specimens was by a ductile mode (steel yielding occurred before concrete crushing). Moreover, a slightly more shear reinforcement is needed if STM model is adopted compared to the other design procedures.

Yang et. al. (2011), proposed a mechanism analysis based on upper-bound theorem (the energy principle) to estimate the critical plane of failure plane and the corresponding shear strength of RC DEBs., Figure. 4. Results were compared with the failure load for 47 specimens obtained in literature. In addition, two techniques have been utilized to analyze the DEBs, which were; PCI design method and the simplified STM model of ACI 318-05. Results revealed that:
1- The two techniques highly underestimated the shear strength of DEBs. In contrast, adequate prediction was found by the proposed mechanism analysis.
2- The PCI design method insufficiently treated effects of a/d ratio, amount of HR and value of horizontal load on the shear strength of DEBs. Whereas, such effects may be included by the mechanism analysis.
3- Both the developed mechanism analysis and STM model obviously determine the critical plane and similar results were obtained from the two methods.

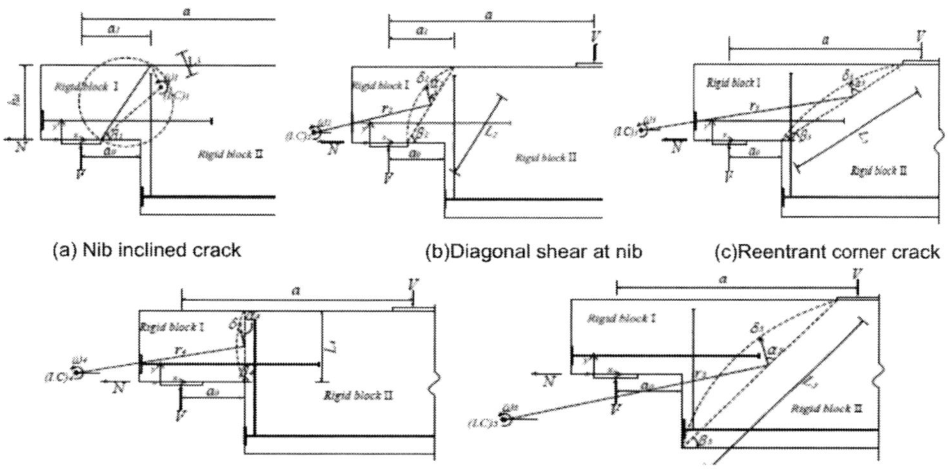

(a) Nib inclined crack (b)Diagonal shear at nib (c)Reentrant corner crack

(d)Direct shear crack (e)Diagonal tension

Figure.4 failure mechanism of dapped-end beams (Yang et al. 2011)

3. Grade of Concrete Effect

In 2003, Lin et al. tested 24 DEB specimens with concrete strengths of (34-69) MPa. The amount of main DE steel, and (a/d) ratio were studied besides the grade of concrete. Results indicated that the shear strength of DEs enhanced with increasing the main nib reinforcement and adopting higher strengths of concrete. Moreover, it relates inversely with a/d ratio. A simplified procedure to predict the shear capacity of RC DEs was proposed and verified against the approach of the PCI Design Handbook. It was found that the proposed method could predict efficiently the shear strength of RC DEBs if compared with the shear friction method.

In 2003, Lu et al. tested twelve RC DEBs considering the same variables of Lin et.al. 2003. A new system of STM elements was proposed to predict the shear strength of the DEBs based on the relationship between vertical and horizontal shears in terms of the ratios of the vertical shear carried by the diagonal and horizontal mechanisms. Several recommendations have been addressed:
1, Small values of shear slenderness ratio, higher amounts of main nib reinforcement may result in a significant enhancement in the shear capacity of DEBs. Furthermore, some improvement in shear resistance was observed when adopting higher grades of concrete
2. The proposed procedure was more convergent than the PCI approach that underestimates the shear capacity.

In 2012, Lu et al. tested 24 RC DEs considering the grade of concrete, a/d ratio and the horizontal load value. Results revealed that the stiffness and the failure load increases with reducing shear slenderness the ratio and ratio of a horizontal load. The softened strut-and-tie model for estimating the shear capacity of DEBs with a shear slenderness ratio greater than unity was extended and calibrated with the STM model of ACI318 code and the PCI method. The result showed that the proposed model and STM model could assess the shear capacity of the RC DEBs more adequately than the PCI approach.

In 2015, Aswin, et al. considered the concrete strength as the main parameter of their study involved testing four large-scale RC-DEB specimens. Three were (NSC) of F'c=27MPa, while the forth was HSC with PVC fibers (79 MPa) at DE only. A value of (a/d) of 0.91 was used to satisfy the PCI requirements. Other variables studied were ratios of the nib and main beam reinforcement. Comparison have been made with the results obtained using three codes PCI, ACI STM model, RELIM (2000). It was reported that:

1- Using HSC or ECC (Engineering Cementitious concrete) with polyvinyl alcohol (PVA) fibers in the dapped area improved the capacity by (51.9%).
2- Increasing the amount of nib and main flexural reinforcements led to an enhancement in the failure load by (62.2%) and (46.7%) respectively.
3- No single code is suitable for analyzing all the beams; Analysis using PCI code was found to become accurate than ACI code and RILEM provision in predicting the failure load of RC-DEBs.

In 2019, Mohammed et al. tested experimentally twenty-seven full-scale DEs with two grades of engineered cementitious composites (ECC), namely were 85 Mpa (ECC-85) and 105 Mpa (ECC-105). The control specimen was designed according to the requirement of standard code and of normal strength concrete of 28 MPa. In some specimens, engineered cementitious composites (ECC-85&ECC-105) have been incorporated in the DE region. Results revealed that the reinforced ECC- DEB showed better performance compared to the NSC-DEBs and higher ECC strength resulting in higher failure capacity. The main points of the study were:
1- The First crack emanated from the re-entrant corner for all DEBs. Moreover, using Diagonal Reinforcement (DR) in DE- zone improved the capacity by 40.79% and deflection of 24.29% compared to the case with vertical HR.
2- It was recommended not to use the diagonal reinforcement as a full substitution for the HR. Because the bottom corner of the full-depth beam experienced severe damage due to the loss in confinement at an un-dapped zone.
3- There was an average enhancement in the ultimate load of 59.97% and maximum deflection of 46.67% for DEB with ECC-85 relative to the RC-DEBs. Meanwhile, DEB utilized ECC-105 exhibited an average improvement on the ultimate load of 11.26% and the ultimate deflection of 11.88% relative to DEB utilized ECC-85.

4. Effect of steel fiber

In 2008, Mohamed and Elliott studied reducing the effect of disturbance of stresses within the dapped end region by incorporating (1%) hooked-end steel fibers as a partial substituent for vertical HR. Twenty self-compacting SFRC DEBs were tested experimentally. The variables that investigated were the detailing of the reinforcement, the grade of concrete as well as the depth of the extended end. A softened STM model was proposed to predict the shear capacity. The proposed model was verified against the PCI method. Several conclusions have been reported:
1. The STM concept is an efficient method to treat the disturbed regions of the DEs and the "compressive diagonal strut" carried all shear forces. In addition, the proposed method yielded results higher than the PCI procedure
2. The mode of failure for most of the tested beams was the "shear-compression" type, which occurred along the diagonal strut and the strength of steel bars has a small influence on the shear capacity of the DEBs.
3-Using steel fibers yielded positive results, in either full addition or partial substitution for HR.

5. Effect of a/d or depth of nib end

In 2005, Wang et al. tested 24 DEB with several configurations of inclined HR. The main variables were the depth of nib end, the type and effective distance to provide the stirrups and the bent detail of the main steel. Furthermore, a Truss Mechanism for shear strength prediction has been proposed. Several conclusions were reported:
1- First cracking initiated at the re-entrant corner of the DE with the angle in range (40-60) degrees from the horizontal. With progress in loading, the diagonal crack developed towards the point of application of the concentrated load of about 30 degrees. The failure occurred whenever the diagonal cracks penetrated through the most thickness of the compression consequently a failure of the shear-compression type occurred.
2- The depths of beams and extended end play a dominant role in developing the diagonal cracks and shear strength. Thus, the depth of DE must not be less than 0.45h (where h: full depth of the beam).
3- The inclined stirrups and longitudinal bent reinforcements had a significant impact on the shear strength than the closed vertical stirrups. Also, The HR should be distributed within a distance of (half

to full depth) of the extended end, provided that the cover not less than 40 mm.

In 2013, Ahmad, et al. studied the effect of nib depth on strength of DEBs. Four DEBs were tested experimentally. Specimens were divided into two groups, one with (nib depth/beam depth) of (279 mm/457 mm) whereas the second has relative values of (178 mm/305 mm). It was concluded that the shear capacity was effected by a/d ratio rather than the overall depth of the beam. In addition, the results revealed that the STM model underestimated results for DEs with high dap depths only, and the design using the STM model was influenced by the strut inclination. Furthermore, it was obtained that DEBs have the ability to resist loads beyond the formation of the diagonal crack due to arch action.

In 2014, Aswin, et al. proposed a simplified STM model considering the study of (Wang et al., 2005). The results were calibrated against the PCI method and STM models of (ACI- 318-08, Euro Code 2 and BS 8110) codes. Grade of concrete, nib geometry, a/d ratio, types and space of distribution of the hanger stirrups, the bent shape of the longitudinal steel, amount of the reinforcements of the extended end and hanger were considered as influential parameters. It was found that Euro Code 2 and BS 8110 were less accurate in comparison with the PCI method and ACI-318-08. Moreover, it was concluded that using inclined stirrups, bent reinforcement less than 90 degrees, higher nib depth, closer and higher number of stirrups yielded higher failure loads.

In 2015, Lu, et al. tested 24-DEB with (a/d) ratio of (1.19-1.51). Concrete strengths of (32.5-70) MPa have been adopted. Parameters that studied included (a/d) ratio, grade of concrete, nib longitudinal reinforcement and hanger stirrups of the DEBs. Results revealed that the shear capacity might be improved when adopting higher strengths of concrete and smaller values of (a/d) ratio. Moreover, a model has been proposed to predict the shear capacity of the dapped ends based on linear bending theory. Such a model was calibrated against the results of 32 specimens available in literature in addition to the specimens tested throughout the study. It was concluded that the proposed model adequately estimated the shear strength of DEs. However, it was reported that the prediction of the STM of the ACI Code is noticeably conservative.

6. Internal detailing of dapped ends

In 2009, Peng studied the effect of HR detailing of the DE- region. Two DEB with four different reinforcing details which are; C-Shaped, closely distributed HR, conventional HR and Two-sided stirrups HR. Such detailing was studied in terms of the anchorage of the HR and the main steel at the nib. It was found that:
1- STM models adopted by ACI318-02 resulted in conservative results. Whereas, using the early version of PCI 1971 and PCI 1999 Design Hand Book resulted in poor detailing requirements. Most of the specimens failed in brittle failure modes without yielding of the entire HR.
2- Proper anchorage reinforcement details have a great effect on ductility and affected the shear capacity of about 44%. However, placing the HR close to the dapped ends developed higher shear capacity.

In 2016, Desnerck, et al. studied of effect of the internal defects in reinforcement detailing within the dapped zone on the capacity of the DEBs. Four specimens with different reinforcements detailing. The control specimen was designed adopting STM model; whereas the other DEBs included intentionally one defect either missing diagonal bars, missing nib steel bars or a reduced HR. Results revealed that:
1-STM models underestimates the failure load and would provide a safe design results. In all beams, cracking initiated at the re-entrant corner with a load level of about 20–33% and angle of about (40–45) degrees except the beam without a U-bar where the angle was more than 75 degrees.
2-Elimination the diagonal or U-bar nib steel bars, a mode of nib failure occurred. A reduction in the failure load has been noticed when omitting the diagonal reinforcement bars by about (39%). In contrast, the reduction of the HR resulted in a shear failure in the full-depth section of the beam with a reduction in capacity of about (10%).

In 2017, the same authors studied the influence caused by local defects in hanger and beam main reinforcement. Large-scale tests on nine DEBs were carried out with low (LSC) and normal strength concrete (NSC) considering four types of defects which were; local corrosion, anchorage cracking, small shear reinforcement ratio, and inconvenient internal reinforcement detailing. It was reported that:

1- A considerable drop in capacity of 35% caused by a local reduction of 50% of the amount of the diagonal, U and vertical steel bars at the extended end of the DEB.
2- The loss of capacity due to inappropriate anchorage of the diagonal bars, a deficiently shear reinforcement or existence of cracking in concrete around the tension steel located at the CTT zone when exist separately, did not exceed 10%. However, simultaneous occurrence of several defects led to drop in capacity in a range of (40%-60%).

In 2018, Desnerck et al. extended the previous study to determine the effects of single and combined defects when exist close to the bottom corner of the full-depth beam (CTT, compression –Tension-Tension nodal point). Twelve DEB were tested with various deficiencies. STM modelling was adopted to design the DEB and simulate deteriorations such as corrosion, insufficient anchorage lengths, and crack formation. It was addressed that:
1- Corrosion in steel can be included by reducing cross-sectional area of the reinforcing bar, and then, the capacity of the tension tie, can be included. Furthermore, insufficient anchorage lengths can be considered using some reduction factor in the tie capacity or by reducing the residual bond capacity. In addition, those STM simulations yielded conservative predictions of loss in capacity within a range of 16-57%.
2- To incorporate the effect of longitudinal cracking around steel into the capacity, it was suggested to use some reduction factor ranging (0.85 to 0.3 or in some cases 0.0) for the bond strength of the concrete.

In 2019, Falcon et al. tested experimental 28 DEBs in which 15 HR reinforcement arrangements, with and without diagonal reinforcement were considered. The ratio and the scheme of reinforcement were discussed. HR either installed as one layer or distributed within hanger zone. Based on the results of the experimental tests, a simplified approach to determine the geometry of the STM models for DEB with one layer- HR was proposed. The nodal point on top of the HR in DEB, which is called compression-compression-tension (CCT) Node, was considered in the study. The main findings that were addressed:

1– The deformation capacity of the DEB arrangements is efficient to produce the full strength of the mechanisms with orthogonal and inclined alignment.
2– The failure mode within the nodal region at the top end of the vertical tie at hanger reinforcement was of "concrete spalling" which controls the failure load of DEB, except for very few amount of reinforcement.
3– The proposed procedure, allows the capacity of the CCT node to be developed including the concrete cover. Furthermore, the HR distribution over wide distance reduces the member's strength due to increasing (a/d) ratio

In (2020), the impact of introducing openings in DEBs without any additional reinforcement was monitored by Shakir et al. Nine specimens were investigated in the experimental study. Different shapes, sizes and locations for opening were examined. It was found that a significant improvement in failure load with range of 23 to 25% was gained when the main reinforcement of the nib (extended end) reinforcement was substituted by inclined reinforcement. In addition, the results revealed that the best location of opening was within regions of high positive bending moment. Moreover, if opening reinforcement was not provided, unexpected modes of failure may occur.

7. Prestressed dapped ends

A new approach to improve the response of the DEB was adopted by making use of prestressing concept. In which prestressing steel is extended to the extended end. Thus flexural and shear stress will be reduced, cracking initiation will be delayed corresponding to non-prestressed DEBs. In 1973, Werner and Dilger tested five posttensioned DEBs. One included no special reinforcement at DE region. Some of the other four DEBs had inclined support at one end to introduce horizontal tensile forces of 50% of the vertical reaction. Some specimens included conventional stirrups at DE-region, others included inclined bars as shown in Figure 5. The configuration and quantity of DE reinforcement, horizontal tensile force and load level to initiate cracking at the RE were discussed. The main conclusions were:
1- For some specimens, premature failures occurred at the anchorage regions of the post-tensioning bars and it was concluded that there is a need for external reinforcement in such regions. One of the beams failed by a flexure mode caused by concrete crushing with the mid-span zone.

2- It was recommended that the main steel bars at extended end to be extended a distance not less that beam depth after the re-entrant corner and to be provided with hooks to provide full bond with concrete.

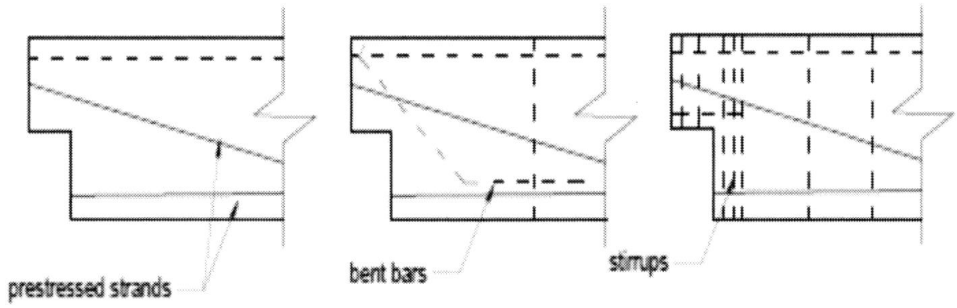

(a) control (b) bent bars (c) HR stirrups

Figure. 5 Reinforcement details for the specimens by Werner and Dilger 1973

In 1974, Hamondi and Phang investigated several variables that affect the response of PC DEBs through testing six prototype-flanged beams. Such variables were shape of the DEs; detailing of reinforcement in D-region; (a/d) ratio; and prestressing level. It was concluded that:

1- Reinforcing the DE zone with prestressed inclined bars can be considered as a good method to satisfy strength requirements thereby develop the required performance.
2- Prestressing may restrict development of shear cracks within a DE at working load stage.

In 1975, Hamoudi et al. tested eight PC T- beams with DEs to check the adequacy of different detailing proposals for shear reinforcement which were the form of Post-tensioned HS rods, U-shaped stirrups, and inclined bars with bent. The design was based on stress functions obtained from the analysis of the DE using the theory of elasticity. The main conclusions were:

1- Shear capacity of PC beams with DEs can be estimated with acceptable accuracy according to the elastic analysis. In addition, shear capacity for DEB enhanced with increasing prestress force and reducing shear slenderness ratio.

2- For DEs reinforced with post-tensioning rods, the load of shear crack was the maximum capacity of the beam. Whereas for DEs included mild steel, the web reinforcement carried the loads beyond shear cracking.

3- In beams with small amount of reinforcement and high prestressing load, flexural cracks initiated first and spread in a high rate causing yielding of the tension steel and overall failure. For beam with high steel area and low prestressing load, shear cracks will develop first and failure may be caused by crushing of concrete at the DE.

In 1986, Mattock and Theryo tested 12- full scale PC T-beams subjected to a combined vertical and horizontal loads with several reinforcement arrangements for the DEs. Inclined loading setup was used to introduce a horizontal force of 20% of the support reaction. Some of the finding of this study were:

1- The horizontal extensions of the inverted U-looped HR need not to be less than 1.7 times the steel development length(l_d). Furthermore, better results were obtained, in terms of crack development control, by using inclined alignment of hanger bars in comparison with the vertical hanger stirrups.

2- Making half of the prestressing strands of the drapped profile through the dapped end contributes in significant improvement in crack restraining.

In 1989, So K. M. P. tested two PC T- beams with DEs under combined vertical and horizontal loads. Each specimen represented one-half of a standard double-T- section and the four dapped ends have different DE detailing of reinforcement. Noticeable enhancement in ductility of the nib end when using two layers of wire fabric, which used as a shear reinforcement in the hanger zone. Also, it was addressed that introducing some inclination in the bottom of a DE yield more adequate than the rectangular one.

This is due to the small stress intensity compared with the rectangular nib end.

Nanni and Huang (2002), proposed a new updated alternative arrangements of HR called as" a Z-bar" that satisfied design requirements of DEs based on the PCI Design approach. Two PC double T- beams with DEs specimens were tested. Three prestressing strands located below the nib and four strands located above the nib. It was reported that:
1- The anchorage of the steel reinforcement is necessary and the bars fillet welded to a two- piece end plate appears to perform satisfactorily".
2- The failure pattern for both beams was of the "shear-flexure" type at the full depth of the beam, due to the small value of (a/d) which was not more than 1.0.

In 2013, Brinkley examined the efficiency of various dapped-end reinforcement arrangements in four full-scale prestressed single tee beams PC T-beams. The loading set-up included the application of vertical and horizontal forces in DE, with eight HR configurations. The study was focused on the detailing of the bend of HR, the use of a heavy U- WWF (Welded Wire Fabric), extending prestressing strands through the nib end. Several conclusions have been reported:
1- The first initiation of cracking occurred at the re-entrant corner and propagated diagonally (at about 45 degrees) towards the top compression fibers of the beam. Meanwhile, the angle of cracks became less steep with the progress in loading and at failure, the angle become close to 30 degrees. Furthermore, it is found that adopting inclined bars as HR significantly restricted widening, rate of propagation of cracks, and contribute in capacity enhancement by about 30% relative to the conventional HR detailing.
2- All the tested DEBs cracked at service loading stage and failed by "diagonal-shear" within the full-depth section type that developed independently from initial cracks emanated from the re-entrant corner.
3- Passing one strand within the extended end resulted in good enhancement in cracking performance if compared with specimens with stands passed below the nib end.

In 2014, Martínez and Meli tested four-PC DEB models at a 1:3.6 scale under vertical loads. Study focused on the efficiency of adopting diagonal reinforcement in lieu vertical stirrups within the DEs with or without some prestressing in the nib, to control the diagonal crack emanating at the RE corner. The important findings were:
1- Prestressing of the dapped end is an adequate method to reduce the re-entrant cracks and can be used in practices in case where very strict measure of corrosions control is necessary.
2-The PCI method predictions were conservatives to determine the loading capacity for the critical failures modes. The STM model that proposed by Mattock has the best method to evaluate the load capacity. Moreover, failure of all the specimens was governed by the diagonal tension at the re-entrant corners.

In 2014, Al- Khazraji studied experimentally and numerically (by ANSYS 9) the response of PC DEBs under static and impact loads. Seventeen SCC (Self-Consolidating Concrete) specimens have been tested. The study was focused on several variables including; the prestress effect, strand profile, and the grade of concrete. Design of specimens was based on PCI method. Results showed an enhancement in the static failure load by (35%) and (97%) in case of straight and draped strand when using prestressing stress of (0.62*ultimate strength of prestressing steel). Furthermore, an enhancement in failure load by 15% and 10 % for the two strand profiles when the grade of concrete increased from (40 MPa) to (60 MPa) respectively was obtained and that the pre-stress reduced deflection by (23%) with negligible difference between the strand profiles.

In 2015, Botros, then in 2017, Botros et al., conducted an extensive studies considering several variable including prestressing force, grade of concrete, detailing of HR, depth of the nib end, length of horizontal extension of HR. Many full scale PC T-beams have been tested with including six different HR configurations. Four of vertical orientation, which were L-, CZ, C- and Z-shapes. The two others were inclined L-shape and Welded Wire Reinforcement (WWR). It was concluded that:
1- The first crack Initiated at the re-entrant corner of most DEs. Cracking at service load could be restrained by passing the strand within the nib end, Moreover, It was reported that the suggested configurations can be adopted in the practice, and PCI method yielded a conservative solution for capacity by (35% to 74%). Furthermore, the inclined L configuration resulted in a good improvement

in strength and serviceability compared with other arrangements because the HR is nearly perpendicular to the direction of the diagonal crack.
2- Adopting higher concrete strengths contribute in improving the strength of the DE region but not directly proportioned with the square root of the concrete strength. Furthermore, it was concluded that increasing amount of HR increases strength and ductility after cracking; however, it does not prevent the sudden nature of failure.

8. Studies on Strengthening of RC Dapped End Beams

Two main techniques were considered by researchers for strengthening DEBs, which are; utilizing Fiber Reinforced Polymers (FRP) and external prestressing. In 2000, Huang, et al., used externally bonded (EB) FRP composites in strengthening PC double Tee DEBs with/without end anchors. Two different strengthening systems were investigated using orthogonal wrapping technique with end-anchors to insure fiber rupture in lieu delamination. It was found that:
1-Using EB FRP arrangements enhanced ductility and strength. Specimens failed by "shear-flexure" rather than "shear at the re-entrant corner". Specimens which were designed by PCI approach yielded very conservative results for the conventional DEBs.
2-The end U-anchor configuration improved the failure capacity of a DEBs, and attain fiber rupture in lieu of delamination of the FRP sheets

In 2001, Tan investigated experimentally several strengthening techniques with different FRP composites on DEs. Upgrading was done with carbon FRP plates (CP), carbon fiber sheets (CS), or glass fiber fabrics (GS). The FRP system was fixed with several anchorage schemes on both ends. Beyond failure, the weaker side was upgraded with additional reinforcement and the beam was tested again to determine the activity of the FRP system at the other end. It was addressed that:
1-An enhancement in failure load of 43%, 75% and 80% for the CP, CS and GS schemes respectively was obtained and STM model was accurate enough to estimate the shear strength of the DEs.
2-Using an anchor bolt in the CP system yielded enhancement in capacity by16 %. Moreover, using of bonded transverse fabric resulted in further increase in strength of 33% and 41% for the CS and GS systems respectively.

In 2005, Taher studied efficiency of strengthening DEs including various defects in steel detailing of the dapped ends. 52 DEs were strengthened with different techniques including bonding of steel angle at the recess corner, unbounded inclined steel bolt anchoring in pre-drilled hole, external steel plate jacketing, exterior carbon fiber wrapping within the beam stem, exterior CFRP stripping and combination of carbon fiber wrapping and strapping. A comparison was conducted with the STM model. It was concluded that the mode of failure was mainly influenced by the introduced defect in the recess zone (extended end zone). EB FRP strengthening systems constituted a viable solution for repairing applications and to provide easy fabrication, horizontal carbon fiber wrapping with inclined CFRP stripping is recommended.

In 2006 Huang and Nanni, studied strengthening of DEBs with two techniques, internal steel reinforcement and external bonding (CFRP) laminates. Three full scale beams with dapped ends have been considered with five tests with several reinforcement arrangements including; two 1-layer CFRP strengthened DEs (one with U-anchor) and one steel reinforced DE with loading at 2.4 m from far support, and one 2-layers FRP strengthened DE and one steel reinforced one at the loading distance of 1.5 m. Two failure scenarios were observed: delamination and rupture failure of CFRP.
Some of the findings of this study were:
1- Using the U-anchor systems to the external debonding (CFRP) laminates increased the ultimate capacity and ensured fiber rupture instead of debonding of the FRP sheets
2- ACI 440.2R-02 predictions for the shear strength of FRP strengthened DEs are found to be highly conservative.

In 2012, Nagy-Gyorgy et al., discussed the adequacy of upgrading RC DEBs using EB CFRPs. Due to errors during assembly stage of several PPC (precast prestressed concrete) beams, some diagonal cracks propagated in the DEs.Thus, on-site retrofitting was done using CFRP plates to control further crack widening. In the empirical phase of the study, four similar specimens were considered: one control specimen, two retrofitted with HS CFRP plates, and one with high-modulus CFRP sheets with layouts 45/90, 0/90 and 0/45/90 respectively. The specimens upgraded with plates had slightly higher failure load than the control specimen, but damaged by debonding, while the specimens strengthened with sheets exhibited no enhancement in capacity and failed by the rupture of fibers. Nonlinear FEA of the specimens demonstrated that debonding is more likely to occur at the inner end of DEs and the capacity have been enhanced by 20% if the plates had been fixed with end anchors.

In 2013, Dăescu extended the previous study (Nagy-Gyorgy et al) by assessing the activity of upgrading RC DEBs using CFRP sheets using non-linear finite element (FEA). The same strengthening configurations were investigated in terms of ultimate capacities and the failure modes. Same parameters have been considered. It was observed that:

1- The strengthening systems with "vertical fibers have been provided marginal increasing for some specimens of (2%)", for others have provided increasing capacities about (6.9% to 37.7%).

2- The optimum solution for strengthening is the fabrics with fibers at horizontal and at (45^0) with increment in capacity of (37.7%).

Sas et al., 2014 used nonlinear FEA modeling to investigate the most appropriate arrangement of (CFRP) composites to upgrade RC DEBs. 24 external bonding and NSM (Near Surface Mounted) reinforcement schemes have been investigated. Parameters that considered included the characteristics of the CFRP, the strengthening scheme and the inclination of the sheets with respect to the longitudinal axis. Two failure modes were observed: rupture and debonding of the CFRP. The results demonstrated that HS (High Strength) NSM FRPs can significantly improve the capacity of DEBs and yielding strains in reinforcement can be effectively reduced by using high modulus CFRPs.

In 2014, Atta and Al-Shafeiy studied experimentally the applications of (FRP) sheets, (FRP) laminates and external prestressing steels under torsional moments on behavior of the DEs. The main conclusions were: using of external prestressing increase cracking loads, failure loads and ductility compared to other techniques. Also, strengthening by CFRP laminates following the external bonding technique produced a severe drop in rotation capacity within the final stages of loading due to debonding failure between concrete surface and CFRP sheets.

In 2016, Atta and Taman tested seven RC DEB that strengthened with different shapes of external prestressing technique directions: horizontal, vertical and inclined, as shown in Figure.6. Results showed that:

1- Vertical external prestressing techniques are a very effective to increase the capacity of DEB up to (82%).

2- Strengthening of dapped-end beams using vertical external prestressing techniques is the best technique from a ductility point of view. Horizontal external prestressing techniques lead to compression failures at the compression zone. The use of two layers of external vertical prestressing strengthening techniques at the undapped zone leads to sudden brittle failure at the extended end.

3- The best technique for strengthening dapped-end beams from both the ultimate load and ductility points of view is using two external prestressing layers, one at the dapped end and the other at the undapped end with two upper separate rigid plates.

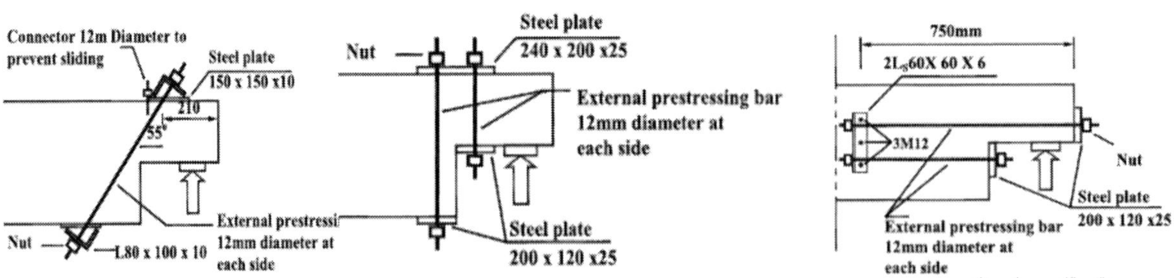

Figure. 6 Configuration of strengthening techniques of Atta and Taman, 2016

Shakir and Abd (2019) conducted an experimental investigation related the response of self-consolidating HSC DEBs retrofitted with CFRP fabrics. A total of 14-specimens have been tested, amongst 12-specimns included intentionally various defects in either nib or HR. Two values of (a/d) ratio namely, 1.5 and 1.0 are considered. It was concluded that using half of the design hanger reinforcements, results in a reduction of capacity by (13%) for both values of (a/d). Moreover, it was noticed that fixing the CFRP sheets with 45 degrees yielded the best performance in comparison with other arrangements.

Shakir and Abd (2020) Investigated experimental the extent of efficiency of using the NSM steel bars in strengthening the SCC RC dapped end regions under short-term static loading with (a/dd) values of 1.0 and 1.5. It was concluded that upgrading the nib region with horizontally aligned NSM steel bars resulted in increasing the load capacity by about 28.7% and 24% for a/d =1 and 1.5 respectively. Furthermore, the enhancement in failure load for strengthened hanger regions vertical and inclined (with 45 degrees) configurations for a/d =1 was 21.4% and 14.2% respectively.

It can be seen that there is no study considered repairing and rehabilitation of damaged dapped ends, behavior of dapped ends under elevated temperature, seismic, cyclic and dynamic loading. Furthermore, incorporating rolled steel sections within the nib end and to be extended to the full depth part. More investigations are needed to study the behavior of DE beams having opening that accommodate cross beam or dapped end beams.

9. Conclusions

Based on the extensive review on dapped end, the following conclusions can be drawn:

1-It was mostly reported that shear friction (SF) method yield less accurate results if compared with the STM models because does not adequately characterize the effect of (a/d) ratio, horizontal load effects, hanger reinforcement. However, STM models of Euro Code 2 and BS 8110 were less accurate in comparison with PCI method and ACI-318-08(STM models).

2-It was recommended the zone of hanger stirrups to be half to full depth (not more than nib depth), cover of the first hanger stirrup of 40 mm, Moreover, it recommended that (nib depth/ depth of beam) not to be less than 0.45.

3- Increasing amount of HR increases strength and ductility after cracking; however, it does not prevent the sudden nature of failure. Furthermore, solutions obtained by ACI318 models seemed to be conservative i.e., the reduction of capacity is less influenced by HR reduction of no more than 50%. However, significant enhancement in capacity can be obtained if the nib main reinforcement made inclined to cross the diagonal cracking from the re-entrant corner.

4-The first cracks usually initiate from the re-entrant corner at a load level of 20-33%, and propagated towards the compression face with an angle (40-60) degrees from the horizontal. Then, the failure crack follows mostly angle of 30 degrees. Thus, the inclination of the strengthening sheets to be orthogonal with the expected path of cracking.

5- The closed vertical stirrups are less effective than of inclined orientation and bent bar configuration. Using inclined hanger or steel bars result in better serviceability expressed by minimizing crack width and crack development rate. An improvement in capacity by about 30% was recorded compared with vertical hanger detail.

6- The concrete strength is not a dominant factor to control the behavior of DEBs as most of the DE zone is under tension. However, incorporating some fibers with suitable aspect ratio may improve tensile strength. but from the economical view, it is recommended to use fibers within DE only.

7- The size, shape and location of such openings affect the response of dapped ends. The optimum location of opening was within regions of high positive bending moment. However, introducing non-preplanned may result in unexpected modes of failure openings.

8-Different schemes used to strengthen the dapped end region, (L-, CZ, C- , Z-, inclined L-shapes and welded Wire Reinforcement (WWR)) However, the inclined L form yielded good performance in comparison to other forms.

9- Prestressing the DE zone with inclined bars of drapped profile may result in significant development in the performance, enhancing shear capacity and restrict propagation (and widening) of diagonal shear cracks at working load stage.

10- Using EB FRP configurations with carbon FRP plates (or sheets) or glass fiber fabrics for strengthening the DE zone improved both ductility and strength and shifted the failure to be as "shear-flexure" rather than "shear at the reentrant corner". Moreover, using the U-anchor systems with the externally bonded laminates improved the strength and resulted in fiber rupture in lieu of deboning of the laminates.
11- The optimum orientation sheet (or plates) used for strengthening the dapped end zone (DE) is 45 degrees from the horizontal.
12-The FEA method can be adopted adequately to study the effect of different variables that may influence the behavior of conventional and prestressed RC dapped ends. Also, It can be utilized to determine the optimum material, quantity and configuration of strengthening (or rehabilitation) need to be used. Some studies showed that the effects of internal defects could be studied effectively.

13- For most specimens, the predicted load carrying capacity by STM was governed by the yielding of the reinforcing bars at the range of (70-95) % for inadequate anchorage lengths, rather than exceeding the capacity of the concrete struts. This was the case even for the low concrete strength specimens. There are three failure modes in DEBs with a /d < 1: flexure failure, diagonal compression failure and tensile failure initiated by the yielding of hanger bars. The failures modes of DEBs with a /d > 1 is dominated by flexure failure

14- Some specimens exhibited ductile behavior in which the steel yielded before the concrete failed in compression. Some studies addressed four failure modes were observed which are diagonal tension cracks within the nib or the full-depth of the beam. Flexure-shear cracking in the full-depth precipitated by longitudinal splitting of the web and strand bond failure, crushing of the diagonal strut near the bottom corner of the stem.

References

[1] Mattock A.H. and Chan T. C. (1979). Design and Behavior of Dapped-End Beams. PCI Journal/November-December.28-45

[2] Torres, F. M. (1998). Theoretic-experimental analysis of reinforced concrete corbels. Dissertation (Master Science in Structural Engineering) - Engineering School of São Carlos, São Paulo University, São Carlos.

[3] Mattock A.H. (1976) Design Proposals for Reinforced Concrete Corbels. PCI J. 21(3), 18-42.

[4] PCI Industry Handbook Committee. (2004). PCI Design Handbook: Precast and Prestressed Concrete. (6th ed). Precast/Prestressed Concrete Institute, Chicago, Illinois, USA.

[5] ACI Committee 318M- 2019. Building Code Requirements for Structural Concrete (ACI 318M-19) and Commentary (ACI 318M R-19). ACI, Farmington Hills, MI, USA.

[6] Liem, S. K. (1983). Maximum shear strength of dapped-end or corbel. M.Sc. Thesis, Concordia university, Montreal, Quebec, Canada,127p

[7] Barton D.L., Anderson R.B., Bouadi, A. Jirsa J.O. and Breen J.E. (1991). An Investigation of Strut-And-Tie Models for Dapped Beam Details. Research Report Number 1127-1,187p

[8] G. Menon, Furlong R. W. (1977) Design of Reinforcement for Notched Ends of Prestressed Concrete Girders. Research Report Number 196-1F, 153p

[9] Yang K.-H., Ashour A.F, And. Lee J.-K, (2011). Shear Strength of Reinforced Concrete Dapped-End Beams using Mechanism Analysis. The University of Bradford Institutional Repository, 1-38.

[10] Lu W.-Y., Lin I.-J., Hwang S.-J. & Lin Y.-H. (2003). Shear strength of high-strength concrete dapped-end Beams. Journal of the Chinese Institute of Engineers,26(5), 671-680.

[11] Lin, I.-J., et al., (2003). Shear Strength of Reinforced Concrete Dapped-End Beams. Structural Engineering and Mechanics, 16(3), 275-294.

[12] Lu, W. Y., I. J. Lin and H. W. Yu (2012). Behaviour of reinforced concrete dapped-end beams. Magazine of Concrete Research 64(9), 793-805.

[13] Aswin, M. Mohammed B. S., Liew M. S., and Syed Z. I., (2015). Shear Failure of RC Dapped-End Beams. Hindawi Publishing Corporation Advances in Materials Science and Engineering Vol. 2015, 1-10.

[14] RILEM TC 162-TDF, (2000). Test and design methods for steel fibre reinforced concrete. Materials and Structures, 36(262),560–567.

[15] Mohammed B. S., Aswin M., Liew M. S. and Zawawi N. (2019). Structural Performance of RC and R-ECC Dapped-End Beams Based on the Role of Hanger or Diagonal Reinforcements Combined by ECC. Int J. Concr Struct Mater 13:44

[16] Mohamed R N, Elliott K S, (2008). Shear Strength of Short Recess Precast Dapped End Beams Made of Steel Fibre Self-Compacting Concrete. 33rd Conference on Our World in Concrete & Structures: 25 – 27, August, Singapore.

[17] Wang Q., Guo Z., Hoogenboom, P. C.J., (2005). Experimental investigation on the shear capacity of RC dapped end beams and design recommendations. Structural Engineering and Mechanics.21(2), 221-235.

[18] Ahmad, S., Elahi A., Hafeez, J., Fawad, M. and Ahsan Z., (2013)," Evaluation of the Shear Strength of Dapped Ended Beam Life Science Journal;10(3),1038-1044.

[19] Aswin, M.,Syed Z. I., Wee T., Liew M.S.,(2014). Prediction of Failure Loads of RC Dapped-End Beams. Applied Mechanics and Materials, Vol. 567, 463-468.

[20] Lu W.-Y., Chen T.-C., and Lin, I.-J. (2015). Shear Strength of Reinforced Concrete Dapped-End Beams with Shear Span-To-Depth Ratios Larger Than Unity. Journal of Marine Science and Technology, 23(4), 431-442.

[21] Peng T. (2009). Influence of Detailing On Response of Dapped End Beams. Mcgill University,Montréal, Canada.103p

[22] Desnerck, P., Lees, J.M. and Morley C. T., (2016). Impact of the reinforcement layout on the load capacity of reinforced concrete half-joints. Engineering Structures, 127, 227–239.

[23] Desnerck, P., Lees, J.M. and Morley C. T., (2017). The effect of local reinforcing bar reductions and anchorage zone cracking on the load capacity of RC half-joints. Engineering Structures 152(1), 865-877.

[24] Desnerck, P., Lees, J.M. and Morley C. T., (2018). Strut-and-tie models for deteriorated reinforced concrete half-joints. Engineering Structures, 161, 41–54.

[25] Mata-Falcón J., Pallarés L. and Miguel P. F. (2019). Proposal and experimental validation of simplified strut-and-tie models on dapped-end beams. Engineering Structures 183,594–609.

[26] Shakir Q. M., Al-Tameemi H. A., Kamonna H.H. H., (2020). Effect of Opening Characteristics on the Performance of High Strength Rc Dapped-End Beams. Journal of Engineering Science and Technology,15(3).693-708

[27] Werner, M. P., and Dilger, W. H., (1973). Shear Design of Prestressed Concrete Stepped Beams. PCI Journal,18(4), 33-49.

[28] Hamoudi, A A and Phang, MKS (1974),"Shear Strength of Post-Tensioned Daps in Prestressed Concrete Beams", American Society of Civil Engineers, 100(ST12).

[29] Hamoudi, A.A., Phang, M.K.S. and Bierweiler, R.A., (1975). Diagonal Shear in Prestressed Concrete Dapped Beams. ACI Journal ,72(7),347-350, cited by Peng,2009

[30] Mattock, A. H. and Theryo, T. S., 1986. Strength of Precast Prestressed Concrete Members with Dapped Ends. PCI Journal, 31(5), 58-75, cited by Peng,2009.

[31] So, K. M. P., (1989). The Behaviour of Thin Stemmed Precast Prestressed Concrete Members with Dapped Ends. M.Sc. thesis, McGill University, Montreal, Canada,155p.

[32] Nanni A.and Huang, P.-C., (2002). Validation of an Alternative Reinforcing Detail for the Dapped Ends of Double Prestressed Tees. PCI Journal, January-February 2002, 38-49

[33] Brinkley F., M., (2013). Behavior of Prestressed, Precast Concrete Thin-Stemmed Members with Dapped Ends. M.Sc. thesis, North Carolina State University, USA. 224p

[34] Martínez J.Y. M. and Meli R., (2014). Experimental study on the structural behavior of concrete dapped-end beams. Engineering Structures, 75,152–163.

[35] Al- Khazraji S.D.M., (2014). Response of Dapped-End Prestressed Concrete Girders to Static and Impact Loads. PhD Thesis, University of Baghdad, Iraq, 219p

[36] "ANSYS Structural Analysis Guide", ANSYS Release 9.0, ANSYS, Inc. South pointe 275 Technology Drive Canonsburg, PA 15317.

[37] Botros, A. W.H. (2015). Behavior and Design of Dapped Ends of Prestressed Concrete Thin-Stemmed Members. Ph. D. Thesis, Raleigh, North Carolina, USA ,397p.

[38] Botros A. W, Klein G. J., Lucier G. W., Rizkalla S. H., and Zia P., (2017). Dapped ends of prestressed concrete thin-stemmed members, Part 1, experimental testing and behavior. PCI Journal, March–April.

[39] Huang, P.C, J.J Myers and A. Nanni, (2000). Dapped-End Strengthening in Precast Prestressed Concrete Double Tee Beams with FRP Composites. Proc., 3rd Inter. Conf. on Advanced Composite Materials in Bridges and Structures, Ottawa, Canada, 15-18 Aug., 545-552.

[40] Tan K. H. (2001). Shear Strengthening of Dapped Beams Using FRP Systems. FRPRCS-5: Vol. 1: Proceedings of the fifth international conference on Fibre-reinforced plastics for reinforced concrete structures, Cambridge, UK, 16–18 July 2001

[41] Taher S. F., (2005). Strengthening of critically designed girders with dapped ends. Structures & Buildings 158(2), 141-152.

[42] Huang, P.C, and A. Nanni, (2006). Dapped-End Strengthening of Full-Scale Prestressed Double Tee Beams with FRP Composites. Advances in Structural Engineering, 9(2), 293-308.

[43] Nagy-György, G. Sas, A.C. Da˘escu , J.A.O. Barros , V. Stoian ,(2012). Experimental and numerical assessment of the effectiveness of FRP-based strengthening configurations for dapped-end RC beams. Engineering Structures, 44, 291-303.

[44] A.C. Dăescu, T. Nagy-György, B.G. Sas, J.A.O. Barros, C. Popescu, (2013). Numerical Assessment of Dapped Beam Ends Retrofitted with FRP Composites. 11th International Symposium on Fiber Reinforced Polymer for Reinforced Concrete Structures (FRPRCS-11).

[45] Gabriel Sas, Cosmin Dăescu, Cosmin Popescu, Támas Nagy-György (2014). Numerical optimization of strengthening disturbed regions of dapped-end beams using NSM and EBR CFRP. Composites: Part B.

[46] Atta, A.M. and Al-shafiey, T. F. (2014). Strengthening of RC dapped end beams under torsional moments. Magazine of concrete research, 66(20), 1065-1072.

[47] Atta A. and Taman M., (2016). Innovative method for strengthening dapped-end beams using an external prestressing technique. Materials and Structures, 49(8), 3005–3019.

[48] Shakir Q. M. and Abd B. B. (2019). Strengthening of Self-Consolidating High Strength RC Dapped Ends with CFRP Fabrics. Journal of materials and Engineering Structures, 6(3), 359–374.

[49] Shakir Q. M. and Alliwe R. (2020). Upgrading of Deficient Disturbed Regions in Precast RC beams with Near Surface Mounted (NSM) Steel Bars. Journal of materials and Engineering Structures, 7(2).

Effect of Chemical Activators on the Compressive Strength of Cement-Slag Mortar

Hersh Faqe

College of Engineering, University of Halabja, Kurdistan Region, Iraq
hersh.faqe@uoh.edu.iq Tel: +9647501573488

Abstract. The reaction of ground granulated blast furnace slag (GGBFS) with water is slower compared to the hydration of ordinary Portland cement. This led researchers to concentrate on increasing the hydration rate of the GGBFS. Previous studies were done by (SCA) showed that GGBFS can be used with activators or a certain amount of Portland cement. Also, they found that alkali hydroxides and GGBFS can be reacted alone without any calcium hydroxide obtained from Portland cement hydration to obtain a strong cement paste in applications. This experimental study conducted by the TS EN 196-1 standard, with the ratio (79:21)% slag to cement and 0%, 1%, 2%, 3% and 4% levels of activators ($CaCO_3$, Ca (OH) 2, Na_2SO_4, K_2SO_4 and $CaSO_4.2H_2O$); results indicates that with above ratio slag can be used with activator of (Na_2SO_4, K_2SO_4 and $CaSO_4.2H_2O$) ,greater compressive strength of 5-6 MPa was obtained at 2 and 7-day curing ages when K_2SO_4 was used as an activator at the dosages of 3% and 4%, while the activator ($CaSO_4.2H_2O$) can improve all ages, and there was not a clear difference among control mixture without activator and slag-mortars activated by $CaCO_3$ and Ca (OH).

Keywords: *Slag-cement Mortar, Compressive Strength, Chemical Activation*

1. Introduction

The most extensively used construction material all over the world is concrete which is mainly composed of Portland cement, several aggregate types, water, and admixtures (if required). Concrete production for each one of the human beings living on Earth accounts for approximately one ton and with this much of consumption all over the world, concrete is the second most used material after water.as a result of the production of individual ingredients of concrete, environmental concerns arise. Production of Portland cement can be given as a striking example for the environmental pollution so that each one ton of cement production is responsible for the release of 1 ton of carbon dioxide to the atmosphere and cement production accounts approximately for the 7% of the total greenhouse emissions all over the world.to consider regulations and limitations on CO2 emissions. In the view of global sustainable development, the utilization of pozzolanic materials as a partial replacement of cement stands as one plausible way to reduce the environmental impact caused due to cement manufacture. Along with the environmental effects, the usage of pozzolans could substantially reduce the final cost of concrete mixtures since these materials are quite cheaper in comparison to Portland cement. According to ACI 116R-90 (1994), pozzolans is "a siliceous or siliceous and aluminous material, which in itself possesses little or no cementitious value but will, in finely divided form and in the presence of moisture, chemically reacts with calcium hydroxide at ordinary temperatures to form compounds possessing cementitious properties." Mineral admixtures that are, or both cementitious and pozzolanic can be used as a partial replacement for Portland cement. Some of the most used materials are ground granulated blast furnace slag (slag hereafter), fly ash, condensed silica fume. When properly used as a portion of the cementitious material, these pozzolanic admixtures can improve the properties of the fresh and hardened concrete. Although several types of supplementary cementitious materials exist, the focus of this review will be mainly on slag. Throughout the world, loads of ground granulated blast furnace slag is manufactured from the iron-steel industry every year. Utilization of this by-product not only has a great impact on the reductions of environmental issues but also contributions to the economy. It has been reported by many researchers that utilization of such by-products in mortars and concrete mixtures has marked influences on durability characteristics which was found to be highly important for infrastructure types built-in or near aggressive environments. Overall, it can be stated that beneficial effects of slag on mechanical and durability properties have started to be widely recognized and this leaded the extensive usage of the material in Portland cement mortar and concrete mixtures by many authorities. Utilization of slag in concrete mixtures is advantageous since the material shows superior cementitious characteristics over ordinary Portland cement and it has both economic and environmental benefits. However, there is a limitation on the usage of slag in concrete mixtures due to the problems arising related to low early-age compressive strength (Bougara et al. 2009). It is generally being accepted that while increased fineness of slag increases the compressive strength of conventional concrete mixtures, increments in slag replacement rates decrease the compressive strength. However, the fineness of slag is not that influential at the later ages when compared to an early age. Major factors that are in relation to the development of early-age strength can be water-binder ratio, sand-binder ratio, the number of pozzolans used in mixtures, environmental conditions, size of the testing specimen and so on. As previously mentioned, ground granulated blast furnace slag in great quantities is being produced each year as a residue from iron-steel factories. Although slag is utilized as aggregate in concrete mixtures and as a filling material in the roadbases, much of these materials are still being environmental burden causing further pollution. Moreover, when slag is used as an aggregate and/or filling material, pozzolanic behavior and cementitious characteristics are being ruled out and an advantage in terms of these properties is not taken. In order to further reduce the environmental burden of by-products and provide assistance in terms of economy, slag was replaced with Portland cement in mortar mixtures in the present research. While doing this, special attention was paid so as not to have decreased compressive strength results from the mortar specimens. A detailed literature review showed that there are limited studies related to the chemical activation of slag. Therefore, in the research concentration was mainly placed upon the chemical activation of slag.an experimental study is undertaken in which five different types of chemical additives were utilized inside of the mortar mixtures. Chemical additives were added to the mixtures at the levels of 1%, 2%, 3%, and 4% of the total binder amount (Portland cement+slag) and influence of chemical addition rates were also investigated in this way. Along with the production of mixtures with chemical additives, control mixtures incorporating no chemical additives were also produced for comparison purposes. A small summary of the detailed experimental program was presented in Figure 1. Overall, the main objective of the present study is to evaluate the effect of type and dose of some chemical activators on cement-slag mortar compressive strength at the highest possible level of slags.

IOP Conf. Series: Materials Science and Engineering **978** (2020) 012004 doi:10.1088/1757-899X/978/1/012004

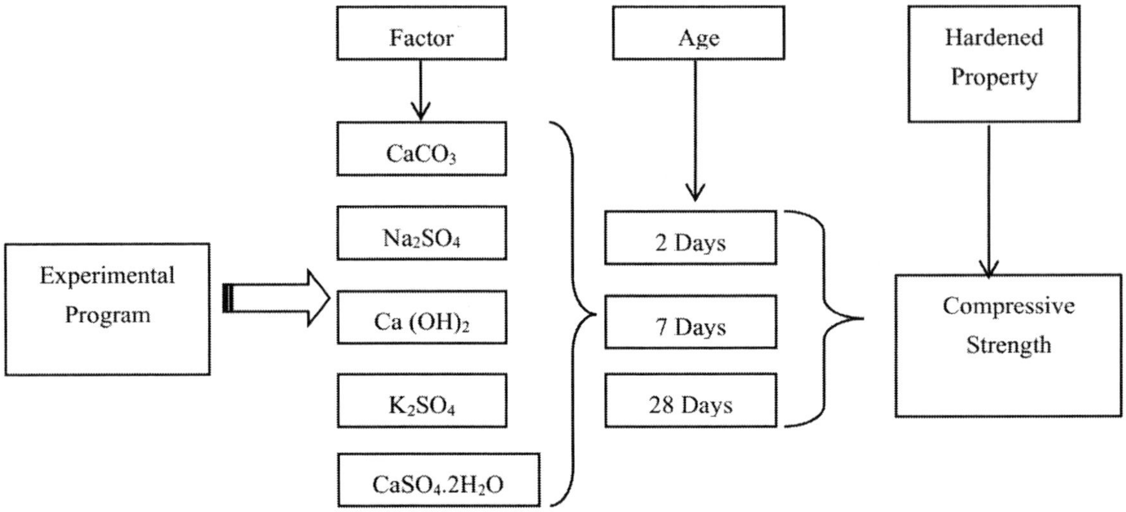

Figure 1. Summary of the experimental program

2-Materials

2-1.Chemical Activation:

Ground granulated blast furnace slag is described as a cementitious material having high capability to show enhanced compressive strength results when alkali activators are used. Throughout this study, slag was tried to be activated by utilizing chemical activators having the formula of $CaCO_3$, $Ca(OH)_2$, Na_2SO_4, K_2SO_4 and $CaSO_4.2H_2O$. The dosage of chemical activators was set to be 1%, 2%, 3% and 4%, by mass of total binders (cement+slag). Addition of the chemical activators in the mixtures was made by first dissolving the chemicals in the mixing water.

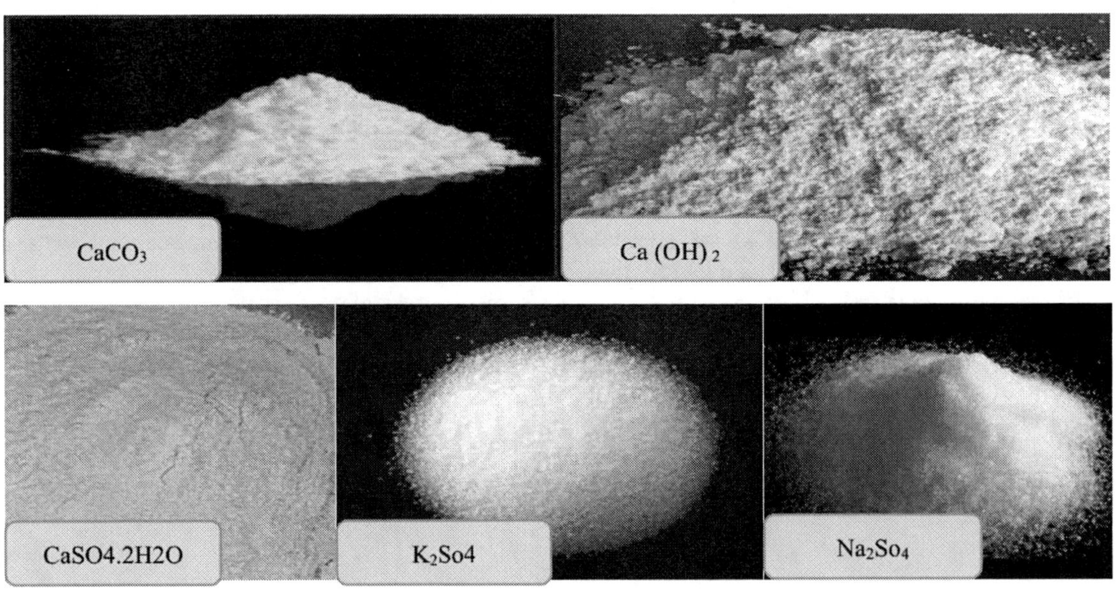

Figure2. chemical activator

2.2. Portland cement:

During the production of mortar specimens CEM I 42.5 N normal Portland cement satisfying the requirements stated by TS EN 197-1 (2012) was used. Portland cement was supplied from Karçimsa Cement Factory which is placed in Karabük, Turkey. The specific gravity of Portland cement was 3.06 and Blaine fineness was 325 m^2/kg. The chemical composition of Portland cement was presented in Table 1.

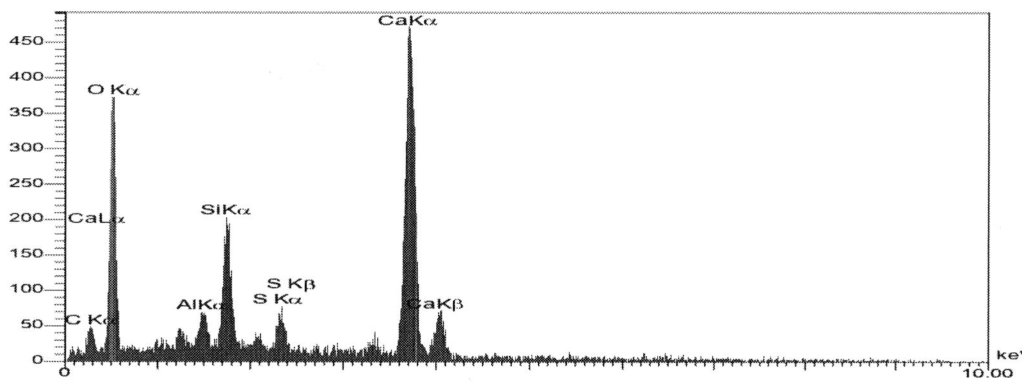

Figure3. X-ray spectroscopy (EDX) of slag

Table 1. The chemical composition of Portland cement and slag

Chemical Composition (%wt.)	Portland Cement	Slag
CaO (%)	61.43	34.16
SiO2 (%)	20.77	39.57
Al2O3 (%)	5.55	10.11
Fe2O3 (%)	3.35	0.74
MgO (%)	2.49	6.40
SO3 (%)	2.49	4.29
K2O (%)	0.77	1.21
Na2O (%)	0.19	0.28
Loss of Ignition (%)	2.20	0.90
SiO2+Al2O3+Fe2O3	29.37	50.42

2-3.Slag:

Slag used throughout the present study was the one obtained from Kardemir iron-steel factory which was then grinded to the desired fineness values. The chemical composition of the slag was given in Table 1. The specific gravity of the slag was 2.79 g/cm^3. The elemental structure of slag used throughout the study was identified by energy dispersive X-ray spectroscopy (EDX) Shown in figure3.

Morphological properties of slag were analyzed by using scanning electron microscopy, shown in Figure 4.

Figure 4. Particle morphology of slag determined by SEM

2-4.Water:

Water was with the requirements of TS EN 1008 (2003) standard.

2-5.Aggregate:

According to the TS EN 196-1 (2002) standard, CEN reference sand is used as a material, as shown in Table 2. Table 2. The sieve analysis of sand

Sieve Size (mm)	Cumulative Sieve remaining (% wt.)
2.0	0
1.6	7 ± 5
1.0	33 ± 5
0.5	67 ± 5
0.2	87 ± 5
0.08	99 ± 5

3-Mixing and Specimen Preparation:

The composition of Portland cements was explained in TS EN 196-1 (2002) in detail. According to the standard, minimum and maximum values for the slag usage ranges between the values the research used the ratio of (79% of slag with 21% of cement) in a total binder material as cementitious material, all of the mortar specimens were produced based on a constant fraction of 1binders (cementitious) x 3(CEN reference sand) x ½(water). Water to binder (cementitious) ratio was set to be 0.5. Which are Twenty-one different mixtures were prepared in total during the research. One of the mixtures was prepared as a control without any chemical activator addition and four mixtures for the five different types of chemical activators (CaCO3, Ca(OH)2, Na2SO4, K2SO4 and CaSO4.2H2O) with the different activator addition levels (1%, 2%, 3% and 4%) in percentage were used.

IOP Conf. Series: Materials Science and Engineering **978** (2020) 012004 doi:10.1088/1757-899X/978/1/012004

4-Test Procedure

Twenty-one different mixtures were prepared, for each mixture, nine (9) samples of 40x40x160 mm prism moulds were prepared, and merged in control temperature water each of three (3) samples was tested at the end of the (2,7,28) days with the Compressive strength testers were performed in accordance with ASTM C 39 (2003).

Figure 5. Compressive strength machine.

Figure 6 (a). Prism moulds.

Figure 6 (b). Automatic water machines.

Figure 7. Mixer.

5-RESULTS AND CONCLUSIONS:

Compressive strength test results of slag-mortars are summarized in charts:

1 - There was not a clear difference among control mixture without activator and slag-mortars activated by $CaCO_3$ and Ca $(OH)_2$.

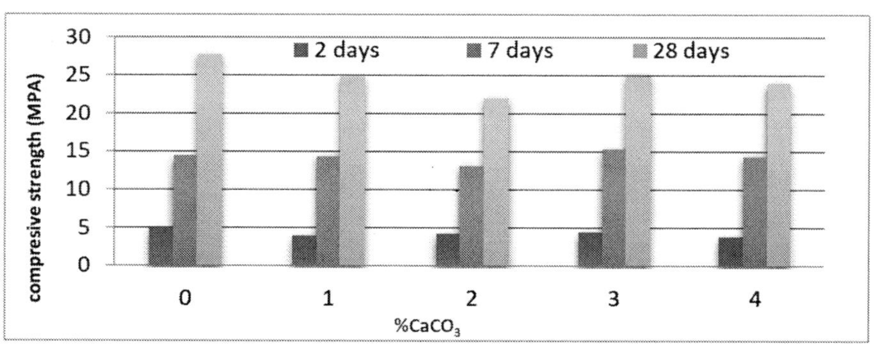

Figure 8. Compressive strength of different amount ($CaCO_3$)

2 - When the slag-mortars were activated by Na_2SO_4 compressive strength results were higher than control specimens by 1.0 and 2.0 MPa for 2 and 7-day of age, respectively. As a result of this, it was found that 7-day compressive strength greater than 16.0 MPa can be easily produced for 32.5 N type cement. However, 28-day compressive strength criteria (≥32.5 MPa) were not satisfied with all dosages of activator of Na_2SO_4.

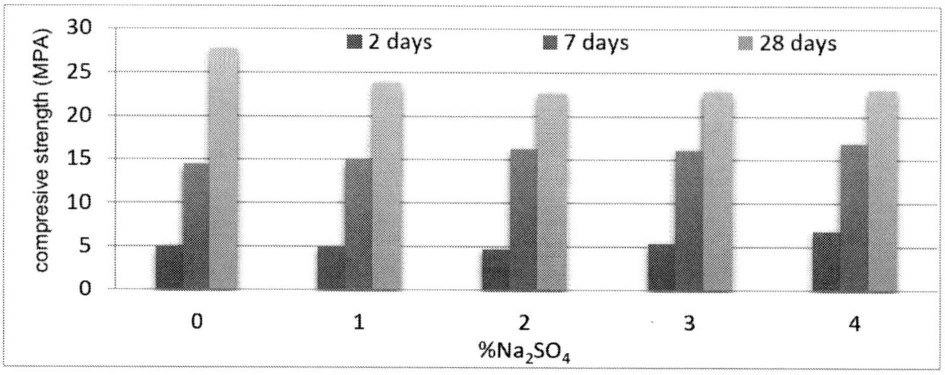

Figure 9. Compressive strength of different amount (Na_2SO4)

3 - In comparison to control mixture, greater compressive strength of 5.0-6.0 MPa was obtained after 2 and 7-day of curing when K_2SO_4 was used as an activator at the dosages of 3% and 4%. 28-day compressive strength values were the same as other activators that were under 32.5 Mpa.

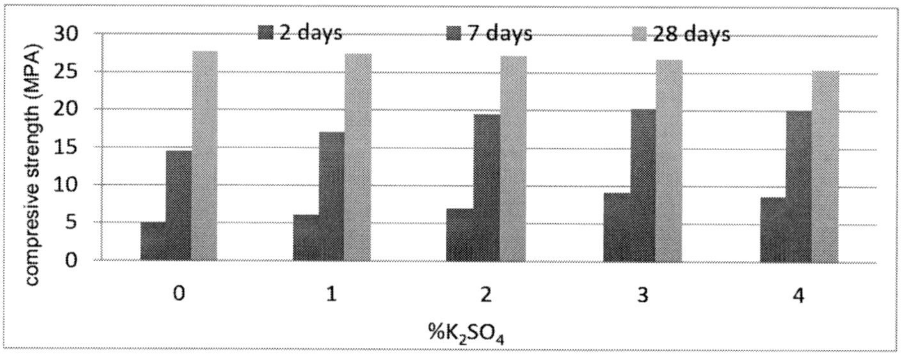

Figure 10. Compressive strength of different amount (K_2SO4)

4 - Compressive strength of 20.0 MPa at 7-day curing age was obtained when gypsum was used as an activator while it was at about 30.0 MPa at 28-day curing age.

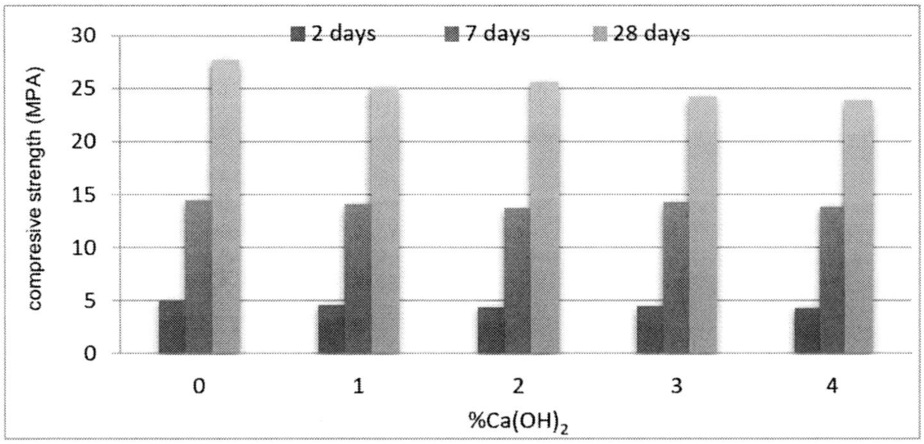

Figure 11. Compressive strength of different amount (Ca (OH) 2

5 - When $CaSO_4.2H_2O$ is used to activate the cementitious materials regardless of the age of testing and amount of activator used. Increments in the results were sometimes 27% more than that of obtained from control specimens with respect to age and dosage of activator used.

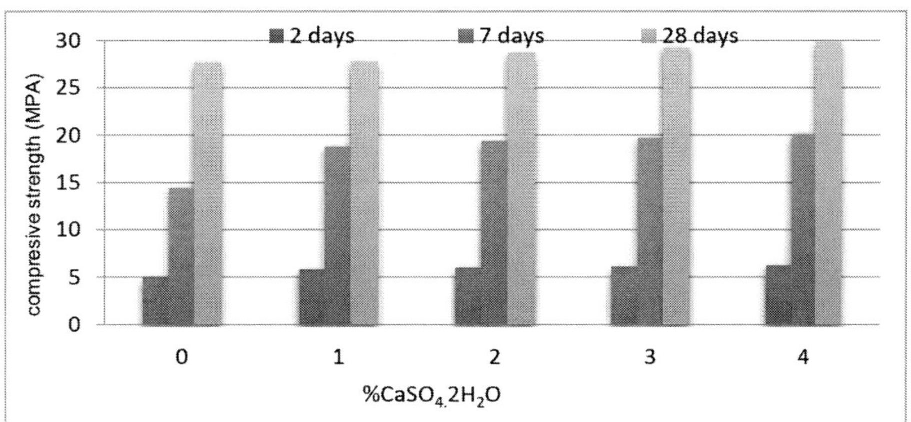

Figure 12. Compressive strength of different amount ($CaSO_4.2H_2O$)

6-Conclusion:

Now a day using ground granulated blast furnace slag (GGBFS) as cementitious material instead of Portland cement in concrete are very common because of environmental issue, and it have slower reaction with water that cause low early strength in mortar and concrete, more than one way to activation slag in this study we used five chemical activators with high range (79%) of replacement of slag, by the test result:

1 - All activator can not to be used as chemical activation of cement –slag mortar.

2 - The activators (Na_2SO_4) and (K_2SO_4) improve the compressive strength at early age when compared with the control, while the 28 days result is less that control.

3 – Activator ($CaSO_4.2H_2O$) can improve and can be used as activator for all age.

IOP Conf. Series: Materials Science and Engineering **978** (2020) 012004 doi:10.1088/1757-899X/978/1/012004

Reference:

[1] ACI Committee 116R. (1994). Cement and concrete terminology. ACI manual of concrete practice, American Concrete Institute.

[2] ACI Committee 233. (1989). Ground granulated blast furnace slag as a cementitious constituent in concrete. USA, American Concrete Institute.

[3] ACI Committee 318R (2002). Building code requirements for structural concrete and commentary. Farmington Hills, Michigan's, American Concrete Institute.

[4] ACI. (2000). Ground granulated blast-furnace slag as a cementitious constituent in concrete, ACI 233R-95. USA.

[5] ASTM C 989. (1994). Standard specification for ground granulated blast-furnace slag for use in concrete and mortars. Annual Book of ASTM Standards, USA.

[6] ASTM Standard C39. (2003) Standard test method for compressive strength of cylindrical concrete specimens. American society for testing and materials. West Conshohocken, PA, USA.

[7] Bougara, A., Lynsdale, C., and Ezziane, K. (2009). Activation of Algerian slag in mortars. Construction and Building Materials, 23(1), pp542-547.

[8] CNA. (2001). Mineral admixture for high performance of concrete. GB18736. China, Chinese national standard.

[9] Mehta, P. K. (1983). Mechanism of sulfate attack on Portland cement concrete- Another look. Cement and Concrete Research, 13 (3), pp401-406

[10] MIA. (2010). the use of ground granulated blast-furnace slag and fly ash in mortar, UK, Mortar Industry Association.

[11] Moranville-Regourd, M. (1998). Cements made from blast furnace slag, Lea's Chemistry of Cement and Concrete, 4th Edition, P. C. Hewlett Ed., and Arnold, pp. 663-674.

[12] Neville, A.M. (1996). Properties of Concrete. Fourth Edition, John Wiley & Sons, USA.

[13] SCA. (2010). Using slag cement in sustainable construction, USA, Suite 349 2516 Waukegan Road Glenview, IL 60025 847-977-6920, Slag Cement Association.

[14] Stark, J. and Ludwig, H. M. (1997). Influence of water quality on the frost resistance of concrete. Freeze-thaw durability of concrete, J. Marchand, M. Pigeon, and M. Setzer Eds., E & FN Spon, pp. 157-164.

[15] Taylor, H. F. W. (1997). Cement chemistry. Second Edition. Thomas Telford Publishing, UK.

[16] TS EN 1008. (2003). Beton karma suyu numune Alma, deneyler ve beton endüstrisindeki işlemlerden geri kazanılan su da dahil olmak üzere suyun, beton karma suyu olarak uygunluğunun tayini kuralları. Turk Standartları Enstitüsü, Turkey.

[17] TS EN 196-1. (2002). Çimento deney metodları - bölüm 1: Dayanım tayini. Turk Standardları Enstitüsü, Turkey.

Fresh Behavior and Hardened Properties of Self-Compacting Concrete Containing Coal Ash and Fly Ash as Partial Replacement of Cement

Nahla N. Hilal[1,*], Noor A. Rajab[2], Rabar H. Faraj[3]

[1] Scholarships & Foreign Relation Department, University of Fallujah, Iraq
E-mail: nahla.naji@uofallujah.edu.iq
[2] College of Engineering, University of Anbar, Anbar, Iraq
E-mail: eng.noor85@uoanbar.edu.iq
[3] Civil Engineering Department, University of Halabja, Halabja, Kurdistan Region, Iraq - E-mail: rabar.faraj@uoh.edu.iq
* Corresponding author: Scholarships & Foreign Relation Department, University of Fallujah, Iraq. Tel: +96447809361161

Abstract. Due to the higher price of cement compared to other concrete ingredients, the problems associated with greenhouse gases emissions during its production and its high popularity have arisen concerns in its substitution with other supplementary cementitious materials (SCMs). In this regard, the current experimental program was established to investigate the fresh and hardened performance of self-compacting concrete (SCC) made with coal ash (CA) and fly ash (FA) as fractional substitution of cement. In this paper an attempt was made to use the remaining CA from the barbecue process of the restaurants from Fallujah city, Iraq. A total number of 7 concrete mixes were prepared with unchanged (w/b) ratio of 0.37 and total cementitious content of 450 kg/m^3. Reference mix was produced by using 100% cement without FA or CA. Then, other mixtures were batched with 10%, 20% and 30% (by weight) replacement of cement by FA and CA, respectively. The results illustrated that, in contrast to the FA, the CA has negative influences on the fresh properties of SCC but the results still meet the criteria for the fresh behavior of SCC. On the other hand, the inclusion of CA was significantly improved the strength and water absorption of SCC. Sustainable high strength SCC can be produced due to the addition of CA.

Keywords: *Self-compacting concrete; Coal ash; Fly ash; Fresh properties; Hardened properties*

1. Introduction

Due to the high demand for cement as the main construction material, its higher prices compared to other concrete constituents and the problems associated with emissions of greenhouse gases during its production made a huge interest to replace them with pozzolanic materials such as SCMs. As mineral admixtures, there have been numerous waste materials used in concrete to protect the environment. However, by using pozzolanic mineral admixtures, mechanical properties and the stability of the concrete have been improved. Besides, they can also be utilized as a way for decreasing the consumption of cement and the emission of carbon dioxide which causes global warming and waste of raw materials. The term green concrete can be used for these types of concretes (Golewski, 2018).

In the previous studies, a huge amount of investigations can be found regarding the incorporation of different pozzolanic materials into the concrete to partially replace Portland cement (Al-Jumaily et al. 2015), the materials used include FA (Güneyisi et al. 2016; Al-Hadithi, & Hilal, 2016; Adak et al. 2017; Faraj et al. 2020), silica fume (Mastali & Dalvand, 2018; Faraj et al. 2019), rice husk ash (Chakraborty & Goswami, 2015; Ahsan & Hossain, 2018), palm oil fuel ash (Mujah, 2016; Zeyad et al. 2016), coconut shell ash (Oyedepo, 2015), shell sunflower ash and shell pumpkin ash (Shahbazpanahi & Faraj, 2020).

From the chemical point of view, the pozzolan materials are siliceous or siliceous-aluminous. They have little inherent cementitious value alone, while during their reaction with saturated calcium hydroxide (C-H) extra calcium silicate hydrate (C-S-H) is produced. Thus, calcium silicate and calcium hydroxide represents the major products of cement hydration (Omrane et al. 2017). Therefore, the pozzolanic materials can have the ability to form additional C-S-H during hydration by which the matrix can be improved. The pozzolanic reactivity of mineral admixtures can significantly impact the strength and durability of concrete. Burning and grinding of waste materials are the best ways to produce pozzolan materials (Shahbazpanahi & Faraj, 2020).

SCC can be regarded as high performance and special concrete type because it can be poured and vibrated inside any form shapes due to its self-weight without any outside efforts, also it assures total filling of forms even when the reinforcement with little spacings are present. Therefore, the cohesiveness along with the flowability of fresh concrete is highly crucial for SCC (Güneyisi & Gesoğlu, 2008). On the other hand, SCC is generally produced by lower w/b ratio and higher cement quantity is required compared to the traditional concrete. Thus, a higher amount of binders such as FA, lime dust, silica fume, and GGBFS can be used as a substitute of cement in order to prevent dissimilar dispersion of larger particles in the fresh concrete. Also, the flowability, passing ability, and stability of SCC can be enhanced due to the addition of viscosity modifying and superplasticizer admixtures, but the total cost of the produced concrete is amplified (Güneyisi et al. 2015; Kurt et al. 2016).

The expenditure of SCC usage can be decreased with the inclusion of mineral admixtures. Moreover, to develop the workability and slump of concrete mixes, generally mineral admixtures such as FA, limestone filler and GGBFS can be easily quantified (Faraj et al. 2020). It is noted that FA, among these materials, enables to enhance the mechanical and resilience features of concrete when it replaces cementitious materials. The inclusion of FA also allows decreasing the need for chemical admixtures to control the viscosity level (Bilodeau et al. 1994).

Global concerns are raised to decrease the consumption of natural resources in concrete production and replace them with alternative waste materials. Therefore, the main aim of this study is to develop sustainable high strength SCC by reducing the cement content through its replacement with mineral waste materials such as FA and CA.

2. Research significance

Due to a high number of restaurants especially in the tourist places and in the city centers a large amount of remaining coal ash (CA) from the barbecue process is generated. This ash can be regarded as waste material and may be hazardous to the environment. Thus, in this paper an attempt was made to use the remaining CA from the barbecue process of the restaurants from Fallujah city, Iraq, which their number is more than 20 restaurants. The researches regarding the influence of the CA on the performance of SCC is limited. Therefore, this article mainly aims to make a comparison between the effects of CA and

IOP Conf. Series: Materials Science and Engineering **978** (2020) 012005 doi:10.1088/1757-899X/978/1/012005

FA as a partial replacement of cement on the fresh and some hardened features of SCC.

3. Experimental program

3.1 Materials

The main material which has been used as a cementitious material in this study was Ordinary Portland cement from Al-MASS Company. While the supplementary cementitious materials used for partially replacing the cement were Class F FA and CA. The chemical composition of CA, FA, and cement are shown in Table 1. Moreover, the physical characteristics of cement are reported in Table 2. Also, figure 1 demonstrates both SEM images analysis of FA and CA respectively. The CA powder which was obtained from different restaurants was sieved through sieve 75 μm. The Grading and physical properties of used coarse aggregate and natural river sand utilized in this study are given in Table 3 and Table 4, respectively. Thus, for the sake of accomplishing the required workability of SCCs, a high-performance superplasticizer known commercially as (ViscoCrete - 5930) was used with a specific gravity of (1.08).

Table 1. Chemical compositions of PC, FA and CA

Chemical analysis (%)	PC	FA	CA
SiO_2	20.9	68.67	2.098
CaO	63.22	1.926	31.29
Fe_2O_3	3.85	1.725	0.807
Al_2O_3	4.89	0.43	0.159
MgO	2.82	2.49	4.37
SO_3	2.73	0.758	5.454
K_2O	0.92	2.488	4.77
Na_2O	0.22	-	-
P_2O_5	-	0.69	2.406

Table 2. Physical properties of cement used in the mixtures

Test type	Specification (the Iraqi standard No.5/1984)	Cement used
Fineness by Blain air permeability apparatus (cm^2/gr)	≥ 2300	2500
Initial setting (min.)	≥ 45	130
Final setting (min.)	≤ 600	250
Soundness	≤ 0.8	0.8
Compressive strength		
7 days (MPa)	≥ 23	27.5
28 days (MPa)	-	32.9

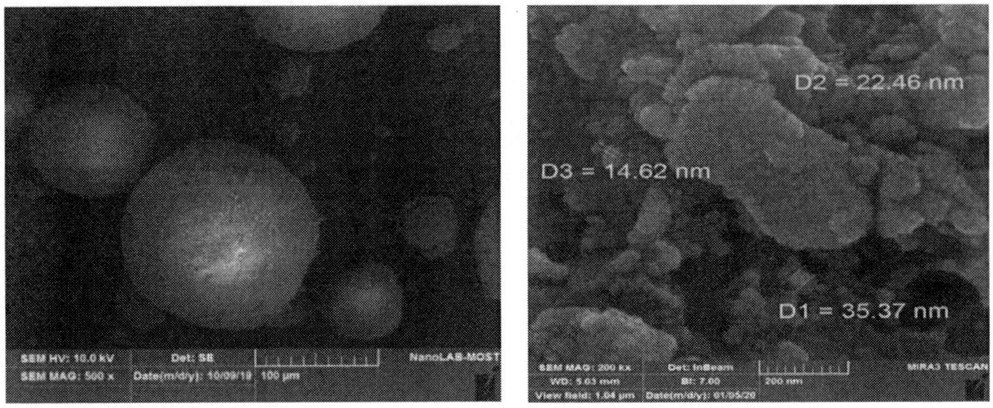

(a) (b)

Figure 1. SEM images of (a) FA and (b) CA

Table 3. Sieve analysis results and physical properties of coarse aggregate

Sieve size (mm)	% Passing	% Passing (ASTM C33)
12.5	100	100
9.5	86.76	85-100
4.75	14.12	10-30
2.36	0.16	0-10
Physical Properties		
Specific gravity	2.7	
Water absorption	0.3%	
Maximum particle size	10 mm	

Table 4. Sieve analysis results and physical properties of fine aggregate (river sand)

Sieve size (mm)	% Passing	% Passing (ASTM C33)
4.75	100	95-100
2.36	85.1	80-100
1.18	70.5	50-85
0.6	58.86	25-60
0.3	28.41	5-30
0.15	7.2	0-10
0.075	1.9	0-3
Physical Properties		
Specific gravity	2.7	
Water absorption	0.8%	
Fineness modulus	2.5	

3.2 Mix proportions

As a methodological procedure for evaluating the performance of SCC containing CA and comparing its effects with FA, an overall number of 7 SCC mixtures were designed with a constant (w/b) ratio of 0.37 and total binder content of 450 kg/m^3. The control mixture was produced by using 100% cement without FA or CA. Then, other mixtures were prepared with 10%, 20%, and 30% (by weight) replacement of cement by FA and CA, respectively. The mixture proportions of SCC are shown in Table 5.

Table 5. Mixture proportions (Kg) of SCC for 1 m^3 of concrete

Mix ID	Cement	Sand	Gravel	Water	SP	FA	CA
C	450	915	885	170	8	0	0
M1	405	915	885	170	8	45	0
M2	360	915	885	170	8	90	0
M3	315	915	885	170	8	135	0
M4	405	915	885	170	8	0	45
M5	360	915	885	170	8	0	90
M6	315	915	885	170	8	0	135

3.3. Mixing procedure and Preparation of the samples

Mixing procedure is an actual significant stage in SCC production that depends on the quantity of elements comprising of binders, coarse and fine aggregates, furthermore the water and admixtures. The procedures set by EFNARC (2005) as a methodology for the production of SCC as well as the direct technique proposed by Kheder et al. (2010) were followed in this study. Mixtures of SCC were prepared by adding a small quantity of water with half of aggregates (sand + gravel) to the mixing pan. The mixer was rotated for 1 minute, then part of the cementitious materials (the cement or cement + FA or CA) was supplemented to the mixture plus partial amounts of sand. The mixture rotated for another 1 minute. Next, quantitatively, half of (water plus superplasticizer) that has been added to the mixture and the process of mixing were continued for about 2 minutes till materials became homogenous, finally remaining constituents including (sand and gravel, superplasticizer and water) added to the mixer. At this stage, the mixing was continued for an extra 3.0 minutes. Afterward mixing, the fresh SCC mixtures were tested for fresh behavior, then for each mixture nine cubic molds with sizes of 100*100*100 mm and three prism molds with a measurement of 500 *100*100 mm were cast for completing tests of compressive strength, density water absorption and flexural strength, respectively. The molds of SCC were enclosed with pieces of polyethylene for 1 day. Finally, the samples were de-molded and maintained in water at 20 C° temperature till the age of the tests.

3.4. Testing procedures

In this section, the procedure used to carry out the tests for fresh and hardened properties of SCC mixtures according to the related standards for each test is explained in detail.

3.4.1 Fresh properties

For measuring slump flow diameter (SFD) and the slump flow time (SFT), a cone with an ordinary slump flow was chosen, and then it was filled up with concrete without being leveled and compacted. After lifting the cone and the concrete is spread, the average diameters of the spread concrete have been subsequently measured as shown in Figure 2. As proposed by EFNARC (2005), three typical classes of slump flow have been regarded to identify the range of applications. The typical application areas and various boundaries lower and upper, of the different classes, are provided in Table 6. However, the test of slump flow was conducted and its time (T_{50}), the determinant that identified the time in which the concrete reached the 500 mm spread circle, was also calculated. EFNARC suggests T_{50} of 2 to 5 seconds for SCC.

Figure 2. Measurement of SFD and SFT for SCC mixtures according to EFNARC (2005).

Figure 3 illustrates the process of evaluating the flowability and the viscosity of the mixture in which the experiment was accomplished. The determination of time has been done by using a simple methodology in which the funnel has been filled with a material of fresh concrete. Additionally, the time flow has been measured during the opening of the orifice until the funnel being drained completely. An adequate and stable concrete would have been consuming a short time for flowing out. Two kinds of viscosity classes in the EFNARC (2005) SCC guide according to the measured V-funnel time (VFT) and T_{50}. Thus, the viscosity classification was also shown in Table 6.

Figure 3. Photographic view of VFT measurement of SCC mixtures according to EFNARC (2005).

The L-box equipment consisted of a rectangular-section box which resembled an L in shape, having horizontal and vertical parts, was used. The horizontal and vertical parts were separated by a movable gate in which vertical bars are fitted (Figure 4). The vertical part was filled out with concrete and subsequently the concrete flew into the horizontal part. After the gate has been lifted and the flow of the concrete being stopped, the height of the concrete settled in the horizontal part was regarded as the ratio of the concrete retained in the vertical section (H_2/H_1). It indicated the slope of the rest concrete. This procedure is deliberated on the passing ability or the degree by which evacuation of concrete via the bars is limited.

Figure 4. Photographic view of L-box apparatus used for the measurement of H_2/H_1 ratio of SCC mixtures according to EFNARC (2005).

Table 6. Passing ability, slump flow and viscosity classes according to EFNARC (2005)

Class	T_{50} (seconds)	VFT (seconds)
Viscosity		
VS1/VF1	≤2	≤8
VS2/VF2	>2	9-25
Passing ability		
PA1	≥ 0.8 with two rebar	
PA2	≥ 0.8 with three rebar	
Class		SFT
Slump flow		
SF1		550-650
SF2		660-750
SF3		760-850

3.4.2 Hardened properties

As for the compression strength, the experiment of compressive strength has been accomplished by using a 100 mm cubes at the age of 7 and 28 days. This experiment was performed by using of 2000 KN hydraulic machine. The stress rate of compression has been measured in terms of dividing the maximum capacity load of the cubes by the face zone and the average of three cubes is reported (Standard, B 1983).

For determining the flexural strength of the specimens the three-point bending test was performed and the loading rate was designated to be 1.0 ($N/mm^2/min$) (ASTM, C78. 2010).

The water absorption test was conducted on the cube specimens following (ASTM, C642. 2008). The average of three cubes was reported for each mix.

The cube samples were also used to determine the dry density of the SCCs. The procedure was conducted according to (ASTM, C138. 1986)

4. Results and discussion

4.1 fresh properties

4.1.1 SFT and T_{50} SFD

The SFT and T_{50} results for SCC made with dissimilar dosages of FA and CA are shown in Figure 5 and Figure 6, respectively. It is noted from Figure 5 that as a result of increasing the FA content, the flowability of SCC has been increased too remarkably. Increasing FA content from 0% to 30% resulted in an increase of SFD from 700 mm to 770 mm. On the other hand, the addition of CA into the SCC resulted in the opposite behavior as it is compared to FA. The addition of 30% CA into the SCC reduced the SFD by about 14% compared to the control mixture. Moreover, the mixture that contains 30% FA has 22% higher SFD as compared to the mixture containing the same percentage of CA. Adding 30% FA shifted the mixture class from SF2 to SF3, while adding the same percentage of CA shifted the mixture class from SF2 to SF1. All produced mixtures containing either FA or CA satisfied the fresh criteria for SCC proposed by EFNARC (2005), but the classes were remarkably different for mixtures made with FA or CA.

Regarding the time of slump flow, as it is shown in Figure 6, the addition of CA also increases the T_{50} time, which means higher time is required for SCC to reach 500 mm diameter, thus the flowability decreased. However, the inclusion of FA slightly decreased T_{50} times. The mixture containing 30% FA has 77% lower T_{50} time compared to the same mixture but contained 30% CA.

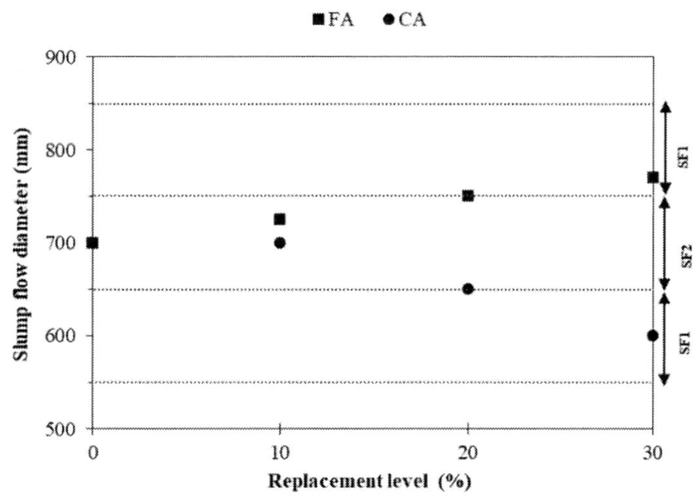

Figure 5. SFD of SCC mixtures containing different percentages of FA and CA

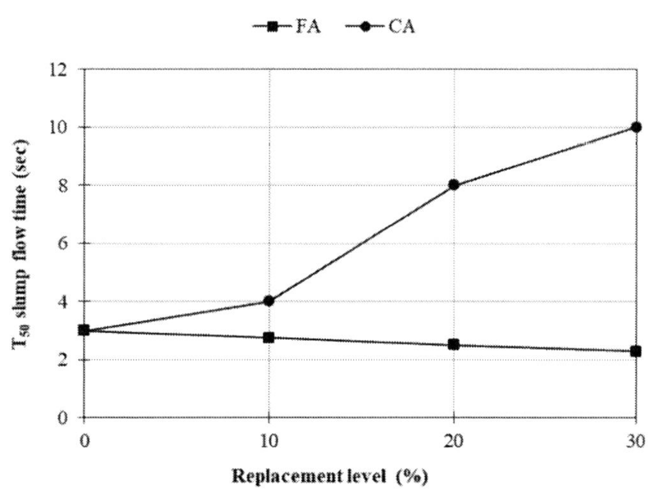

Figure 6. SFT of SCC mixtures containing different percentages of FA and CA

4.1.2 V-funnel flow time

Figure 7 shows the VFT versus the partial replacement of FA and CA with cement. It is obvious from the results that the addition of CA significantly amplified the VFT, thus the filling ability of SCC has been decreased. Increasing the CA content from 0% to 30% amplified the VFT form 8 seconds to 17 seconds. In contrast, the addition of FA gradually decreased the VFT. The mixtures containing CA can be categorized as VF2, while the mixtures containing FA can be classified as VF1 according to EFNARC (2005).

On the other hand, the VFT versus T_{50} flow time for SCC made with FA and CA as it is shown in Figure 8. This figure can be used to identify the viscosity classes of SCC. As can be seen from Figure 9 mixtures containing FA can be regarded as VS1/VF1 class whereas the mixture that contains CA should be regarded as VS2/VF2 class.

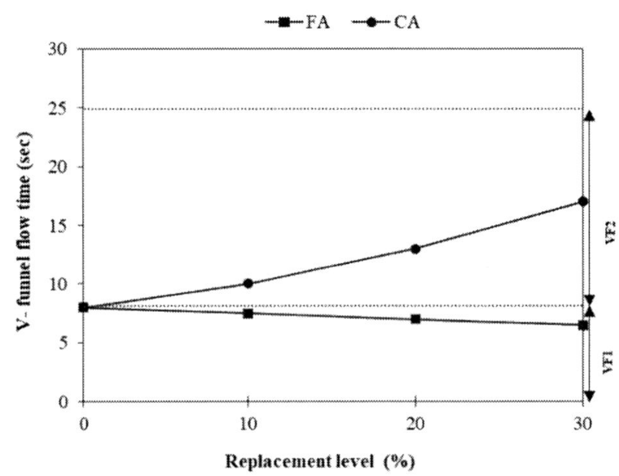

Figure 7. VFT of SCC mixtures containing different percentages of FA and CA

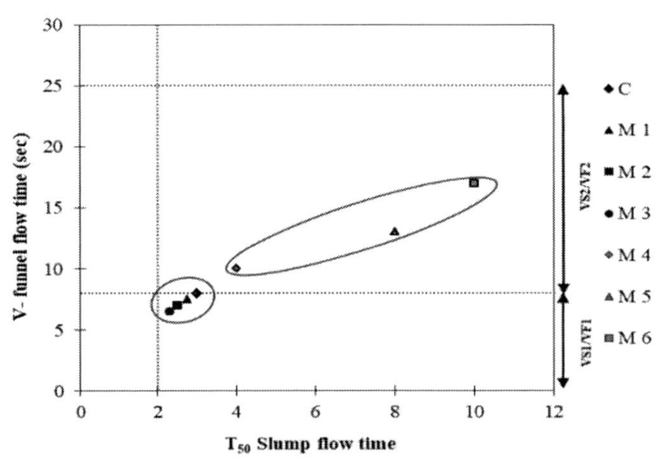

Figure 8. VFT versus T_{50} flow time of SCC mixtures containing different percentages of FA and CA

4.1.3 L-box ratio

As for the passing ability of SCC in terms of the L-box height ratio, it was significantly reduced by amplifying the CA content as demonstrated in Figure 9. Moreover, contrasting other fresh properties the addition of FA also decreased the passing ability of SCC but with a small range compared to CA. The inclusion of 30% CA resulted in a 37% reduction in the L-box height ratio. The mixtures containing CA cannot meet the fresh criteria for SCC regarding the passing ability

The overall fresh results illustrated that the CA involves a negative impact on the fresh behavior of SCC, but the majority of the results still meet the criteria required for the fresh behavior of SCC. The main reason behind decreasing the fresh properties with increasing the CA content is assigned to the fact that those particles of CA was in the nano-scale compared with particles of cement and FA which in the micro-scale, thus the CA particles have higher surface area and needs more quantity of water to wet all surface area of the particles (Shahbazpanahi & Faraj, 2020).

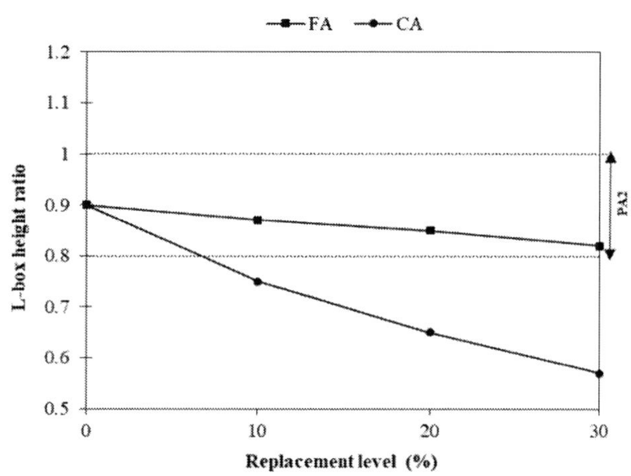

Figure 9. L-box height ratio of SCC mixtures containing different percentages of FA and CA

4.2 Hardened properties

4.2.1 Dry density

The dry density of SCC containing different percentages of FA and CA is presented in Figure 10. Mixtures with CA had higher dry density compared to those containing FA at all testing ages. The inclusion of CA leads to the surge of the dry density of SCC at 7 and 28 days. This may be accredited to the fact that, because the CA particles had greater surface area when compared with cement and FA particles, thus during the hydration process more water was converted to the hydration products, made the micro structure less porous, more packed and increased the density of the produced concrete. The mixture containing 20% FA had a dry density of about 7% and 9.5% lower than that containing the same percentage of CA at 7 and 28 days, respectively.

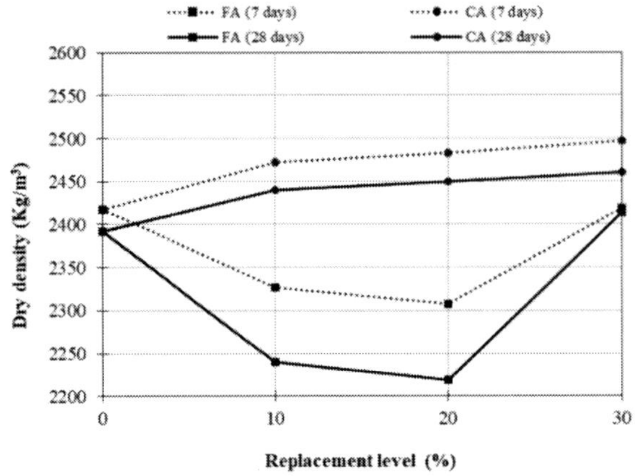

Figure 10. Dry density of SCC mixtures containing different percentages of FA and CA

4.2.2 Compression strength

The compressive strength of SCC has remarkably been increased with increasing CA content up to 20% at 7 and 28 days, as shown in Figure 11. Nevertheless, this behavior was different for SCC containing FA. The compressive strength was decreased with the inclusion of 10% FA, and then it was increased with the addition of 20% and 30% of FA at both testing ages. The mixtures containing 10% and 20% of CA and 30% of FA might be considered as self-compacting high strength concrete (SCHSC), because the strengths higher than 60 MPa were obtained at 28 days (Faraj et al. 2019; ACI 363R-10). The compressive strength of SCC was increased by about 30% and 46% with the inclusion of 20% and 30% of CA and FA, respectively. Increasing the compressive strength with the addition of CA can be assigned to the highly lower diameters and higher surface area of the particles that enhanced the ITZ and microstructure of the matrix (Al-Jumaily et al. 2015). The results also demonstrated that the compression failure mode of SCC mixtures containing FA were more brittle than the reference mixture and mixtures made with CA as shown in Figure 12.

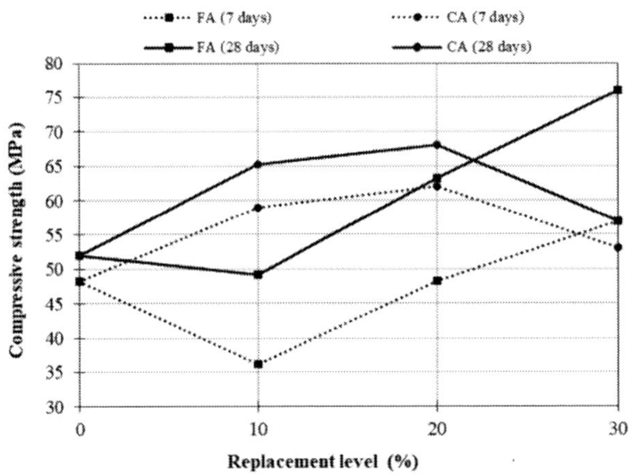

Figure 11. Compressive strength of SCC mixtures containing different percentages of FA and CA at 7

and 28 days

(a) (b) (c)

Figure 12. Compression failure mode of different samples at 28 days (a) reference SCC mixture (b) Mixture containing 10% FA, and (c) Mixture contains 10% CA.

4.2.3 Flexural strength

The flexural strength of SCC containing various percentages of FA and CA at 28 days is illustrated in Figure 13. The flexure failure mode of the different SCC mixtures is also presented in Figure 14. It is obvious from the findings that the flexural strength of SCCs involving different percentages of FA and CA were lower than the control sample. However, the lowest flexural strength of SCC was achieved for the mixture containing 10% FA which was 46% lower than that of the control mixture. On the other hand, the highest flexural strength was gained for the mixture that contains 30% CA which was about 8% higher than the reference sample. Moreover, the mixtures containing CA had a higher flexural strength of the mixes containing FA, because the microstructure of the matrix and ITZ was better improved with the addition of CA compared to the FA.

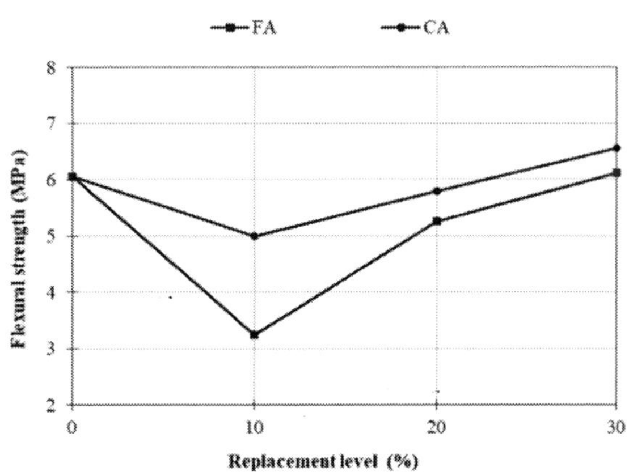

Figure 13. Flexural strength of SCC mixtures containing different percentages of FA and CA at 28 days

(a) (b) (c)

Figure 14. Flexure failure mode of different samples at 28 days (a) reference SCC mixture (b) Mixture containing 10% FA, and (c) Mixture contains 10% CA.

4.2.4 Absorption

The water absorption test results of SCC comprising different percentages of FA and CA are shown in Figure 15. Generally, the water absorption of SCC was remarkably decreased with the inclusion of FA and CA. The positive impact of CA on the water absorption was more obvious especially at a 30% replacement level compared to FA. The water absorption of SCC was reduced by about 29% and 58% by the addition of 30% FA and 30% CA, respectively. A main reason behind decreasing the water absorption of SCC is that, due to the pozzolanic reaction of FA and CA the microstructure of the matrix was significantly improved (Faraj et al. 2020). Moreover, the CA particles were much smaller than cement particles, thus during hydration the compactness of the hydrated structure was considerably enhanced, which resulted in the disconnected pore structure of the matrix.

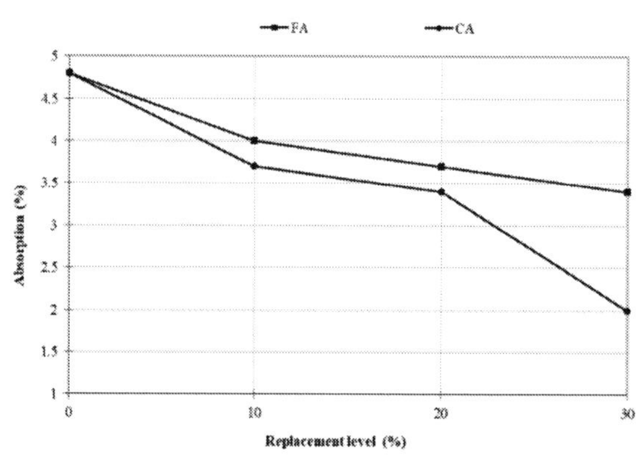

Figure 15. Water absorption ratio of SCC mixtures containing different percentages of FA and CA

5. Conclusions

According to the outcomes reached through this study, some conclusions can be drawn:

1- The waste material resulted from the barbecue process was obtained and used as a partial replacement with cement to produce sustainable SCC. Moreover, the influence of CA on the SCC fresh as well as the hardened properties was studied and compared with the FA effects.
2- Generally, the fresh behavior of SCC has been negatively impacted by the CA but the majority of the results still meet the criteria required for the fresh behavior of SCC.
3- The fresh properties of SCC have been improved by the FA addition in terms of flowability, filling ability, and passing ability.
4- The dry density of SCC has been increased along with the increase of CA content starting from 0% to 30% and it was reduced with increasing the FA content.
5- There has been a significant enhancement in the compressive strength of SCC that has been significantly improved when the CA content up increased to 20% and then decreased with a further increase of the CA content.
6- The CA can be regarded as a significant pozzolanic material for producing SCHSC with compressive strength higher than 60 MPa.
7- Both CA and FA involve a negative effect on the flexural strength of SCC .at. 28 days, but it's worth mentioning that the influence of CA has been insignificant compared to the effect of FA.
8- The water absorption of SCC was increased gradually by raising the (CA) content, and it was also dramatically amplified with the addition of FA.

Abbreviations and definitions

CA	coal ash
C-H	calcium hydroxide
C-S-H	calcium silicate hydrate
EFNARC	European Federation of National Associations Representing for Concrete
FA	fly ash
GGBFS	ground granulated blast furnace slag
LHR	L-box height ratio
PC	Portland cement
SEM	scanning electron microscopy
SCC	self-compacting concrete
SCHSC	self-compacting high strength concrete
SCM	supplementary cementitious materials
SFD	slump flow diameter
SFT	slump flow time
SP	superplasticizer
VFT	V-Funnel flow time
W/b	water/binder

References

[1] ACI Committee 363. (2010). *Report on High-Strength Concrete (ACI 363R-10)*. ACI.

[2] Adak, D., Sarkar, M., & Mandal, S. (2017). Structural performance of nano-silica modified fly-ash based geopolymer concrete. *Construction and Building Materials*, *135*, 430-439. https://doi.org/10.1016/j.conbuildmat.2016.12.111

[3] Ahsan, M. B., & Hossain, Z. (2018). Supplemental use of rice husk ash (RHA) as a cementitious material in concrete industry. *Construction and Building Materials*, *178*, 1-9. https://doi.org/10.1016/j.conbuildmat.2018.05.101

[4] Al-Hadithi, A. I., & Hilal, N. N. (2016). The possibility of enhancing some properties of self-compacting concrete by adding waste plastic fibers. *Journal of Building Engineering*, *8*, 20-28. https://doi.org/10.1016/j.jobe.2016.06.011

[5] Al-Jumaily, I. A. S., Naji, N., & Kareem, Q. (2015). An overview on the Influence of Pozzolanic Materials on Properties of Concrete. International Journal of Enhanced Research in Science Technology & Engineering, 4(3), 81-92.

[6] ASTM, C78. (2010). Standard test method for flexural strength of concrete (using simple beam with third-point loading). In *American Society for Testing and Materials* (Vol. 100, pp. 19428-2959).

[7] ASTM, C138. (1986). Test Method for Unit Weight of Concrete Specimens, Annual Book of ASTM, Standards, Section 4, Vol. 02, pp. 96–98. American Society for Testing and Materials, Philadelphia.

[8] ASTM, C642. (2008) Standard Test Method for Density, Absorption, and Voids in Hardened Concrete, United States Am. Soc. Test. Mater., pp 11–13 (2008).

[9] Bilodeau, A., Sivasundaram, V., Painter, K. E., & Malhotra, V. M. (1994). Durability of concrete incorporating high volumes of fly ash from sources in the USA. Materials Journal, 91(1), 3-12.

[10] Chakraborty, A., & Goswami, A. (2015). Conservation of environment by using fly ash and rice husk ash as a partial cement replacement in concrete. *J. Energ. Res. Environ. Technol*, *2*(1), 9-11. http://www.krishisanskriti.org/jeret.html

[11] Concrete, S. C. (2005). The European Guidelines for Self-Compacting Concrete. BIBM, et al, 22., http://www.efnarc.org.

[12] Faraj, R. H., Sherwani, A. F. H., & Daraei, A. (2019). Mechanical, fracture and durability properties of self-compacting high strength concrete containing recycled polypropylene plastic particles. *Journal Of Building Engineering*, *25*, 100808. https://doi.org/10.1016/j.jobe.2019.100808

[13] Faraj, R. H., Ali, H. F. H., Sherwani, A. F. H., Hassan, B. R., & Karim, H. (2020). Use of recycled plastic in self-compacting concrete: A comprehensive review on fresh and mechanical properties. *Journal of Building Engineering*, 101283. https://doi.org/10.1016/j.jobe.2020.101283

[14] Golewski, G. L. (2018). Green concrete composite incorporating fly ash with high strength and fracture toughness. *Journal of Cleaner Production*, *172*, 218-226. https://doi.org/10.1016/j.jclepro.2017.10.065

[15] Güneyisi, E., & Gesoğlu, M. (2008). Properties of self-compacting mortars with binary and ternary cementitious blends of fly ash and metakaolin. Materials and Structures, 41(9), 1519-1531. https://doi.org/10.1617/s11527-007-9345-7

[16] Güneyisi, E., Gesoğlu, M., Booya, E., & Mermerdaş, K. (2015). Strength and permeability properties of self-compacting concrete with cold bonded fly ash lightweight aggregate. *Construction and Building Materials*, *74*, 17-24. https://doi.org/10.1016/j.conbuildmat.2014.10.032

[17] Güneyisi, E., Gesoglu, M., Naji, N., & İpek, S. (2016). Evaluation of the rheological behavior of

fresh self-compacting rubberized concrete by using the Herschel–Bulkley and modified Bingham models. *Archives of civil and mechanical engineering*, *16*(1), 9-19. https://doi.org/10.1016/j.acme.2015.09.003

[18] Kheder, G. F., & Al Jadiri, R. S. (2010). New Method for Proportioning Self-Consolidating Concrete Based on Compressive Strength Requirements. ACI Materials Journal, 107(5).

[19] Kurt, M., Gül, M. S., Gül, R., Aydin, A. C., & Kotan, T. (2016). The effect of pumice powder on the self-compactability of pumice aggregate lightweight concrete. Construction and Building Materials, 103, 36-46. https://doi.org/10.1016/j.conbuildmat.2015.11.043

[20] Mastali, M., & Dalvand, A. (2018). The impact resistance and mechanical properties of fiber reinforced self-compacting concrete (SCC) containing nano-SiO2 and silica fume. *European Journal of Environmental and Civil Engineering*, *22*(1), 1-27. https://doi.org/10.1080/19648189.2016.1177604

[21] Mujah, D. (2016). Compressive strength and chloride resistance of grout containing ground palm oil fuel ash. Journal of cleaner production, 112, 712-722. https://doi.org/10.1016/j.jclepro.2015.07.066

[22] Omrane, M., Kenai, S., Kadri, E. H., & Aït-Mokhtar, A. (2017). Performance and durability of self-compacting concrete using recycled concrete aggregates and natural pozzolan. Journal of Cleaner Production, 165, 415-430. https://doi.org/10.1016/j.jclepro.2017.07.139

[23] Oyedepo, O. J., Olanitori, L. M., & Akande, S. P. (2015). Performance of coconut shell ash and palm kernel shell ash as partial replacement for cement in concrete. *Journal of Building Materials and Structures*, *2*(1), 18-24. http://dx.doi.org/10.5281/zenodo.241986

[24] Shahbazpanahi, S., & Faraj, R. H. (2020). Feasibility study on the use of shell sunflower ash and shell pumpkin ash as supplementary cementitious materials in concrete. *Journal of Building Engineering*, 101271. https://doi.org/10.1016/j.jobe.2020.101271

[25] Standard, B. (1881). Part-116 (1983) Method for Determination of Compressive Strength of Concrete Cubes, London. *British Standard Institution*.

[26] Zeyad, A. M., Johari, M. A. M., Tayeh, B. A., & Yusuf, M. O. (2017). Pozzolanic reactivity of ultrafine palm oil fuel ash waste on strength and durability performances of high strength concrete. Journal of Cleaner Production, 144, 511-522. https://doi.org/10.1016/j.jclepro.2016.12.121

Reducing Rutting in Flexible Pavement Using Specified Polymers with HMA

Mohammed Abbas Al-Jumaili[1,*] and **Hamid A. Al-Jameel[2]**

[1]Faculty of Engineering, University of Kufa, Iraq
E-mail: mohammedah.aljumaili@uokufa.edu.iq
[2]Faculty of Engineering, University of Kufa, Iraq
E-mail: hamid.aljameel@uokufa.edu.iq
*Corresponding author: Faculty of Engineering, University of Kufa, Najaf, Iraq. Tel: +9647801341152.

Abstract. Iraqi roads mainly suffer from rutting distresses which resulting from extremely axle loading and high difference in temperature between seasons and even in day and night for the same day. Different types of modifier polymer have been inserted into asphalt materials to enhance the characteristics of asphalt concrete mixture as a whole. Thus, this paper has generally been dedicated to investigate the effect of using two types of polymers; Styrene-Butadiene-Styrene (SBS) and High-Density Polyethylene (HDPE) on the rutting distress. Accordingly, five ratios (4%, 6%, 8% and 10%) of SBS and HDPE have been used as a percentage of total mix in order to get the optimum influence of each ratio. Hence, the best ratios for surface (Type III A) layer are 8% HDPE and 6% SBS; whereas these ratios for binder layer are 6% HDPE and 2% SBS. The results indicate that the reduction in rutting percent are (72% and 65%) for HDPE and SBS for each type of additives at 40°C. Accordingly, this means these types of polymers have a significant role in reduction of rutting in surfacing and binder layers.

Keywords: *Hot Mix Asphalt, High-Density Polyethylene, Styrene-Butadiene-Styrene, rutting distress.*

1. Introduction

The reviewing of appropriate existing literature revealed that several detrimental types of distresses may influence on flexible pavement within its service life. Traffic load, weather and construction practices are one of the causes that lead to such distresses. Some of these distresses are rutting, shoving, fatigue and stripping; which may cause the functional failure and even the structural failure of the flexible pavement. The influence of these distresses may reflect uncomfortable ride to drivers (Hafeez et al., 2013; Sorum et al., 2014; White, 2018; and Bala et al., 2018).

In this era, using modified asphalt binders in constructing flexible pavement is widely spreading.

Modifiers, such as styrene-butadiene-rubber (SBR) and high density polyethylene (HDPE), can enhance the physical and rheological properties of asphalt materials (Lu and Isacsson, 2001). HDPE improves the elastic properties of asphalt cement at high temperature wherever permanent deformation influences the pavement. As the percentage of HDPE increases, the mechanical behavior improves (Perez-Lepe, et al., 2005 and Razak, 2015).

In fact, rutting decreases the beneficial service life of the pavement and generates severe risks for road operators by influencing car handling characteristics. In Iraq, rutting problems can be witnessed at some sites in the roads because of the heavy axle loading and high temperature at summer seasons (Yassoub, 2005). Utilizing recent technologies to enhance pavement effectiveness and increase its service life which is a significant measure of pavement designing and rehabilitating systems. Several technologies may offer benefits such as improving pavement effectiveness, one of these technologies is additives (Ai-Qadi and Appea, 2003). The stripping distresses are triggered by the weak bond of the bitumen binder to specific aggregates, particularly in the existence of water. Therefore, a serious need for recent or modified binders is required (Fernando and Guirguis, 1983). The usage of Polymer-Modified Asphalt (PMA) to attain improved flexible pavement performance has been noticed for a long term. The enhanced functional properties contain permanent deformation, fatigue, low temperature cracking, and aging (Raad et al., 1996). Recent studies indicated that the charactersitics of PMA are mainly based on the polymer properties, polymer content and binder origin in addition to the mixing process. Regardless of the high polymeric products, there are somewhat few kinds which are appropriate to asphalt binder modification, when utilized as asphalt modifiers. Particular polymers have to be appropriate to asphalt binder, be able to be processed by conventional blending and laying apparatus, and be capable of maintaining their superior properties during blending and application in service. utilizing a modifier has to also be cost-effective (Al-Bana'a, 2010). Polymers can be categorized as elastomers and plastomers. Elastomers, at primary low strain level, can add just slight strength to the binder; whereas these polymers could be elongated out and become stronger at high strain and return as the applied load is left. Plastomers is a rigid three-dimensional network and indicates tensile strength under high load whereas crack at higher strains. The roads have indicated initial marks of distresses because of the severe weathering conditions and heavy traffic loads. The high temperature on the pavement surface may be 60°C in Iraq in summer which decreases the stiffness of paving mix and accordingly leads to pavement deformation (Al-Abdul Wahhab et al., 1996). Al-Khateeb, et. al., (2011) used ABAQUS program to inspect the static repetitive load influence on rutting of asphalt pavement structure. They confirmed that the depth of rut rises with growing temperature and tire pressure and reducing subgrade strength. Hossain et al., (2019) used SBS, HDPE, polyphosphoric acid (PPA), polyethylene (PE) and crumb rubber (CR) to improve a neat bitumen binder (PG 64-22). They noticed a significant improvement in asphalt characteristics such as increase in SBS percentage rises the ductility and an increase in HDPE percentage reduces the ductility in the studied rates.

Ibrahim (2019) investigated the effect of 5% HDPE on dense-graded Hot Mix Asphalt (HMA) and stone mastic asphalt mixtures. These mixes were exposed to aging process at 100°C with different levels of aging for 48 and 96h. Then, these effects have been characterized by Marshall stability, indirect tensile strength, tensile strength ratio, and resilient modulus. Moreover, other characteristics, such as penetration, ductility, softening point, loss of heat and air (aging) have been evaluated, too. The results reveals that the 5%HDPE for dense-graded asphalt and stone mastic asphalt mixtures are higher 1.4 for indirect tensile strength under 48h and 96 h aging than the control mixes. Whereas, the same behavior has been observed for tensile strength ratio, and resilient modulus (Ibrahim, 2019). On the other hand, some studies found that using 5% of HDPE improved asphalt mix effectiveness (Al-Hadidy and Yi-qiu, 2009; Al-Hadidy, 2018a and 2018b; and Attaelmanan et al., 2011). Moreover, even using reclaimed PE can improve the characteristics of HMA such as increasing the tensile strength of the asphalt mixtures at low temperatures (5°C) which may reduce the cracking potential of pavements at low temperatures (Punith and Veeraragavan, 2007). The authors found that the optimum PE content required for improving the performance of asphalt concrete mixtures is 5% by weight of asphalt for 80/100-grade asphalt cement.

Recently, Abed and Bahia (2020) found out the effect of using HDPE and SBS modifiers on the asphalt binder and asphalt concrete mixtures. The used asphalt was PG (64-16) with 3% and 5% of nano-HDPE and SBS, respectively, by weight of total asphalt binder. The results of their research indicated that HDPE could be utilized as a significant modifier with lower blending and compaction temperature in addition to equal or considerably better rutting and strength performance than traditional SBS Modifier.

The permanent deformation of flexible pavements has a main effect on the performance of pavement. The accumulation of plastic deformations within asphaltic layers has been recognized as one of the main reasons of rutting in asphaltic pavements. Rutting was the chief distress in Iraq resulting from the excess axle loads and high local summer temperature. The objective of this paper is to investigate the effect of using HDPE and SBS with five different ratios of both of these modifiers at 40°C to reduce the rutting in asphalt mixture for both binder and wearing layers.

2. Research methodology

It is noteworthy to highlight that the current study has used two modifiers (i.e., HDPE and SBS) with five percentages (2%, 4%, 6%, 8% and 10%) for wearing and binder courses to reduce rutting failure. Then, wheel tracking test has been implemented to assess the rutting distress.

3. Materials and testing methods

The used materials in the current study are locally existing and commonly utilized for constructing the pavement in Iraq. The asphalt cement used here is (40-50) penetration grade gained from Al-Daurah refinery in Bagdad. Table 1 reveals the physical properties of the asphalt cement that have been tested in civil engineering department lab at university of kufa.

Table 1. Physical properties of the used asphalt cement.

Test	Unit	Tested results	SCRB
Penetration ((ASTM-D5))	0.1mm	46	40 -50
Ductility ((ASTM- D113))	cm	112	>100
Flash point ((ASTM-D92)	°C	261	>232
Solubility ((ASTM-D2042))	%	99.2	>99
Specific Gravity 25°C ((ASTM-D70))	---	1.01	---
After Thin-Film Oven Test ((ASTM-D1754))			
%Retained penetration ((ASTM-D5))	%	66	>55%
Ductility of residue ((ASTM-D113))	cm	>100.00	>25
Loss in weight ((ASTM-D1754))	%	0.31	≤ 0.75

As indicated in Table 1, all physical characteristics such as the penetration, ductility, flash point, solubility are satisfied with the Iraqi specifications. These tests have been implemented according to ASTM as shown in Table 1.

Two maximum sizes of 25mm and 19mm were used to aggregate gradation adopted by SCRB specifications for wearing and binder courses and two categories of additives (HDPE, SBS) as modifies to use in HMA. The selected gradation is in line with the mid-limits gradation of the SCRB specification (2003). Table 2 demonstrates the gradation for wearing layer and Table 3 indicates the gradation for binder layer. The gradation for both layers is fallen within the required limits.

Table 2. Gradation of aggregate for wearing layer.

Sieve size	Sieve opening (mm)	Percentage passing by weight of total aggregate	
		Finer wearing course Type III A (%Passing by weight of total aggregate + Filler)	
		Specification limits (SCRB, 2003)	Mid-point gradation
3/4"	19.00	100.00	100.00
1/2"	12.50	90-100	95
3/8"	9.50	76-90	83.00
No.4	4.75	44-74	59.00
No.8	2.36	28-58	43.00
No.50	0.30	5-21	13.00
No.200	0.075	4-10	7.00
Asphalt content (%)		4-6	-----

The HDPE is obtained from Iran, which is a white color with density greater than 0.941 and less than 0.965. It is a plastomers thermoplastic polymer material which consists of carbon and hydrogen atoms forming a product with a high molecular weight changing into ethylene. Subsequently, applying heat and pressure converts the HDPE into polyethylene. This material consists of 500,000 to 1,000,000 carbon units long. Consequently, the longer chain molecules have the large the number of atoms which has the heavier the molecular weight. This could be a measure of several mechanical and chemical properties of the resulted product.

The physical properties of optimum asphalt binder with the addition of polymers are indicated in Table 4. The results of tests indicate that adding 8% HDPE leads to reduction in some characteristics from the specifications such as penetration, ductility and ductility of residue. This could be attributed to its effect of HDPE on the asphalt nature.

The SBS is widely used as an asphalt polymer modifier. This modifier was obtained from Baghdad, Iraq. The modifier is used as an asphalt modification. The SBS can be found in various lengths with different arrangement of the molecules. These differences could mainly be due to the effect of modification produced by the SBS, in addition to the ease of mixing and the storage stability.

Table 3. Gradation of Aggregate to Level (binder) Course.

Sieve Size	Sieve Opening (mm)	Percentage passing by Weight of total Aggregate	
		Finer leveling (binder) Course Type II (%Passing by Weight of Total Aggregate + Filler)	
		Specification Limits (SCRB, 2003)	Mid-point Gradation
1"	25.00		100.00
3/4"	19.00	100.00	95.00
1/2"	12.50	90.00-100.00	80.00
3/8"	9.50	76.00-90.00	68.00
No.4	4.75	44.00-74.00	50.00
No.8	2.36	28.00-58.00	36.00
No.50	0.30	5.00-21.00	12.00
No.200	0.075	4.00-10.00	6.00
Asphalt content (%)		4.00-6.00	-----

Table 4. Some properties of modified asphalt with (8%) HDPE.

Test	Unit	The results	SCRB
Penetration ((ASTM-D5))	0.1mm	23.00	40.00 -50.00
Ductility((ASTM-D113))	cm	35.00	> 100.00
Flash point ((ASTM-D92))	'C	251.00	≥ 232.00
Solubility ((ASTM-D2042))	%	99.70	> 99.00
After Thin-Film Oven Test ((ASTM-D1754))			
Retained penetration ((ASTM-D5))	%	84.00	> 55.00%
Ductility of residue ((ASTM-D113))	cm	16.00	> 25.00
Loss in weight ((ASTM-D1754))	%	0.35	≤ 0.75

A copolymer has two various kinds of repeating molecular units, the most widespread co-monomer utilized with styrene is butadiene. Block copolymers, such as SBS, have these repeating molecular units in a frequently happening block pattern. The SBS binders have the capability to withstand long-term deformation and to reduce fatigue and cracking due to low temperature. The physical properties of optimum asphalt cement when with the addition of polymers as indicated in Table 5. The addition of SBS, as indicated Table 5, has approximately the same behavior in reduction some of the physical characteristics (penetration and ductility) of the modified asphalt.

Table 5. Some properties of modified asphalt with (6%)SBS by weight of asphalt.

Test	Unit	Tested results	SCRB
Penetration ((ASTM-D5))	0.1mm	34.00	40.00 -50.00
Ductility ((ASTM-D113))	cm	53.00	> 100.00
Flash point ((ASTM-D92))	'C	251.00	≥ 232.00
Solubility ((ASTM-D2042))	%	99.50	> 99.00
After Thin-Film Oven Test ASTM D 1754			
Retained penetration ((ASTM-D5))	%	94.00	> 55.00%
Ductility of residue ((ASTM-D113))	cm	38.00	> 25.00
Loss in weight ((ASTM-D1754))	%	0.35	≤ 0.75

4. Permanent deformation test

There were two main categories of rutting: HMA rutting and subbase rutting. Asphalt mix rutting occurred as a lower layer did not rut until the pavement surface reveals wheel path distresses because of compaction and insufficient mixture design. Granular material/subgrade rutting happens as the layers displayed wheel path depressions because of extreme loading. The asphaltic slab specimens used for rutting test. The roller compacter device could compact asphaltic layers to a required density utilizing loads per unit roll width, which have been satisfied with those pavements rollers utilized in the road construction. The roller compactor gives a pneumatically powered means of compacting layers of bitumen material in the laboratory with conditions, which simulate in-situ compaction as demonstrated in Figure 1.

Figure 2 demonstrates the dimensions of samples (400 mm × 300 mm × 40 mm) for surface layer and (400 mm × 300 mm × 60 mm) which is assigned for binder layer. Rutting testing were achieved at 4% AV by roller compactor device at National Central of Construction Laboratories (NCCL) consistent with (EN12697, 2003). The mass of the asphaltic mix is a function of several factors as demonstrated in Eq. 1 (EN12697, 2003):

$$M = 10^{-6} \times L \times I \times e \times \rho m \times \left(\frac{100-v}{100}\right) \quad \text{......} \quad (1)$$

Where:

M: is the weight of slab (kg);

L: is the interior length of mold (mm);

I: is the interior width of mold (mm);

e: is the compacted thickness of slab, in millimeters (mm);

ρm: is the max. theoretical density of an asphalt mix, (kg/m³);

v: is the air voids in slab (%);

Figure 1. Wheel Tracking Test for Compacting Slab for at NCCLR in Baghdad.

The weight of sample to surface and binder layers as in Eq. (1) are 10.828 kg and 16.243 kg; respectively. Based on the test scale, the samples were loaded ranging from full size to small tries. The sample in wheel-tracking tests was loaded frequently by moving the wheel on the sample. Whereas, in the case of small scale wheel tracking tests, by moving the specimen under the load.

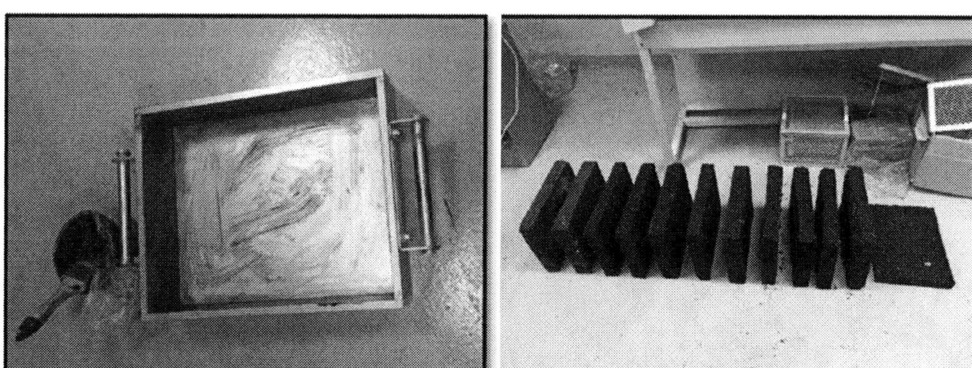

Figure 2. Steel Mold and Specimen.

The samples were usually loaded in one direction. Furthermore, some wheel tracking tests have the facility of crossing to simulate wheels roaming. The experiment was done at specified temperatures or in representative climatic conditions by making pavement sections outside. The sample was loaded at determined passes and the central depth of rut for the loaded region and/or transverse section profile observed with time or the number of load passes as demonstrated in Figure 3. The test was implemented consistent with (EN 12697-22, 2003) where it provided with the rate of permanent deformation from moving and concentrated loads. It used a Linear Value Displacement Transducer (LVDT) to determine the deformation of the sample. The loaded wheel used about (700±10) N of load at contact points and passed repetitively over the specimen up to 10,000 cycles. The center line of the tire path (0.5cm) was from the center of the sample. The center contact area of the frame represents a simple harmonic motion related to the center of the upper surface of the test sample having a total transmission distance of (23.0 ± 1.0) cm. Continuous loading frequency of (26.5 ± 1.0) loading cycle per 1 min of the test device in about 10,000 loading cycles or 2.0 cm with a maximum of acceptable deformation gotten. The design of a permanent deformation test experiment is a full factorial with two asphalt contents, one test temperature (40° C), one type of asphalt which is either native or modified with best percent of (SBS and/ HDPE) and two types of aggregate grading (surface and binder). The compressed samples (400 mm x 300 mm x 40 mm) for the surface layer and (400mm x 300mm x 60mm) for the binder layer were cooled to room temperature for 24 hours in line with (EN-12697-22, 2003).

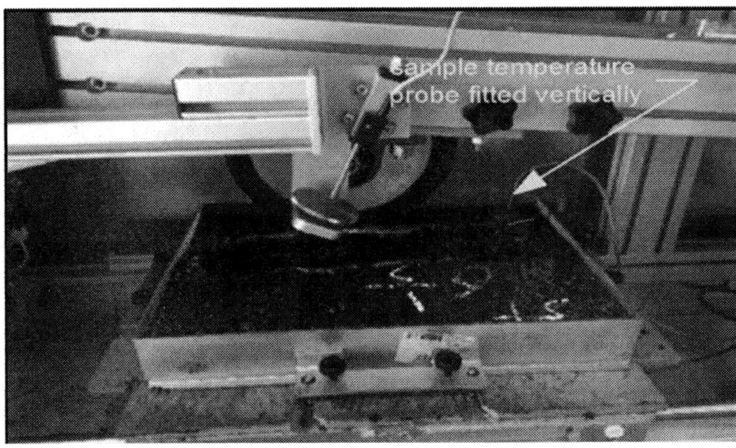

Figure 3. Wheel truck apparatus.

5. Results and discussion

The test results are gained including the impact the polymer types on modified asphalt concrete mixtures used in construction of wearing and binder courses. Table 6 indicates the results of Marshall tests which are conducted in this study such as stability, bulk density voids in total mix (VTM), and voids in mineral aggregate (VMA) and the optimum asphalt content. The Marshall properties of (wearing and binder) mixtures are consistent with the Iraqi specifications limits (SCRB, 2003).

Table 6. Asphalt Mixtures Properties and Specification limit S. C.B.R for two layers.

Properties	Wearing Layer	S.C.R.B Specification	Binder Layer	S.C.R.B Specification
Marshall Stability (KN)	11.21	8 min	9.17	7 min
Marshal Flow (mm)	3.05	2-4	3.21	2-4
Bulk Density	2.327	----	2.283	----
VTM. (%)	4.02	3-5	4.05	3-5
VMA (%)	15.67	14 min	16.31	13 min
Optimum AC (%)	4.90	4-6	4.60	4-6

The permanent rutting resistant for surface layer has been assessed utilizing wheel track test. The permanent deformation test is an experimental with two asphalt contents, one testing temperature, two kinds of additive, the details of testing variables could be clarified, as follows:

The optimum asphalt content obtained has two values 4.6% and 4.9% for binder and surface (Type IIIA) layers, respectively, by the weight of total mixture. In this study, the testing temperature is 40 C°. Generally, two types and two ratios of polymers have been used; (8%HDPE,5% SBS) for surface (Type IIIA) layer and (6%HDPE, 2% SBS) for binder layer. Cycles Rate: samples have been tested with one cycling rate; 26.5 cycle/minute. The limitation of the test is in line with EN 12697-22(2003) (put the machine in motion and determine readings at the vertical displacement of the wheel, primarily, thereafter as a minimum 6 or 7 times in the first hours, and as a minimum one reading every 500 load cycles thereafter. The perpendicular position of the wheel has been identified as the average value of the profile of the sample n a length of ± 50 mm about the center of the loading area at the center of traverse, determined in as a minimum 25 points almost correspondingly spaced. The wheel vertical position is determined without bringing it to a stop. Tracking until being continued 10000 load cycles were applied or until a depth of rutting 20 mm was (whatever is the smaller). Table 7 demonstrates the depth of rutting for modified asphalt mixes. Figures (4, 5 and 6) indicate the influence modified mixes on the depth of rutting at 40°C testing temperature for the used additives (HDPE, SBS) at 10000 cycle. Table 8 demonstrates the rut depth (RD) for the optimum asphalt content (4.6 and 4.9) % for the two layers with different percent of HDPE and SBS. The value of RD is much less in the modified mixes than unmodified ones.

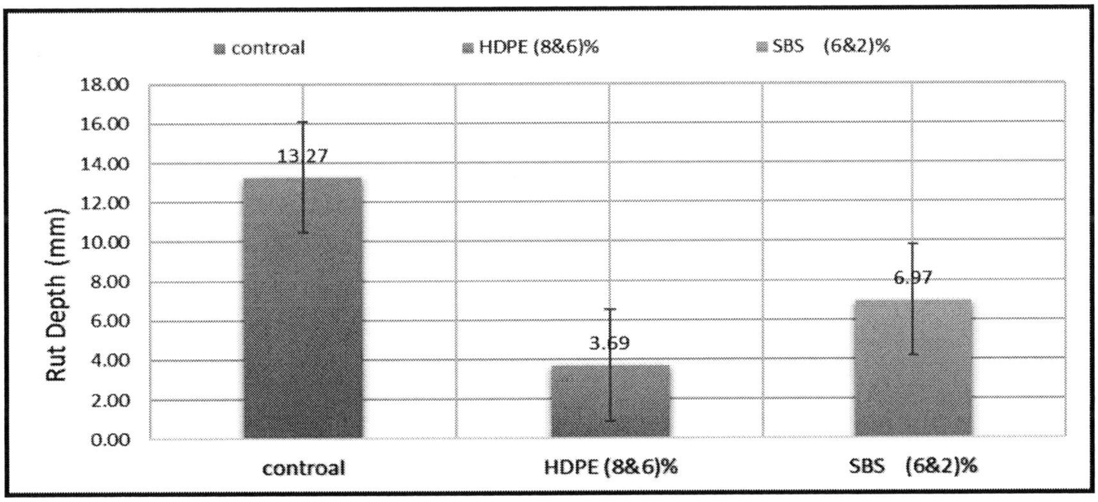

Figure 4. Influence of modified mixes on permanent deformation for two layers.

IOP Conf. Series: Materials Science and Engineering **978** (2020) 012006 doi:10.1088/1757-899X/978/1/012006

Table 7. The depth of rutting for wearing (Type III A) and binder Layers

Type of additives	AC (%)	Additive (%)	Thickness (mm)	IR (%) modified asphalt samples
control	4.9	0	10	
	4.6	0		0
HDPE	4.9	8	10	
	4.6	6		72
SBS	4.9	6	10	
	4.6	2		48

Table 8. Modified asphalt layers with improved ratio comparing with unmodified asphalt at 10000 cycle

Optimum Asphalt Content(%)	Thickness (mm)	Polymer percent (%)	Temperate (C°)	Unmodified mixtures RD(mm)	modified mixtures RD(mm)
(4.6 & 4.9)%	10	8% HDPE 6% HDPE	40	13.27	3.69
(4.6 & 4.9)%	10	6% SBS 2% SBS	40	13.27	6.97

Figure 5 indicates the rutting depth values for modified and unmodified asphalt layers at 10000 cycle and two categories of modifiers. While the number of cycles increases from 0.0 to 10000 for the control mix, the depth of rutting increases from approximately zero value to approximately 13mm. This demonstrates how the number of cycles has the significant effect on developing the rutting distress. Whereas, using SBS reduces the depth of rutting to approximately 7mm which is approximately the half for the control mix. Similarly, using HDPE reduce the value of rutting depth up to half value for the SBS modifier.

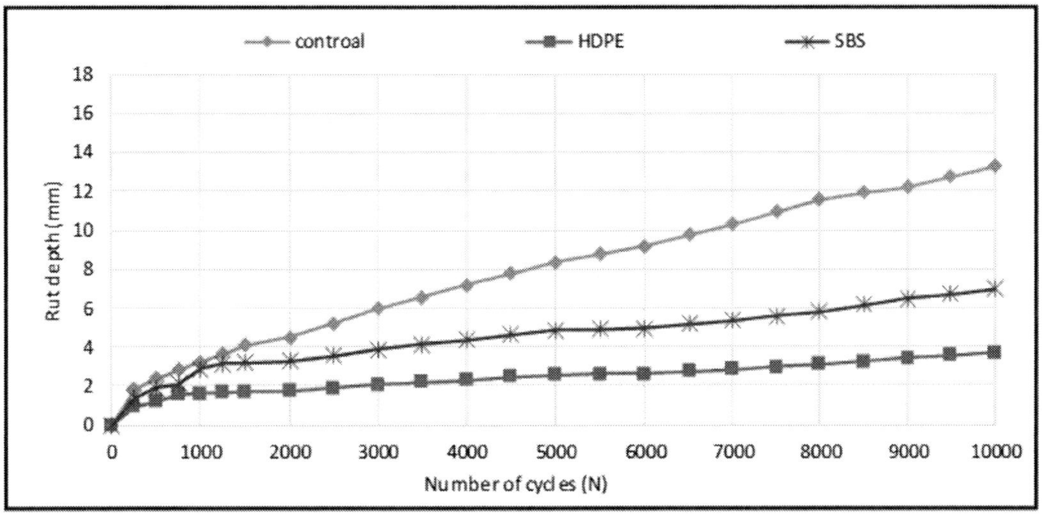

Figure 5. The depth of rutting for each polymer at temperature of 40°c.

Figure 6 demonstrates the difference in the Improvement Ratio (IR) of the RD at 40°C for the two layers. The results reveal that also modified mixes give higher values of IR than unmodified mixes.

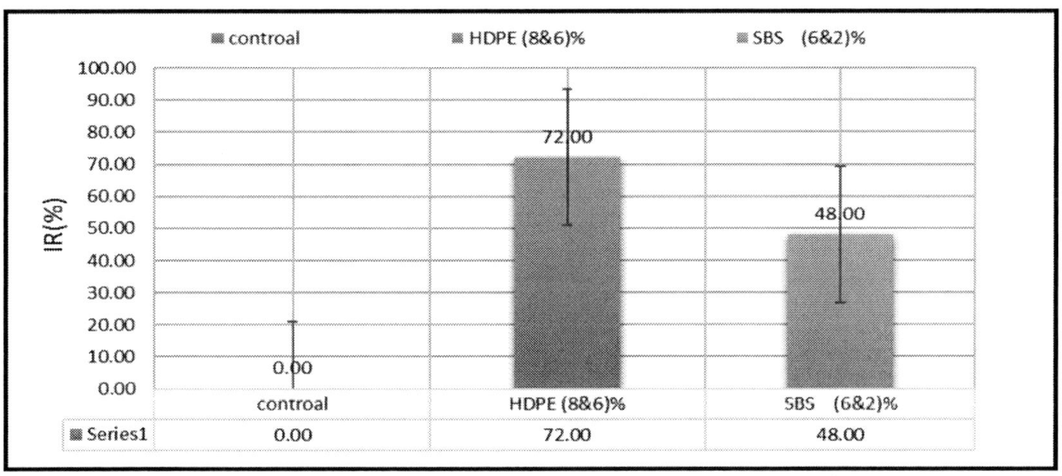

Figure 6. Difference in the IR (%) with modified asphalt mixes for two layers.

6. Conclusions and recommendations

All conclusions for modified mixtures (improvement mixtures performance) as percentages for optimum content of polymers compared with control asphalt mixtures. The modified mixes with Polyethylene and SBS have higher performance properties rather than the other types of polymers, therefore the best plastomers polymer is HDPE and the best elastomers polymer is SBS due to HDPE have higher stiffness related to its molecule weight and SBS-copolymers have different arrangement in molecule structure from other types of polymers. The results from wheel track test indicated that the depth of rutting for the modified specimens are lower than control mixtures, therefore the resistance to permanent deformation for modified mixtures is higher than control mixtures under the test temperatures (40 C°). It's about (72-48) % for (40 C°) times higher than control asphalt mixture. This study contributed that the fact is SBS polymer improves the rutting resistance of asphalt mixture. The recommendation of this study is to conduct fatigue test on modified asphalt mixture to study the effect of polymer and repetition loading on fatigue life.

References

[1] Abed, A. and Bahia, H. (2020). Enhancement of permanent deformation resistance of modified asphalt concrete mixtures with nano-high density polyethylene. Construction and Building Materials, 236(10).

[2] Ai-Qadi, I. L., & Appea, A. K. (2003). Eight-year field performance of secondary road incorporating geosynthetics at subgrade-based interface. *Transportation research record*, *1849*(1), 212-220.

[3] Airey, G. D. (2004). Styrene butadiene styrene polymer modification of road bitumens. *Journal of Materials Science*, *39*(3), 951-959.

[4] Al-Bana'a, J. (2010). *Effect of Polymer Type on the Performance of Modified Asphalt Paving Mixture* (Doctoral dissertation, M. Sc. Thesis, University of Babylon, College of Engineering).

[5] Al-Hadidy, A. I. (2018 a). Engineering behavior of aged polypropylene-modified asphalt pavements. *Construction and Building Materials*, *191*, 187-192.

[6] Al-Hadidy, A. I. (2018 b). Effect of laboratory aging on moisture susceptibility and resilient modulus of asphalt concrete mixes containing PE and PP polymers. *Karbala International Journal of Modern Science*, *4*(4), 377-381.

[7] Al-Hadidy, A. I., & Yi-qiu, T. (2009). Effect of polyethylene on life of flexible pavements. *Construction and Building Materials*, *23*(3), 1456-1464.

[8] Al-Khateeb, L. A., Saoud, A., & Al-Msouti, M. F. (2011). Rutting prediction of flexible pavements using finite element modeling. *Jordan Journal of Civil Engineering*, *159*(2980), 1-18.

[9] ASTM. (2012). Annual book of ASTM standards, Road and Paving Materials, Vol.04.03.

[10] Attaelmanan, M., Feng, C. P., & Al-Hadidy, A. I. (2011). Laboratory evaluation of HMA with high density polyethylene as a modifier. *Construction and Building Materials*, *25*(5), 2764-2770.

[11] Bala, N., Napiah, M., & Kamaruddin, I. (2018). Effect of nanosilica particles on polypropylene polymer modified asphalt mixture performance. *Case studies in construction materials*, *8*, 447-454.

[12] EN 12697-33, Bituminous mixtures - Test method for hot mix asphalt - Part 33: Specimen prepared by rolling compactor.

[13] EN 12697-33, Bituminous mixtures - Test method for hot mix asphalt - Part 22: Wheel tracking.

[14] Fernando, M. J., & Guirguis, H. R. (1983, August). Rubberized asphalt in pavement technology for hot climates. In *Road Engineering Association of Asia and Australasia, Conference, 4th, 1983, Jakarta, Indonesia* (Vol. 3).

[15] Gabriel, L. H. (2011). History and Physical Chemistry of HDPE Plastic Pipe Institute. *Corrugated Polyethylene Pipe Design Manual and Installation Guide.*

[16] Hafeez, I., Kamal, M. A., Mirza, M. W., & Bilal, S. (2013). Laboratory fatigue performance evaluation of different field laid asphalt mixtures. *Construction and Building Materials*, *44*, 792-797.

[17] Hossain, R., Arafat, S., & Wasiuddin, N. M. (2019). Identification and Evaluation of the Elastomeric, Plastomeric, and Chemical Modifiers in Asphalt Binders. In *Airfield and Highway Pavements 2019: Testing and Characterization of Pavement Materials* (pp. 255-265). Reston, VA: American Society of Civil Engineers.

[18] Ibrahim, A. H. A. (2019). Laboratory investigation of aged HDPE-modified asphalt mixes. *International Journal of Pavement Research and Technology*, *12*(4), 364-369.

[19] King, G., et al., (1999) " Additives in Asphalt ", J. Assoc Asphalt Paving Technol A, (75th Anniversary Historical Review and Index of Journals, Proceedings of the Association of Asphalt Paving Technologists).

[20] Lu, X., & Isacsson, U. (2000). Modification of road bitumens with thermoplastic polymers. *Polymer testing*, *20*(1), 77-86.

[21] Punith, V. S., & Veeraragavan, A. (2007). Behavior of asphalt concrete mixtures with reclaimed polyethylene as additive. *Journal of materials in civil engineering*, *19*(6), 500-507.

[22] Raad, L., Sebaaly, J., Epps, J., Coetzee, N., & Camilli, R. (1996). Low temperature cracking of polymer-modified AC mixes. *Alaska Cooperative Transportation and Public Facilities Research Program Project No. SPR-95-14 Task Rep.*

[23] Razak, N. A. (2015). *Utilization of Waste High Density Polyethylene (HDPE) as a Course Aggregate Replacement with Fly Ash Filler on Hot Mix Asphalt* (Doctoral dissertation, UMP).

[24] Sorum, N. G., Guite, T., & Martina, N. (2014). Pavement distress: a case study. *Vol. 3*, 274-284.

[25] State Cooperative of Roads and Bridges (SCRB). (2003). Standard Specifications for Roads and Bridges. Baghdad, Republic of Iraq: Ministry of Housing and Construction, Department of Design and Studies, Addendum No.3.

[26] Taha, R., Ali, G., & Delwar, M. (1998). Evaluation of coke dust-modified asphalt using Superpave. *Journal of materials in civil engineering*, *10*(3), 174-179.

[27] Wahhab, H. A. A., Ali, M. F., Asi, I. M., & Al-Dubabe, I. A. (1995). Adaptation of shrp performance based binder specifications to the gulf countries. Final Report, KACST Project AR-14-60, King Abdulaziz City for Science and Technology, Riyadh, Saudi Arabia.

[28] White, G. (2018). State of the art: Asphalt for airport pavement surfacing. *International Journal of Pavement Research and Technology*, *11*(1), 77-98.

[29] Yassoub, N. (2005). *Prediction Model of Rutting Potential in Flexible Pavements* (Doctoral dissertation, Ph. D Thesis, Civil Engineering, University of Baghdad).

Smart Parking System based on Improved OCR Model

Rami Bassam[1,*] and Fars Samann[2]

[1]BSc Student at College of Engineering, University of Duhok, Kurdistan Region, Iraq- E-mail: Ramibassam393@gmail.com

[2] College of Engineering, University of Duhok, Kurdistan Region, Iraq - E-mail: Fars.samann@uod.ac

*Corresponding author: BSc Student at College of Engineering, University of Duhok, Kurdistan Region, Iraq. Tel: +9647507435988

Abstract. Smart parking system becomes essential nowadays specially in urban area because it can reduce the time and the fuel wasted in searching for an empty parking slot. The aim of this paper is to develop smart parking system based on improved Optical Character Recognition (OCR) model. The proposed system consists of three stages: 1) OCR based parking slot detection, 2) User notification based on IoT approach, 3) Smart parking meter based on Simple Mail Transfer Protocol (SMTP). In this system, the vacant parking slots are detected using improved OCR model by labeling the parking slots with specific characters. The system identifies empty parking slots by detecting these characters with installed camera above the parking slots, otherwise the parking slots are occupied. The performance of the proposed OCR model in detecting these characters is enhanced by considering two stages of advanced morphology filter to remove unwanted small/large objects from the image. The accuracy of the proposed OCR model is tested by eight images having different parking street textures as a background and characters written with seven font styles. The idea of detecting empty parking spaces is to share the number of vacancies on a website for the drivers searching for a parking space. However, a time-controlled parking system with different charge rate each hour can be considered to make the best utilization of parking spaces available in urban areas. Therefore, a smart parking meter is proposed to notify the user through Email about the time left that allowed to be parked and the change in charge rate. The proposed system is developed and implemented using vision assistant of LabView.

Keywords: *Smart parking system; Optical Character Recognition (OCR); Parking slot detection.*

1. Introduction

The increase in number of vehicles especially in urban areas makes finding an empty parking slot very difficult nowadays. The analysis of traffic congestion shows that the main reason of traffic congestion in city center is due to 70 % of drivers are slowing down their speed while they are searching for empty parking slot (Athira, Lekshmi, Vijayan, & Kurian, 2019). This will increase the wastage of fuel, air pollution caused by vehicles and the time wasted in finding vacant parking space. Besides, in most of big cities, more than 31% of the land were wasted for parking coverage due to the large population (Lin et al., 2017). Several research papers are working on developing smart parking system to overcome these problems. Most of the current solutions are focused on developing empty parking lot detection using different type of electronic sensors or image processing methods with IoT approach to inform the drivers about vacant parking spaces through website or smart phone application. These research papers will be discussed in related work section.

The proposed approach of solving parking space problems in urban areas consists of three stages. The first stage is focused on developing empty parking slot detection based on improved optical character recognition (OCR). The second stage is to share the number of vacant parking slots on a website for drivers looking for empty parking space using IoT approach. The last stage deals with the smart parking meter system which notify the user about time-limit left of parking through an Email. This system is based on Simple Mail Transfer Protocol (SMTP).

2. Related Works

Most of the recent smart parking systems are based on detecting the vacant parking space using IR sensors or ultrasound sensors. In the work (Seth, Ashritha, & Namith, 2019), IR sensors with IoT approach are considered to develop smart parking system. The data from IR sensors about the status of parking spaces are sent to the cloud web services to inform the users about the number of available parking spaces that updated continuously. This approach has the disadvantage of complexity in installation since each parking slot requires IR sensor. In (Patil, Kadam, & Kaushal, 2018), smart parking system based on reservation service is proposed to allow the driver to pre-reserve the parking space online with mobile application. The drawback of reservation system is not efficiently utilizing parking lots due to the wasted time of empty reserved parking space because the driver is able to reserve the slot online from long distance.

Smart parking systems based on image processing method proposed by (Yusnita, Norbaya, & Basharuddin, 2012)(Athira et al., 2019) (Jain, Choudhury, & Kashyap, 2017). Different image processing techniques were used to develop parking slots detection instead of electronic sensors. Parking space detection using image processing was proposed by (Yusnita et al., 2012) that based on detecting brown rounded circles which are not hidden by parked car. In (Athira et al., 2019), OCR based smart parking system was developed that identify the empty parking spaces by detection the numbers which are not covered by parked car. However, most of these proposed OCR models were developed using one street texture background and one font style of park slot label. Since each garage may consider different street texture and different font style for labeling parking slots, it is important to consider the impact of the background texture of the garage on the performance of the OCR model.

An improved OCR model is considered in this paper because it is very efficient and simple method to identify empty spaces, beside it requires couple of installed cameras to cover the parking lot. Sharing the number of vacant spaces to drivers through IoT approach is also considered in this work. Moreover, a smart parking meter based on SMTP protocol is proposed to inform the user through Email about the time-limit left for parking. Moreover, a time-controlled parking system is proposed that based on increasing the charge rate each hour. This approach may make the utilization of parking lot more efficient.

3. Proposed Smart Parking System

The proposed smart parking system consists of three stages as shown in figure 1. In the first stage, the empty slots are detected by image processing technique known as OCR. The second stage shares the number of vacant parking spaces on a website. This will allow the user to track the nearest parking lot with the number of available spaces. In the last stage, a smart parking meter is considered to alert the user about the time-limit left of parking and the changes in the charge rate. This approach will help the user to define the limit of money to be spent by getting an Email about the time-limit left for parking. Figure 2 illustrates schematic model of the proposed system.

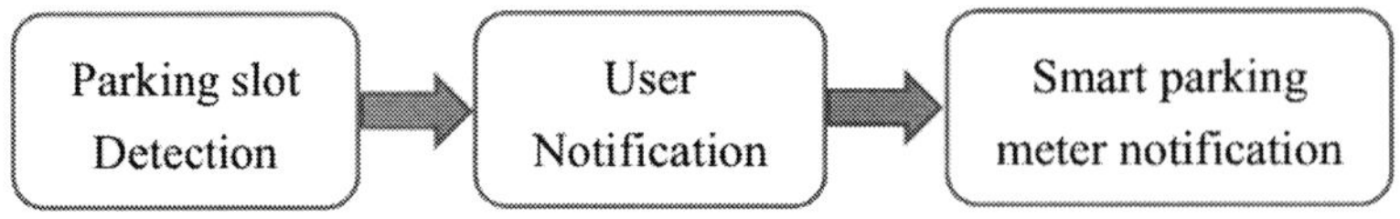

Figure 1. Block diagram of the proposed smart parking system

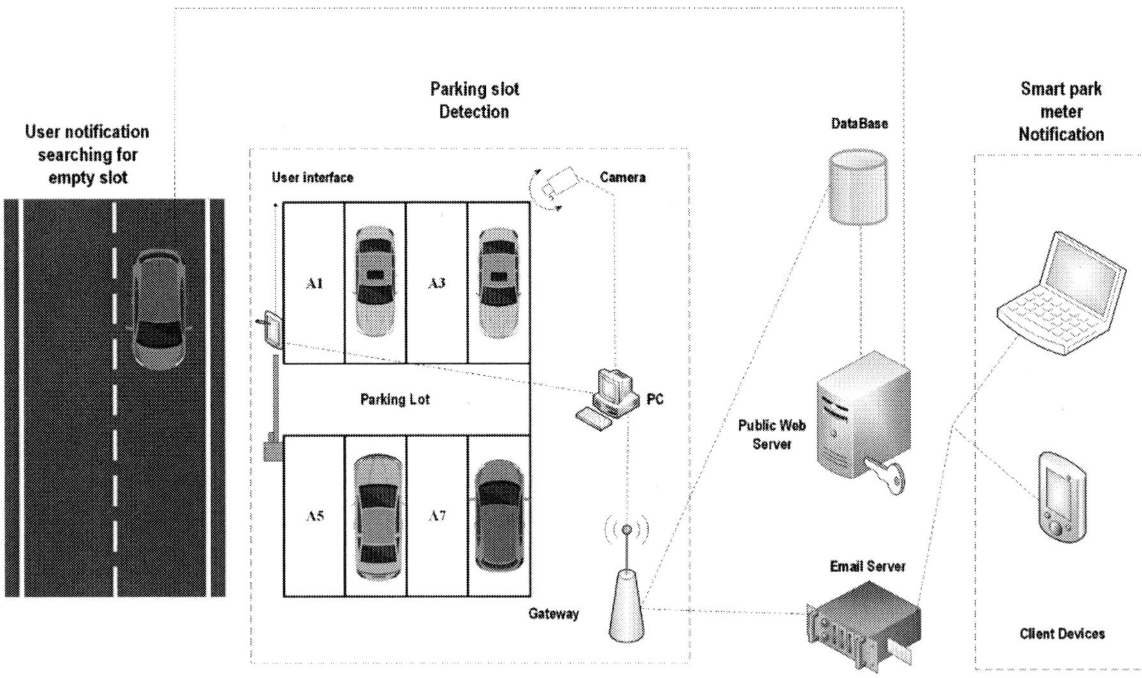

Figure 2. Proposed model of a smart parking system

3.1 OCR based Parking Slot Detection

Optical Character Recognition (OCR) is an advance image processing technique that able to convert the characters in the images captured by digital camera to digital characters. In this paper, the vacant slots are identified using modified OCR model. This can be achieved by labeling all parking slots with unique code in white color which consists of two digits as shown in figure 4 (the first digit is a character, and the second digit is a number). Considering such a complicated labeling of two digits is to avoid the wrong detection of a number printed over the vehicle. The camera is fixed in a position to capture the maximum number of slots in one frame. Depending on the design of the parking lot, it is possible to make the camera rotating and continuously capturing images of the entire garage with one camera only. The proposed OCR model is based on several of image processing phases as illustrated in figure 3. The first phase is capturing images of the whole parking garage, then the captured images are converted to grayscale image and improved by Math-lookup processing based on exponential operation. The process of Math-lookup is based on replacing the pixel values with defined values from lookup table (National Instruments, 2003). The use of exponential operation is to extend the contrast of large pixel values and reduce the contrast of small pixel values. The histogram phase is used to estimate the threshold value which is the sum of the standard deviation and the mean value. This threshold value will be considered in the segmentation phase. The image segmentation phase is based on thresholding the images to separate the characters. Before identifying the characters with OCR model, two stages of advanced morphology filters are considered to remove the unwanted large and small objects. The two advanced morphology filters are basically high-pass and low-pass filter which remove the particles with width greater $K1$ pixels or less than $K2$ pixels respectively (National Instruments, 2003). These two stages of advanced morphology filters have removed most of the large and small particles resulted from parking texture and the parked vehicles (see figure 5). Moreover, a lookup table with equalize function is considered to make the image having evenly distributed grayscale levels. Before applying OCR algorithm, a median filter with size of 2x2 is considered for further smoothing from salt and pepper noise. Then, the OCR algorithm is applied to identify the characters. This algorithm is based on template matching approach. Therefore, the algorithm of OCR model should be trained with a data set of numbers and characters. The detected characters from the processed image are compared with predefined vector of all unique codes of the parking slots, then missing code is denoted as an occupied slot. This is because the parked vehicle covered the code of that slot as shown in figure 5D.

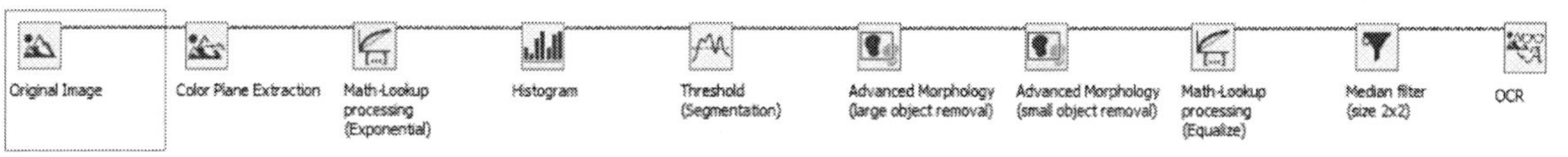

Figure 3. phases of OCR based parking slot detection

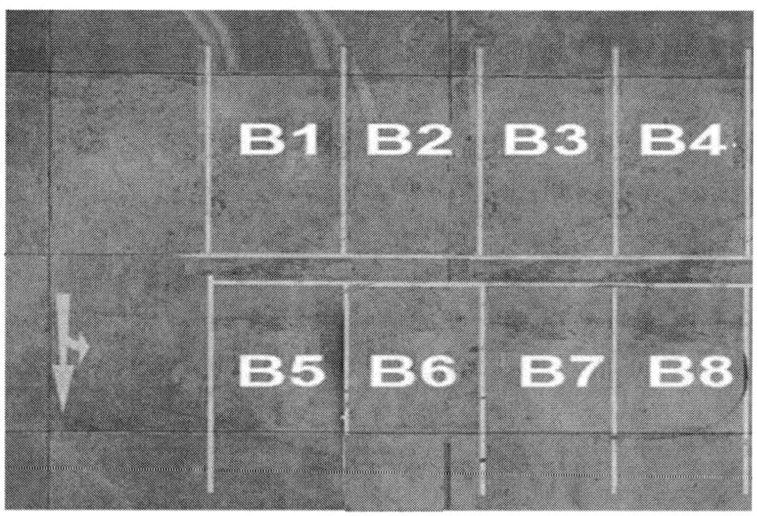

Figure 4. Labelling the slots of parking garage with white color unique code

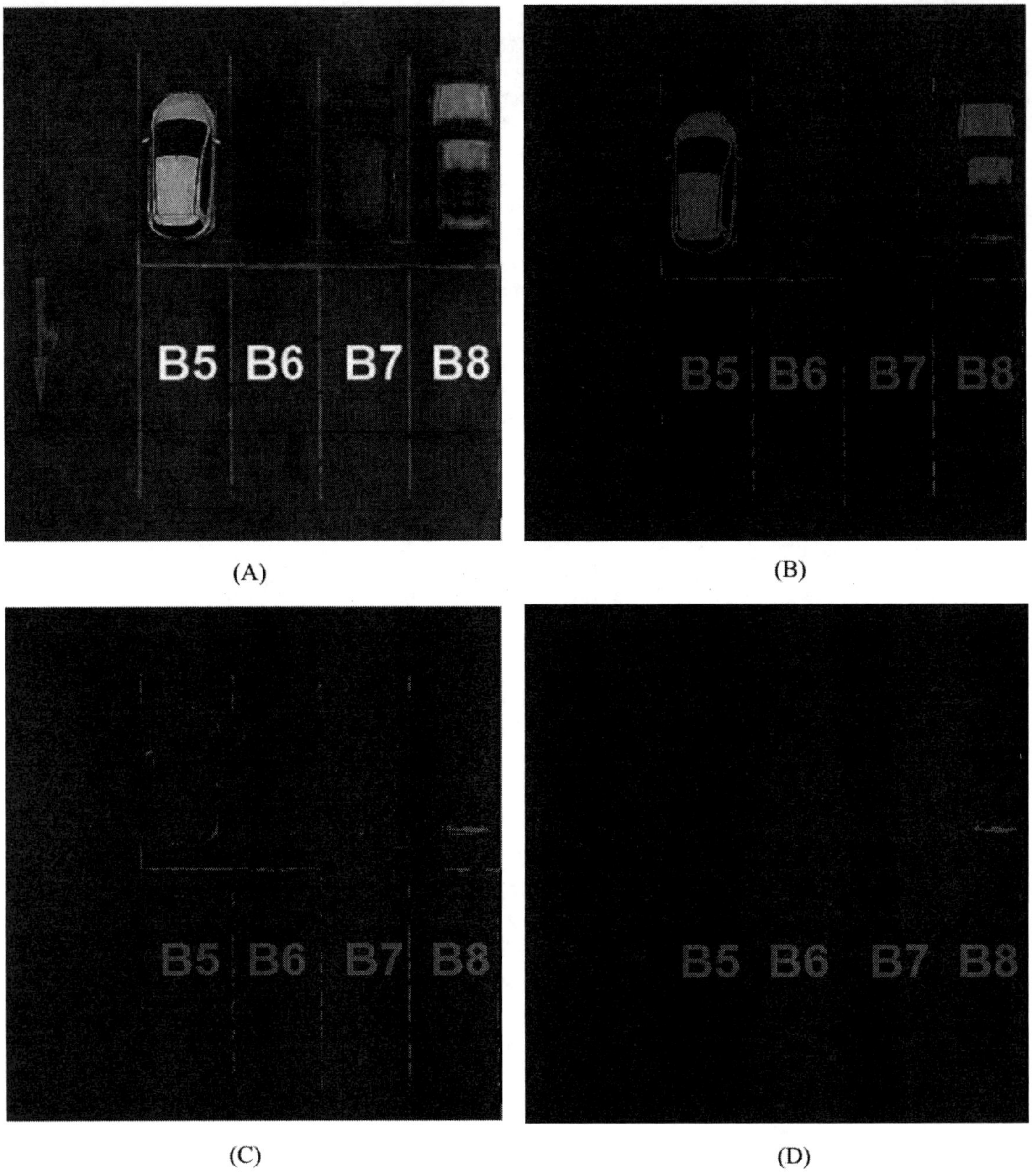

Figure 5. the processed image of different phase, (A) Math-lookup using exponential operation phase, (B) Segmentation phase, (C) advanced morphology filter for removing large objects, (D) advanced morphology filter for removing small objects

3.2 User Notification System

The main aim of smart parking system is to share the number of empty slots for the drivers searching for the nearest garage with vacant space. The approach of internet-of-thing (IoT) is considered to send the data acquired from parking slot detection method to web-enabled without any human intervention. The 'ThinkSpeak' website which is free access for IoT project is used. Two channels for two different

IOP Conf. Series: Materials Science and Engineering **978** (2020) 012007 doi:10.1088/1757-899X/978/1/012007

garages are designed, and each channel has the number of empty slots and the location of the garage at google map as shown in figure 6. This will help the drivers to find the nearby garage which has empty slots.

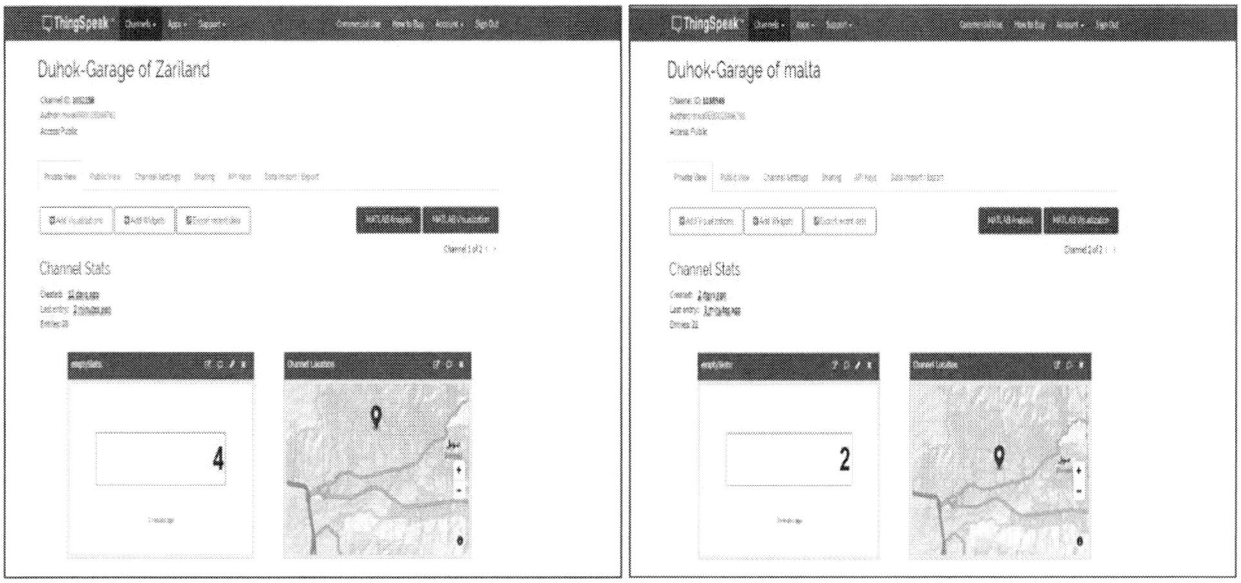

Figure 6. User notification website interface that shows number of vacant and the location of the park on google map.(Left) first channel "Duhok-Garage of Zariland", (Right) second channel "Duhok-Garage of Malta"

3.3 Smart Parking Meter

A time-controlled parking system is proposed in this paper, where the charging rate is increasing with respect to the parking duration as illustrated in table 1. Therefore, a smart parking meter based on SMTP protocol is developed to alert the user through Email about the time-limit left for parking and the change of charge rate (see figure 7). The system allows the users to set their time-limit of parking duration for Email notification by writing their email, the parking duration and the unique code of the slot that is occupied by the user as shown in figure 7 A. The technical interface helps the user to see the actual location of the empty slot. This can reduce the time wasted searching for the location of empty slot inside the garage as shown in figure 7 B. This system notifies the client with an Email ten minutes before the end of the time-limit parking duration, and each time the charge rate is changed (see figure 8). This approach will improve the utilization of parking lots specially in city center of urban areas by controlling the average parking duration through increasing charge rate with respect to parking duration. The proposed charge rate in this paper can be adopted with respect to the regulations of that city and it can be limited to specific number of hours.

Table 1. Charge rate with respect to parking duration

Parking Duration	Charge rate
< 60 minutes	0.25$ per 10 minutes
> 60 minutes & < 120 minutes	0.5$ per 10 minutes
>120 minutes & <180 minutes	1$ per 10 minutes
>180 minutes	1.5$ per 10 minutes

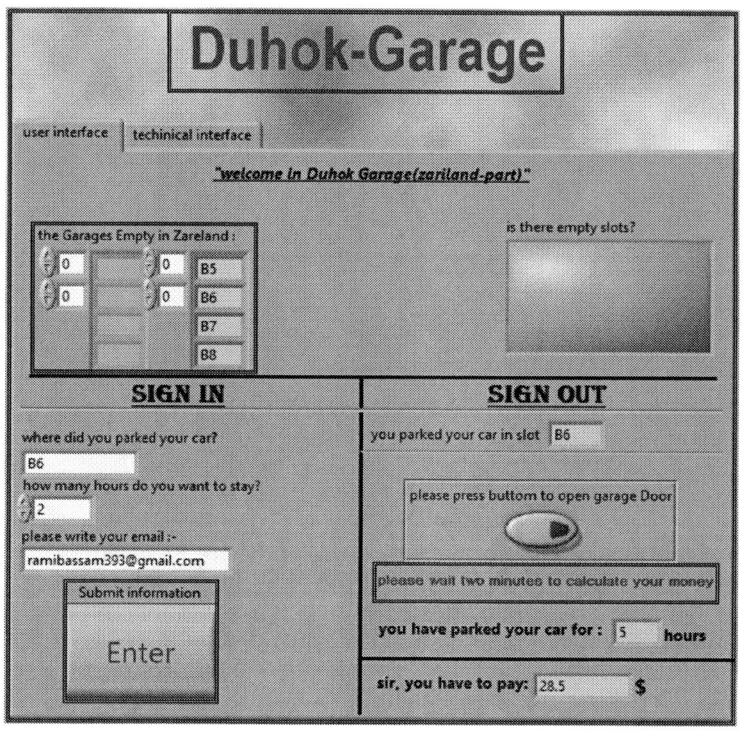

(A)

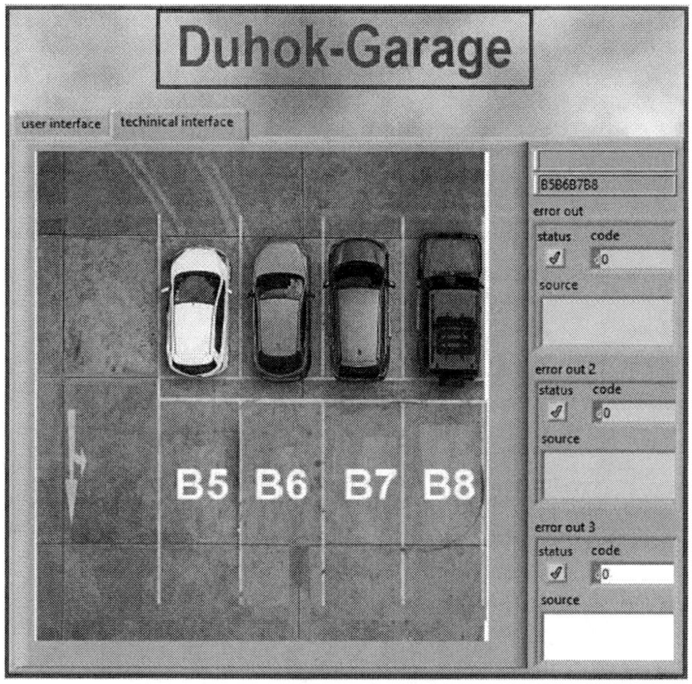

(B)

Figure 7. User interfaces at main gate of the garage. (A) this interface shows Sign in/out of smart parking meter system and the code of the vacant slots (B) the technical interface gives the real-time images of the actual location of the vacant slots

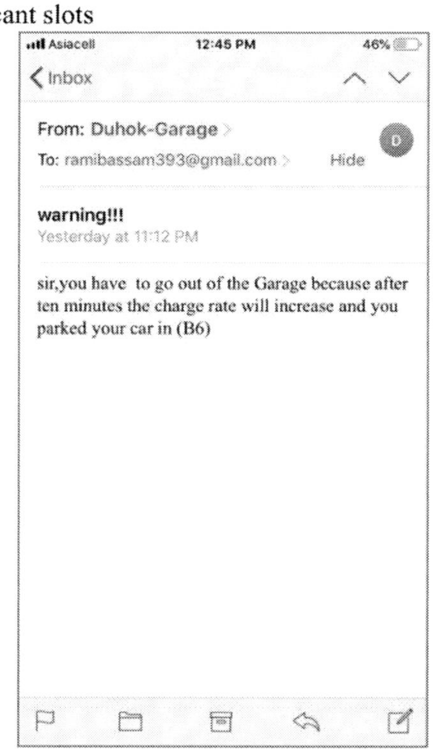

Figure 8. Email notification from smart parking meter

4. Results and Discussion

The proposed OCR based parking slot detection is developed using vision assistant tool in LabView 2018. A data set of images is generated using Microsoft office to train and test the proposed OCR model. The training image consists of eight numbers and eight English characters of seven different font styles and white color font with black background as shown in figure 9. The testing images are similar to the training image, but each testing image has a specific parking street texture as a background (see figure 10). Besides, the images of parking slots with/without parked vehicles from (Athira et al., 2019) are also used. The accuracy of OCR model is calculated by following formula (Samann, 2018):

$$accuracy\% = \frac{No.of \ \ correctly \ detected \ characters}{Total \ No.of \ characters} \quad (1)$$

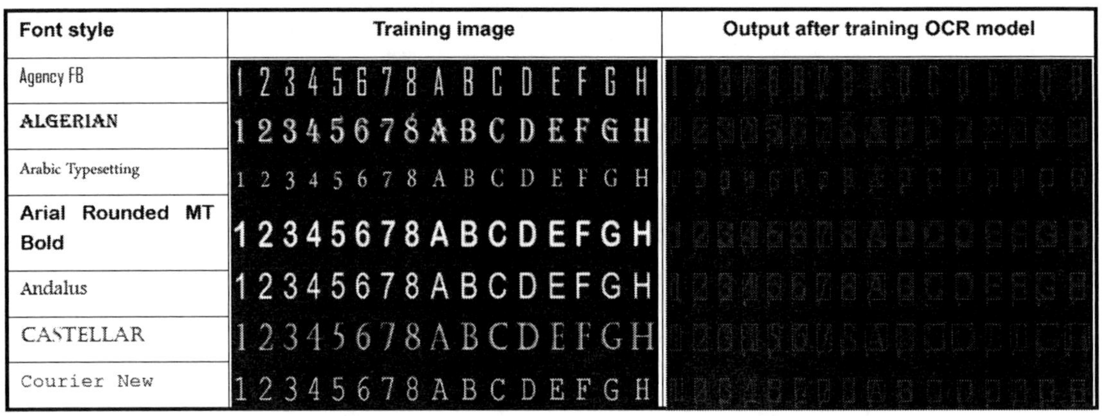

Figure 9. Training image and the output of OCR model after training

The most common parking street textures (see figure 11) are considered to measure the performance of the proposed OCR model in different environment. The accuracy in detecting the characters for different parking street textures are illustrated in table 2. In general, the results show that the smooth parking street textures (see texture 6 & 7 in figure 11) has excellent accuracy in detecting characters. Moreover, the results in table 2 indicate that font style plays significant role in the accuracy of the proposed OCR model. For example, the following font styles, Arial bold and Courier New, have excellent accuracy with most of parking textures.

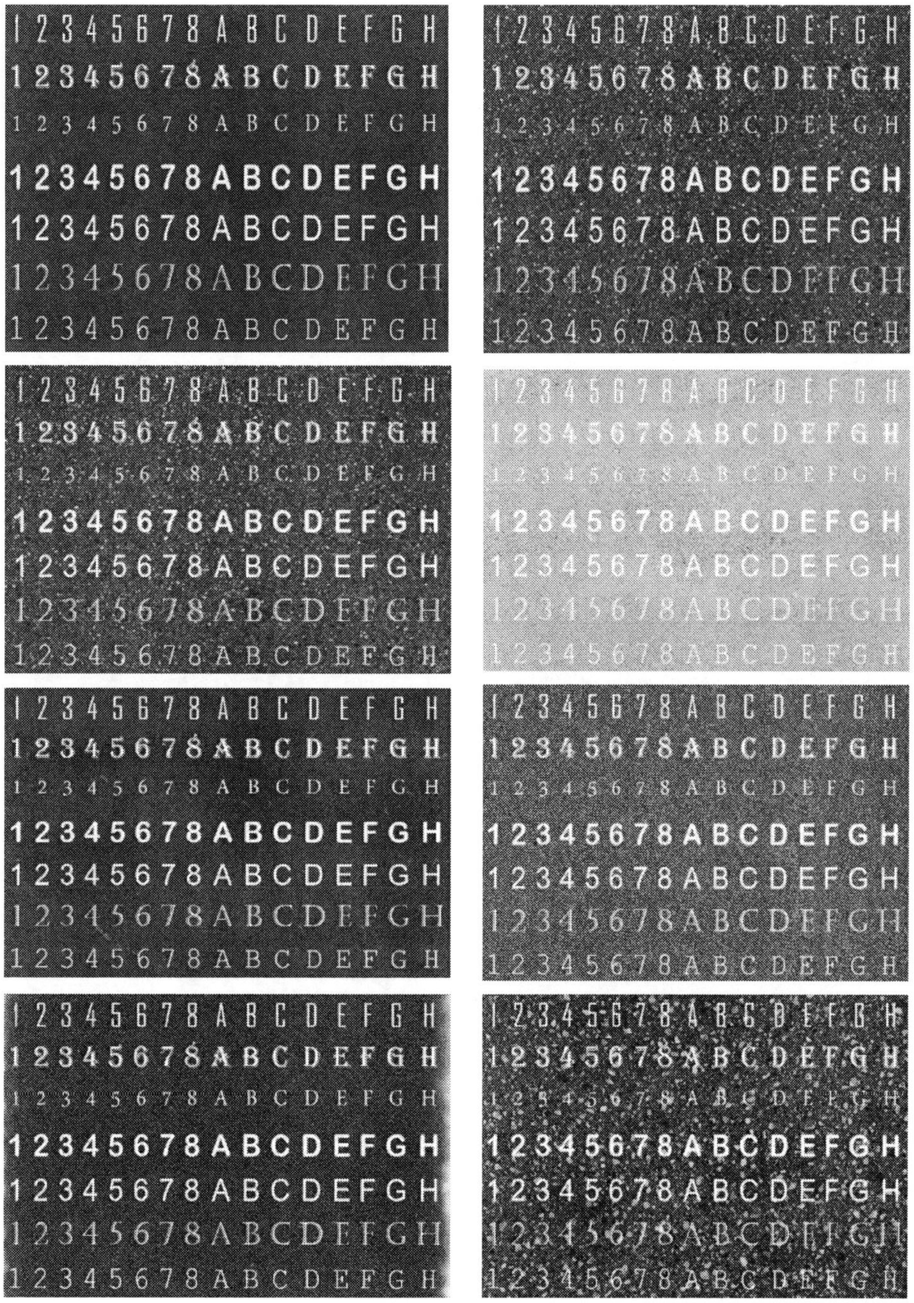

Figure 10. Testing images are used to test the performance of the proposed OCR model

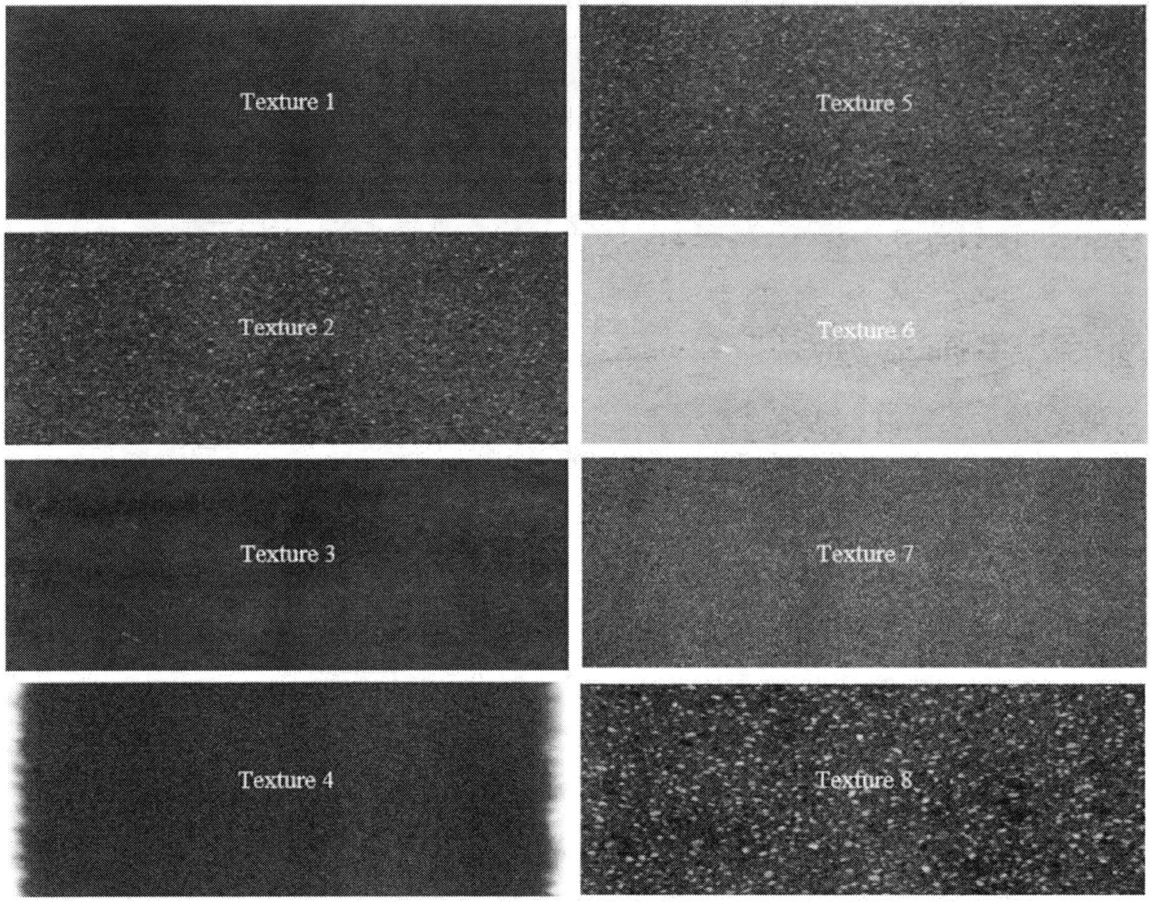

Figure 11. Eight different parking street textures

Table 2. Accuracy of the proposed OCR model

Front style	Texture 1	Texture 2	Texture 3	Texture 4	Texture 5	Texture 6	Texture 7	Texture 8
Agency FB	100%	100 %	100 %	100 %	93.75	100 %	100 %	93.75 %
Algerian	100 %	93.75 %	100 %	100 %	93.75	100 %	100 %	75 %
Arabic Typesetting	100 %	81.25 %	100 %	100 %	87.5 %	100 %	100 %	93.75 %
Arial Rounded MT Bold	100 %	100 %	100 %	100 %	100 %	100 %	100 %	93.75 %
Andalus	100 %	100 %	100 %	100	93.75 %	100 %	100 %	81.25 %
Castellar	87.5 %	87.5 %	87.5 %	87.5	87.5 %	100 %	100 %	87.5 %
Courier New	100 %	100 %	100 %	100 %	93.75 %	100 %	100 %	100 %

While not smooth textures like texture 8 & 5 have the worst accuracy due to the high salt and pepper noise resulted from the remain of the textures itself. This can affect the performance of OCR algorithm in identifying the characters. On the other hand, Castellar font style has the lowest accuracy comparing with other fonts. This is because their characters consist of two sides with line in the middle. Table 3 shows the average accuracy of proposed OCR model in comparison with OCR model developed in LabView by (Zhang, Cui, & Yang, 2015)

Table 3. Comparison in term of average accuracy between two OCR models developed in LabView

OCR model developed in LabView	Average Accuracy
OCR model proposed by this paper	>90%
OCR model proposed by (Zhang et al., 2015)	>80%

The following scenarios are considered as the main drawbacks of the proposed smart parking system: 1) The label of slot is covered by an obstacle, so the slot is considered occupied, but smart parking meter will be inactive because it requires Sign-In, 2) Label is not covered by the vehicle, the slot is considered not occupied and the user will be notified by User interface at main gate of the garage, when the user is trying to Sign-In.

5. Conclusion

Smart parking slot detection based on improved OCR model is proposed with efficient utilization approach of available slots through smart parking meter notification system. The information about the number of vacant parking slots and the location of the park on google map are available on a website which can be accessed by drivers. In this paper, the accuracy OCR model is improved by considering several image processing phases. Where the threshold value for segmentation phase can be calculated automatically by adding standard deviation and mean values from histogram phase, then two stages of advanced morphology filter are used to remove the unwanted small and large particles from the image. A median filter with 2x2 kernel is also used for further salt and pepper noise reduction before applying OCR algorithm. To evaluate the performance of the proposed OCR model, eight parking textures and seven font styles are considered as a testing data set. The results indicate that the accuracy of the OCR model is reliable specially in case of having smooth texture and clear and bold font style.

References

[1] Athira, A., Lekshmi, S., Vijayan, P., & Kurian, B. (2019). Smart Parking System Based On Optical Character Recognition. *2019 3rd International Conference on Trends in Electronics and Informatics (ICOEI)*, 1184–1188. Tirunelveli, India: IEEE.

[2] Jain, K., Choudhury, T., & Kashyap, N. (2017). Smart vehicle identification system using OCR. *2017 3rd International Conference on Computational Intelligence & Communication Technology (CICT)*, 1–6. Ghaziabad, India.

[3] Lin, T., Rivano, H., Mouël, F. Le, Lin, T., Rivano, H., Le, F., … Lin, T. (2017). A Survey of Smart Parking. *IEEE Transactions on Intelligent Transportation Systems*, *18*(12), 1–24.

[4] National Instruments. (2003). *IMAQ Vision Concepts Manual.*

[5] Patil, L., Kadam, S., & Kaushal, R. (2018). Reservation based Smart Parking System. *International Journal of Scientific & Engineering Research*, *9*(3), 194–197.

[6] Samann, F. (2018). Real-time Liquid Level and color Detection system using Image Processing. *Academic Journal of Nawroz University (AJNU)*, *7*(4), 223–227.

[7] Seth, J., Ashritha, P., & Namith, R. (2019). Smart Parking System using IoT. *International Journal of Engineering and Advanced Technology (IJEAT)*, *9*(1), 6091–6095.

https://doi.org/10.35940/ijeat.A1963.109119

[8] Yusnita, R., Norbaya, F., & Basharuddin, N. (2012). Intelligent Parking Space Detection System Based on Image Processing. *International Journal of Innovation, Management and Technology, 3*(3), 232–235.

[9] Zhang, T., Cui, Y., & Yang, Y. (2015). Design of a Character Recognition System Based on LabVIEW. *4th International Conference on Mechatronics, Materials, Chemistry and Computer Engineering (ICMMCCE 2015)*, (1), 347–350.

Using AHP Methods in Maintenance to Improve Reliability and Equipment Performance

Kadhum A. Abed[1,*], Sattar A. Mutlag[2]

Department of Mech. Eng., College of Engineering, University of Anbar, Ramadi, Anbar, Iraq - E-mail: kadhum1968@uoanbar.edu.iq
[2]College of Engineering, University of Anbar, Department of Mech. Eng. Iraq-Anbar-Ramadi- E-mail: satmutt1961@uoanbar.edu.iq
* Corresponding author: College of Engineering, University of Anbar, Iraq-Anbar-Ramadi Tel: +9647832949085

Abstract. The engineers are usually expected to consider several influential criteria when selecting as a single best alternative for water maintenance system (WMS). This research discusses the approach based on the Analytic Hierarchy Process (AHP) as a decision making tool for selecting WMS. Supported with information of equipment breakdown, the questionnaire method as well as personal interviews as a main tool to collect data and information in the practical aspect of the study. In this study, the AHP decision support tool was applied in a maintenance systems field in the town of AL- Fallujah Water Treatment Project, Iraq. The AHP was applied to evaluate and an alternative is chosen from a set of maintenance alternatives: [Preventive Maintenance (M1), Proactive Maintenance (M2), Total Productive Maintenance (M3), Predictive maintenance (M4) and Corrective Maintenance (M5)] vs eight of Maintenance pillars. Results indicate the advantages over traditional Decision Making selection methods. By using this information with AHP the best rank of alternatives was M3(11.85), M4(9.67) and so on. Moreover, the relationship between M3 and the pillars of maintenance: indicates a very high correlation 78%. The study reached a number of conclusions, the most important of which is the existing of a very strong impact for the proposed maintenance system in improving the performance and reliability of FWTP project operations.

***Keywords**: maintenance systems; decision making; Analytic Hierarchy Process (AHP); Pareto analysis; Reliability function.*

1. Introduction

It has become clear that the continuation of industrial and service projects depends on the efficiency of the operation of the machinery and equipment. so, it is necessary to rely on maintenance work, which includes all activities that industrial projects need to maintain machinery, equipment and buildings and make them conform to the standard operating condition. Saxena et al [1] As a result of this role that maintenance plays, scientific and practical measures must be taken to apply maintenance activities at certain times. Vishnu & Regikumar, [2] By means of which the operational processes continue with a high efficiency and reduce the occurrence of sudden failures with at the lowest possible cost, so for the purpose of achieving this goal. Ahmad et al. [3] a modern scientific methods are adopted, which have become that can be relied upon in choosing the appropriate time for implementation and at the lowest possible cost. Harrison, et al. [4] the function used to evaluate the functional performance of systems and machines is the reliability function. There are some auxiliary tools that are used in this evaluation such as failure rate, Pareto analysis, maintenance function, and others tools [5,6]. The research problem: through field visits to the reality of the FWTP in Fallujah, it was found that the project suffers from a problem of high maintenance costs and fluctuation in its implementation times, which in turn created a gap between the current performance of the project and the expected performance of it, and it cannot bridging this gap except by developing existing maintenance policies and processes, by applying a modern and advanced maintenance system (predictive maintenance system based on Preventive maintenance system)[7]. This system is considered one of the most important factors for the success of contemporary production systems [1,7]. Depending on the above, the research aims to improve the reliability and performance of FWTP equipment using modern scientific methods of maintenance. Thus, despite many researchers have already discussed about criteria, sub-criteria and criteria and allocation of resources, the criteria and criteria selection of each scenario is particularly unique.

2. Material and Methods

The choice of an optimal maintenance strategy for a system is a complex and multidimensional decision-making problem because of the data collection phase, numerous, conflicting criteria and decision makers from various fields, as well as a variety of components and functions that need to be addressed as a systematic approach Mohammad et al. [5]. The MCDM approach has received great interest in the last half of the 2000s. MCDM methodology is applied to select the most appropriate maintenance strategies in various industries [7]. Many MCDM approaches have been proposed in the literature. Bevilacqua & Marcello [8] in an Italian petrol company, Zaim [9] et al. and Wang et al [10] and Mutlag and Kassam [11] in a water treatment plant, and Lizot et al, [7] in the water treatment system in a Brazilian sanitation company, have chosen the most appropriate maintenance strategy using the Analytic Hierarchy Process (AHP) method. Giada [12] evaluated cost, safety, added value, equipment and technology criteria, and CM, time-based preventive maintenance (TBPM) and condition-based maintenance (CBM) strategies with the AHP method. Triantaphyllou et al. [13] selected the most appropriate maintenance strategy for a hospital with the AHP method. The study evaluates Preventive Maintenance (M1), Proactive Maintenance (M2), Total Productive Maintenance (TPM)(M3), Predictive maintenance (M4) and Corrective Maintenance (M5) as a maintenance strategies alternatives.

2-1. Total Productive Maintenance (TPM):

Melesse & Ajit [14] is a maintaining system and improving the integrity of production, safety and quality systems through the machines, equipment, processes, and employees. The objective of TPM is maximization of equipment effectiveness. TPM aims at maximization of machine utilization and not merely machine availability maximization. As one of the pillars of TPM activities, Kaizen pursues efficient equipment, operator and material and energy utilization, that is extremes of productivity and aims at achieving substantial effects.

Ikuobase et al. [15] collected two traditional maintenances (predictive preventive maintenances) were developed by Seiichi Nakajima of Japan. The results of his work on the subject led to the Total Productive Maintenance TPM process in the late 1960s and early 1970s. Nippon Denso (now Denso), a company that created parts for Toyota, was one of the first organizations to implement a TPM program. Abbas et al [16] This resulted in an internationally accepted benchmark for how to implement TPM. Incorporating lean manufacturing techniques, TPM is built on eight pillars. The eight pillars of total productive maintenance focus on proactive and preventive techniques to help improve equipment reliability. The eight pillars are [13-18]: Supporting top management P1, focused improvement P2, Educating and Training P3, Maintenance Information Support Systems P4, Supporting and Assistance all Departments P5, The Total Cost of Maintenance P6, Quality Maintenance P7, Participation of all employees P8.

2-2. Statistical Probability Distributions in Reliability Analysis:

Exponential Distribution: It is the most widely used distribution in the field of reliability analyzes and is used to describe equipment with a fixed fault rate. As in the following equation [5]:

$$R_{(t)} = e^{-\lambda t} \quad \text{where: } \lambda \text{, Distribution prameter and t, time} \tag{1}$$

Weibull Distribution: it comprehensively describes all stages of the life cycle of the stomach, as it describes the phenomenon of decreasing and increasing failure rate, in addition to the phenomenon of stability. Reliability function is measured [5]:

$$R_{(t)} = e^{\left(-\frac{t}{\alpha}\right)^{\beta}} \quad \text{where: } \beta, \alpha \text{ Distribution prameters and t, time.} \tag{2}$$

Pareto Chart Analysis: One of the seven basic tools for quality control data analysis is the Brito chart. Pareto notes that "the majority of the holidays are due to the small (few) reasons." Accordingly, he divided the elements to be studied into two parts: "vital few" and "trivial many" [3]. Then he put the following rule (90% -80%) from the results that occur due to (20% -10%) of the causes. The method was developed by (Juran 1990) to become its present form [9].

$$\text{The percentage of failure repeated Fr\%} = \frac{\text{Number of occurrences of Failure NF}}{\text{Total Number of Failures TNF}} \tag{3}$$

The impact of TPM on reliability: Reliability improvement due to maintenance, Improving the reliability of a complex system can be achieved through TPM that aims to keep the system in good condition with a certain level of performance and reliability, as it maintains a constant malfunction rate and increases the actual operating period of the TBF system [5]. The expected goal of maintenance involves reducing the costs of unreliability, which is the loss of production as a result of the waiting period for components to be repaired or maintained to return to service [2]. To reach this goal by using the Integrated Approach methods within management, operation and maintenance as activities for the reliability improvement program for customer satisfaction [5]. The system reliability with preventive maintenance Rm (t) is [2,5]:

$$R_m(t) = R(T)^n \, R(t - nT) \quad for \quad \begin{matrix} nT \le t < (n+1)T \\ n = 0,1,2,3 \ldots \end{matrix} \tag{4}$$

Where: Rm(t): system reliability with predictive maintenance. R(t): reliability of the system without maintenance. T: is the time between two preventive maintenance cycles. $R(T)^n$: The possibility of the system remaining running up to n maintenance cycles. $R(t - nT)$: The probability that the system will remain up to (t-nT) from the time after the last maintenance cycle.

Figure (1) Shown reliability R(t) without preventive maintenance, reliability Rm(t) with preventive maintenance. Therefore, the reliability improvement over time by preventive maintenance, and the curve at the top represents preventive maintenance and TPM maintenance returns the system as good as new at the end of each preventive maintenance cycle. the reliability function Rm(t) appears as a decreasing function over time [5].

Therefore, the reliability Equation (5) with maintenance when Tp = nT can be [2]:

$$R_m(T_P) = exp\left[-n\left(\frac{T_P}{n\alpha}\right)^{P}\right] = \exp\left[-n^{1-\beta}\left(\frac{T_P}{\alpha}\right)^{P}\right] \tag{5}$$

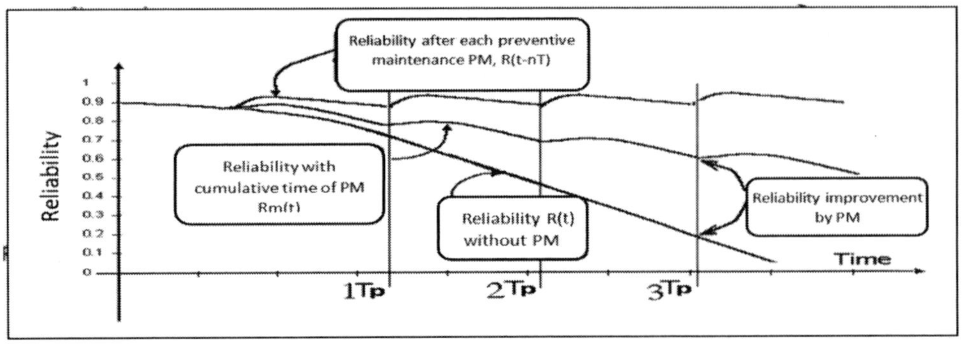

Figure 1. Periodic maintenance reliability curve for an increasing failure rate

2-3. The Mathematical model to calculate the total cost of maintenance:

The expected total cost per unit of working time represented by the symbol $C(t_P)$[5]:

$$C(t_P) = \frac{C_P + C_F * H\ (T_P)}{T_P - B * H\ (T_P)} \tag{6}$$

where: C_P: the cost of a preventive replacement before a breakdown occurs, and includes the cost of the idle segment and the cost of stopping production. C_F: replacement cost after the breakdown, and includes the cost of spear parts and cost of production loss, as that $C_F > C_P$ B: breakdown period to repair the failure when it occurs. $H\ (T_P)$: hazard rate during the period $(T_P, 0)$ and as in the following formula:

$$H\ (T_P) = \int_0^{T^P} \frac{\beta}{\alpha}\left(\frac{t}{\alpha}\right)^{\beta-1} dt \qquad where: (T_P, \beta, t, \alpha)\text{: explained earlier} \tag{7}$$

3. Case Study

3-1. The city of Al-Fallujah is one of the most important and largest cities in Anbar Governorate. it is witnessing rapid population growth and growth as the population has reached approximately (200,000 people), so improving the infrastructure, especially drinking water purification plants, will lead to improve the social level of the health for them. The study of the new Fallujah Water Treatment Project (FWTP), which is the largest and most important drinking water supply station in the city of Al-Fallujah, where this station supplies nearly 80% of the city's population with drinking water. it is design capacity 2000 cubic meters / hour and its working system is 24 hours. The project consists of a group of systems [23]:

1- Intake System (A1,2,3): The intake system consists of three Pumps. The design capacity of each pump is 720 m3 / hour.
2- Deposition station(B1,2,3). It consists of a mud sweeper containing on the motor and a gear box and a median mixer for the tub, and valve to enter and other to exit water for sedimentation process.
3- Alum system(C1,2,3,4): It consists of four alum injector.
4- Chlorination system(D1,2): It consists of two chlorine devices and two posters. Liquid chlorine is added from large cylinders to the water by means of a chlorine injection device.
5- Clay station(F1,2): It consists of three pumps.
6- Filters station(G1,2). It consists of ten filters for each entry and exit gate filter; in addition to that it has two reverse wash pumps, and also contains two air compressors.
7- Ground tank(H): It store the water that all the treatment operations have been made to be ready to be pumped into the water networks through main Lifting system.
8- The high Lifting system (E1,2,3,4,5,6): consists of two lines (two tubes) with a diameter of 500 mm. Each line contains three pumps with a capacity of 540 m3 / hour.

3-2. Analysis of the actual reality of maintenance operations: in the FWTP Based on personal interviews with the officials (top management, engineers, Expert technicians, workers) of FWTP Project, or observations of researchers. The maintenance unit is one of the main units in FWTP Project. so, the maintenance main partner that production operations in the project cannot be completed successfully unless they are based on efficient and effective maintenance actives. The reason is that the production operations in the water project are largely based on (equipment and machinery) that consists Including project equipment. Therefore, top management took a pioneering step towards the maintenance unit by giving it a distinctive role in the decision-making process by occupying a vital position in the project organizational structure, as is the case with other units such as (management and operation) and this means obtaining financial support sufficient moral from top management.

3-3 Data Analysis Steps: To study and evaluate the project reliability, some auxiliary tools were used to analyze and measure reliability, which are explained in paragraph (2-3).

3-3-1 Calculation of reliability analysis:

Calculating the number of frequently of failure: From the preliminary data of the project failures, which were taken from the project records and after their classification, and calculating the number of recurring failures for each part of the project and arranged.

Table 1. Represents the project's failure times after sorting its data.

A1		Intake Pump A1			
	date	FROM(hr)	TO(hr)	TBF(days)	TTR(days)
1	01/01/2018	2.5	3.5	103.7	
103	14/04/2018	19.2	10.5	169.0	0.4
169	30/09/2018	11.0	19.5	44.5	0.4
45	14/11/2018	7.0	12.5	28.9	0.2
29	13/12/2018	10.0	1.0		0.4

A2		Intake Pump A2			
	date	FROM(hr)	TO(hr)	TBF(days)	TTR(days)
31	01/02/2018	9.0	7.0	104.8	
104	16/05/2018	25.7	14.0	78.1	0.5
78	02/08/2018	17.5	23.0	94.6	0.2
95	05/11/2018	13.5	16.0		0.1

E5	High Lifting Pumping E5				
	date	FROM(hr)	TO(hr)	TBF(days)	TTR(days)
14	15/03/2018	3.9	10.6	15.4	
15	14/08/2018	20.6	17.6	58.8	0.1
59	17/11/2018	12.4	26.6		0.6

A3		Intake Pump A3			
	date	FROM(hr)	TO(hr)	TBF(days)	TTR(days)
14	15/01/2018	7.8	9.7	15.5	
15	30/01/2018	22.3	16.7	59.0	0.2
59	30/03/2018	16.3	25.7	75.4	0.4
76	14/06/2018	12.3	18.7	151.9	0.3
152	13/11/2018	15.3	7.2	20.9	0.3
21	04/12/2018	5.3	6.2		0.0

E6	High Lifting Pumping E6				
	date	FROM(hr)	TO(hr)	TBF(days)	TTR(days)
14	15/03/2018	3.9	10.6	15.4	
15	30/01/2018	22.3	16.7	59.0	0.2
59	17/11/2018	12.4	26.6		0.6

E1		High Lifting Pumping E1			
	date	FROM(hr)	TO(hr)	TBF(days)	TTR(days)
10	11/01/2018	13.2	23.9	100.5	
101	22/04/2018	11.0	19.9	133.5	0.4
134	03/09/2018	8.8	15.9	50.6	0.3
51	24/10/2018	6.6	11.9	59.7	0.2
60	23/12/2018	4.4	8.0		0.1

E2	High Lifting Pumping E2				
	date	FROM(hr)	TO(hr)	TBF(days)	TTR(days)
41	11/02/2018	19.7	27.4	100.6	
101	23/05/2018	17.5	23.4	133.7	0.25
134	04/10/2018	15.3	19.4	50.7	0.17
51	24/11/2018	13.1	15.4		0.10

E3		High Lifting Pumping E3			
	date	FROM(hr)	TO(hr)	TBF(days)	TTR(days)
24	25/01/2018	18.5	30.1	100.4	
101	06/05/2018	16.3	26.1	133.5	0.4
134	17/09/2018	14.1	22.1	50.6	0.3
51	07/11/2018	11.9	18.1	0.8	0.3

E4		High Lifting Pumping E4			
	date	FROM(hr)	TO(hr)	TBF(days)	TTR(days)
14	15/03/2018	14.6	31.0	14.2	
15	16/10/2018	12.4	27.0	58.3	0.61
59	06/12/2018	10.2	23.0	1.0	0.53

Calculating the value of TTR: Calculating the value of time to repair (TTR), by calculating the difference between the time of the start of the break and the time of completion of the maintenance, inspection and operation, in hourly and minute and converted into days.

Calculating TBF: the value of Time Between Failure (TBF) by calculating the difference between the start time of the equipment after a failure and the next failure in hour and minute and convert it into days. The popular statistical software (Minitab-19) was used to process these basic and subsequent basic and complementary accounts in the research. By relied on Table (1) as input for calculating TBF and TTR.

3-3-2. Calculating Pareto Analytics:

From Table (1), and using Equation (7), the frequency of the failure of each part was calculated, the results are listed in Table (2). The results analyzed using the Pareto rule to determine which parts have the most impact on project performance. From figure (2) conclude that the intake system A and high lifting system E are the most influential as they cause more than 75% of the project's failures from work, so we will focus the study on these two systems.

Table 2. Repetition of failure for each system (station) of the FWTP

system	Frequency	Fr%	Cumulative Fr%
Intake System	18	0.36	36%
High Lifting system	16	0.32	68%
Alum system	7	0.14	82%
Deposition station	5	0.1	92%
Filters station	2	0.04	96%
Clay station	1	0.02	98%
Chlorination system	1	0.02	100%
Ground tank	0	0	100%

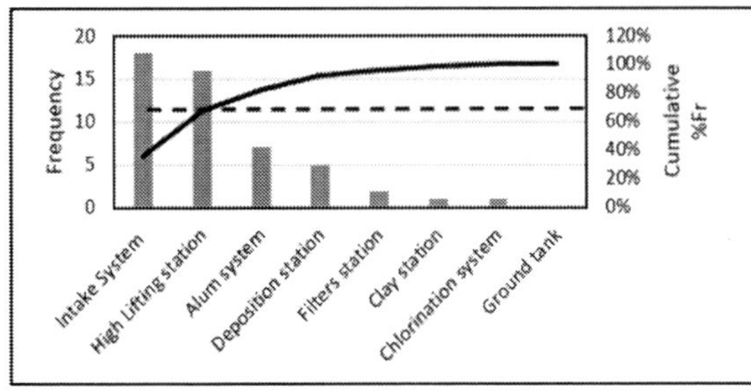

Figure 2. Pareto's analysis

3-3-3. Calculation of Parameters of Project Parts:

The computational steps for estimating distribution parameters are done by using the Minitab 19 software and listed in Table (3). The most parts of the project have an increased rate of failure $\beta > 1$, especially the two systems of Intake system and High Lifting system.

Table 3. Elements of system parameters α β and for project equipment

system	β value	α days
Intake System	2.58	122.33
High Lifting system	2.77	183.40
Alum system	2.57	27.98
Deposition station	1.40	37.90
Filters station	1.21	53.55
Clay station	0.78	22.30
Chlorination system	0.59	42.45

3-3-4. Reliability analyzes: the impact of TPM on improving project reliability:

1- One of the most important elements of the total cost of maintenance is the cost of repair time and loss in production. Table (4) shows TTR_{plan} and TTR_{unplan} project, maintenance cost MC, spare parts price SPP, cost of maintenance workers MW, which were collected from the FWTP project data.

2-Estimate the cost of failure maintenance: the total maintenance costs are divided into two main types:

A- To estimate the total maintenance cost for the intake system, for example: From Table (4):

Cf = cost of production loss + cost SPP + cost MW to complete fault repair tasks when they happen (8)

= (Production Capacity * Unit Price * $TTR_{unplanned}$) + SPP + (Cost of MW * $TTR_{unplanned}$).

= (2000 m3 / hour * 24 * [2 *10^3Dr] *0.34 days) + (250* 10^3) + (170* 10^3 Dr/ day * 0.34 days).

= 18259.8 *10^3Dr

B- Estimate the total cost of predictive maintenance of the failure:

Cp = cost of production loss + cost SPP + cost MW to complete PM tasks (9)

= (Production Capacity * Unit Price * $TTR_{planned}$) + cost SPP + (Cost MW * $TTR_{planned}$).

= (2000 m3 / hour * 24 * [2 *10^3Dr] * 0.2 days) + (250* 10^3) + (170* 10^3 Dr/ day * 0.2 days).

= 10844* 10^3Dr

The above steps can be repeated to calculate the maintenance cost of the high lifting system. Table 4 shows a summary of the results of maintenance cost calculations for intake and high lifting systems, which will rely on in calculating the optimal time period for preventive maintenance.

Table 4. Summary of the results of maintenance cost calculations for intake and high lifting systems

Average β	Average α day	value β	valueα day	C_p	C_f	$TTR_{unplaned}$ day	TTR_{planed} day	symbol	system
1.58	122.33	2.58	122	10844	18259.8	0.34	0.20	A1	Intake system
		2.51	120	10844	19848.9	0.37	0.20	A2	Intake system
		2.5	125	10844	17764.8	0.33	0.20	A3	Intake system
1.77	183.4	2.79	1 87	16141	24086.5	0.45	0.30	E1	High lifting pump
		2.82	182	16141	25145.9	0.47	0.30	E2	High lifting pump
		2.75	190	16141	24616.2	0.46	0.30	E3	High lifting pump
		2.72	176	16141	30509.1	0.57	0.30	E4	High lifting pump
		2.77	182	16141	20378.6	0.38	0.30	E5	High lifting pump
		2.73	184	16141	22376.2	0.45	0.30	E6	High lifting pump

3-3-5. Calculating the optimum time period for preventive maintenance based on TPM: cost and method of calculation are described in the paragraph (2-3) by using the Equation (6,7) to calculate the total cost against the expected period of maintenance. By using Table (4) the expected total cost of maintenance for high lifting and intake pumps summarized in Table (5), which can be represented by the maintenance cost vs operation period in scheme, from the Table (5) & Figure (3) shows the optimal cycle time of preventive maintenance for high lifting pumps (86) days and intake pumps (121) days.

Table 5. Optimum cost and optimum time maintenance by TPM method

lifting pump		intake station	work time day
354.13		371.80	31
306.84		328.32	36
192.03		234.36	61
180.69		227.49	66
171.37		222.71	71
163.71		219.66	76
157.40		218.03	81
152.25		217.60	86
148.07		218.21	91
144.73		219.71	96
142.12		221.99	101
140.16		224.95	106
138.77		228.53	111
137.89		232.65	116
137.47		237.27	121
137.47		242.34	126
137.85		247.82	131
138.58		253.68	136
139.64		259.89	141
141.00		266.43	146
142.64		273.27	151

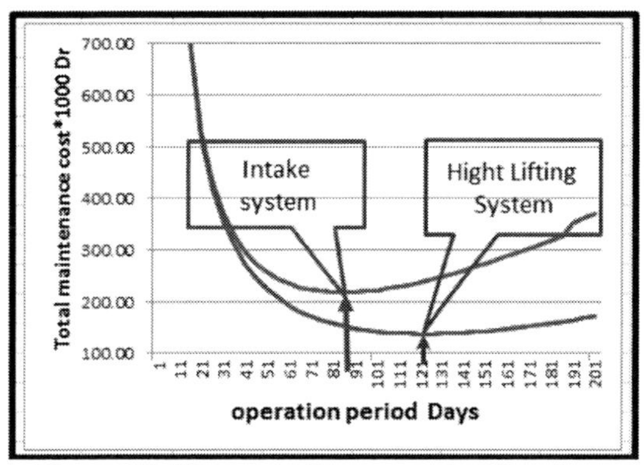

Figure 3. Shown optimum cost and optimum time maintenance for high lifting and intake pumps

3-3-6. Calculating a Reliability Improvement Ratio using TPM:

To calculate the reliability improvement ratio of the intake system, for example: The time period for the predictive preventive maintenance cycle was calculated in the previous paragraph and it was (121 days). Therefore, the improvement rate is calculated using the two Equations (4,5) as follows:

Reliability with corrective (sudden) maintenance, that is, without preventive maintenance:

$$R_{(nT)} = e^{-\left(\frac{t}{\alpha}\right)^{\beta}} = e^{-\left(\frac{121}{183.4}\right)^{2.77}} = 0.73 \qquad (10)$$

Reliability with preventive maintenance:

$$R_{m(nT)} = e^{\left[-(n^{(1-\beta)})\left(\frac{T(p)}{\alpha}\right)^{\beta}\right]} = e^{-\left[(2^{(1-2.77)})*\left(\frac{121}{183.4}\right)^{2.77}\right]} = 0.89 \quad (11)$$

The reliability improvement ratio of the intake system

$$\frac{R_{m(nT)}}{R_{(nT)}} = \frac{0.89}{0.73} = 0.17 = 17\% \quad (12)$$

The reliability improvement ratio of the intake system is 17%. With the same steps above the reliability improvement ratio of the high lifting system is 14%.

3-4 The Hierarchical Analysis Process AHP method: AHP is a way to classify solutions and work out the best solutions when providing more than one alternative. The study was made up of several alternatives. The proposed methodology for choosing a maintenance system method consists of two basic stages: (1) the hierarchical of problem, and (2) the process of calculating AHP.

3-4-1. The first step: defining the problem and then formulating the problem as a hierarchical structure. Where the hierarchy of the FWTP decision is created as explained in Figure 4.

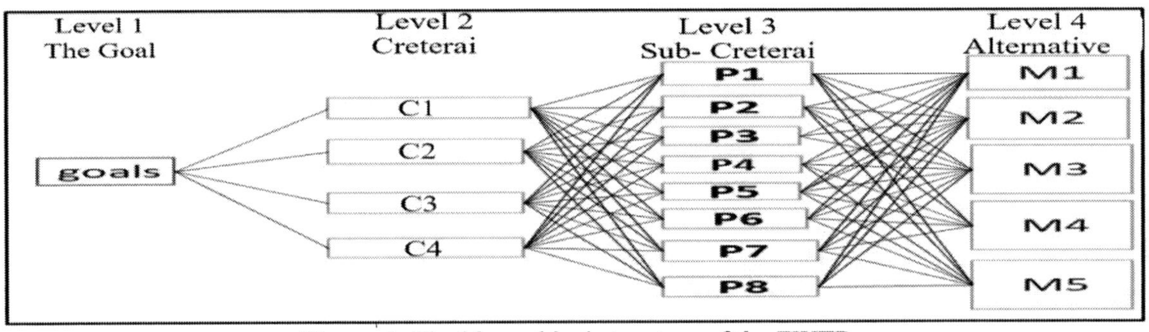

Figure 4. The hierarchical structure of the FWTP

3-4-2. The second step: building binary comparison matrices [11]. used to determine the scores achieved by each alternative according to each criterion (P_{ij}), where the decision maker compares each of the two alternatives according to each criterion individually using the preference scale which consists of nine degrees (1-9), where (9 very high important and 1 low important), to build the comparison matrix as in Equation (13)

$$P = (p_{ij})_{nxn} = \begin{pmatrix} p_{11} & p_{12} & \cdots & p_{1n} \\ p_{21} & p_{22} & \cdots & p_{2n} \\ \vdots & \vdots & \vdots & \vdots \\ p_{n1} & p_{n2} & \cdots & p_{m} \end{pmatrix} \quad (13)$$

3-4-3. The third step: 1- Prioritization calculation: Weights are estimated to obtain a rough estimate of relative priorities by the following equations [11]:

$$P_{ij\ Norm} = \frac{P_{ij}}{\sum_{k=1}^{n} P_{kj}}, \quad i,j = 1,2 \ldots.,n \quad (14)$$

$$P_{Norm} = (P_{ij\ Norm})_{n\times n} \quad (15)$$

$$W_{i\ Norm} = \sum_{j=1}^{n\ Norm} P_{ij} \quad ,I = 1,2 \ldots,n \quad (16)$$

$$Wi = \frac{w_i\ Norm}{\sum_{k=1}^{nNorm} w_k} \quad , \ i = 1,2,n \quad (17)$$

$$\lambda_{max} = \frac{1}{n}\sum_{i=1}^{n}\frac{(PW)i}{Wi} \quad (18)$$

2- Finding the consistency ratio by the following equations:

$$CR = \frac{CI}{RI} \qquad where{:}\ CI = \frac{(\lambda_{max} - n)}{(n-1)} \tag{19}$$

3- RCI from the table (6):

Table 6. Random consistency index (RC) values [10]

N	2	3	4	5	6	7	8	9	10
RCI	0	0.58	0.90	1.12	1024	1.32	1.41	1.45	1.51

$$\lambda_{max} = \frac{1}{n}\sum_{i=1}^{n}(P.W)i/Wi = \frac{1}{8}\left[\frac{3.4399}{0.3540} + \frac{1.3122}{0.2183} + \frac{0.8123}{0.1061} + \frac{0.5064}{0.0730} + \frac{0.4913}{0.0659} + \frac{0.4854}{0.0331} + \frac{0.5302}{0.0808} + \frac{0.4216}{0.0689}\right] =$$

$$\frac{65.1220}{8} = \mathbf{8.1403} \ldots\ldots\ldots\ldots\ldots\ldots\ldots\ldots\ldots\ldots\ldots (20)$$

$$CI = \frac{8.1403-8}{8-1} = 0.020$$

Create a binary comparison matrix for the main criteria and the set of alternatives within each standard and as in tables (7-9).

Table 7. $P_{ij\ Norm}$ Matrix with Weight of each W_{max} Standard

										Normalization	
0.06	0.24	0.07	0.30	0.07	0.10	0.03	0.01	-	-	0.1227	0.116
0.01	0.03	0.04	0.10	0.29	0.05	0.01	0.02	-	-	0.0641	0.060
0.28	0.31	0.33	0.39	0.29	0.19	0.18	0.18	-	-	0.2592	0.273
0.02	0.03	0.08	0.10	0.14	0.24	0.44	0.37	-	-	0.1554	0.168
0.06	0.01	0.08	0.05	0.07	0.10	0.18	0.09	-	-	0.0669	0.075
0.03	0.03	0.08	0.02	0.04	0.05	0.02	0.05	-	-	0.0351	0.035
0.17	0.21	0.16	0.02	0.04	0.19	0.09	0.18	-	-	0.1438	0.131
0.39	0.14	0.16	0.02	0.07	0.10	0.04	0.09	-	-	0.1359	0.142
-	-	-	-	-	-	-	-	-	-	0.0084	0.000
-	-	-	-	-	-	-	-	-	-	0.0084	0.000
										Eigenvalue:	11.18552

Table 8. (P* w_{max}) for each subcriteria

Subcriteria	P1	P2	P3	P4	P5	P6	P7	P8
P*W_{max}	0.150	0.083	0.200	0.107	0.159	0.165	0.059	0.078

Table 9. Result of matrix for subcriteria P1

Matrix	M1 (1)	M2 (2)	M3 (3)	M4 (4)	M5 (5)	normalized principal Eigenvector
M1 (1)	1	1	1/6	1/8	1/7	6.17%
M2 (2)	1	1	1/2	1/4	1/2	7.80%
M3 (3)	6	2	1	2	1	17.70%
M4 (4)	8	4	1/2	1	1	17.84%
M5 (5)	7	2	1	1	1	17.55%

					Normalization	
0.04	0.08	0.03	0.02	0.02	0.0634	0.062
0.04	0.08	0.08	0.03	0.08	0.0729	0.078
0.23	0.15	0.16	0.27	0.16	0.1751	0.177
0.31	0.31	0.08	0.14	0.16	0.2042	0.178
0.27	0.15	0.16	0.14	0.16	0.1867	0.175
-	-	-	-	-	0.0117	0.000
					Eigenvalue:	8.842764

Table 10. AHP method results for choosing the best maintenance strategy alternative

Matrix P1	M1 (1)	M2 (2)	M3 (3)	M4 (4)	M5 (5)	normalized principal Eigenvector
M1 (1)	1	1	1/6	1/8	1/7	6.17%
M2 (2)	1	1	1/2	1/4	1/2	7.80%
M3 (3)	6	2	1	2	1	17.70%
M4 (4)	8	4	1/2	1	1	17.84%
M5 (5)	7	2	1	1	1	17.55%

Matrix P2	M1 (1)	M2 (2)	M3 (3)	M4 (4)	M5 (5)	normalized principal Eigenvector
M1 (1)	1	1	1/6	1/7	1/5	6.64%
M2 (2)	1	1	1	1/3	1/2	9.27%
M3 (3)	6	1	1	2	2	18.69%
M4 (4)	7	3	1/2	1	1	16.93%
M5 (5)	5	2	1/2	1	1	14.37%

Matrix P3	M1 (1)	M2 (2)	M3 (3)	M4 (4)	M5 (5)	normalized principal Eigenvector
M1 (1)	1	1	1/5	1/8	1/6	6.24%
M2 (2)	1	1	1/2	1/4	1/5	7.18%
M3 (3)	5	2	1	2	2	18.65%
M4 (4)	8	4	1/2	1	1	17.46%
M5 (5)	6	5	1/2	1	1	16.87%

Matrix P4	M1 (1)	M2 (2)	M3 (3)	M4 (4)	M5 (5)	normalized principal Eigenvector
M1 (1)	1	2	1/2	1/4	1/4	9.42%
M2 (2)	1/2	1	1/2	1	1	10.44%
M3 (3)	2	2	1	1	1/2	13.12%
M4 (4)	4	1	1	1	1	15.09%
M5 (5)	4	1	2	1	1	16.63%

Matrix P5	M1 (1)	M2 (2)	M3 (3)	M4 (4)	M5 (5)	normalized principal Eigenvector
M1 (1)	1	3	1/4	1/7	1/6	8.69%
M2 (2)	1/3	1	1	1/2	1	9.43%
M3 (3)	4	1	1	2	2	17.32%
M4 (4)	7	2	1/2	1	1	16.80%
M5 (5)	6	1	1/2	1	1	14.81%

Matrix P6	M1 (1)	M2 (2)	M3 (3)	M4 (4)	M5 (5)	normalized principal Eigenvector
M1 (1)	1	1	1/2	1/8	1/4	7.47%
M2 (2)	1	1	1	1/2	1/2	9.55%
M3 (3)	2	1	1	2	2	16.01%
M4 (4)	8	2	1/2	1	1/2	16.65%
M5 (5)	4	2	1/2	2	1	16.06%

Matrix P7	M1 (1)	M2 (2)	M3 (3)	M4 (4)	M5 (5)	normalized principal Eigenvector
M1 (1)	1	2	1/3	1/7	1/3	8.39%
M2 (2)	1/2	1	1	1/2	1/2	9.10%
M3 (3)	3	1	1	2	2	16.87%
M4 (4)	7	2	1/2	1	1/2	16.31%
M5 (5)	3	2	1/2	2	1	15.21%

Matrix P8	M1 (1)	M2 (2)	M3 (3)	M4 (4)	M5 (5)	normalized principal Eigenvector
M1 (1)	1	1	1	1/3	1/2	10.11%
M2 (2)	1	1	1	1/2	1	11.27%
M3 (3)	1	1	1	1	1	12.25%
M4 (4)	3	2	1	1	1	16.11%
M5 (5)	2	1	1	1	1	13.49%

Table 11. Results of calculations for all subcriteria

Alternatives	Score of AHP prosess								R^2	$\sum w_i/n$	score
	P1	P2	P3	P4	P5	P6	P7	P8			
M1	7.80	9.27	7.18	10.44	9.43	9.55	9.10	11.27	68.00%	0.389	8.69
M2	17.84	16.93	17.46	15.09	16.80	16.65	16.31	16.11	55.00%	0.209	7.75
M3	17.70	8.68	18.65	13.12	17.32	16.01	16.87	12.25	78.00%	0.098	11.85
M4	17.55	14.37	16.47	16.63	14.81	16.06	15.21	13.49	60.00%	0.069	9.67
M5	6.17	6.64	6.24	19.42	8.69	7.47	8.39	10.11	19.00%	0.063	7.42
										0.032	
										0.077	
										0.064	

4. Analyzing the results of field research in the FWTP:

Through theoretical and analytical study by adopting modern methods of operations research and advanced tools for data analysis and by using simple and available information in the project, a set of valuable conclusions can be reached that enables the administration to raise the level of performance of the project as follows:

1- The adoption of the method of predictive maintenance system based on Preventive maintenance system improved and raised the reliability and performance of the intake and high lifting systems in theory. high lifting (17%) and intake system (14%).

2- Predictive maintenance based on preventive maintenance (TPM) from recent innovations to assist top management in making maintenance decisions. From Table (5), the optimum maintenance time for both the intake system (85 days) and the high lifting system (121 days).

3- Top management, middle management, and experts realized that the TPM maintenance system (proposed) could advance the reality of production in the project through its focus on technical, engineering and administrative aspects of maintenance operations, and these results were consistent with the results reached by the study of Zaim and colleagues [9].

4- Table (11) provided a summary of the study results from the impact ratio of TPM pillars {P1-P8} on the alternative available maintenance system alternatives {M1-M5}. Analysis of the AHP method refers to the relationship between the alternative of maintenance system and the pillars of maintenance in FWTP, individually and as follows:

   1- The relationship between Corrective Maintenance M5 and the pillars of maintenance, indicates the weak relationship between the sudden maintenance and pillars 19%. The correlation score for top Management Support P1 (6.17), Focused Improvement P2 (6.64), Educating and Training P3 (6.24), Maintenance, Supporting and Assistance all Departments P5 (8.69), and Total Cost of Maintenance P6 (7.47) (weak relationship). This is a logical conclusion, as in the sudden maintenance, management support, the expected performance of it, and the training of workers are very weak, and this results in a lack of quality of the product in addition to that it did not take into account maintenance costs (a very small relationship), which causes significant losses that cost the project management through stop losses. Production and high maintenance cost. Since the score is 7.42.

   2- The relationship between TPM M3 and the pillars of maintenance: indicates a very high correlation 78%. Where high levels of correlation refer to P1 (17.7), P3 (18.6), P4 (13.1), P5 (17.3), P6 (16.0), P7 (16.8), P8 (12.3) all high Relationship. This indicates that there is a need for high support for top management and other departments of the Maintenance Division, in addition to training workers to the required level and the participation of all workers in the maintenance process. The table indicates a major point, which is the high relationship between TPM and its cost, which reduces large amounts and losses of project that were caused by the other of maintenance if used.

3- From Table (11), the last column indicates a summary of the results of the AHP where it was pointed out that the highest result from the effect of the eight pillars on the maintenance process is on the TPM (11.85).
4- By using AHP helped us and top management to come up with a set of options that enable the top management to take the appropriate decision based on logical and scientific considerations in choosing one of the set of options available to solve the problem of selecting the type of appropriate maintenance at the optimal time that takes place against the lowest possible cost in carrying out maintenance work to raise the performance and reliability of the project equipment.

5. conclusions

Considering the existence of a many number of maintenance criteria selection for WTP, the research highlights the importance of using MCDM tools in water treatment, since selection a maintenance system based on inadequate criteria can have serious long-term consequences. The novelty is related to the use of specific criteria by a multi-criteria analysis applied to a real case study. For an overall analysis, the AHP was applied to evaluate and an alternative is chosen from a set of maintenance alternatives vs eight of Maintenance pillars. In this research investigated improved and raised, in theory, the reliability and performance of the intake and high lifting systems high lifting (17%) and intake system (14%) by adoption of the predictive maintenance system based on Preventive maintenance.

AHP provides redundancy for preference assignment of criteria, alternatives and mechanism to validate consistency. As with all complex system problems, selecting the most suitable maintenance system for WTP, where showed the results obtained from (AHP) that the best rank of the alternative was M3(11.85), M4(9.67) and so on for the other alternatives. Moreover, the relationship between M3 and the pillars of maintenance: indicates a very high correlation 78%. from these results we conclude that the existing of a very strong impact for the proposed maintenance system in improving the performance and reliability of FWTP project operations. These results indicate the advantages over traditional Decision Making selection methods. The AHP allows diction makers of WTP to make these decisions in a clear and straightforward manner, taking into account during DM all the variables that directly influence the outcome of this process. Fuzzy AHP can be used for further study with other systems

Acknowledgments: The authors wish to thank AL- Fallujah Water Treatment Project (FWTP) and all employee of it for support of this research.

References

[1] Saxena, A., Singh, S. K., Kumar, A., & Ram, M. (2019). Water treatment system performance evaluation under maintenance policies. In System Performance and Management Analytics (pp. 147-165). Springer, Singapore.

[2] Harrison E. M., Saroj K. Sh., Kalanithy V. (2011). Multi-criteria Decision Analysis: A Strategic Planning Tool for Water Management. Water Resource Manage, Springer Netherlands, 25, 3947-3959.

[3] Vishnu, C. R., & Regikumar, V. (2016). Reliability based maintenance strategy selection in process plants: a case study. Procedia technology, 25, 1080-1087.

[4] Ahmad, R., Kamaruddin, S., Azid, I. & Almanar, I. (2011). Maintenance management decision model for preventive maintenance strategy on production equipment. Journal Industrial. Engineering. Inteter, 7, 22-34.

[5] Mohammad M., Abdolreza Y., Ali L., Edmundas K. & Zenonas T. (2018). Maintenance Strategy Selection using AHP and COPRAS under fuzzy environment. International Journal of Strategic Property Management, 16 (1), 85-104.

[6] Mutlag, S. A. (2019). Decision Making in Materials Selection: An Integrated Approach with AHP. Anbar Journal of Engineering Sciences, 7(4), 399-407.

[7] Lizot, M., Goffi, A. S., Thesari, S. S., Trojan, F., Afonso, P. S., & Ferreira, P. F. (2020). Multi-criteria methodology for selection of water treatment systems with maintenance, economic, social, technical and environmental aspects. Environment, Development and Sustainability, 1-25.

[8] Bevilacqua, M. & Marcello B. (2017). "The analytic hierarchy process applied to maintenance strategy selection." Reliability Engineering & System Safety, 70(1), 71-83.

[9] Zaim, S., Turkyılmaz, A., Acar, M. F., Al-Turki, U., & Demirel, O. F. (2012). Maintenance strategy selection using AHP & ANP algorithms: a case study. Journal of Quality in Maintenance Engineering, 18(1),16-29.

[10] Wang, L. Chu, J. & Wu May J. (2017). Selection of optimum maintenance strategies based on a fuzzy analytic hierarchy process. International Journal of Production Economics, 107(1), 151-163.

[11] Mutlag, S, A. and Kassam A. H. (2019) "Fuzzy Reliability-Vulnerability for Evaluation of Water Supply System Performance" Anbar Journal of Engineering Science, 8, 72 – 82.

[12] Giada F., Ilaria L.,Valentina C. & Vittorio D. (2015). A Strategic Planning Tool for Water Supply Systems Management" Water Resour Manage, Springer, 2(11), 21-36.

[13] Triantaphyllou, E., Kovalerchuk, B., Mann, L. & Knapp, GM. (2017). Determining the most important criteria in maintenance decision making. Journal of Quality in Maintenance Engineering, 3(1). 16–24.

[14] Melesse, W. & Ajit, S. (2019). Total Productive Maintenance: A Case Study in Manufacturing", Global Journal of Researches in Engineering, 12(1), 12-25.

[15] Ikuobase E., Rosemary A. and Alan J. (2016). Elements of Maintenance System and Tools for Implementation Within the Framework of Reliability Centered Maintenance- A Review. 8(2).

[16] Abbas M., Ahmad J., Edmundas K., Zavads k. (2015). multiple criteria decision-making techniques and applications – Two decades review from 1994 to 2014. Expert Systems with Applications, Elsevier, 42 (8), 4126-4148.

[17] Mutlag S. A., H. K. Dawood (2020) "A Maintenance Management -based Decision Support System for Multicriteria using AHP Methodology for Power Plants" 82(5) 10146-10153

[18] Mutlag, S. A., & Dawood, H. K. (2019, October). Optimal Refrigerant by Integrated Operational and Environmental Criteria in Air Conditioner. In 2019 12th International Conference on Developments in eSystems Engineering (DeSE) (pp. 309-314). IEEE.

[19] Dawood, H. K., Mutlag, S. A., Abed, K. A. Abed, (2018, September). Priority-Based Decision Support System (PBDSS) by Genetic Algorithm as a Tool for Network Problem. In 2018 11th International Conference on Developments in eSystems Engineering (DeSE) (pp. 306-311). IEEE.

[20] The documents of the AL- Fallujah Water Treatment Project (FWTP).

Teeth Model Reconstruction Based on Multiple View Image Capture

Faten Imad Ali [1,*] and Ziad Tarik Al-dahan[2]

[1]College of Engineering, Al-Nahrain University, Iraq
E-mail: fatenemadali@gmail.com
[2]College of Engineering, Al-Nahrain University, Iraq
E-mail: ziadrmt1959@yahoo.co.uk
* Corresponding author: College of Engineering, Al-Nahrain University, Iraq. Tel: +9647709630264.

Abstract. Various dentistry fields such as teeth treatment, dental restoration, and denture production require the application of three-dimensional intraoral scanners to create digital impressions. This study presents a method to reconstruct a three-dimensional teeth crown by utilizing a prototype intraoral custom design with publicly available software. The system consists of custom containing an endoscope camera and two light sources (white LED and blue laser), a personal control computer, and photogrammetric open source software packages (VisualSFM, CMVS/PMVS, and MeshLab). Photogrammetry methods were used to acquire the three-dimensional teeth model with high precision. Multi images have been captured of the interesting teeth using the custom system from different angles. The capturing process was done in two phases using one illumination source each time. The captured images are loading to Photogrammetry software to generate the three-dimensional dense point cloud. A comparison has been made of the resulted dense point cloud at each phase. Finally, the dense point cloud is loading to MeshLab software to generate the three-dimensional mesh that can be utilized in CAD/CAM software. The results have shown the superiority of using blue illumination with the photogrammetry software that shows good accuracy of teeth crown details and measurements that are related to the reconstruction algorithm which yields in more number of 3D points of the object to generate good three-dimensional meshes of the teeth crown. This study helps to offer low cost, simple design, and a user-friendly system to generate a three-dimensional teeth crown.

Keywords: *3D reconstruction, Teeth reconstruction, Photogrammetry, Laser, Intraoral camera.*

1. Introduction

Computer vision is a field that includes methods for recognizing, processing, analyzing, and understanding images. It is known as image analysis or scene analysis. It imitates the capability of human vision by the means of electronically recognizing and analyzing the image. Through the computer vision, image data can be processed in abundant forms like video sequences, depth images, images captured from several cameras, medical scanner, satellite sensors etc. Three-dimensional (3D) reconstruction is a significant application of computer vision [1].

Three-dimensional reconstruction is a rife expression of multiple processes that target obtaining 3D information of scanned objects. This 3D image-based method processes utilize two dimensional cameras to obtain the 3D reconstruction. The 3D reconstruction methods can be divided into active and passive methods [2].

Active methods, also called range data methods, give the depth-map and reconstruct the 3D profile either by controlling the illumination of the scene like Kinect or via measuring the return time and phase shift of reflected waves like laser scanner [3]. These methods do not proceed effectively in sunlight or a poorly controlled environment. Generally, these methods are expensive and need the consultation of experts [4].

Passive methods determine the reflected rays emitted from the surface of objects, using a sensor to derive the 3D structure by image analysis. These methods are commonly based on multiple capturing of a certain object with a video camera or multiple cameras [5]. These methods can be utilized in wider fields compared with the active methods. In these methods light is used only for illumination and not in the reconstruction points' triangulation which utilizing stereo algorithm to match the known similar points of image sequences [6].

Intraoral Digital Impression is a method that allows a dental professional to make a 3D reconstruction of a patient's teeth, using lasers or other optical scanning devices and generate the image on a dedicated computer screen. Digital impression scanner can capture clear and highly accurate impression data very quickly, easier and more comfortable for both patients and dentist than the conventional impression methods [7]. Generally, intraoral scanners based on confocal technology need massive and expensive hardware. Most scanners devices used visible wavelengths rang (400-700 nm) since sound dental enamel has high light scattering, that prevent imaging through the tooth layers and improve the surface scanning process [8].

Over the century there has been many alternative options examined to improve digital tooth Impression methods and reduce its cost. In 2000 Sameh M. et. al. constructs a contact system for 3D model of patient's jaw utilizing an intraoral video camera. Modified method of shape from shading (SFS) was used in the system. The suggested system used camera calibration and perspective projection to extracts the 3D data from a series of two dimension (2D) jaw images. Triangulation method utilized for solid 3D model reconstruction. The result showed acceptable reconstruction accuracy [9]. In 2016 Knyaz V. and Gaboutchian A. utilize photogrammetric system to generate high resolution 3D dental models. The proposed system included two high resolution digital cameras, a high resolution structure light projector and computer to control the rotation position stage. Odontometric methods wad used on the final 3D model to estimate maximal teeth sizes in different dimensions [10]. Also, in 2017 Xiaoming F. et. al. discusses utilizing multi-baseline digital close-range photogrammetry (MBDCRP) method with single-lens reflex camera to reconstruct 3D orthodontic model of plaster cast. The result showed that MBDCRP with inexpensive hardware are alternative mode to save and scale dental models without requiring a special room to keep plaster casts [11].

In this study Photogrammetry process with inexpensive hardware are used to estimate the 3D coordinates of the teeth. photographs are used as the essential medium of measuring shapes and sizes by taking and analyzing images from multiple angles. The captured images are used as an input for a specialized software which search and match the visible 2D images features (a pixel) using the data collected from the camera's positions and dimensions. The software output is 3D mesh of the teeth crown.

IOP Conf. Series: Materials Science and Engineering **978** (2020) 012009 doi:10.1088/1757-899X/978/1/012009

Initially, the total pipeline used to create the digital 3D teeth models is characterized. The pipeline displays the steps used in our experiment. First, the Structure from Motion software, called (VisualSFM), is used for images calibration estimation or cameras parameters, which will be saved as (.txt). The software also creates a 3D sparse point cloud that will be saved as (. PLY). The VisualSFM output is used as an input to the 'Clustering Multi View Stereo' (CMVS) software package to generate the 3D dense point cloud. Finally, MeshLab software is used to generate the teeth 3D mesh which could be exported to several 3D formats in compliance with the user's requirements.

2. METHODS AND PROCEDURES

A. System outline

The proposed system is a computer-based system uses photogrammetric programs for 3D surface reconstruction of the teeth crown. As illustrated in Figure 1, the system consists of the following components:

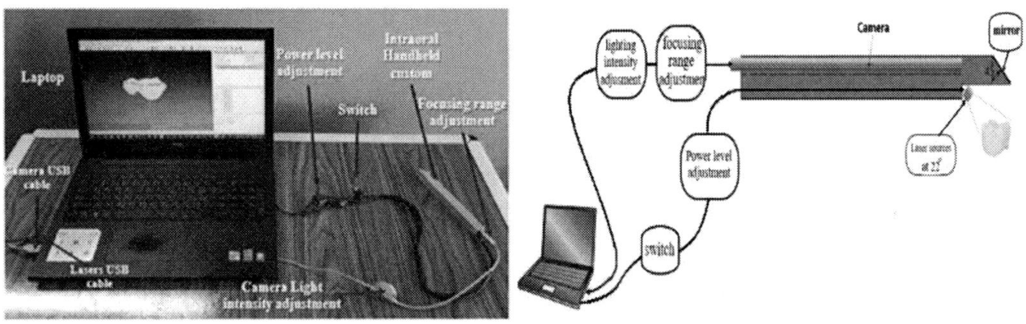

Figure 1. Complete system structure

1. An intraoral custom designed using AutoCAD and solidwork software and printed using 3D printer to allow easy imaging access to the top and bottom teeth. It has a section of about 15 mm and length of 150 mm as illustrated in Figure 2.

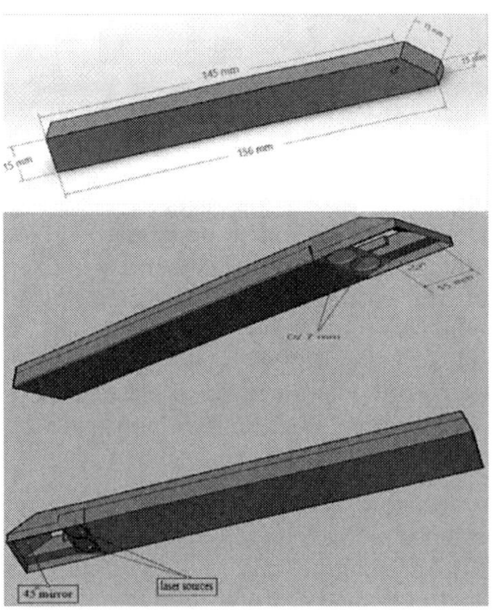

Figure 2. Proposed intraoral custom design

2. A low-cost digital camera type (1800 endoscope) with resolution up to (1600x1200) pixels and can be manually adjusted in both lighting intensity and focus rang as show in Figure 3. A mirror at 45-degree angle is used for the horizontal viewing.

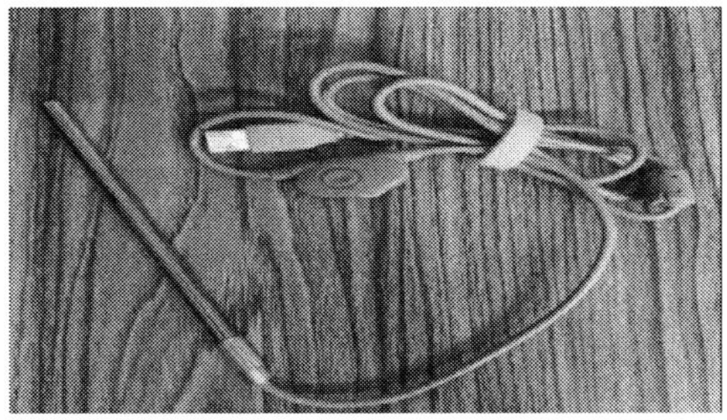

Figure 3. Endoscope (model 1800)

3. Two light sources
   a) Light emitting diode (LED).
   b) A blue diode laser with 405 nm wavelength and 5mW maximum output power. It is angled at 22 degrees horizontally to illuminate the teeth surface.

The capturing system can save images as JPEGs or any other format needed. The images are viewing and storing in the computer without the need for external storage devices. A USB power supply output cable connects each component to the computer. A suitable power level is used with the laser component to decrease the optical speckle noise and enhance the image contrast. The images are taken by positioning the intraoral system parallel to the teeth surface to illuminate the designated region.

B- 3D Imaging Using Photogrammetry Technologies

Photogrammetry is a passive process that estimates 3D coordinates of objects by using photographs as the essential medium of measuring shapes and sizes of objects by taking and analyzing images from multiple angles. Objects are measured with a remote-sensing technique [12]. In this study, 3D reconstruction of objects, observed by Photogrammetry Technologies made with Structure from Motion (SfM) method in which using a set of 2D images to produce 3D points cloud structure of an object. SfM methods are usually pipelined algorithms that each subtask processed sequentially. SfM commonly includes two phases; the correspondence phase which calculates the relations between geometrical input images through three sequential subtasks (feature extraction, feature matching and geometric verification) , and the iterative incremental reconstruction phase which estimates the camera poses for each image and reconstruct the 3D sparse point cloud through three sequential subtasks (image registration, triangulation and bundle adjustment) [13].

detection processes based on searching for features points which define the object within the input picture such as corners, edges, interest points, ridges and etc. many methods used for detection process such as SIFT Algorithm that was used in this study to detect features and compare the images in the collection that produce a high number of correspondences and therefore good candidate for the SfM process.

After detection of the feature points, the next step was the matching process in which the keypoints acquired through feature extraction process are utilized to decide which images describe common parts of the object and are subsequently at least partially overlapping. In this study RANSAC Algorithm (RANdom SAmple Consensus) was used to automatic computation of F matrix during the geometry verification. RANSAC allows the calculation of the mathematical model parameters via random sampling. The algorithm based on that the data formed of inliers, which are the data of the model, and outliers, which represent the data that does not set to the model. When two keypoints in various images have the same description, consequently those keypoints can be considered to be the same in the object. The output of this step is a group of images overlapping minimally in pairs and groups of correspondences among features. After the matching process, the 3D points (comprise the 3D coordinate information of the relate points) were measured using the matching results; these were known as point clouds. Triangulation method was used of predictable a point location in the 3D space from its projections into two or more images to produce the sparse point cloud which performs the images and features that connect the images pair obtained from the geometrical verify.

Once the camera poses have been determining as sparse point cloud, the next process is to determine the surface of the object represented as a dense point clouds. This is an essential step to get a denser point cloud which is very useful for creating a precise 3D model with the objective of correcting occlusions in the object facades. The integration of SfM and MVS methods submit an automated workflow for the production of high-precision 3D dense point cloud [14]. Clustering Views for multi view stereo (CMVS) algorithm using the camera parameters and the 3D sparse point cloud resulting from VisualSFM as input data and then applying MVS process by grouping images with similar views to enhance determination performance. Patch-based Multi-View Stereo (PMVS) algorithm used to produce extra cloud points in to the 3D model based on the groups images specified via CMVS algorithm.

Due to the limited field of view, the prior step of 3D reconstruction does not produce the information for the proximal contact area between the adjusting teeth. So the reconstructed 3D view is incomplete. In this study, the Iterative Closest Point algorithm was used to merge two partial 3D models with applying the rotation matrix and the translation vector to the data set to get a full 3D view of the model [15]. Finally, using the 3D dense point cloud to create the 3D mesh by used the free MeshLab software.

C- Teeth 3D model generation software pipeline

After obtaining informed consent, a volunteer was a participant in the middle age of 25 from the patient population. Overhead lights were turned off while a dentist imaged sites of interest. Multi images of the interesting area were captured using the handheld IOS from different angles of approximal and occlusal views in two steps each time using one of the illumination sources (white and blue). Figure 4 illustrates the full system architecture.

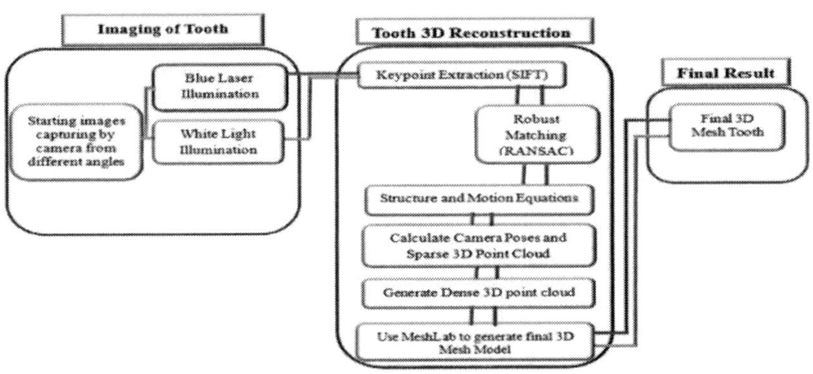

Figure 4. Full system architecture

As mentioned previously, Xiaoming Fu discussed using photogrammetry software to 3D reconstruct models on an orthodontic plaster cast [11]. In this study, the system software was designed to reconstruct the tooth crown. The tooth crown reconstruction algorithm is based on getting point cloud from the captured images then generate poly mesh on these points. The process requires a couple of programs: VisualSFM with CMVS/PMVS and MeshLab. Figure 5 shows the total pipeline utilized to build the 3D model.

IOP Conf. Series: Materials Science and Engineering **978** (2020) 012009 doi:10.1088/1757-899X/978/1/012009

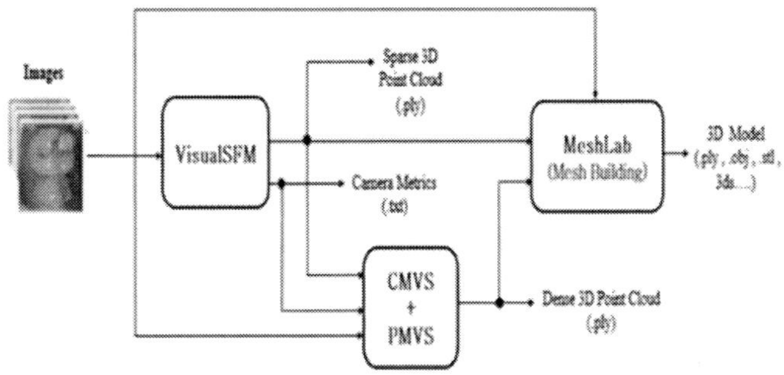

Figure 5. Flowchart of the system design

The images are uploaded as VisualSFM input to estimate the structure from motion (SFM). As indicated, VisualSFM outputs are (.txt) and (. PLY) files which are the matrices of camera and a 3D sparse point cloud, respectively. The algorithm of Multiple Dense View Stereo (MVDS) utilizes images and the camera matrices to produce the dense 3D points cloud. The final mesh model with multi-formats is generated using the captured images and the dense points.

VisualSFM is a Graphics Processing Unit (GUI) which is implemented to produce 3D reconstructions for objects by means of SFM. The system uses many algorithms such as an incremental SFM system, Scale Invariant Feature Transform (SIFT) and the Multicore Bundle Adjustment, which is used for bundle adjustment and detecting matching features. Addendum of sparse reconstruction is used as well. This process offers to run the algorithm of Yasutaka Furukawa's PMVS/CMVS which is utilized in dense reconstruction [16].

CMVS; refers to Clustering Views of Multi-View Stereo (MVS), which is a development of Multi-View Stereo (MVS) software. MVS is a software which uses a collection of photos and their matrices parameters as an input to generate the 3D object structure in dense points cloud form. MVS algorithms are not scaled well with massive numbers of input images due to the lack of computational and memory resources. CMVS is used to analysis the photos within small sized image clusters. This lets the MVS treat each cluster autonomously and in parallel, to ensure the reconstructions unity of all the clusters would not lose any data which can be otherwise acquired from all the image groups [17].

PMVS; refers to Patch-based Multi-View Stereo. It is also a development of MVS and is included in the CMVS package. It is a software which uses a collection of photos and their camera parameters to produce 3D structure of a photographed object. It is used to reconstruct a rigid structure only and automatically deny non-rigid objects. The output is an organized series of a point's model, which calculates both the 3D coordinates and the surface normal at each oriented point. CMVS/PMVS are used in conjugation with the VisualSFM software (Bundler) [18]. Figure 6 shows the VisualSFM/ PCMVS program flow chat.

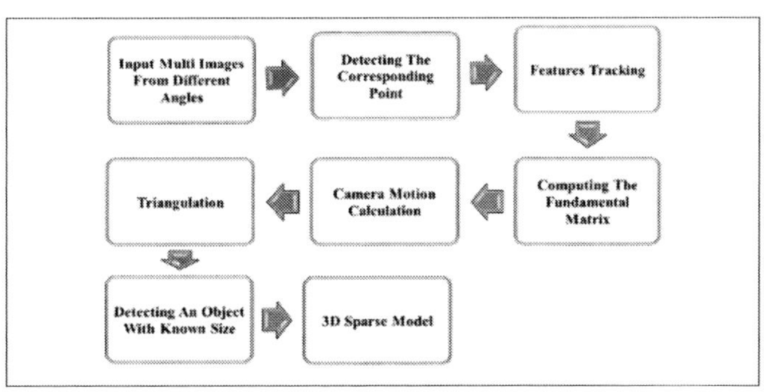

Figure 6. VisualSFM / PCMVS flow chart

MeshLab is an open source program used to process and edit the 3D triangular meshes. It submits a group of tools for 3D mesh processing and models preparation for 3D printing. These tools such as edit, clean, heal, inspect, render, texture and convert meshes [19].

The tooth reconstruction algorithm that include the 3D dense reconstruction program and final mesh model described in follows as shown in figure 7.

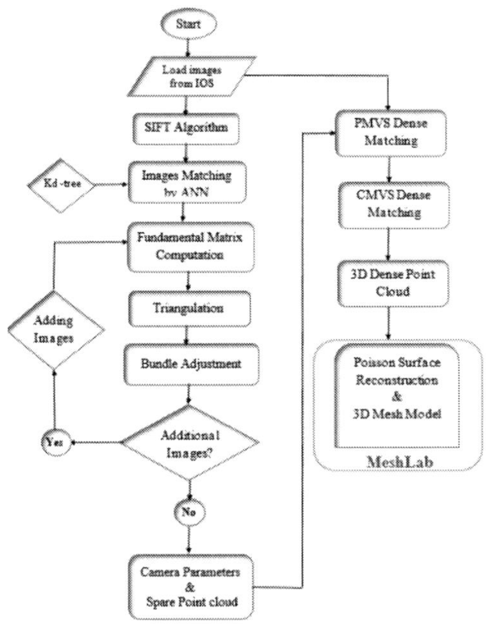

Figure 7. Flow chart for tooth reconstruction algorithm

3. Results

This section presents the experimental results that measures the effectiveness of the submitted method using a group of images. This work runs on computer laptop with the following specifications (CPU: 2.2 GHz, core i5, 8GB RAM, Windows10 64bit). In this method, real images of human teeth were scanned through handled IOS so as to reconstruct 3D teeth model as illustrated in Figure 8.

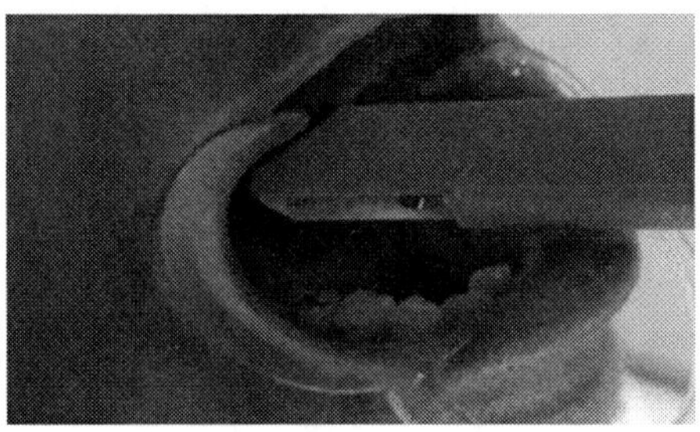

Figure 8. The teeth scanned to reconstruct 3D module

In order to do the reconstruction process, multi images of the molars are captured using the handheld system from different angles at two steps, each time using one of the illumination sources (white and blue). Figure 9 shows some of the images captured.

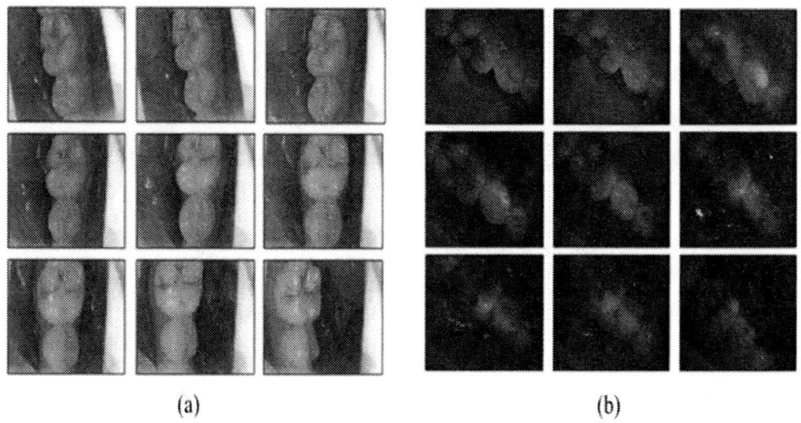

Figure 9. Some of the images used for the reconstruction with (a) white light (b) blue light

Then, the reconstruction process done following these steps:

i. Load the images to the VisualSFM software.

ii. Employ SIFT algorithm to detect object feature points in the multiple images. Figure 10 displays the results of estimated features detection from the captured images for white LED and blue laser light. As shown, blue illumination extracts more points than white illumination for the same images. In this work, images with similar size of (1600*1200) pixels was used. Also, uncelebrated camera was used, thus VisualSFM was used to elicit EXIF file data and focal length of all images. Focal lengths are evaluated per camera set with formula (1) which corresponds to a medium viewing angle.

$$\text{Focal length} = 1.2 * \max(\text{width}, \text{hight}) \quad (1)$$

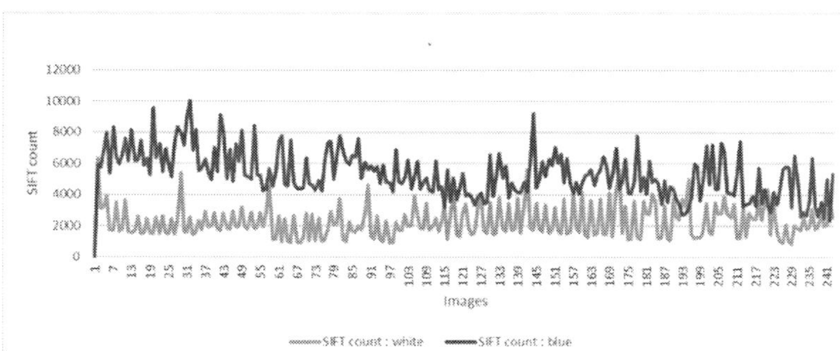

Figure 10. Estimated features detection results

iii. Perform matching. After detecting the SIFT features in every image, VisualSFM starts to match the feature points against each other to find the object points which are visible in multiple pictures. Figures 11a and 11b show the number of matches for a couple images.

IOP Conf. Series: Materials Science and Engineering **978** (2020) 012009 doi:10.1088/1757-899X/978/1/012009

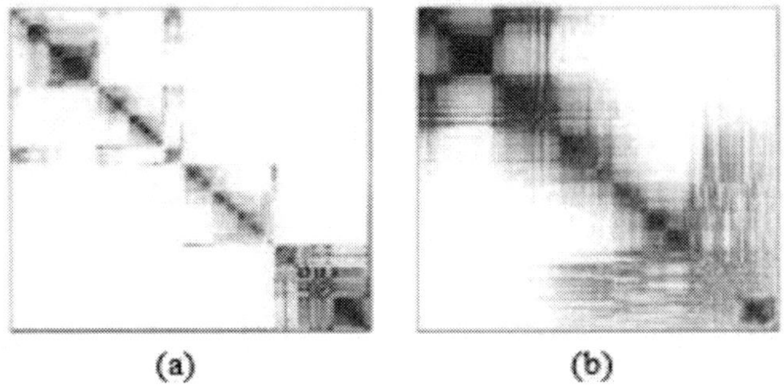

Figure 11. Number of matches for a couples of images with (a) white LED light, (b) blue light

As shown in Figure 11, matching matrix of blue illumination images was superior on white illumination images. Where the white color denotes loss of matches or little created matches, while, dark red color denotes a high number of matching feature points whereas yellow and green mean a lesser number of matching feature points. Figures 12a and 12b show the enlargement of part of Figures 11a and 11b.

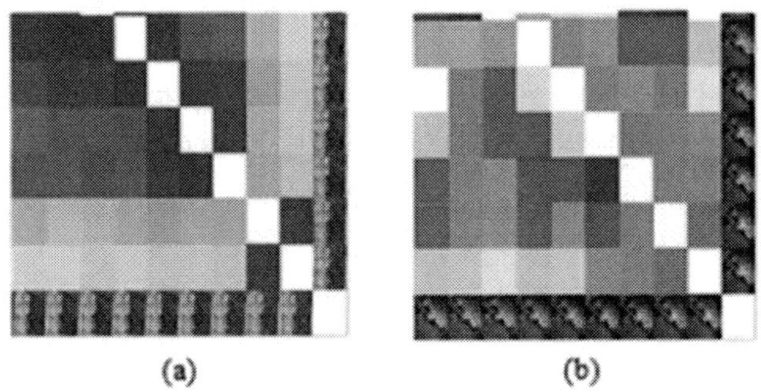

Figure 12. Magnification of part of images features matches with (a) white LED light, (b) blue light

iv. Sparse reconstruction, in this step the dimensions and points cloud was calculated. Once the features matching is accomplished, this process runs the bundler software and generates a sparse point cloud. The automated election of good image pairs starts the 3D reconstruction process. Thus, camera matrices are computed from the main matrix, and the 3D points are estimated utilizing triangulation method of matched points. As a result, an initial 3D point cloud of (1126) 3D points was obtained with white LED light and (8490) 3D points with blue light. The remaining images will be added sequentially. The additional images matrices are set and the produced 3D points are combined to the premier point cloud. Figures 13a and 13b show the achieved results for the 3D point cloud with the primary cameras locations.

(a) (b)

Figure 13. 3D sparse scene with initial position cameras with (a) white LED light (b) blue light

As observed, blue illumination generates more number of 3D points of the object than white illumination. VisualSFM output was saved as (bundle.rd.out) file that holds the 3D sparse of extension (. PLY), and the camera's matrices (.txt).

v. Dense reconstruction using Yasutaka Furukawa's CMVS/PMVS was used to produce the output of dense point cloud. CMVS used the point cloud created in VisualSFM and applies MVS algorithm. CMVS use the images, cameras parameters (.txt) and the sparse 3D point could (. PLY) to estimate the depth of the scene using stereo view. Finally, denser point cloud was formed using the existing points cloud and the depth map of the scene. Figures 14a and 14b display the CMVS output displayed through VisualSFM. As observed, blue illumination images generate denser point cloud than white illumination.

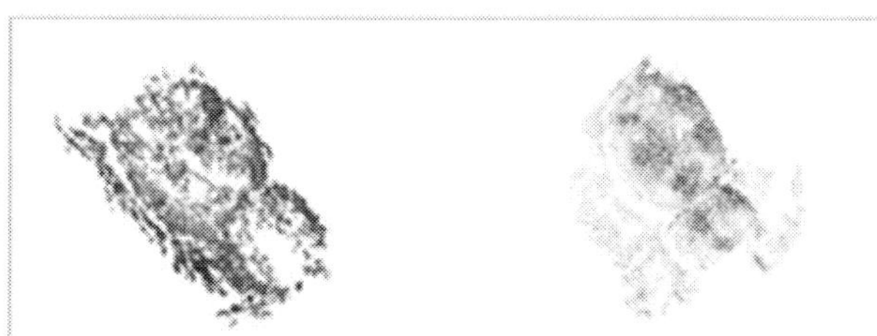

(a) (b)

Figure 14. 3D Dense scene with (a) white LED light, (b) blue light

vi. 3D mesh model was generated using MeshLab. Poisson's surface reconstruction filter was utilized to generate the 3D mesh from the VisualSFM point cloud. Figures 15a and 15b illustrate the mesh created with Poisson's filter and the acquired 3D model.

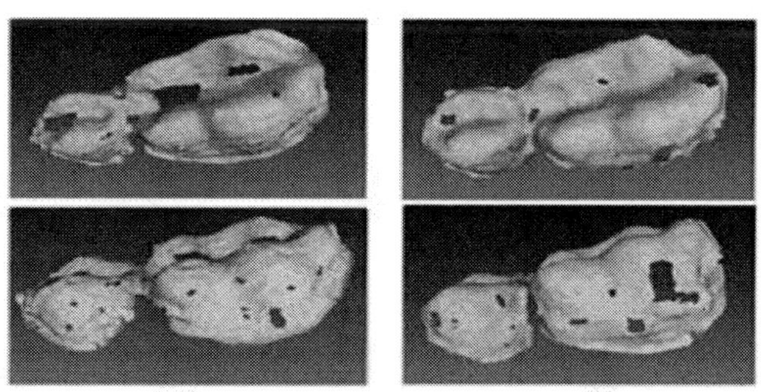

(a) (b)

Figure 15. MeshLab mesh generation with Poisson's surface reconstruction with (a) white LED light (b) blue light

vii. Close Holes filter was applied to verify the digital 3D models shown in Figure 16. To get a full 3D view of teeth by closing the holes and spaces which produce from the absence of feature points in the places where the view is obstructed by other teeth in the mouth. Then the model scale was converted from an arbitrary to a real-world scale using MeshLab formula (2).

$$\text{Scale Factor} = \frac{\text{referencemeasure}}{\text{mesh measure}} \tag{2}$$

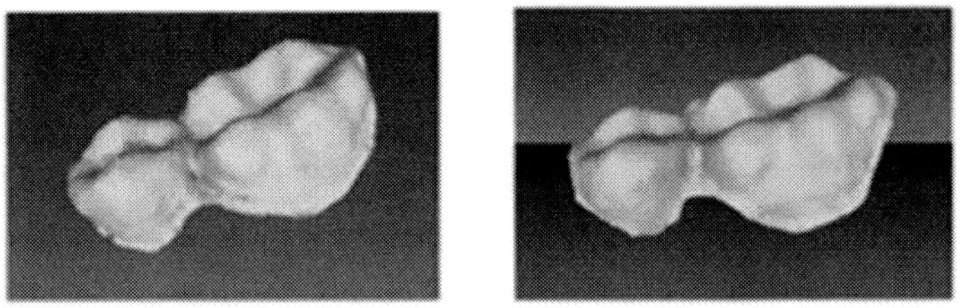

(a) (b)

Figure 16. Final 3D module generation with (a) white LED light, (b) blue light

Finally, the CAD model was printed using 3D printer as shown in Figure 17.

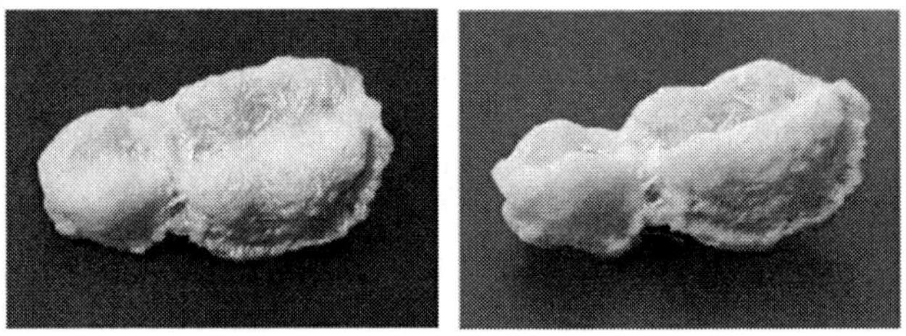

Figure 17. Final 3D printed model

Table 1. represents the main teeth parameters with the result of this testing. To confirm accuracy, the dimension of each parameter by this algorithm was compared to the dimension of real teeth. The difference was measured using Euclidean distance; then accumulated error distances were measured.

Table 1. Dimensions for real and model of the teeth

Teeth parameter	Dimension from teeth model (cm)	Dimension from real teeth (cm)	Accumulated Error (cm)
Tooth surface length(Mesiodistal Diameter of crown)	1.003	1	0.003
Tooth surface width	0.711	0.7	0.011
cervical occlusal length of Tooth crown(crown height)	0.669	0.8	0.131

The measurement errors varied in a wide range from 0.003 cm to 0.131 cm. The results indicate that the errors in the Mesiodistal diameter of the crown of individual model were slightly higher than that in the real teeth. This was due to the shape of the teeth surface at which the view is obstructed by other teeth in the mouth, which makes the automated reconstruction software to consider additional points from the adjacent teeth which consider as noise points. Margin points and noises were removed when processing and calculating measurements using Mesh Lab, leading to control of these errors. Also, this problem can be avoided by using higher resolution images.

Measurement errors in the cervical occlusal length of the crown were slightly lower than that in the real teeth. This was due to the gingivitis among the teeth that causing missing and unmatching points in the images captured from the side. These were fixed when processing and calculating measurements using Mesh Lab, leading to control of these errors. Also, this problem can be avoided by using higher resolution images.

The comparison of the outcome of this study with other literature reviews [9, 10, 11] implies that the recommended system is reliable that shows good accuracy of teeth crown details and measurements, but that there are nevertheless some sections which can influence the accuracy of the reconstruction. These sections related to the difficulty of images capturing and the efficiency and resolution of the used camera. As well as errors resulting from subject angulation errors in 3D and the physical hand and arm trembling.

4. Conclusion

3D reconstruction occupies an important method to create objects in a photorealistic way essentially in biomedical application. For 3D teeth scanner applications, the precision in teeth details and dimensions and facilitated use are the fundamental points that must be considered [20]. This work presents an experimental method to design low-cost movable system used for 3D scanning of the teeth. The proposed system is based on utilizing photogrammetric open source software packages (VisualSFM, CMVS/PMVS, and MeshLab) with two sources of illumination (405 nm blue laser and white LED) for 3D teeth reconstruction in order to get high accuracy 3D model with details and measurements close to reality.

From the reconstruction results, the following can be observed:

1. The VisualSFM features process reconstruction runs better within texture surfaces. This makes VisualSFM less efficient if insufficient features are revealed such as white and unified color objects. Thus using the appropriate blue illumination gives additional features to the teeth surface that enhance the performance of VisualSFM and give better results than white illumination due to the fact that teeth are white and poor textured.
2. By using sequence video images, the 3D reconstruction measuring casual error was avoided since it raises with increasing distance between captured images of the scanned object.
3. By using the low power laser instead of the LEDs light it will save the power and heating produce from the many numbers of the LEDs, in present work, the 405 nm low power laser diode was used with its minimum power, while six LEDs visible light were used to obtain the appropriate illumination.

The results demonstrate elevated precision of teeth details which are concerned with reconstruction algorithm and software which produce good 3D mesh of human teeth.

5. Acknowledgements

We are very grateful to experts for their appropriate and constructive suggestions to improve this template.

References

[1] G. Donald, G. Stuart, H. Neil, & Y. Laurent (2015). *Visual Turing test for computer vision systems.* National Academy of Sciences of the United State of America, 112 (12), 3618-3623. doi: 10.1073/pnas.1422953112.

[2] B. Peter, D. Frantisek, T. Michal, H. Peter, & K. Martin (2014). *3D map reconstruction with sensor Kinect.* In: Institute of Robotics and Cybernetics (RAAD) Conference, Slovak University of Technology, Smolenice, Slovak republic. doi: 10.1109/RAAD.2014.7002252.

[3] S. Mariusz, & R. Przemyslaw (2014). *A Survey of Passive 3D Reconstruction Methods on the Basis of More than One Image.* Machine GRAPHICS and VISION, 23 (3/4), 57–117.

[4] B. Gianfranco, G. Alessandro, B. Fabio, & M. Maurizio (2013). *A Comparative Analysis between Active and Passive Techniques for Underwater 3D Reconstruction of Close-Range Objects.* Sensors, 13 (8), 11007-11031.

[5] M. James, J. Chandler, A. Eltner, C. Fraser, & P. Miller (2019). *Guidelines on the use of Structure from Motion Photogrammetry in Geomorphic Research.* Earth Surface Processes and Landforms, Wiley. doi:10.1002/esp.4637.

[6] U. Petra, H. Petr, & J. Mikola (2015). *Testing photogrammetry-based techniques for three-dimensional surface documentation in forensic pathology.* Forensic Science International, 250, 77–86.

[7] S. Tingshu, & J. Sun (2015). *Intraoral Digital Impression Technique: A Review.* Wiley Online Library, Journal of Prosthodontics by the American College of Prosthodontists, 24, 313–321.

[8] E. Faten, & T. Ziad (2019). *Imaging of occlusal dental decay with 780 nm NIR light.* International Journal of Advanced Technology and Engineering Exploration, 6(55), 175-179.

[9] M. Sameh, A. Aly,T. David ,& G. Allan (2000). *A 3D Reconstruction System for the Human Jaw using a sequence of optical images.* IEEE Transactions on Medical Imaging, 19(5), 538- 547.

[10] V. Knyaz, & A. Gaboutchian (2016). *Photogrammetry-Based Automated Measurements for Tooth Shape and Occlusion Analysis.* The International Archives of the Photogrammetry, Remote Sensing and Spatial Information Sciences, XLI-B5, 849- 855, doi: 10.5194/isprsarchives-XLI-B5

[11] F. Xiaoming, P. Chun, L. Zan, L. Shan, T. Minmin, & S. Jinlin (2017). *The application of multi-baseline digital close range photogrammetry in three-dimensional imaging and measurement of dental casts.* PLOS ONE, 12(6), 1-13.

[12] H. Aldis, S. Ashley, A. Miguel, B. Paul, & S. Sakti (2018). *Photogrammetry of Human Specimens: An Innovation in Anatomy Education.* Journal of Medical Education and Curricular Development, 5(1), 1-10.

[13] L.Po-Han, H. Jui-Wen, & L. Huei-Yung (2012). *3D model reconstruction based on multiple view image capture.* IEEE International Symposium on Intelligent Signal Processing and Communication Systems (ISPACS 2012), 978(1), 4673-5082.

[14] R. Lagisetty, N. Philip, R. Padhi, & M. Bhat (2013). *Object Detection and Obstacle Avoidance for Mobile Robot using Stereo Camera.* IEEE International Conference on Control Applications (CCA), Hyderabad, India.

[15] Z. Zhang, S. Antipolis, E. Automatique, & E. Recherche (1992). *Iterative point matching for registration of free-form curves.* International Journal of Computer Vision 13.

[16] T. Kersten, & M. Lindstaedt (2012). *Automatic 3D Object Reconstruction from Multiple Images for Architectural, Cultural Heritage and Archaeological Applications Using Open-Source Software and Web Services.* Photogrammetrie Fernerkundung Geoinformation (PFG), Stuttgart, Germany, (6), 727-740. doi: 10.1127/1432-8364/2012/0152.

[17] B. Derek, B. Tamy, & H. Wolfgang (2008). *Accurate multi-view reconstruction using robust binocular stereo and surface meshing.* IEEE Conference on Computer Vision and Pattern Recognition, Anchorage, AK, USA. doi: 10.1109/CVPR.2008.4587792.

[18] M. Westoby, J. Brasington, N. Glasser, M. Hambrey, & J. Reynolds (2012). *Structure from Motion photogrammetry: A low-cost effective tool for geoscience applications.* ELSEVIER, Geomorphology, 179/300–3146. doi: 10.1016/j.geomorph.2012.08.021.

[19] P. Cignoni, M. Callieri, M. Corsini, M. Dellepiane, & F. Ganovelli (2008). *MeshLab: An Open-Source Mesh Processing Tool.* Eurographics Italian Chapter Conference, Salerno, Italy. doi:10.2312/LocalChapterEvents/ItalChap/ItalianChapConf2008/129-136.

[20] P. Amit, B. Despoina, & G. Jorge (2017). *Dental Impression Materials and Techniques.* Dental Clinics of North America, 779–796, doi: 10.1016/j.cden.2017.06.004.

Compressive Strength-Ultrasonic Pulse Velocity Relationship of Concrete Containing Plastic Waste

Azad A. Mohammed

College of Engineering, University of Sulimani, Kurdistan Region, Iraq
azad.mohammed@univsul.edu.iq Tel: +9647504735890.

Abstract. In this paper the relationship between compressive strength and non- destructive behavior of ultrasonic pulse velocity (UPV) of concrete containing different plastic wastes has been investigated. For this purpose, a total of 102 data observations were collected from nine past works on properties of concrete with plastic waste and analyzed. Relationship between both compressive strength and UPV with plastic waste aggregate was found weak. Also, relationship between compressive and UPV is not strong, and in contrast, there is a relatively strong correlation between compressive strength ratio (concrete with plastic waste/control concrete) and UPV ratio. Using regression analysis, an equation was developed for calculating compressive strength of concrete with plastic waste depending on the UPV test data. The proposed equation is helpful to assess the residual compressive strength and consequently the structural health monitoring of those structures made of concrete with aggregate partially replaced with plastic wastes, which depends essentially on test data of UPV and compressive strength of control concrete.

Keywords: Compressive strength, Plastic waste, Recycled concrete, Regression analysis, Ultrasonic pulse velocity

1. Introduction

Different kinds of plastics are nowadays in use and they have become an inseparable and integral part of our life. Post consuming plastics can contribute to an ever increasing volume in the solid waste stream. Among the important ones there are Polyethylene terephthalate (PET), High density polyethylene (HDPE), Unplasticized Polyvinyl chloride (UPVC), Plasticized polyvinyl chloride (PPVC), Low density polyethylene (LDPE), Polypropylene (PP), Polystyrene (PS) and Expanded polystyrene (EPS) (Pacheco-Torgal, 2019, Siddique, 2008). A large volume of plastics consumed by the citizens worldwide usually accompanied with a large amount of wastes. These wastes need to be managed correctly in order to avoid land and water pollutions. Such waste has been thrown in the form of landfilling or been subjected to an incineration (Siddique, 2008, Sadat-Shojai and Bakhshandeh, 2011). Unfortunately, these methods are classified as the non-desired solution in the developed countries, mainly because the plastic is not biodegradable and on burning poisonous gases will release to the air. One can say that on recycling plastic wastes, additional expenses are required for reprocessing, accompanied with the color change and been extensively investigated by the past researchers and a relatively large amount of test data is now available in the published literature (Gu and Ozbakkaloglu, 2016, Siddique et al., 2008). Among the most important properties of concrete essentially modified with the existence of plastic waste one can mention the compressive strength. Strength and durability of different concrete structures are mostly dependent on this concrete property. Compressive strength gives an overall picture of quality of concrete, because this property is directly related to the structure of the hydrated cement paste. Also, compressive strength of concrete is almost invariably a vital element of structural design (Neville, 2004). Ultrasonic pulse velocity (UPV) test is a long established non-destructive method which determines the velocity of longitudinal waves. The determination consists of measurement of a time taken by a pulse to travel a measured distance (Neville, 2004). The test method is prescribed by ASTM C597-16 and by BS 1881: Part 203: 1986. The velocity of the ultrasonic pulse through concrete is the outcome of the time taken by the pulse to travel through the hardened cement paste and through the aggregate. Neville (2004) reported that there is no physical relation between the compressive strength and UPV. He noted that within limitations the UPV test can be used to asses the strength of concrete. There are many factors governing UPV including richness of the mix with cement, w/c ratio, aggregate content, moisture condition and the existence of steel reinforcement. With regard concrete containing plastic waste which it the issue of the current study, there is another parameter affecting UPV resistance of concrete which is the existence of plastic instead of natural aggregate. Ultrasonic pulse velocity travels inside concrete mass may vary with the existence of plastic because of the fact that UPV resistance of plastic is different from that of normal concrete materials beside the relevant cracks and flaws between plastic particles surface and the hardened cement. Akcaozog˘lu et al. (2013) reported that UPV value increased when the density of concrete was light, causing a reduction in the transmission of ultrasonic speed. Also, the addition of plastic materials to concrete mix accounts for the porosity and running dispersal of the transmitted ultrasonic speed (Rahmani, 2013). The previous studies show that plastic contained concrete has lower UPV due to development of pores. Rahmani et al. (2013) observed the reduction in UPV with an addition and increase of PET particle content in concrete. SEM analysis shows that the addition of PET particles makes concrete porous (Senhadji et al, 2015). Therefore, lower ultrasonic pulse velocity was recorded for concrete containing PET contents. Ultrasonic pulse velocity of concrete with PVC aggregate was also studied by Senhadji et al. (2015). The ultrasonic wave velocity values of concrete were found to decrease as the amount of PVC granules in the mixture increase. This reduction was attributed to the fact that the elastic properties of concrete and plastic are different, in addition to increasing porosity of concrete and cavities formed by the PVC particles. These factors are able to attenuate the ultrasonic wave by acoustic impedance. Based on their results there is a linear relationship between compressive strength and ultrasonic wave velocity of concrete. This will lead to the fact that the ultrasonic wave velocities can potentially be used to predict the compressive strength of concrete containing PVC aggregate. The following relationship was obtained for the relation between UPV and compressive strength of concrete at the ages of 28 and 90 days.

$$UPV = 76.08 f_c' + 1807 \quad (1)$$

In which UPV is the ultrasonic pulse velocity measured in m/s and f_c' is compressive strength measured in MPa. For the above relationship R^2 is equal to 0.9337. Properties of concrete containing plastic waste aggregate (PWA) subjected to high temperature were studied by Lima (2012). This researcher found that the incorporation of PWA can cause a degradation of UPV, which found to depend on the replacement ratio of the natural aggregate with PWA. This is explained by the higher porosity of concrete due to the decomposition of PWA. The relationship between residual UPV and compressive strength was found to be nearly linear (see Fig. 1), which confirms the suitability of measuring the UPV as an expeditious and non-destructive test to measure the residual compressive strength of concrete containing PWA exposed to fire.

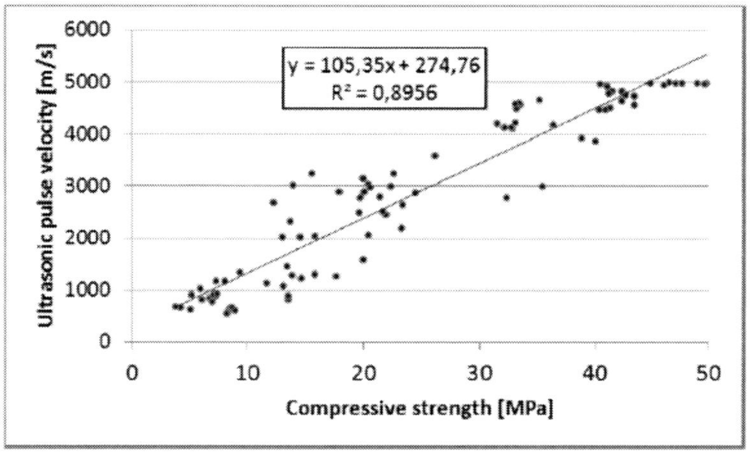

Figure 1. Relationship between UPV and compressive strength (tests by Lima, 2012)

Based on test data of compressive strength and UPV of normal concrete with PVC aggregate, the following two relationships were obtained by Mohammed et al. (2019)

For fine aggregate replaced with PVC aggregate

$$UPV = 1521\,(f_c')^{0.3014} \quad (2)$$

For coarse aggregate replaced with PVC aggregate

$$UPV = 59.43\,f_c' + 2280.9 \quad (3)$$

R^2 is equal to 0.83 and 0.908 for the relationships (2) and (3), respectively, applied for aggregate replacement with PVC ratio between 5% and 85%.

The author thinks that there is a relatively large amount of test data on compressive strength and ultrasonic pulse velocity of concrete with different plastic wastes. However, a general relationship for the compressive and UPV is not available to be applied on concrete with different plastic wastes for the purpose of structural health monitoring of those structures made of concrete with plastic waste aggregate. The available proposed relations is applied for one type of concrete and mainly proposed based on a few number of test data. In the current study, a relatively large number of test data has been collected and analyzed for the purpose of developing a more general equation for the relation between compressive strength and UPV.

2. Available Test Data Inspection

In this study, test data obtained by past nine published works (Akcaozog˘lu et al., 2013, Albano et al., 2009, Rahmani et al, 2013, Senhadji et al., 2015, Mohammed et al., 2019, Safi et al., 2013, Correia et al., 2014, Latroch et al., 2018, Al-Hadithi and Al-Ani, 2018) on the properties of concrete with different plastic wastes have been utilized. Test variables attempted by the researchers are given in Table 1. Plastic wastes used by the researchers were found to be PET, PVC from scrapped pipes or sheets, and expanded PVC. Plastic ratios used were varied from 2.5 to a maximum of 85%, used mostly as fine aggregate replacement. As given in Table 1, totally, 102 data points were taken for the regression analysis.

IOP Conf. Series: Materials Science and Engineering **978** (2020) 012010 doi:10.1088/1757-899X/978/1/012010

3. Proposed Equation

It is required here to show variations of compressive strength (f_c') and UPV with plastic waste content in addition to the relation between f_c' and UPV, from which the quality of the relation can be assessed. Fig. 2 shows variation of compressive strength with the plastic waste aggregate ratio. In general, there is a compressive strength loss, being increased with increasing plastic waste content, and the trend seems to be nonlinear. There is a relatively weak correlation between f_c' and the plastic waste ratio because the low coefficient of determination (R^2) which is only 0.3. Fig. 3 shows variation of UPV with the plastic waste ratio. There is a continuous loss in the UPV when the aggregate is replaced with plastic waste, and there is no chance for this property to be increased. Again, there is a relatively weak correlation between UPV and the plastic waste ratio because the highest R^2 was found to be 0.5. The reason of this low correlation of the two properties of concrete with PW ratio is attributed to the fact that different concrete mixes were used by the researchers. So, there is a vital need to construct a correlation based of the compressive strength ratio (concrete with plastic waste/ control concrete) and UPV ratio. Fig. 4 shows variation of f_c' ratio with plastic waste ratio and Fig. 5 shows variation of UPV ratio with plastic waste ratio. Highest R^2 for the two relationships are 0.67 and 0.75 respectively. Because of lower R^2 value, there is a high discrepancy of test data related to the compressive strength variation. This may be attributed to the sensitivity of concrete under compression and changing failure mode attributed to the change of internal structure of the hardened mass, accompanied with the normal aggregate replacement with plastic waste. This effect is well observed at low plastic waste contents as shown in Fig. 4. However, existence of plastic waste aggregate in concrete has a regular effect on UPV value, as observed from Fig. 5. Accordingly, there is chance to calculate UPV ratio of concrete for any plastic waste value.

Table 1. Test variables attempted by the past investigators

Reference	Compressive strength (MPa)	Mix proportion, kg/m^3 (Control mix) (C:FA:CA) (W/C ratio)	Type of plastic	Plastic ratio (% volume)	Number of test data	Remark
Akcaozog˘lu et al. (2013)	43.2, 55.9	500:877:575 (250/500)	PET aggregate (MS=4 mm)	30,40,50, 60	10	-
Albano et al. (2009)	27.9,28.9, 21.3,23	24.1:64.9:41.2 kg (0.5) 19.1:68.6:43.6 kg (0.6)	PET aggregate (Average size =0.26 cm, 1.14 cm)	10,20	28	-
Rahmani et al. (2013)	35.36, 38.71	488.1:654.9:976.1 (0.42) 379.6:745.9:976.1 (0.54)	PET aggregate (MS=7 mm)	5,10,15	8	There is an increase in f_c' at 5% and 10% aggregate replacement
Senhadji et al. (2015)	32.84, 34.1	400:470:1320 (0.48)	Scrapped PVC pipe, Graded 0–3 mm and 3– 8mm	30,50,70	8	-
Mohammed et al. (2019)	41.51	455.4:569.3:1138.5 (236.8/455.4)	PVC aggregate (Shredded sheet MS=10 mm)	5,15,30,45, 65,85	13	There is an increase in f_c' at 5% CA replacement
Safi et al. (2013)	61.9	664.1:1372.4 (Mortar) (W/B = 0.38)	Rejected plastic bags (MS=5 mm)	10,20,30, 50	15	-
Correia et al. (2014)	44	350:800.7:1003.2 (189/350)	Three types of PET waste: coarse (size between 2 and	7.5,15	6	-

			11.2 mm), fine (size between 1 and 4 mm) and nearly-spherical pellets obtained from a post heat treatment.			
Latroch et al. (2018)	40, 42	450:1350 g (Mortar) (W/C = 0.5)	Expanded PVC aggregate	15,25,50, 75	10	-
Al-Hadithi and Al-Ani (2018)	69	450:900:750 (W/B = 150/500)	PET aggregate (MS=4.75 mm)	2.5,5,7.5	4	-

* MS = Maximum size of aggregate, f_c'= Compressive strength, C=Cement, B=Binder, W=Water, CA= Coarse aggregate, FA= Fine aggregate.

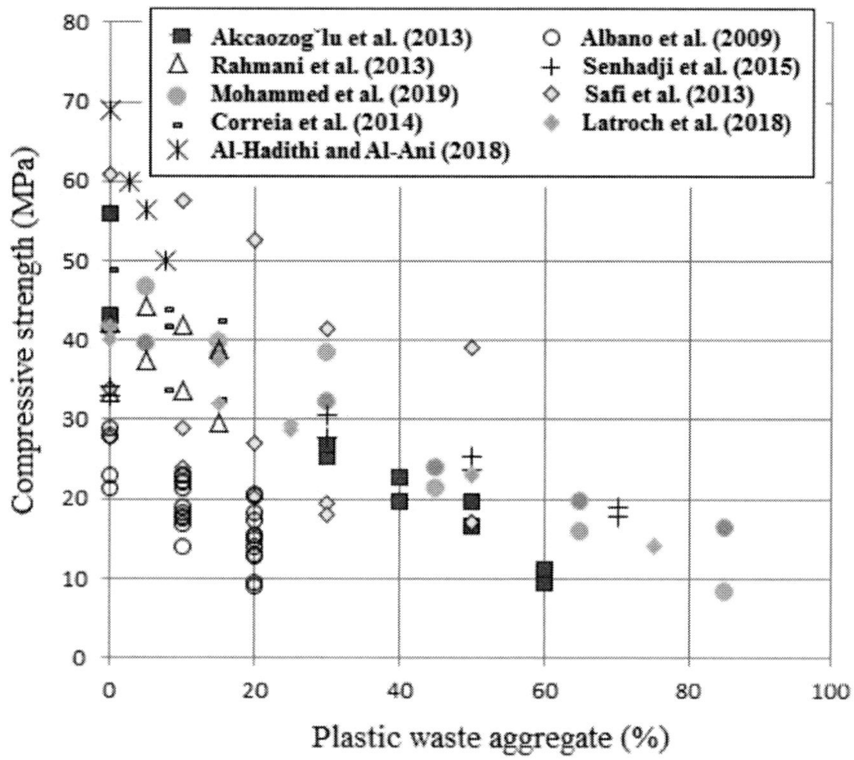

Figure 2. Variation of f_c'with plastic waste ratio

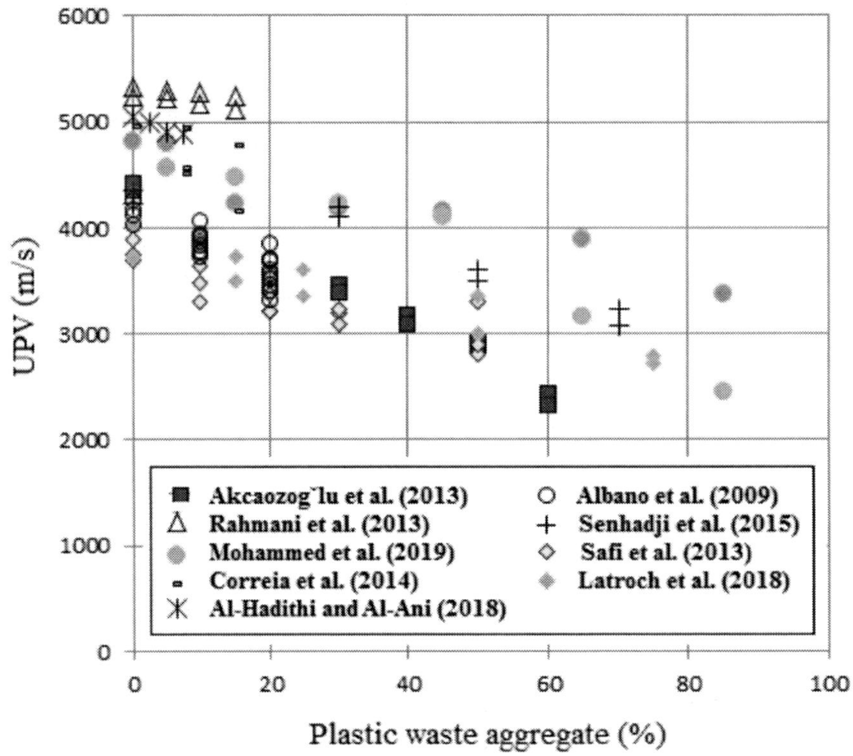

Figure 3. Variation of UPV with plastic waste ratio

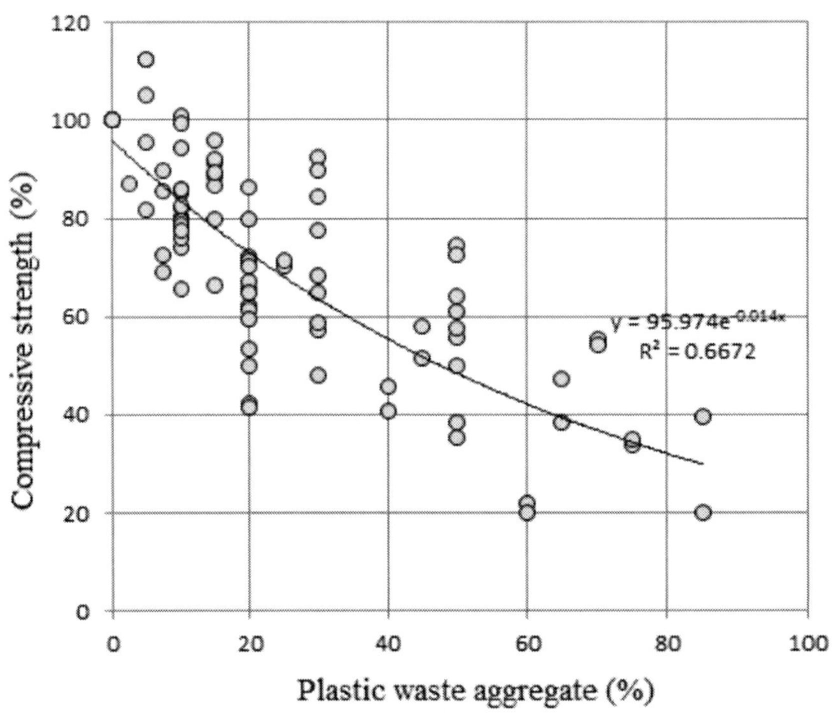

Figure 4. Variation of f_c' ratio with plastic waste ratio

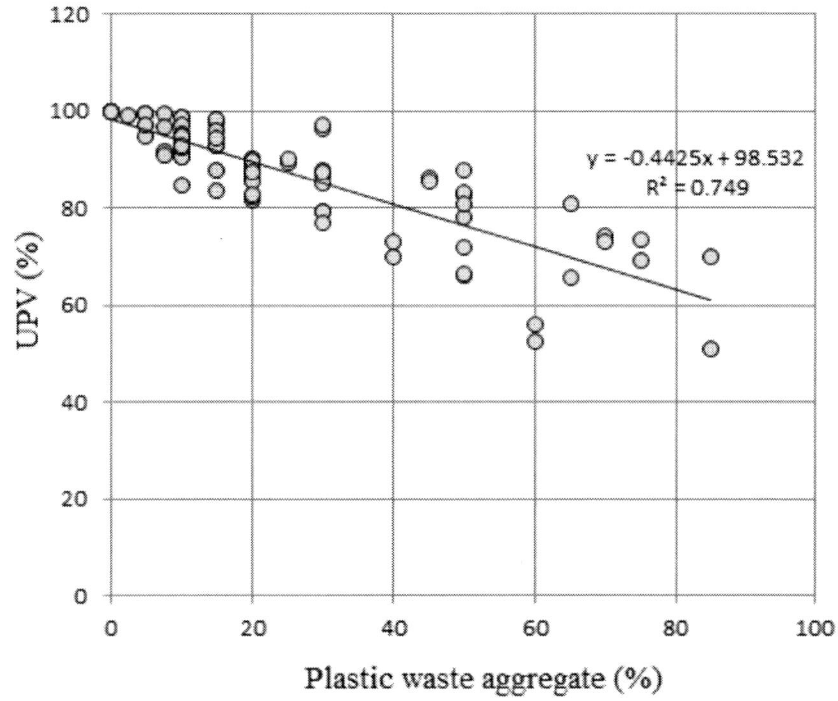

Figure 5. Variation of UPV ratio with plastic waste ratio

The following linear relationship was obtained from regression analysis

$$UPV\% = 98.532 - 0.4425(PW) \quad (4)$$

In which PW is the plastic waste percentage varied from 2.5% to 85%, R^2 for the above relationships is 0.75.

Fig. 6 shows variation of f_c' with UPV. One can find a poor correlation between the two properties, of R^2 equal to 0.48, mainly because different types of concrete and plastic waste have been used as stated before. Consequently, the equations proposed by the past investigators (i.e. Eqs. 1,2 and 3 and that given by Lima (2012) has a limited application and could not be used for concrete with different plastic wastes. For this purpose, it is necessary to draw the variation of compressive strength ratio with UPV ratio. Fig. 7 shows the correlation between the two percentages. In contrast to the previous relations, there is a relatively strong correlation between the two percentages because of high R^2 which is equal to 0.91. The relation between compressive strength ratio and UPV ratio was found to be nonlinear represented by the following equation obtained from regression analysis

$$f_c'\% = 0.0011(UPV\%)^{2.4652} \quad (5)$$

Based on the above equation, the residual compressive strength of concrete containing was is calculated as follows

$$f_{cp}' = f_c' \, [0.0011(UPV_p / UPV_c)^{2.4652}] \quad (6)$$

In which f_{cp}' is compressive strength of concrete with plastic waste, f_c' is compressive strength of control concrete, UPV_p is ultrasonic pulse velocity of concrete with plastic waste, and UPV_c is ultrasonic pulse velocity of control concrete.

Regarding of health monitoring of concrete structure, the applicability of Eq. 4 may not be of great importance, because there is a chance to perform the UPV test easily in all locations. However, Eq. 6 has a vital application to assess the variation of compressive strength with time and to diagnose any related problem. Eq. 6 can be utilized to assess the structural health against any damage via calculating the residual compressive strength f_{cp}'. To perform this, there is a need to measure UPV_p experimentally. UPV_c and f_c' are related to control concrete without plastic waste. On substitution in Eq. 6, the residual compressive strength (f_{cp}') is calculated. The obtained compressive strength is a tool to assess the infrastructures performance against any damage and to decide if there is a need for strengthening or not. Based on the discussion given above, the author recommends that there is a need to perform UPV test on a concrete member at the early lifetime of the structure. The existence of the early test data of f_c' and *UPV* for control concrete is helpful for the application of proposed equations for the purpose of structural health monitoring, such as Eq. 6 proposed in this study.

4. Conclusions

From the following research study, the following conclusions can be drawn

1- Analysis of a large number of test data indicates a high discrepancy of concrete compressive strength variation with plastic aggregate ratio, in which there is a high strength loss or increase at the same plastic waste content, especially at low plastic contents. Accordingly, there is a poor correlation between compressive strength of concrete and plastic aggregate content. Similarly, a weak correlation between UPV and the plastic waste content has been found. The reason of this is attributed to the fact that different plastic wastes have been tested by the past researchers.

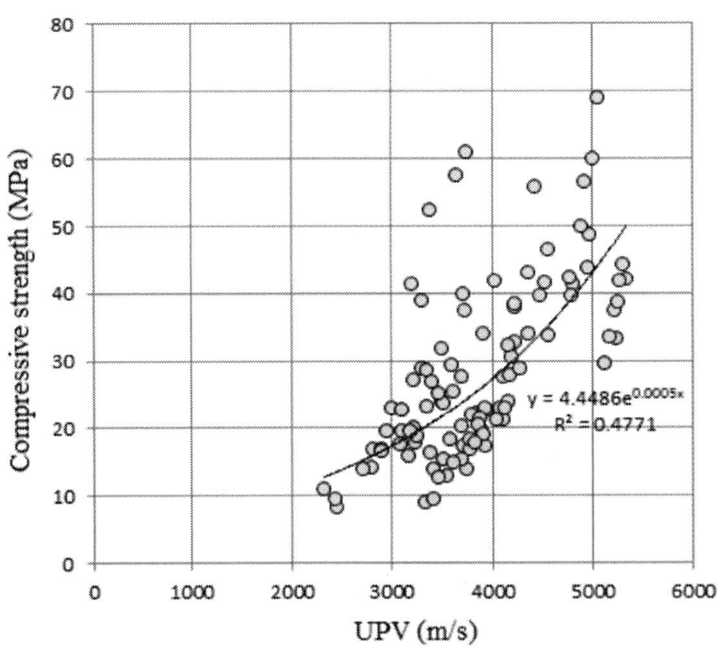

Figure 6. Variation of f_c' with UPV

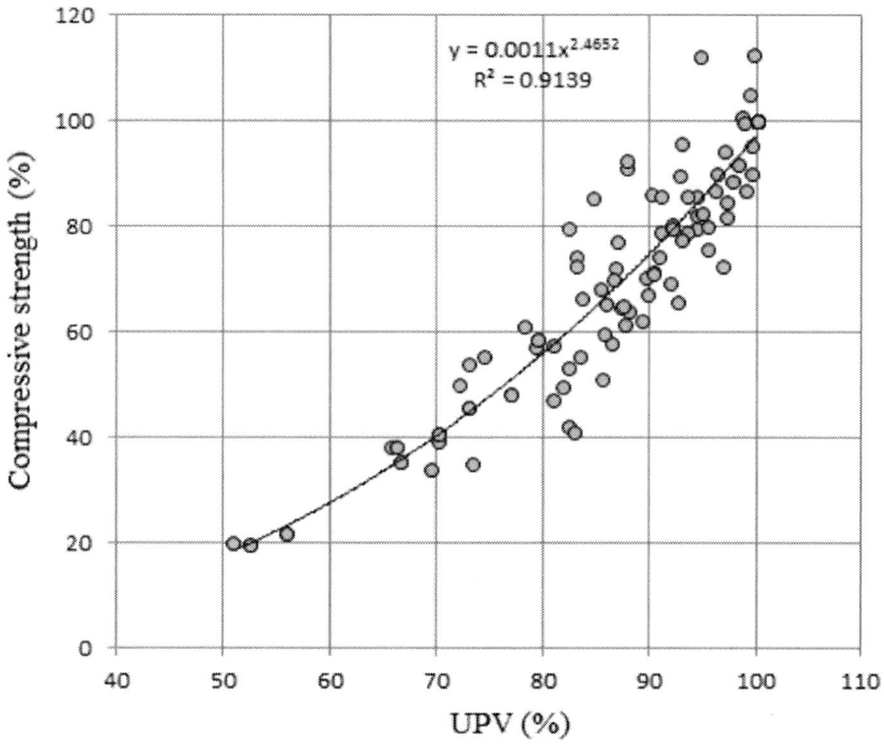

Figure 7. Variation of f_c' ratio with UPV ratio

2- There is a relatively strong correlation between UPV ratio (concrete with PW / control concrete) and plastic waste aggregate content. Eq. 4 was obtained from regression analysis to calculate this property based on plastic waste ratio.

3- It is more convenient to use the data of compressive strength and UPV ratios (concrete with PW / control concrete) to develop an equation for the relation between the two properties. Using regression analysis an equation was proposed (Eq. 6) to assess compressive strength of concrete based on the measurements of UPV. The proposed equation can be used to assess the residual compressive strength and to know the performance of concrete member.

4- It is important to perform UPV test and to have test data on control concrete at the early lifetime of the structure. The available test data is helpful for future applications of those equations proposed for the relation between compressive strength and UPV, for the purpose of structural health monitoring.

References

[1] Akcaozog˘lu, S. Akcaozog˘lu, K. and Atis, C.D. (2013). Thermal conductivity, compressive strength and ultrasonic wave velocity of cementitious composite containing waste PET lightweight aggregate (WPLA), *Composites: Part B* 45, 721–726.

[2] Albano,C., Camacho, N., Hernandez, N., Matheus, A., Gutierrez, A. (2009). Influence of content and particle size of waste pet bottles on concrete behavior at different w/c ratios, *Waste Management*, 209, 2707-2716.

[3] Al-Hadithi, A.I. and Al-Ani, M.F. (2018), Effects of Adding Waste Plastics on Some Properties of High Performance Concrete, Paper presented at the 11th International Conference on Developments in eSystems Engineering (DeSE), 273-279.

[4] Correia, J.R., Lima, J.S. and de Brito, J. (2014), Post-fire mechanical performance of concrete made with selected plastic waste aggregates, *Cement and Concrete Composites,* 53, 187–199.

[5] Gu, L. and Ozbakkaloglu, T. (2008). Use of recycled plastics in concrete: a critical review. *Waste Management*, 51(2016): 19-42.

[6] Latroch, N., Benosman, A.S., Bouhamou, N. and Senhadji, Y. (2018), Physico-mechanical and thermal properties of composite mortars containing lightweight aggregates of expanded polyvinyl chloride. *Construction and Building Materials,* 175, 77–87.

[7] Lima, J.S. (2012), *Fire behavior of concrete produced with selected plastic waste aggregate*, IST Tecnico Lisboa, Extended Abstract, pp. 19.

[8] Mohammed, A.A., Mohammed, I.I., Mohammed, S.A. (2019), Some properties of concrete with plastic aggregate derived from shredded PVC sheets, *Construction and Building Materials,* 201, 232-245.

[9] Neville, A. M. (2004). *Properties of Concrete,* 4th edition, Pearson Prentice Hall.

[10] Pacheco-Torgal, F. (2019). *Introduction to the use of recycled plastics in eco-efficient concrete, in Use of recycled plastics in eco-efficient concrete*, edited by: F. Pacheco-Torgal et al, Woodhead Publishing- Elsevier, 1-8.

[11] Rahmani, E. Dehestani, M., Beygi, M.H.A., Allahyari, H., Nikbin, I.M. (2013), On the mechanical properties of concrete containing waste PET particles. *Construction and Building Materials* 47, 1302–1308.

[12] Rebeiz, K.S. Flower, D.W. and Paul, D.R. (1993), Recycling plastics in polymer concrete for construction applications. *Journal of Materials in Civil Engineering,* 5(2), 237–48.

[13] Sadat-Shojai, M. and Bakhshandeh, G. (2011). Recycling of PVC wastes. *Polymer Degradation and Stability* 96, 404-415.

[14] Safi, B., Saidi, M., Aboutaleb, D., Maallem, M. (2013), The use of plastic waste as fine aggregate in the self-compacting mortars: Effect on physical and mechanical properties. *Construction and Building Materials* 43 (2013): 436-442.

[15] Senhadji, Y., Escadeillas, G., Benosman, A.S., Mouli, M., Khelafi, H. Ould Kaci,S. (2015), Effect of incorporating PVC waste as aggregate on the physical, mechanical, and chloride ion penetration behavior of concrete, *Journal of Adhesion Science and Technology,* 29 (7) .

[16] Siddique, R. (2008). *Waste Materials and By-Products in Concrete.* Springer.

[17] Siddique, R. Khatib, J. and Kaur, I. (2008). Use of recycled plastic in concrete: A review, *Waste Management*, 28, 1835–1852.

IOP Conf. Series: Materials Science and Engineering **978** (2020) 012011 doi:10.1088/1757-899X/978/1/012011

Possible Scenarios of Iraqi Marshland Restoration for Future Water Resources Management

Alaa H. Alshami[1], Mohammed A. Ibrahim[2], Haitham A. Hussein[3*], Hassan Hameed Gatiea[4]

[1]Civil engineering department, Al-FARABI University, Baghdad, Iraq
E-mail: alshami_alaa@yahoo.com
[2]Civil engineering department, Al-Nahrain University, Baghdad, Iraq
E-mail: moh_env@yahoo.com
[3]Civil engineering department, Al-Nahrain University, Baghdad, Iraq
E-mail: haithamalshami@yahoo.com
[4]Ministry of Water Resources-National centre of water resources management, Baghdad, Iraq- E-mail: haramuma4@gmail.com
*Corresponding author Haitham A. Hussein Civil engineering department, Al-Nahrain University, Baghdad, Iraq. Tel: +9647702662997

Abstract. Agriculture is the only source of the Tigris and Euphrates Rivers in Iraq. Annually more than 70% of the available water is used for irrigation. The total amount of land that can be cultivated in Iraq is 12 million hec. Only 2.2 million hec in Tigris basin and 1.37 million hec in the Euphrates basin were cultivated in 2001. According to the development of the projects in Turkey and development irrigation in Syria for the Euphrates basin, Iraq will face reduction in the amount of water entering Iraq with 50% for Tigris basin and 80% for Euphrates basin and with deficit of 21-22 billion cubic meters for the coming year 2030. The existing area of marshland restoration equal to 58%. The study suggests in case to increase the area of restoration Marshland to 75%, it should be need water amount to 2.5 BCM or 70 m^3/s continuously with existing discharge.

***Keywords**: Annual Flow; Flooded Area; Marshland Scenarios; Water management; Water resources of Iraq;*

1. Introduction

More than 70% of the available water of Iraq enters the country that originates in Turkey and Iran (Italian et al., 2006) . Most of water comes from the Tigris and Euphrates that originates in Turkey (Davis & Hanbali, 2005). Before the development of the hydraulic project on both river basins in the upstream countries, the average annual flow of Euphrates as it enters Iraq was estimated at 30 Billion Cubic Meters (BCM) out of which about 10 BCM per year drained in to the Marshes (Al-Ansari, 2013). On the Tigris prior to the construction of new dams and since records were kept the average annual runoff as it enters Iraq was estimated to be 21.2 BCM additional tributaries in the east added an average annual total of about 15 BCM from inside Iraq and 6 BCM from Iran. The huge water body exists where Tigris and Euphrates meet is the location of the Mesopotamian Marshlands. These marshes lie at the southern part of Iraq and are divided in to major areas (Hawaizah, Central, and Hammar) and nine to eleven smaller

IOP Conf. Series: Materials Science and Engineering **978** (2020) 012011 doi:10.1088/1757-899X/978/1/012011

ones. Due the political reason in the 1990s had a devastating impact on the ecosystem indigenous population and their economy the marshland once estimated to be between 15000 to 20000 Km^2 were reduced to about 2000 to 3000 Km^2 by 2004 (Ayied & Reiss, 2019). If we consider the year 1973 is the maximum permissible flooded then the area will be equal to 10000 square kilometers which are distributed in to three main marshlands which are the Central Marshlands of 3121 km^2 , Hammer Marshes with areas estimated 2729 km^2 and Hawaizeh Marshes with area equal to 3717 km^2 with additional part insides Iran estimates 640 km^2 called Alathem marshes (USAID, 2006). Fig (1) and Fig (2) represent the location of the marshes land in the south region of Iraq and the compression of marshes between maximum for year 1973 and minimum for year 2000 respectively.

2. The Southern Marshes

Before the drying of the southern marshes the area including lakes and reservoir covered about 4% of Iraq. The largest of marshes as known as the Mesopotamian marshes that include Huwaizah, central (also known as Qurna) and Hammar, flooding usually took place at the end of February or beginning of March and coincide with beginning of spring flood of Tigris and Euphrates Rivers. Central marsh used to receive Tigris flood water from Al- Musendak (downstream Kut) in addition to Butaira , Orieth , and Great Major (Ewaid & Abed, 2017). Huwaizah marsh receives water through the canals of Kahla, Mushrah and Micheriya as well as an additional flood water came from Iran through Tib, Dewerridj and Karkheh rivers that emerge from the mountainous region of Iran. The source of flood water to Hammar were from the Euphrates, Central marsh and Graff Canals (Makki, Abood, & Al-Umar, 2019). The restoration of the marshes is particularly challenging for many reasons, perhaps the most concern the current and the future delivery of water to Iraq via the Tigris and Euphrates. Given the volume of water that would require to meet the high evaporation losses in southern Iraq (2600-3000 mm/year) and accounting for the amount of water required to maintain through consume a significant amount to the total available supply in the catchments and would represent are unrealistic given other sector needs in the country, clearly there are decision to be made by the government of Iraq regarding the priority locations and extent areas to be restored (NWRC, 2010).

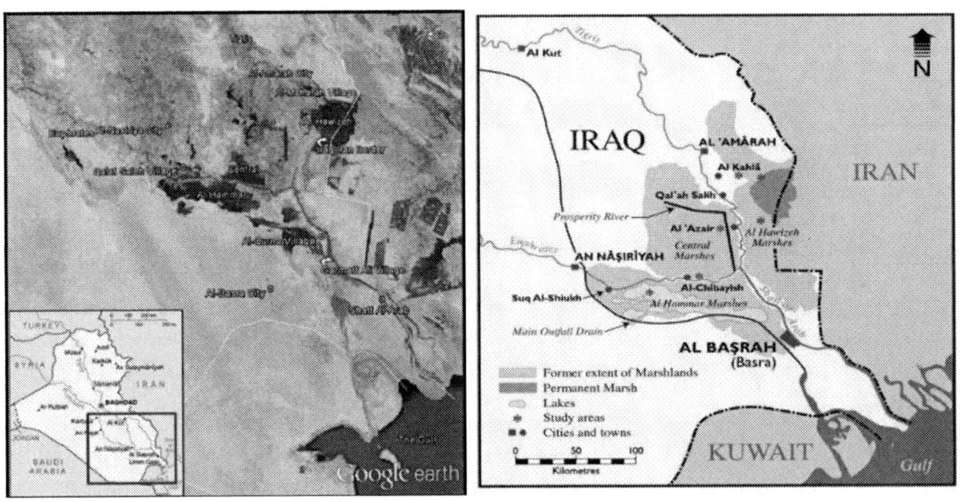

Figure 1. Location of marshes land in south of Iraq (USAID, 2006)

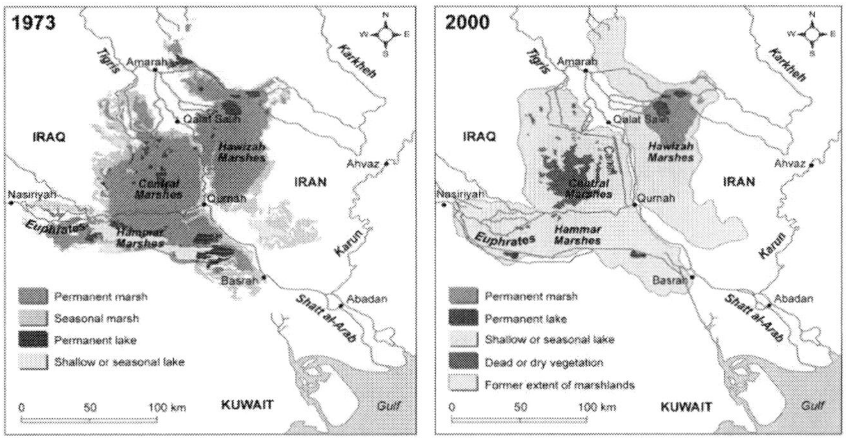

Figure 2. Map of the Marshes of southern Iraq 1973 compared to 2000 (Mahdi & Fawzi, 2014)

3. The Southern Marshes

Before the drying of the southern marshes the area including lakes and reservoir covered about 4% of Iraq. The largest of marshes as known as the Mesopotamian marshes that include Huwaizah, central (also known as Qurna) and Hammar, flooding usually took place at the end of February or beginning of March and coincide with beginning of spring flood of Tigris and Euphrates Rivers. Central marsh used to receive Tigris flood water from Al- Musendak (downstream Kut) in addition to Butaira , Orieth , and Great Major (Ewaid & Abed, 2017). Huwaizah marsh receives water through the canals of Kahla, Mushrah and Micheriya as well as an additional flood water came from Iran through Tib, Dewerridj and Karkheh rivers that emerge from the mountainous region of Iran. The source of flood water to Hammar were from the Euphrates, Central marsh and Graff Canals (Makki et al., 2019). The restoration of the marshes is particularly challenging for many reasons, perhaps the most concern the current and the future delivery of water to Iraq via the Tigris and Euphrates. Given the volume of water that would require to meet the high evaporation losses in southern Iraq (2600-3000 mm/year) and accounting for the amount of water required to maintain through consume a significant amount to the total available supply in the catchments and would represent are unrealistic given other sector needs in the country, clearly there are decision to be made by the government of Iraq regarding the priority locations and extent areas to be restored (NWRC, 2010).

4. Turkish and Syrian Infrastructures projects

Iraq is heavily dependent on the flow water from Tigris and Euphrates Rivers (Al-Ansari, 2013). Turkey, in particular and-to lesser extent Syria, controls headwaters of the Euphrates and a major portion of Tigris and intends to fully exploit these reservoirs. Many dams with huge storage capacities have been constructed or are planned by the two countries on the main rivers channel or their Tributaries. Table (1) represent the status of these reservoir and dams

Table 1. Status of Reservoir in Turkey and Syria

Name of the project	Country	Designed Storage capacity BCM	Current Status
Kiban	Turkey	30.9	completed
Karkaia	Turkey	9,58	completed
Ataturk	Turkey	48.7	completed
Perigick	Turkey	1.22	completed
Karakamish	Turkey	0.157	completed
Tishreen	Syria	1.90	completed
Tabaka	Syria	14.16	completed
Bath	Syria	0.09	completed
Kiralkezi	Turkey	1.92	completed

Tigris	Turkey	0.59	completed
Debokgedi	Turkey	0,2	completed
Solivan	Turkey	8.7	Planned
Ceaser	Turkey	-----	planned
Batman	Turkey	1.18	completed
Gezran	Turkey	0.5	planned
Illisu	Turkey	10.41	completed
Cezar		----	planned

4.1 Turkish Projects

The projects currently being constructed, planned by Turkey on the Tigris and Euphrates are carry the name South East Anatolia project or GAP. The project consists of 22 Dams and 19 hydroelectric stations and number of irrigation Tunnels, Canals and projects. Turkey completed the first stage of the project for one of the largest irrigation canal in the world (Orfa tunnel) which takes the water from Ataturk dam to land situated beyond the Euphrates basin (USAID, 2006). Two new dams Brrecik and Karkamis, are being constructed on the Euphrates near the Turkish-Syrian border, enabling Turkey to be increase full control of the water of the river. In 1997 Turkey plan to irrigate 1.091 million hectares' in the Euphrates basin and 0.602 million hectares' in the Tigris basin. The total area prescribed for irrigation (within the framework of the completed development of the storage and irrigation project in the Euphrates basin) will be more than 1.6 million hectares (USAID, 2006).

4.2 Syrian Projects

The World Bank estimate that the agriculture lands cultivated in Syria on the Euphrates in 1965 was 213000 hectars. According to the information presented by the Syrian technical delegation during the negation of the joint technical committee on water in 1983, the area irrigated on the Euphrates basin at that year were 258000 hectares and their water requirement in the region were 3.87 BCM.

The irrigation areas within the Euphrates basin in 1998 are estimated at 258000 hectares with plans to expand to about 778000 hectares. The current areas or the irrigated lands along Tigris basin is unknown but there are planned to irrigates up to 227000 hectares (USAID, 2006).

5. Data Collection and Results for Existing and Future Water Resources Management

5.1 Existing Water Resources in Tigris River and its Tributaries

According to the study (USAID, 2006) the following quantity of water resources in Tigris and Tributaries rivers are distributed in the Table 2. The water resources as an annual average from the year 1990 to 2006 are recorded from National Water Resources Center/ ministry of water resources and the annual mean flow is equal to 42.8 BCM (Italian et al., 2006).

Table 2. Existing water resources in Tigris River and Tributaries (USAID, 2006)

Rivers	Annual flow BCM	Outside Iraq %	Inside Iraq %
Main Tigris	19.43	100	-----
Khabor	2.1	58	42
Greater zab	14.32	42	58
Lesser - zab	7.07	36	64
Al- Audhaim	0.7	---	100
diyala	5.86	59	41
Total	49.48	56% for Turkey	32% for Iraq
		12% for Iran	

5.2 Future Water Resources Requirement in Tigris River and its Tributaries

The average annual flow for Tigris and Tributaries rivers before and after 2030 as shown in Table 3 (USAID, 2006). The annual average flow required for Turkey, Syria and Iraq up to 2040 are presented in the Table 4.

Table 3. Amount of average annual flow before and after 2030 in BCM for Tigris and Tributaries (USAID, 2006)

Item	Iraq-Turkey Border		Tributaries of Tigris	
	Before	after	before	after
Mean Annual BCM	19.43	9.16	30.05	23.53
Quality of Water in PPM	250	375	250	> 375

Table 4. Expected average annual flow needed up to 2040 for Turkey, Syria and Iraq in BCM (USAID, 2006)

Rivers	Turkey	Syria	Iraq	Total
Tigris	6.50	3.0	41.8	51.30
Euphrates	18.50	11.50	23.0	53.0

From Table 2 which shows the annual average flow required for Tigris equal to 49.48 BCM and form National Water Resources Center the actual mean annual flow (1990-2006) equal to 42.8 BCM. In other words, there is deficit about 6.68 BCM. In future water resources in Tigris River, the water amount after 2030 equal to 32.69 BCM as mentioned in Table 3, while the quantity required after 2040 equal to 41.8 BCM as shown in Table 4, so it will give a deficit equal to 9.11 BCM. From final discussion, it clearly shows that Iraq will face with deficit equal to 6.68BCM before 2030 and this deficit will increase to reach 9.0 BCM up to 2040.

5.3 Existing Water Resources in Euphrates River

Iraq will receive only 25-30 percent of what used to enter the country three decades ago lagging behind the Euphrates river basin development. According to the previous study (Italian et al., 2006), the Euphrates River basin for Turkey (GAP) and development of Syrian Euphrates basin which propose three alternative scenarios which depend on the basis of entering water to Hit station in Iraq which is equal to 33.46 BCM.

5.4 Future Water Resources Requirement in Euphrates River

The water required in BCM in Turkey, Syria and Iraq for the period between 2010 to 2040 after developing the GAP in Turkey and developed agriculture in Syria (Middle alternative) it will get the following result shown in the Table 5.

Table 5. Water required for the year 2010 to 2040 in BCM (Bank, 2006)

Year	Turkey	Syria	Iraq
2010	10.814	9.405	13.241
2020	12.385	9.405	11.67
2030	14.365	9.405	9.69
2040	16.908	9.405	7.197

If refer to the Table 4, the amount of water required up to 2040 equal to 23.0 BCM, while the water entering Hit station up to year 2040 equal 7.197 BCM as shown in the Table 5. Hence Iraq will face a deficit of water equal to 16.0 BCM in Euphrates river. The information from the study (USAID, 2006) which was submitted to the ministry of water resources/Restoration of Marshland Center and prepared by (USAID), the following information of water quantity and quality is shown in the Table 6.

Table 6. Quantity and Quality of the Euphrates River before and the expected after 2030 (USAID, 2006)

Item	Turkey and Syria boarder		Iraqi and Syria boarder	
Year	Before 2030	After 2030	Before 2030	After 2030
Average annual flow BCM	32	14.2	30.37	8.48
PPM	250	500	457	1250-1350

To make the comparison for water entering to Iraq, we will find the quantity of water after 2030 equal 8.45 BCM. Also quantity required for irrigation and other uses equal to 23.0 BCM. Therefore, the total deficit will be face equal to 14.55 BCM. The actual data collected from the control department/National Water Resources Center between water year 1931- 1932 to 2005- 2006 for Hit station at Euphrates river are divided in to three results parts depending three time periods. First period (1931-1932) -(1972-1973) the mean annual equal 30.70 BCM.Second period (1972-1973) -(1997-1998) the mean annual equal to 22.8 BCM. And Third period (1997-1998) -(2005-2006) the mean annual equal to 18.5 BCM.

The total average for the three periods equal to 25.0 BCM while the required amount equal to 30.37 BCM as mention in table (6) and this will give deficit equal to 5.77BCM before 2030 and it will increase after 2030 to reach between 15-16 BCM

According to the data analysis, the results clearly shows that Iraq will face deficit equal to 21-22 BCM for both Tigris and Euphrates Rivers and this amount will increase if we consider the average flow in Hit station equal to 25.0BCM instead of 33.46 BCM.

6. Scenario of the Restoration of the Marshland

In order to make the restoration it is necessary to supply with great amount of water while Iraq face with deficit of water due to the development of the irrigation projects in the riparian countries.

The dried area of the marshes has been used by constructing houses and as cultivation areas but in spite of that it could be possible to restore these marshes and bring it back to the previous condition. The following information are collected from the Marshland Center/MOWR and (Italian et al., 2006) which consist these facts

a) The flooded area of the Marshland considered equal to 100% as it was existing in 1973.

b) According to the above the area flooded of the Marshlands equal to 9567 Km^2 inside Iraq and 640 Km^2 inside Iran.

c) The Marshland distributed in the three parts, Central Marshes (Qurna marshes) with area equal to 2729 Km^2, Hammar Marshes with area equal to 3121 km^2, and Huwaza marshes with area equal to 3717 km^2

It is necessary to consider that the area of the Restoration will be increased or decreased according to the change in the hydrological cycle (wet season, normal water year and dry season) and the area to be flooded cannot interfere with existing projects such as the agriculture area constructed in the dried area of the marshes and field oil project by Mathematical programing (Hec-RES-SIM).

6.1 Scenario (1) by taking the flooded area between 0-25%

It is similar to year 2000-2002 and the area flooded equal to 2400 km^2 as shown in Figure 3a (Italian et al., 2006)

6.2 Scenario (2) by taking the flooded area between 25-50%

This alternative will be similar to year 2005 and the area will be flooded equal to 2400- 4800 km^2 and this will have achieved in normal water year as shown in Figure 3b

6.3 Scenario (3) the percent of flooded area between 50-75%

In this alternative is similar to year 1990 and the total area will be flooded between 4800- 7200 km^2 and this can have achieved during the wet water year and also will not crossed with the existing project but it needs to help from the riparian countries to supply with sufficient water Figure 3c

6.4 Scenario (4) the percent of flooded area between 75-100%

This alternative is similar to water year 1973and the area will be flooded between 7200-10000 km^2 as shown in Figure 3d this proposal will crossed with the existing projects and need to supply with water from the riparian courtiers.

From these alternative Scenarios number (3) is recommended to flood the marshland area with 50 – 75% and can covered an area between 4800-7200 Km^2. Figure 4 indicate the flooded area for short and medium term in order to restore the marshes (Italian et al., 2006).

7. Conclusion

1- According to the development of the projects in Turkey for GAP and the development of irrigation projects in Syria for the Euphrates basin, Iraq will face a reduction in the amount of water entering with 80% for Euphrates basin and 50% for Tigris basin and with deficit between 21-22 BCM for the coming years (2030), the restoration will face reduction of water and then will affect the area of the restoration.

2- The study analyzed the area of the existing flooded that is equal to 58% and it will need to increase to 75% of flooded area which will require additional amount of water to 2.5BCM or 70 m^3/sec

3- In this study if flooded are 75%, then the total area of the marshes to be flooded equal to 7200 km^2 hence the evaporation will be great amount so it need to discharge continuously.

4- The operation of the restoration of the marshlands needs to cooperate between Iraq and the riparian courtiers in order to supply with sufficient water and the indication show a reduction in the marshlands due to the deficit in the water in the near future.

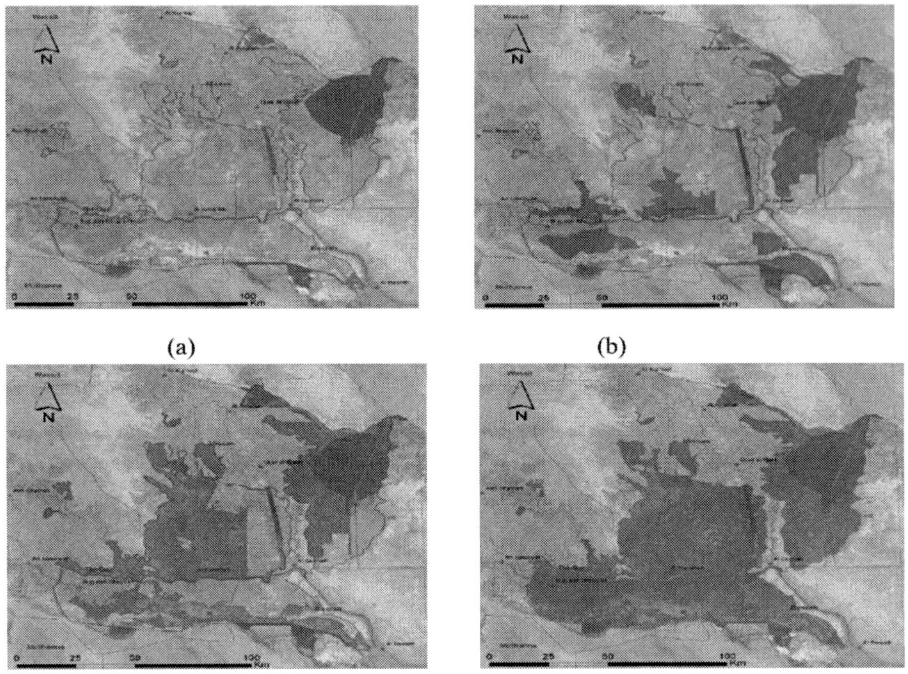

(a) (b)

(c) (d)

Figure 3. Different Scenarios of Marshland a. flooded area 0-25%,b. flooded area between 25-50%, c. flooded area 50-75% and d. flooded area 75-100%

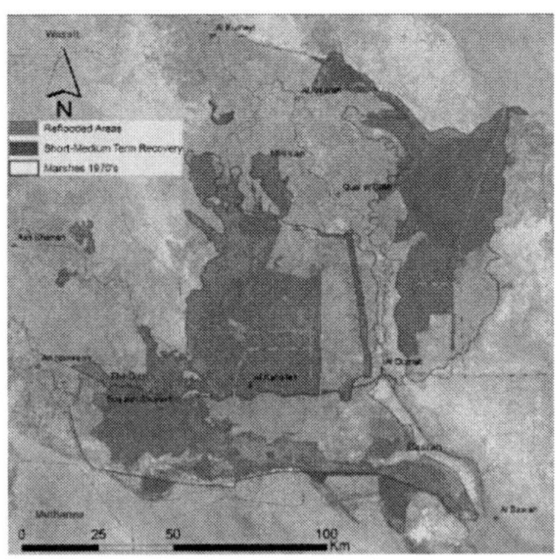

Figure 4. Proposed 50-75% Area Marshland Restoration

References

[1] Al-Ansari, N. (2013). Management of water resources in Iraq: perspectives and prognoses. *Engineering, 5*(6), 667-684.

[2] Ayied, A. Y., & Reiss, P. (2019). Impact of the Iraq Marshlands Restoration Program on Livestock Population and Production in the Southern Marshes of Iraq. *Journal of Buffalo Science, 8*(2), 25-33.

[3] Bank, W. (2006). Water Resources Interim Assistance Strategy. *Water, Enviroment, Social and Rural Development Department Middle East and North Africa Region.*

[4] Davis, D. W., & Hanbali, F. (2005). Water management in Iraq—Capability restoration and implications for historic marsh restoration *Impacts of Global Climate Change* (pp. 1-9).

[5] Ewaid, S. H., & Abed, S. A. (2017). Water quality index for Al-Gharraf river, southern Iraq. *The Egyptian Journal of Aquatic Research, 43*(2), 117-122.

[6] Italian, Ministry, Environment, Territory, Free, Iraq, & Foundation. (2006). New Eden master plan for the integrated water resources management in the Marshland area, Main Report, Iraqi Ministries of Environment. *Water Resources Municipalities and Public Works with cooperation of the Italian Ministry for the Environment and Territory and Free Iraq Foundation, 20.*

[7] Mahdi, B. A., & Fawzi, N. A.-M. (2014). Iraq's inland Water quality and their impact on the North-Western Arabian Gulf. *Marsh Bulletin, 9*(1), 1-22.

[8] Makki, J. S., Abood, A. R., & Al-Umar, M. H. (2019). Restoration of Al-Sanaf Marsh under Various Operation Conditions. *Journal of Ecological Engineering, 20*(11), 178-189. doi: 10.12911/22998993/113038

[9] NWRC. (2010). Ministry of water resources *National Water Resources Center.*

[10] USAID. (2006). Iraq Marshland Restoration program. *(U.S. Agency for International Development).*

IOP Conf. Series: Materials Science and Engineering **978** (2020) 012012 doi:10.1088/1757-899X/978/1/012012

Using PAVER Software to Evaluate Pavement Condition Index of Highway Segment in the North Sector of Najaf City (Case Study)

Khawla H. H. Shubber[1*], Mohammed Noamman Mahdi[2]

[1]Faculty of Engineering, University of Kufa, Najaf, Iraq- E-mail: khawla.shubber@uokufa.edu.iq
[2]Ministry of Transport, Public company for managing Private transport, Babylon- Iraq E-mail: psbbakdo@gmail.com
* Corresponding author: Faculty of Engineering, University of Kufa, Iraq. Tel: +9647705888525

Abstract. Current paper present part of comprehensive case study for urban highway networks of north sector of Najaf city located 161 km Southwest of Baghdad the capital of Iraq. The main goals of current paper are to evaluation pavement condition index denoted as (PCI). Values of PCI acting a main factor on depending maintenance of pavement, selecting economic alternative, increasing development of engineers and technician's capabilities, in order to reach the best maintenance management process and achieve the best riding level with lowest costs and the shortest travel time. PAVER software uses surveying data to calculate pavement condition index which should be within rating from zero, which mean pavement in failed condition, to 100, that mean excellent pavement condition. Pavement's condition is providing predicting its maintenance and Rehabilitation requirements. Free issue of PAVER 5.2 software was used to calculate values of PCI for a minor arterial urban highway in north sector of Najaf city, that is the Najaf-Karbala highway located from Al-Askarain Tunnel intersection to the end of Al-Nidaa district, with length (8011m). Manual geometric and types of distresses survey was depended instead of using assessment technologies to consider the type, degree of severity, of distresses for sample unites along the highway. As a results of implementing PAVER 5.2 software depending on field survey data collection, PCI value is equal to 80. That's mean this part of the selected study section was in good situation.

Keywords: *Pavement maintenance Management System (PMMS), PAVER 5.2, flexible pavement, pavement condition index (PCI).*

1. Introduction:

Each activity aimed to improve pavement structure performance and made any pavement segment in acceptable level of service represented as Highway maintenance. Highway maintenance should give a respectable contribution between environmental condition and transportation requirements. The growing number of traffic load, increasing road mileage, increasing commercial vehicles weight, needs for advanced specifications for maintenance and performance illustrate the main factors controlling the need for maintenance and rehabilitation, (Almuhanna, et al, 2018).

IOP Conf. Series: Materials Science and Engineering **978** (2020) 012012 doi:10.1088/1757-899X/978/1/012012

Road construction projects are always costly, especially if implemented with high quality standards. In order to achieve the functional purpose for which it was established for the best and long design life, a scientific program for maintenance work must be provided by the competent authorities.

It is an inaccuracy to follow traditional survey methods, which rely on the visual scanning of the engineer and his cumulative experience as this method offers a wide range of personal explanations and a variety of opinions which gives an unclear picture to make the right decision to perform the necessary maintenance of the road. Pavement management system (PMS) contain many engineering activities related to pavement design, construction, and maintenance. Figure 1, shows Pavement maintenance management systems (PMMS). Pavement maintenance operations represent an effective approach in process strategies and fund resource keeping (R Haas et al, 1994). Evaluate current pavement condition, forecast future conditions, analysis process of alternatives, in addition to select proper rehabilitation strategy represent consequent PMMS process stages for any given pavement section, (Garber& Hoel, 2008).

Typically, pavement condition index (PCI) is pavement rating system employs by PMMS as the source of assessment of present and expected pavement performance. Figure 2 demonstrate traditional curve characterizes pavement's expected performance relationship with pavement life in years. Usually, future performance prediction can be obtained from this curve. The estimated pavement condition after time period, multiple economical approach and maintenance methodologies can be done according to optimum cost effective maintenance treatment, (TRB, 2005).

In other words, (PMS) is all analytical tools and methods that contribute to finding the best strategies for maintenance usable berths over a specified period time. Maintenance strategies include all set of different activities identified for each highway system to upgrade the pavement classification to a higher level than the minimum requirement (LU& Lyton, 1976).

Highways in north sector of Najaf city (as all the city highways network) suffers from bad management or lack of using such system in order to put the priority of maintenance process and selecting the best alternatives for its road network. Although extensive research has been carried out on specific cities in Iraq, no such single study exists which deals with evaluation pavement condition in north sector of Najaf city. Therefore, the aim of the present study is to shed light on the type of distress spreading in part of selecting study area and to evaluate Pavement Condition Index (PCI) using PAVER 5.2 software. Pavement Maintenance Management System that illustrated in PAVER software, initially was established in the late 1970s to help in the pavements maintenance and rehabilitation process.

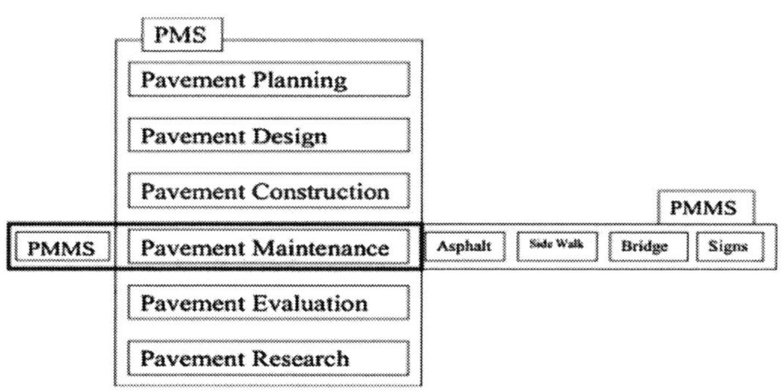

Figure 1. PMMS as a components of PMS (Garber& Hoel, 2008), (Abo-Hashema et al, 2006).

IOP Conf. Series: Materials Science and Engineering **978** (2020) 012012 doi:10.1088/1757-899X/978/1/012012

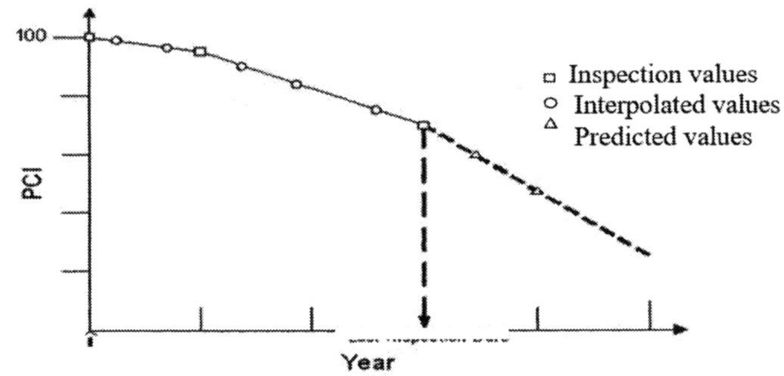

Figure 2. Typical performance curve (TRB, 2005), (Shahin& Smith, 2005).

AASHTO has identified the maintenance, rehabilitation and reconstruction strategies of the pavement as follows (AASHTO, 1997)

- Routine maintenance: The application of daily activities includes maintenance by cadres to control all defects in early times and maintain the functional status of the road. Examples of such activities are drill trenching and filling or sealing cracks.
- Rehabilitation of the pavement: It is divided in to two part.
  - Simple rehabilitation is intended to maintain the current pavement.
  - The main rehabilitation involves improving the durability of the pavement by increasing the thickness of the pavement and as a result, highway service life extended.
- Reconstruction of the pier: This option is used when the functional performance of the road is no longer available. Requiring the removal of the current pavement structure and replacing it with the equivalent structure.

2. Distress of flexible pavement;

According to many references, failures in pavement take two essential different types. these failures are, (Yoder& Witczak, 1976):

- Functional distress: the functional distress all that disturbed pavement surface and caused decrease in comfort for drivers in vehicle along the highway,
- Structural distress: any breakdown or collapse in structure components of pavement in one layer or more, and come to be in un capable to transmit load passes over its surface. functional distress always couples/ or not with structural distress. While, structural distress always accompanied with frailer in pavement surface (functional distress).

Hot mix asphalt pavement represent the high percent of highways constructed in Iraq, as usual in Najaf city. Hence, in this paper, distresses in flexible pavement types had been proposed. Flexible pavement distress evaluated by occasion preservation treatments. It well known, the procedure of pavement distresses evaluation has been known in term of inspection survey (LTPP, 2003). Weak of structural capacity, using low material quality, insufficient design, bad construction techniques in addition (or not) absence of preventive maintenance process are represent key factors or a grouping of factors affected Pavement distress. PAVER 5.2 software categorize 5 major classes in hot mix asphalt pavement surface distresses (Miller& Bellinger, 2003). These categories are:

A- Pavement cracking such as "alligator cracking or known as fatigue cracking, longitudinal and transverse cracking, block cracking, slippage cracking, joint reflective cracking, and edge cracking",

IOP Conf. Series: Materials Science and Engineering **978** (2020) 012012 doi:10.1088/1757-899X/978/1/012012

B- Surface Deformation take one form of "rutting, corrugations, shoving, depressions, swell, bumps and sags",

C- Disintegration defects which take place from surface to the lower layers such as "potholes, patching and utility cut patching",

D- Surface defects like effect of "weathering and raveling, bleeding, polishing",

E- Miscellaneous distresses which take different forms such as "lane/shoulder drop off, railroad crossing".

Below, table 1 shows flexible pavement distress groups.

Table 1 PAVER 5.2 Classification Distress for Asphalt Pavements According to its help file (Shahin& Smith- 2005, Yoder& Witczak-1976, Miller& Bellinger-2003, Almuhanna, et al-2018, PAVER5.2 helps index).

Code	Distress	Measure Unit	Defined Severity Levels?	Type of Distress	Main Cause
1	Alligator Cracking	m^2	yes	Structural	Load
2	Bleeding	m^2	yes	Functional	Other
3	Block Cracking	m^2	yes	Structural	Climate
4	Bumps And Sags	m^2	yes	Structural & Functional	Other
5	Corrugation	m^2	yes	Functional	Other
6	Depression	m^2	yes	Functional	Other
7	Edge Cracking	m^2	yes	Functional	Load
8	Joint Reflection	m^2	yes	Structural	Climate
9	Lane/Shoulder Drop-Off	m^2	yes	Functional	Other
10	Longitudinal And Transverse Cracking	m^2	yes	Structural	Climate
11	Patching And Utility Cut Patching	m^2	yes	Structural & Functional	Other
12	Polished Aggregate	m^2	no	Functional	Other
13	Potholes	Number	yes	Structural & Functional	Load
14	Railroad Crossings	m^2	yes	Functional	Other
15	Rutting	m^2	yes	Functional	Load
16	Shoving	m^2	yes	Functional	Load
17	Slippage Cracking	m^2	yes	Structural	Other
18	Swell	m^2	yes	Functional	Other
19	Weathering And Raveling	m^2	yes	Functional	Climate

3. Selected Study Area Description

As shown in Figure 3, the selected highway section represents part of north sector of Najaf city, located in AL-Najaf governorate located 161 km Southwest of Baghdad the capital of Iraq. Name of section was Najaf-Karbala highway, extended from Tunnel intersection Al-Askarain to Al-Nidaa district. Length of selected highway section was (8011m). Geometric design of selected section was two-way divided highway with three lanes in each direction have total width of (10.5 m) and two side walk of 2.0 meter, width of median was (3 m).

4. PAVER 5.2 Software Program:

The pavement condition as mentioned earlier, is usually assessed by (PCI). The system developed by United States Army groups of engineers and approved by several associations such as the American Public Works Association (APWA) and the American Society for Testing and Materials (ASTM). According to PAVER5.2, PCI ranging from 0 to 100 as issued in Illinois center for Transportation research report (APWA, 2012). Currently, the PAVER 5.2 software and its new versions are widely using by at least 600 cities, countries, corporations, private foundations and airport, (Obead, 2012). PAVER represents only standard (ASTM) certified platform standard in (ASTM, 1999B) standards. The PAVER software is interesting and providing user support. Positive feedback on program performance was one of the most important elements that helped (PMS) modify and improve the program (APWA, 2012).

Selected road divided to suitable numbers of segments. These segment parts are defined boundaries within which pavement components in terms of geometrical and functional porperties. According to PAVER 5.2, highway section divided into sample units, each segment should have an area approximately 2500 ± 1000 sqf define a asphalt surfaced sample unit of roads, (Shahin& Smith, 2005). The collected data needed for calculating Pavement Condition Index including:

1- Distress in terms of category, severity, and density, in addition to other data of section type, location, and extension (Obaidat& Al-kheder, 2006),
2- Roadway Geometry and types if "arterial, collector or local", details of road if its divided with median or non-divided, title name, segment length (m), highway width (m), and directional classification (one or two-way).

PAVER 5.2 recommended a methodology depend on inspection survey data results to find PCI:

1. Types of distresses found in Surface of segment unit boundary which should be measured in terms of total area, segment length or width and evaluation process depend on these distresses type, frequency and severity, (ASTM, 2011), (ERDC-CERL, 2009).

2. A photograph for each type of sample distresses with different severity in section of roadway necessary to evaluate the pavement condition index.

According to PAVER 5.2 Pavement condition index have rating as guide shows present condition of pavement as compared to standard PCI uses a scale of 7 different kinds started from excellent, to good, to satisfactory…etc. until reach to failed; on the other hand, there is a specific version of PCI scale depend on 3 classes. These are good, fair, and poor, in order to easy discrimination of these categories, they take different colors assigned by PAVER 5.2 software as shown in Figure 4 (Shahin& Smith, 2005).

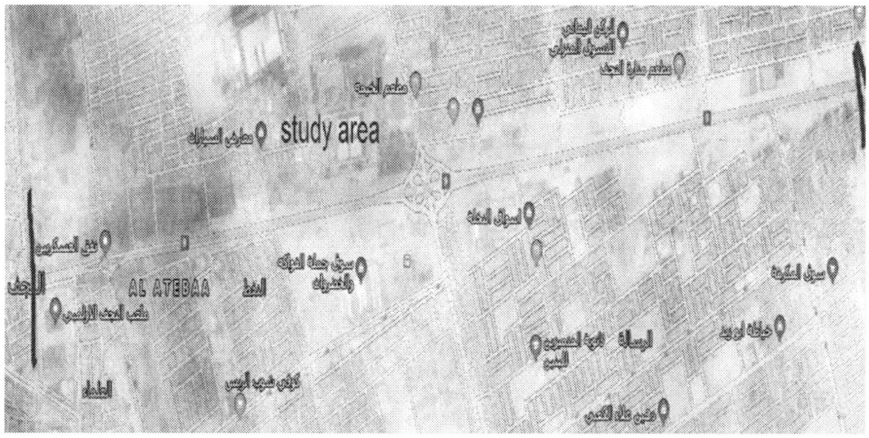

Figure 3 Aerial Photograph for Selected Highway

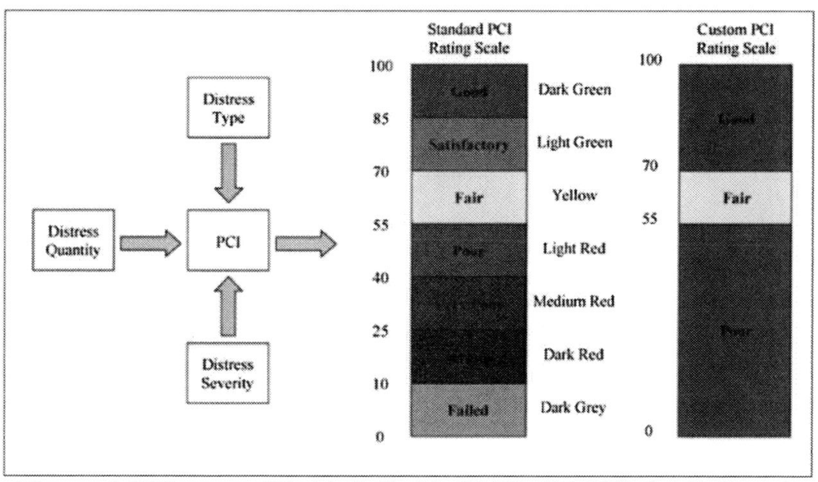

Figure 4. PCI ranges (Shahin& Smith, 2005). (PAVER 5.2 Software).

5. Field Measurements:

Field measurement in the current study area consists of distress types. Distresses shown in present study are classified into the following types below and shown in the Table 2:

1- Alligator Cracking: Also called fatigue cracking, this distress type severity is figured according to development of cracks withen network. Measured by tape, in area unit(m^2).

2- Potholes: quantity of distress is examined by number, while severity is based on diameter and depth of the potholes.

3- Rutting: Quantity is measured in area units (m^2), and the severity is determined in terms of average depth of each rut.

4- Longitudinal cracking: is measured by using tape in (m. L).

IOP Conf. Series: Materials Science and Engineering **978** (2020) 012012 doi:10.1088/1757-899X/978/1/012012

5- Edge cracking: that looks like alligator cracking which is found within (0.3048) to (0.609) m, from edge of the pavement.

6- Transverse cracking: Is measured by tape in (m. L).

7- Roughness: Is caused by weathering and raveling distresses, measured by tape in area units (m^2).

Table 2 types of flexible pavements distresses listed in the study from data collection survey

No.	Distress type	Photo taken from studied section
1-	Alligator Cracking	
2-	Potholes	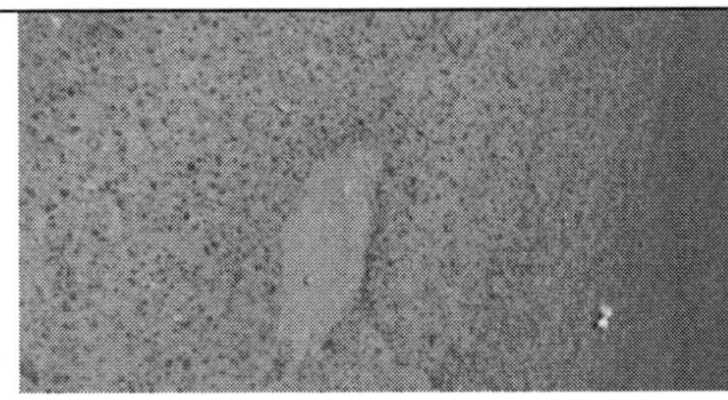

Table 2 types of flexible pavements distresses listed in the study from data collection survey **(continued)**

3-	Rutting	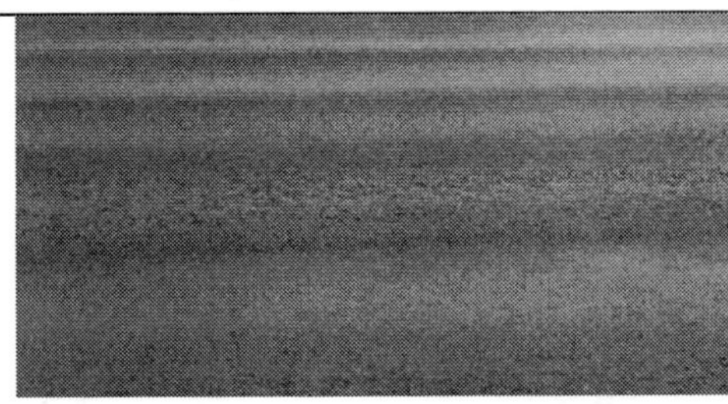
4-	Longitudinal cracking	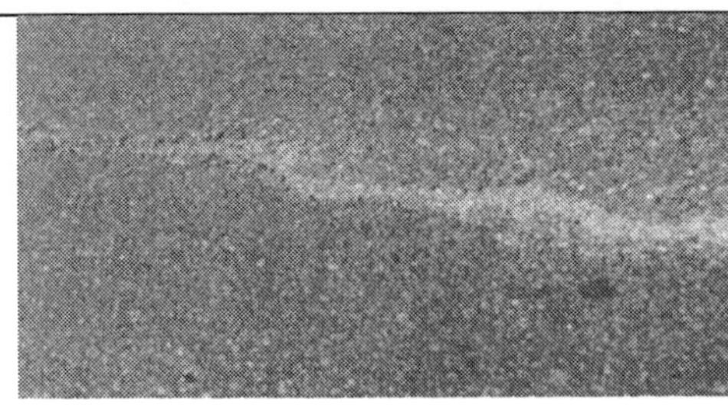

5-	**Edge cracking**	
6-	**Transverse cracking**	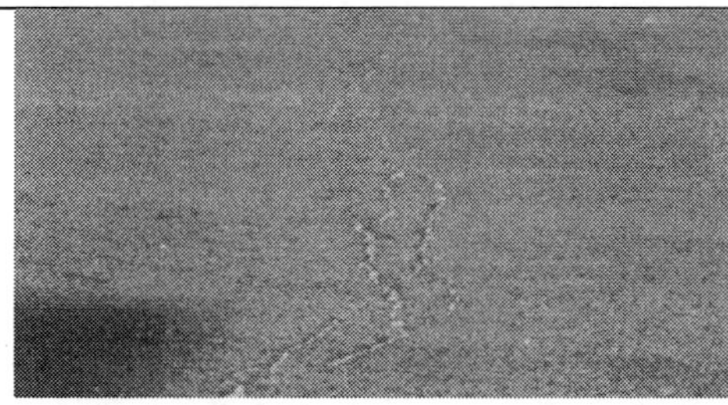
7-	**Roughness**	
8-	**Patching**	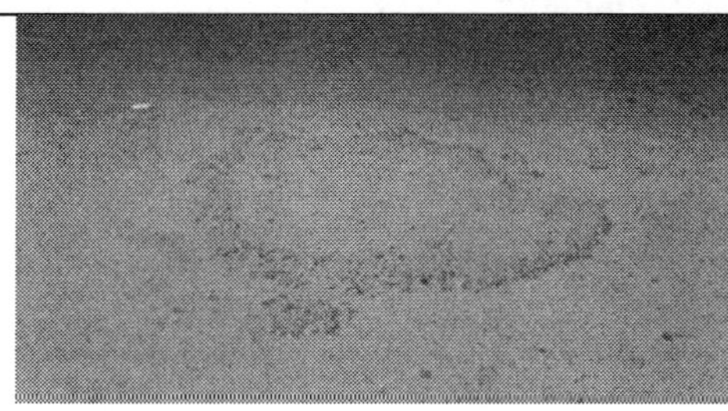

6. Result of Application PAVER 5.2 Software:

The PAVER 5.2 methodology is building on assessment by determining distress type, extent and severity on pavement surface condition. Hand calculations was using in estimation values of PCI for selected section of the study area. As mentioned earlier, quantity of each type distress had been measured in terms of square feet (square meters) for area, linear feet (meters) for length and depth, or number of occurrences in other wise.

First to all, the pavement section should be divided into number of segments represent sample units. Each unit should have an area within the range of 225± 90 m^2, (Shahin, 2005). So, the study area was divided into (267) inspection units, each unit segment was 315 m^2 in area. Type of distress, level of severity and quantity, needed for estimate PCI value for each sample units. As mentioned earlier, PCI values of the highway sections calculated by PAVER 5.2 software procedure. Sampling procedure was beginning by calculating smallest sample units' number that called (n). According to "Shahin& Smith, 2005 et al., (1976-84)". this number was given an estimation reasonably for the PCI mean value of the unit, within ±5 point of the true mean PCI that means there is 95% confidence, to obtained correct PCI must be inspected all sample units. Equation 1 listed in ASTM D6433-07, shown below:

$$n= \frac{(N*s^2)}{\left(\frac{e^2}{4}\right)(N-1)+s^2} \quad (1)$$

n: represent whole sample units number of selected pavement section.

N: section area / sample area

e: acceptable error in estimated value of PCI (usually e equal to 5)

s: PCI value standard deviation (usually s equal to 10)

Area of section=10.5 × 8011=84115.5 m^2

Area of sample = 315 m^2

According to all above values and by using equation 1:

N=84115/315=267 (complete number of sample)

n =15 (smallest number of samples)

The first sample should be chosen randomly by technique named as" systematic random" to determine spaces between sample units, the following steps should be observing (Shahin& Smith, 2005).

1- Selection sampling interval (i): Can be considered (i = N/n=17) where i is approximated smaller integer number value
2- Random star (s): random chose between sample unit No. 1 and interval sampling (i).
3- Determined sample unit as (s, s+i, s+2i, s+3i, etc.) for the designated pavement section:

Since i =17 then sample 1 could be the first sample to be inspected, then the successive samples are as

1, 1+17, 1+17*2, etc. that is mean the selected sample take the sequence of: 1, 18, 35, 52, 69, 86, 103, 120, 137, 154, 171, 188, 205, 222, and finally 239

After input the information of all sample unite of selected section, PCI value for all tested sample units had been calculated. Calculation of PCI value relies on subtract values; weighting factors extending from 0 to 100 represent degree of every distress on pavement. It well knowing, 100 deduct value gives an indication the operational condition on pavement surface is Excellent, on the other hand, a value of zero indicates very poor (Shahin& Smith, 2005). Through data entry and processing by PAVER 5.2 software, calculated PCI value of selected highway was found to be equal to 80. Therefor the pavement condition is Good conferring to evaluation of highway surface distress type, density and severity.

PAVER 5.2 allows the pavement manager to select suitable process for maintenance and rehabilitation for any highways sections depend upon regular of calculated PCI value, and then priorities of field processing become easy to decided. Concerning maintenance action for studied sections need preventive maintenance such as crack sealing and patching. It well knowing that the preventive maintenance always necessary for keep the pavement condition in good quality, and stop progressing the distresses to moderate or high severity. Preventive Maintenance applies lower-cost treatments to retard a road's deterioration, maintain or improve the functional condition, and extend the pavement service life.

7. Conclusions

Based on the outcomes of the present study, the following conclusion were formed:

1- Most of the distress of selected section are cracks from different types.
2- The condition of the pavement was good (PCI=80).

3- Preventive maintenance is the best treatment to improve pavement condition by filling and sealing the cracks as well as using batching techniques.

References:

[1] American Association of State Highway and Transportation officials, D.C (1997) (AASHTO): AASHTO Guide for Design of Pavement Structures, Washington.

[2] ASTM Standard Practices for Roads and Parking Lots Pavement Condition Index Surveys. D6433, American Society for Testing and Materials, West Conshohocken. PA, United States, 2011.

[3] Distress Identification Manual for the LTPP (Fourth Revised Edition), (2003) (Appendix B, Manual for Fault Meter Measurements, Publication Number: FHWARD- 03-031, Washington, D.C., June).

[4] E.J. Yoder, M.W. Witczak, Principles of Pavement Design, 2nd ed., John Will y & Sons, Canada, 1976.

[5] ERDC-CERL, PAVER™ Distress Identification Manual: Asphalt Surfaced Roads and Parking Lots, U.S. Army Corps of Engineers, USA, 2009.

[6] J.S. Miller, W.Y. Bellinger, Distress Identification Manual for the Long-Term Pavement Performance Program (Fourth Revised Edition), (2003) (Report No. FHWA-RD-03-031, McLean VA).

[7] LU &Lytton R/L. (1976), "Strategic Planning for Pavement Rehabilitation and Maintenance Management System" Transportation Research Record,589: 29-35

[8] M.A. Abo-Hashema, A.M. Abdel Samad, Y.A. Al-Zaroni, M.M.S. Hawwary, Integrating pavement maintenance management practices and geographic information system in Al ain city, UAE, Third Gulf Conference on Roads (TGCR06), Muscat, March 6–8, 2006, pp. 279–287 (ISSN 1817-4310).

[9] M.Y. Shahin, R.E. Smith (Ed.), Pavement Management for Airports. Roads, and Parking Lots, second edition, Springer Science Business Media, Inc., New York. NY. U.S.A, 2005.

[10] Micro PAVER 5.2 User Manual, U.S. Army Corps of Engineering-USACE, USA, 2014.

[11] Mohammed Taleb Obaidat, Sharaf A. Al-kheder, Integration of geographic information systems and computer vision systems for pavement distress class, Constr.Build. Mater. (November) (2006).

[12] N.J. Garber, L.A. Hoel, Traffic and Highway Engineering, fourth edition, Cengage Learning Publishing Company, Toronto, Canada, 2008.

[13] Obead F. Y., (2012). Development of Pavement Maintenance Alternatives Based on Multi-Criteria System, M.Sc. Thesis, Highway &Transportation Department. University of Mustansiriyah.

[14] R. Haas, W.R. Hudson, J.P. Zaniewski, Modern Pavement Management, Krieger Publishing, Florida, 1994.

[15] Raid R.A. Almuhanna, Hussein Ali Ewadh, Saja J.M. Alasadia, Using PAVER 6.5.7 and GIS program for pavement maintenance, Case Studies in Construction Materials vol. 8 pp 323–332, 2018 Published by Elsevier Ltd

[16] Transportation Research Board (TRB), Pavement management system based on financial consequence, in pavement management, transportation research record, J. TRB Wash. DC, USA (1940) (2005) 196 (TRB).

New Efficient Algorithm for the Discrete Hartley Transform

Mounir T. Hamood

Electrical Engineering Department, College of Engineering, University of Tikrit,
Salah Al-Deen Governorate, P O Box 42, Tikrit, Iraq
m.t.hamood@tu.edu.iq Tel: +9647714748928.

Abstract. In real-time signal processing applications, the discrete version of the transform introduced by Hartley (DHT) has been proved to be an efficient substitute to the discrete Fourier transform (DFT). A new algorithm for fast calculations of the DHT (FHT) based on radix$-2/4/8$ method is introduced in this paper. In comparison with the split radix FHT algorithm, the proposed algorithm has a comparable arithmetic complexity, but preserves the regularity and the simple butterfly structure of the radix-2 algorithm. The development of the algorithm is motivated by firstly deriving a new radix$-2/8$ FHT algorithm and then cascading it with the radix-4 and radix-2 FHT algorithms. The arithmetic complexity of the developed algorithm has been implemented and analyzed by calculating the number of real additions and multiplications for different transform lengths. Comparisons with the existing FHT algorithms have shown that this algorithm can be considered as a good compromise between structural and arithmetic complexity.

Keywords: *Fast transform algorithms; Discrete Hartley Transform (DHT); Discrete Fourier Transform (DFT); radix-based algorithms,*

1. Introduction

Over the decades, fast orthogonal transforms have been playing a major role in signal processing and communication fields. The result of the research carried out in this discipline over the last few years has been inspiring, producing in a wide range of fast algorithms and new transforms (Madisetti, 2017; Poularikas, 2009). In fact, transforms are continuously evolving and new ones are needed to solve many new or current challenges created by scientific developments or the request of new applications and tasks such as modern communication systems, signal/image processing and just to name a few. One the most important orthogonal transforms that was introduced by Bracewell (Bracewell, 1990; K. Jones, 2010), known as the discrete Hartley transform (DHT) become a progressively common as a substitute to the real valued discrete Fourier transform (RDFT). The DHT is particularly attractive in the applications involving real sequences, as it is likely that all the properties related with the DFT such the shifting and circular convolution theorems (Bi & Zeng, 2012; Oppenheim, Schafer, & Buck, 1999) can be applied to the DHT as well. As most spectrum analysis problems in real world involve real valued data only, therefore, substantial computational improvements can be gained using DHT. This is achieved by the fact that the computation of N-point real valued data samples using the DHT results N real-valued samples. However, the DFT will produce $2N$ real valued sequence, where half of them are redundant. Thus, the arithmetic operations and the memory storage will be halved when using the DHT, as it requires real data arrays rather than complex as compared with the DFT. Furthermore, the DHT has the self-inverse property, i.e. there is no need to distinguish between the forward and inverse except from the scale factor (K. Jones, 2010). This feature is of most significance and in specific for the multi-dimensions (MD-DHTs) for large amounts of data is to be processed.

In (Lihong, Yonghong, Jouni, & Hannu, 1998), a VLSI oriented FFT algorithm was presented, for pipeline implementation of the discrete Fourier transform (DFT) by developing a radix$-2/4/8$ FFT algorithm. This algorithm is beneficial that it provides an improved efficiency over existing radix-8 and radix-4 algorithms and it offers versatility in that it is not restricted to transform lengths based on powers of 8 and 4 respectively, and that it is significantly easier to implement than the well-known split-radix algorithm. This is achieved by cascading radix-2, radix$-2/8$ and radix$-2/4$ butterflies together into a single butterfly.

A new decimation-in-frequency (DIF) algorithm for the fast calculation of the DHT is presented in this paper, by developing a new radix$-2/8$ FHT algorithm. This algorithm is further developed by incorporating the methodology given in (Lihong, Bingxin, Yonghong, & Tenhunen, 1998; Lihong, Yonghong, et al., 1998) and cascading it with radix-2 and radix-2/4 algorithms into a single algorithm. It provides a complexity that is comparable to the split radix FHT algorithm, but with the benefit of an implementation that is significantly easier. The real-time succeeded applications of the DHT depend mainly on the existent of the efficient algorithms, since the well- known DHT fast algorithms (FHTs) such as radix-2 (M. Hamood, 2016; Mounir Taha, 2013; Shah & Rathore, 2010), radix-4 (M. T. Hamood & Boussakta, 2011; K. J. Jones, 2006) and split-radix (Bi, 1994, 1997; Bouguezel, Ahmad, & Swamy, 2004; Longyu, Huazhong, Jiasong, Lu, & Lotfi, 2010) algorithms are developed. Consequently, the aim of this paper is to develop an efficient new algorithm using radix$-2/4/8$ method.

The paper is organized as follows; section 2 presents the complete development of the proposed radix$-2/4/8$ FHT DIF algorithm. Section 3 describes analysis of the algorithm's performance in contrast to existing FHT algorithms. Lastly, the conclusion is given in section 4.

2. Algorithm Derivation

This section develops and analyses a new radix$-2/4/8$ FHT algorithm using decimation-in frequency (DIF) approach. This has been achieved firstly by developing the radix-2/8 FHT algorithm. The derivation has been accomplished by using radix-8 decompositions to the odd-indexed sequence and the radix-2 decompositions to the even-indexed sequence and of $X(k)$. The desired algorithm will then be obtained by further incorporating it with the radix$-2/4$ and radix-2 FHT algorithms.

2.1 Development of the new FHT Algorithm

Conventionally, the radix$-2/4/8$ algorithm is developed using a combination of different fixed-radix decompositions (Lihong, Yonghong, et al., 1998). It is therefore convenient to describe these decompositions by generalizing them as a radix$-q$ FHT algorithm, where q is assumed to be a power-of-two. The DHT $X(k)$ of an real valued sample $x(n)$ with transform size N is defined as

$$X(k)=\sum_{n=0}^{N-1}x(n)cas(\theta nk) \qquad 0\le k\le N-1 \tag{1}$$

where $cas(\theta nk) = cos(\theta nk) + sin(\theta nk)$ and $\theta = 2\pi/N$.

The development begins by dividing (1) into q $-$inner summations and substituting n with $(n + \lambda N/q)$ for $n = 0, 1, \dots\ N/q - 1$

$$X(k)=\sum_{n=0}^{N/q-1}\sum_{\lambda=0}^{q-1}x(n+\lambda\tfrac{N}{q})cas(\theta(n+\lambda\tfrac{N}{q})k) \tag{2}$$

In Eq. (2), the input samples $x(n)$ is divided into $q = 2^s$ sets (s positive integer) so that each inner sum denotes DHT of size N/q. The output samples $X(k)$ is calculated as q −distinct parts, and each of them represented by $X(qk + l)$ has (N/q) successive components indexed by k for $k = 0,1, \dots, N/q - 1$ and $l = 0,1, \dots, q - 1$. Therefore, (2) becomes

$$X(qk+l) = \sum_{n=0}^{N/q-1}\sum_{\lambda=0}^{q-1} x(n+\lambda \tfrac{N}{q}) cas(\theta(n+\lambda \tfrac{N}{q})(qk+l))$$
$$= \sum_{n=0}^{N/q-1}\sum_{\lambda=0}^{q-1} x(n+\lambda \tfrac{N}{q}) cas\left(\theta(qnk+nl+\lambda l \tfrac{N}{q})\right) \quad (3)$$

Equation (3) is the general radix$-q$ formula for the DIF FHT algorithm; decomposing it gives the preferred output sequence. Accordingly, the radix$-2/8$ algorithm can be calculated using (3) by using both radix -8 decimation (considering $q = 8$ and $l = 0, 1, .., 7$) and radix -2 decimation (considering $q = 2$ and $l = 0, 1$). Consequently, the radix-2 decimation for even-indexed sequences ($q = 2$ and $l = 0$), becomes

$$X(2k) = \sum_{n=0}^{N/2-1}\sum_{\lambda=0}^{1} x(n+\lambda \tfrac{N}{2}) cas(2\theta nk)$$
$$= \sum_{n=0}^{N/2-1}\left[x(n) + x(n+\tfrac{N}{2})\right] cas(2\theta nk) \quad (4)$$

and radix-8 decimation for the first odd-indexed sequences ($q = 8$ and $l = 1$), is given by:

$$X(8k+1) = \sum_{n=0}^{N/8-1}\sum_{\lambda=0}^{7} x(n+\lambda \tfrac{N}{8}) cas\left(\theta(8nk+n+\lambda \tfrac{N}{8})\right) \quad (5)$$

Expanding the partial summation of (5) yields:

$$X(8k+1) = \sum_{n=0}^{N/8-1}\left\{\begin{array}{l} x(n)cas\left(\theta n(8k+1)\right) + x(n+\tfrac{n}{8})cas\left(\theta n(8k+1)+\tfrac{\pi}{4}\right) + x(n+\tfrac{n}{4})cas\left(\theta n(8k+1)+\tfrac{\pi}{2}\right) + \\ x(n+\tfrac{3n}{8})cas\left(\theta n(8k+1)+\tfrac{3\pi}{4}\right) + x(n+\tfrac{n}{2})cas\left(\theta n(8k+1)+\pi\right) + x(n+\tfrac{5n}{8})cas\left(\theta n(8k+1)+\tfrac{5\pi}{4}\right) \\ +x(n+\tfrac{7n}{8})cas\left(\theta n(8k+1)+\tfrac{7\pi}{4}\right) \end{array}\right\} \quad (6)$$

Using the following cas identity

$$cas(a+b) = cas(a)cos(b) + cas(-a)sin(b) \quad (7)$$

the following relationship can be obtained:

$$cas\left(\theta n + \theta \tfrac{N}{2}\right) = cas(\theta n)cos(\pi) + cas(-\theta n)sin(\pi) = -cas(\theta n) \quad (8)$$

$$cas\left(\theta n + \theta \tfrac{N}{4}\right) = cas(\theta n)cos(\tfrac{\pi}{2}) + cas(-\theta n)sin(\tfrac{\pi}{2}) = cas(-\theta n)$$
$$= -cas\left(\theta n + \theta \tfrac{3N}{4}\right) \quad (9)$$

$$cas\left(\theta n + \theta \tfrac{N}{8}\right) = cas(\theta n)cos(\tfrac{\pi}{4}) + cas(-\theta n)sin(\tfrac{\pi}{4}) = \tfrac{1}{\sqrt{2}}\left[cas(\theta n) + cas(-\theta n)\right]$$
$$= -cas\left(\theta n + \theta \tfrac{5N}{8}\right) \quad (10)$$

$$cas\left(\theta n + \theta \tfrac{7N}{8}\right) = cas(\theta n)cos(\tfrac{7\pi}{4}) + cas(-\theta n)sin(\tfrac{7\pi}{4}) = \tfrac{1}{\sqrt{2}}\left[cas(\theta n) - cas(-\theta n)\right]$$
$$= -cas\left(\theta n + \theta \tfrac{3N}{8}\right) \quad (11)$$

$$cas(8\theta nk+\theta n)=cas(8\theta nk)cos(\theta n)+cas(-8\theta nk)sin(\theta n) \tag{12}$$

$$cas(-8\theta nk-\theta n)=cas(-8\theta nk)cos(\theta n)-cas(8\theta nk)sin(\theta n) \tag{13}$$

Substituting (8)-(13) into (6), we get:

$$X(8k+1)=\sum_{n=0}^{N/8-1}\left\{\begin{array}{l}\left[x(n)-x(n+\frac{N}{2})\right]cas\left(\theta n(8k+1)\right)+\left[x(n+\frac{N}{4})-x(n+\frac{3N}{4})\right]cas\left(-\theta n(8k+1)\right)\\ +\frac{1}{\sqrt{2}}\left[x(n+\frac{N}{8})-x(n+\frac{5N}{8})\right]\left[cas\left(\theta n(8k+1)\right)+cas\left(-\theta n(8k+1)\right)\right]\\ -\frac{1}{\sqrt{2}}\left[x(n+\frac{3N}{8})-x(n+\frac{7N}{8})\right]\left[cas\left(\theta n(8k+1)\right)-cas\left(-\theta n(8k+1)\right)\right]\end{array}\right\} \tag{14}$$

Replacing $[x(n)-x(n+N/2)]$ by $u(n)$ into (14) we get,

$$X(8k+1)=\sum_{n=0}^{N/8-1}\left\{\begin{array}{l}\left[u(n)+\frac{1}{\sqrt{2}}u(n+\frac{N}{8})-\frac{1}{\sqrt{2}}u(n+\frac{3N}{8})\right]\left(\theta n(8k+1)\right)+\\ \left[u(n+\frac{N}{4})+\frac{1}{\sqrt{2}}u(n+\frac{N}{8})+\frac{1}{\sqrt{2}}u(n+\frac{3N}{8})\right]\left(-\theta n(8k+1)\right)\end{array}\right. \tag{15}$$

Using the relation:

$$\sum_{n=0}^{N/8-1}u(n)cas(-8\theta nk-\theta n)=\sum_{n=0}^{N/8-1}u(-n)cas(8\theta nk+\theta n) \tag{16}$$

Substituting (16) into (15) and after some manipulations, the expression for $X(8k+1)$ can be written as:

$$X(8k+1)=\sum_{n=0}^{N/8-1}\left\{\begin{array}{l}\left[u(n)+\frac{1}{\sqrt{2}}\left(u(\frac{N}{8}-n)+u(n+\frac{N}{8})\right)+u(\frac{N}{4}-n)+\frac{1}{\sqrt{2}}\left(u(\frac{3N}{8}-n)-u(n+\frac{3N}{8})\right)\right]cos(\theta n)-\\ \left[u(-n)+\frac{1}{\sqrt{2}}\left(u(n+\frac{N}{8})+u(\frac{N}{8}-n)\right)+u(n+\frac{N}{4})+\frac{1}{\sqrt{2}}\left(u(n+\frac{3N}{8})-u(\frac{3N}{8}-n)\right)\right]sin(\theta n)\end{array}\right\}cas(8\theta nk) \tag{17}$$

Succeeding the same procedure given by (5)-(17), other odd-indexed sequences $X(8k+3), X(8k+5)$ and $X(8k+7)$ can be written as:

$$X(8k+3)=\sum_{n=0}^{N/8-1}\left\{\begin{array}{l}\left[u(n)+\frac{1}{\sqrt{2}}\left(u(\frac{N}{8}-n)-u(n+\frac{N}{8})\right)-u(\frac{N}{4}-n)+\frac{1}{\sqrt{2}}\left(u(\frac{3N}{8}-n)+u(n+\frac{3N}{8})\right)\right]cos(3\theta n)-\\ \left[u(-n)+\frac{1}{\sqrt{2}}\left(u(n+\frac{N}{8})-u(\frac{N}{8}-n)\right)-u(n+\frac{N}{4})+\frac{1}{\sqrt{2}}\left(u(n+\frac{3N}{8})+u(\frac{3N}{8}-n)\right)\right]sin(3\theta n)\end{array}\right\}cas(8\theta nk) \tag{18}$$

$$X(8k+5)=\sum_{n=0}^{N/8-1}\left\{\begin{array}{l}\left[u(n)-\frac{1}{\sqrt{2}}\left(u(\frac{N}{8}-n)+u(n+\frac{N}{8})\right)+u(\frac{N}{4}-n)-\frac{1}{\sqrt{2}}\left(u(\frac{3N}{8}-n)-u(n+\frac{3N}{8})\right)\right]cos(5\theta n)-\\ \left[u(-n)-\frac{1}{\sqrt{2}}\left(u(n+\frac{N}{8})+u(\frac{N}{8}-n)\right)+u(n+\frac{N}{4})-\frac{1}{\sqrt{2}}\left(u(n+\frac{3N}{8})-u(\frac{3N}{8}-n)\right)\right]sin(5\theta n)\end{array}\right\}cas(8\theta nk) \tag{19}$$

$$X(8k+7)=\sum_{n=0}^{N/8-1}\left\{\begin{array}{l}\left[u(n)-\frac{1}{\sqrt{2}}\left(u(\frac{N}{8}-n)+u(n+\frac{N}{8})\right)-u(\frac{N}{4}-n)-\frac{1}{\sqrt{2}}\left(u(\frac{3N}{8}-n)-u(n+\frac{3N}{8})\right)\right]cos(7\theta n)-\\ \left[u(-n)-\frac{1}{\sqrt{2}}\left(u(n+\frac{N}{8})+u(\frac{N}{8}-n)\right)-u(n+\frac{N}{4})-\frac{1}{\sqrt{2}}\left(u(n+\frac{3N}{8})-u(\frac{3N}{8}-n)\right)\right]sin(7\theta n)\end{array}\right\}cas(8\theta nk) \tag{20}$$

Therefore, radix$-2/8$ FHT algorithm divides an N-point DHT given by (1) into four $N/8$-point DHTs defined by (17)-(20) and one $N/2$-point DHT defined by (4).The next step in deriving the algorithm, is by applying further the radix-4 and radix-2 FHT algorithms to the odd- and even-indexed samples of (4) respectively. Therefore, we get the following decompositions

i. For even-indexed samples

$$X(4k) == \sum_{n=0}^{N/4-1}\left[x(n)+x(n+\tfrac{N}{4})+x(n+\tfrac{N}{2})+x(n+\tfrac{3N}{4})\right]cas(2\theta nk) \tag{21}$$

ii. For odd-indexed samples

$$X(8k+2) = \sum_{n=0}^{N/8-1}\left\{\begin{array}{l}\left[x(n)+x(\tfrac{N}{8}-n)-x(n+\tfrac{N}{4})-x(\tfrac{3N}{8}-n)+x(n+\tfrac{N}{2})+x(\tfrac{5N}{8}-n)-x(n+\tfrac{3N}{4})-x(\tfrac{7N}{8}-n)\right]cos(2\theta n)- \\ \left[x(N-n)+x(n+\tfrac{N}{8})-x(\tfrac{N}{4}-n)-x(n+\tfrac{3N}{8})+x(\tfrac{N}{2}-n)+x(n+\tfrac{5N}{8})-x(\tfrac{3N}{4}-n)-x(n+\tfrac{7N}{8})\right]sin(2\theta n)\end{array}\right\}cas(8\theta nk) \tag{22}$$

$$X(8k+6) = \sum_{n=0}^{N/8-1}\left\{\begin{array}{l}\left[x(n)-x(\tfrac{N}{8}-n)-x(n+\tfrac{N}{4})+x(\tfrac{3N}{8}-n)+x(n+\tfrac{N}{2})-x(\tfrac{5N}{8}-n)-x(n+\tfrac{3N}{4})+x(\tfrac{7N}{8}-n)\right]cos(6\theta n)- \\ \left[x(N-n)-x(n+\tfrac{N}{8})-x(\tfrac{N}{4}-n)+x(n+\tfrac{3N}{8})+x(\tfrac{N}{2}-n)-x(n+\tfrac{5N}{8})-x(\tfrac{3N}{4}-n)+x(n+\tfrac{7N}{8})\right]sin(6\theta n)\end{array}\right\}cas(8\theta nk) \tag{23}$$

Finally, the preferred algorithm (radix$-2/4/8$ FHT) can be obtained by applying radix-2 FHT algorithm into (21), resulting in the remaining decompositions:

$$X(8k) = \sum_{n=0}^{N/8-1}\left[x(n)+x(n+\tfrac{N}{8})+x(n+\tfrac{N}{4})+x(n+\tfrac{3N}{8})+x(n+\tfrac{N}{2})+x(n+\tfrac{5N}{8})+x(n+\tfrac{3N}{4})+x(n+\tfrac{7N}{8})\right]cas(8\theta nk) \tag{24}$$

$$X(8k+4) = \sum_{n=0}^{N/8-1}\left\{\begin{array}{l}\left[x(n)-x(n+\tfrac{N}{8})+x(n+\tfrac{N}{4})-x(n+\tfrac{3N}{8})+x(n+\tfrac{N}{2})-x(n+\tfrac{5N}{8})+x(n+\tfrac{3N}{4})-x(n+\tfrac{7N}{8})\right]cos(4\theta n)- \\ \left[x(N-n)-x(\tfrac{N}{8}-n)+x(\tfrac{N}{4}-n)-x(\tfrac{3N}{8}-n)+x(\tfrac{N}{2}-n)-x(\tfrac{5N}{8}-n)+x(\tfrac{3N}{4}-n)-x(\tfrac{7N}{8}-n)\right]sin(4\theta n)\end{array}\right\}cas(8\theta nk) \tag{25}$$

Performing the computation in-place using double butterfly with 16-points provides an in-place butterfly of the proposed radix$-2/4/8$ DIF algorithm as depicted in Figure. 1.

IOP Conf. Series: Materials Science and Engineering **978** (2020) 012013 doi:10.1088/1757-899X/978/1/012013

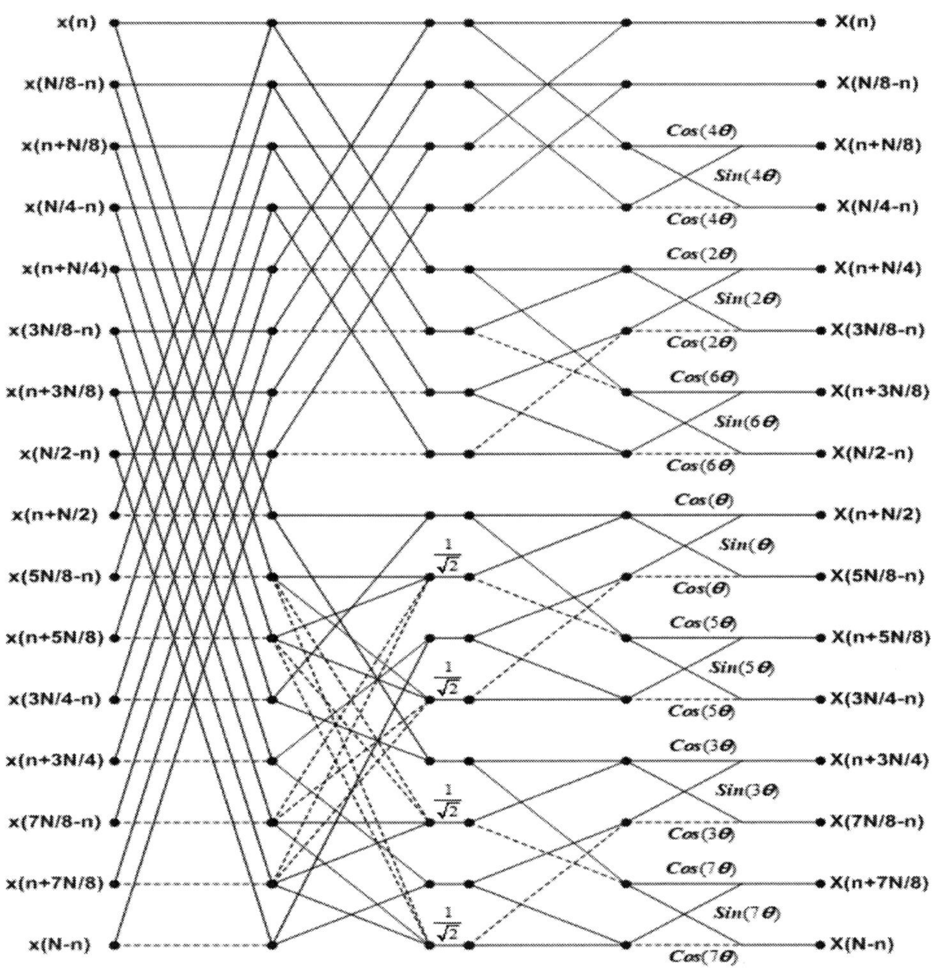

Figure 1. Radix$-2/4/8$ butterfly for the FHT algorithm, where dotted and solid lines stand for subtractions and additions respectively.

3. Computational Complexity

Equations (17)-(20) and (22)-(25) represent the radix$-2/4/8$ DIF FHT algorithm. By merging two butterflies together will remove the trivial arithmetic operations (multiplying by $(-1, 0, 1)$ as shown in Figure 1. The developed in-place butterfly computes sixteen points and needs 66 additions and 32 multiplications. The whole transform fulfills the following relations.

$$M(N) = 8M(\tfrac{N}{8}) + 2N - 22$$

$$A(N) = 8A(\tfrac{N}{8}) + \tfrac{31N}{8} - 10 \qquad (26)$$

where $A(N)$ and $M(N)$ represents the total number of real additions and multiplications respectively.

The computational complexity given by (26) is iterative. To get the closed form complexity at various transform size, the fundamental complexities values are required, which are the number of real additions and real multiplications that are required by size two and size four DHTs, that are given as $A(2) =$

IOP Conf. Series: Materials Science and Engineering **978** (2020) 012013 doi:10.1088/1757-899X/978/1/012013

$2, A(4) = 8$ $A(8) = 24$ and $M(2) = 0, M(4) = 0, M(8) = 2$. The comparison in total number of operations for different FHT algorithms is depicted in Table 1.

Table 1. Operations counts for different FHTs algorithms

Transform Size N	Radix−2/4/8 Algorithm	Radix−4 Algorithm	Radix−2 Algorithm
8	24	24	30
16	74	78	94
32	210	212	262
64	520	562	678
128	1312	1362	1670
256	3088	3314	3974
512	7344	7602	9222
1024	16624	17586	20998

An assessment has been done between radix-2, radix-4 and the developed radix$-2/4/8$ FHT algorithms, in terms of the total number of additions and multiplications as depicted in Figures. 2 and 3 respectively. As shown from these figures, a substantial saving can be achieved by using the newly developed radix$-2/4/8$ algorithm over other radix based FHT algorithms such as radix-4 and radix-2.

Furthermore, in order to verify the results based on arithmetic complexity for the developed radix$-2/4/8$ FHT algorithm. This algorithm and the corresponding radix-2 and radix-4 algorithms are implemented using MATLAB, and tested based on computer run-times on a Core i5 computer with speed of 1.8 GHz and 4 GB RAM. The results of the run-times are shown in Figure 4. The results of the computer run-time comparison are better than those obtained for the arithmetic operations comparison. This confirms the fact that the butterfly of the proposed algorithm computes triple stages concurrently as compared with the computation of single and double stages for radix-2 and radix-4 butterflies respectively.

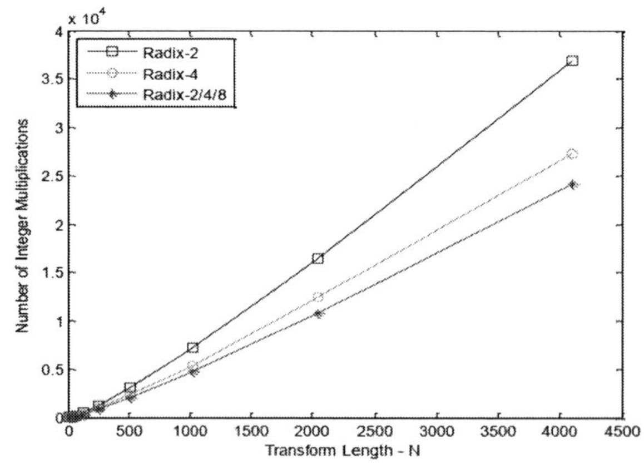

Figure 2. Total number of multiplications comparison for different FHT algorithms

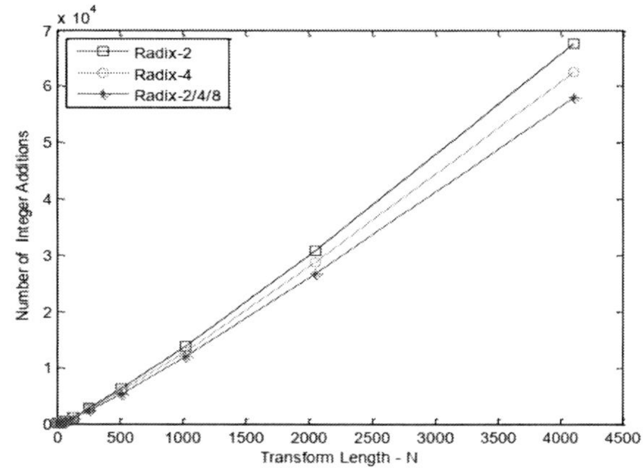

Figure 3. Total number of additions comparison for different FHT algorithms

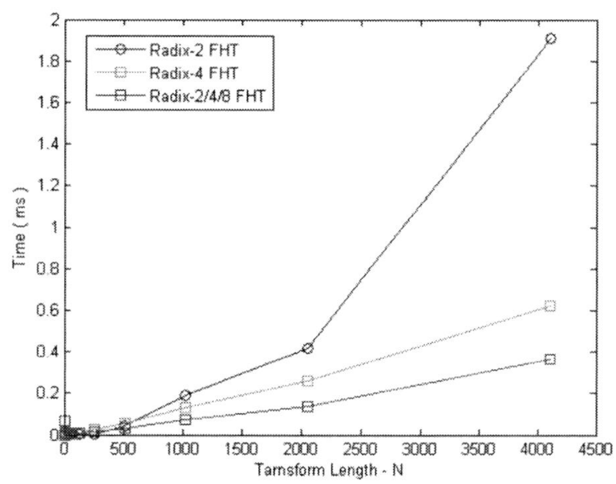

Figure 4. The run-time for different FHT algorithms using MATLAB runs on Core i5 computer with speed of 1.8 GHz

4. Conclusion

The paper has been presented a newly developed radix$-2/4/8$ FHT algorithm using the decimation-in- frequency method. The efficiency of the proposed algorithm is comparable with the well-known split radix FHT algorithm in terms of arithmetical complexity, but it out-performs the split-radix by means of simple structure and regular indexing scheme which is offers a significant ease in both hardware and software implementations, in retrospect to the split-radix FHT algorithm. The new algorithm has been developed by introducing the radix$-2/8$ FHT DIF algorithm. The proposed radix$-2/4/8$ algorithm has been developed further by incorporating both radix-2 and radix$-2/4$ decimations. The resulting algorithm contains a number of radix-based FHT algorithms in a cascading configuration, by generalizing them as radix$-q$ FHT algorithms, where q is assumed to be a power-of-two. The algorithm has been examined in respect to its arithmetical complexities of numerous transform sizes, in comparison to existing FHT algorithms based on radix-2 and radix-4 approaches.

References

[1] Bi, G. (1994). Split radix algorithm for the discrete Hartley transform. *Electronics Letters, 30* (22), 1833-1835.

[2] Bi, G. (1997). New split-radix algorithm for the discrete Hartley transform. *IEEE Transactions on Signal Processing, 45*(2), 297-302.

[3] Bi, G., & Zeng, Y. (2012). *Transforms and Fast Algorithms for Signal Analysis and Representations*: Birkhäuser Boston.

[4] Bouguezel, S., Ahmad, M. O., & Swamy, M. N. S. (2004). *An efficient split-radix FHT algorithm.* Paper presented at the Proceedings of the International Symposium on Circuits and Systems, ISCAS '04.

[5] Bracewell, R. N. (1990). Assessing the Hartley transform. *IEEE Transactions on Acoustics, Speech and Signal Processing, 38*(12), 2174-2176.

[6] Hamood, M. (2016). New Decimation-In-Time Fast Hartley Transform Algorithm. *International Journal of Electrical and Computer Engineering (IJECE), 6*, 1654-1661.

[7] Hamood, M. T., & Boussakta, S. (2011). *New radix-based FHT algorithm for computing the discrete Hartley transform.* Paper presented at the IEEE International Conference on Acoustics, Speech and Signal Processing (ICASSP).

[8] Jones, K. (2010). *The Regularized Fast Hartley Transform*: Springer Netherlands.

[9] Jones, K. J. (2006). Design and parallel computation of regularised fast Hartley transform. *IEE Proceedings -Vision, Image and Signal Processing., 153*(1), 70-78.

[10] Lihong, J., Bingxin, L., Yonghong, G., & Tenhunen, H. (1998). *Implementation of a low power 128-point FFT.* Paper presented at the International Conference on Solid-State and Integrated Circuit Technology. .

[11] Lihong, J., Yonghong, G., Jouni, I., & Hannu, T. (1998). *A new VLSI-oriented FFT algorithm and implementation.* Paper presented at the IEEE International ASIC Conference.

[12] Longyu, J., Huazhong, S., Jiasong, W., Lu, W., & Lotfi, S. (2010). A novel split-radix fast algorithm for 2-D discrete Hartley transform. *Trans. Circuits and Systems Part I, 57*(4), 911-924.

[13] Madisetti, V. (2017). *Digital Signal Processing Fundamentals*: CRC Press.

[14] Mounir Taha, H. (2013). Improved FHT Algorithms for Fast Computation of the Discrete Hartley Transform. *Tikrit Journal of Engineering Sciences 20*(1), 62-69.

[15] Oppenheim, A. V., Schafer, R. W., & Buck, J. R. (1999). *Discrete-time signal processing*: Prentice Hall.

[16] Poularikas, A. D. (2009). *Transforms and applications primer for engineers with examples and MATLAB*: Taylor and Francis.

[17] Shah, G. A., & Rathore, T. S. (2010). *A New Architecture for Radix-2 DHT.* Paper presented at the International Conference on Computational Intelligence and Communication Networks.

Evaluation of Al-Kufa City Road Networks Using GIS

Ahmed Yahya Abdlabass [1,*] **and Hamid Athab Al-Jameel** [2]

[1]Dep. of Geographic, University of Kufa, Iraq-E-mail: feyada@uokufa.edu.iq
[2]College of Engineering, University of Kufa, Iraq
E-mail: hamid.aljameel@uokufa.edu.iq
* Corresponding author: College of Engineering, University of Kufa, Najaf, Iraq. Tel: +9647801341152.

Abstract. One of the important points to develop any region is the suitable road network and understanding the relationship between land use and transportation system. To effectively develop any network, there is an urgent need to obtain proper connectivity and orientation. However, Al-Kufa road network has received significantly limited attention and studies to evaluate this network. Therefore, this study has focused on analyzing the present road network connectivity in Al-Kufa city. Furthermore, the GIS program has been adopted in this study to represent the main features in the city. Several measures have been adopted in this study to perform the evaluation for the road network such as beta index, gamma index, alpha index, spreading index, correlation index, etc. The main objective of this study is to assess how the road network is successful to containment the present road traffic and suggestions for future traffic movement performance to accommodate the growth. The results demonstrate that urgent improvements are required such as adding new roads. Finally, the network density could give better another important prediction of the road network measures.

Keywords: *Al-Kufa road network; Connectivity; GIS; Network density; Spreading index.*

1. Introduction

The urban road network is one of the traditional transport networks that has drawn a great deal of interest from the various research communities (Bando et al., 1995, Wang et al., 2012). The main objective of the study of urban road traffic network is to prevent or relieve traffic congestion, which has become an urgent problem to solve with the continuous growth of urban population and automobiles.

Khazaal (2009) conducted geographical analysis for paved road network in Erbil city, Iraq in order to determine the connectivity of the network. In addition, the study focused on the analysis of detour. The results mainly indicated that there are low connectivity and high detour index because of nature of the hilly nature of the area.

Aduory (2010) studied the geographical analysis for Al-Dur road network, in Salahaddin city, Iraq to know the routes, lengths, density and the efficiency of the under study road network. The author concluded that the road density was low comparing with other cities there. The roads in the under study area were influenced by natural factors such as geographical location, water resources and also were affected by agricultural, industrial activities and population distribution.

Ogunleye (2011) studied the spatial of road networks in Ekiti city in Nigeria. The results indicated that low connectivity of roads in this city. Therefore, the recommendations of this study were to maintain the existing roads and construct the new ones for increasing the accessibility of people and goods.

Taran and Makhamra (2015) did a spatial analysis for a rural road network in Al-Mafraq city in Jordan using GIS in order to measure the connectivity and circuit. The study found that the percent of connectivity was 42% as indicated by Gamma index. Whereas, the circuit is very low as indicated by 11% of alpha index.

Taran et al., (2016) conducted structural analyses road network study in Karak city in Jordan using graphs theory and some quantitative methods to identify the reality of the road network in terms of the degree of connectivity and circuitry, accessibility and density. As well as the relationship between the lengths of the network and population distribution analysis. The authors found that the road network suffers from the low level of connectivity and circuitry in general. The study also concluded that there is a disparity in the road network distribution within the city and that the lengths of roads ratio does fit with population ratios. Therefore, the study recommended the need to develop the connectivity to ensure redistribution of the population and meet their development plans.

Al-Jameel and Abdulabas, (2017), indicated that the city network suffers from poor mobility and accessibility. Limited studies discussed the assessment of Al-Kufa road network. Therefore, the main purpose of this study is to evaluate the road network in Al-Kufa city using several indices such as connectivity and coverage.

2. Historical background

Al-Kufa is an ancient, historical and religious city which has several historical places such as Al-Kufa Mosque, Al-Sahila Mosque and Muslim Ibn Aqeel holy shrine. This city lies north east of Al-Najaf city, as indicated in Figure 1, with an area of 49.44 Km^2 and population of 185232 persons in 2018.

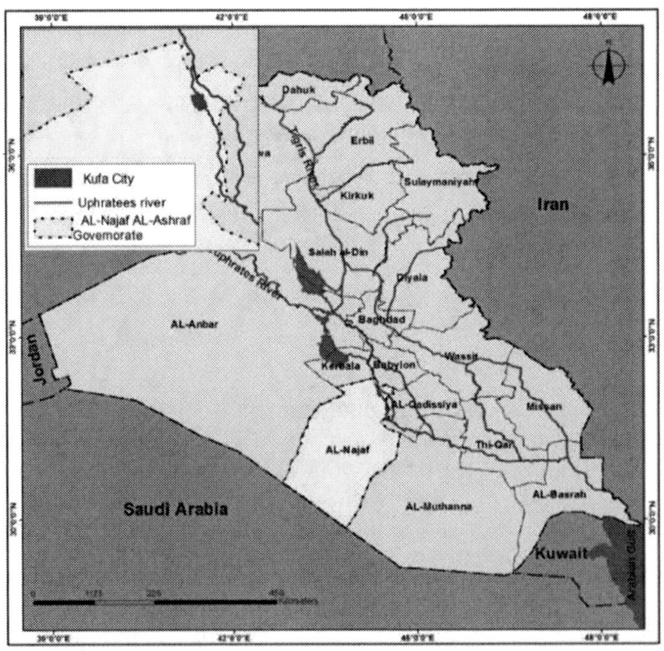

Figure 1. The Site of Al-Kufa city from Al-Najaf governorate and Iraq.

Road network display both topologic and geometric variations in their structure. There are variety of indices, proposed in earlier studies applicable for evaluating road network properties. The main role of network is to connect local facilities and people. Therefore, the efficient network is an essential part of sustainable improvement in the quality of life and development of the city (Sreelekha et al., 2016).

Basically, a road network consists of links (roads) and nodes (intersections, civic center, community centers, or any natural separators such as river ...etc.). The most important indicators to evaluate the network are connectivity and efficiency (Sreelekha et al., 2016). The graph theory has been developed to quantify the spatial structure of road network. Among the geographers who understand the network structure, the transportations engineers whom understanding the topology of transport network has received considerable interest. This could be attributed to different applications such as heterogeneity patterns and continuity [Xie, 2005, Gastner and Newman, 2006; and Xie and Levinson, 2007).

3. Al-Kufa network

The first appearance of modern means of route used for transportation in Al-Kufa city was in 1907. This was the carriages drawn by horses as reported by Musabee (1978). Then, the first paved road network appeared in 1954 (Al-Jaboory, 2007). The number of population was increased from 13700 capita in 1947 to 157845 capita in 2013 as reported by Al-Kalaabee (2017). This increment means more area to be served by roads. According to the Directorate of planning in Iraq, the criteria of transportation and land occupied road for each person is 25m^2; however, this value was 15.8m^2 per capita in 2007 and 16.03m^2 per capita in 2013, respectively (Al-Kalaabee, 2017). These values are less than the standard value. Therefore, one could conclude that there is a lack in the roads which may lead to traffic congestion.

4. Methodology

The methodology involves applying GIS for evaluating the road network of the study area as indicated in Figure 2. Urban road network evaluation based on GIS involves collecting the data resources, digitizing the network, building road network database, extracting the network structure etc. These data have been collected from different sources such as descriptive data relating to the names of roads and intersections, number of population, number of vehicles and special data such as sector maps for Al-Kufa city with scale of 1:1000, maps of fundamental design with scale of 1:60000 and satellite image (Quick Bird US Satellite) with accuracy of 60cm.

ArcGIS9.3 was used in this study. All the roads including arterial, collector streets and local streets were digitized from the satellite image. Moreover, all required lengths have been measured as indicated in Figure 3.

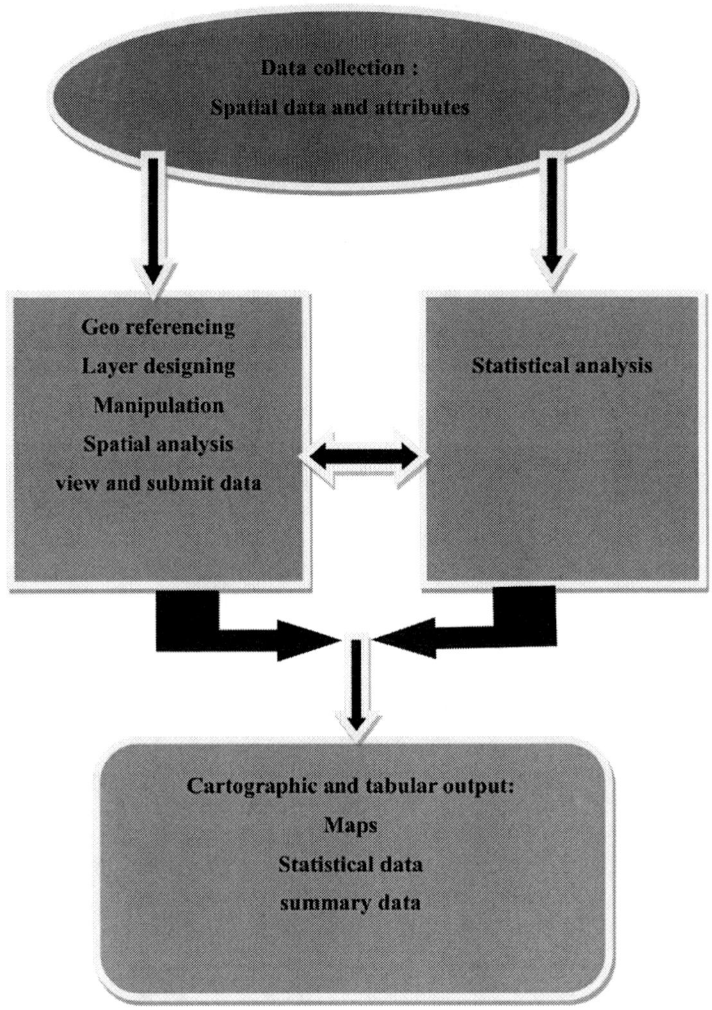

Figure 2. Flow chart describing GIS part of study.

The required ESRI shape file format has been produced from the GIS. In this study, the GIS has an important role in representing and analyzing the road network in the city. This process of the GIS has been used to develop a map for Al-Kufa city with scale 1:1000 and 1:2000. These digital maps have been produced depending on maps from Ministry of Municipality and Public Works, Directorate of Al-Najaf Municipality and Planning Unit. Moreover, maps from fundamental design prepared by the company of Architecture Design (ADEC), Liewelyen Davies Yeang and TRIBAL Urban Studio in 2010 with scale 1:60000 have been also adopted to develop digital map produced by GIS. Finally, the satellite image for Al-Kufa city using Satellite called Quik Bird with accuracy 60cm has been used.

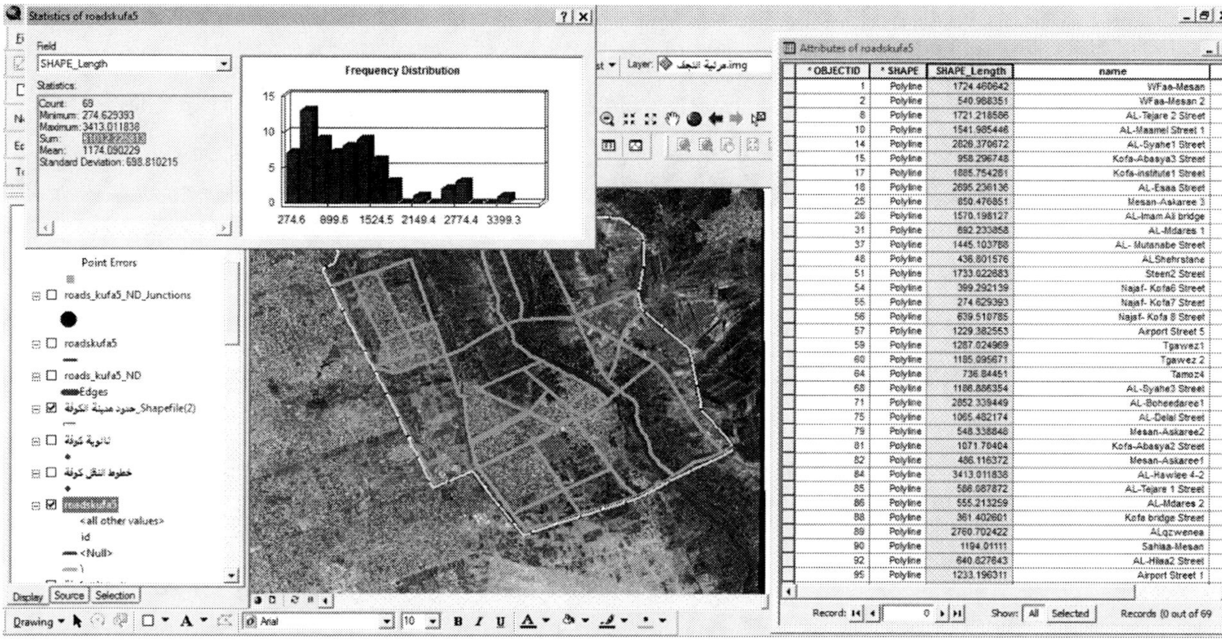

Figure 3. Application interface for how to calculate the actual lengths of Al-Kufa road networks using (Arc Map 9.3).

5. Characteristics of road network

To evaluate the road network, there are several indices which have been proposed in previous studies such as connectivity, coverage and spatial pattern. These measures have different applications in both planning and transportation practices.

6. Calculations:

6.1 Connectivity:

Connectivity refers to the directness of travel between origins and destinations. A well-connected network has many short links, numerous intersections, and minimal dead-ends providing continuous, direct routes to destinations (Sreelekha et al., 2016). To determine the connectivity for transportation network, the real network should be changed to simple network which is called topological diagram (Sreelekha et al., 2016). In the diagram, a road could be represented as a link which connects nodes. The graph theory deals with links and nodes according to network theory and data. The application of network theory in transportation network has influential effect in directing the analysis of road networks. Various indices used for evaluating connectivity pattern of road transport network are Alpha Index, Beta Index, Gamma Index, Eta Index and Grid Tree Pattern Index (Kansky, 1963; and Noda, 1996). To apply such as indications, the network has been changed to topological map as in Figure 4 using ArcGIS 10.5.

6.1.1 Beta index

This index consists of dividing the number of links by nodes. The value of this index ranges between 0 and 1 (Buyong, 2007). If this value is zero this represents no links and the value is one this means complete graph. The greater the value of β, greater is the connectivity. As transport networks develop and become more efficient, the value of β should be high (Buyong, 2007).

$$\beta = e/v \qquad (1)$$

where e - number of edges in the network, v - number of vertices in the network, and $\beta = 66/43 = 1.53$

Applying such index on Al-Kufa network indicates that value is 1.53. This means that there is more

IOP Conf. Series: Materials Science and Engineering **978** (2020) 012014 doi:10.1088/1757-899X/978/1/012014

than one complete network in the city.

6.1.2. Gamma index (γ)

This is the best measurements to know the connectivity degree (Davey,1971). This index could be determined according to Equation 2.

$$\gamma = \frac{e}{3(v-2)} \quad \text{......(2)}$$

where; e - number of edges in the network and v – is the number of vertices in the network.

If there is no connectivity the value approaches from zero but if the value is one this means fully connected.

$\gamma = (69/\,(3(43\text{-}2))) = 0.56$. The resulted value indicates that connectivity of the network is at low level. To improve the level of connectivity, there is a need to add 54 links.

6.1.3 Alpha index (α):

This is the ratio of actual number of circuits to the maximum number of circuits in the network (Dill, 2003). It ranges from 0 (no circuits) to 1 (completely interconnected network).

$$\alpha = \frac{e-v+p}{2v-5} \quad \text{......(3)}$$

This measure should be determining the connectivity degree. The value of (α) for Al-Kufa city is calculated as follows:

$$\alpha = \frac{69-43+3}{2X43-5} = 0.36.$$

This means that the connectivity is very low and there is a need for 58 links to satisfy the required connectivity degree.

Figure 4. The topological map of the roads of A-Kufa city.

IOP Conf. Series: Materials Science and Engineering **978** (2020) 012014 doi:10.1088/1757-899X/978/1/012014

6.2 Coverage

Coverage measures describe the density aspect of the elements of a network, as intersections and links (Buyong, 2007). The Coverage measures are useful in determining compactness and development. Higher the value, more the network is developed. Network density and Intersection density attribute road network coverage in an area.

Spreading degree: This index indicates how the lines of network spreading and level of access. To measure the closeness and divergence among the nodes, will be done according to the following indices:

6.2.1 Eta index(η)

This index uses for measuring the real lengths of links between the nodes as indicated below (Buyong, 2007):

$$\eta = M/e \qquad (4)$$

where M - total network length in km

This value starts from zero. If this value is zero, this means that no spreading for the networks in spatial urban (i.e. this network is without access).

For Al-Kufa city, $\eta = 81.012/43 = 1.884$

According to this value of index, the degree of spreading is 1.884 which represents good spreading degree and there is no closeness between nodes.

6.2.2 First Betty index

This index is used to measure the level of spreading in its region. The value of this index ranges from one and more. If this value is zero, this means no spreading. This index determines from the following equation (Buyong, 2007):

Betty index=no. of links- no. of nodes +1...(5)

By applying this index for Al-Kufa city, its value = 69-43+1=27. This means a huge spreading for the networks in its region.

6.2.3 Detour index

The shortest distance is the straight one. So, the best roads which are straightly extended; however, most roads deviate either right or left and ascend or descend under different conditions. This leads to increase the lengths of these roads which is called detour index (Buyong, 2007).

This index is used to evaluate the efficiency of roads by knowing the lengths of roads in transportation network in addition to the straight lengths. Consequently, this will help to know where you have to add or cancel some links in the network understudy. The value of this index could be determined according to the following equation:

Detour index= (the actual length of the road/the straight length of the same road) *100........(6)

The value of this index is always equal or higher than 100%. If this value approaches from 100% this means that roads are straight with high efficiency. If the value exceeds 100% this refers to inadequate cases. The detour index for Al-Kufa roads is indicated in Table1 and Figure 2.

Table 1. Detour index for The main streets in the Kufa City

no. of links	Street	Actual length (m)	Theoretical length (m)	Detour index%	Increase (m)
1	Airport Street 1	1233.196	1233.196	100	0

2	Airport Street 2	1341.988	1341.988	100	0
3	Airport Street 3	692.482	692.482	100	0
4	Airport Street 4	329.457	329.457	100	0
5	Airport Street 5	1229.382	1216.559	101.054	12.823
6	AL-Hawlee 4-2	3413.011	3413.011	100	0
7	AL-Hawlee 4-3	2331.596	2331.596	100	0
8	AL-Maamel Street 1	1541.985	1541.985	100	0
9	AL-Maamel Street 2	928.729	923.724	100.541	5.005
10	Najaf- Kufa3 Street	969.647	969.647	100	0
11	Najaf- Kufa5 Street	710.743	709.437	100.184	1.306
12	Najaf- Kufa6 Street	399.292	390.328	102.2965	8.964
13	Najaf- Kufa4 Street	1446.331	1446.331	100	0
14	Najaf- Kufa8 Street	639.510	638.966	100.0851	0.544
15	Najaf- Kufa7 Street	274.629	274.629	100	0
16	Kofa bridge Street	361.402	361.402	100	0
17	AL-Delal Street	1065.482	1064.707	100.0728	0.775
18	AL-Syahe1 Street	2826.370	2808.028	100.6532	18.342
19	AL-Esaa Street	2695.236	2549.690	105.7084	145.546
20	AL-Hilaa1 Street	1476. 325	1476.325	100	0
21	AL-Hilaa2 Street	640.827	573.027	111.8319	67.8
22	Kufa-Abasya1 Street	320.503	320.503	100	0
23	Kufa-Abasya2 Street	1071.704	1053.007	101.7756	18.697
24	Kufa-Abasya3 Street	958.296	940.912	101.8476	17.384
25	AL-Syahe2 Street	939.960	939.960	100	0
26	AL-Syahe3 Street	1186.886	1181.043	100.4947	5.843
27	Kufa-institute1 Street	1885.754	1879.170	100.3504	6.584
28	Kufa-institute2 Street	2778.588	2778.588	100	0
29	AL-Imam Ali bridge	1570.198	1563.304	100.440	6.894
30	Mesan- ALImam Ali bridge	1584.183	1579.504	100.296	4.679
31	WFaa-Mesan	1724.460	1630.207	105.7817	94.253
32	WFaa-Mesan 1	434.909	434.909	100	0
33	WFaa-Mesan 2	540.988	540.988	100	0
34	WFaa-Mesan 3	572.736	572.736	100	0
35	AL-Mocktar 1	931.830	931.830	100	0
36	AL-Mocktar 2	1710.089	1710.089	100	0
37	AL-Boheedaree1	2852.339	2801.797	101.803	50.542
38	AL-Tejare 1 Street	586.087	586.087	100	0
39	AL-Tejare 2 Street	1721.218	1721.218	100	0
40	Steen1 Street	586.783	586.783	100	0
41	Steen2 Street	1733.022	1733.022	100	0
42	Tamoz1	1455.767	1455.767	100	0
43	Tamoz4	736.844	736.844	100	0
44	Tgawez1	1287.108	1287.108	100	0
45	AL- Taqaa Street	1518.455	1518.455	100	0

46	AL- Mutanabe Street	1445.103	1445.103	100	0
47	Sahlaa- kufa 1	720.875	720.875	100	0
48	Sahlaa- kufa 2	1324.868	1324.868	100	0
49	AL-Sahalea Street	735.216	735.216	100	0
50	Medicine college street 1	651.326	651.326	100	0
51	Medicine college street 2	796.516	796.516	100	0
52	Sahlaa-SutAL kufa 1	1157.454	1157.454	100	0
53	Sahlaa-SutAL kufa 2	485.668	485.668	100	0
54	Sahlaa-Mesan	1194.011	1130.351	105.631	63.66
55	Ibn Blal 1 Street	642.034	642.034	100	0
56	Ibn Blal 2 Street	1255.248	1255.248	100	0
57	ALSekaa	752.551	670.288	112.272	82.263
58	ALShehrstane	436.801	436.801	100	0
59	AL-Mdares 1	692.233	692.233	100	0
60	AL-Mdares 2	555.213	555.213	100	0
61	ALqzwenea	2760.702	2736.989	100.866	23.713
62	Mesan-Askaree 1	486.116	486.116	100	0
63	Mesan-Askaree 2	548.338	548.338	100	0
64	Mesan-Askaree 3	850.476	850.476	100	0
65	Tgawez 2	1185.095	1185.095	100	0
66	AL-Senaee 1	1438.993	1438.993	100	0
67	AL-Senaee 2	563.270	563.270	100	0
68	Kornesh street 1	1774.815	1765.770	100.512	9.045
69	AL-Kornesh street 2	1323.030	1252.614	105.621	70.416

Source: Depending on the spatial database

A. The average detour index is 100.8713%. The increment of 0.8713% from the ideal value represents that Al-Kufa city has an efficient network which could be attributed for level terrain.

B. It was found that 47 roads have the value of 100% for this index whereas the rest roads have the value of this index ranges from 112.27%-100.072%.

6.3 Correlation index

This index represents the correlation degree between links in the network. This index is calculated as follows (Buyong, 2007):

Correlation index= no. of links / max. no. of possible links. Where; The max possible links

$= 1/2(n^2-n)$ ....(7)

n= no. of nodes.

If the value for this index is zero which represents there is no link in the network. Whereas, when the value is one this represents that the number of links is the max possible number. By applying the above equation, the value of this index is:

The max possible links= $1/2\ (43^2-43)$ =903

Correlation index= 66/903=0.073

This value is so low and the degree of correlation is so weak.

IOP Conf. Series: Materials Science and Engineering **978** (2020) 012014 doi:10.1088/1757-899X/978/1/012014

6.4 Grid Trip Proportion GTP

The grid-tree proportion (GTP) (Sreelekha et al., 2016) index for road connectedness and the crossing factor (CF) (Xie and Levinson, 2007) for road density are selected from existing pattern indexes. Various indexes describing the relationship between the number of roads and the number of intersections has been proposed in the field of quantitative geography (Gastner and Newman, 2006). An index for connectedness is expanded to make it possible to assess the relationship in a square grid road network, and is defined as the GTP index as

$$\text{GTP index}=\frac{e-v+p}{(\sqrt{v}-1)^2}=0.841 \quad (8)$$

The relationship between the total length of road and the number of intersections has been studied mainly by using a random line model based on integral geometry, in which random straight lines are treated as a road network.

6.5 Density of roads to the area and the number of population

This indication represents the economic development for the area and it also gives an idea about the number of roads whether it is sufficient with the geographical area in order to evaluate the performance of these roads in the area understudy (Buyong, 2007).

The density of roads to the area= lengths of the roads (km)*1000/ (the area served by roads (km^2)) ....(9)

The density of roads in Al-Kufa City = 117.3965*1000/49.4422=2374.419 km/1000km^2.

Therefore, this density in the city is higher than the standard specification (105km/1000km^2)

The density of roads to the population could be determined as:

The density of roads to population=lengths of roads (km)*100,000/population.................................(10)

=117.3965*100,000/185,232

So, the density for 2018 is 63.3 km/100,000 people. This value is less than the standard specification which is (496km/100000people) (Buyong, 2007).

7. Discussion of the results

Having determined the results of connectivity, the Beta index is 1.53 which means that more than one network in the city. Whereas, Gamma index was found to be 0.53 which indicate there is a need for 26 links. However, the alpha indicates there is a need for 4 links. Therefore, all indices of connectivity demonstrate that there is a need for more links in the network.

On the other hand, the spreading degree indicates that there is a wide spread in the network and nodes are away from each other because Eta index is 14.30 while first betty index is 27. In addition, detour index is approximately demonstrating that most roads are straight. Among other measures, correlation index shows very weak correlation of the network. On the contrary, the density of roads in the city to the area and the number of population is 6772.435km/1000km2 which is too much higher than standard value (105km/1000km2) whereas, the density of roads to population (182.628km/100000) is less than standard value (496km/100000people). Finally, the ratio of roads to vehicles is 8.80m which is inadequate.

8. Conclusion and recommendation

The main conclusions of this study could be summarized as:

1. The connectivity indications for Al-Kufa city show weak connectivity according to its indices. The values of α, β and γ are 0.25, 1.47 and 0.5, respectively.
2. The coverage indication of the network shows, according to Eta index, widely spread and there is no closeness among nodes. This indication, according to first Betty index, demonstrates a huge spreading for the network in its region.
3. The detour index states that Al-Kufa roads are efficient.

4. The correlation index indicates that the degree of correlation among roads (links) so weak.
5. The density of roads to the area of the city is higher than ideal value; whereas the density of roads to population is less than the ideal value.
6. There is a need for putting a plan for the city to avoid the lack of the network which in return will help increasing the efficiency of road network.

References

[1] Aduory, R. (2010). Geographical Analysis of the Network of Roads in the Duor for 2008". Journal of Tikrit University for the humanities, University of Tikrit, Iraq, 17(3).

[2] Al-Jaboory, K. (2007). Kufa in Othman era. A-Safeer Journal, 5th issue, Al-Dar Al-beatha for printing and publishing, Al-Najaf.

[3] Al-Jameel, H. A., and AbdAbas, A. Y. (2017). Assessment of Urban Structure for The Holy City of Al-Najaf in light of Sustainability. *Journal of University of Babylon*, *25*(4), 1088-1100

[4] Al-Kalaabee, G. (2017). Spatial Analysis: Urban land uses in Kufa City. 1st edition, Dar. Alkari, Iraq.

[5] Bando, M., Hasebe, K., Nakayama, A., Shibata, A., & Sugiyama, Y. (1995). Dynamical model of traffic congestion and numerical simulation. *Physical review E,* 51(2), 1035.

[6] Buyong, T. (2007). *Spatial data analysis* for *geographic information science*. Penerbit UTM.

[7] Davey, J. A. (1971). Peter Haggett and Richard J. Chorley, Network Analysis in Geography, Edward Arnold, London, 1969, 348 pages. *Pacific Viewpoint*, *12*(2), 200-200.

[8] Dill, J. (2003). Joint Congress of ACSP-AESOP. *Leuven, Belgium.*

[9] Gastner, M. T., & Newman, M. E. (2006). The spatial structure of networks. *The European Physical Journal B-Condensed Matter and Complex Systems*, *49*(2), 247-252.

[10] Kansky, K. J. (1963). Structure of transportation networks: relationships between network geometry and regional characteristics. Khazaal, S. (2009). A geographical Analysis of Road Network System in Erbil governorate". Journal of Diyala, (34). University of Salahaddin, Iraq.

[11] Musaab, M. (1978). Translating the book entitled: Kufa planning and discussing its map by Louis Massignon.

[12] Noda, H. (1996). A quantitative analysis on the patterns of street networks using mesh data system. *City Planning Review*, *202*, 64-72.

[13] OGUNLEYE, O. S. (2011). Spatial Structure of Road Infrastructure in Ekiti State, Nigeria: Options for Transformation. *Advances in Natural Science*, *4*(2), 138-142.

[14] Sreelekha, M. G., Krishnamurthy, K., & Anjaneyulu, M. V. L. R. (2016). Interaction between road network connectivity and spatial pattern. *Procedia technology*, *24*, 131-139.

[15] Taran, A. and Makhamra, Z. (2015). Quantitative Analysis of Road Network in the Mafraq Governorate, the Dirasat journal (Humanities and social sciences), University of Jordan, (42).

[16] Taran, A., Aldayaflah, O., & Nawaiseh, S. (2016). Structural Analysis of the Road Network in the Governorate of Karak, Southern Jordan.

[17] Wang, Y., Han, Q., & Pan, H. (2012). A clustering scheme for trajectories in road networks. In *Advanced Technology in Teaching-Proceedings of the 2009 3rd* International *Conference on Teaching and Computational Science (WTCS 2009)* (pp. 11-18). Springer, Berlin, Heidelberg.

[18] Xie, F. (2005). *The evolution of road networks: A simulation study based on network degeneration* (Doctoral dissertation, University of Minnesota).

[19] Xie, F., & Levinson, D. (2007). *Measuring* the structure of road networks. *Geographical analysis*, *39*(3), 336-356.

Innovation of Analytical Software for Financing Construction Projects: Infrastructure Projects in Iraq as a Case Study

Saja Hadi Raheem[1,*] **Faiq M. S. Al-Zwainy**[2]

[1] College of Engineering, University of Baghdad, Iraq
E-mail: s.rahim1001@coeng.uobaghdad.edu.iq
[2] College of Engineering, Al-Nahrain University, Iraq
E-mail: faiq.al-zwainy@eng.nahrainuniv.edu.iq
* Corresponding author: College of Engineering, University of Baghdad, Iraq. Tel: +9647717021830.

Abstract. This study includes application of an Analytical Software system to explain the financing methods applied in infrastructure projects in Iraq and highlight the importance of the tax in support the Iraqi economy by knowing the percentage of its contribution in supporting the financing of infrastructure projects. The proposed program is flexible, and easy to use. It enables the users adding and managing unlimited number of projects as well as any related information. The information obtained from this research can be used by the government policy maker and regulators and thus develop better understanding of the application and implementation of tax system in Iraq. It is concluded that the tax system leads to make tax revenues contribute to solving 70% of financing problems if applied in infrastructure projects, and therefore the contribution of government funding be 30%.

Keywords: *Taxation, Financing, Infrastructure Projects, Computer Software.*

1. Introduction

The tax system in any nation is the financial deanship. The tax system reflects all the tax laws and regulations that exist in it and these laws are often affected by the economic thinking of the state or the social system that prevails in the state, leading to a different tax system from one country to another, such factors affect the tax system from one culture to another (Oberlender Garold, 2000).

The tax system is a collection of taxes that the government imposes on citizens at a certain time, and is an effective tool in the state's hands to achieve the goals they want, however financial, political and social factors influence this process (Omidvar & Bordbar, 2013).

The most important component of the tax system is the legislative authority responsible for enforcing tax laws, as well as the administrative executive authority administering those laws and the judicial system responsible for settling disputes that may occur between taxpayers and the tax administration while implementing those laws. The tax defined as a monetary duty paid by the individual to the State or one of its public bodies (Alavipour & Arditi, 2019).

The tax is also defined as a monetary amount to which the state is obliged to repay by individuals according to its ability to contribute with the state to bear the public burdens in order to achieve political, social and economic objectives within the state(Bird, 2011).

2. Types of taxes

Taxes are classified according to the method of taxation(Chase Bank, 2016);

1) Direct taxes: These are taxes that are imposed directly on income and capital and settle on the taxpayer and cannot transfer their burden, including taxes on profits and taxes on real estate sales.

2) Indirect taxes: A tax paid by the taxpayer and then transferred its burden to another person, it collected from the producer, exporter or importer in some way and its burden is ultimately on the consumer indirectly through high prices for goods produced or imported such as production tax, customs taxes and trading taxes such fees. Types of taxes are illustrated in figure1 below:

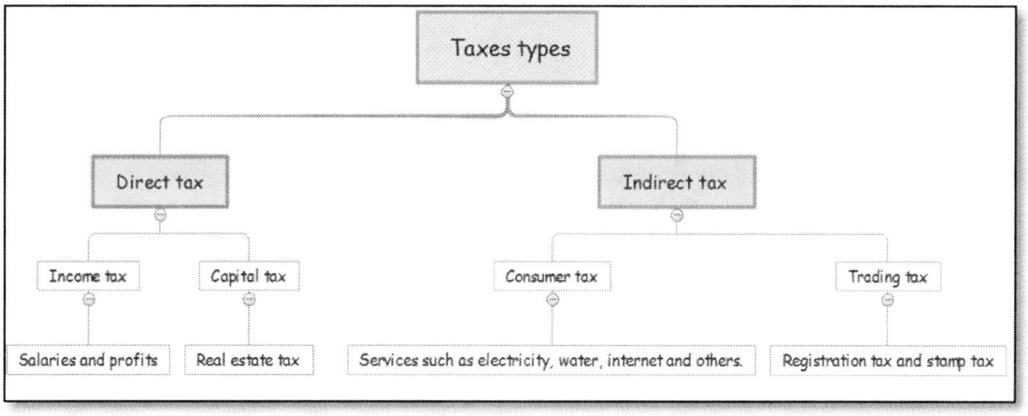

Figure 1. Types of taxes (Researcher based on Bird, Richard and Zolt, 2003)

2.1 Attributes of tax system

The tax system differs between growing and developed countries in terms of both the extent of reliance on taxes and the importance of taxes.

What concerns us here is the tax system in developing countries, as its economic system is the closest to Iraq's economic system, distinguished by the following characteristics:

1) A reduction in final tax revenue until it does not reach, on average, more than 15% of national income, while more than 30% of national income is reached in other developing countries. Perhaps the reason for the low tax revenue is due to the following:

a. Decreased national income from these countries due to inadequate growth in the production sector, which is reflected in lower individual income and a rise in the proportion of tax exemptions.
b. The autonomy of the agricultural sector, which in fact suffers from numerous problems over other productive sectors, ensures that it is not possible to impose taxes on this sector and therefore reduce the total tax revenue.
c. The restricted and primitive industrial sector, in addition to the influence of foreign companies, which receive significant tax exemptions to enable them to invest.

2) The tax system in these countries is characterized by an imbalance in which indirect taxes dominate, with a collection rate of 60%-80% of the general tax revenue.

3) Low significance of direct taxation, such as taxes on income and capital, due to low wages and widespread deprivation, and the lack of big investment projects that carry a significant proportion of the withholding tax.
4) The difficulty of tax collection and the prevalence of tax evasion due to the lack of advanced tax systems addressing tax evasion issues and the low efficiency of tax administration.

2.2 Types of taxes in the Iraqi tax system

Taxes in Iraq can be divided into two basic types(Cox & McLure, 2019):

1) Direct Tax: direct tax includes two varieties:

a. Income tax: It is the widest type of tax and is imposed on the largest segment of taxpayers engaged in various economic activities and generating annual revenue for them. This tax is regulated under Iraqi Income Tax Law No. 113 of 1982.

b. Real estate tax: It is the tax that is imposed on the total annual taxable income of his real estate income other than his residence, and its provisions are laid down in Real Estate Tax Act No. 162 of 1959.

2) Indirect Tax: includes several types;

a. Consumer tax: this type consists of:

- *Customs taxes*: taxes levied on anything related to the flow of goods and services from or to the state or across borders. It is considered to be one of the most significant indirect taxes, because it levies large amounts of money on the state, and most developing countries depend on it for easy collection and increased resources. In Iraq, its provisions are governed by Iraqi Customs Tariff Law No. 77 of 1975, as amended, and the customs tax has been repealed and replaced by the Iraq Reconstruction Tax levied by CPA Order No. 28 of 2003. The previous two laws have been repealed and replaced by Customs Tariff Law No. 22 of 2010.

- Taxes on production: these are taxes imposed on certain locally produced materials, and the state often uses exemptions for this tax, particularly in developing countries, to encourage national industries and compete with foreign products. This tax regulates the different laws. For example, the tax imposed on alcoholic beverages by Law No. 17 of 1973, the tax on cigarette production No. 64 of 1966 and the tax on oil production No. 9 of 1939 as revised.

b. Trade taxes: they are cash sums for the service provided by the State to the taxpayer and are divided into several types:

- Tax on stamps: This tax is imposed on financial stamps affixed to the document relating to lawful. The rules of this tax are laid down in Stamping Law No. 16 of 1974. This tax is characterized by the optimistic economic condition of the Iraqi tax system.

- Registration tax: In Iraq, they are referred to as registration fees and are charged while registering or documenting legal changes of property. Examples include tax on property registration and tax on vehicle registration. Its requirements are regulated by many laws, including the Law on Real Estate Registration No. 43 of 1971 and the Law on Judicial Fees No. 114 of 1981. This tax is considered to be small in price because it does not exceed 2% of the basic value of the transaction, as it is primarily concerned with the service rendered by the State to individuals.

3. Infrastructure projects

Usually an infrastructure is considered as the set of all the physical components which are needed to be build and operated to supply customers with a goods, e.g.: materials, energy, information, communication, etc. When the extension of these services involves a whole country, or a community of countries, the possible problems of an infrastructure may become problems for the country, or many countries (Cox & McLure, 2019).

Actually, an infrastructure is a complex organization structured in multilevel hierarchies. Its components must work and be operated in an integrated way. Examples of infrastructures are the electric power system, cold chain, fuels supply, communications networks, transportation, social services, health care, military defense, etc.as illustrated in figure 2. Each infrastructure uses the services and the resources provided by other infrastructures and this can create the conditions for a mutual dependence. It is enough to think how many services today depend on electricity and telecommunications. Recently, this dependence increased quickly and it evolved into a strong interaction between the infrastructures. The mentioned blackout events showed how these interactions become critical and now it is usual to mention the main infrastructures as critical. The problem is already at economic and political levels. So, institutions started to consider, evaluate and plan the protection criteria for critical infrastructures and some reference definitions has been prepared(Cox & McLure, 2019)(Mohammed, Al-Zwainy, & Sameer, 2014).

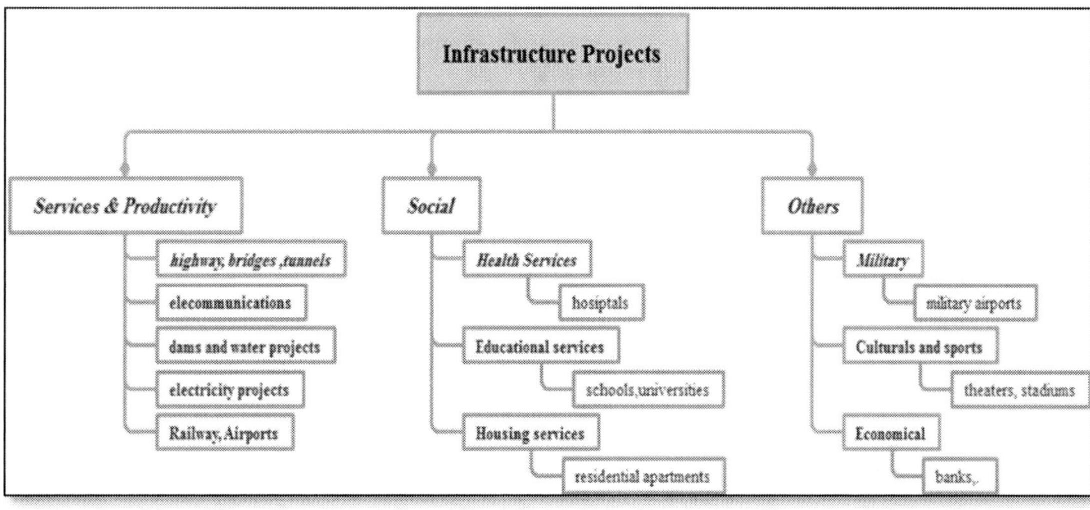

Figure 2. Infrastructure projects Types (Researcher)

3.1 Developing of the computer software

The main objective of this computer software was to introduction an active tax system, which also supports projects financing. This may be justified according to the following points;

1) The importance of such software is due to its high accuracy in tax system innovation since the recent error that has occurred by the use of such software is so small that it can be neglected.
2) The lack of effective documentation software in Iraq that can be used to recover the tax system innovation, using them to predict the requirements for financing in infrastructure projects.
3) The tax system affects the state revenues on the one hand and the money of citizens in the country on the other hand, as well as the national economy, especially with the adoption of financial and tax legislation to protect the national economy, and social through the issuance of tax legislation to protect public health or the environment.
4) Lack of the importance of direct taxes such as income and capital taxes, due to low incomes and the spread of poverty and the absence of large investment projects that bear a large proportion of tax deduction.
5) Difficult tax collection and the prevalence of tax evasion, due to the lack of sophisticated tax systems to treat the issues of tax evasion and low efficiency of tax administration.
6) There is no similar software currently in use by contractor construction companies or any other organization related to infrastructure projects in Iraq. All these authorities and organizations have agreed on the importance of the software as indicated by the questionnaire and personal interviews.
7) The infrastructure projects in Iraq still suffers from the lack of scientific software and expert for calculating accurate infrastructure estimated financing. The total financing estimation is important

IOP Conf. Series: Materials Science and Engineering **978** (2020) 012015 doi:10.1088/1757-899X/978/1/012015

in the bidding stage; therefore, the most popular reason for the failure to awarding construction contracts is the weakness in cost estimation process.

3.2 The Program specification

The system was designed and programmed using the following languages:

1) The programming language used for the design of the interface is HTML, CSS and JQuery.
2) The programming language used to analyze and connect databases and solve mathematical equations is PHP and JavaScript.
3) The database used is MySQL Server.
4) The local server that was used to open the program is Apache Server.
5) Responsive Web App technology has been used to simulate mobile devices, which ensures that the program can be run and used on any mobile device.

The system was programmed using Web Application technique; It ensures that the program can be run on an internal local network and that the program can be run on all devices connected to that internal network. The program can also be uploaded to a global network (Cloud). The program can be run anywhere in the world through an Internet service.

4. The Dialog of the computer program

The layout of the proposed tax system program is outlined in the flow chart as shown in figure3.

After installation of the system and pressing its icon the interface will appear, and consist of five options: *Case study, Start New Infrastructure Project, My Case Study, Information and Contact Information.*

Case study: previously created and stored in the system database.

When the user presses Case Study, the application is working, and the project case study window appears with three options: information about case study, show case study, and review of case study images. For *Show Case Study*: This window contains two options; *Overall View Flow Chart* and *Start Case Study.*

Overall View Flow Chart: Provides an outline of the flow chart that was prepared by the researcher, which reflects steps that demonstrate the methods of financing, including taxes, in order to know the magnitude of the tax contributions to the financing of infrastructure projects. Al-Nahrawan water treatment plant project is presented as a case study in this section.

Start Case Study: In this option, the information and data of the case study are entered and saved. as shown in Figure 4.

IOP Conf. Series: Materials Science and Engineering **978** (2020) 012015 doi:10.1088/1757-899X/978/1/012015

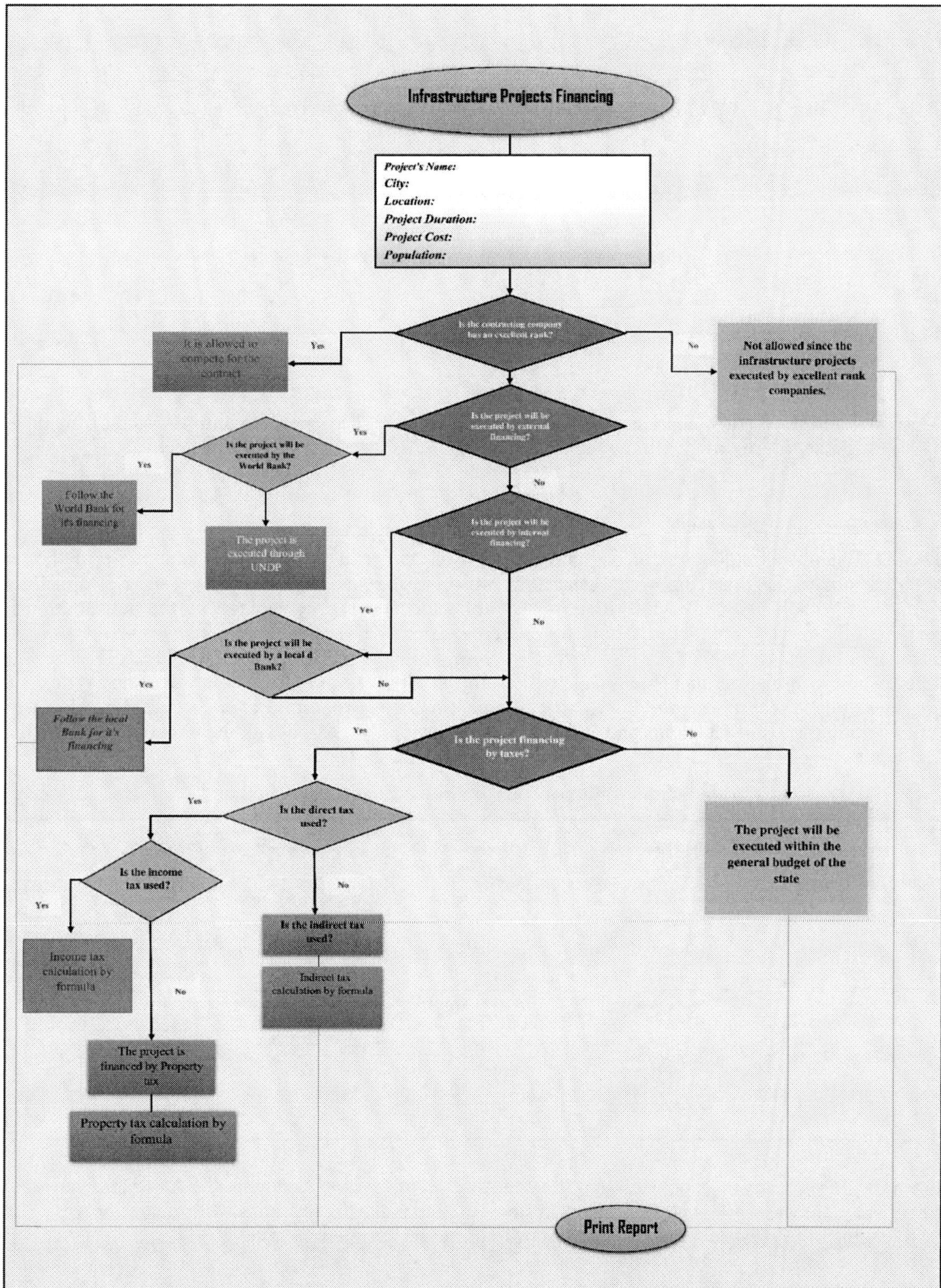

Figure 3. Flow-chart for Infrastructure Projects Finance

Figure 4. Screen of case study option

As shown in figure3 (flow chart steps), the researcher chooses the option No until the step of financing by taxes is reached, in order to know the contribution of taxes in financing the project.

5. Case study analysis result

The AL- Nahrawan water treatment plant project was analyzed to assess the magnitude of the tax contributions to the financing of infrastructure projects. The steps that were used to accomplish the financing through taxes. The information was collected and the percentage of the contribution in financing the project was found. The laws and conditions of living in Al- Nahrawan were taken into consideration, as were the rates of taxes that were imposed and applied in the city. The researcher was able to find information about the taxes applied in the city, which are, income tax, property tax and consumer tax. After entering the information as equations, the proportion of each type was calculated. After getting the result, the ratios and amounts available were compared with the remained, as shown in Table 1.

Table 1. Ratio of taxes contribution in financing (Researcher)

Financing method by taxes	*Available amount ID [1]*	*Projects cost ID [2]*	*Remained amount ID [2-1]*	*Ratio % [(2-1)/2]*
Income tax	28,800,000,000	82,810,506,791	54,610,506,791	66%
Property tax	5,000,000,000		77,810,506,791	94%
Consumer tax	25,200,000,000		57,610,506,791	69.5%
Total	59000000000	82,810,506,791	23,810,506791	29%

From the Table 1, it was concluded that the contribution of taxes in financing the Al- Nahrawan water treatment plant was 71%, such that the tax system leads to make tax revenues contribute to solving 70% of financing problems, and therefore the contribution of government funding is 30%.

Start New Infrastructure Project: The user selects this option to create a new infrastructure project. The user must enter the data and information required to start the next step; all the required information is shown in Figure 5. The operation of the computer software is represented by the overall flow chart, which shows the logic that was used and the components of the software-user interface.

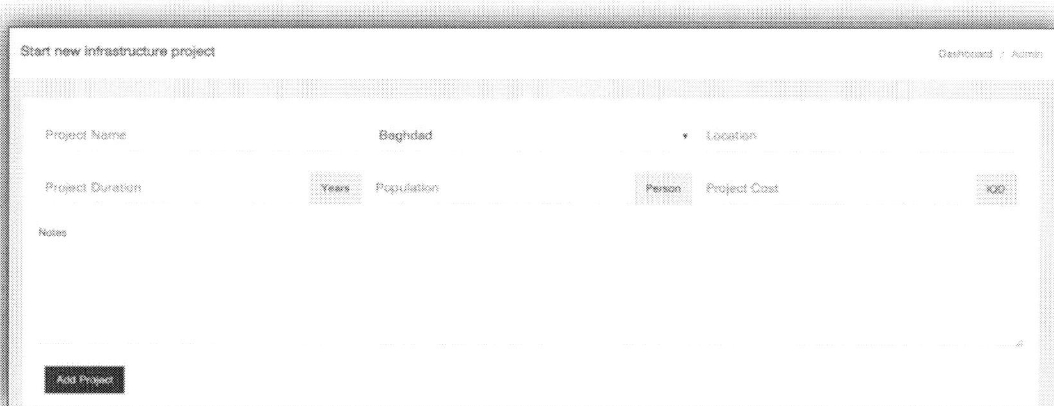

Figure 5. First window to start new infrastructure project

My Case Study: This option shows a new project that has been entered into the system and saved by the by the user.

Information: This option displays general information about the designer.

Contact us: This option allows Communication via e-mail.

6. Evaluation of the Proposed Program

The proposed program was presented to a group of experts for evaluating it, and in a style similar to the method used in the study of (Moatasem et al, 2018). The evaluation process had made by (10) experts, who have experience in construction management and in financial investment. The evaluation form distributed to individuals for obtain their opinion, comments, and recommendations about the operation and feasibility of proposed program. The questions related to evaluating the proposed program and the answers of respondents are summarized in Table 2.

The following sequences of steps were undertaken to complete the computer software evaluation interviews:

1) The evaluation documents were prepared to include a summary of the prepared software such as software dialog.

2) The prepared evaluation documents were handed over to the selected concerned people with a general overall preliminary presentation of its contents.

3) Sufficient time period was allowed to study these documents.

4) A prepared sort of questionnaire form Table2 was filled for each evaluation characteristics or questions.

5) Five answers as choices (Very Low, Low, Medium, High, Very High) were given.

6) Selection of the appropriate answer choice was left up to the satisfaction of the concerned person for estimation.

7) Upon completion of the interview, evaluation replies were personally collected.

Table 2. The Respondents' Answers on the Questions Related to Program Evaluation

No	Questions	V. Low (1)	Low (2)	Medium (3)	High (4)	V. High (5)	A.M	Rating level
1	What is your opinion about the assistance provided by the proposed program to the project manager to manage and implement the infrastructure project?	0	0	0	3	7	4.7	V. High
2	What do you think about the effectiveness of the proposed program at the project funding?	0	1	2	5	2	3.8	High
3	What do you think about the importance of the proposed program for your workplace?	0	0	3	5	2	3.9	High
5	How easily is use of the proposed program by the users?	0	0	0	2	8	4.8	V. High
6	What extent the proposed program has achieved its purpose and applicable?	0	0	1	5	4	4.3	V. High
7	What is your opinion about the accuracy of the information provided by the proposed system?	0	0	2	3	5	4.3	V. High
8	What is the effect of the proposed program in developing process of managing the construction projects?	0	0	2	5	3	4.1	V. High

The results indicate that the proposed development of the computer software is acceptable and can achieve the required financing of infrastructure projects. This indicates that the proposed software is efficient and complete and it can be used with more effectiveness if the decision-making authorities support it.

7. Conclusions

This study includes the application of an Analytical Software system to explain the financing methods applied in infrastructure projects in Iraq and highlight the importance of taxes in supporting the Iraqi economy, by finding the percentage of its contribution to the financing of infrastructure projects. The proposed program enables the users to add and manage an unlimited number of projects and any related information. The information obtained from this research can be used by government policy makers and regulators to develop better understanding of the application and the implementation of the tax system in Iraq. The study concludes that the tax system leads to make tax revenues contribute to solving 70% of financing problems if applied in infrastructure projects, and therefore the contribution of government funding would be 30%.

Acknowledgements

Authors would like to thank those respected persons who have helped and guided us to achieve this paper; they really deserve all respect and gratitude. We also want to thank and give our gratitude to everyone who has guided or advised us in submitting or writing this paper.

Author Contributions

Saja H. Al-Dhamad developed the Program System, wrote the manuscript and performed the computations; Faiq M. S. Al-Zwainy devised the project, the main conceptual ideas and proof outline and worked out almost all of the technical details, proposed the Analytical Software and involved in planning and supervised the current study. All authors discussed the results and contributed to the final manuscript.

References

[1] Alavipour, S. M. R., & Arditi, D. (2019). Maximizing expected contractor profit using an integrated model. *Engineering, Construction and Architectural Management*, *26*(1), 118–138. https://doi.org/10.1108/ECAM-04-2018-0149.

[2] Bird, R. M. (2011). Tax Challenges Facing Developing Countries: A Perspective from Outside the Policy Arena. *SSRN Electronic Journal*, (November). https://doi.org/10.2139/ssrn.1393991.

[3] Cox, M., & McLure, C. (2019). Taxation. *Encyclopædia Britannica, Inc.Retrieved from https://www.britannica.com/topic/taxation.*

[4] Mohammed, I. A., Al-Zwainy, F. M. S., & Sameer, D. (2014). INTERNATIONAL. *INTERNATIONAL JOURNAL OF CIVIL ENGINEERING AND TECHNOLOGY (IJCIET)*, *5*(9), 145–155.

[5] Oberlender Garold, D. (2000). *PROJECT MANAGEMENT FOR ENGINEERING AND CONSTRUCTION" McGraw-Hill Higher Education.*

[6] Omidvar, M., & Bordbar, F. (2013). *Advanced decision support systems for managers*. *2*(3), 700–708.

[7] Moatasem I. A., Ibrahim A. Mohammed, Faiq M. S. AL-Zwainy, Huda F. Ibraheem, 2018. Development of an Analytical Software for Cost Estimation for Highway Project, International Journal of Applied Engineering Research, 13(9), 6944-6951.

Traffic Assignment of Al-Kufa City Using TransCAD

Zainab Salam Al-Duhaidahawi[1,*], Raid R. A. Almuhanna[2], Ahmed Yahya Abdabas[3] and Hamid A. Al-Jameel[4]

[1]College of Engineering, University of Kerbala, Iraq
E-mail: Zoazzoaz1986@gmail.com
[2]College of Engineering, University of Kerbala, Iraq
E-mail: raidalmuhanna@yahoo.com
[3]College of Arts, University of Kufa, Iraq - E-mail: ahmed.feyada@uokufa.edu.iq
[4]College of Engineering, University of Kufa, Iraq
E-mail: hamid.aljameel@uokufa.edu.iq
* Corresponding author: College of Engineering, University of Kufa, Najaf, Iraq. Tel: +9647801341152.

Abstract. Modern urban cities around the world suffer from high congestions due to high daily trips for different purposes. Al-Kufa city has a lot of congested roads in its road network. The main objective of this study is to evaluate the current traffic situation. Firstly, the number of links and nodes has been counted using GIS software. These links represent major and minor arterials. The collected data include flow, speed and geometric characteristics for each link and node in the network. The traffic assignment model (Stochastic User Equilibrium) in TransCAD software was adopted in this study. The results indicated that the value (v/c) for most of these roads is more than one which indicated congested roads.

Keywords: *Al-Kufa city, traffic assignment, TransCAD, stochastic user equilibrium.*

1. Introduction

In 1907, the first transportation mean appeared in Al-Kufa city. This was the tramway operated by horses as reported by Musabee (1978). Then, the first paved road network appeared in 1954 (Al-Jaboory, 2007). The number of population was increased from 13700 capita in 1947 to 157845 capita in 2013 as reported by Al-Kalaabee (2017). This increment means more area to be served by roads. According to the Directorate of planning in Iraq, the criteria of transportation and road area for each person is 25m^2; however, this value is 15.8m^2 per capita in 2007 and 16.03m^2 per capita in 2013 (Al-Kalaabee, 2017). These values are less than the standard value. Therefore, it can be concluded that there is a lack in the roads which may lead to traffic congestion.

In addition, the rate of car ownership per household increases from 0.43 in 2003 to 0.92 in 2013 for middle regions in Iraq such as Najaf, Babylon and Al-Qadisiyah cities which includes Al-Kufa city, in addition to the annual traffic growth rate during the last 10 years is found about 11.5% which was higher corresponding the population rate 3.14% (Al-Obaedi, 2019).

Al-Jameel and Yahya (2017) reported that Al-Najaf city suffers from a lack in the road hierarchy. They found that the number of arterials and ring roads is insufficient to provide both accessibility and mobility in the city. It means that the land uses are spread over a wide area and the population density is distributed in an irregular manner, and thus the movement and access is difficult in the city. This leads to interconnected traffic with public streets within the residential areas that are not protected from the heavy traffic.

The objective of this study is to evaluate the current traffic status at the Al-Kufa network using TransCAD and GIS software in addition to field data collected by different means.

2. Literature review

In most Iraqi cities there is limited or no comprehensive published planning transportation studies or even traffic management (Qasim, 2015). Abedali and Qasim (2017) used TransCAD version 4.5 to model traffic assignment in Al-Amarah city using stochastic user equilibrium with exclusion of the local roads. The results indicated that most roads in the city with Level of Service LOS (D) in the base year (2015). However, the performance of the city network indicated that LOS (F) was found for most of the roads in the city center in 2035.

Qasim et al., (2017) evaluated the network of Nasiriyah city which the capital of Thi-Qar province in south of Iraq. This evaluation was achieved using several softwares such as TransCAD, GPS and GIS. The analyzed results demonstrated that few roads in the city centers suffered from high congestion with value of v/c was equal or more than 1.0 whereas other roads in the city with LOS (B) which represented good performance.

Alwash (2019) implemented an evaluation of Al-Hilla road networks just for major and minor arterials using GIS and TransCAd programs. The results reveal that there is a need for new roads to tackle the congestion in both current case (2018) and future case (2030).

Having reported the aforementioned studies in few Iraqi cities; Al-Kufa city has no such comprehensive study. Therefore, the current study aims to assign the flow of traffic for each major road in the city and investigate the performance of these roads.

3. Methodology

The main steps of this study could be summarized as: firstly, describe and determine the boundaries of the study area. Secondly, identify the required links and nodes in the network using GIS tool. Thirdly, collect the required traffic data and general information using field surveys and questionnaires. Finally, evaluate the current traffic condition using stochastic user equilibrium in TransCAD.

4. Study area

The city of Al-Kufa is located at the intersection of longitude (44-19)° east and latitude (31-59)° north. It is located in the southwestern part of Iraq as shown in Figure 1, the city of Al-Kufa is located in the eastern side of the province of Najaf, and is located to the western side of the Al-Kufa river, which is an important branch of the Euphrates River. The city of Al-Kufa is an important urban center for the province of Najaf. It is bordered to the north by the Kafel region (Babil Governorate), with a distance of (20 km), from the east by the Abbasiyya area, by a distance of (5 km), and from the west by the city center of Najaf, and from the south by the Manadhrah district, by a distance of (12 km). Al-Kufa city is (160 km) from Baghdad.

Figure 1. Location of A-Kufa City (Source: Directorate of Statistics Najaf Governorate, 2015).

Given the importance of the historical and religious center within it, the city has become a popular gathering place for a large number for visitors, which gave it a local importance that contributed directly to the city's development and urban growth.

The city of Al-kufa is one of the attractive cities for visitors because of the religious factor where the city contains a number of religious shrines, including the mosque of Al-Kufa and shrine of Muslim Ibn Aqil and Hani Ibn Erwa, as well as containing the house of Imam Ali and AL-Sahla mosque and shrine of Ibrahim AL-Gammer. The inhabitants' growth rate is 2.5% according to (Directorate of AL-Najaf Statistic). Table 1 shows the growth in population for the last 47 years.

Table1. Population Growth of AL-Kufa city as Directorate of AL-Kufa Statistic.

Year	1965	1977	1987	1997	2009	2013	2016	2018
Population	77813	304832	77279	97626	162452	157845	161711	185232

5. Distribution of Arterials at Al-Kufa City

This study focuses on just major roads and excludes local streets and minor collectors. These main streets are:

The main streets in the city link the old city of Al-Kufa with its provinces in several directions, including:

(Kufa –Najaf Street): which connects the city of Al-Kufa and the city of Najaf, andAl-Maemal Street: which starts from the bridge to the direction of the cement factory Kufa and from there to Al-Manathira.

Airport Street: it is one of the main streets in Najaf, where it was built a few years ago and starts from the airport towards the Al-Kufa river, passing through Al-Ma'mal Street, Al-Malaab Street and Maitham Al-Tamar Street.

Hospital Al-Sadir Street: it is the street that extends from Sadir Hospital to the Palace of Culture, passing through the University of Kufa (Engineering Gate), and this makes it is always crowded, especially during peak times.

Main Maysan Street: it is one of the main entrances to the city of Al-Kufa, through the Mukhtar tunnel.

Whereas, the minor streets are the second phase in the city of Al-Kufa through the traffic density include:

1. Al-Jsir Street: It passes the old city in the northeast of Al-Kufa Mosque and limits the Waqf and Al-Judidat and the bridge runs through the Euphrates and has a high traffic density.
2. Sikka Street: It passes through the Al-Saray and Al-Jadidat and is linked to Muslim Bin Aqil street where it travels the Al-Jumhuria and Al-Waqf.
3. Al-Sahla Street: Consists of four lines leads from the Mosque of Al-Kufa towards the Mosque Al-Sahla.
4. Cornish Street: Starting from the southern sections of the city center along the side of the river, passes Al-Judidat and Al-Saray to the north off Al- Kufa.
5. Al-Jumhuria Street: It is characterized by less traffic density, there are main streets within the neighborhoods, often separating the neighborhoods.
6. Maitham Al-Tamar Street: it is the road between Al-Matar Street and Kufa-Najaf Street, passing through Maitham Al-Tamar neighborhood.

6. Data Collection

The main aim of collecting data is to gather all data that would be necessary to investigate the circumstances of traffic flow at the study area. The main types of these data are:

1. Geometric features (e.g., lane width for each approach, no. of lane per approach and so on).
2. Data of traffic flow which included; traffic volume of vehicles at peak period and vehicles classification (passenger car, bus, motor cycle).
3. Traffic regulator information like; the type of signal control that used in the intersection and the type of traffic control, signal cycle length and phases information.
4. The speed data at each internal link in the network.

7. GIS Analysis

Arc GIS 10.5 has been used to prepare three types of shape files. These files are links, nodes and sectors which were developed by observing the satellite image and the master plan as shown in Figure 2. Al-Kufa municipality has divided the city into four sectors according to the land use as demonstrated in Figure 2. In addition, different characteristics have been obtained from the Arc GIS such as geometric deign such as length, width and number of roads as illustrated in Figure 3.

Beta (β) index

The main part of spatial analysis of road network is to develop graph theories depending on links (edges), nodes (vertices), etc. It is possible to compute the degree of complexity of road network as follows:

- Number of edges (e) =69
- Number of vertices (v)= 45
- Number of separate non-connecting sub-graphs (p) =3

The beta β index, which is also called as the link-node ratio, determines the "completeness" of a graph.

The index determines the level of connectivity of the transport network. The β index is found as follows: (Buyong, 2007)

$$\beta = e/v \quad \text{.......................................(1)}$$

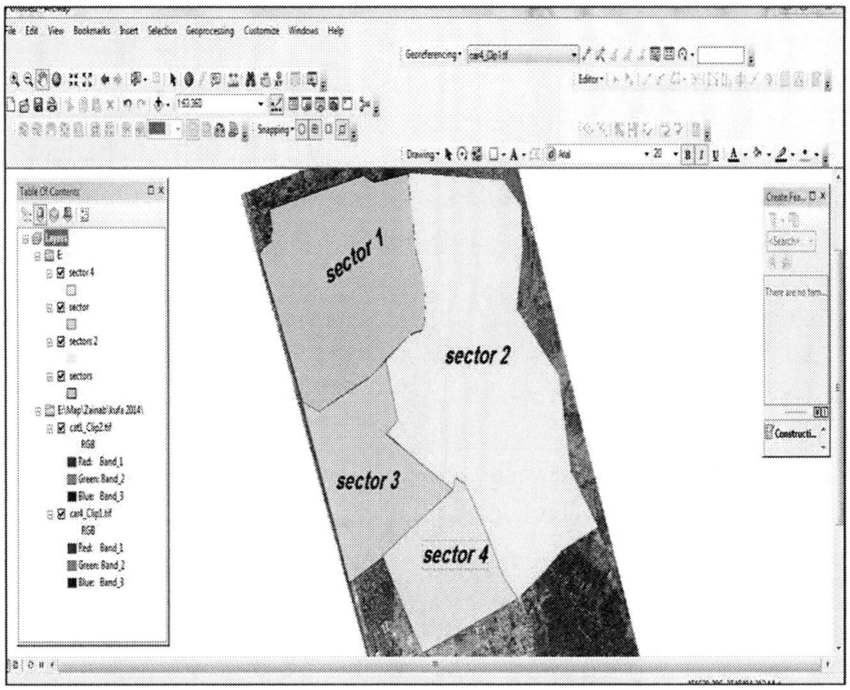

Figure 2. Sectors Divisions of Al- Kufa City

The β index is in fact 1/2 of the average number of links servicing each node and it represents the circulation of road network. It varieties between (0) and (3) for most route networks. If a value of β is less than one, this means a branching like network. If a value of β is greater than one, this indicates an increasing complication (very well connected network). Hence; β index discriminates simple topological structures from complex topological structures. The value of β =e/v= 69/45= 1.533

α (Alpha) Index

The alpha index utilizes the conception of a circuit- a finite, closed path starting and ending at a single node (Dill, 2003). This index for connectivity is the ratio of the number of fundamental circuits to the maximum possible number of circuits which may be in a network. This alpha index is a measure of the redundancy or duplication in the system. This index determines the ratio of the number of observed fundamental circuits in a graph to the maximum number of fundamental circuits which may occur (Morlock, 1967). This index is calculated as follows (Buyong, 2007):

$$\alpha = \frac{e-v+p}{2v-5} \quad \text{............(2)}$$

The α index compares the number of closed circuits in a network (e-v+p) against the maximum possible number of circuits which it might have (2v-5) and it is measuring of rotation (density of road network in specific area). The value ranges between zero and unity. A higher value of α Indicates a higher possible level of connectivity. α= 69-45+3/ (2*45)-5=0.317. These values will be used in the TransCAD software. This means that the connectivity is very low and there is a need for (61) links to satisfy the required connectivity degree.

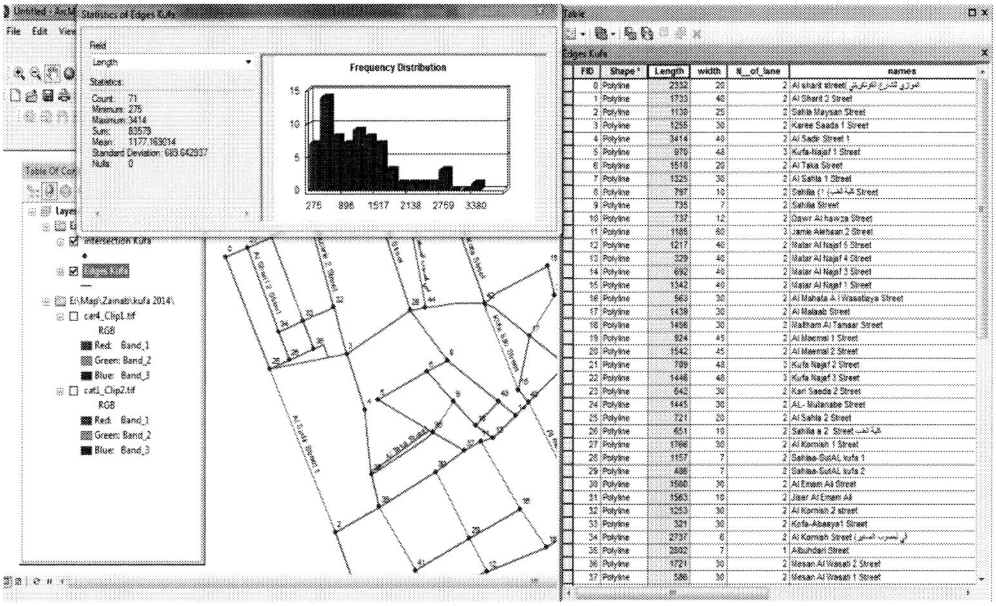

Figure 3. The links, nodes and other characteristic from ArcGIS.

8. Determination of Peak Hour Periods

The most congested period for traffic data collection predicted by various individual observations and survey, using video camera, was implemented in the selected area. Many personal interviews; interviewing people face to face, have also been achieved with many concerned people such as; policemen, owners of shops and offices at the study area and different road users. These observation and survey showed that there were two peak periods at (7:30-9:30 A.M.) and (1:30-4:30 P.M.) for typical weekdays (Sunday, Tuesday, Thursday and Friday). Intersections were collected by field measurement that data was collected during morning peak hour (7:30-8:30 A.M.).

It could be noticed that the traffic volumes passing through selected intersections (that located in entertainment places and shopping centers) were approximately equal during all the weekdays. Whereas there were big differences in the traffic volumes between the workdays and the weekend in the intersections located at the school zones and governmental buildings.

9. Data Collection and Implementation

In order to achieve the aims of the study, different types of data from different sources were needed. There are:

1. Official data:

These data collected from different organizations and they are:

- The number of household and population and growth factor for each year obtained from Directorate of AL-Najaf statistics.
- The number of vehicles and classification according to type which is obtained from AL-Najaf Traffic Police Directorate.
- Maps for AL-Kufa city provided by Municipality Directorate.

IOP Conf. Series: Materials Science and Engineering **978** (2020) 012016 doi:10.1088/1757-899X/978/1/012016

2. Socioeconomic data:

These data can be collected using design questionnaire forms which include socioeconomic characteristics such as income, household size, vehicle ownership.

3. Manual count

The manual count method utilized in the traffic data collection reflects the existing traffic characteristics of the study area. Traffic volume studies are implemented to find the degree, movements and classification of road vehicles in the significant location.

These data can help in determining critical flow times, or determining the effect of large vehicles or the flow of pedestrian traffic, or document traffic trends. The length of the sampling period bases on the kind of count taken and the intended use of the recorded data. For example, the interactive count is performed during the peak flow period where the manual count can be used with 5-minute intervals to get traffic flow data.

Traffic count was achieved on 1/12/2019 for the period from 7:00 am to 9:00 am. The count was implemented on normal working days: Sunday, Monday and Tuesday, which are the typical count days. The data were categorized according to the class of vehicle and according to the direction. For each movement, traffic volume was categorized into five kinds by the vehicle type. The first category refers to private cars. The second kind represents buses, inclusive of all vehicles with passenger capacity of 22 persons. The third kind refers to minibuses. A truck is any vehicle with two or more axles and it represents the fourth kind. The fifth kind refers to bikes and motorcycles; this is shown in Table 2.

The traffic volumes accounted per hour and per maximum 15 minute for each link in both directions in the network for peak hour. As shown in Table 3 and Figure 4 shown number of each link.

10. Traffic Volume Data for Intersections and roundabout

These data include counting the traffic volumes extracted from a video recording for each leg at the intersections and roundabout in the chosen network, also, traffic composition and the volume of turning movements. The data are required for this software for each intersection alone. Many runs were implemented to exclude the bias data. SIDRA INTERSECTION 5.1 software was used to analyze the existing traffic flow patterns for intersections roundabouts as isolated at the study area. Table 4 indicate the output results of simulation runs include the degree of saturation, total delay, and level of service for each intersection in the study area for peak hour. All the output results are categorized by medium to high total delays, thus the level of service for these intersections are shows in Figure 5 (a, b and c).

Table 2. Part of Numbers and Percentages of Transportation used in the Study Area.

Investigated Street	Dir.	PC	%	BUS	%	MB	%	HV	%	MT	%
Kufa-Najaf1	AB*	1831	77.2	43	1.8	411	17.3	12	0.5	75	3.2
	BA*	2914	79	37	1	457	12.3	90	2.4	194	5.3
Kufa-Najaf2	AB	964	67	40	2.8	250	17.4	100	7	150	10.4
	BA	864	63.3	30	2.2	250	18.3	100	7.33	120	8.8
Kufa-Najaf3	AB	964	67	40	2.8	250	17.4	100	7	150	10.4
	BA	864	63.3	30	2.2	250	18.3	100	7.33	120	8.8
Al-Kornich	AB	945	82.7	1	0.08	56	4.9	82	7.2	59	5.2
	BA	969	85.6	1	0.08	45	3.9	72	6.4	44	3.9
Al-Sikaa	AB	250	74	0	0	12	3.5	20	6	56	16.5
	BA	185	70.6	0	0	38	14.5	16	6.1	23	8.7
Al-Sahla	AB	457	66.6	2	0.3	103	15	49	7.1	75	11
	BA	436	73	1	0.2	88	14.7	38	6.4	34	5.7
Al Meamal	AB	894	66.9	5	0.4	197	14.7	106	7.9	133	9.9
	BA	862	72.5	4	0.3	122	10.3	111	9.3	89	7.5
Al-Takaa and AL-Mutanabe	AB	382	82.5	3	0.65	27	5.8	35	7.5	16	3.45
	BA	301	84.5	1	0.3	18	5.1	14	3.9	22	6.2
Kare Saada	AB	1996	82.6	5	0.2	120	4.96	164	6.8	131	5.4
	BA	1309	82.8	8	0.5	99	6.3	91	5.7	74	4.6
Maysan Al Rayiysi	AB	1942	77.8	5	0.2	322	13	71	2.8	154	6.2
	BA	855	62	6	0.43	283	20.5	119	8.6	115	8.3
Hospital Al-Sadir	AB	1628	84.8	9	0.47	48	2.5	112	5.8	121	6.3
	BA	1473	92.7	12	0.75	47	2.9	18	1.1	39	2.4
Al Sharit	AB	104	70.2	2	1.4	13	8.7	12	8.1	17	11.5
	BA	104	70.2	2	1.4	13	8.7	12	8.1	17	11.5
Muslim bin Aqeel	AB	760	65	5	0.42	285	24.4	68	5.8	51	4.3
	BA	979	61.3	2	0.12	340	21.3	113	7.1	162	10.2
Al-Jiser	AB	1004	76	1	0.07	135	10.2	78	6	103	7.8
	BA	955	70.5	0	0	198	14.6	92	6.8	108	7.98
Imam Ali and bridge	AB	2190	81.4	4	0.15	187	7	128	4.7	181	6.7
	BA	1326	78.5	3	0.2	60	3.5	119	7	180	10.6
Almazarie 1,2	AB	440	61.6	4	0.56	100	14	90	12.6	80	11.2
	BA	380	63	4	0.66	86	14.3	53	8.77	81	13.4
Sahla Maysan	AB	466	50	5	0.5	200	21.4	112	12	150	16.1
	BA	1215	66.1	4	0.22	328	17.8	139	7.55	153	8.32
Sahilia	AB	443	62.1	4	0.56	112	15.7	78	10.9	76	10.5
	BA	392	64	3	0.48	84	13.7	53	13.5	81	20.6

*pc= passenger car, **MB= mini bus, TR=Truck, ****MT= motorcycle),, *AB and *BA mean trips in direction AB and BA

IOP Conf. Series: Materials Science and Engineering **978** (2020) 012016 doi:10.1088/1757-899X/978/1/012016

Table 3. Samples of links flows of Al-Kufa network.

Investigated Street	Direction	flow (veh/hr)	Link Number
Kufa-Najaf1	AB*	3044	L6
	BA*	4092	
Kufa-Najaf2	AB	1464	L22
	BA	1392	
Kufa-Najaf3	AB	1464	L23
	BA	1392	
Al-Kornich	AB	1256	L28, L33
	BA	1264	
Al-Sikaa	AB	420	L67
	BA	332	
Al-Sahla	AB	776	L8, L26
	BA	720	
Al Meamal	AB	1488	L20, L21
	BA	1324	
Al-Takaa and AL- Mutanabe	AB	652	L7,L25
	BA	560	
Kare Saada	AB	2648	L4,L24
	BA	1744	
Maysan Al Rayiysi	AB	2756	L66
	BA	1720	
Hospital Al-Sadir t	AB	2072	L5
	BA	1960	
Al Sharit	AB	150	L1
	BA	150	
Muslim bin Aqeel	AB	1332	L49,L51
	BA	1724	
Al-Jiser	AB	1416	L52
	BA	1420	
Imam Ali and bridge	AB	2756	L31,L32
	BA	1720	
Kofa-Abasya	AB	550	L34, L55, L62, L64
	BA	500	
Almazarie 1,2	AB	800	L48,L65
	BA	700	
Sahla Maysan	AB	1904	L3
Sahilia	AB	800	L10
	BA	700	

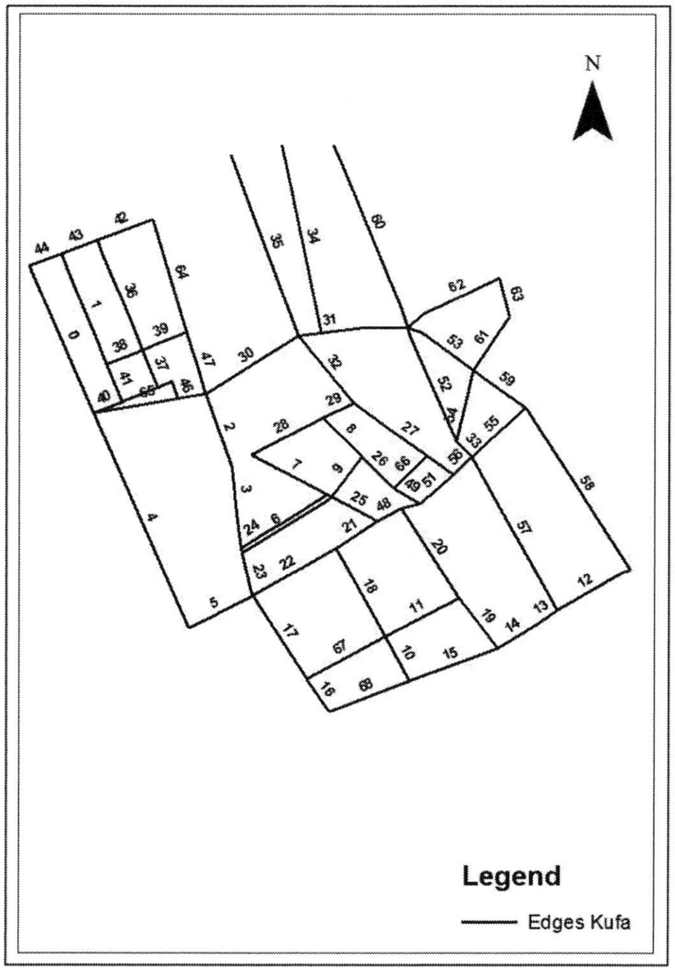

Figure 4. Number of links in Al-Kufa city.

Table 4. Results of Analysis Intersections and Roundabouts by SIDRA 5.1 Intersections for Existing Condition.

Intersection Code	Intersection Name	Max. Degree of saturation (V/C)	Average Delay (sec / veh)	LOS
1	Al-Najaf Hospital Intersection	6.085	463.7	F
2	Al Shahrstani Roundabout	1.654	278.1	F
3	Habib Al Raash Roundabout	1.087	46.9	D

IOP Conf. Series: Materials Science and Engineering **978** (2020) 012016 doi:10.1088/1757-899X/978/1/012016

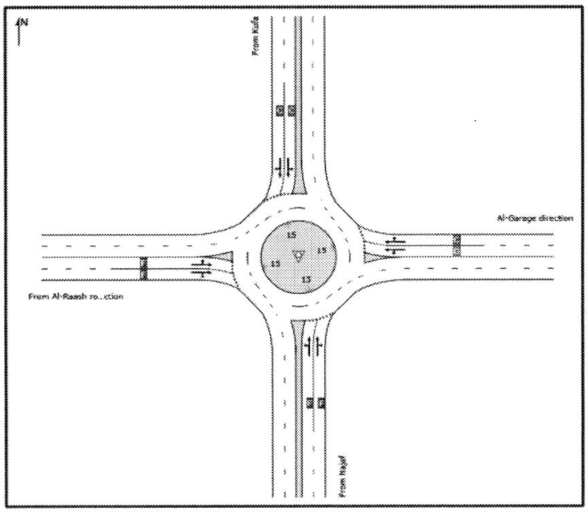

a. Al Shahrstani Roundabout

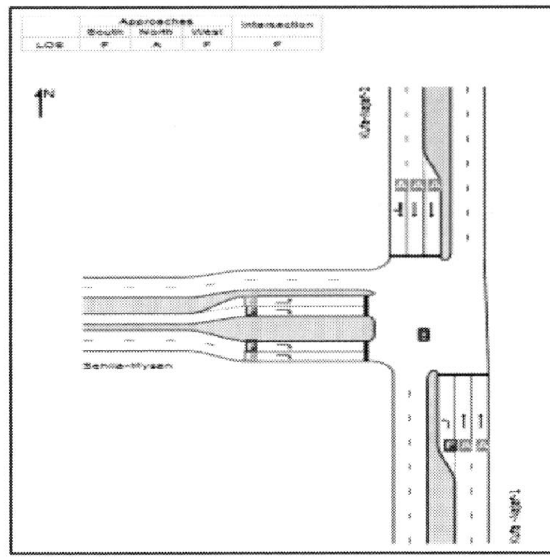

b. Al-Najaf Hospital Intersection

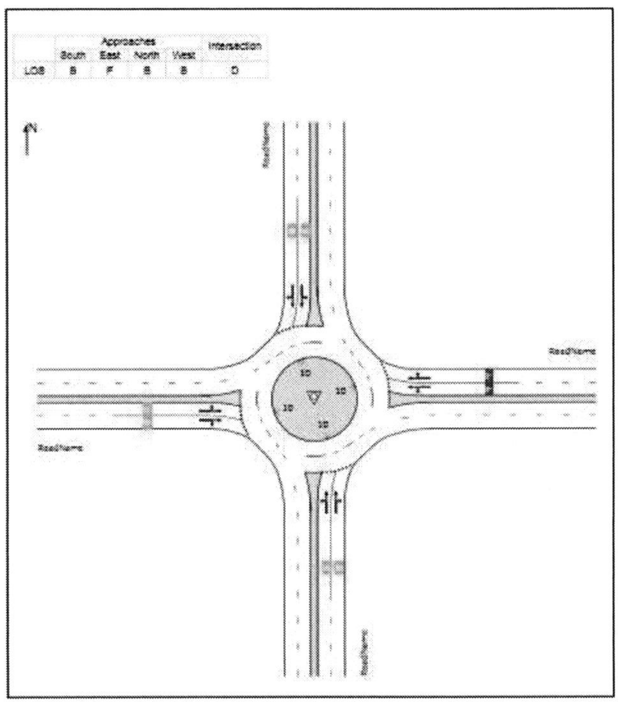

c. Habib Al Raash Roundabout

Figure 5. LOS analysis of intersection (T-intersection) and Roundabouts.

11. Public Transport at Al-Kufa City

The structure of public transportation varies from place to place and according to the transportation infrastructure and means of transportation that the population can travel. As for the city of Al-Kufa, public transportation depends on minibuses. whose passenger numbers range from 11-14 passengers. These buses do not have private parking spaces, but according to the desire of the passengers. Other significant things are absence of timetable for these public transportation lines. Actually, no suitable stations are available through the city. Furthermore, all these vehicles utilized as public transportation are for private sector (each vehicle has private owner not belong to the government) therefore, no uniform label (such as bus No.) for these buses even the distributed lines and routes of these buses are out of control. Subsequently, no control on the public transportation could be forced. Moreover, the characteristics enjoyed by minibuses of capacity between 11 to 14 passengers are the deficiency of stops, speed and freedom to maneuver in traffic at traffic jams and exceeding the time taken to complete the number of passengers provided it many sources of attraction. These minibuses provide flexibility for public transportation than buses because the number of passengers is less than buses. Moreover, they can stop anywhere without depending on regular stops

12. Existing problems of Al-Kufa Roads Network

The existing conditions issues concerning transportation system in Al-Kufa included the following:

- Extreme obstruction of traffic at peak hours due to lack of traffic management;
- Illegal parking and lack of rules on parking.

The main road to Al-Kufa-Najaf, there are a lot of vehicles standing parallel parking on road side, these vehicles leads to reducing the street capacity from three lanes to two lanes. Therefore, just one lane is used for traffic movement in the following sections:

1. Section from Al-Najaf hospital to the holy shrine Maitham Al-Tamar.
2. Section close to Muslim-Tunnel to the Al-Kufa old bridge.
3. Al-Sikah street.
4. The section between Al-Raash Roundabout to Al-Shahristani Roundabout.

Therefore, on-street parking (using strict traffic enforcement) along the main road network must be prevented parking vehicles and adequate off-street parking spaces (parking and garages) must be installed.

13. Traffic assignment model

The traffic assignment model, which is available in the TransCAD Software 4.5, stochastic user equilibrium model was adopted in this study.

A. Input parameters

The main inputs required for this model are road network file and origin-destination (O-D) matrix file. Firstly, the road network file can have imported from GIS software as in this study which has the characteristic of the road network such as Identify Data (ID), length, width, number of lanes, direction of flow, capacity, flow and speed for each link. Furthermore, the characteristics of each node, which represents intersection, such as ID and x-y coordinates.

Secondly, the O-D matrix is the simplest way to represent the travel demand between the selected sectors which are Sectors 1, 2, 3 and 4. Without such matrix, TransCAD can't operate in addition to starting and ending trips at some discrete points in the city network. That is why centroid nodes have been employed to overcome this problem. The centroid point is the geographic center of a sector. Therefore, these centroids will not be connected to the road network, a new set of links, call centroid connectors are created to make connection. Table 5 represents Al-Kufa city peak hour O-D matrix among sectors.

Another type of data is required for traffic assignment, the external- external traffic volume; trips start and end outside the study area; which should be located on the road network links. Then, the external and internal matrixes are combined together into one matrix.

Table 5. Al-Kufa City Peak Houre O-D matrix (vph)

	Sector 1	Sector 2	Sector 3	Sector 4
Sector 1	0	4367	5615	1787
Sector 2	421	0	10029	1137
Sector 3	2046	4806	0	1828
Sector 4	1081	3559	2746	0

B. Performing a traffic assignment

After preparing the O-D matrix with the suitable ESRI shape filesfiles (with all required data as mentioned above, the execution of the program using stochastic user equilibrium with a default number of iterations which are 20 per assignment is performed. The outputs of the program are exported to specific files.

Mainly, trips for each O-D pair are then allocated to the links in the minimum path and the trips are added up for each link as indicated in Figure 6. The allocated flow is then compared with the capacity of each link to find out is it congested or not. If a link is congested the travel time is adjusted to result in a longer travel time on that link. Consequently, changes in travel time means that the shortest path may change. This process is replicated several iterations until there is an equilibrium between travel demand and travel supply. The results of stochastic user equilibrium model indicated the LOS for each link in terms of (v/c) according to the value demonstrated in Table 6.

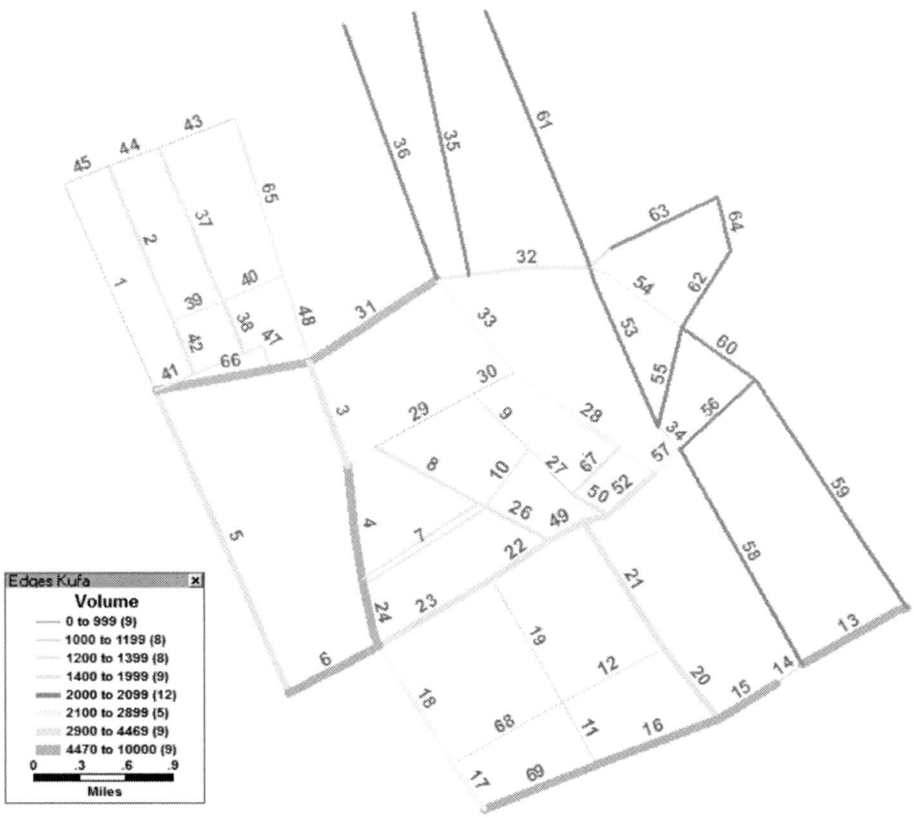

Figure 6. Flow map for Kufa network morning peak.

Table 6. Relationship between v/c ratio and LOS for Urban Roads (HCM, 2010)

LOS	V/C Ratio	Operating Condition
A	0.00-0.50	Free flow conditions
B	0-51-0.70	Reasonably unimpeded operations
C	0.71-0.80	Generally stable operations
D	0.81-0.90	Approaching unstable conditions
E	0.91-0.99	Significant delays and low average speed
F	≥1	Severe congestion and delays

Therefore, a lot of links with (v/c>1.0) LOS F have been observed as indicated in Figure 7. These roads with (v/c>1.0) are always for roads carrying trips towards University of Kufa. These links need more improvements to mitigate traffic congestion such as preventing parking and convert some of these intersections discussed above to interchange.

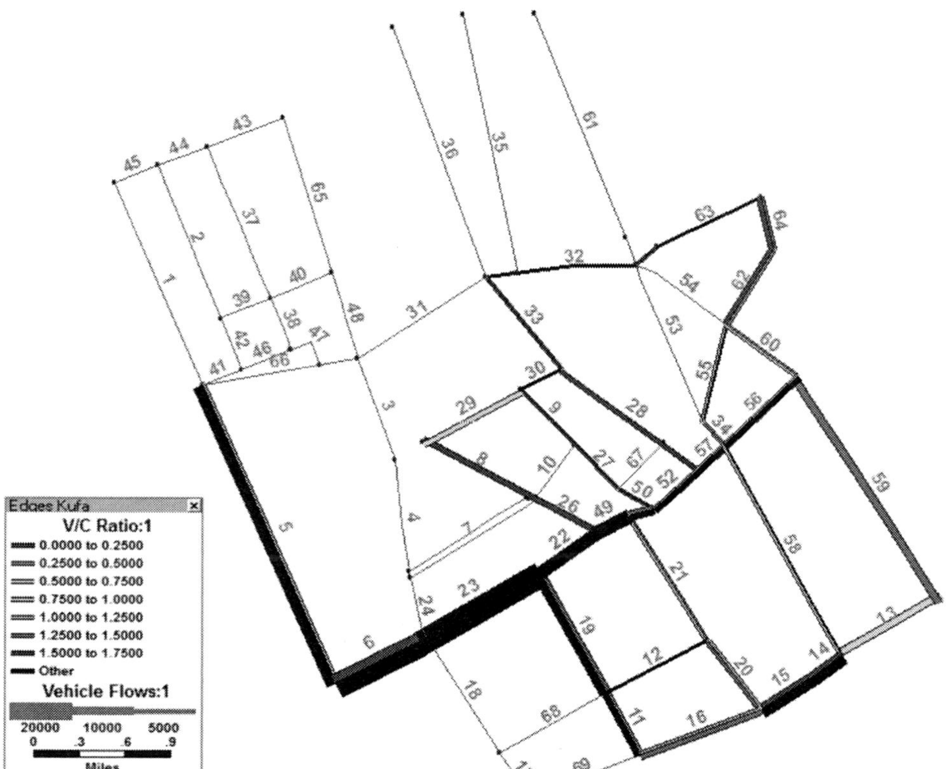

Figure 7. Stochastic User Equilibrium Method Assignment Model Result.

14. Conclusions

The main significant points of this study could be summarized as following:

1. The activities of the network have two main centers: the first is the University of Kufa in Sector 1 and the second is the city center of Al-Kufa which is close to the river.
2. The network suffers from high congestion with links of (v/c) close and higher 1.0 especially Al-Kufa-Najaf and Najaf-Miasan Roads.
3. There is a need for adding more road to satisfy the network connectivity along with reducing the congestion.

References

[1] ABEDALI, A. H., & QASIM, G. J. (2017). TRAFFIC ASSIGNMENT MODEL OF AL-AMARAH CITY. *Journal of Duhok University*, 319-334.

[2] Al-Jaboory, K. (2007). Kufa in Othman era. A-Safeer Journal, 5th issue, Al-Dar Al-beatha for printing and publishing, Al-Najaf.

[3] Al-Jameel, H. A. E. (2016). Evaluation of Al-Najaf Hospital Intersection Performance Using Simulation model: Case Study. *Journal of University of Babylon*, *24*(1), 136-143.

[4] Al-Jameel, H. A., & AbdAbas, A. Y. (2017). Assessment of Urban Structure for The Holy City of Al-Najaf in light of Sustainability. *Journal of University of Babylon*, *25*(4), 1088-1100.

[5] Al-Kalaabee, G. (2017). Spatial Analysis: Urban land uses in Kufa City. 1st edition, Dar. Alkari, Iraq. College,University of Baghdad.

[6] Al-Obaedit, J. T. S. (2019). Highlighting the Ongoing Roads Infrastructure Challenges: Car Ownership Rates in Iraq. *University of Thi-Qar Journal for Engineering Sciences*, *10*(1), 1-4.

[7] Buyong, T. (2007). *Spatial data analysis for geographic information science*. Penerbit UTM.

[8] Dill, J. (2003). Joint Congress of ACSP-AESOP. *Leuven, Belgium*.

[9] Lima, and Associates, (2006). Lincoln MPO Travel Demand Model. Draft Model Documentation, Prepared for Lincoln Metropolitan Planning Organization, Transportation- G.I.S.

[10] Morlock, E. K. (1967). An analysis of transport technology and network structure. *Evanston. Illinois, the Transportation Center Northwestern University.*

[11] Musaab, M. (1978). Translating the book entitled: Kufa planning and discussing its map by Louis Massignon.

[12] Qasim, G. (2015). Travel demand modeling: AL-Amarah city as a case study. *Doctor thesis submitted to the Engineering College, University of Baghdad.*

[13] Qasim, Z., Ziboon, A. R., & Falih, K. (2018). TransCad analysis and GIS techniques to evaluate transportation network in Nasiriyah city. In *MATEC Web of Conferences* (Vol. 162, p. 03029). EDP Sciences.

[14] Transportation Research Board, National Research Council, "Highway Capacity Manuual" Washington, DC, 2010

Severity of Damage Due to 2017 Kurdistan Region of Iraq-Iran border Earthquake (Halabja Earthquake)

Wrya Abdullah[1,*], SerwanKhorsheedRafiq[2]and Younis M. Alshkane[3]

[1]College of Engineering, University of Sulaimani, Kurdistan Region, Iraq
E-mail: Wrya.faraj@univsul.edu.iq
[2]College of Engineering, University of Sulaimani, Kurdistan Region, Iraq
E-mail: Serwan.rafiq@univsul.edu.iq
[3]College of Engineering, University of Sulaimani, Kurdistan Region, Iraq
E-mail: Younis.ali@univsul.edu.iq
* Corresponding author: College of Engineering, University of Sulaimani, Kurdistan Region, Iraq. Tel: +96477736989217.

Abstract. On 12th of November 2017, a major earthquake of a magnitude of M 7.3 hit Halabja and caused severe structural damage and collapse in Iran, however, only a few areas of Iraq were mildly affected by it. It was considered the deadliest earthquake of the year as its casualties surpassed 400 people most of which were in Iran. Halabja, was only 32 km away from the epicenter, yet the quake caused no noticeable structural damage to the buildings and no casualties or serious injuries were reported from there. This paper presents the possible engineering reasons behind the gigantic gap of severity of the damage between Kurdistan region of Iraq and Iran. The analysis is done after a series of site visit to Darbandikhan and Halabja and taking more than 300 photographs of the damaged and collapsed buildings. Several crucial factors are discussed in detail, the major of which is directionality effects of the quake as its epicenter was in Iran and the quake caused a meter of slip towards south and half a meter of displacement towards west of the epicenter. Due to the directionality effect, the damage was more in the south and the west of the epicenter knowing that most parts of Kurdistan region of Iraq were located at the north of the epicenter. It is also found that the geological location of the buildings was insignificant in increasing the damage level of the buildings.

***Keywords:** Earthquake effects; Structural damage levels; Halabja Earthquake; Directionality*

1. Introduction

Approximately 10 to 20 strong earthquakes of a magnitude of M 7 and above hit the earth every year, one of which occurred in Iraq-Iran region on 12th of November 2017 at 18:18 GMT. The magnitude of the earthquake was M 7.3 according to (USGS, 2017). The name was attributed to Halabjah by (USGS, 2017) as its epicenter was only 32 km south of Halabjah, a city in Kurdistan region of Iraq, with the local depth of 23 km, even though the epicenter was in Iran near Ezgelecity in Kermanshah province at 34.905°N latitude 45.956°E longitude. Various agencies specified different exact locations of the epicenters, however, all of them confirmed that it was near Ezgele, therefore, some other names are attributed to it such as Ezgele earthquake(Alavi et al., 2018).The quake had some foreshocks of different magnitudes ranging from M 1.9 to M 4.5 and 529 aftershocks with magnitudes ranging from M 1.8 to M 4.7 until 28th of November 2017(Alavi et al., 2018).

The mechanism of faulting of the tremor was thrust faulting as a result of movement of Arabian as it is being pushed away by Eurasia Plate. There are two main active faults which are Zagros Mountain Front Fault (MFF) and High Zagros Fault (HZF) which are considered in interpreting the cause of the main quake with a peak recorded ground acceleration of 681 cm/s^2N-S and 585 cm/s^2W-Ein Sarpol-eZahab earthquake station(Yekrangnia et al., 2017). More than 1,800,000 people live with within 100 kilometers of the epicenter of the quake, 10000 of which were injured and approximately 500 of them were in Iraq. It was considered as the deadliest earthquake of the year with more than 400 people died, 10 of the dead were reported to be in Iraq and the rest were in Iran(IFRC, 2018). In terms of structural damage, it caused damage to more than 12000 buildings in Iran and some building collapse and some buildings were severely damaged in Kurdistan Region of Iraq (Taha & Hasan, 2018; Yekrangnia et al., 2017). The affected areas in Kurdistan region of Iraq were Darbandikhan,Halabja, Khanaqin, Kalar, Sulaymaniyah and Erbil knowing that most of the destructions were in Darbandikhan city in Sulaymaniyah governorate with a distance of approximately 40 km from the epicenter. Halabjah was approximately 30 km away from the epicenter and the name of the earthquake was attributed to it, yet, no serious damages or causalities reported from there. While, the affected areas in Iran were Sarpol-eZahab, Qasr-e Shirin and Eslamabade-e gharb. Sarpol-e Zahab with a population of more than 80000 people was 42.35 km away from the epicenter. However, it was the epicenter of the destruction in Iran where more than 85% of the causalities occurred in there with 387 deaths. Qasr-e-Shirin with population of 24000 experienced 16 deaths (Yekrangnia et al., 2017). It was considered as one of the most catastrophic earthquakes in the area that has had influences on reviewing the designing buildings in that vicinity. Comparing Halabjah and Sarpol-e Zahab, it can be seen that Halabjah is closer to the epicenter than Sarpol-e Zahab yet the level of damage and causality is incomparable as Sarpol-e Zahab lost 387 lives of its inhabitants and none was reported in Halabjah. This triggered a question amongst the engineers and designers as to why this huge gap? To answer this question, some of them speculated wrong theories such as lateral forces were counted for in the design stage of the buildings in Kurdistan of Iraq, knowing that most of the buildings were built in an era where the residents built their own houses without consulting with an engineer or the design codes used in calculating earthquake loads are out of date now.Therefore, the major scientific reasons for effects of earthquakes on different areas of a region is discussed in this section to find possible reasons behind the huge gap between level of damage in Kurdistan region of Iraq and Iran due to the same earthquake.

2. Methodology of the work

A research investigation is carried out to study the reasons behind the gap of level of damage between Kurdistan region of Iraq and Iran. For that reason, a series of site investigations were carried out to all the damaged buildings in Kurdistan region of Iraq particularly to Halabjah and Darbandikhan. Firstly, the authors took a trip to Halabjah right after the earthquake on 14th of November, 2017. A thorough investigation was carried out by asking the officials and the locals for the places where they heard of damages or building collapses. In Halabjah, the total of 8 damaged buildings were investigated. Then two site visits were made to Darbandikhan on 14th of November 2017 and on 1st of December 2017. The total of 16 damaged buildings were investigated with the location of the buildings. The damage analysis is carried out using engineering judgement and visual inspection of the buildings.

A thorough investigation is carried out on the impacted places in Iran. Literature was the source of the research including all the reports on the earthquake and the type of buildings in the area. The investigation included reading and taking satellite data of the area hit with the earthquake to find out all possible reasons behind the huge difference of level of damage between Kurdistan region of Iraq and Iran. Some of the papers that are used as references were preliminary official reports on the earthquake and then later on updated and the rest were investigation performed by humanitarian agencies and local searchers.

In this paper the effects of Halabjah earthquake is investigated in both Kurdistan regions of Iraq and Iran to comprehend the reasons leading to a huge gap between the structural damage in the buildings.

3. Earthquake Effects

Halabjah earthquake had a huge impact on Iran rather than Iraq for numerous of reasons. As described by Thomas Paulay and M. J. N. Priestley (1992), the earthquakes' effects on a specified area depend on several crucial factors that can vary from one place to another. The main factor of destruction is the fault rupture that occurs during earthquakes which can extend to be larger than a meter therefore, any structure built on those fault lines experience major damage if their foundation is not unified as a raft or mat foundation. Although, this dislocation of earth surface is considered as the most destructive factor for the structures but because it affects only a specific area therefore, it does not establish a major thread. The other factor which is significant and interests most of the structural designers, is the structural response of the structures during releasing energy and the fault slip which depends on various variables and factors. These factors contribute in enhancing structural response for a particular earthquake which causes more damage and sometimes collapse, which are magnitude, the vertical component of the earthquake, intensity, soil type, directionality effects, geographical effects and the quality of the material used in the constructions. In this section each of those details are discussed in detail.

3.1 Magnitude:

The higher the magnitude of an earthquake, the more energy release would be. Each magnitude higher means, the amount of energy released by that earthquake will be 32 times higher than the previous one. It is estimated that an area of 10000 km^2 will be affected and structures in that vicinity may be damaged by an earthquake of a magnitude M 7 to M 8. For Halabja earthquake, the magnitude was estimated to be M 7.3 therefore, the whole Iraq- Iran boarder affected by it but the epicenter was in Iran near Ezgele, in Kermanshah Province therefore, the consequences of the quake was more obvious in the surrounding area of the epicenter and the fault lines. Different magnitudes cause different fault rupture lengths for instance an earthquake of a magnitude of M 5+ may cause of a fault slip in a few kilometers while that length can be extended to 400 km if the magnitude increased to M 8.The epicenter of the quake was located in between two active fault lines of Zagros Mountain Front Fault (MFF) of a length of 1368 km and High Zagros Fault (HZF) which its fault length is 1375 km, where the movement of MFF caused the earthquake(Roudbari et al., 2020). Both of these faults are located in Iran. HZF extends from north-west of the epicenter towards south-east in a linear pattern. Whereas the fault line of MFF passes along north-west of the epicenter extending towards west of it on Iraq- Iran border sweeping to the south of the epicenter. Most of the damaged areas were a long those areas where the MFF is passing.

The duration of the earthquake can increase the length of the fault line which can increase the damage as the shock waves arrives latter. The duration of Halabja earthquake was 30 seconds which was felt in Caucasus, Turkey and Syria(Yekrangnia et al., 2017).

To show the relationship between the damage caused by an increase in the magnitude of the earthquake, (Nievas et al., 2020) collected a database of 1062 earthquakes with different magnitudes of 4. To 5.5 and plotted on a graph, shown in Figure 1, to show the relationship between the magnitude and the amount number of building destroyed. As it is shown in Figure 1, the earthquakes that caused a destruction of the lower and/ or upper bound of 1000 buildings are the ones which their magnitudes are greater than 4.5.

IOP Conf. Series: Materials Science and Engineering **978** (2020) 012017 doi:10.1088/1757-899X/978/1/012017

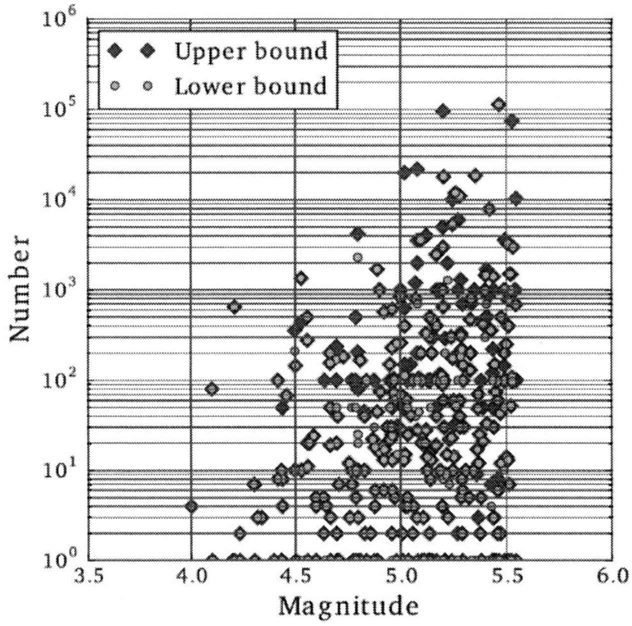

Figure 1. Relationship between the number of destroyed buildings with the magnitude of earthquakes (Nievas et al., 2020)

3.2 Vertical acceleration:

The vertical component of the earthquake is usually neglected by the structural designers and it is usually taken as the two-thirds of the peak horizontal acceleration, however, it is proven that sometimes that the value of the vertical acceleration is as high as the horizontal acceleration especially close to the epicenter. The ground acceleration of the Halabjah earthquake was recorded by 101 stations inside Iran and the strongest peak accelerations were recorded by Sarpol-e Zahab station. Figure 2 shows the vertical acceleration of the quake was 358 cm/s^{2}which considered to be high as the peak was 585 cm/s^{2}in W-E and highest was along N-S with 681 cm/s^2(Yekrangnia et al., 2017). This might be adding reasons for the destructiveness of the earthquake.

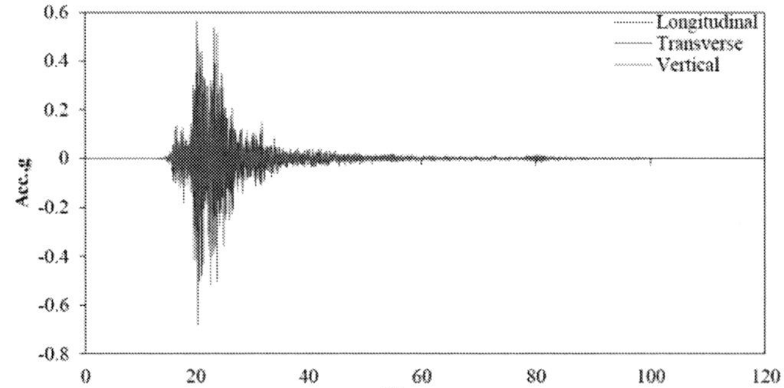

Figure2. Peak ground accelerations of Halabjah earthquake (Yekrangnia et al., 2017).

To emphasize the effects of the vertical acceleration on the structures in general. A study on its effects on spacious underground R.C structures is performed by (He & Chen, 2019). They took two other variables into account which are the performance level of the structure varying from fully operational (FO), operational (OP), life safety (LS), and near collapse (NC) and the type of the soil it is built on which varies from stiff site (site I), medium stiff site (site II), medium soft site (site III), and soft site (site IV). They found out that the effects of the vertical acceleration would be significant on the underground structures especially if the type of soil was soft clay regardless of the performance level as it is shown in Figure 3 which represents the results of (FO) on the left and (OP) performance level on the right.

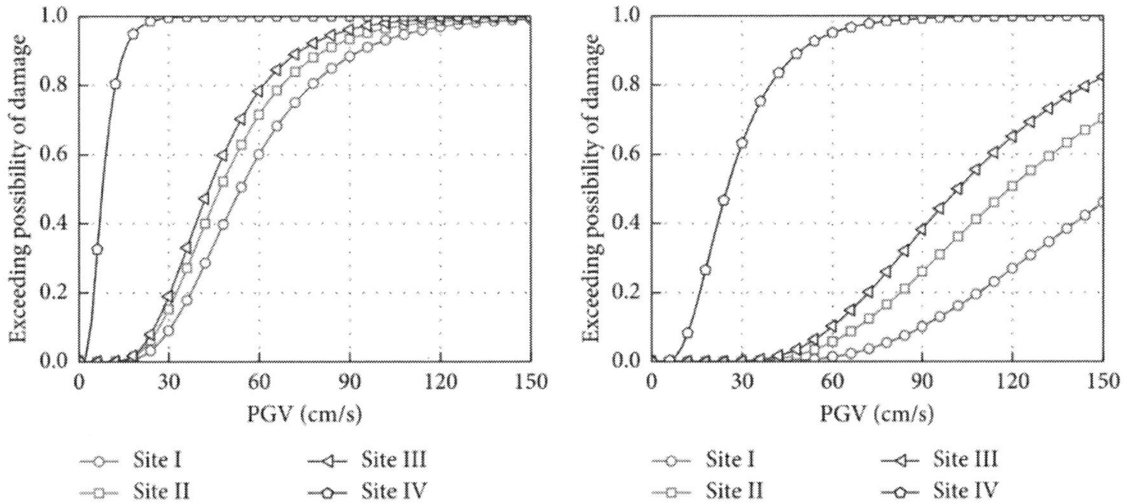

Figure 3. Peak ground vertical accelerations with the possibility of damage (He & Chen, 2019)

3.3 Intensity:

It is the factor which measures the level of damage and danger on the hit areas which means the effects of the earthquake on a specific region and how the inhabitants felt it. Therefore, it is a subjective assessment based on the level of damage felt by the inhabitants so it has no direct relationship with the magnitude of the earthquake. For instance, (Paulay & Priestley, 1992) stated that both 1986 M 5.4 in San Salvador and the 1985 M 7.8 Chilean earthquake were having the same Peak Horizontal Ground Acceleration of 0.7g. Other factors may contribute in increasing the intensity of quake such as intensity of the population, distance form epicenter and focal depth. The depth of the focus affects the intensity of the quake as the shallower it is, the more intense it would be. The depth of the quake was only 23 km in case of Halabjah earthquake so it has increased the intensity of the quake. Density of population around the area is another factor. According to the United Nations Institute for Training and Research (UNITAR- UNOSAT), who took the survey of a population of 79,000,000 in Iran. From that number, only 1,222,000 people reported to have felt moderate shaking, 178,000 people reported to have felt strong shaking, and 6,800 people reported to have felt very strong shaking. While in Iraq they surveyed the population of 36,000,000 form which only 2,669,000 people reported a moderate shaking effects, 582,000 people testified a strong shaking and 110000 people verified a very strong shaking effect. This variation is due to the density of the population in the affected area which does not indicate the level of damage of the buildings because even though more people in Iraq reported very strong shaking, the number of damaged and collapsed buildings were more in Iran which is an indication that intensity cannot be taken as effective factor but the factors affecting it, can be taken.

3.4 Influence of the soil type:

It is another factor contributing to the amount of structural Damage. As it can be seen in Figure 4, the softer the soil is, the stronger the ground acceleration reading will be because of wave reflection and amplifying in soft soil.

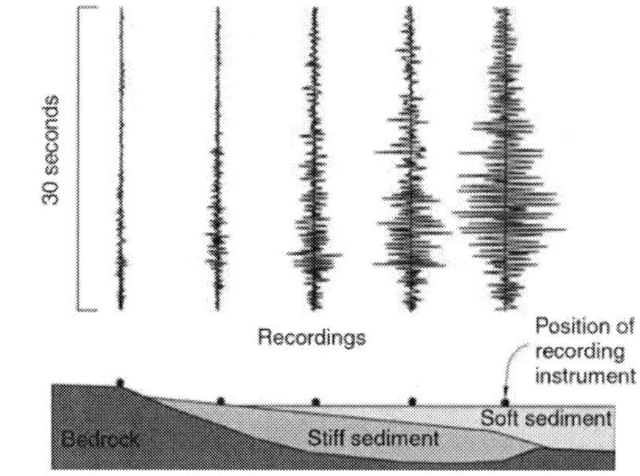

Figure 4. Soil type effects on amplifying the waves (Charleson, 2012).

According to (Roudbari et al., 2020) in the Halabjah earthquake most of the damage occurred in the areas where built on soft soil. For example, two complex residential buildings in Maskan-e Mehr in Sarpol-eZahabare compared namely Karmandi complex which was built on a mountainous area on a foothill with stiff soil or soft rock at 34°28′22″N, 45°50′59″E and ShahidShiroodi complex which was built on a soft soil which was originally farmland and it had a high ground water table at 34°28′1″N, 45°50′39″E (Roudbari et al., 2020). The distance between the two sites is only 1500 m and both sites are approximately 6 to 8 km away from MFF and nearly 45 km away from HZF. After analyzing the damaged buildings, it was concluded that only 4 buildings were damaged on Karmandi complex site while that number was 30 in ShahidShiroodi site. Figure 5 shows the damage consequence that can be observed form a damaged map drawn by (UNOSAT, 2017).

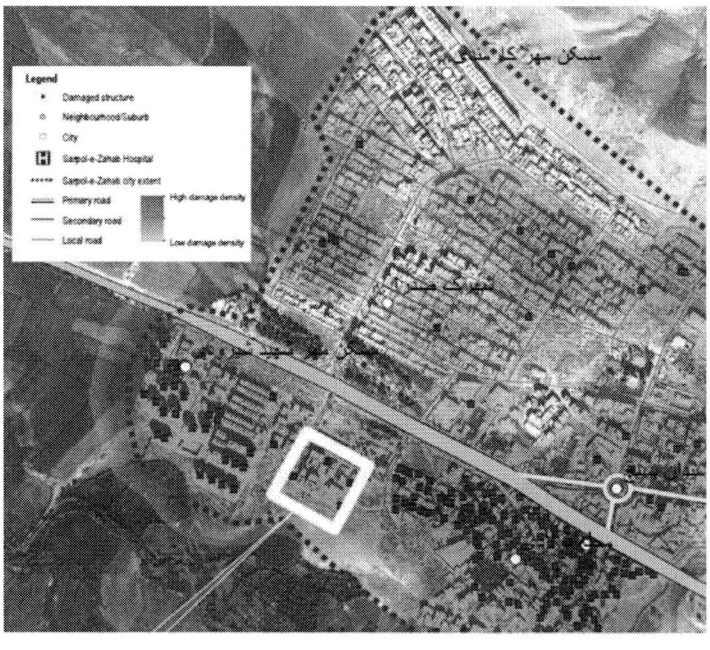

Figure 5. The difference of level of damage due to soil effects in Sarpol-e Zahab (UNOSAT, 2017).

In Kurdistan Region of Iraq, however, generally, a layer of brown clay with lumps of carbonate covers most of the cities. According to the geotechnical conducted by Consulting Engineering Bureau of College of Engineering at Sulaimani University (Rashid, 2008 and Rashid 2012), the mean values of the Unconfined Compression Strength (UCS) for Darbandikhan and Halabja cities were 355.568 kPa and 613.62kPa, respectively. The UCS of Darbandikhan varies from 158.59 kPa to 885.57 kPa while the UCS of Halabja varies from 167.87 kPa to 1024.88 kPa. Table 1 presents the approximate consistencies of clays on the basis of their unconfined compression strength (Das, 2007). According to Table 1, the consistence of brown clay with lumps of carbonates of the mentioned regions varies from stiff to hard clay. As explained in Figure 4, the stiffer the soil is, the less the amplification of the waves would be leading to less damage as a consequence.

Table 1. General relationship of consistency and unconfined compression strength of clays (Das, 2007)

Consistency	Unconfined Compression Strength (q_u)	
	Kg/cm^2	kN/m^2
Very Soft	<0.25	<25
Soft	0.25-0.50	25-50
Medium	0.5-1.0	50-100
Stiff	1.0-2.0	100-200
Very Stiff	2.0-4.0	200-400
Hard	>4.0	>400

3.5 Directionality effects:

Contrary to the popular believe, it is proven that during a fault eruption, most of the energy propagates in one direction most of the times and not in all directions of the epicenter. Rather, it will release its energy in most of the cases unidirectionally. Further, as it can be seen in the Figure 6, the peak ground acceleration at a downstream, station A, is enhanced by the Doppler effects which caused by the interaction of the shock wave with the waves released from the fault rupture along the fault line. This results in high-amplitude and shorter period of earthquake. However, at the upstream, station B, the amplitude of the ground shaking is low with extended period of time.

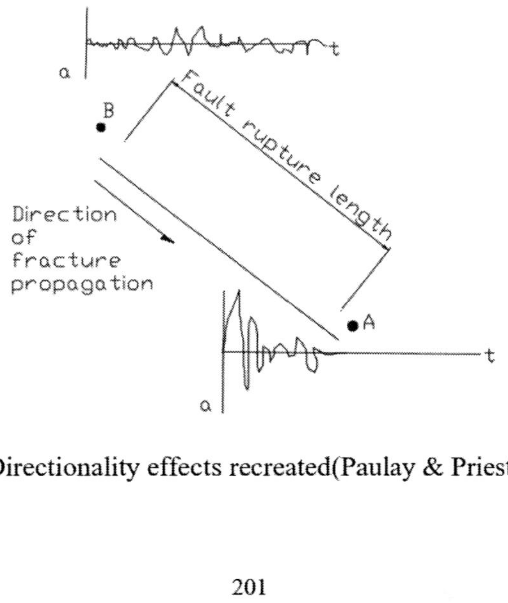

Figure 6. Directionality effects recreated(Paulay & Priestley, 1992)

According to (Roudbari et al., 2020) directionality effects had inputs in amplifying the localized or directional ground motions in Halabjah earthquake as it did and caused the high level of damage in 1994 Northridge earthquake in California and 1995 Kobe earthquake in Japan due to the wave propagation within the site and direction of the fault rupture both of which affected the near fault rupture directivity effects. The damage due to directionality effects would be higher if the weaker axis of the building is aligned with the stronger horizontal motion of the quake.

To understand the propagation of the waves in Halabjah earthquake, a deeper understanding is needed in the way the ground motion has propagated and the direction of the waves released from the focus. For which, some of the researchers have taken the data from different satellites to understand the amount of the ground displacement in each direction. For example, (Zare et al., 2017) took the data from Sentinel-1 satellite in two different dates namely 30/10/ 2017which is before the earthquake and 17 /11/2017which is after it. They used a SNAP software to simulate the motion and as it can be seen from Figure 7, there is an intense propagation of the waves on Arabia plate where most of the structural damage occurred. They used color code to understand the phases each of which represents 2π. They discovered that the Arabia plate has moved nearly a meter while Eurasia plate had moved only 10 cm.

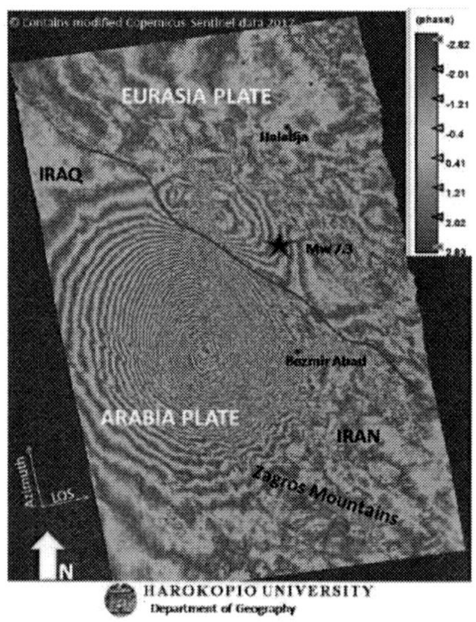

Figure 7. Ground displacement of the Halabjah earthquake based on Synthetic Aperture Radar (SAR) data(Zare et al., 2017)

Moreover, some other researchers of Geospatial Information Authority of Japan (GIAJ, 2017) observed the crustal deformation by using ALOS-2/PALSAR-2 data. They have observed the substantial displacement of nearly half a meter westward, as it can be seen in Figure 8 and more than 90 cm upwards, as it can be seen in Figure9, in approximately 20 km NNW of Sarpol-e Zahab. It can be seen from Figure 10 that the propagation of the ground motion waves was from 25 km depth of the focus towards south-west where most of the structural damages occurred. In Halabjah and Kurdistan Region of Iraq, on the other side, the displacement was nearly zero as they were located at the north of the epicenter as it can be seen from Figure 8 and 9. Therefore, the level of damage was not as severe as it was in Iran.

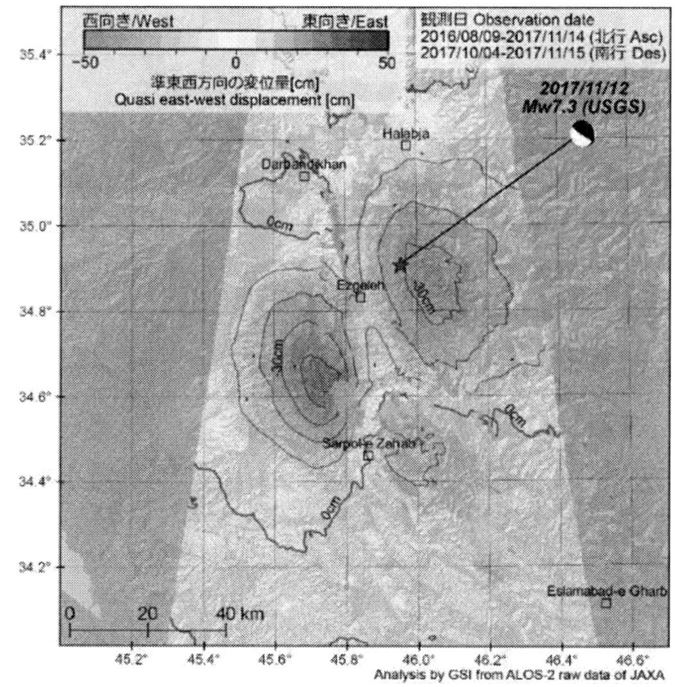

Figure 8. East- West Displacement of Serpol el Zahab (GIAJ, 2017)

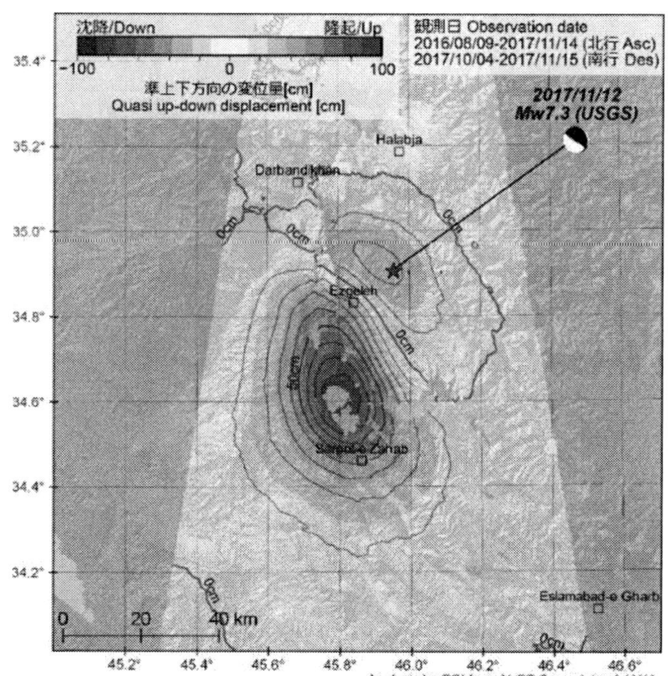

Figure 9. Up-Down Displacement of Serpol el Zahab (GIAJ, 2017)

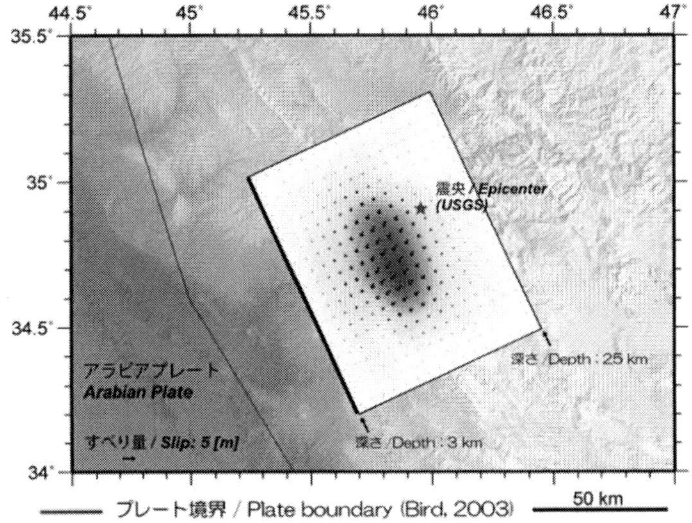

Figure 10. Construction of the faults (Model) (GIAJ, 2017)

3.6 Secondary effects:

It is believed that the movement of MMF in Sarpol-e Zahab region triggered the earthquake. The quake was strong enough to cause huge landslides, rockslide and liquefactionthat caused damage in the direction of the ground motion. For example, in MelaKabod a huge land slide occurred, as shown in Figures 11 and 12, in an area of 4 $km^2$where the maximum horizontal displacement was 34 m and maximum vertical displacement was 10 m(Vajedian et al., 2018).This landslide caused damage to some buildings as reported by (Masakatsu & Setiawan, 2018) observed during their site visit to the location.

Figure 11: Land sliding in MelaKabod(Vajedian et al., 2018)

Figure 12:The scale of the land slide(Vajedian et al., 2018)

However, no land sliding has been reported in Kurdistan region of Iraq due to the quake. Additionally, rock sliding was reported in both area of Kurdistan region of Iraq and Iran without mentioning any damage to it, which is reasonable due to vibration at the instance of the earthquake and lack of friction between the rocks. based on a comprehensive paper on the Darbandikhan dam written by (Yousif et al.,

2019), no damage was attributed to rock falling even though the size of the rocks vary between a few centimeters to 2.5m. Liquefaction is reported to have caused damage and collapse of some buildings in Sarpol-e Zahab due to construction in an area with high level of ground water table in that area and the soil type was originally a farmland (Masakatsu & Setiawan, 2018; Roudbari et al., 2020).

Figure 13 shows the distribution of the collapsed buildings in Darbandikhan which suggests that most of the damage took place in the south and east of the city which is in line with the attenuation of the earthquake wave as they propagated in the direction of south- west. The type of fault slips is thrust faulting. The slip of the fault was only 50 cm towards west which can be considered as the main factor in collapsing the buildings in Darbandikhan while the slip to south was nearly a meter which caused catastrophic damages in Sarpol-e Zahab as it is located to the south of the epicenter.

Figure 13. Collapsed buildings locations in Darbandikhan from satellite view(https://www.bing.com/maps)

3.7 Geographical Effects:

It is true that the response acceleration of structures would be higher with increase of height. Similar effect might be true when it comes to topography of the sites as it is shown in Figure 14 where (Paulay & Priestley, 1992) observed different level of damage for the same location with the same earthquake but different heights. This effect is local, affecting a specific site, therefore, comparison between two cities and ignoring the above factors would not be reasonable.

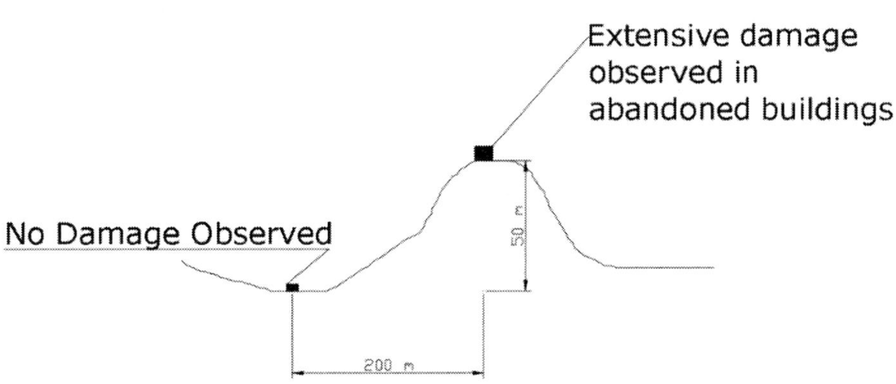

Figure 14: Geographical effects(Paulay & Priestley, 1992)

The altitude of the epicenter was not trivial to find as the epicenter had different locations based on the agencies who calculated them. It was estimated to vary from 763m to 833.5 m. Otherwise, both Darbandikhan and Sarpol-e Zahab have nearly the same altitude of 535m and yet the damages are incomparable.

3.8 Quality of material:

It is another effective factor that causes server injuries and sometimes causalities. Figure 15 shows the relationship between the percentage of damage with quality of material being used in construction(Stein & Wysession, 2009). As it can be seen from the Figure 15, the level of damage of earthquakes of ground acceleration of 0.65 g is nearly a 100% when adobe and brick are used in construction. The ground acceleration of the Halabjah earthquake was nearly 0.7 g and most of the damaged structures in Iran had brick as the main material of the construction.

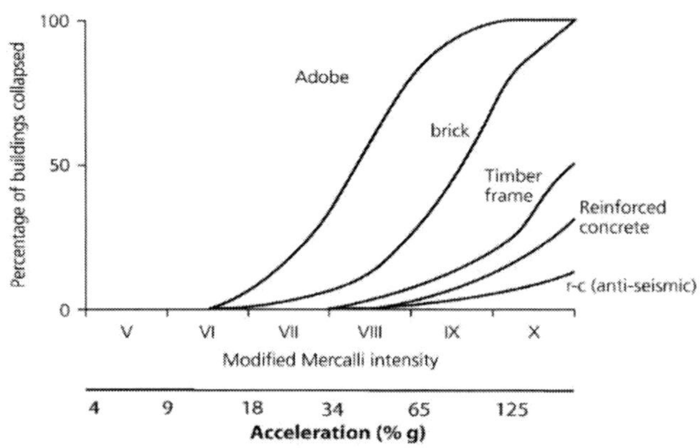

Figure 15: Effects of quality of materials and level of damage(Stein & Wysession, 2009)

The styles of housing and constructing them are different from Kurdistan region of Iraq and Iran. Based on reports by (Derakhshan, 2017) published in World Housing Encyclopedia (WHE), a large part of the buildings (47.5%)which is nearly a half of the buildings in that area werebuilt with a combination between brick and steel section as it can be seen in Figure 16. Of this portion, 67% was in Sarpol-e Zahab which is considered to be the maximum percentage of this type of the building in the area. As discussed previously, the highest fatality rate was also in Sar-Pol-e-Zahab with 387 deaths.

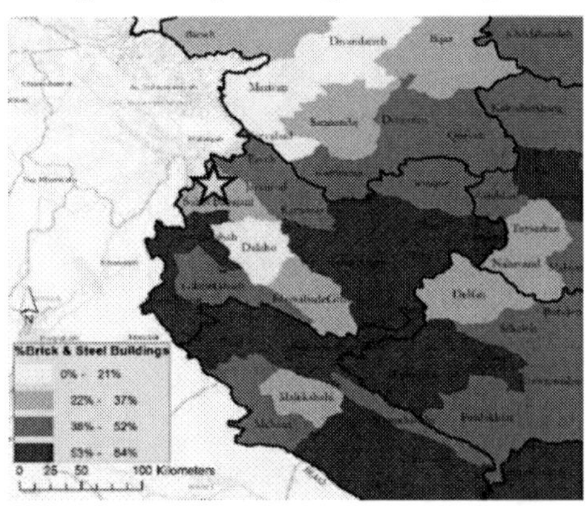

Figure 4. Percentage of Brick and Steel buildings near the epicenter (Derakhshan, 2017)

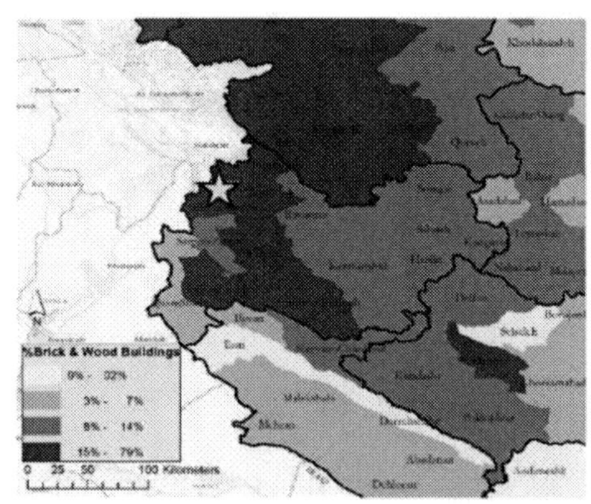

Figure 5. Percentage of brick and wood buildings near epicenter (Derakhshan, 2017)

(Alavi et al., 2018) discussed the reasons of this type of building's failure and mentioned several factors which are improper design of the steel sections, buckling of the compression members, as it can be seen in Figure 18 and improper design and construction in general as it can be seen in Figure 19.

Figure 6: Buckling of the braces

(Alavi et al., 2018)

Figure7: Buckling of the braces

(Alavi et al., 2018)

The other common types of buildings in the area were steel frame buildings shown in Figure 20 with the average of 17%, brick and wood building, shown in Figure 17, with the average of 13.8%, concrete buildings shown in Figure 21 with the average of 8%, and adobe and wood buildings with the average of 7%.

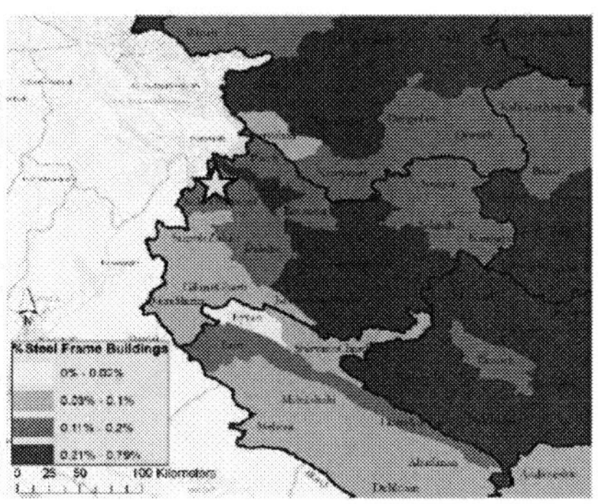

Figure 20. Percentage Steel buildings near the epicenter

(Derakhshan, 2017)

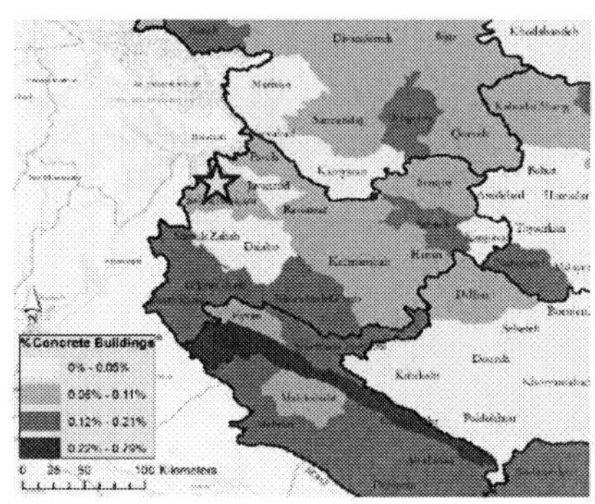

Figure 21. Percentage of Concrete buildings near epicenter

(Derakhshan, 2017)

However, in Kurdistan region of Iraq, most of the residential houses are masonry buildings built as load bearing walls and constructed with concrete block walls with a concrete slab on top. Also, most of the masonry building consist of one floor or two, therefore, they are not tall structures(Yaseen & Ahmed, 2018).A typical masonry building in Kurdistan region of Iraq is shown in Figure 22. In an analysis, (Yaseen & Ahmed, 2018) took an example like the one shown in Figure 22, and analytically tested it for 59 different ground motions. They deduced that this type of building is only susceptible to light and moderate level of damage and no server damage was observed in their investigation even though they interpreted it as a "questionable" phenomenon. This type of building is strong enough to survive the moderate quakes as it mainly consists of concrete which is stronger than brick and adobe. Similar conclusion can be drawn as the fatality rate was very low in the regions of Iran where the concrete buildings rate was high and the steel buildings rate were medium as it can be seen in figure 20 and 21. While the damage rate was very high in the regions where brick and wood and brick with steel were used in construction as in Figures 16 and 17.

Figure 22: Masonry building example in Kurdistan region of Iraq (Yaseen & Ahmed, 2018)

4. Collapse of the buildings in both regions

There were different levels of damage in both Kurdistan regions of Iraq and Iran but when it comes to collapse, some buildings collapsed in Darbandikhan in Kurdistan region of Iraq as did in Iran. The reason of the collapse of these buildings related to structural design, construction and detailing and it is out of the scope of this paper. A collapsed building in Darbandikhan is shown as a satellite image in Figure 23 as in its actual state and the collapsed state. It was an R.C frame of which was three stories high and after the collapse most of the members were did not lose their shapes. However, satellite images from an area in Sarpol-e Zahab is shown in Figure 24 and it looks like an explosion aftermath as brittle materials used in their construction.

(a) (b)

Figure 23: Collapsed R.C frame building in Darbandikhan (35.105141, 45.687696) (a) undamaged building (http://bing.com/maps) (2014) (b) collapsed building (2017)

Figure 24: Collapse of some masonry buildings in Kermanshah, Iran (UNOSAT, 2017)

5. Remarks

Finally, it is worth mentioning that in this paper the major reasons behind the difference of the levels of damage and levels of causality discussed between Kurdistan region of Iraq and Iran based on scientific reasons proved by scientists and expertise in the area of seismology and structural engineering. Only the major reasons were discussed and some of the reasons related structural detailing and analyzing, the design code used in designing the structures and details of their foundations are not discussed in this paper.

Although, the level of damage in Kurdistan region was not as high as it was in Iran, still, the hazards and risks related to earthquakes have to be acknowledged to be prepared for future quakes. Because of this reason, (Issa, 2017) urged the engineers to collaborate on establishing and designing a national code to rely on as a step of preparedness in minimizing the risks attributed to future earthquake loads. Therefore, more thorough investigations needed which cannot be done individually as it takes time, effort and money so there should be a national plan in achieving that goal.

The detail of structural design and the philosophy of design are not discussed in this paper even though they have influences on the destination of a building during an earthquake. However, the paper tried to discuss the main reasons behind that gap between the damage levels in general and the details of the construction and design have not been discussed. Otherwise there might be other factors including the soil, altitude, location, etc. that might have had inputs in the damages. Therefore, there is a strong the need for a thorough investigation emphasizing the mistakes in the design and construction of the buildings that led to their collapse. Based on that, a unified national code can be established for all the designers and site engineers which will represent the reality of the buildings as it would have been established based on data collection and site investigations.

6. Conclusion

The level of damage due to an earthquake varies between a place to another. During Halabjah earthquake, the structural damage and causality rates had a huge gap between Kurdistan region of Iraq and Iran, so this paper investigated the reasons in an attempt to interpret that gap. The reasons can be summarized as:

1- Magnitude, intensity, and vertical component of the ground acceleration of the earthquake had their footprints in echoing the damage level in the buildings in Iran. Because the focus and epicenter of the quake was in Iran, so the impact would be logically higher in the vicinity of the epicenter. The vertical component of the acceleration was as high as the horizontal component which might have added reasons of building damages around the epicenter as it will, generally, be ignored in the design.

2- Stiffness of soil plays a crucial role in the buildings that collapsed as most of which were on farmland soils with high level of ground water tables. The soil stiffness in Kurdistan Region, however, can be regarded as stiff which will not amplify the earthquake waves as the soft soils do.

3- Directionality effects can be considered as the main and most effective reason in creating the huge gap in the level of damage between the two areas of the study as the waves traveled south-west of the epicenter while Kurdistan region of Iraq is located at north of the epicenter. The quake had a lateral displacement of 50 cm towards west and 90 cm to a meter of displacement towards south, the two places with most of the damage. The fault movement of MFF which caused the earthquake is located in Iran and most of the damaged area are located in the direction of the wave attenuation.

4- Secondary effects such as land sliding, liquefaction and rock sliding occurred in Iran adding more reasons to increase the damage.

5- Building material quality can be considered another factor in increasing the impacts of the earthquake in Iran as most of the buildings in that area constructed using adobe and brick which are considered as weakest material in resisting lateral loads. Nearly 48% of all the building in the affected area in Iran are made out of brick with steel which is vulnerable to earthquake loads. However, wall bearing system and hybrid or combination between R.C frame and masonry system are mainly used in Kurdistan region of Iraq which can resist earthquake to a satisfying degree if they build without tolerating principles of earthquake design.

6- The effects of intensity of the quake and geographical location of the buildings were not significant as the former is a subjective decision that changes with intensity of the population and the latter has local effects only.

References

[1] Alavi, E., Mahootchian, A., Yadegari, S., Shamsodin, M., Nouri, M. B., & Ordoubadi, B. (2018). Report of: M7.3 Ezgele, Kermanshah, Iran Earthquake on November 12, 2017. *Earthquake Engineering Research Institute, February*, 1–22. https://www.eeri.org/wp-content/uploads/Kermanshah-Earhquake-Report-E.A

[2] Charleson, A. (2012). Seismic design for architects: Outwitting the quake. In *Seismic Design for Architects: Outwitting the Quake* (Vol. 9780080888). https://doi.org/10.4324/9780080888255

[3] Das, Braja M. (2007), Principles of Foundation Engineering 6th Edition. Thomson, 2007.

[4] Derakhshan, S. (2017). Distribution of building types in the earthquake affected regions of Iran. *EERI*, 2881.

[5] GIAJ. (2017). *Crustal Deformation Observed by Synthetic Aperture Radar (SAR)*. Global Engineers & Technologists Review. http://www.gsi.go.jp/cais/topic171115-index-e.html

[6] IFRC. (2018). Emergency Plan of Action (EPoA) Iraq : Earthquake A . Situation analysis. *International Federation of Red Cross and Red Cresent Societies, December 2017*, 1–11.

[7] Issa, H. (2017). *Earthquakes and building regulations in Kurdistan: a turning point*. Kurdistan 24. https://www.kurdistan24.net/en/news/8d8dfd15-caac-4ff3-b521-8439b19c49fd

[8] Masakatsu, P. M., & Setiawan, H. (2018). *Geotechnical Damage on the 2017 Iran – Iraq Earthquake*.

[9] Paulay, T., & Priestley, M. J. . (1992). *Seismic design of reinforced concrete and masonry building* (T. Paulay & M. J. . Priestley (eds.); first). Wiley interscience publication.

[10] Rashid, Kamal Ahmed (2008), Geotechnical report for Halabja Chemical Diseases Hospital in Halabja city-Halabja Governorate, Unpublished report, Cosulting Engineering Bureau-College of Engineering-University of Sulaimani.

[11] Rashid, Kamal Ahmed (2012), Geotechnical report for Kindergaten Building in Darbandikhan city/Sulaimani Governorate, Unpublished report, Cosulting Engineering Bureau-College of Engineering-University of Sulaimani.

[12] Roudbari, S., Heris, M., Hakhamaneshi, M., & Dashti, S. (2020). Mediating Design Claims: The Social Media and Housing Disaster of the 2017 Halabja Earthquake. *Natural Hazards Review*, *21*(2), 1–10. https://doi.org/10.1061/(ASCE)NH.1527-6996.0000352

[13] Stein, S., & Wysession, M. (2009). *An Introduction to Seismology, Earthquakes, and Earth Structure (Google eBook)*. http://books.google.com/books?hl=en&lr=&id=-z80yrwFsqoC&pgis=1

[14] Taha, B. O., & Hasan, S. A. (2018). A Comparative Study of the Seismic Provisions between Iraqi Seismic Codes 2014 and 1997 for Kurdistan Region/Iraq. *Eurasian Journal of Science and Engineering*, *4*(1), 180–192. https://doi.org/10.23918/eajse.v4i1sip180

[15] UNOSAT. (2017). *Damaged structures and related density map in Sarpol-e-Zahab, Kermanshah, Iran*.

[16] USGS. (2017). *Magnitude 7.3 Earthquake Iran/Iraq Border*. Website. https://www.usgs.gov/news/magnitude-73-earthquake-iraniraq-border

[17] Vajedian, S., Motagh, M., Mousavi, Z., Motaghi, K., Fielding, E. J., Akbari, B., Wetzel, H. U., & Darabi, A. (2018). Coseismic deformation field of the Mw 7.3 12 November 2017 Sarpol-e Zahab (Iran) earthquake: A decoupling horizon in the Northern Zagros Mountains inferred from InSAR observations. *Remote Sensing*, *10*(10). https://doi.org/10.3390/rs10101589

[18] Yaseen, A. A., & Ahmed, M. S. (2018). Seismic Vulnerability Assessment of Single-Story Masonry Buildings in Kurdistan Region - Iraq. *Academic Journal of Nawroz University*, *7*(4), 18.

https://doi.org/10.25007/ajnu.v7n4a267

[19] Yekrangnia, M., Eghbali, M., Seyri, H., Panahi, M., Zanganeh, S. Y., Beyti, M., Hayatgheybi, D. S. V., Nazarpour, M., & Amiri, G. G. (2017). *A preliminary report on school buildings performance during M 7.3 Ezgeleh, Iran earthquake of Participants Project Team Steering Committee Advisory Committee. November.* https://www.eeri.org/wp-content/uploads/Report-Kermanshah-DRES-Ver2-2.pdf

[20] Yousif, O. S. Q., Zaidn, K., Alshkane, Y., Khani, A., & Hama, S. (2019). *Performance of Darbandikhan Dam during a major earthquake on November 12, 2017. May.*

[21] Zare, M., Kamranzad, F., Parcharidis, I., & Tsironi, V. (2017). Preliminary report of Mw7. 3 Sarpol-e Zahab, Iran earthquake on November 12, 2017. *EMSC Report, 2017*(Fig 4), 1–10.

The Hydraulic Jump Formed Downstream a Stepped Gabion Weir: An Experimental Study

Ali Mekki Al-Fawzy[1,*], Kadhim Naief Al-Taee[2], Fadhil Mohammed Al-Mohammed[3], and Ali Hassan Hommadi[4]

[1]Directorate of Water Resources in Kerbala City, State Commission on Operation of Irrigation and Drainage Projects, Ministry of Water Resources, Iraq
E-mail:ali_alfawzy85@yahoo.com
[2]College of Engineering, University of Babylon, Iraq
E-mail: altaee_kadhim@yahoo.com
[3]Kerbala Technical Institute, Al-Furat Al-Awsat Technical University, Iraq
E-mail: fmhmjme@yahoo.com
[4]Al-Hindyiah Barrage Project -Babylon, State Commission of Dams, Ministry of Water Resources, Iraq - E-mail: alihassan197950@yahoo.com
* Corresponding author: Directorate of Water Resources in Kerbala City, State Commission on Operation of Irrigation and Drainage Projects, Ministry of Water Resources, Iraq. Tel: +9647723715510.

Abstract. The current study focus on the effect of using stepped gabion weir in a laboratory channel on the hydraulic jump distance which form at its toe. A series of 175 operation tests and 25 laboratory experiments were conducted by using a laboratory flume 10 m long, 0.3 m wide, and 0.5 m deep. The tested gabions had different five lengths 0.72 m, 0.84 m, 0.96 m, 1.08 m, and 1.20 m respectively, and the material used to fill the gabions was natural quarry mono graded gravel in five different sample sizes of diameters ranged between (09.5-14.0) mm, (14.0-19.0) mm, (19.0-25.0) mm, (25.0-37.5) mm, and (37.5-50.0) mm. The operation flow rate values ranged between $2.33*10^{-3}$ to $50.00*10^{-3}$ $m^3/s/m$. Dimensional analysis was used to generate dimensionless parameters, and correlated them using the Buckingham Pi-Theorem. The results of this study showed that the hydraulic jump distance increases by increasing the flow rate value, but increasing the values of both of the gravel sample used and the total length of the weir have an undular effect on the hydraulic jump distance.

Keywords: *dimensional analysis; gabions; hydraulic jump; hydraulic models; stepped gabion weirs.*

1. Introduction

Occurrence of hydraulic jumps in open channels take many forms according to the nature of channel, the tested hydraulic structure, and the target study. Generally, this phenomenon can be defined as a transformation of flow state from high-velocity, super-critical flow to low-velocity, sub-critical flow (Abady and Akhtari, 2018; Abbas *et al.* 2018; Abbaspour *et al.* 2013; Abbaspour *et al.* 2019; Al-Fawzy *et al.* 2020; Anandraj, 2012; De Padova *et al.* 2018; Fisher *et al.* 2018; Gavhane *et al.* 2017; Ghose *et al.* 2019; Jafar, 2016; Maatooq *et al.* 2013; Movahed *et al.* 2018; Palermo and Pagliara, 2018; Torkamanzad *et al.* 2019; and Yousefi *et al.* 2018). The hydraulic jump characteristics have been studied to explain and comprehend a lot of problems related to the management of water resources. (Abady and Akhtari, 2018) performed a numerical simulation of the hydraulic jump in divergence rectangular sections, and compared the laboratory results with that made using Flow-3D program and some standard turbulence models to evaluate the effect of vertical and curved blocks on the specification of hydraulic jump, whereas the study results showed that creating of vertical blocks reduces the length of the jump as much as 17.64%, and in case of curved blocks the jump reduces to 35.29%. (Habib, 2013) carried out an experimental study to investigate the effect of relative pipe diameter on the characteristics of the free hydraulic jump that formed downstream a weir controlled by a sluice gate, by comparing the hydraulic characteristics of the jump using a weir with circular openings and weir without openings, and using the dimensional analysis for developing theoretical model to predict the relative depth of water and relative energy loss of the hydraulic jump. (Maatooq *et al.*2013) proposed a treatment for the hydraulic jump formation in Diyala weir structure, whereas the scour occurs due to the position of the jump because its sequent depth is higher than exist tail water depth, by present a suitable stilling basin or use of weir with low height at the end of the basin to ensure stability of the jump. While (Jafar, 2016) used a flow partitioning structure, FPS, downstream a sluice gate to raise the tail water level to move the jump into a closer distance from the gate's outlet. He found that using this structure increases the tail water level due friction resistance, and force the hydraulic jump back toward the gate. (Fisher *et al.* 2018) tried to control the hydraulic jump using the electro-magnetic force in a liquid metal flow, whereas this force could be generated using electrical currents and magnetic fields injected externally to control the jump. They found that applying this force has a repeated and predicted impact on the hydraulic jump, and could be useful in solving future practical problems related to it. (De Padova *et al.* 2018) modeled the smooth particle hydrodynamics of the hydraulic jump at an abrupt drop. The results of this modeling showed that particularly both of the vorticity, the turbulent kinetic energy fields, the velocity of flow, depths of water, and pressure spectra downstream of the jump can computed and analyzed. (Abbas *et al.* 2018) investigated the characteristics of the hydraulic jump on smooth adverse slope in three different states without appurtenances, and they illustrated that the jump length ratio was reduced by 22.1%. (Yousefi *et al.* 2018) tried to estimate the hydraulic jump length in a rectangular channel, and propose roughness for the used stilling basin. From their study they found that all the upstream and downstream water depths, upstream Froude number, and bed roughness are independent variables affect the hydraulic jump formation, and accordingly the jump length reduces as the stilling basin roughness increases. Also, (Ghose *et al.* 2019) tested the hydraulic jump formation in a laboratory flume has inclination, and they illustrated that the jump length dimensionless parameter depend significantly on the bad of the channel. (Abbaspour *et al.* 2019) investigated the operation of porous screens in super-critical flow condition experimentally, by installing them on two reverse slopes to evaluate the losses of energy of flow. Their investigation showed that the distance of screen from the hydraulic jump toe has insignificant effect on the energy dissipation. (Ljubičić *et al.* 2018) proposed the use of reverse slope stilling basin with stepped chutes instead of smooth chutes to study the characteristics of flow aeration, energy dissipation and the hydraulic jump, whereas they found that the proposed state has a better agreement with the experimental data, as compared with standard state, and useful in primary design. (Jalil, 2018) modeled the hydraulic jump that generated partially on a sloped apron, using multi-slope bed to generate the hydraulic jump at different locations by controlling the tail water depth. Results of study showed that the energy exchange process depends on the Froude number and the location of the hydraulic jump. The main purpose of this study is to investigate the effect of use the stepped shape of gabion weir on the distance of hydraulic jump experimentally.

2. The laboratory work

The experimental works conducted at the hydraulic laboratory of College of Engineering, University of Babylon in Iraq. The laboratory has a tilting flume of 10 m length, 0.3 m width and 0.5 m depth. The bed of flume was fabricated from iron plates and the flume side walls were made of anti-crush glass supported by stainless steel bars. The tested physical models have total lengths of 0.72, 0.84, 0.96, 1.08, and 1.20 m, and these models are assigned as SGW.A, (Stepped Gabion Weir A), SGW.B, SGW.C, SGW.D, and SGW.E respectively. Figure 1 shows the sketch of general shape of the tested gabion weirs. All physical models have constant cross-section width of 0.3 m and maximum weir height of 0.4 m. Table 1 shows the dimensions of the models used in this study, whereas h1, h2, and h3 present the effective distance of step height, L1, L2, and L3 are divided according to maximum weir length which is equal to 1.20 m; where L1 has a percentage of 33.3% of the maximum weir length, L2 has a percentage of 16.7% of the maximum weir length, and L3 has a percentage of 10% of the maximum weir length, for SGW.A, and the percentage of L3 increases accumulatively by 10% of maximum weir length with every physical model tested according to their alphabetical arrangement above. All the tested models were installed at distance of 4 m from the beginning of the flume. The gravel samples used as filling material for the weir models were five mono-sized gravel samples with diameters ranged between (9.5-14.0), (14.0-19.0), (19.0-25.0), (25.0-37.5), and (37.5-50.0) mm, and numbered as GS.I, (gravel sample number I), GS.II, GS.III, GS.IV, and GS.V respectively. The frame of the weirs was made of thin steel plate bars, covered by a wire mesh, and fixed inside the flume by silicone glue. A photo of SGW.B with GS.III is shown in Figure 2. A centrifugal pump having a capacity of 40 l/s was used to deliver flow to the flume. Two movable carriages with point gages were mounted on brass rail at the top of flume sides, which have accuracy of 0.1mm to measure the depths of water. The first was located at the upstream side of the weir to measure the upstream water depth at equal distances starting from 0.0 m to 1.0 m before the weir during the single test run. The second was at the downstream side of the weir to measure the downstream water depths before and after the hydraulic jump location during the single test run. The total number of test runs was 175, and their flow rate values were varied between $2.33*10^{-3}$ - $50.00*10^{-3}$ m3/s/m.

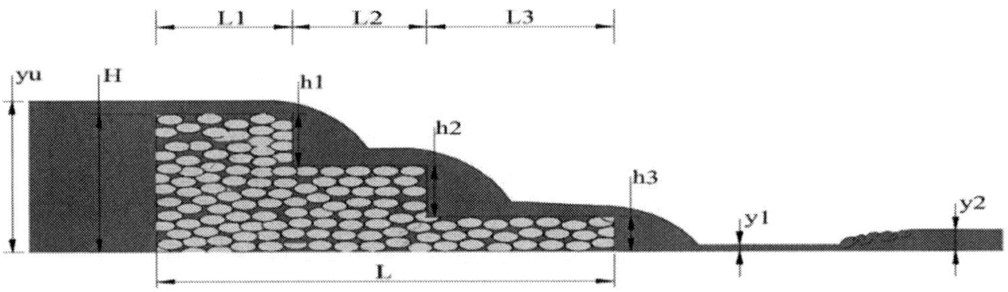

Figure 1. The sketch of the general shape stepped gravel gabion weir tested in this study.

Table 1. Details of step dimensions for all steps of the tested gabion weir.

Stepped gabion weir	Step dimensions					
	1st step dimensions		2nd step dimensions		3rd step dimensions	
	h1	L1	h2	L2	h3	L3
	cm	cm	cm	cm	cm	cm
A	15	40	15	20	10	12
B	15	40	15	20	10	24
C	15	40	15	20	10	36
D	15	40	15	20	10	48
E	15	40	15	20	10	60

Figure 2. Stepped gabion weir B with gravel sample number III.

3. Theoretical background and dimensional analysis

Generally, the relationships which combine multi-components can be represented in many different ways as the direct relations or (x-y) relation, empirical equations, standard formulas, or correlated by using the dimensional analysis (e.g. Abbas et al. 2018; Abbaspour et al. 2013; Abbaspour et al. 2019; Abozeid et al. 2010; Al-Fawzy et al. 2020; Anandraj, 2012; Dhahir and Mohammed-Hasan, 2017; Habib, 2013; Jafar, 2016; Jalut and El-Baaja, 2014; Ljubičić *et al.* 2018; Movahed *et al.* 2018; Obaed *et al.* 2016; and Sharif and Kabiri-Samani, 2018). In this study, the geometric and hydraulic components of the stepped gabion weir upon which the hydraulic jump distance may depend can by as:

$$DHJ = f_1\ (q, d_G, L_3, L, \rho_w, g) \tag{1}$$

whereas DHJ is the hydraulic jump distance formed at the weir toe (m), q is the unit discharge or flow rate, ($m^3/s/m$), d_G is the possible diameter of the used gravel sample (mm), L_3 is the third step length of the weir (m), L is the total length of the weir (m), ρ_w is water mass density (kg/m^3), and g is the gravitational acceleration (m/s^2).

The Buckingham Pi – Theorem is an effective method used for representing of relationships of components functionally in dimensionless form (e.g. Abbas et al. 2018; Abbaspour et al. 2013; Abbaspour et al. 2019; Abozeid et al. 2010; Al-Fawzy et al. 2020; Dhahir and Mohammed-Hasan, 2017;

IOP Conf. Series: Materials Science and Engineering **978** (2020) 012018 doi:10.1088/1757-899X/978/1/012018

Habib, 2013; Jafar, 2016; Movahed et al. 2018; Sharif and Kabiri-Samani, 2018; Torkamanzad et al. 2019; and Yousefi et al. 2018). Thus, the general equation of the hydraulic jump by the other components, within the limitations of this study, can be expressed dimensionally as follows:

$$\frac{\text{DHJ}}{\text{Chosen component}} = f_1\left(\frac{\text{Component 1}}{\text{The same chosen component}}, \frac{\text{Component 2}}{\text{The same chosen component}}, \frac{\text{Component 3}}{\text{The same chosen component}}, \ldots\ldots, \text{etc}\right) \quad (2)$$

Where the chosen component is one of the components in the right side of equation (1), that chosen for specific study, and components 1, 2, 3, …. , etc are other components on the same side of the chosen one.

4. Results and discussion

4.1 Effect of discharge on the hydraulic jump distance

The relationship between the discharge and the distance of the hydraulic jump has been represented by using of the direct mode, the x-y relationship, (Al-Fawzy *et al.* 2020), (Jafar, 2016), (Jalut and El-Baaja, 2014), and (Obaed *et al.* 2016). The figures 3, 4, 5, 6, and 7 show the drawn relationship between the discharge and the hydraulic jump distance.

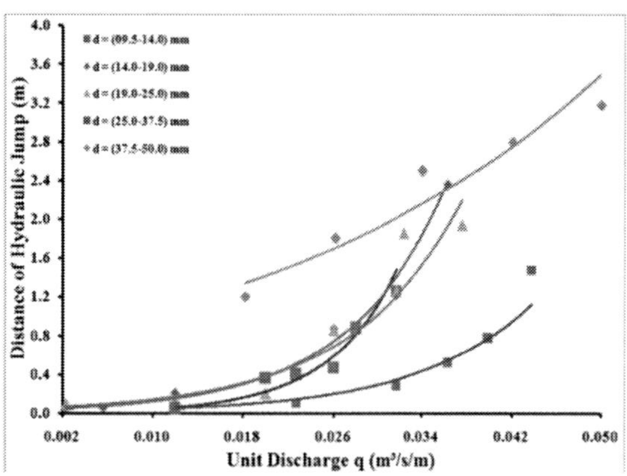

Figure 3. The discharge-hydraulic jump distance relationships for all gravel samples used with SGW.A.

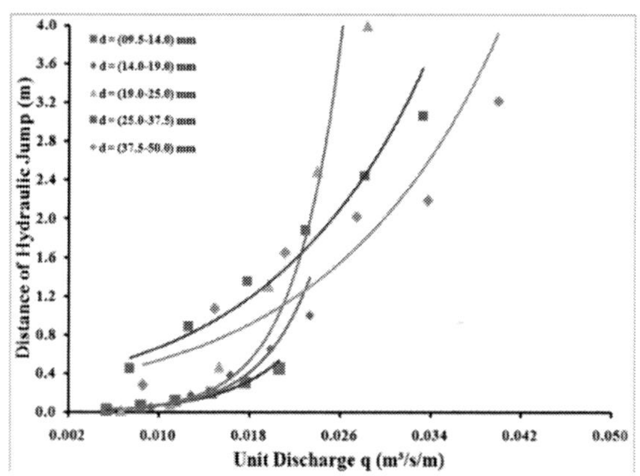

Figure 4. The discharge-hydraulic jump distance relationships for all gravel samples used with SGW.B.

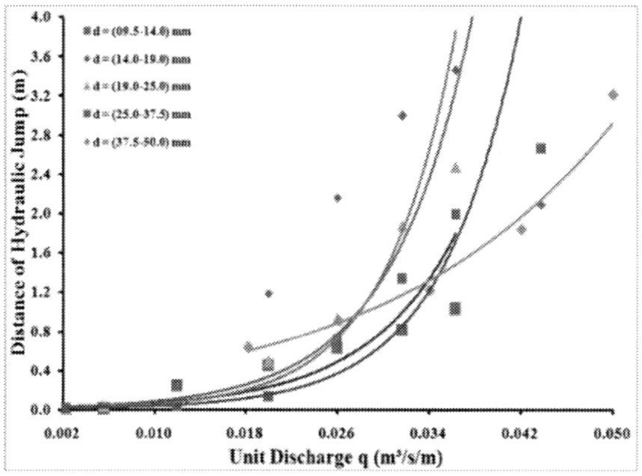

Figure 5. The discharge-hydraulic jump distance relationships for all gravel samples used with SGW.C.

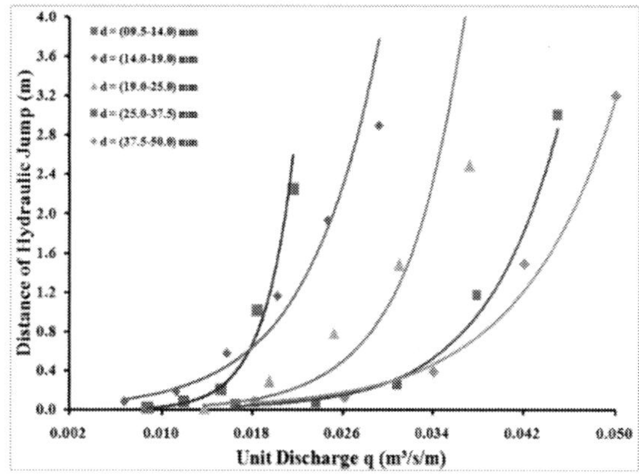

Figure 6. The discharge-hydraulic jump distance relationships for all gravel samples used with SGW.D.

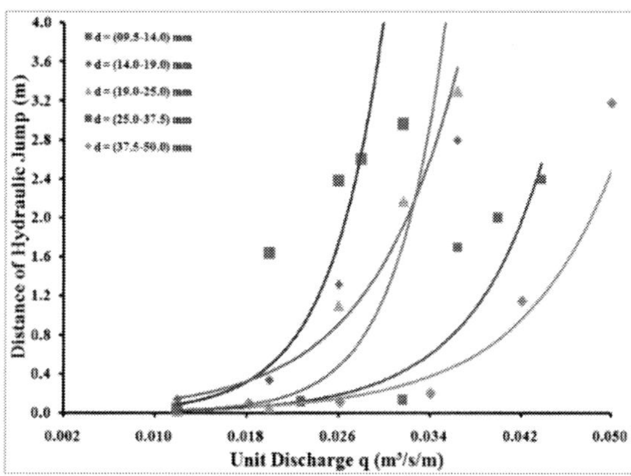

Figure 7. The discharge-hydraulic jump distance relationships for all gravel samples used with SGW.E.

In these figures, the equation of trend lines that represent this relationship varies between the exponential and power forms.

$$DHJ = c_1 e^{c_2\ q} \tag{3}$$

$$DHJ = c_3\ q^{c_4} \tag{4}$$

whereas c_1, c_2, c_3 and c_4 are constants, and table 2 presents their values. The larger value of discharge listed in table 2 represent the maximum, and safe value of discharge recorded for every series of test runs of a specified physical model and gravel sample, taking into consideration prevention of water from pass over the flume sided walls, and keeping the hydraulic jump at a safe distance about 0.75 m from the flume end. Generally, for the figures 3, 4, 5, 6, and 7 the hydraulic jump distance increases as the discharge value increases, and the proportion is direct.

Table 2. Values of constants c_1 and c_2 in equation (3), and c_3 and c_4 in equation (4).

SGW	GS	q (m³/s/m)		DHJ (m)		c_1	c_2	c_3	c_4	R^2
		From	To	From	To					
A	I	0.01200	0.03168	0.05	1.26	-	-	11320.00	3.312	0.969
	II	0.00233	0.03633	0.05	2.36	0.0390	113.0	-	-	0.984
	III	0.00233	0.03767	0.09	1.94	0.0500	100.4	-	-	0.924
	IV	0.01200	0.04380	0.07	1.84	0.0160	097.0	-	-	0.953
	V	0.01823	0.05000	1.20	3.80	-	-	63.35	0.976	0.981
B	I	0.00540	0.02067	0.03	0.45	-	-	1760.00	2.145	0.998
	II	0.00583	0.02333	0.02	1.00	-	-	65202.00	2.947	0.988
	III	0.00670	0.02833	0.02	4.00	-	-	4E+06	3.815	0.992
	IV	0.00750	0.03333	0.45	3.07	-	-	240.50	1.284	0.999
	V	0.00860	0.04000	0.28	3.22	-	-	384.80	1.466	0.938
C	I	0.00567	0.03633	0.01	1.03	-	-	3352.00	2.346	0.912
	II	0.00233	0.04380	0.01	2.09	-	-	3225.00	2.140	0.908
	III	0.00233	0.03633	0.01	2.48	0.0011	160.7	-	-	0.962
	IV	0.00233	0.04380	0.01	2.66	0.0070	149.4	-	-	0.954
	V	0.01823	0.05000	0.63	3.21	0.2420	49.86	-	-	0.985
D	I	0.00877	0.02167	0.02	2.24	0.0001	137.5	-	-	0.991
	II	0.00680	0.02917	0.09	2.90	-	-	18394.00	2.490	0.984
	III	0.01377	0.03727	0.02	2.49	-	-	2E+07	4.739	0.947
	IV	0.01657	0.04500	0.05	3.00	0.0020	154.7	-	-	0.970
	V	0.01823	0.05000	0.09	3.20	0.0070	120.6	-	-	0.973
E	I	0.01200	0.03168	0.04	2.96	-	-	3E+07	4.495	0.894
	II	0.01200	0.03633	0.14	2.80	0.0029	131.5	-	-	0.961
	III	0.01200	0.03633	0.02	3.30	0.0010	230.1	-	-	0.931
	IV	0.01200	0.04380	0.03	2.40	0.0030	148.3	-	-	0.877
	V	0.01823	0.05000	0.09	3.18	0.0060	119.0	-	-	0.920

4.2 Effect of the gravel sample on the hydraulic jump distance

The figures 3-7, in each figure it's obvious that there is no clear effect of increasing the diameter of the used gravel sample on the hydraulic jump distance at low values of discharge. This leads to the possibility of use any available material within the limitations of this study in field works, whereas the water flow through the pores of gravel particles in laminar state, and there is no formation of hydraulic jump in most of case. However, increasing of discharge value, this behavior transforms to an undular effect on the distance of the hydraulic jump. Also, using the dimensional analysis after combining the data of the figures 3-7, and draw the relationship between the parameters ($\frac{DHJ}{d_G}$) and ($\frac{q}{g^{0.5} * d_G^{1.5}}$) as shown in figure 8, where every single trend line represent a gravel sample used with all physical models, gives the same result of undulation. From figure 8 it is clear that the smallest gravel samples may be useful in detention of water, but it pushes the hydraulic jump away from the weir toe, while the largest samples of gravel do the opposite behavior by making the hydraulic jump come closer, whereas it makes the weir downstream portion behave as a stilling basin and reduce the length of the formed hydraulic jump. Furthermore, the accuracy of equation of trend line reduces with increasing the diameter of the used gravel sample due dispersion in their data, and the equation of these trend lines varies between the exponential and power forms too. Equations (5) and (6) may be used for design purposes.

$$\frac{DHJ}{d_G} = c_5 e^{c_6\left(\frac{q}{g^{0.5} * d_G^{1.5}}\right)} \tag{5}$$

$$\frac{DHJ}{d_G} = c_7\left(\frac{q}{g^{0.5} * d_G^{1.5}}\right)^{C_8} \tag{6}$$

Whereas c_5, c_6, c_7 and c_8 are constants, and table 3 presents their values.

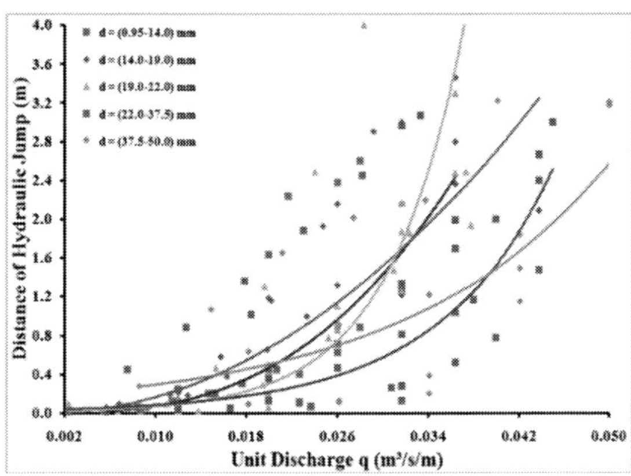

Figure 8. The gravel sample-hydraulic jump distance non-dimensional relationship.

Table 3. Values of constants c_5 and c_6 in equation (5), and c_7 and c_8 in equation (6).

SGW	GS	c_5	c_6	c_7	c_8	R^2
All	I	-	-	23567	2.769	0.810
All	II	-	-	01718	2.005	0.858
All	III	0.014	151.30	-	-	0.781
All	IV	0.029	098.84	-	-	0.518
All	V	0.169	054.47	-	-	0.318

4.3 Effect of the gabion total length on the hydraulic jump distance

To view the effect of weir length on the hydraulic jump distance, two values of applied discharges, minimum $2.33*10^{-3}$ $m^3/s/m$ and $12.00*10^{-3}$ $m^3/s/m$, were used for comparison purposes as shown in table 4. From this table it's clear that the gabion total length has slight effect on the hydraulic jump distance, and this result was verified by using of dimensional analysis to draw the relationship between the parameters ($\frac{DHJ}{L}$) and ($\frac{q}{g^{0.5} * L^{1.5}}$) as shown in figure 9. Whereas the smallest lengths may be beneficial in dissipating the energy of flow more than reducing the hydraulic jump length, while the largest values of L_3 works as a part of stilling basin. Furthermore, the accuracy of trend line equation reduces with increasing diameter of the used gravel sample due to the clear dispersion of data, and the equation of these trend lines varied between the exponential and power forms too. Equations (5) and (6) probably used for design purposes.

$$\frac{DHJ}{L} = c_9 e^{c_{10}\left(\frac{q}{g^{0.5} * L^{1.5}}\right)} \qquad (7)$$

$$\frac{DHJ}{L} = c_{11}\left(\frac{q}{g^{0.5} * L^{1.5}}\right)^{c_{12}} \qquad (8)$$

Whereas c_9, c_{10}, c_{11} and c_{12} are constants, and table 5 shows their values.

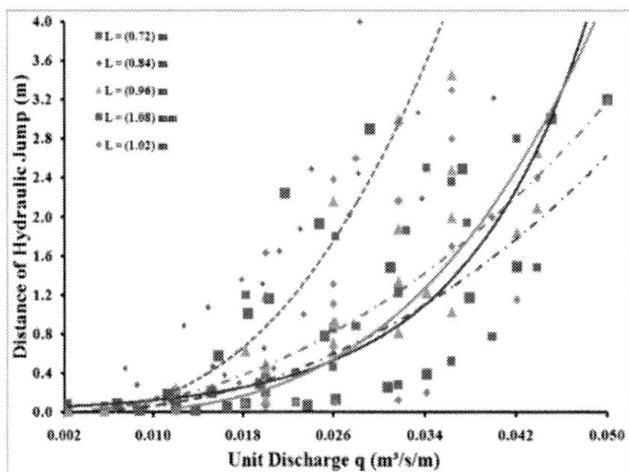

Figure 9. The gabion length-hydraulic jump distance non-dimensional relationship.

Table 4. Values of distance of hydraulic jump at $2.33*10^{-3}$ and $12.00*10^{-3}$ $m^3/s/m$ applied discharges.

GS	Applied discharges * 10^{-3} ($m^3/s/m$)	Distance of hydraulic jump (m)				
		Physical models				
		SGW.A	SGW.B	SGW.C	SGW.D	SGW.E
I	2.33	0.00	0.00	0.00	0.00	0.00
	12.00	0.20	0.12	0.26	0.43	0.11
II	2.33	0.00	0.00	0.00	0.00	0.00
	12.00	0.17	0.19	0.32	0.60	0.05
III	2.33	0.00	0.00	0.00	0.00	0.00
	12.00	0.08	0.48	0.07	0.22	0.12
IV	2.33	0.00	0.00	0.00	0.00	0.00
	12.00	0.09	0.18	0.09	0.16	0.06
V	2.33	0.00	0.00	0.00	0.00	0.00
	12.00	0.09	0.07	0.05	0.03	0.00

Table 5. Values of constants c_9 and c_{10} in equation (7), and c_{11} and c_{12} in equation (8).

SGW	GS	c_9	c_{10}	c_{11}	c_{12}	R^2
A	All	0.049	91.12	-	-	0.722
B	All	-	-	26747	2.638	0.806
C	All	-	-	01592	2.076	0.886
D	All	-	-	02272	2.257	0.517
E	All	-	-	53696	3.149	0.599

Using multi-linear regression for the data set to correlate the dependent and independent components (e.g. Abbas *et al.* 2018; Al-Fawzy *et al.* 2020; Dhahir and Mohammed-Hasan, 2017; Jalil, 2018; Sharif and Kabiri-Samani, 2018; Torkamanzad *et al.* 2019; and Yousefi *et al.* 2018), 60% of data were used in regression process, (data recorded of test runs of the 1st, 3rd, and 5th total lengths of the weir), and 40% of data were used in verification of the resulted equation, (data recorded of test runs of 2nd and 4th total lengths of the weir). After many trials, equation (2) can be re-written as,

$$\left[\frac{DHJ}{L}\right]^{0.005} = (0.8267) + 0.1668 * \left[\frac{q}{g^{0.5} * L^{1.5}}\right]^{0.1} + 0.0716 * \left(\frac{L_3}{L}\right)^{0.05} + 3044.4112 * \left(\frac{d_G}{L}\right)^{5} \qquad R^2 = 0.67 \tag{9}$$

The figures 10 and 11 show the relationship between the calculated and the measured hydraulic jump distance (e.g. Abbaspour *et al.* 2013; Abozeid *et al.* 2010; Ljubičić et al. 2018; Movahed *et al.* 2018; Obaed *et al.* 2016; Palermo and Pagliara, 2018; Torkamanzad *et al.* 2019; and Yousefi *et al.* 2018). Equations (10) and (11) express this relationship,

$$DHJ_{Calculated} = 0.467\,(DHJ_{Measured})^{0.61} \qquad R^2 = 0.82 \tag{10}$$

$$DHJ_{Calculated} = 0.682\,(DHJ_{Measured}) + 0.232 \qquad R^2 = 0.73 \tag{11}$$

According to these figures, an acceptable agreement has been achieved between the values. Both values of hydraulic jump distance, calculated and measured, has a percentage of errors, that was calculated for this study by equation (12) with an average value of -49.1%. May be this value becomes better and more accurate if a separation has been made between the data of through flow and the over flow.

$$E\% = \frac{(DHJ_{Calculated} - DHJ_{Measured})}{DHJ_{Calculated}} \tag{12}$$

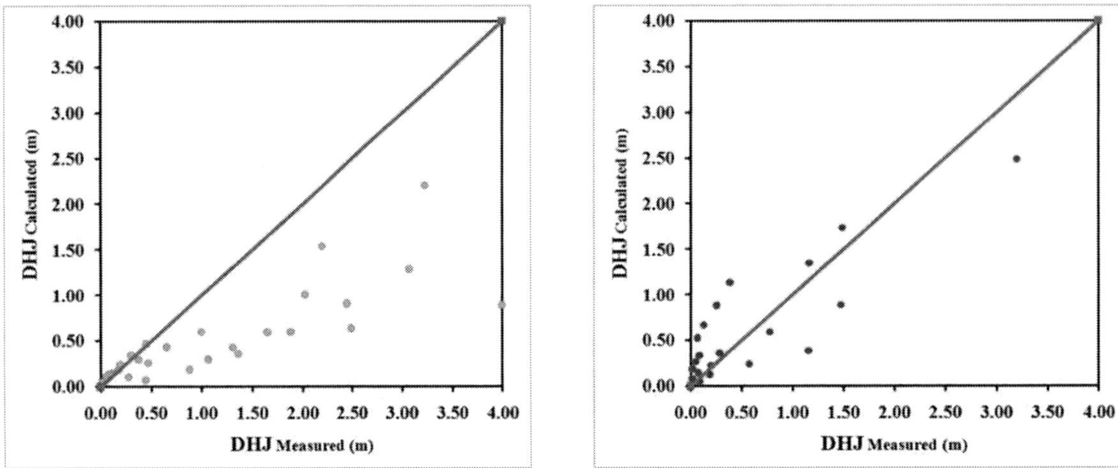

Figures 10 and 11. The relation between the calculated and measured values of distance of hydraulic jump for the 2nd length, (on the right side), and the 4th length of, (on the left side) of the gabion

5. Conclusions

The current study aimed to investigate how the distance of hydraulic jump affected by placing a stepped gabion weir of three steps in a rectangular Laboratory channel. Relationships between components were made either directly or in non-dimensions' style, while the regression process was used to correlate all components. Within the limitations of this study, it can have concluded that:

- The hydraulic jump distance increases by increasing the discharge.
- The hydraulic jump distance behaves undular with increasing both the diameter of the used gravel sample and the total length of the weir.
- The formulated dimensionless equation to combine all components of the current study has R^2 equal to 0.67.
- The calculated-measured relationship of the hydraulic jump distance has been drawn and the fitness equations were with R^2 equal to 0.82 and 0.73 respectively.
- The percentage of errors was -49.1% for the relationship between the calculated and measured distance of hydraulic jump.

Acknowledgements

A great full thanks and gratitude for all people who have a share in supporting and completing this work, and specially for the technical team, and their leader in fluid laboratory of college of engineering in Babylon university.

References

[1] Abady, Kh., & Akhtari, A. (2018). Effect of vertical and curved blocks on hydraulic jump characteristic in diverging rectangular sections with Flow-3D software. *Modares Civil Engineering Journal*, 17, 6, 269-279.

[2] Abbas, A. S., Alwash, H., & Mahmood, A. A. (2018). Experimental study of hydraulic jump in adverse stilling basin at smooth bed.*Diyala Journal of Engineering Sciences*, 11, 3, 7-13.

[3] Abbaspour, A., Farsadizadeh, D., & Ghorbani, M. A. (2013). Estimation of hydraulic jump on corrugated bed using artificial neural networks and genetic programming. *Water Science and Engineering Journal*, 6, 2, 189-198.

[4] Abbaspour, A., Taghavianpour, T., & Arvanaghi, H. (2019). Experimental study of the hydraulic jump on reverse bed with porous screens. *Applied Water Science*,https://doi.org/10.1007/s13201-019-1032-7

[5] Abozeid, G., Mohamed, H. I., & Shehata, S. M. (2010). Hydraulics of clear overfall weirs with bottom openings. *Journal of Engineering Sciences, Assiut University*, 38, 1, 19-28.

[6] Al-Fawzy, A. M., Al-Mohammed, F. M., & Alwan, H. H. (2020). Energy dissipation in Gabion Weirs. *IOP Conference Series: Material Sciences and* Engineering, 671, 012068, 1-12.

[7] Anandraj, A. (2012). Investigational study on self aeration characteristics of hydraulic jump. *Journal of Mechanical and Civil Engineering*, 4, 2, 27-31.

[8] De Padova, D., Mossa, M., & Sibilla, S. (2018). SPH numerical investigation of the characteristics of an oscillating hydraulic jump at an abrupt drop. *journal of hydrodynamics*, 30, 1, 106-113.

[9] Dhahir, F. M., & Mohammed-Hasan, A. M. (2017). Using of hydraulic jump characteristics as criterion for energy dissipation in rectangular gabion weir. *International Journal of Advanced Research*, 5, 7, 268-276.

[10] Fisher, A. E., Kolemen, E., &Hvasta, M. G. (2018). Experimental demonstration of hydraulic jump control in liquid metal channel flow using Lorentz force. *Physics and Fluids Journal*, 30, 6, 067104, 1-9.

[11] Gavhane, A. T., Sathe, N. J., Hinge, G. A. & Jain, S. S. (2017). Studies on design of hydraulic jump type stilling basin in laboratory for Gunjawani dam, Maharashtra, India. *journal of water resources engineering and management*, 4, 3, 15-22.

[12] Ghose, D. K., Mandal, P., & Samantaray, S. (2019). Experimental study of hydraulic jumps in an inclined rectangular flume. *Pertanika Journal of Science and Technology*, 27, 1, 397-407.

[13] Habib, A. A. (2013). Effect of weir opening on hydraulic jump characteristics d. s. of the weir. *Minia Journal of Engineering and Technology*, 32, 1, 49-61.

[14] Jafar, M. S. (2016). The friction resistance effect of the hydraulic jump location and energy dissipation, a laboratory study. *Kufa Journal of Engineering*, 7, 2, 90-103.

[15] Jalil, Sh. A. (2018). Modeling of Hydraulic Jump Generated Partially on Sloping Apron. *Journal of University of Babylon, Engineering Sciences*, 26, 1, 81-93.

[16] Jalut, Q. H., & El-Baaja, N. F. (2014). Experimental study for energy dissipation using stilling basin with one and two consecutive drops. *Diyala Journal of Engineering Sciences*, 7, 2, 61-82.

[17] Ljubičić, R., Zindović, B., Vojt, P., Pavlović, D., Kapor, R., & Savić, L. (2018). Hydraulic Jumps in Adverse-Slope Stilling Basins for Stepped Spillways. *WATER, Hydraulics-Setion*, 10, 460-477.

[18] Maatooq, J. S., Al.Adili, A. Sh., & Sameen, S. S. (2013). Relevant problems of a hydraulic jump at Diyala weir and the proposed remedy. *Eng. and Tech. Journal, University of Technology*, 31, A, 2504-2512.

[19] Movahed, S. A. M., Mozaffari, J., Davood-Magami, D., & Akbari, M. (2018). A semi-Analytical equation to estimate hydraulic jump length. *PeriodicaPolytechnica Civil Engineering*, 62, 4, 1001-1006.

[20] Obaed, I. H., Al-Salim, N. H. A., & Al-Fatlawy. Th. J. M. (2016). Experimental study for hydraulic characteristics of flow over compound regular notches. *Journal of Babylon University, Engineering Sciences*, 24, 1, 38-55.

[21] Palermo, M. & Pagliara, S. (2018). Semi-theoretical approach for energy dissipation estimation at hydraulic jumps in rough sloped channels. *Journal of Hydraulic Research*, 56, 786-795.

[22] Sharif, M., &Kabiri-Samani, A. (2018). Flow regimes at grid drop-type dissipators caused by changes in tail-water depth. *Journal of Hydraulic Research*, 56, 1-12.

[23] Torkamanzad, N., Dalir, A. H., Salmasi, F., & Abbaspour, A. (2019). Hydraulic jump below abrupt asymmetric expanding stilling basin on rough bed. *Water*, doi:10.3390/w11091756

[24] Yousefi, F., Mozaffari, J., & Movahed, S. A. M. (2018). Developing a hydraulic jump length model on horizontal rough beds. *Journal of the South African Institution of Civil Engineering*, 61, 3, 2-6.

The Influence of Various Factors on the Drying Shrinkage of Basalt Fibre Reinforced Cement-Based Composites Designed by the Taguchi Method

Adeyemi Adesina[1*], Emad Booya[2], Gorospe Karla[3] and Sreekanta Das[4]

[1]Department of Civil and Environmental Engineering, University of Windsor, Canada
E-mail: adesina1@uwindsor.ca
[2]Department of Civil and Environmental Engineering, University of Windsor, Canada
E-mail: booya@uwindsor.ca
[3]Department of Civil and Environmental Engineering, University of Windsor, Canada
E-mail: gorospek@uwindsor.ca
[4]Department of Civil and Environmental Engineering, University of Windsor, Canada
E-mail: sdas@uwindsor.ca
* Corresponding author: Department of Civil and Environmental Engineering, University of Windsor, Canada.

Abstract. Cement-based composites are widely used for various construction applications due to their outstanding mechanical and durability performance. However, the high drying shrinkage of cement-based composites has resulted in detrimental effects on its long-term performance. Drying shrinkage in cement-based composites causes cracks that serve as a pathway for the ingression of deleterious materials into the composites. Therefore, improving the drying shrinkage resistance of cement-based composites will result in an enhancement of the overall performance of the composites. The incorporation of short fibres and supplementary cementitious materials are some of the effective ways to improve the resistance of cement-based composites against drying shrinkage. Therefore, this study aims to evaluate the effect of the incorporation of basalt fibres and fly ash on the drying shrinkage of cement-based composites. This paper presents the results from the experimental and numerical investigation on the influence of various factors on the drying shrinkage of basalt fibre reinforced cement-based composites. The mixtures evaluated in this study were designed using the Taguchi method and a total of nine mixtures were made. The influence of the fly ash, water, sand and basalt content on the drying shrinkage of the mixtures were investigated. The findings from this study showed that the use of fly ash to cement ratio of 4, sand to binder ratio of 1, water to binder ratio of 0.25 and basalt fibre dosage of 2% is optimum to reduce drying shrinkage.

***Keywords**: Cement-based Composites; drying shrinkage; Fly Ash; Basalt fibre; Taguchi method.*

1. Introduction

Cement-based composites are the most used building material for various construction purposes (Han, Yu and Ou, 2014). The preference in cement-based composites for construction is due to its excellent strength performance, versatility and availability of materials locally (Neville and Brooks, 2010). However, the application of cement-based composites is plagued by its shrinkage which is consequential to its performance and service life. One of the common shrinkages that occur in cement-based composites is drying shrinkage with is caused by the rapid loss of moisture from the composites.

In order to mitigate the shrinkage in cement-based composites, several materials can be incorporated. It has been found out the use of supplementary cementitious materials (SCMs), short fibres, low water content can be used to mitigate the drying shrinkage in cement-based composites. The presence of coarse aggregates in cement-based composites is also known to reduce drying shrinkage as a result of the restraint they provide within the matrix. However, the majority of the current studies have only explored the effect of fibres such as polyvinyl alcohol (Adesina and Das, 2020c), carbon (Alcaide *et al.*, 2007), natural fibres (Booya *et al.*, 2018) on the drying shrinkage. The use of basalt fibres which is a sustainable alternative in cement-based composites made without coarse aggregate is limited creating a huge knowledge gap (Duic et al. 2020; Branston et al., 2019).

Hence, it is critical and imminent to determine the effect of various factors alongside basalt fibre dosage on the drying shrinkage of cement-based composites. Due to the high level of various factors required in order to effectively evaluate the effect of various factors; the design of experiments was made using the Taguchi method. Taguchi method eliminates the need to prepare a high number of mixtures thereby reducing the time and cost of experimental studies (Manjunath and Narasimhan, 2018).

Taguchi method was used recently to assess the influence of various factors on the mechanical performance of basalt fibre-reinforced composites (Adesina and Das, 2020a). However, there exists no use of this method to investigate the influence of various factors on the drying shrinkage of the composites. Hence, in this study; the Taguchi method was used to design the experiments and assess the impact of various factors on the drying shrinkage of the basalt fibre reinforced composites. Fly ash and basalt fibres which are environmentally friendly materials were used as SCM and reinforcement, respectively in all mixtures. The factors investigated in this study are the ratio of FA to PC, sand to binder, water to binder. The impact of three different dosages of basalt fibres was also considered as one of the factors.

2. Experimental Program

2.1 Materials

Fly ash and Portland cement satisfying the requirements of ASTM C 618 (2010) and ASTM C 150 (2012), respectively were used as the binder in this study. Fine aggregates with a specific gravity of 2.64. Basalt fibre with a diameter and length of 16 μm and 18 mm, respectively were used.

2.2 Mixture design

In order to assess the influence of different factors effectively, the design of the experiment was developed using three levels of each factor based on the practical composition of conventional cement-based composites. The factors and corresponding levels are presented in Table 1. Nine mixtures were developed based on the Taguchi method and the composition presented in Table 2.

Table 1. Factors and levels used

Factor ID	Factor	Levels		
		1	2	3
A	Fly ash to cement ratio	1	2	4
B	Sand to Binder ratio	0.25	0.45	0.35
C	Water to binder ratio	0.25	0.30	0.35
D	Fibre dosage (%)	1	2	5

Table 2. Composition of mixtures

Run	Factor A	Factor B	Factor C	Factor D
1	1	0.25	0.25	1
2	2	0.25	0.35	5
3	2	1	0.25	2
4	4	0.45	0.25	5
5	4	0.45	0.35	1
6	1	1	0.30	2
7	1	1	0.30	5
8	4	0.25	0.30	2
9	2	0.45	0.35	1

2.3 Sample preparation and curing

For each mixture presented in Table 2, the binder and aggregates were mixed initially before adding water to the mixtures while the mixing continued. This process was followed by the slow addition of the basalt fibres. The mixture was mixed for an additional two minutes after all the fibres were added to achieve a homogenous mixture. The fresh mixture was poured into bar moulds after mixing was completed and samples demoulded after 24 hours. After demoulding the samples were left open in laboratory conditions to allow drying of the samples to progress.

2.4 Test method

The drying shrinkage of each mixture was measured up to 28 days by measuring the length change of the samples per ASTM C 157 (2016). For each mixture, a total of four samples were made and measured. The average drying shrinkage obtained from four samples for each mixture was used to obtain the effect of each factor on the drying shrinkage.

3. Results and Discussion

The results and discussion made in this section are based on the drying shrinkage of the composites after 28 days of subjecting to drying conditions. Based on the results, Equation 1 was formulated to predict the drying shrinkage (*DS*) of the mixtures. From the analysis of the results, the influence of different factors on the drying shrinkage of the cement-based composites are presented in Figure 1 to Figure 4.

$$DS = 0.45 + 0.082A + 0.029B - 0.064C - 0.14D \quad (1)$$

3.1 Effect of fly ash to cement ratio

It is evident from Figure 1 that increasing the content of FA as replacement of the PC resulted in a decrease in the drying shrinkage with a more significant reduction made at FA to PC ratio greater than 2. The reduction in the drying shrinkage with a higher content of FA can be ascribed to the pore filling effect of FA particles and the formation of pozzolanic products which results in the densification of the microstructure. This observation corresponds to other studies where FA was used to replace PC in cement-based composites (Yang, Yang, and Li 2007; Adesina and Awoyera 2019; Awoyera et al. 2019).

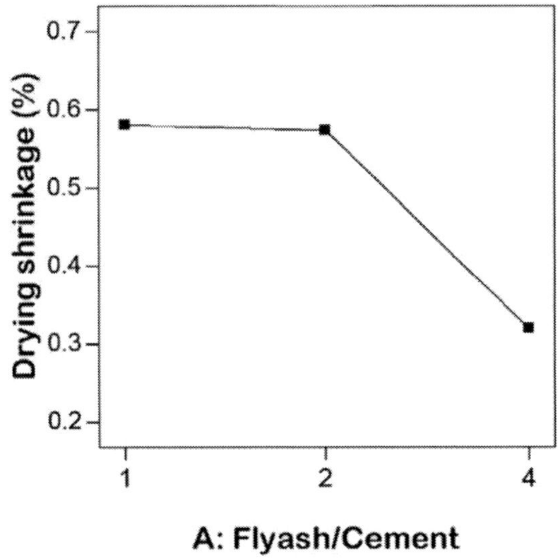

Figure 1. Effect of fly ash to cement ratio on drying shrinkage

3.2 Effect of sand to binder ratio

Figure 2 presents the effect of drying shrinkage of the cement-based composites. The results obtained are anticipated as the increase in the sand content is expected to provide more restraint in the matrix, hence, lower change in the length of the composites. The results showed that there is no considerable effect on the drying shrinkage when the sand to binder was increased from 0.25 to 0.45. However, a considerable reduction in the drying shrinkage ensued when the sand to binder increase to 1. These results agree with the observation made by Ye et al. (2019) where it was observed that increasing the sand to binder ratio results in lower drying shrinkage of engineered cement-based composites. Hence, it can be deduced that increasing the content of aggregates in cement-based composites will mitigate drying shrinkage. Nonetheless, it is worth mentioning that increasing the sand to binder content will result in more reduction in the strength capacity. Hence, it is recommended to ensure that other requirements such as strength and durability are within acceptable limits when higher sand to binder ratio is employed to mitigate drying shrinkage.

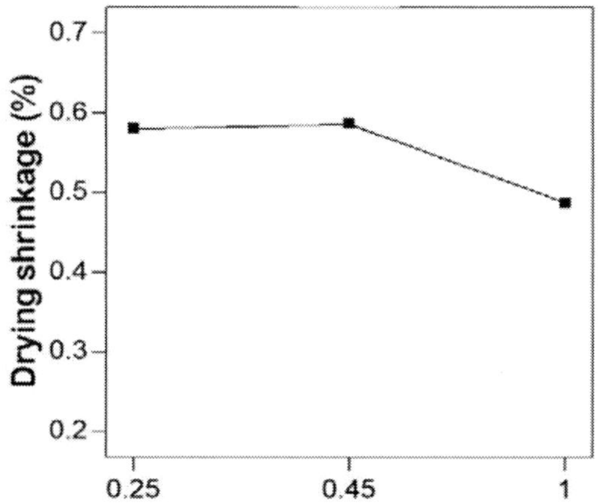

Figure 2. Effect of sand to binder ratio on drying shrinkage

3.3 Effect of water to binder ratio

Generally, increasing the water to binder ratio is detrimental to the performance of cement-based composites. A similar effect on drying shrinkage was observed when the water to binder ratio in cement-based composites was increased from 0.25 to 0.30. However, a slight reduction (i.e. less than 0.05%) was observed when the water to binder ratio was further increased to 0.35. These results confirm the observation made by other studies where the use of lower water to binder ratio in cement-based composites was found to yield lower drying shrinkage (Adesina and Das, 2020c, 2020b, 2020d).

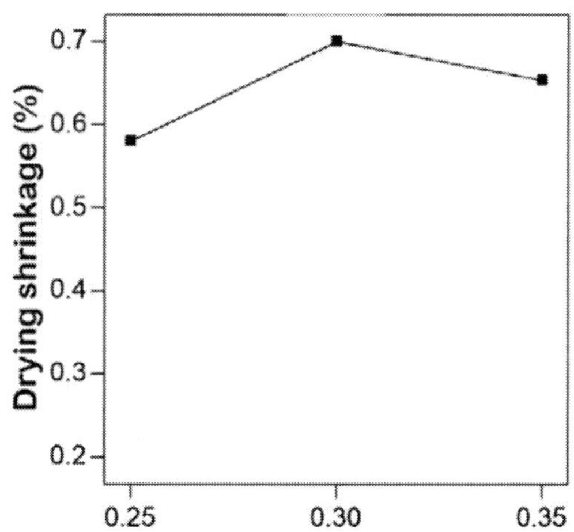

Figure 3. Effect of fly ash to cement ratio on drying shrinkage

3.4 Effect of basalt fibre dosage

The incorporation of fibres into cement-based composites is expected to mitigate drying shrinkage as the fibres provide restraints within the matrix. Figure 4 presents the effect of basalt fibre dosage. It is clear from Figure 4 that 2% is the optimum basalt fibre dosage to reduce drying shrinkage. At higher content of basalt fibre (i.e. 5%), the drying shrinkage of the composites was higher (i.e. greater than when 2% was used). The reduction in the drying shrinkage at 2% basalt fibre can be ascribed to the effective bridging effect of the basalt fibres in the matrix. The increase in the drying shrinkage when the fibre dosage was increased to 5% can be attributed to the agglomeration of the fibres due to improper dispersion in the mixture (Branston *et al.*, 2016). These accumulated fibres create a weak point and void in the matrix resulting in higher drying shrinkage due to their inability to offer restraint within the matrix. It is worth mentioning that the presence of fibres has been found to reduce the compressive strength of cement-based composites. Hence, it is critical to ensure that there is no significant deterioration in the strength capacity when basalt fibres are used to mitigate drying shrinkage. Also, this optimum dosage is applicable to these mixtures without coarse aggregate. Therefore, a different behaviour or optimum dosage might be applicable to concrete mixtures.

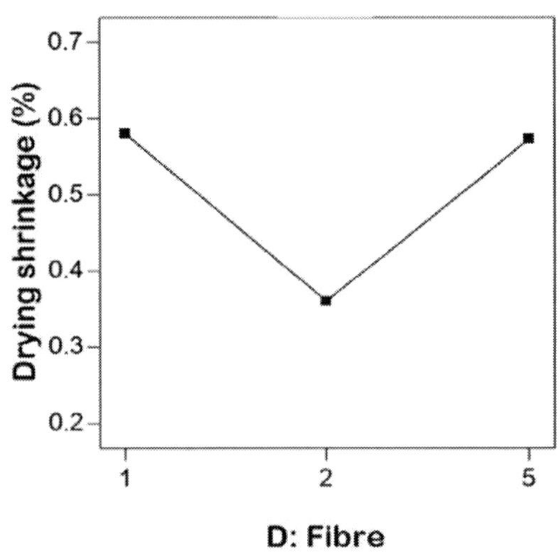

Figure 4. Effect of basalt fibre dosage on drying shrinkage

4. Conclusions

The effect of various factors on the drying shrinkage of cement-based composites reinforced with basalt fibre was investigated in this study. Based on the outcome from this experimental and numerical investigation it was found out that replacing Portland cement with fly ash results in lower drying shrinkage. Similarly, increasing the sand content of the mixture reduces the drying shrinkage of the composites as more restraints are provided. In terms of the basalt fibre dosage, the optimum was found to be 2% as a higher content of the basalt fibre resulted in higher drying shrinkage.

In general, the drying shrinkage of cement-based composites can be reduced with the use of basalt fibre at a dosage of 2% alongside higher content of fly ash and sand in the mixtures. However, the use of fly ash and basalt fibres are more effective in mitigating drying shrinkage compared to higher sand to binder ratio and lower water to binder ratio. The use of fly ash and basalt fibres in cement-based composites not only help to mitigate the drying shrinkage but also improves the sustainability of the composites.

Acknowledgement

The authors acknowledge the Natural Sciences and Engineering Research Council of Canada for the financial support provided.

References

[1] Adesina, A. and Awoyera, P. (2019) 'Overview of trends in the application of waste materials in self-compacting concrete production', *SN Applied Sciences*. doi: 10.1007/s42452-019-1012-4.
[2] Adesina, A. and Das, S. (2020a) 'Development of Eco-Friendly Basalt Fibre Reinforced Cementitious Composites Using Taguchi Method', *Silicon*. Springer, pp. 1–10. doi: 10.1007/s12633-020-00532-z.
[3] Adesina, A. and Das, S. (2020b) 'Drying shrinkage and permeability properties of fibre reinforced alkali-activated composites', *Construction and Building Materials*. doi: 10.1016/j.conbuildmat.2020.119076.
[4] Adesina, A. and Das, S. (2020c) 'Influence of glass powder on the durability properties of engineered cementitious composites', *Construction and Building Materials*. doi: 10.1016/j.conbuildmat.2020.118199.
[5] Adesina, A. and Das, S. (2020d) 'Performance of green fibre-reinforced composite made with sodium-carbonate-activated slag as a binder', *Innovative Infrastructure Solutions*. doi: 10.1007/s41062-020-00296-w.
[6] Alcaide, J. S. *et al.* (2007) 'Carbon fibre-reinforced, alkali-activated slag mortars', *Materiales de Construccion*.
[7] ASTM (2012) 'Standard Specification for Portland Cement - ASTM C-150', *ASTM D-150*.
[8] ASTM C 618 (2010) 'Standard Specification for Coal Fly Ash and Raw or Calcined Natural Pozzolan for Use', *Annual Book of ASTM Standards*.
[9] ASTM C157 (2016) 'StandardTest Method for Length Change of Hardened Hydraulic-Cement Mortar and', *Annual Book of ASTM Standards*. doi: 10.1520/C0157.
[10] Awoyera, P. O., Adesina, A. and Gobinath, R. (2019) 'Role of recycling fine materials as filler for improving performance of concrete - a review', *Australian Journal of Civil Engineering*. doi: 10.1080/14488353.2019.1626692.
[11] Booya, E. *et al.* (2018) 'Durability of cementitious materials reinforced with various Kraft pulp fibers', *Construction and Building Materials*. doi: 10.1016/j.conbuildmat.2018.10.139.
[12] Branston, J. *et al.* (2016) 'Influence of basalt fibres on free and restrained plastic shrinkage', *Cement and Concrete Composites*, 74, pp. 182–190. doi: 10.1016/j.cemconcomp.2016.10.004.
[13] Branston, J. *et al.* (2019) 'Microstructure and mechanical properties of basalt fibre reinforced concrete', in *17th Euroseminar on Microscopy Applied to Building Materials*. Toronto, Canada, pp. 121–127.
[14] Duic, J. *et al.* (2020) 'Field Applications of Basalt Fiber Materials for Rehabilitation of Deteriorated Concrete Structures', *Journal of Performance of Constructed Facilities*, 34(5), pp. 1–10. doi: 10.1061/(ASCE)CF.1943-5509.0001500.
[15] Han, B., Yu, X. and Ou, J. (2014) *Self-Sensing Concrete in Smart Structures, Self-Sensing Concrete in Smart Structures*. doi: 10.1016/C2013-0-14456-X.
[16] Manjunath, R. and Narasimhan, M. C. (2018) 'An experimental investigation on self-compacting alkali activated slag concrete mixes', *Journal of Building Engineering*. doi:

10.1016/j.jobe.2018.01.009.
[17] Neville, A. M. M. and Brooks, J. J. J. (2010) *Concrete Technology Second Edition, Building and Environment*. doi: 10.1016/0360-1323(76)90009-3.
[18] Yang, E. H., Yang, Y. and Li, V. C. (2007) 'Use of high volumes of fly ash to improve ECC mechanical properties and material greenness', *ACI Materials Journal*. doi: 10.14359/18966.
[19] Ye, B. *et al.* (2019) 'Effect of water to binder ratio and sand to binder ratio on shrinkage and mechanical properties of High-strength Engineered Cementitious Composite', *Construction and Building Materials*. doi: 10.1016/j.conbuildmat.2019.07.303.

IOP Conf. Series: Materials Science and Engineering **978** (2020) 012020 doi:10.1088/1757-899X/978/1/012020

Performance of High Strength Cementitious Composites with High Volume Supplementary Cementitious Materials

Adeyemi Adesina[1*], Emad Booya[2], Gorospe Karla[3] and Sreekanta Das[4]

[1]Department of Civil and Environmental Engineering, University of Windsor, Canada
E-mail: adesina1@uwindsor.ca
[2]Department of Civil and Environmental Engineering, University of Windsor, Canada
E-mail: booya@uwindsor.ca
[3]Department of Civil and Environmental Engineering, University of Windsor, Canada
E-mail: gorospek@uwindsor.ca
[4]Department of Civil and Environmental Engineering, University of Windsor, Canada
E-mail: sdas@uwindsor.ca
* Corresponding author: Department of Civil and Environmental Engineering, University of Windsor, Canada. Tel: +14386806399.

Abstract. Portland cement (PC) is the major binder used for cementitious composites. However, due to the increase in the demand for cementitious composites for the construction of infrastructures, there's a consequential effect of the production of this binder (i.e. PC) on the sustainability of our environment. The production of PC emits huge amounts of carbon dioxide into the environment and posses a huge strain on the natural deposits of its raw material. In order to create a sustainable environment while meeting the high demand for cementitious composites; it is paramount to replace the PC with locally available waste materials. This study incorporates a high volume of slag alongside silica fume at a ratio of 2.2 to that of PC to produce high strength cementitious composites. The effects of these compositions were determined experimentally on its fresh and hardened properties. Results from this study showed that high strength cementitious composites can be produced with a high volume of supplementary cementitious composites up to 80%. The use of slag as 80% replacement of slag resulted in a 9.3% increase in the compressive strength and a 48.1% decrease in sorption.

***Keywords**: Cementitious Composites; Slag; Strength; Durability; Sustainability.*

1. Introduction

The primary binder used in the production of various cementitious composites is Portland cement (PC). However, the production of PC causes a huge menace in the environment due to the high carbon dioxide emissions and the high consumption of raw materials (Purnell, 2013; Sivakrishna *et al.*, 2019). Hence, PC is the major contributor to the embodied carbon and energy in cementitious composites.

High strength cementitious composites with a compressive strength greater than 50 MPa are in huge demand due to their outstanding performance and the viability to reduce the size of structures made with those composites while conserving the performance. However, the high content of PC required in such mixtures has resulted in high strength cementitious composites having higher embodied energy and carbon (Thomas and Chandra Gupta, 2016). In order to promote more sustainable construction practices; alternative materials with lower environmental impact must be used as components in cementitious composites. Of such promising and viable sustainable alternatives that can be used to produce high strength cementitious composites are blast furnace slag (BFS) and silica fume (SF) which are waste products from the production of metals. These materials can be used to partially replace the PC, hence, classified as supplementary cementitious materials (SCMs).

BFS and SF have been incorporated over the years to improve the sustainability and performance of cementitious composites (Dimitriou, Savva and Petrou, 2018; Awoyera and Adesina, 2019a; Booya *et al.*, 2019). However, there is limited study on its use in high volume to produce high strength cementitious composites. Also, studies on the use of these SCMs have incorporated them at low volumes (i.e. less than 50%) and only focused on either the mechanical or durability performance. Therefore, there is a limited comprehensive study on the influence of the high-volume use of these SCMs on the fresh, mechanical and durability properties.

Hence, this study was carried out to investigate the influence of using high volume BFS and SF on the properties of high strength cementitious composites. The BFS was used to replace the PC up to 80% while the SF was used as a replacement of the BFS up to 20%. The performance of the composites was evaluated in terms of slump, air content, density, compressive strength, porosity, and water absorption.

2. Experimental Program

2.1 Materials

The primary binder used in this study is Portland cement (PC) with an average particle size of 15 μm and satisfying the requirements of ASTM C 150 (2012). The supplementary cementitious material (SCM) used to replace the PC are the Slag and silica fume with an average size of 10 μm and 0.4 μm, respectively. The specific gravity of the PC, BFS and SF used is 3.13, 2.84, and 2.21, respectively. The chemical composition of the PC and SCMs are presented in Table 1. Only fine aggregate with a specific gravity of 2.63 and a maximum size of 3 mm was used as aggregate. Superplasticizer (SP) was used as the chemical admixture to improve the workability of the mixtures.

Table 1. Binder composition

	PC	BFS	SF
CaO	62.43	37.6	0.1
SiO_2	19.78	36.9	96.3
Al_2O_3	5.38	9.08	0.2
Fe_2O_3	2.67	0.61	0.3
SO_3	3.47	1.96	0.2
Na_2O	0.12	0.25	0.1

2.2 Mixture Design, Sample Preparation and Curing

The composition of all mixtures was kept constant except for the composition of the binder. SCMs (i.e. combined BFS and SF) were used to replace up to 80% the content of PC while the SF was used as a replacement of BFS up to 20%. The water to binder ratio was kept at 0.35 for all mixtures and the binder to sand ratio was fixed at 0.40. SP was premixed with the mixing solution at a dosage of 0.8 kg/m^3 for all mixtures. Table 2 presents a detailed composition of the mixtures evaluated in terms of ratio to the binder content. The mixture ID used represents the composition of the binder.

Each mixture was produced by first mixing the dry component (i.e. binders and aggregates) before adding water to the mixture as the mixing continues. When the mixing was completed, the fresh mixtures were poured into moulds and covered with a plastic sheet to avoid loss of water from the samples. All samples were demoulded after one day of casting and cured in water for 28 days before testing.

Table 2. Composition of mixtures

Mixture ID	PC	BFS	SF	Fine aggregate	Water
100PC	1	0	0	0.40	0.35
80BFS	0.2	0.80	0	0.40	0.35
75BFS5SF	0.2	0.75	0.05	0.40	0.35
BFS10SF	0.2	0.70	0.10	0.40	0.35
BFS15SF	0.2	0.65	0.15	0.40	0.35
60BFS20SF	0.2	0.60	0.20	0.40	0.35

2.3 Test Methods

The air content of each mixture was measured following the test procedures in ASTM C 231 (2014). The air content test was carried out immediately after the mixing was completed. The compressive strength was evaluated using cubic samples with a dimension of 50 mm x 50 mm x 50 mm and following the test procedure in ASTM C 109 (2012). For each mixture, a total of four samples were tested and the average results presented.

Cylindrical samples with a height and diameter of 50 mm and 100 mm, respectively were used to evaluate the porosity (i.e. permeable voids), absorption, and sorption of the mixtures. The permeable voids and absorption were measured as per the test procedure in ASTM C 642 (2006)while the test procedures in ASTM C 1757 (2013) were used to assess the sorption of the mixture. Four samples were used for each test, and the result section presents the average obtained from the four samples for each mixture.

3. Results and Discussion

3.1 Air content

The effect of BFS and SF content on fresh content is shown in Figure 1. It can be observed that the replacement of 80% PC with BFS resulted in a 33.4% reduction in the air content. This reduction in the air content can be associated with the smaller particle size of the BFS which reduced the air voids in the fresh mixture. However, replacing BFS with SF in the range of 5% to 20% resulted in an increase in the air content 14.2% to 28.6%.

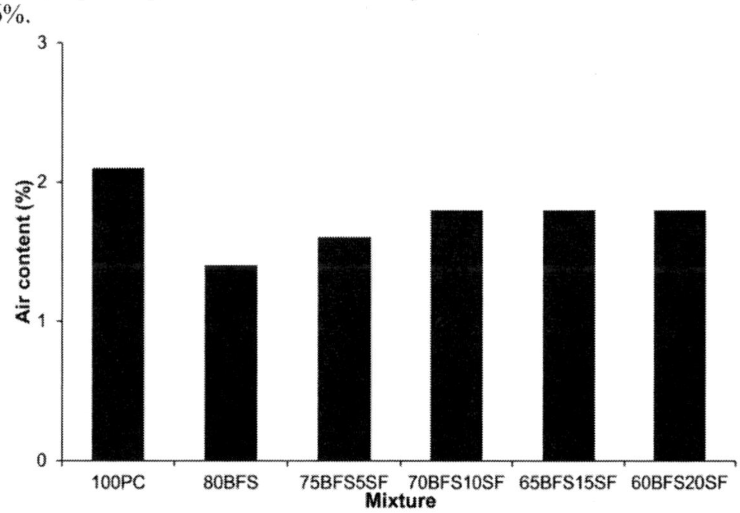

Figure 1. Air content of mixtures

3.2 Density

Figure 2 shows the effect of the SCMs on the dry density of the cementitious composite. It is being observed that the replacement of 80% PC with BFS resulted in a significant reduction in the density. This reduction can be associated with the lower specific gravity of the BFS. However, the replacement of BFS with SF up to 20% resulted in a slight increase in the density by up to 3.4%. This observation was not expected as SF particles have lower specific gravity compared to PC and BFS. This phenomenon could be attributed to the SF particles acting as a nucleation site resulting in the formation of more products and a corresponding increase in the density (Adesina, 2019, 2020). Nonetheless, the density of composites incorporating SF is lesser than that made with only PC as the binder (i.e. 100PC). However, it is recommended that more studies should be carried out in order to fully understand the results.

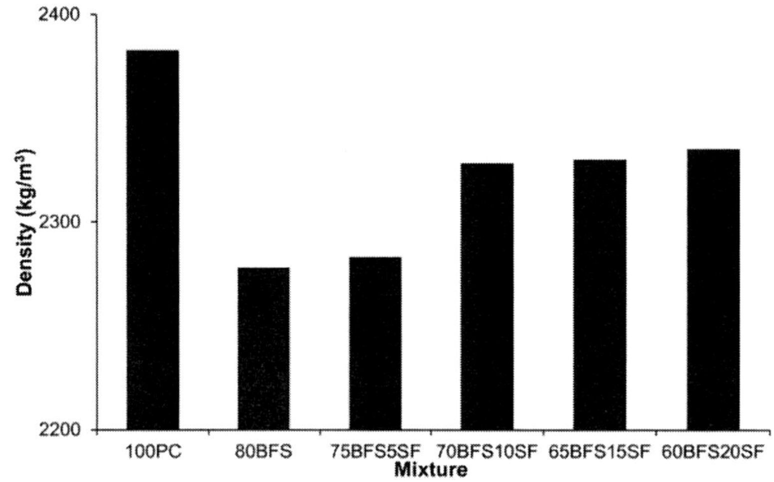

Figure 2. Density of mixtures

3.3 Compressive strength

Compressive strength is an essential property of cementitious composites and a good pointer of the overall mechanical performance. It can be observed from Figure 3 that the replacement of high content of PC with SCMs (i.e. BFS and SF) improved the compressive strength. The compressive strength of cementitious composite made with BFS as 80% replacement of the binder is 9.3% higher than that made with PC as the binder. However, the replacement of BFS with SF does not result in any significant improvement in the compressive strength with the compressive strength reducing at SF content greater than 15%. The increase in compressive strength with the use of up to 80% BFS and SF up to 15% can be associated with the possible refinement of the microstructure coupled with the formation of additional pozzolanic reaction products. The effect of 20% replacement of BFS with SF (i.e. mixture 60BFS20SF) contradicts the hypothesis made for the density results due to the decrease in compressive strength observed. This reduction in compressive strength when SF was used as a 20% replacement of BFS can be attributed to the possible excess silica in the matrix.

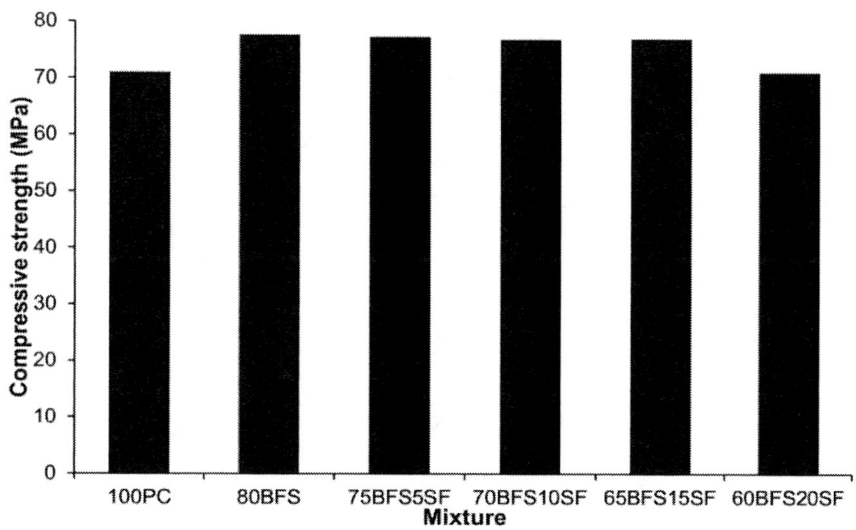

Figure 3. Compressive strength of mixtures

3.4 Porosity

The effect of the SCMs content on the porosity (i.e. permeable void) of the mixtures is shown in Figure 4. It is evident from Figure 4 that the incorporation of BFS and/or BFS as replacement of the PC resulted in a lower the porosity of the composites. The reduction in the porosity of the high strength cementitious composite with the incorporation of the SCMs can be associated with the pore refinement and pozzolanic capability of BFS and SF.

It can also be observed that more reduction in the porosity of the mixture was evident when the content of SF used as the replacement of BFS increased. This phenomenon can be attributed to the very fine particles of SF which proffer more refinement to the microstructure. The pozzolanic products from the pozzolanic reaction of the SF particles would also contribute significantly to the pore refinement of the microstructure of the composites. This observation corresponds with that of other studies where the use of BFS and/or SF was found to result in lower porosity of cementitious composites (Türkmen, Gül and Çelik, 2008; Chahal and Siddique, 2013; Awoyera and Adesina, 2019b; Adesina and Das, 2020). It will also be noticed that in contrast to the fresh air content, the incorporation of SF resulted in lower porosity. This can be ascribed to the ability of the SF to refine the microstructure of cementitious composites after it has reacted with other components in the composite.

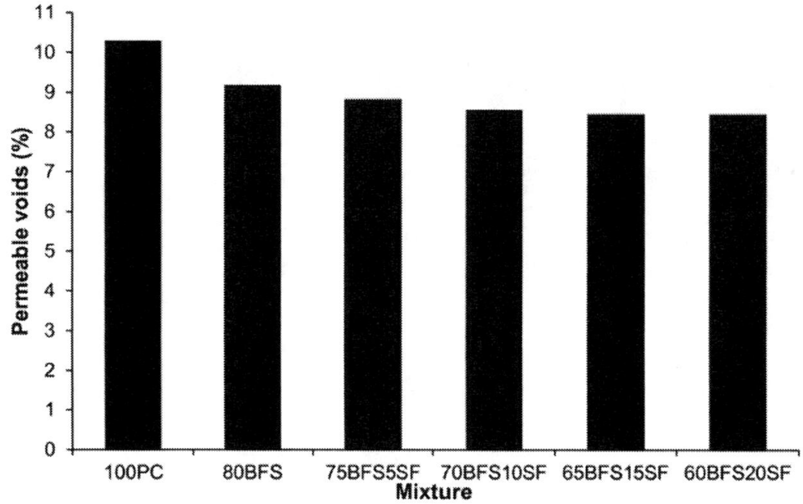

Figure 4. Porosity of mixtures

3.5 Absorption and Sorption

The permeability properties of cementitious composites reflect their overall durability as these properties are dependent on the ease of penetration and amount of water that can get into it. Figure 5 and Figure 6 presents the absorption and sorption of the mixtures, respectively. It can be observed that the incorporation of BFS and SF as replacement of PC in the mixtures resulted in a significant decrease in the permeability (i.e. absorption and sorption) of the composites. These reductions in permeability can be associated with the ability of the BFS and SF to refine the microstructure limiting the ingress of moisture into the composites. The correlation between the absorption, sorption and permeable voids confirmed this as there is a strong correlation between these properties as shown in Figure 7.

In contrast to the effect of SF on the compressive strength, there is a decreasing trend observed in the reduction in the absorption and sorption of the mixtures with increasing SF content. This shows that the more refinement capability of the SF plays a more significant role in the refinement of the microstructure compared to the effect of the products formed.

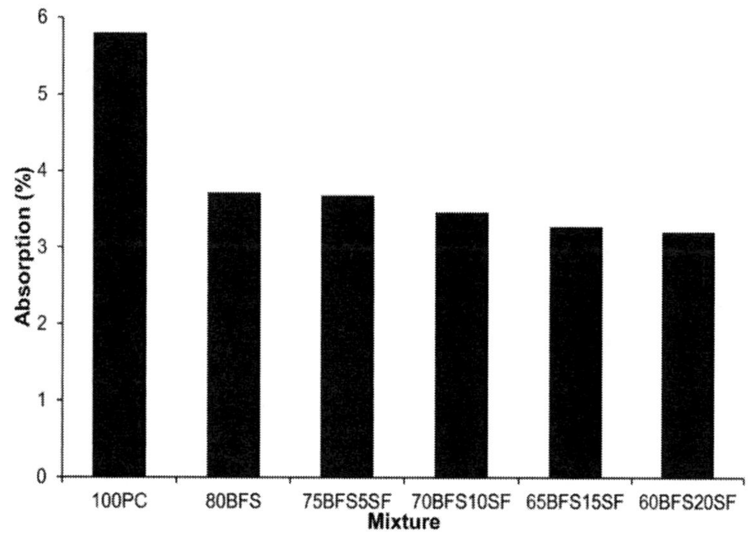

Figure 5. Absorption of mixtures

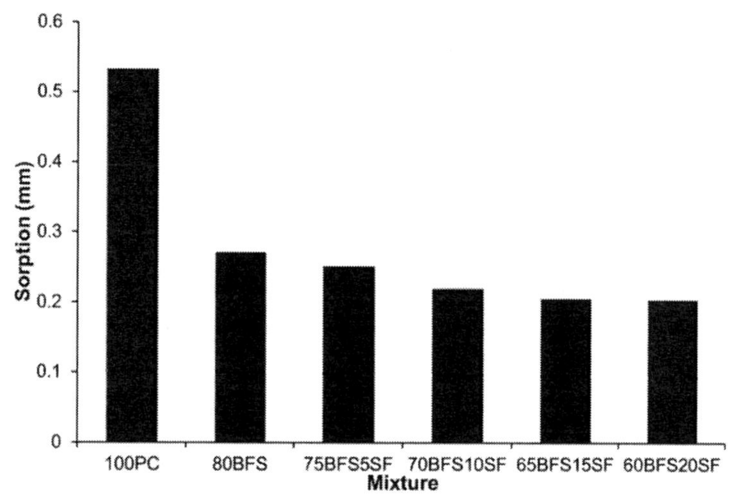

Figure 6. Sorption of mixtures

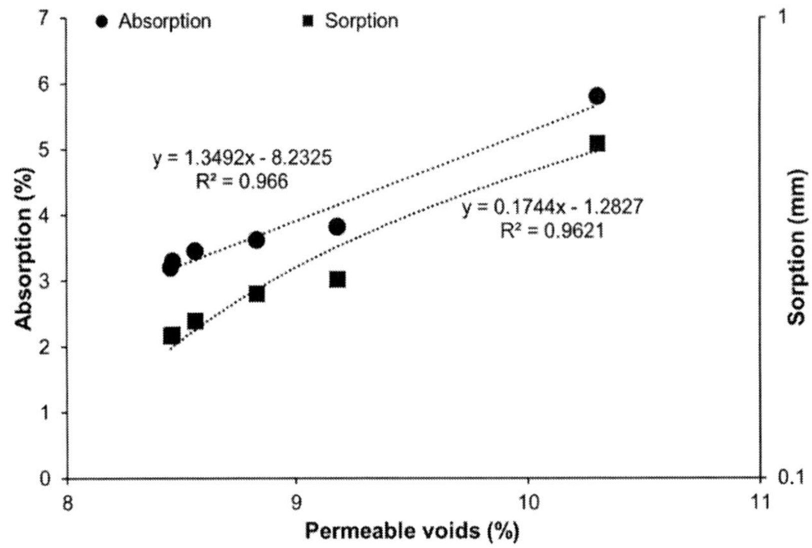

Figure 7. Correlation between absorption, sorption and permeable voids

4. Conclusions

In the current paper, the performance of high strength cementitious composites made with high volume supplementary cementitious material (i.e. slag and silica fume) was evaluated experimentally. The performance of the composites in terms of the fresh and hardened properties of the composites was investigated. The following conclusions can be made based on the results of this study:

1. Supplementary cementitious materials such as slag and silica fume can be used to replace Portland cement in cementitious composites without any significant effect on the corresponding performance.
2. The incorporation of blast furnace slag and silica fume resulted in a decrease in the density of the composites due to their lower specific gravity. Hence, in addition to the sustainability advantage with the use of these SCMs as replacement of Portland cement; they can be incorporated to reduce the weight of cementitious composites.
3. The incorporation of blast furnace slag and silica fume as replacement of Portland cement improved the performance of the composites in terms of increasing the compressive strength and reducing the permeability of the composites. Hence, these SCMs are sustainable alternatives to PC to produce high strength cementitious composites.

Acknowledgement

The authors acknowledge the Natural Sciences and Engineering Research Council of Canada for the financial support provided.

References

[1] Adesina, A. (2019) 'Durability Enhancement of Concrete Using Nanomaterials: An Overview', *Materials Science Forum*. doi: 10.4028/www.scientific.net/msf.967.221.

[2] Adesina, A. (2020) 'Nanomaterials in cementitious composites: review of durability performance', *Journal of Building Pathology and Rehabilitation*. Springer, 5(1), p. 21. doi: 10.1007/s41024-020-00089-9.

[3] Adesina, A. and Das, S. (2020) 'Drying shrinkage and permeability properties of fibre reinforced alkali-activated composites', *Construction and Building Materials*. doi: 10.1016/j.conbuildmat.2020.119076.

[4] ASTM (2012) 'Standard Test Method for Compressive Strength of Hydraulic Cement Mortars', *ASTM International*. doi: 10.1520/C0109.

[5] ASTM C 642-2006 (2006) 'Standard test method for density, absorption, and voids in hardened concrete', *ASTM International, West Conshohocken, PA*, (4), pp. 1–3. doi: 10.1520/C0642-13.5.

[6] ASTM International (2012) 'Standard Specification for Portland Cement', *ASTM International*', 552(d), pp. 1–9. doi: 10.1520/C0150.

[7] ASTM International (2013) 'C1757 – 13 Standard Test Method for Determination of One-Point, Bulk Water Sorption of Dried Concrete', *ASTM Standards*. doi: 10.1520/C1757-13.

[8] ASTM Standar C 231 (2014) *'Test Method for Air Content of Freshly Mixed Concrete by the Pressure Method', ASTM International*. doi: 10.1520/C0231_C0231M-14.

[9] Awoyera, P. and Adesina, A. (2019a) 'A critical review on application of alkali activated slag as a sustainable composite binder', *Case Studies in Construction Materials*. doi: 10.1016/j.cscm.2019.e00268.

[10] Awoyera, P. and Adesina, A. (2019b) 'Durability Properties of Alkali Activated Slag Composites: Short Overview', *Silicon*. doi: 10.1007/s12633-019-00199-1.

[11] Booya, E. *et al.* (2019) 'Durability properties of engineered pulp fibre reinforced concretes made with and without supplementary cementitious materials', *Composites Part B: Engineering*. doi: 10.1016/j.compositesb.2019.05.070.

[12] Chahal, N. and Siddique, R. (2013) 'Permeation properties of concrete made with fly ash and silica fume: Influence of ureolytic bacteria', *Construction and Building Materials*. doi: 10.1016/j.conbuildmat.2013.08.023.

[13] Dimitriou, G., Savva, P. and Petrou, M. F. (2018) 'Enhancing mechanical and durability properties of recycled aggregate concrete', *Construction and Building Materials*. doi: 10.1016/j.conbuildmat.2017.09.137.

[14] Purnell, P. (2013) 'The carbon footprint of reinforced concrete', *Advances in Cement Research*. doi: 10.1680/adcr.13.00013.

[15] Sivakrishna, A. *et al.* (2019) 'Green concrete: A review of recent developments', *Materials Today: Proceedings*. Elsevier BV. doi: 10.1016/j.matpr.2019.08.202.

[16] Thomas, B. S. and Chandra Gupta, R. (2016) 'Properties of high strength concrete containing scrap tire rubber', *Journal of Cleaner Production*. doi: 10.1016/j.jclepro.2015.11.019.

[17] Türkmen, I., Gül, R. and Çelik, C. (2008) 'A Taguchi approach for investigation of some physical properties of concrete produced from mineral admixtures', *Building and Environment*. doi: 10.1016/j.buildenv.2007.02.005.

Solar Integrated Unit to Reduce Power and Cooling Penalty of a Natural Gas Combined Cycle Plant integrated with a Carbon Capture System

Saif W. Mohammed Ali[1, *], Sahib S. Ahmed[2] and Nabeel A. Ghayadh[3]

[1] Ph. D, and Lecturer at the Mechanical Department, Kufa University, Najaf, Iraq
E-mail: saifw.ali@uokufa.edu.iq saifw.alturaihi@uokufa.edu.iq
[2] Ph. D, and Lecturer at the Mechanical Department, Kufa University, Najaf, Iraq
E-mail: sahib.aldulaimi@uokufa.edu.iq
[3] Ph. D, and Lecturer at the Mechanical Department, Kufa University, Najaf, Iraq
E-mail: nabeel.ghayadh@uokufa.edu.iq
* Corresponding author: College of Engineering, Kufa University, Najaf, Iraq - Tel: +9647733910416.

Abstract. Integrating solar assisted plant with a carbon capture unit to provide the required reboiler heat duty has been recently investigated extensively to reduce the negative impact of carbon capture unit on plant performance. However, using solar assisted plant to provide the required energy in the reboiler of the stripper leads to increase plant's condenser duty unfavorably. In the present study, two novel configurations have been proposed for a concentrated solar assisted power plant to generate the required reboiler duty in the stripper of a post-combustion carbon capture unit (PCCC-SAPP) in a (630 MWe) natural gas combined cycle (NGCC) power plant to reduce both the condenser duty and power penalty of the plant simultaneously. In the present study, both the solar plant and plant's low-pressure steam (LPS) are contributing to deliver the required reboiler thermal energy in the first method while the extraction is from the cooling system load in addition to the solar plant in the second method. Solar contribution factor (SCF) is the main criteria which has been used to evaluate the findings. Condenser duty with its associated water requirements, power penalty, and the required solar field land area are the parameters which have been investigated in the preset study. It has been found that the second configuration method has better impact on the performance of the plant and its associated cooling system than the first configuration method where zero power penalty is consumed in the second configuration as well as reducing the condenser duty with its related water usage amounts. In addition, it was shown that raw water withdrawal and consumption amounts in term of (gal/min)/MW_{net} decrease with increasing SCF in the first configuration where raw water withdrawal and consumption amounts are 7.23 (gal/min)/ MW_{net}, and 5.17 (gal/min)/ MW_{net} at SCF equals to 0 while these amounts reach to 6.74 (gal/min)/ MW_{net} and 4.82 (gal/min)/ MW_{net} at SCF of 0.95 respectively. On the other hand, these amounts increase with increasing SCF in the second configuration where water withdrawal and consumption values approach 6.75 (gal/min)/ MW_{net} , and 4.92 (gal/min)/ MW_{net} at SCF equals to 0 while the amounts are equal to 7.93 (gal/min)/ MW_{net} and 5.81 (gal/min)/ MW_{net} at SCF equals to 0.95 respectively despite that condenser duty is increasing in the both methods. This comes from the fact that the net produced power is increasing in the first scenario with SCF while it keeps constant in the second scenario. Furthermore, the relationship between the solar field's required land area and solar direct normal irradiation (DNI) was investigated in the present paper under four different SCF; 0.2, 0.4, 0.6, and 0.8. It was found that reducing the solar field's required land area requires increase in the DNI and decrease in the SCF where the reboiler heat duty is constant in the PCCC and the relationship is almost exponential. The validation was performed with an American governmental report (NETL, 2015) showing that the presented results has a significant agreement with the report.

Keywords: *Carbon capture unit, Solar assisted power plant, Natural gas combined cycle, Carbon capture reboiler duty, Water energy nexus in power plant, Cooling efficiency in thermal power plants.*

1. Introduction

Adding carbon capture unit to thermoelectric power plant almost doubles water requirements in the unit (Zhai and Rubin, 2011). In addition, integrating this unit causes power penalty because of the extracted LP steam from the entrance of the LP turbine to be used for solvent regeneration in the stripper reboiler of the capture unit. On the other hand, this extraction from the LP steam leads to reduce the main plant's condenser duty by about 30% (Saif W. M. et al., 2019). Although the net waste heat duty is increased because of the additional carbon capture duty, adding the carbon capture unit has an advantage on the main plant's condenser duty by reducing it. To avoid a such penalty in the net produced power, solar power assisted plant would be integrated with the carbon capture unit to be used for solvent regeneration in the stripper's reboiler instead of extracting from the LP steam and causing the penalty in the produced power. However, using this solar assisted power plant has two disadvantages; first, large area is required for the solar plant and its related accessories, and this area is involved directly with the net produced thermal power by the solar plant. This means that increasing or decreasing the produced thermal energy by the solar plant leads to increase or reduce plant's area directly. The second disadvantage is as there is no required extraction from the LP steam, the main plant's condenser duty would not be reduced by which the net cooling duty of the plant, including the effect of the carbon capture, is increased remarkably leading to increase in the water usage amounts of the cooling system. Therefore, new approaches should be investigated to reduce both plant's area and plant's main condenser duty properly and simultaneously in parallel with reducing the penalty in the turbine's produced power. The present study investigates two approaches to reduce the condenser duty with its related water requirements in a NGCC integrated with a PCCC and a solar assisted power plant to regenerate the solvent in the reboiler of the capture unit.

The concept of using carbon capture solar assisted power plant has been studied extensively in the literature. Cau et al. (2014) studied the economic and technical effect of producing three types of superheated steam by a concentrated solar power plant on a performance of ultra-supercritical (USC) steam power plant integrated with a carbon capture unit. These streams are LP steam for the solvent regeneration in carbon capture unit, IP steam and HP steam to be used in the steam cycle of the (USC) steam plant. It was shown that the LP steam production has the worst impact on the performance of the (USC) plant than the IP and HP steams production. In term of condenser duty, producing the IP and HP steam to the (USC) steam cycle shouldn't go over a certain limit because any undesirable increase in steam production may lead to increase the condenser duty over the favorite limits because of increasing in the steam mass flow rate causing plant power shortage. Liu et al. (2016) conducted a research involves to a technical and economic investigations for using two different solar assisted power plants to be used for solvent regeneration of ammonia-based carbon capture unit; vacuum tube (VT), and Parabolic Trough Collector (PTC). Studying the effect of different regions in China corresponding to its DNI (Direct Normal Irradiation) and weather data has implemented in this work too. It was demonstrated that using the (VT) is the most feasible approach technically than the (PTC). Regarding the economic aspect, the prices of the solar assisted plants should be reduced to a specific limit per area to be viable economically. Mokhtar et al. (2012) investigated an economic feasibility method to utilize solar thermal energy in contributing with LP steam extraction for solvent regeneration in post-combustion carbon capture unit which being integrated with a 300 MWe pulverized coal-fired power plant installed in New South Wales, Australia by using actual weather data and electricity prices. It was found that solar thermal energy can be viable for solar collector price equals to 100 \$/m2 when the optimal solar load fraction equals to 22%. Zhai et al. (2018) used thermo-economic theory to evaluate two configurations of solar thermal power plants to be integrated with 1000 MWe coal-fired power plant equipped with MEA based-carbon capture unit. The first design of the solar thermal plant was configurated by utilizing the solar power energy to superheat the HP feed water stream in the steam cycle of the coal-fired power plant where this characterized by (SA-PG-CC) while LP steam extraction was used for solvent regeneration in the carbon capture reboiler. The second one was designed to be used in the stripper reboiler directly for solvent regeneration and characterized by (SA-CC-PG). Both cases were compared with the referenced case which was not integrated with the solar power plant (PG-CC). Overall, the first scenario (SA-PG-CC) is the better choice from the performance point of view and economic point of view.

Number of studies have been conducted to evaluate the net water usage in solar power plants. Erica Suzanne (2015) has developed an optimization model to minimize water and land requirement in power plants by replacing it with three renewable energy technologies; wind, Photo Voltic (PV) solar cell, and Collector Solar Parabolic (CSP). She found that dry CSP consumes higher water without considering land constraint and less with considering it. Saria Bukhary et al. (2018) has performed an evaluation for water and land usage in a solar development system in the U.S in addition to analyzing its associated CO2 emissions. STELLA dynamic system tool, SAM, and Solar DS tools were used to evaluate the cost effectiveness and the efficiency of the system. It was found that PV technology takes the smallest water requirement comparing to CSP system. Regarding CSP systems, it was demonstrated that CSP tower takes the largest land demand, and CSP wet cooling takes the largest water requirement with a good agreement with the literature. X.D. Wu, and G.Q. Chen (2017) had a novel investigation by including water usage through construction stages of solar plant infrastructure in the total water amount by using Input-Output data. It was found that by including this water, solar plant will not be feasible in term of water saving. Integrating solar assisted plant with a pre-combustion and oxy combustion was not absent from the literature scientific works and researches. Li Y. et al., (2013) investigated using solar energy sources to be employed for producing syngas/hydrogen through the pre-combustion CO_2 capture processes. Heliostat collector was used as a medium energy source to separate the Methane from the hydrogen in a membrane while parabolic trough collector was used to regenerate the steam as a low solar energy source. Using solar energy as an external energy source to increase the efficiency of the chemical looping system in the oxy-combustion carbon capture was investigated by Hong and Jin (2005). It was found that using the solar energy in oxy-combustion system gives the possibility to store the solar energy chemically where it can be used during the absence of solar radiation at night.

In the present study, two novel configurations of a post-combustion carbon Capture-Solar assisted (PCCC-SAPP) power plant have been investigated regarding its impact on plant performance and condenser duty of a NGCC power plant. New parameters have been studied through the present work which have not been studied yet including the effect of the integrated solar units on the water usage amounts in term of (gal/min)/MWnet and their effect on the required land area of the solar field. The first configuration is installed by contributing two LP steam streams to be used for solvent regeneration in the stripper; the extracted steam from the LP turbine, and the solar superheated LP stream. The second design is a novel design in term of carbon capture - solar assisted study. This design is configurated by extracting heat from the condenser unit to be used and contributed for solvent regeneration beside the solar part contribution by which the net condenser duty of the main plant's condenser would be reduced. The solar field contribution factor (SCF) has been taken in this study as a criterion for the comparison study. The technical impact of these two configurations on condenser duty with its related water usage, plant power penalty, and solar field area has been studied in the present work. COCO software has conducted in this research to perform all the simulations with enabling Excel Workbook Sheet to implement all the governing equations which be used in the solar field calculations. Concentrated solar collector was used as the solar assisted power plant in the present work.

2. The Effect of post-combustion carbon capture on the plant and cooling system performance.

Carbon capture unit is very necessary to reduce CO_2 emissions from thermoelectric power plants by which its negative impact on global warming would be mitigated. On the other hand, adding this unit requires penalty on the net waste heat duty and the net produced power. Fuel gas stream reacts with MEA solvent in the absorber to absorb most of the CO_2 content in the gas stream to generate a rich CO_2-Sorbent component which would be separated in the stripper by applying a massive heat amount. A high heating value comes from a reboiler which is integrated in the capture unit's stripper to regenerate the solvent and separate the CO_2 with a vapor content. LP steam is extracted from the entrance of the LP turbine to be exploited in the stripper's reboiler by which the net produced power is reduced. However, extracting from the LP turbine leads to reduce the cooling duty of the main condenser by about 30% according to Saif W. M. et al. (2019).

3. Post-combustion carbon capture solar assisted plant (PCCC-SAPP).

To avoid such a penalty in the produced power, solar assisted power plant would be integrated to generate the required thermal energy for solvent regeneration in the stripper instead of extracting from the LP steam so that there will be no penalty in the produced power. Although using solar assisted power plant reduces plant power penalty, there will be no reduction in the plant's main condenser duty beside the required large land area for the solar plant and its accessories. To reduce both the penalty and condenser duty simultaneously in the NGCC plant which is equipped with the carbon capture unit, two novel configurations have been investigated in the present study by integrating a concentrated solar collector assisted plant with a post-combustion carbon capture (PCCC) unit as being detailed in the next two sections.

3.1. LP steam turbine extraction configuration (First Configuration).

Thermal energy provided by the concentrated solar collector plant beside thermal energy provided by the extracted LP steam from the LP turbine are contributing to regenerate the solvent and separate the CO_2 in the stripper's reboiler of the PCCC unit. Solar contribution factor (SCF) is the criterion to find the most viable value by which power penalty, condenser duty with its related water usage, and solar plant required area are reduced in a proper way. The distributed heat between the extracted LP steam and solar thermal power plant can be achieved by controlling the net steam mass flow rate. To control solar thermal energy simply and in a proper manor, it has been assumed that the temperature and pressure of the steam at the solar plant outlet are equal to these values for the extracted LP steam from the LP turbine. Thus, the only factor which would control the solar thermal energy is the collector's medium fluid mass flow rate. Figure 3.1 shows the first configuration of the concentrated solar collector which being integrated with the PCCC's reboiler and the extracted LP steam from the LP turbine.

3.2. Condenser duty extraction configuration (Second Configuration).

In this scenario, there would be no LP steam extraction form the LP turbine. Therefore, there would be no penalty in the LP turbine's produced power. On the other hand, the net waste heat duty in this scenario is less than the net waste heat duty in the first scenario as a result of reducing the main condenser duty by about 30% although that the water requirements per power unit is increasing in this system in contradiction with the first scenario where net produced power is constant in the present configuration while is changing in the first one. To reduce both the condenser duty and the required solar land area with consuming zero penalty, a novel design would be proposed by sharing part of the condenser duty to superheat the medium fluid stream to be contributing beside the solar superheated steam for solvent regeneration in the reboiler by which plant's main condenser duty can be reduced for the same level of the first scenario or less in addition to reducing the solar required land area because of reducing its thermal energy contribution factor (SCF) as being shown in figure 3.2. In this configuration, the total thermal energy duty in the reboiler would be distributed between the gained heat duty from the condenser and the solar thermal energy part.

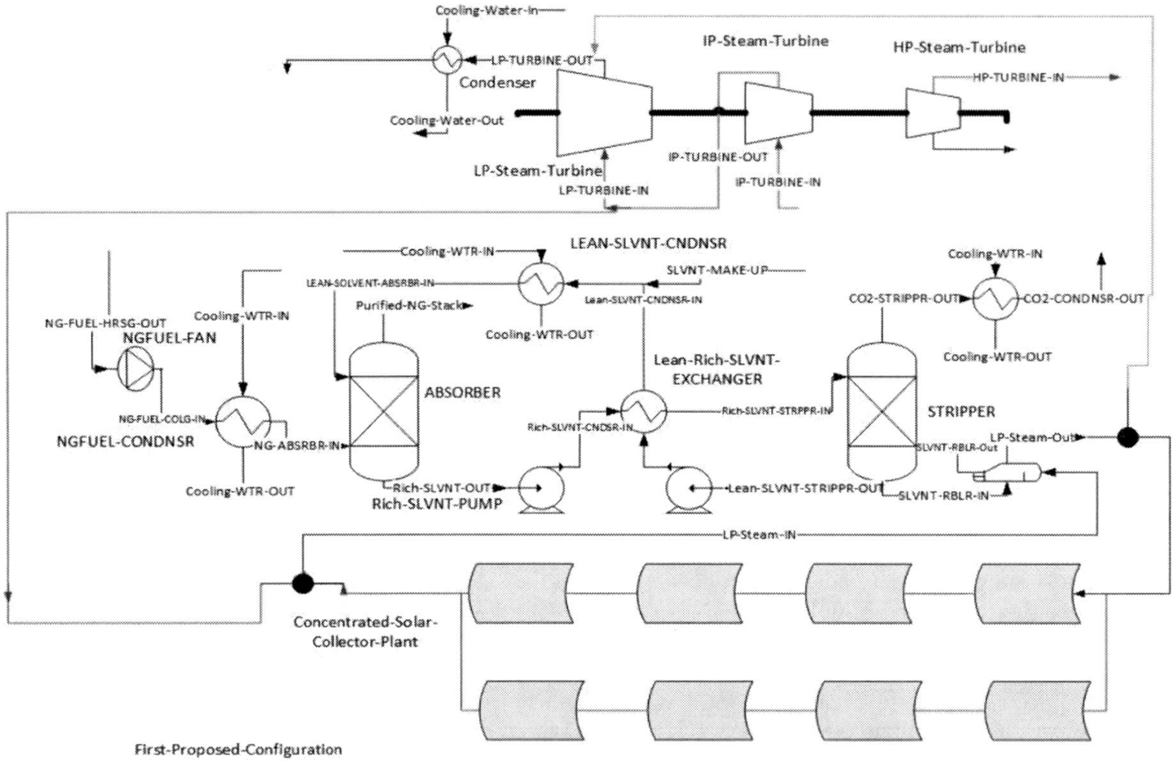

Figure 3.1. shows the first proposed PCCC-solar assisted plant (First Configuration).

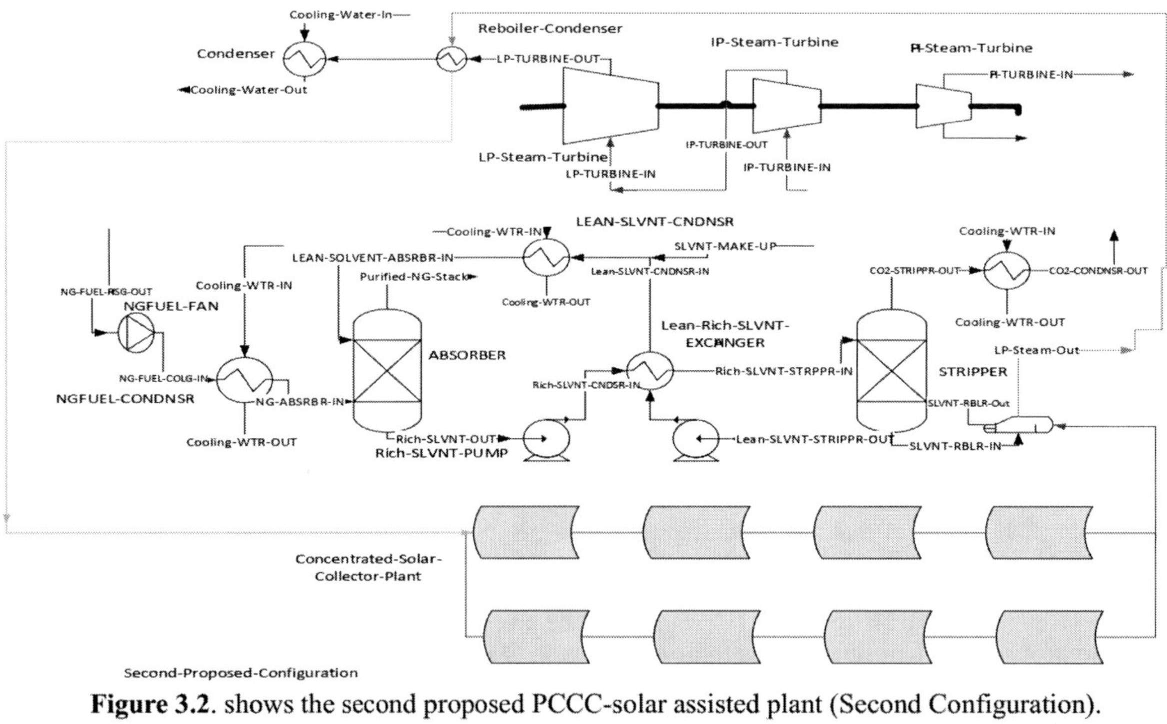

Figure 3.2. shows the second proposed PCCC-solar assisted plant (Second Configuration).

4. Cooling water model including the effect of the carbon capture.

The net cooling water mass flow rate in the plant's cooling system including the effect of carbon capture unit can be calculated according to Zhai and Versteeg (2011) as:

$$m_c^{\cdot} = \frac{(Hr-3600).MWg.1000.(1+\eta_{aux})-q_r^{CCS}}{C_P\Delta T_w.1000} + m_c^{CCS} \quad (1)$$

Where Hr is the steam cycle heat rate (kJ/kWh), 3600 is a unit conversion factor, $m_c^{\cdot}$ is the total recirculated cooling water (kg/h), m_c^{CCS} is the used cooling water in the CCS unit, MWg is the plant gross electric power, q_r^{CCS} is the extracted heat for solvent regeneration in the cooling system (kJ/h) where (Zhai and Versteeg, 2011):

$$q_r^{CCS} = q_{reboiler}^{Duty} + q_{Auxiliary}^{CCS} \quad (2)$$

As can be seen, most of the carbon capture duty comes from the reboiler. ΔT_w in Eq. (1) is the temperature difference between the inlet and the outlet of the condenser (oC), η_{aux} is the auxiliary loading (%), C_P is the water specific heat (kJ/kg.oC). In addition, as the used cooling system in the present work is wet cooling system, raw water withdrawal and consumption can be determined according to Saif W. M. et al. (2019) as:

$$Raw\ water\ consumption \left(\frac{\frac{gal}{min}}{MW_{net}}\right) = \frac{Evaporated\ losses}{Net\ Produced\ Power} \quad (3)$$

$$Evaporated\ losses = 0.0008 \times \Delta T_{tower} \times m_c^{\cdot} \times 1.8 \quad (4)$$

Where ΔT_{tower} is the temperature difference across the cooling tower.

$$Blowdown\ losses = \frac{Evaporated\ losses}{COC-1} \quad (5)$$

Where COC is the number of cycles of concentration where four (4) COC are considered in the present work. Furthermore, draft loss is assumed to be equal to 0.1% of the recirculated mass flow $(m_c^{\cdot})$ in the present study.

Finally, raw water withdrawal can be calculated as (Saif W. M. et al. 2019):

$$Raw\ water\ withdrawal \left(\frac{\frac{gal}{min}}{MW_{net}}\right) = \frac{Evaporated\ losses + Blowdown\ losses + Draft\ Losses}{Net\ Produced\ Power} \quad (6)$$

As it can be seen from equation (1) that extracting heat from the LP turbine entrance has an advantage on plant condenser heat duty and consequently on net water requirements of the system in spite of its impact on reducing turbine's net produced power. Despite utilizing solar assisted plant thermal energy for solvent regeneration in the stripper leads to reduce power penalty, condenser duty and its associated cooling water mass flow rate would be increased according to Eq. (1). In addition, the related solar land area is involved directly with the solar plant's produced thermal energy.

5. Solar field and solar contribution factor (SCF) calculations.

To make using solar assisted plant more viable in term of water and condenser duty saving, penalty decreasing, and solar land area reducing, two proposed configurations have been investigated in the present study to share the reboiler required heat duty between the solar thermal energy and the extracted heat form the LP turbine entrance one time and from the condenser heat duty another time as have been mentioned in the previous two sections. Therefore, reboiler duty $q_{reboiler}^{duty}$ (W) from Eq. (2) can be calculated as:

$$q_{reboiler}^{duty} = q_{solr}^{therm.enrg} + q_{extracted}^{therm.enrg} \quad (7)$$

Where $q_{solr}^{therm.enrg}$ is the solar produced thermal energy (W) while $q_{extracted}^{therm.enrg}$ is the extracted thermal energy whether from the LP turbine entrance or from the condenser heat duty (W).

Now, to calculate the solar plant's required land area, the below steps should be followed (Zhai and

Versteeg, 2011):

$$q_{solr}^{therm.enrg} = m_{fluid}^{\cdot} \times (h_{out} - h_{in}) \quad (8)$$

$$q_{solr}^{therm.enrg} = q_{rec} - q_{losses} \quad (9)$$

$$q_{losses} = F \times Ac \quad (10)$$

Where $m_{fluid}^{\cdot}$, h_{out} and h_{in} are the mass flow rate and the inlet and the outlet enthalpy of the solar superheated fluid stream, q_{rec} is the heat transferred to the collector receiver, F is losses factor. In the present study, it has been assumed that F is equal to (0.008 W/m^2 °C).
On the other side,

$$q_{rec} = Q_{Solr} \times \eta_{opt} \times IAM(\theta) \times \eta_{END} \times \eta_{CLN} \quad (11)$$

$$Q_{solr} = DNI \times Ac \quad (12)$$

Where η_{opt}, and η_{CLN} are the optical and cleanliness efficiencies provided by the manufacturer, *DNI* is the direct normal irradiation (W/m^2). $IAM(\theta)$ is the incident angle modification, η_{END} is the end-losses efficiency, and this can be calculated as (Zhai and Versteeg, 2011):

$$\eta_{END} = 1 - \frac{F_L}{M_L}\tan(\theta_L) \quad (13)$$

F_L is the collector focal length while M_L is the collector module length. θ_L is the longitudinal angle in degree between line perpendicular on the horizontal surface and solar rays' projection on the collector longitudinal plane. This angle can be calculated as (Zhai and Versteeg, 2011):

$$\theta_L = sin^{-1}(\cos(\gamma).\cos(\alpha)) \quad (14)$$

Where γ is the azimuth angle between solar ray's horizontal plane projection and the south side while $\boldsymbol{\alpha}$ is the elevation angle between solar rays and their projection on a horizontal plane. Figure 5.1. shows a concentrated solar collector configuration showing all the relevant angles and dimensions.

To calculate solar contribution factor (SCF) two different models be would be derived from the two different proposed configurations according to the previous equations. For the first method, the net steam mass flow rate can be calculated as:

$$m_{rebilr.strm}^{\cdot} = m_{LP-TRBN.EX}^{\cdot} + m_{LP-Solar}^{\cdot} \quad (15)$$

Where $m_{rebilr.strm}^{\cdot}$ is the net steam mass flow rate which comes from both the LP extracted steam and the solar superheated steam to be used in the reboiler while $m_{LP-TRBN.EX}^{\cdot}$, and $m_{LP-Solar}^{\cdot}$ are the mass flow rates of steam streams for both the extracted LP steam and the solar superheated steam part respectively. Thus, solar contribution factor (SCF) would be calculated in the present configuration as:

$$SCF_{first.Conf.} = \frac{m_{LP-Solar}^{\cdot}}{m_{rebilr.strm}^{\cdot}} \quad (16)$$

While in the second configuration, the total thermal energy duty in the reboiler would be distributed between the gained heat duty from the condenser and the solar thermal energy part as below: -

$$Q_{reboiler-duty} = Q_{Condenser-Duty} + Q_{Solar-Thermal} \quad (17)$$

Q in Eq. 17 represents the heat duty of the system. Thus, solar contribution factor (SCF) can be calculated in the current design as: -

$$SCF_{second.Conf.} = \frac{Q_{Solar-Thermal}}{Q_{reboiler-duty}} \quad (18)$$

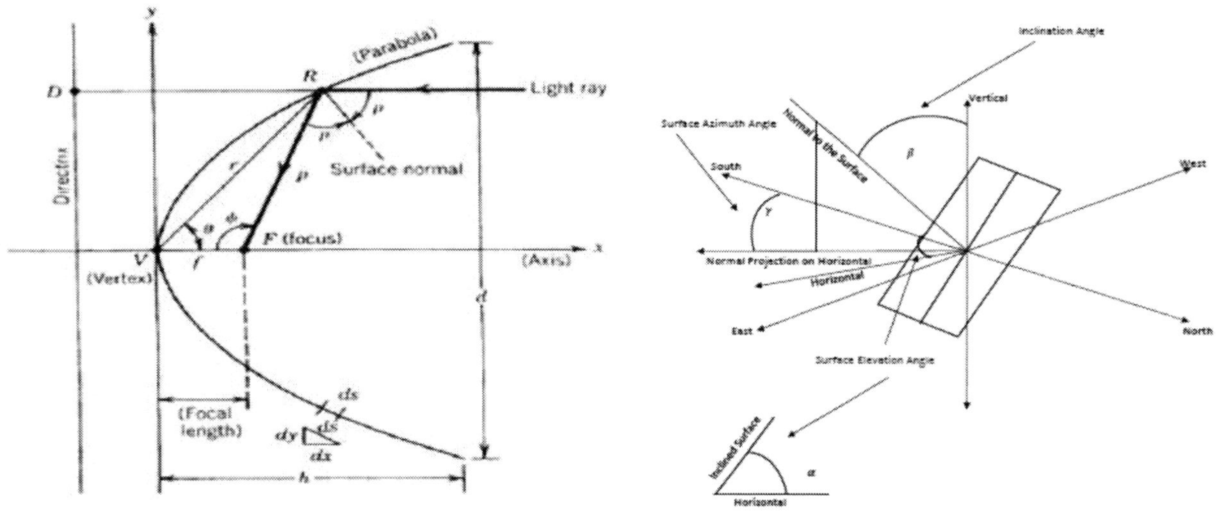

Figure 5.1. Shows all the relevant angles and dimensions for the concentrated solar collector (Sup A. B. et al., 2015).

6. Results and Discussion.

The effect of solar contribution factor (SCF) on plant's main condenser duty with its associated water usage, plant's power penalty, and the required solar field area at different direct normal irradiation (DNI) values under four different SCF; 0.2, 0.4, 0.6, 0.8 would be discussed in this section.

6.1. Result Validation.

The result has been validated for a 559 MWe NGCC power plant by comparing the based plant simulation outputs with a data provided by NETL report (2015). As being shown in the table below.

Table 6.1. Result validation with the NETL report (2015) for the based NGCC equipped with a PCCC unit.

Parameters for Validation	NETL report Result	Present Study Result
Plant Gross Power with PCCC (MWe)	601	601
Steam LHV efficiency with PCCC	33.5	33.5
Net LHV Efficiency with PCCC	50.6	50.7
Condenser Duty Without PCCC (MW_{th})	355.83	373.90
Raw Water Withdrawal without PCCC (gal/min)/MWe_{net}	4.20	4.26
Raw Water Consumption without PCCC (gal/min)/MWe_{net}	3.30	3.32
Condenser Duty with PCCC (MW_{th})	246.66	260.98
Raw Water Withdrawal with PCCC (gal/min)/MWe_{net}	7.20	7.34
Raw Water Consumption with PCCC (gal/min)/MWe_{net}	5.40	5.72

It can be seen from table 6.1 that all the main parameters of the simulated NGCC plant, which is integrated with the PCCC unit, have been validated properly with the corresponding data from the report except a minor difference in the values related to the condenser duty and its involved water requirement. This likely happens because of using different thermophysical property method for the saturated steam calculations than the method which was used by the NETL report to calculate the thermophysical properties of the saturated steam.

Table 6.2. shows the main parameters which would be used in the solar field calculations according to Cau et al. (2014).

Table 6.2. represents the main the parameters used in the solar field calculations of the present sturdy.

The parameters	The corresponding value
Elevation Angle / Azimuth Angle (degree)	75.0°, 0.0 °
Focal Length (m)	7.4
Module Length (m)	45
IAM (θ)	0.19
η_{opt}	0.66
η_{CLN}	0.98

6.2. The effect of SCF on plant performance and cooling system performance.

The increase in the solar contribution factor (SCF) means that solar thermal energy is increasing as a part of the reboiler duty in the carbon capture unit. Any raise in the solar part thermal energy has consequences on plant performance where the net produced power would be changed in addition to its effect on the performance of the cooling system as a result according to Eq. (1). Thus, the increase in the SCF leads to a reduction in the power penalty or increase in the net produced power and increase in the net condenser duty. Figure 6.1-a and Figure 6.1-b demonstrate the effect of increasing the solar contribution factor (SCF) on both the plant's net condenser duty and plant's power penalty respectively for the proposed scenarios. Figure 6.1-a shows that the increase in the SCF leads to increase in the condenser heat duty for both scenarios. However, the increase in the cooling load for the second scenario is less than the increase for the first one for SCF less than 0.7. This comes from the fact that net produced power and mass flow rate of the LP-steam would not be changed in the second scenario where the extraction will be from the condenser heat duty directly. On the other hand, the increase in the SCF in the first scenario would be accompanied with a change in the net produced power and in the mass flow rate of the LP-steam leading to more increase in the cooling load. While after SCF equals to 0.7, the change in the net produced power and mass flow rate of the LP-steam in the first configuration will not be significant comparing to the second case and the difference in the condenser heat duty would be minor between both designs.

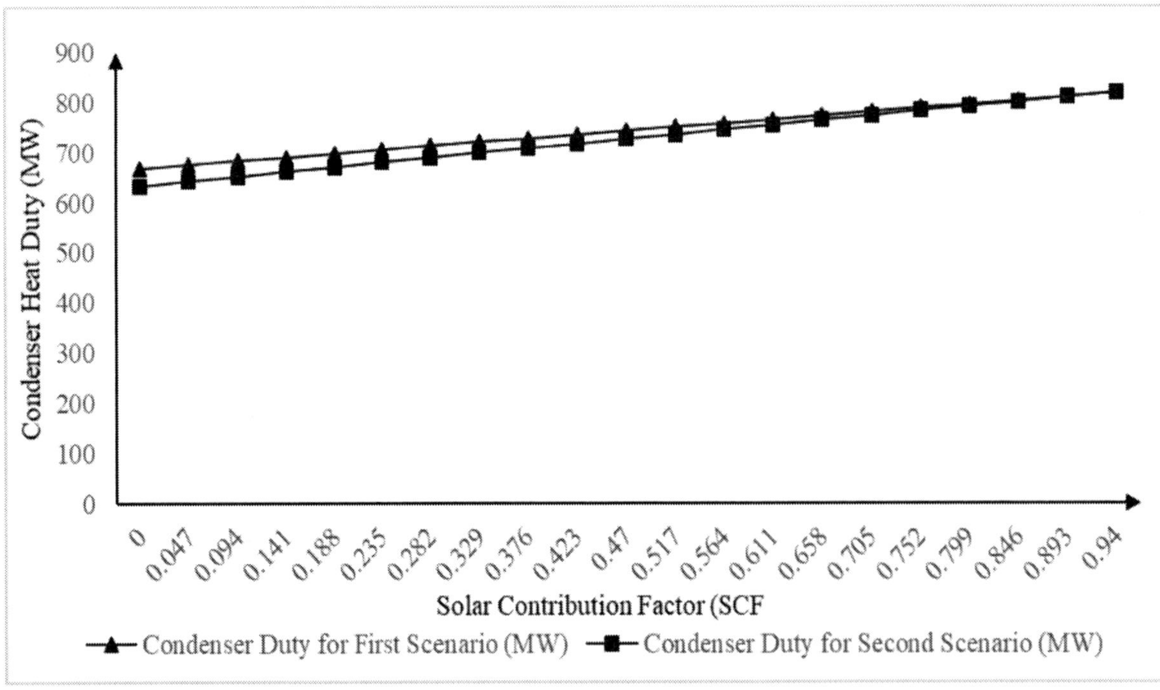

Figure 6.1.-a Shows the effect of solar contribution factor (SCF) on the Condenser Duty for both scenarios.

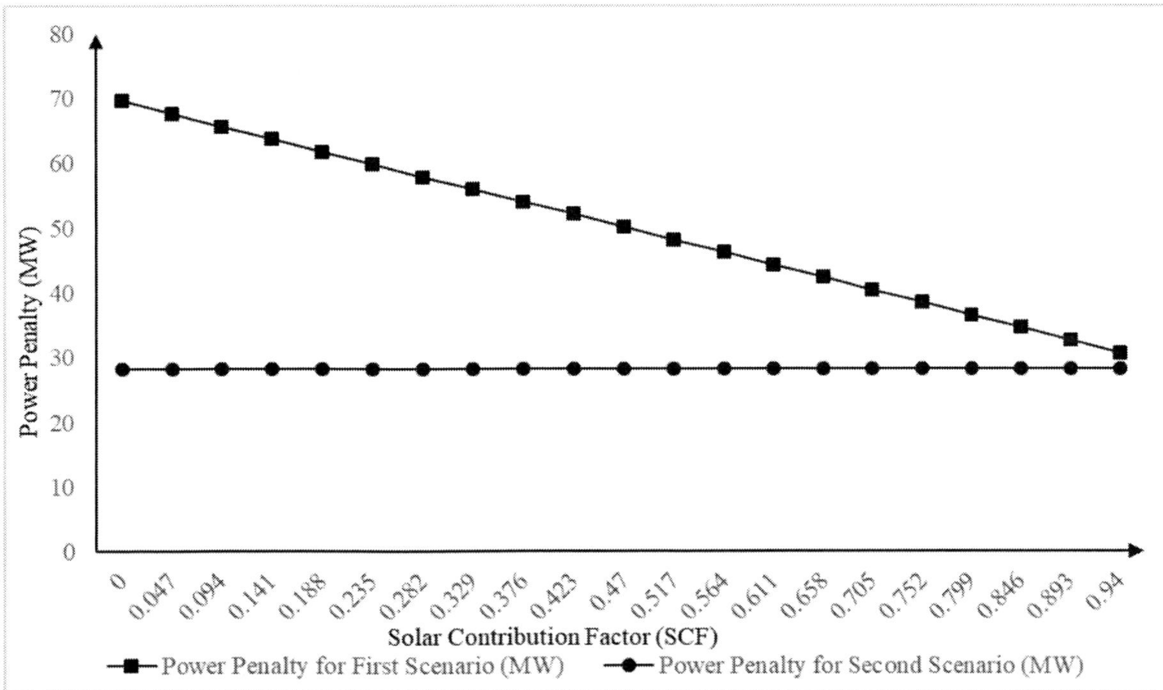

Figure 6.1.-b Shows the effect of solar contribution factor (SCF) on the power penalty for both scenarios.

Figure 6.1.-b shows that power penalty is reduced with increasing SCF in the first scenario while it keeps constant in the second scenario. As has been mentioned before, the reason behind this is that the extraction from the LP-turbine is reduced gradually with increasing the SCF leading to decrease in the power penalty or increase in the net produced power consequently. In contrast, the extraction would be from the condenser heat duty directly in the second scenario without need to extract from the LP-turbine causing zero penalty in the net produced power by neglecting the effect of the back pressure in the condenser.

Figure 6.2. demonstrates the effect of SCF on the net raw water withdrawal and consumption in the integrated wet cooling system for both the first and the second scenarios in term of flow rate unit per power unit. It is shown that increasing SCF leads to reduce these amounts in the first scenario in contrast with the second scenario where these amounts would be increased. This is caused by the increase in the net produced power with increasing the SCF in the first scenario while net produced power keeps constant in the second scenario. Therefore, because the water amounts in figure 6.2 is in term of flow rate unit per power unit where power amount is increased in the denominator for the first scenario and keeps constant in the second scenario, the increase in the SCF leads to a reduction in the water amount per unit power in the first scenario and increase in this amount in the second scenario. This comes as a result of two reasons; first, there is no power penalty in the second scenario leading to a constant produced power with increasing SCF while the increase in the SCF of the first scenario causes reduction in the penalty leading to increase the net produced power as has been demonstrated in figure 6.1-b. However, as increasing SCF leads to increase in the net condenser duty for both scenarios as was shown in figure 6.1-a, the increase in the SCF leads to increase in the water usage amounts without considering the effect of the power unit in the denominator because the consumed water amount relates directly with the net condenser duty.

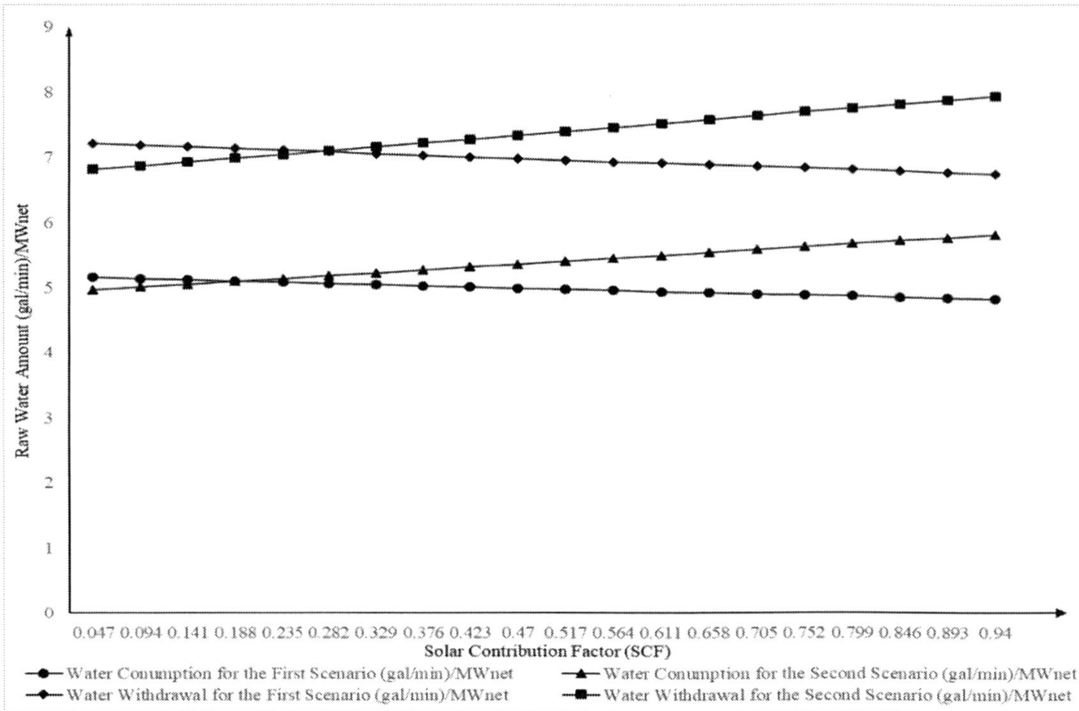

Figure 6.2. Shows the effect of solar contribution factor (SCF) on water withdrawal and consumption.

6.3. The effect of the SCF on the solar field required area.
In addition to its effect on the plant and cooling system performance, SCF has a direct and significant impact on the required solar field land area.

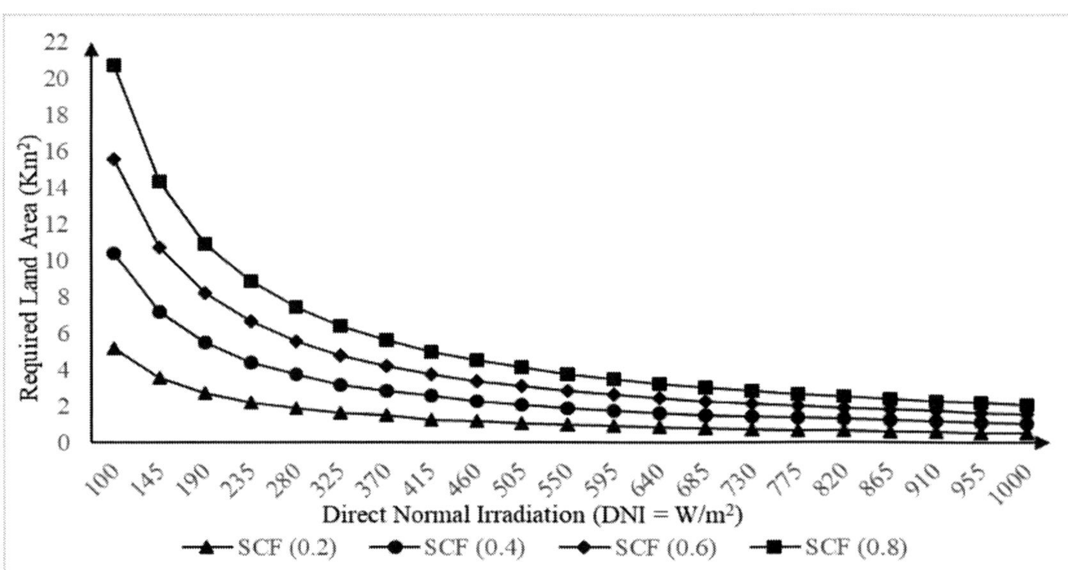

Figure 6.3. Shows the effect of solar contribution factor (SCF) on the solar field land area at different DNI (W/m^2) for any of the above studied cases because the required solar energy is constant for both methods.

Figure 6.3. shows the relationship between the required solar field land area and the direct normal irradiation (DNI) at four different SCF; 0.2, 0.4, 0.6, and 0.8. As the figure demonstrates, as DNI increases, the required solar field land area decreases exponentially when the reboiler heat duty keeps constant. This behavior is expected because large DNI, small land area is required when the produced thermal energy keeps constant. However, this relation has been discussed to get an indication about the required solar land area at different SCF and different DNI values. An interesting finding is that a significant reduction in the solar field land area at large DNI values was observed from figure 6.3 as a result of the exponential relationship between the land area and the DNI. In addition, it is noticed from figure 6.3 that as the increase in the SCF leads to increase in the solar field land area expectedly, the differences in the land area values under different SCF at large DNI is much lower than these differences at small DNI values. This likely happens because of the exponential relationship between the solar field land area and the DNI values.

7. Conclusions and recommendation.

Two novel configurations have been proposed in the present work for a solar assisted power plant to be integrated to a post combustion carbon capture unit (PCCC) of a (630 MWe) natural gas combined cycle (NGCC) power plant. The effect of these configurations on the performance of the plant produced power and the cooling system load was investigated in this study. In addition, the effect of the proposed configurations on the required solar field land area has been impeded in the present study too. The results were evaluated in the present work using the SCF as the main criteria. It was shown that the second configuration method is better than the first one regarding condenser cooling load and power penalty reduction. However, water usage amounts in term of flow rate unit per power unit are reduced with increasing SCF in the first method rather than the second one in spite of the increasing in the condenser heat duty with SCF in both scenarios because the net produced power would be reduced with increasing the SCF in the first configuration while it keeps constant in the second configuration as the second configuration consumes zero penalty from the plant produced power. Another interesting finding is that at high DNI values, the required solar land area would be reduced significantly where the smallest area was 0.518 Km^2 at SCF equals to 0.2 when the maximum solar radiation value reaches 1000 W/m^2 concluding that the solar assisted plant is more viable in regions provided higher concentrated solar radiation. Finally, the validation with the NETL report (2015) proves the reliability of the results presented in the current paper. For the future work, it is recommended to use different configuration for the second method where this method represents the most feasible case as was proven above. The recommendation is that controlling the SFC factor would be by controlling the mass flow rate of the LP-stream instead of controlling the SCF by the partial heating between the solar and cooling systems.

Acknowledgments

The Author is very pleased to thank HCED (Higher Committee of Education Development in Iraq) and Kufa university for providing the required financial support to perform this research in addition to his thanks to Lehigh University for allowing him doing this research in Lehigh's intellectual property.

References

[1] Bukhary, S., Ahmad, S., & Batista, J. (2018). Analyzing Land and Water Requirements for Solar Deployment in the Southwestern United States. *Renewable and Sustainable Energy Reviews*, *82*, 3288-3305, https://doi.org/ 10.1016/j.rser.2017.10.016.

[2] Cau G., Cocco D., & Tola V., (2014), Performance Assessment of USC Power Plants Integrated with CCS and Concentrated Solar Collectors, Energy Conversion and Management, (88) 973-984, https://doi.org/ 10.1016/j.enconman.2014.09.040.

[3] Hong H., and Jin H., (2005), A novel solar thermal cycle with chemical looping combustion, Int. J. Green Energy, 2(4), 397-407, https://doi.org/10.1080/01971520500288022.

[4] Li Y., Zhang N., and Cai R., (2013), Low CO_2-emissions hybrid solar combined-cycle power system with methane membrane reforming, Energy, (58) 36-44, https://doi.org/10.1016/j.energy.2013.02.005.

[5] Lui L., Zhao J., Deng S., & An Q., (2016), A Technical and Economic Study on Solar-Assisted Ammonia-Based Post-Combustion CO2 Capture of Power Plant, Applied Thermal Energy, (102) 412-422, https://doi.org/ 10.1016/j.applthermaleng.2016.03.154.

[6] Mukhtar M., Ali T. M., Khalilpour R., Abbas A., Shah N., Hajaj A. A., Armstrong P., Chiesa M., & Sgouridis S., (2012), Solar-Assisted Post-Combustion Carbon Capture Feasibility Study, Applied Energy, (92) 668-676, https://doi.org/ 10.1016/j.apenergy.2011.07.032.

[7] Mohammed Ali SW, Vahedi N, Romero C, Banerjee A. Water usage and energy penalty of different hybrid cooling system configurations for a natural gas combined cycle power plant—Effect of carbon capture unit integration. Int J Energy Res. 2019;1–18. https://doi.org/10.1002/er.4694.

[8] Stults, E. S. (2015). Minimizing Water Requirement for Electricity Generation in Water Scarce Areas (Ph.D.'s Dissertation, Worcester Polytechnic Institute).

[9] Sup A. B., Zainudin F. M, ZanariaShamsirAli T., Bakar A. R, & Ming L. G., (2015), Effect of Rim Angle to the Flux Distribution Diameter in Solar Parabolic Dish Collector, Energy Procedia, (86) 45-52. https://doi.org/ 10.1016/j.egypro.2015.03.231.

[10] Wu, X., & Chen, G. (2017). Energy and Water Nexus in Power Generation: The Surprisingly High Amount of Industrial Water Use Induced by Solar Power Infrastructure in China. Applied Energy, 195, 125-136. https://doi.org/ 10.1016/j.apenergy.2017.03.029.

[11] Zhai H., and Rubin E.S., (2011), Carbon Capture Effect on Water Usage at Pulverized Coal Power Plant, *Energy Procedia*, 4:2238-2244, https://doi.org/10.1016/j.egypro.2011.02.112.

[12] Zhai, H., Rubin, E. S., & Versteeg, P. L. (2011). Water Use at Pulverized Coal Power Plants with Post combustion Carbon Capture and Storage. *Environmental Science & Technology*, *45*(6), 2479-2485. https://doi.org/10.1021/es1034443.

[13] Zhai R., Liu H., Wu H., Yu H., & Yang Y., (2018), Analysis of Integration of MEA-Based CO2 Capture and Solar Energy System for Coal-Based Power Plants Based on Thermo-Economic Structural Theory, Energy, (11) 1284.

[14] Zoelle, A., Keairns, D., Pinkerton, L. L., Turner, M. J., Woods, M., Kuehn, N., Chou, V. (2015). Cost and Performance Baseline for Fossil Energy Plants Volume 1a: Bituminous Coal (PC) and Natural Gas to Electricity Revision 3, https://doi.org/10.2172/1480987.

IOP Conf. Series: Materials Science and Engineering **978** (2020) 012022 doi:10.1088/1757-899X/978/1/012022

Mechanical Properties of Lightweight Concrete Made with Shredded Waste Polypropylene Yogurt Containers

Ahmed Mudhafar Hashim[1,*], Aseel Mahmood Abdullah [2] Amar Al-Baaji[3] and Niveen Jamal Abdulkader[4]

[1]College of Engineering, University of Al-Qadisiyah, Iraq
E-mail: ahmed.hashim@qu.edu.iq
[2]College of Engineering, University of Technology, Iraq
E-mail: 130027@yotechnology.edu.iq
[3]College of Engineering, University of Al-Qadisiyah, Iraq
E-mail: amar.jabar@qu.edu.iq
[4]College of Engineering, University of Technology, Iraq
E-mail: 130012@yotechnology.edu.iq
* Corresponding author: College of Engineering, University of Al-Qadisiyah, Diwaniyah, Iraq. Tel: +9647724633444.

Abstract. The partial replacement of natural coarse (gravel) aggregate by waste plastic aggregate to produce lightweight concrete has become one of the major challenges in recent years and stresses on the need for a constant and authoritative development for the betterment of the environment. This work was carried out to investigation the possibilities of using waste polypropylene yogurt containers as polymer aggregate in concrete application. The concrete samples with 0%, 5%, 10%, 15%, 20% and 25% recycled waste plastic aggregate as partial replacement by mass of natural coarse aggregate are prepared after a curing period of 28 days. The effect of recycled plastic aggregate on compressive strength, density and water absorption of natural concrete are studied and compared with the results of specimen without plastic aggregate. Both compressive strength and density decreased with increasing the amount of plastic aggregate. The highest reduction in compressive strength and density were 50% and 19% respectively of concrete cubes containing 25% wt polypropylene waste aggregate in comparison of traditional concrete. It was found that water absorption increased as the plastic aggregate content increased up to 15% wt and then decreased when replaced coarse aggregate by 20% wt and 25% wt by recycled polypropylene aggregate. This study is recommended to use recycled polypropylene yogurt containers as a partial substance of traditional coarse (gravel) concrete mix up to 15% wt for applications include those where low compressive strengths and low cost production are desirable.

Keywords: *waste plastic, lightweight concrete, compressive strength, physical properties.*

1. Introduction

All over the world, plastic materials have become an essential and integral part of our modern lives by the variety of its features have been conducted to its recurrent use in wide range sectors. The year by year the rate of global plastic production has increased drastically during the few past years from 300 million tons in 2011 to 335 million tons in 2016 (Richard, P *et al* 2018 and Ossi Martikka *et al* 2019) . A quantity of plastic materials is used in food industry (containers, bottles, packaging etc.). The success of plastics as packaging material has been substantial, they has offer many features over metals and ceramics such as light weight, ease of design, good weight to strength ratio, good chemical resistance, tough and low cost (M., Sulyman *et al* 2016 and M. B. Hossain *et al* 2016) . Unfortunately, these plastics have serious drawbacks. They non – degradable and increasing the amount of solid waste have been making significant contribution to the increase in environmental pollution Mohd Syahrul Hisyam Mohd Sani *et al* 2019.

One of the feasible application of waste plastics as shredded aggregate is in construction fields for produce lightweight concrete in order to reduce the total weight of concrete. Various types of the common waste thermoplastics are used as lightweight aggregate in the preparation of lightweight concrete such as ethylene vinyl acetate EVA E. Q. R. Santiago *et al* 2009, low-density polyethylene LDPE (A. O. Sojobi *et al* 2014 Manish Chaudhary *et al* 2014), high- density polyethylene HDPE (Deividas Rumšysa *et al* 2017 and Gopal Swarup Sangal 2018), polypropylene PP Martina Záleská *et al* 2018, polyvinylchloride PVC (S.C. Kou *et al* 2009, Y. Senhadji *et al* 2015, N. Haghighatnejad *et al* 2016 and H. H. Hussein *et a*l 2017) and polystyrene PS Andrea Petrella *et al* 2020.

Polypropylene PP is one the most conventional widely thermoplastics used in the field of dairy milk packaging due to its good barrier properties, poor gas barriers, tough, inert, clear, good surface finish and low cost Fahad K. Alqahtani *et al* 2015. The lowest value of density making the polypropylene the lightest plastic materials known and this feature makes it the best candidate for lightweight concrete. In this present work, we are aimed the study the effectiveness of waste polypropylene yogurt containers as partial substances for coarse (gravel) aggregates and then compared the attained results to the traditional concrete mix. (cement, fine (sand) aggregates and coarse (gravel) aggregates.

2. Experimental

2.1 Materials

Waste polypropylene yogurt containers used in this study has been collected from various local supermarkets located in Diwaniyah city – south of Iraq. Some of the PP waste containers is shown in Figure 1.

Figure 1. Waste polypropylene yogurt containers

High sulphate resistance Portland cement (Al-Jesr trade mark) type V. CEM I 42.5 SR, coarse aggregates (gravel), and fine aggregates (sand) were supplied from one the local building materials store in Diwaniyah city – south of Iraq. The cement source is Lafarge Iraq and manufactured according to the Iraqi standard number 5-1984. The properties of cement used are summarized in Table 1.

Table 1. Cement resistant sulfate-properties typical.

Parameters		Range
Lime Saturation Coefficient		0.9 - 1.0
Magnesium Oxide (MgO) %		1.5 - 2.5
SO_3 Content %		2 - 2.3
Loss on Ignition %		2.2 - 3.2
Insoluble Residue %		0.5 - 1.2
Tricalcium Aluminates (C_3A) %		2.4 - 3.1
Fineness (m^2/kg)		320 - 335
Initial Setting Time (min)		125 - 175
Final Setting Time (Hr)		3 - 3.7
Soundness Lechatelier (mm)		0 -2
Specific Gravity (kg/litter)		3.10 - 3.18
Bulk Density (kg/m^3)		1.440 ± 0.1
Compressive Strength as per Iraqi Standard (mN/m^2)	3d	30 - 36
	7d	40 - 45
	28d	55 - 60

2.2 Preparation of low weight concrete samples:

The processing steps that were followed to preparation of recycled PP aggregate as follows:

Step one washed the waste plastic after collected with water to remove all remaining products inside the containers and then cleaning the outer surface from contaminants (coating materials) with aprotic solvent (acetone).

Step two shredded and crushed in a plastic crushed machine which has blades that break the plastic into smaller flake sizes (Figure 2).

Figure 2. Waste polypropylene shredded.

Step three the waste plastic crushed were processed in an extruder – molding machine. The extruded PP recycling sheets cutting to small size (8 – 12 mm angular) and then used to replace coarse aggregate (gravel) for making the concrete cubic samples.

Figure 3. Waste Polypropylene aggregates used in concrete.

2.3 Preparation of low weight concrete samples

A steel cube molds of dimension 100 mm × 100 mm × 100 mm respectively was prepared. For the preparation procedure of concrete is done using hand lay-up technique. Calculated of concrete samples were produced using cement: fine aggregates (sand): coarse aggregates (gravel and recycled PP) ration of 1:2:4 and water to cement ratio of 0.5 and cured for 28 days according to ASTM C192.

Figure 4. Casting of concrete.

Five different concentrations of recycled PP aggregates (5%, 10%, 15%, 20% and 25%) of total weight of gravel aggregates were added to the concrete mix proportions. Table 2 illustrated the compositions of the concrete mixtures.

Table 2. Composition of examined concrete mixtures.

code	Waste PP weight fraction %	No. of samples
C0	0	3
CP5	5	3
CP10	10	3
CP15	15	3
CP20	20	3
CP25	25	3

2.4 Testing

2.4.1 Compressive Test

The compressive strength test of cubes samples of concrete with and without PP aggregates for that three cubes were prepared for each mix. The test was performed according to ASTM C109 using machine. The compressive load was applied continuously until the cube sample failed.

2.4.2 Density

In this work, the density of different concrete cube samples is the theoretical density. The density was tested according to ASTM C 642 on the same samples, which utilized for compressive test and calculated by dividing the weight on the volume of each cube sample.

2.4.3 Water Absorption

The water absorption was tested according to ASTM C642 by drying each of cube sample in an oven at 105 C for 24 hours after curried and then weighting it. The samples were immersed in water for 24 hours. The cube samples were taken out from the water, dried and reweighed.

3. Results and Discussion

3.1 Compression Test

Results of summarizes compressive strength of all different of cube concrete samples after 28 days of curing was illustrated in Table 3.

Table 3. Results from compressive test of concrete cubes after 28 days.

code	Waste PP weight fraction %	Average weight kg	Average compressive strength MPa
C0	0	2.370	19.885
CP5	5	2.356	16.835
CP10	10	2.160	14.94
CP15	15	2.095	13.885
CP20	20	2.075	11.52
CP25	25	1.925	10.385

The experimental results indicated that the average value of the compressive strength of traditional concrete slightly decrease to 15 % when coarse aggregate (gravel) replaced by 5% wt with the recycled PP aggregate. The values of compressive strength are more decreasing when the concentrations of recycled plastic aggregate increased to 10%, 20% and 25% wt. the highest reduction was 50% of cube samples with 25% wt waste PP aggregate as shown in Figure 5.

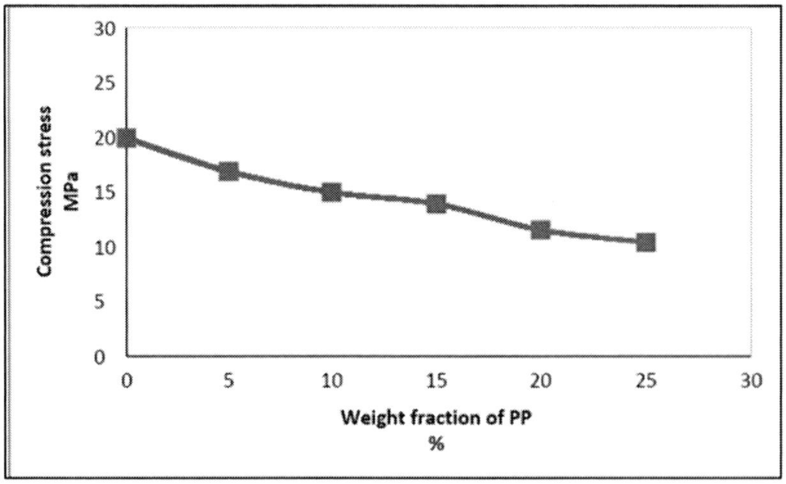

Figure 5. Compression strength of concrete cubes after 28 days.

The lower values of average compressive strength of cube concrete samples containing waste plastic aggregate than samples containing natural aggregates is due to reduction in bonding between plastics pieces with cement matric Raghatate Atul M. 2012 and the lower compressive strength value of PP aggregate.

3.2 Density

The results reported that the density of cube concrete samples decreased with the increase of plastic aggregate ratio in composition by respectively when compared to the traditional concrete as shown in Figure 6. The reduction in densities is related to the lower density of waste PP aggregate than the coarse (gravel) aggregate. The largest reduction in density about 19 % for cube samples containing 25% wt waste PP aggregate.

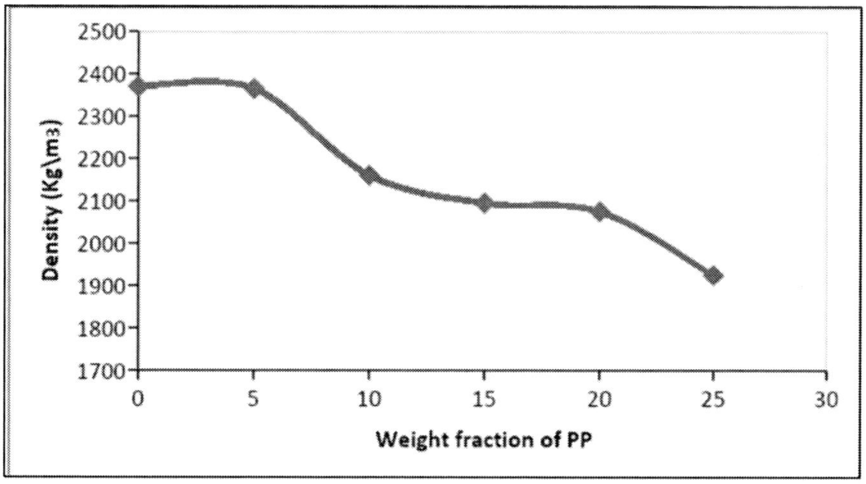

Figure 6. Density of concrete cubes after 28 days.

3.3 Water Absorption

In the case of water absorption, the relationship between the percentages of water absorption and waste plastic content are illustrated in Figure 7. It was observed that the percentage of water absorption of concrete increased the double with increasing the amount of waste PP aggregate up to 15% wt. The percentage of water absorption was 3% for natural coarse aggregate and 6% for concrete with 15% wt waste PP aggregate. Then the percentage of water absorption starts to decrease as polymer content in concrete mix increases, reaching a minimum value of 4.6% for concrete cube sample containing 25% wt of plastic aggregate higher than the value of the concrete sample without any plastic aggregate.

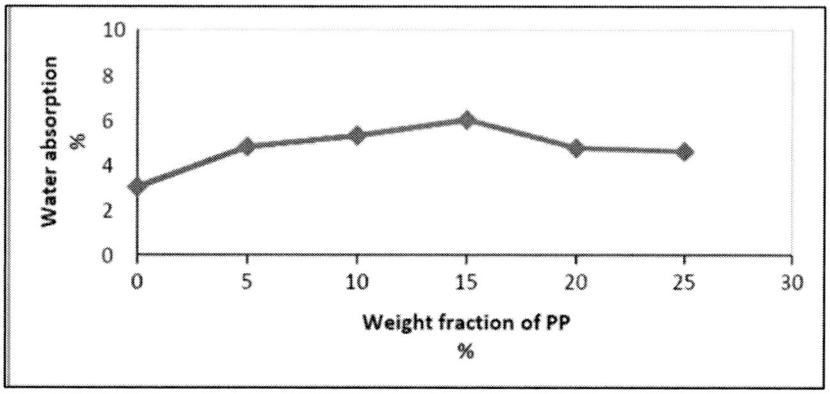

Figure 7. Percentage of water absorption of concrete cubes after 28 days.

IOP Conf. Series: Materials Science and Engineering **978** (2020) 012022 doi:10.1088/1757-899X/978/1/012022

The results of values of water absorption of all concrete mixes is very low under 10 %. This displays the good quality of all concrete mixes in this work (Neville, A.M. 2011)

4. Conclusions

It may be possible to use recycled PP yogurt containers as a partial substance of traditional coarse (gravel) concrete mix up to 15% wt for applications include those where low compressive strengths are desirable.

The density of all types of recycled PP concrete was lesser than that of traditional concrete (only sand and gravel aggregate).

It has been noticed that compressive strength values of concrete containing different concentrations of plastic aggregate decreased between and when compared to the traditional concrete.

Water absorption increased as the plastic aggregate content increased up to 15% wt and then decreased when replaced coarse aggregate by 20% wt and 25% wt by recycled PP aggregate.

References

[1] Richard, P., Catherine, A., Rod Fox & Bill Hopkins (2018). Re-Formative Polymer Composites from Plastic Waste: Novel Infrastructural Product Application. *Recycling*, *3*(54), doi:10.3399/recycling3040054

[2] Ossi Martikka & Timo Karki (2019). Promoting Recycling of Mixed Waste Polymers in Wood-Polymer Compoeites Using Compatibilites. *Recycling*, *4*(6), doi:10.3399/recycling4010006

[3] M., Sulyman, J., Haponiuk & K., Formela (2016). Utilization of Recycled Polyethylene Terephthalate (PET) in Engineering Materials: A Review. *International Journal of Environment Science and Development*, *7*(2), doi:10.7763/IJESD. 2016.V7.749

[4] M. B. Hossain, P., Bhowmik, & K. M. Shaad (2016). Use of Waste Plastic Aggregation in Concrete as a Constituent Material. *Progressive Agriculture*, 27(3), 383-391.

[5] Mohd Syahrul Hisyam Mohd Sani, Muhammad Isha Ismail, Marzuki Ab Rahman & Fadhluhartini Muftah (2019). Waste Paper Ash Pellets as Coarse Aggregate Replacement in Concrete. *International Journal of Engineering and Advanced Technology (IJEAT)*, 8(4), 1112-1117.

[6] E. Q. R. Santiago, P. R. L. Lima, M. B. Leite & R. D. Toledo Filho (2009). Mechanical Behavior of Recycled Lightweight Concrete Using EVA Waste and CDW under Moderate Temperature. *IBRACON Structures and Materials Journal*, 2(3), 211-221.

[7] A. O. Sojobi and H. I. Owamah (2014), Evaluation of the Suitability of Low Density Polyethylene (LDPE) as Fine Aggregate in Concrete, *Nigerian Journal of Technology (NIJOTECH)*, 33(4), 409-425, dx.doi.org/10.4314/njt. v33i4.1.

[8] Manish Chaudhary, Vikas Srivastava and V. C. Agarwal (2014), Effect of Waste Low Density Polyethylene on Mechanical Properties of Concrete, *Journal of Academia and Industrial Research (JAIR)*, 3(3), 123-126.

[9] Deividas Rumšysa, Darius Bačinskasa, Edmundas Spudulisa & Adas Meškėnasa (2017), Comparison of material properties of lightweight concrete with recycled polyethylene and expanded clay aggregates, *Procedia Engineering*, 172, 937 – 944

[10] Gopal Swarup Sangal (2018). Study the Effect of Plastic Waste on Strength of Concrete. *International Journal of Advance Research and Development*, 3(7), 36-39.

[11] Martina Záleská, Milena Pavlíková, Ondřej Jankovský, Jaroslav Pokorný &Zbyšek Pavlík (2018), Lightweight Concrete Made with Waste Expanded Polypropylene-Based Aggregate and Synthetic Coagulated Amorphous Silica, *Ceramics-Silikáty* 62 (3), 221-232.

[12] S.C. Kou, G. Lee, C.S. Poon & W.L. Lai (2009), Properties of lightweight aggregate concrete prepared with PVC granules derived from scraped PVC pipes, *Waste Management* 29, 621– 628.

[13] N. Haghighatnejad, S.Y. Mousavi, S.J. Khaleghi, A.T. Tabarsa & S. Yousefi (2016), Properties of recycled PVC aggregate concrete under different curing conditions, *Construction and Building Materials* 126 943–950.

[14] Y. Senhadji, G. Escadeillas, A.S. Benosman, M. Mouli, H. Khelafi & S. Ould Kaci (2015), Effect of incorporating PVC waste as aggregate on the physical, mechanical, and chloride ion penetration behavior of concrete, *Journal of Adhesion Science and Technology*, 29(7), 625–640.

[15] H. H. Hussein, O.A. Edan & M.K. Ahmed (2017), Mechanical, thermal and acoustical properties of concrete with fine polyvinyl chloride (PVC), *Iraqi Journal of Civil Engineering*, 12(2), 81-91.

[16] Andrea Petrella, Rosa Di Mundo & Michele Notarnicola (2020). Recycled Expanded Polystyrene as Lightweight Aggregate for Environmental Sustainable Cement Conglomerates. *Materials*, 13(988), doi:10.3390/ma13040988

[17] Fahad K. Alqahtani, Gurmel Ghataora, M. Iqbal Khan, Samir Dirar, Azzedine Kioulk & Mansour Al-Otaibi (2015). Light Weight Concrete Recycled Plastic Aggregates *International Conference on Electromechanical Control Technology and Transportation (ICECTT).*

[18] Raghatate Atul M. (2012), Use of Plastic in A Concrete to Improve Its Properties, *International Journal of Advanced Engineering Research and Studies (IJAERS)*, 1(3), 109-111.

[19] Neville, A.M. (2011), "Properties of Concrete", Longman Group Ltd., United Kingdom.

Interfacial Characteristics of Engineered Pulp Fibre Reinforced Concretes Containing Various Pozzolans

Emad Booya[1*], Adeyemi Adesina[2], Karla Gorospe[3] and Sreekanta Das[4]

[1]Department of Civil and Environmental Engineering, University of Windsor, Canada
E-mail: booya@uwindsor.ca
[2]Department of Civil and Environmental Engineering, University of Windsor, Canada
E-mail: adesina1@uwindsor.ca
[3]Department of Civil and Environmental Engineering, University of Windsor, Canada
E-mail: gorospek@uwindsor.ca
[4]Department of Civil and Environmental Engineering, University of Windsor, Canada
E-mail: sdas@uwindsor.ca
* Corresponding author: Department of Civil and Environmental Engineering, University of Windsor, Canada. Tel: (1) (519) 992 1086

Abstract. Increasing durability threats on existing structures made with cementitious composites has prompted a need to find ways to improve the durability of new structures. One of the effective ways to improve the durability properties of cementitious composites is by the incorporation of fibres. A type of modified Kraft pulp fibres was employed in this study to improve the durability properties of cementitious composites. A microstructural investigation was also conducted to study the fibre morphology and its effect on the performance of the cementitious composites. Moreover, the durability tests in terms of water absorption and rapid chloride penetration were carried out. Overall results showed higher improvement of the durability of the cementitious composites with composites incorporated metakaolin, silica fume, and slag. This was further confirmed by microstructural analysis showing that these supplementary cementitious materials densified the cementitious matrix.

***Keywords**: Engineered Fibres; Pulp; SEM; Strength; Durability.*

1. Introduction

More sustainability awareness in the construction industry in recent times has led to the use of various environmentally friendly materials as components in concrete. Of such eco-friendly materials that can be incorporated into concrete to improve the performance are natural fibres such as cellulose fibre. These fibres are sustainable alternatives to the conventional petrochemical-based and synthetic fibres as they exist naturally in the environment and does not require and complex production process (Ardanuy et al. 2015). Similarly, the natural fibres are available worldwide, renewable and do not pose any hazardous threat to the environment. Hence, these materials have several promising potentials to be used in producing environmentally friendly concrete.

The penetration of deleterious materials into concrete is known to pose a durability threat and a corresponding degradation of mechanical performance (Booya et al. 2018 and Booya et al. 2020). These deleterious materials get into concrete can occur by diffusion or by using water as a pathway. On the other hand, concrete could also experience higher shrinkage, low ductility, etc. which might hinder the performance of the composite. The incorporation of cellulose fibres in concrete can be used to improve some of the detrimental properties by enhancing the microstructural permeable pores. The study by Banthia et al. (2012) showed that the use of virgin, fully purified softwood cellulose fibre at dosages of 0.1% and 0.3% by volume of the mixture resulted in a reduction in the pore volumes of cement. In addition, the cryoporometry and mercury intrusion porosimetry (MIP) observations showed that higher dosages of cellulose fibre in cement paste decreased the pore size to approximately 5 nm from approximately 70 nm. Hence, it was concluded from the study that cellulose fibres can the incorporation of cellulose fibres can be used to mitigate the penetration of deleterious materials into the concrete. In a similar study, Sappakittipakorn and Banthia (2012) evaluated the effect of virgin and fully purified softwood cellulose fibre with a length of 2.3 mm on the chloride transport characteristics of fibre reinforced concretes. The findings from the study showed a reduction in the effective chloride (free) diffusion coefficients can be achieved due to the ability of the cellulose fibres to chemically bind the chlorides. Another study carried out at the University of Florida also explored the effect of cellulose fibres on the permeability and durability properties of concrete (Roque 2009). In the study, an unmodified cellulose fibre with a length and aspect ratio of 2.1 mm and 117, respectively was used. Results from the study showed that the incorporation of the cellulose fibres also resulted in an improvement in the durability performance of concrete.

Despite the promising advantage of incorporation of cellulose fibres as evident in the laboratory results, the use of these fibres on large scale is limited as a result of some major challenges. One of the challenges with the use of cellulose fibres in concrete is the high alkalinity of the concrete pore solution which might result in degradation of the fibres in the long-term (Savastano 2009). Another major challenge is the possible migration of hydration products into the lumen of the fibres which can result in the mineralization of the fibre and consequently weaken the fibres.

The degradation of natural fibres in concrete can be prevented or minimized by modifying the fibre or modifying the composition of the concrete (Mohr et al. 2007, Toledo Filho 2003, Mohr et al. 2006). The fibres can be modified chemically or my physical methods (Mohr et al. 2007). These methods have been found to be effective in enhancing the bonding of fibre in the composite and reducing the moisture transfer around and through the fibres. Claramunt et al. (2011) studied the effect of hornificated Kraft pulp fibres on the durability properties of cement mortar composites. It was concluded from the study that the hornification of the fibres enhanced the performance of the composites. On the other hand, composites made with Kraft pulp without any modification exhibited lower performance. Several studies have shown that these natural fibres can be modified by processes such as fibre beating, initial drying age, bleaching, and pulp treatments (thermomechanical or Kraft) (Mohr et al. 2006 and Mohr et al. 2005). It was also concluded from these studies that the drying state and beating did not result in any significant influence on the mechanical performance of the composites when subjected to wet/dry cycles.

The performance of concrete reinforced with natural fibres can also be improved with the use of materials such as supplementary cementitious materials (SCMs). Some studies have explored the use of SCMs as partial replacement of cement in order to refine the pore structure and reduce the alkalinity of the cement matrix. Mohr et al. (2005) evaluated the effect of SCMs in mitigating the fibre degradation resulting from wet/dry cycles. Results from the study showed that the use of 30% silica fume, 90% slag and 30% metakaolin as replacement of cement prevented the degradation of the concrete samples subjected to wet and dry cycles. Similarly, the study by Toledo Filho et al. (2003) showed that the pore solution alkalinity reduced to 12.9 from 13.2 when silica fume was used as a 17% replacement of the cement. Findings from Silva et al. (2010) study also showed that the ultimate bending strength of sisal-

IOP Conf. Series: Materials Science and Engineering **978** (2020) 012023 doi:10.1088/1757-899X/978/1/012023

fibre composites made with metakaolin and calcined waste as replacement of the cement is four times more than those made with only cement as the binder.

Hence, this current study aims to investigate the influence of highly refined pulp fibre and SCMs on the interfacial characteristics and durability properties of concretes. An experimental program was carried out to assess the influence of engineered pulp fibre, namely mechanically modified fibre (MMF). MMF is a modified softwood fibre developed by Domtar Inc and the surface of MMF is modified externally. The compressive strength, chloride permeability, absorption alongside the interfacial properties of concrete with MMF and different types of SCM was evaluated.

2. Experimental Program

2.1 Materials

A Kraft pulp fibre (fully bleached) type like the one used by Booya et al. (2018) was used as reinforcement. The fibre used is a type of softwood fibres that is mechanically modified and engineered to have increased fibril counts and increased surface area. The average fibre length is 1.8 mm, and the average width (or diameter) of the fibre is 18 μm. The fibre had a density of 0.746 gm/cc and negligible lignin content.

General use limestone type cement (GULC) was used in all mixtures. Silica fume (SF), type C Fly ash (FA), metakaolin (MK) and grade 80 slag (SG), which are commercially available, were utilized as supplementary cementitious materials (SCMs). The chemical and physical characteristics of the cement and SCMs satisfies the requirements and recommendations of CSA A3001, ASTM C1240, ASTM C989, and ASTM C618 standards. Oven dried sand with maximum particle size of 5 mm and coarse aggregate that have a maximum particle size of 19 mm were used in all mixtures. A water reducer agent (high range super plasticizer) conforming to ASTM C494, was used to attain workability in the ECC mixtures.

2.2 Mixture Design and Proportioning

In order to determine the effect of SCMs on the fibre reinforced concretes, five mixtures were designed and proportioned with 1.3% by volume fibre content. One of the mixtures had no SCM in it and hence it was considered as a control mixture. The other four mixtures included binary blends in which a 10 percent of cement by weight was replaced with SCM. Varied amounts of HRWRA (superplasticizer) was added to all mixtures, to achieve required workability (slump range between of from 70 to 100 mm). Table 1 lists the mixtures and testing matrix of the tested concretes.

A specific mixing procedure was followed in this study to fabricate the concrete specimens. Moreover, the fibre preparations and conditioning followed similar procedure used by Booya et al. (2019). From each mixture, cylinders having a diameter of Ø100 and height of 200 mm, were cast and compacted. After 24 hours, the specimens were demoulded. Then, the specimens were let to cure in limewater until the time of testing (28 or 90 days). The mixture IDs are designated to reflect the type of SCM used in the concrete for example, mixture M-SF refers to a concrete mixture that included silica fume bends, while M-SG refers to concrete mixtures that included slag blends.

Table 1: Mixture proportions and testing matrix

Property \ Mixture ID	Control	M-SF	M-FA	M-MK	M-SG
Cement (Kg/m^3)	485	436.5	436.5	436.5	436.5
Gravel (Kg/m^3)	935	935	935	935	935
Sand (Kg/m^3)	759	759	759	759	759
Water/binder	0.35	0.35	0.35	0.35	0.35
SCM (Kg/m^3)	N/A	48.5	48.5	48.5	48.5
Fibre volume (%)	1.3	1.3	1.3	1.3	1.3
Fibre mass per cubic meters	N/A	9.7	9.7	9.7	9.7
Cement/SCM type	General use lime	Silica fume	Type C Fly Ash	Metakaolin	Grade 80 Slag
Cement/SCM Supplier Name	St Marys GUL	Euclid Chemicals	Lafarge Canada	Whitemud Resource	Lafarge Canada

2.3 Testing Method

Scanning Electron Microscopy (SEM) technique was used to study the interfacial properties and the fibre fracture properties of the composites. The specimens for SEM testing were made of mortar-fibre composite that is reinforced with 2% MMF fibres by the weight of cement. The mortars also included ten percent cement replacement by a SCM type (i.e., fly ash). The chemical reaction was accelerated by subjecting the specimens to heat at 50°C for 90 days.

ASTM C39 was followed to determine the compressive strength of the produced mixtures. Three cylinders of Ø100 × 200 mm were tested, and the average value was reported at each testing age of 28, and 90 days.

The resistance chloride ion penetration of the mixtures was assessed in accordance with ASTM C1202, and the charge passed through the disk specimen was monitored and calculated. For each mixture, two 50 mm thick disks (cut from the mid-portion of a Ø100 mm diameter by 200 mm high cylinder) were prepared for testing ages of 28 and 90 days. After following the conditioning procedure specified in ASTM C1202, the disk specimens were attached to the testing cell with one face in contact with3% NaCl solution and the other face in contact with 0.3N sodium hydroxide (NaOH) solution. A data acquisition system (DAQ) was employed to log the current (DC) of 60 ± 0.1 volts, and time over a period of six hour. The passed charge through the specimen was then calculated. The chloride permeability rating of concrete was then determined (ranging from "negligible" to "high") per ASTM C1202.

ASTM C642 recommendations were followed to determine the water absorption percentages. Two 50 mm thick cylindrical specimens were cut from Ø100 × 200 mm cylinder specimens. The specimens were conditioned, and oven dried for 24 hours, and then cooled down to room temperature. The disk specimens were then totally immersed in 20°C water for a period of two days. The change in specimen mass was calculated. The two-day immersion water absorption was selected for each testing age of 28, and 90 days.

3. Results and Discussions

3.1 Interfacial properties

Composites containing MMF fibres were examined and investigated for interfacial characteristics. Scanning electron microscope (SEM) image of the control mixture specimen is shown in Figure 1. From this figure, it is obvious that a few fibres were pull-out from the composite matrix. The SEM examinations also showed empty pockets that indicates fibre de-bonding failure mode. Moreover, a gap in the fibre-matrix zone from specimen drying was observed. The existence of such gap contributes to increasing the matrix porosity. Hence, these observations can be associated to increasing permeability values reported for the control concrete mixture.

Figure 2 shows SEM images of silica fume (SF) containing specimen. The figure shows that the hydration products are precipitated on the outside surfaces of the fibres. Moreover, the composite matrix seems to have rough texture. This suggests improved bonding and reduced moisture transfer. Mohr et al. (Mohr et al. 2005) reported similar observations on the effect of including SCMs in composites reinforced with pulp fibre and exposed to wet/dry cycle.

Figure 3 shows an image for fly ash (FA) containing specimen. The SEM image shows a breakage in the fibre tip which suggests a brittle failure mode. Moreover, the image revealed that the composite exhibited fibre detachment from its matrix. Linking these observations with water permeability characteristics of the composites, it can be concluded that the fibre debonding is a predominant failure mode.

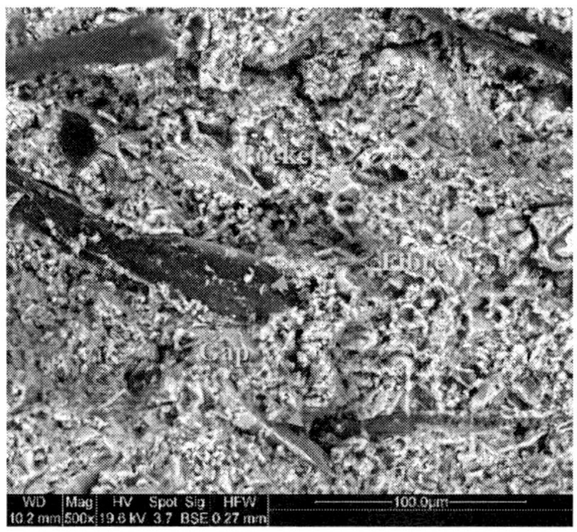

Figure 1 Control SEM image at x500

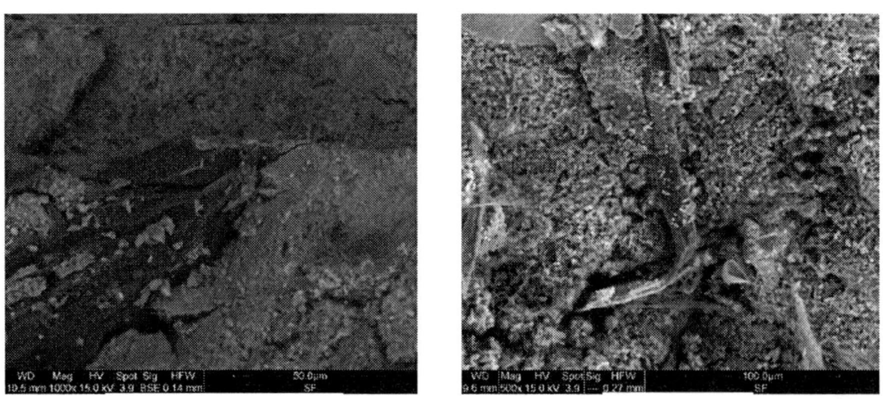

(a) x1000 (b) x500

Figure 2 M-SF mixture SEM image at x500

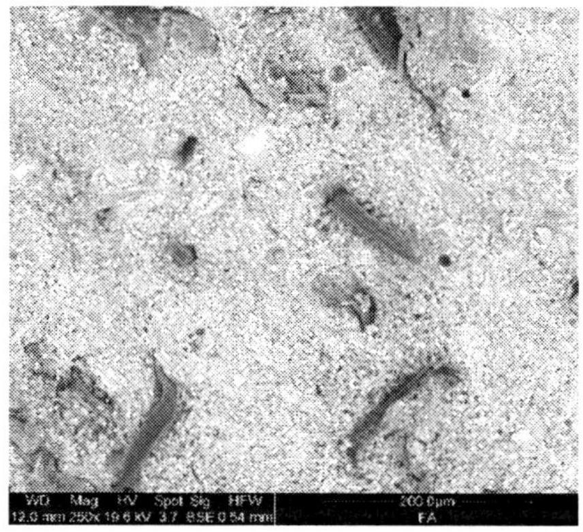

Figure 3 M-FA mixture SEM image at x250

On the other hand, mortar composites with 10% metakaolin blends (Figure 4) had the fibre tips fractured. The improved fibre-matrix bond was enough to break the fibres. SEM image also revealed that the matrix has dense surface which contributes to the lowered water and chloride permeations. No spaces were visible at the fibre-matrix interfacial zones.

Composites containing slag seemed to enhance the fibre-matrix bond and hence had fibre fractured as shown in Figure 5. The behaviour of slag and metakaolin, in terms of permeability and absorption, seems to be similar. However, metakaolin blends are more effective at lower replacement levels.

In all mixtures, SEM investigation did not have any evidence of ettringite formation around the fibre or near the interfacial zone. The research work herein concluded that the MK, SF, and SG blends, are beneficial in modifying the cementitious matrix which resulted in enhanced durability. However, FA blends had insignificant effect on permeability reduction.

Figure 4 M-MK mixture SEM image at x500

(b) x1000 (b) x500

Figure 5 M-SG mixture SEM image (Booya et al. 2019)

3.2 Compressive Strength

The compressive strength values of all mixtures are shown in Figure 6. From this figure, it is obvious that mixtures with fly ash blends had the lowest compressive strength (M-FA), regardless of the testing age. Metakaolin blends have higher specific surface area than cement which optimizes the particle distribution of the mixtures. These fine metakaolin particles reduce the pore sizes that result in reducing the permeability of the concretes (Booya et al. 2019). Mixture M-MK had the highest compressive strength of 53.8 MPa, at 90 days. This is a 32% increase in strength when compared to the control mixture (Control). Silica fume and slag blends (M-SF, and M-SG) slightly increased the compressive strength in comparison to the control mixture. However, the mixture incorporated fly ash (M-FA) blends had 21.2%, and 9.6 % reduction in strength at 28, and 90, respectively, as compared to the control mixture.

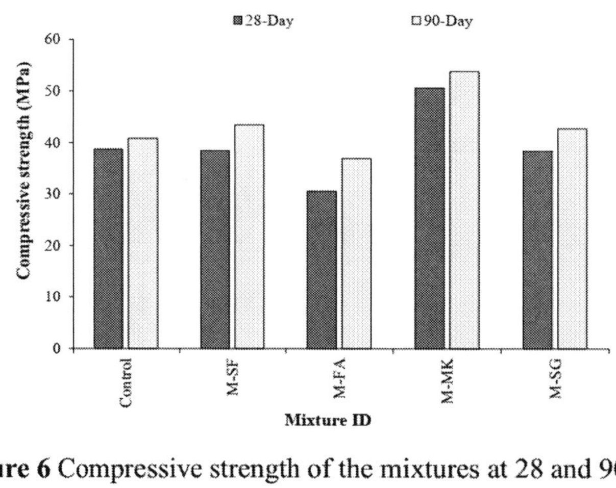

Figure 6 Compressive strength of the mixtures at 28 and 90 days

IOP Conf. Series: Materials Science and Engineering **978** (2020) 012023 doi:10.1088/1757-899X/978/1/012023

3.3 Rapid Chloride Penetration

Chloride transfer through cementitious materials often led to major defects, particularly when chloride ions reach embedded steel reinforcements. The penetration of chloride ions causes the formation of rust products, resulting in reduced durability and service life of cement-based structures (Booya et al. 2018 and Booya et al. 2019). The rapid chloride ion permeability test (RCPT) is a way of assessing and evaluating the long-term durability of concrete (Booya et al. 2020). Table 2 lists the chloride penetration charges in coulombs that passed through the specimen after a period of 6 hours. Incorporating SCM materials in the mixtures decreased the charge passed through the specimens and hence had reduced ion permeability ratings as compare to the control mixture, regardless of testing age. The reduced permeability reported for SCM containing concretes may be attributed to pore refinement or transformation of large pores into fine pores (Güneyisi et al. 2015, Chia and Zhang 2002).

Table 2: RCPT Charges and ratings as per ASTM C1202

Age \ Mixture ID	Control	M-SF	M-FA	M-MK	M-SG
28 day	<5500	<3100	<4700	<2600	<4100
Rating	High	Moderate	High	Moderate	high
90 day	<3500	<1600	<2400	<1700	<2800
Rating	Moderate	Low	Moderate	Low	Moderate

3.4 Immersion Absorption

The immersion water absorption test is a way of assessing and estimating the total interconnected pores that water can reach in cementitious composites. The seventh days absorption test results are shown in Figure 7. Mixture M-FA had the largest absorption of 4.2% at 28.

It is evident from this figure that metakaolin and slag blends reduce water absorption, regardless the testing age. Hence, M-MK had the highest reduction in water absorption among all mixtures, at both testing ages.

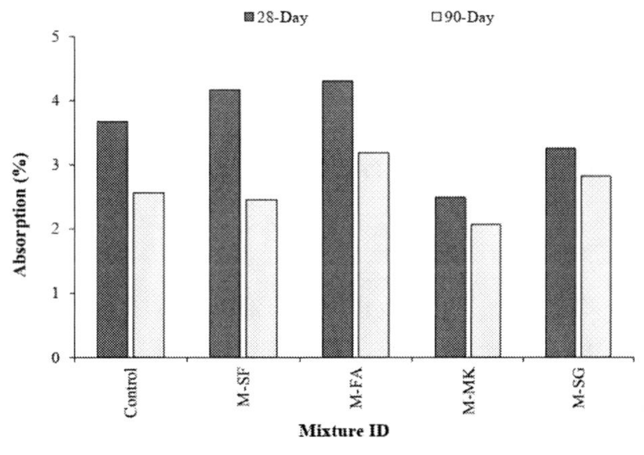

Figure 7 Immersion Absorption of the mixtures at 28 and 90 days

4. Conclusions

Below are the main conclusions that can be drawn based on the experimental results presented in this paper. The conclusions made herein are limited to the specimens, and conditions used in the study.

- Microstructural investigation showed that few fibres were de-bonded from the matrix and that fibres had spaces at the fibre interfacial zones. The presence of these voids at the interfacial zone is expected to increase the permeability of the composites. In contrast, the use of SF, MK and SG resulted in a very dense microstructure.
- The use of MK as a replacement of cement resulted in better enhancement of the compressive strength compared to FA used.
- All concrete mixtures incorporating SCMs exhibited better durability performance compared to the control mixtures.

5. Acknowledgement

This research work was completed with the financial assistance of Natural Sciences and Engineering Research Council of Canada (NSERC) located in Ottawa, Canada.

References

[1] Ardanuy M, Claramunt J, Toledo Filho RD. Cellulosic fibre reinforced cement-based composites: A review of recent research. Constr Build Mater 2015; (79): 115–128.

[2] ASTM 618. (2010). Standard Specification for Coal Fly Ash and Raw or Calcined Natural Pozzolan for Use, ASTM Standards.

[3] ASTM C1202. (2018). Standard Test Method for Electrical Indication of Concrete's Ability to Resist Chloride Ion, ASTM Standards.

[4] ASTM C143. Standard test method for slump of hydraulic-cement concrete, ASTM International, West Conshocken, PA 2015; 1–4.

[5] ASTM C1585. Standard test method for measurement of rate of absorption of water by hydraulic cement concretes, ASTM International, West Conshocken, PA 2013; 1–6.

[6] ASTM C39. Standard test method for compressive strength of cylindrical concrete specimens, ASTM International, West Conshocken, PA 2016; 1–7.

[7] ASTM C494, (2015). Standard Specification for Chemical Admixtures for Concrete, ASTM Standards. 1–9.

[8] ASTM C642. (2013). Standard Test Method for Density, Absorption, and Voids in Hardened Concrete. ASTM Standards, 04.

[9] ASTM C989, (2013). Standard Specification for Slag Cement for Use in Concrete and Mortars, ASTM Standards. 44, 1–8.

[10] Banthia N, Sappakittipakorn M, Jiang Z. On permeable porosity in bio-inspired fibre reinforced cementitious composites. Int J Sustain Mater Struct Syst 2012; (1) 29–41.

[11] Boel V, Audenaert K, De Schutter G. Gas permeability and capillary porosity of self-compacting concrete. Mater Struct 2008; 1283–1290.

[12] Booya E., Ghaednia H., Das S., Pande H. (2018) Durability of cementitious materials reinforced

with various Kraft pulp fibers, Construction and Building Materials, 19.

[13] Booya E., Gorospe K., Das S., Loh P. (2020). The influence of utilizing slag in lieu of fly ash on the performance of engineered cementitious composites. Construction and Building Materials, 256.

[14] Booya E., Gorospe K., Ghaednia H., Das S. (2019). Durability properties of engineered pulp fibre reinforced concretes made with and without supplementary cementitious materials, Composites Part B, 172, 376–386.

[15] Chia KS, Zhang MH. Water permeability and chloride penetrability of high-strength lightweight aggregate concrete. Cem Concr Res 2002; (32): 639–645.

[16] Claramunt J, Ardanuy M, García-Hortal JA, Filho RDT. The hornification of vegetable fibres to improve the durability of cement mortar composites. Cem Concr Compos 2011; (33): 586–595.

[17] CSA A3001, (2013). Cementitious materials used in concrete, Canadian Standards Associations. Mississuiga, Ontario.

[18] Filho RDT, Ghavami K, England GL, Scrivener K. Development of vegetable fibre-mortar composites of improved durability. Cem Concr Compos 2003; (25): 185–196.

[19] Güneyisi E, Gesoğlu M, Booya E, Mermerdaş K. Strength and permeability properties of self-compacting concrete with cold bonded fly ash lightweight aggregate. Constr Build Mater 2015; (74): 17–24.

[20] Mohr BJ, Biernacki JJ, Kurtis KE. Microstructural and chemical effects of wet/dry cycling on pulp fibre-cement composites. Cem Concr Res 2006; (36): 1240–1251.

[21] Mohr BJ, Nanko H, Kurtis KE. Durability of kraft pulp fibre – cement composites to wet / dry cyclingCem Concr Compos 2005; (27): 435–448.

[22] Mohr BJ, Nanko H, Kurtis KE. Durability of thermomechanical pulp fibre-cement composites to wet/dry cycling. Cem Concr Res 2005; (35): 1646–1649.

[23] Roque LG, Kim N, Kim B. Durability of fibre-reinforced concrete in Florida environments. Technical report submitted to Florida department of transportation. 2009.

[24] Sappakittipakorn M, Banthia N. Corrosion of Rebar and Role of Fibre Reinforced Concrete. J Test Eval 2012; (40): 103873.

[25] Sappakittipakorn M, Banthia N. Corrosion of rebar and role of fibre reinforced concrete. Journal of Testing and Evaluation 2012; (40): 127–136.

[26] Savastano H, Santos SF, Radonjic M, Soboyejo WO. Fracture and fatigue of natural fibre-reinforced cementitious composites. Cem Concr Compos 2009 (31): 232–243.

[27] Silva F de A, Filho RDT, Filho J de AM, Fairbairn E de MR. Physical and mechanical properties of durable sisal fibre-cement composites, Constr Build Mater 2010; (24): 777–785.

The Performance of Slag Containing Engineered Cementitious Composites

Emad Booya[1*], Adeyemi Adesina[2], Karla Gorospe[3] and Sreekanta Das[4]

[1]Department of Civil and Environmental Engineering, University of Windsor, Canada
E-mail: booya@uwindsor.ca
[2]Department of Civil and Environmental Engineering, University of Windsor, Canada
E-mail: adesina1@uwindsor.ca
[3]Department of Civil and Environmental Engineering, University of Windsor, Canada
E-mail: gorospek@uwindsor.ca
[4]Department of Civil and Environmental Engineering, University of Windsor, Canada
E-mail: sdas@uwindsor.ca
* Corresponding author: Department of Civil and Environmental Engineering, University of Windsor, Canada. Tel: +1 519 992 1086.

Abstract. Engineered Cementitious Composite (ECC) is a type of fibre reinforced cement-based composite that is used in different construction applications. The use of fly ash in these composites improves tensile strain capacity, multiple cracking behaviour, and durability. However, fly ash is in short supply as many countries are in an effort to phase-out fossil fuels, and are eliminating coal plants. An alternative to fly ash is slag. In contrast to fly ash, the supply of slag is expected to be sustained. Hence, many industries have considered substituting fly ash with slag in the production of ECC. This study was designed to determine the mechanical, and permeability of ECC mixtures containing fly ash and slag. The study contained a total of four mixtures. Two of the mixtures had the ratio of supplementary cementitious materials to cement of 2.2 and two more mixtures had supplementary cementitious materials to cement ratio of 1.2. The water to binder ratio was the same for all mixtures. Test results showed that slag blends improved the mechanical and permeability characteristics.

***Keywords**: Engineered Cementitious Composites; Slag; Fly Ash; Strength; Durability.*

1. Introduction

In recent years, there have been significant advancements in the research and development of high-performance fibre-reinforced cementitious composites (HPFRCC). Superior compressive strength can be achieved with HPFRCC due to its low water-to-cementitious materials ratio (W/CM). In addition, studies have shown that HPFRC has high bending characteristics, improved durability, and good resistance to weathering conditions (Booya et al. 2020 and Li et al. 2018). Engineered cementitious composite (ECC) is a type of HPFRCC that is designed with micromechanical principles (Li 2003 and Li et al. 2001). These micromechanical principles enable the design of high performing composites with minimal amounts of reinforcing fibres. With a typical fibre amount of less than 2% by volume, it has been determined that tensile strain-hardening behaviour of ECC occurs after first cracking, and that strain capacities of ECC are 300 to 500 times higher than normal concrete. Furthermore, it is characteristic of ECC to exhibit self-controlled crack widths under increased loading. Effective fibre bridging and controlled matrix fracture lead to tight cracks in hardened ECC (less than 60 μm) (Şahmaran and Li 2009 and Zhu 2012).

Typical ECC consists of Type F fly ash as a supplementary cementitious material (SCM). However, various concrete associations around the world have reported a shortage in the supply of Class C and Class F fly ash (National Precast Association 2020). The shortage in fly ash supply is a result of the closure of coal-based plants as countries around the world move towards sustainable and green power sources (Booya et al. 2020).

Alternatively, the supply of slag blends, which is a by-product of the iron-making process, is expected to be sustained. Like fly ash, utilizing slag blends to produce ECC decreases environmental impacts. Slag blends enhances the composites' performance and has more consistent chemical and physical characteristics compared to fly ash blends (Slag Cement Association). Therefore, concrete suppliers and construction companies often consider slag as a more reliable SCM.

Few studies on ECC containing slag blends have been undertaken. This study aims to expand the literature by investigating the influence of fly ash and slag, as a binary blend, on the durability and mechanical performance of ECC. The effects of replacing higher SCM volume on the durability and mechanical performance of the ECC mixtures were also determined. The strength and permeability of the ECC mixtures were assessed for compressive strength, four-point bending strength, chloride permeability, and immersion absorption.

2. Experimental Program

2.1 Materials

General use type cement (C) was used in all ECC mixtures. Type F Fly ash (FA) and slag (SG), which are commercially available, were utilized as supplementary cementitious materials (SCMs). The chemical and physical characteristics of the cement and SCMs satisfies the requirements and recommendations of CSA A3001, ASTM C989, and ASTM C618 standards. Oven dried sand with maximum particle size of 600 μm was used in all ECC mixtures. A water reducer agent (high range super plasticizer) conforming to ASTM C494, was used to attain workability in the ECC mixtures. Polyvinyl alcohol type fibres (PVA) of 39 μm diameter and 8 mm length were used. The specific gravity and tensile strength of these PVA fibres were 1.3 and 1600 MPa, respectively.

2.2 Mixture Design and Proportioning

All ECC mixtures had a water to binder (W/B) ratio of 0.25. Two mixtures had SCM/C ratio of 1.2 and sand to binder (S/B) ratio of 0.34. Another two mixtures were prepared with SCM/C ratio of 2.2 and sand to binder ratio of 0.36. Table 1 lists the mixture proportions of all ECC mixtures. From this table, the mixture IDs were selected to reflect the SCM type and the SCM/C ratio used in the ECC mixtures. For example, mixture FA2.2C refers to an ECC mixture that has zero slag content (only fly ash as SCM) and SCM/C ratio of 2.2, while SG1.2C refers to an ECC mixture that has slag (no fly ash) and SCM/C ratio of 1.2.

Table 1: ECC Mixture Proportioning

Mixture ID	SCM/C	S/B	W/B	Cement Content (Kg/m^3)
FA2.2C	2.2	0.36	0.25	380
SG2.2C	2.2	0.36	0.25	420
FA1.2C	1.2	0.34	0.25	600
SG1.2C	1.2	0.34	0.25	600

2.3 Testing Method

ASTM C109 was followed to determine the compressive strength of the produced ECC mixtures. Five cube specimens of 50 mm were tested, and the average value was reported at each testing age of 28, and 90 days.

Four-point bending test was performed on 355 mm x 75 mm x 75 mm prisms. ASTM C1609 recommendations were followed to conduct the testing. The prism deflection was obtained via a 25 mm linear variable displacement transducer (LVDT) that was positioned at the mid-span. The average flexural strength of three specimens was reported for the testing age of 28 and 90 days. Figure 1 shows the test setup for the four-point bending test.

The resistance chloride ion penetration of the ECC mixtures was assessed in accordance with ASTM C1202, and the charge passed through the disk specimen was monitored and calculated. For each ECC mixture, two 50 mm thick disks (cut from the mid-portion of a Ø100 mm diameter by 200 mm high cylinder) were prepared for testing ages of 28 and 90 days. After following the conditioning procedure specified in ASTM C1202, the disk specimens were attached to the testing cell with one face in contact with3% NaCl solution and the other face in contact with 0.3N sodium hydroxide (NaOH) solution. A data acquisition system (DAQ) was employed to log the current (DC) of 60 ± 0.1 volts, and time over a period of six hours. The passed charge through the specimen was then calculated. The chloride permeability rating of concrete was then determined (ranging from "negligible" to "high") per ASTM C1202.

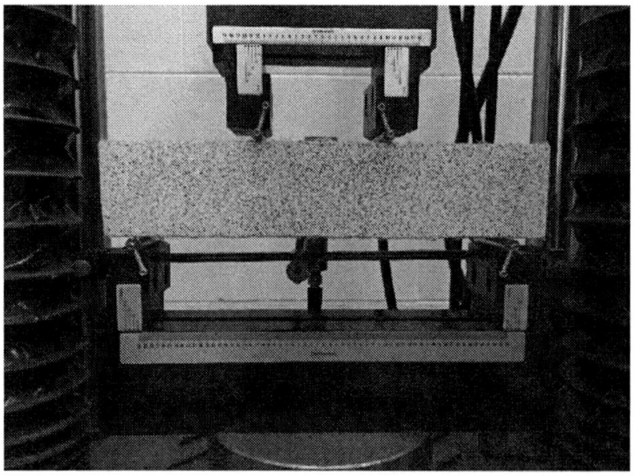

Figure 1 – Flexural Strength Setup

ASTM C642 recommendations were followed to determine the water absorption percentages. Two 50 mm thick cylindrical specimens were cut from Ø100 × 200 mm cylinder specimens. The specimens were conditioned, and oven dried for 24 hours, and then cooled to room temperature. The disk specimens were then immersed in 20°C water for a period of two days. The change in specimen mass was calculated. The two-day immersion water absorption was selected for each testing age of 28, and 90 days.

3. Results and Discussions

3.1 Compressive Strength

The compressive strength values of all ECC mixtures are shown in Figure 2. From this figure, it is obvious that mixtures with fly ash blends had the lowest compressive strength (FA1.2C and FA2.2C), regardless of the testing age. Among all mixtures, FA2.2C had the lowest compressive strength.

ECC mixture SG1.2C had the highest compressive strength of 97.9 MPa, at 90 days. This is a 23% increase in strength when compared to ECC mixture FA1.2C. Slag has pozzolanic properties that are known to be advantageous for early age strength gain (Zhu et al. 2012). Therefore, it was expected that ECC mixtures with slag would exhibit higher strength gain at 3 and 7 days compared to ECC mixtures with fly ash only. Increasing the SCM/C ratio from 1.2 to 2.2 led to compressive strength reduction, as shown in Figure 2. This can be attributed to the lower cement content per cubic meter. Nonetheless, all ECC mixtures had compressive strengths higher than 51.1 MPa, at 28 days. Such value may satisfy the requirement of most construction and engineering applications.

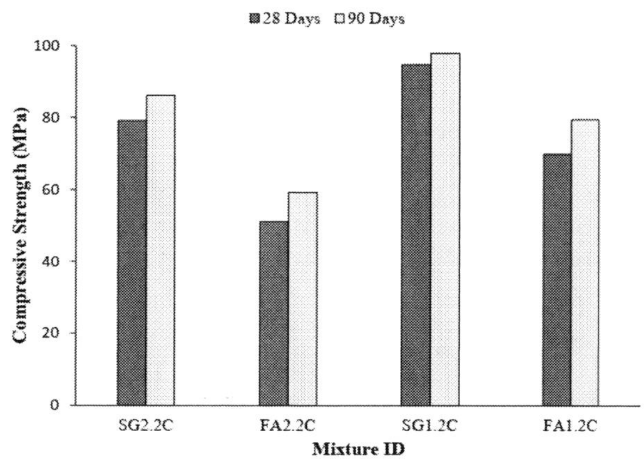

Figure 2 – Compressive strength of the mixtures at 28 and 90 days

3.2 Flexural Strength

Figure 3 lists the ultimate flexural values for the ECC mixtures. From this table, the lowest flexural strength of 7.1 MPa was recorded for mixture SG2.2C, at 28 days. However, mixture SG1.2C had the highest flexural strength of 11.5 MPa and 12.2 MPa at 28 and 90 days, respectively. As can be seen in Figure 3, increasing SCM/C from 1.2 to 2.2 did not contribute to the increase in flexural strength for mixtures FA2.2C and SG2.2C, at 28 days, as compared to mixtures FA1.2C and SG1.2C. For mixtures containing fly ash, FA1.2C, and FA2.2C had a maximum flexural strength of 9.2 MPa and 8.5 MPa, respectively. The flexural strength of mixtures containing fly ash improved at the age of 90 days with FA1.2C and FA2.2C reaching flexural strengths of 10.5 MPa and 9.1 MPa, respectively. It is worth mentioning that at both testing ages, and in all ECC mixtures, very tiny and multiple cracking behaviour was observed.

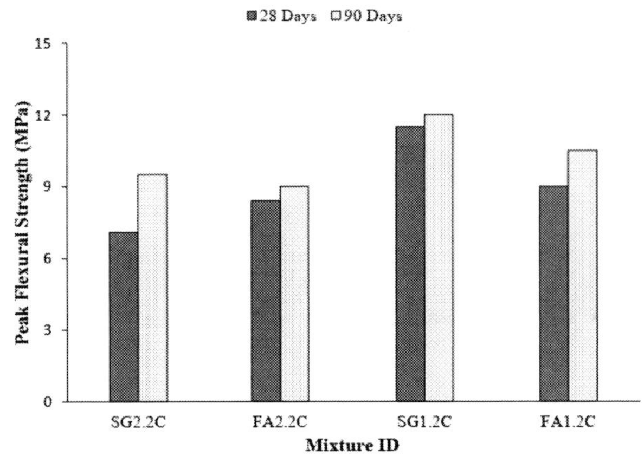

Figure 3 – Flexural Strength of the mixtures at 28 and 90 days

3.3 Rapid Chloride Penetration

Chloride transfer through cementitious materials often led to major defects, particularly when chloride ions reach embedded steel reinforcements. The penetration of chloride ions causes the formation of rust products, resulting in reduced durability and service life of cement-based structures (Booya et al. 2018 and Booya et al. 2019). The rapid chloride ion permeability test (RCPT) is a way of assessing and evaluating the long-term durability of concrete (Booya et al. 2020). Table 2 lists the chloride penetration charges in coulombs that passed through the specimen after a period of 6 hours. ECC mixtures containing fly ash minerals had higher charges compared to mixtures with slag only blends, regardless of the SCM/C replacement ratios. However, the extended curing periods showed a reduction in the charges passed for all mixtures. At 90 days, all mixtures had less than 1000 Coulombs which can be categorized as "very low", per ASTM C1202.

Table 2: RCPT Charges and ratings as per ASTM C1202

Mixture ID	SG2.2C	FA2.2C	SG1.2C	FA1.2C
28 day	<1000	<1500	<800	<1000
Rating	Very Low	Low	Very Low	Very Low
90 day	<700	<800	<500	<400
Rating	Very Low	Very Low	Very Low	Very Low

3.4 Immersion Absorption

The immersion water absorption test is a way of assessing and estimating the total interconnected pores that water can reach in cementitious composites. The two-day absorption test results are shown in Figure 4. Mixture FA1.2C had the largest absorption of 1.92% and 1.73% at 28 and 90 days, respectively. It is evident from this table that higher slag volumes reduce water absorption. Hence, SG2.2C had the highest reduction in water absorption among all ECC mixtures, at both testing ages. The reported absorption value for mixture SG1.2C is 1.75% and 1.62% at 28 and 90 days, respectively. Furthermore, the absorption percent value for mixture SG2.2C was reduced by 37.1% and 41.4% at 28 and 90 days, respectively, when compared to mixture SG1.2C.

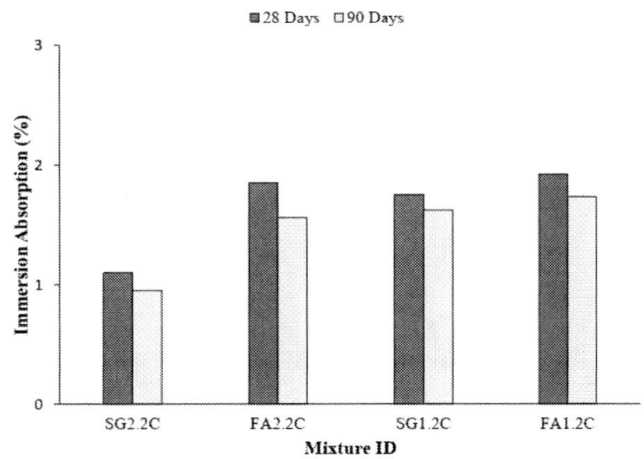

Figure 4 – Immersion Absorption of the mixtures at 28 and 90 days

4. Conclusions

Below are the main conclusions that can be drawn based on the experimental results presented in this paper. The conclusions made herein are limited to the specimens, and conditions used in the study.

- Mixtures incorporating slag had increased compressive strength values in comparison to the fly ash only mixtures, at 28 and 90 days. Slag improved early age strength. Increasing the SCM/C amount from 1.2 to 2.2, led to lower compressive strengths. Nonetheless, all ECC mixtures had compressive strength value of at least 51.1 MPa at 28 days. Extended the curing period (up to 90 days) was effective in increasing the compressive strength.
- Slag addition improved the flexural strength. However, higher fly ash amount (i.e., SCM/C = 2.2) replacement in mixture FA2.2C provided better flexural characteristics.
- The RCPT results indicated that slag blends are more effective in reducing chloride ion transfer compared to fly ash blends, at 28 days. However, prolonged curing period resulted in low chloride ions penetration rating for all ECC mixtures.
- The immersion absorption percent values revealed that slag blends are more effective than fly ash blends in reducing the water absorption. Therefore, incorporating higher slag amount resulted in greater reductions. Increased SCM/C amount from 1.2 to 2.2 reduced the water absorption.

5. Acknowledgement

This research work was completed with the financial assistance of Natural Sciences and Engineering Research Council of Canada (NSERC) located in Ottawa, Canada.

References

[1] ASTM 618. (2010). Standard Specification for Coal Fly Ash and Raw or Calcined Natural Pozzolan for Use, ASTM Standards.

[2] ASTM C1202. (2018). Standard Test Method for Electrical Indication of Concrete's Ability to Resist Chloride Ion, ASTM Standards.

[3] ASTM C109, (2020). Standard Test Method for Compressive Strength of Hydraulic Cement Mortars (Using 2-in. or [50 mm] Cube Specimens), 1–12.

[4] ASTM C494, (2015). Standard Specification for Chemical Admixtures for Concrete, ASTM Standards. 1–9.

[5] ASTM C642. (2013). Standard Test Method for Density, Absorption, and Voids in Hardened Concrete. ASTM Standards, 04.

[6] ASTM C989, (2013). Standard Specification for Slag Cement for Use in Concrete and Mortars, ASTM Standards. 44, 1–8.

[7] Booya E., Gorospe K., Das S., Loh P. (2020). The influence of utilizing slag in lieu of fly ash on the performance of engineered cementitious composites. Construction and Building Materials, 256.

[8] Booya E., Gorospe K., Ghaednia H., Das S. (2019). Durability properties of engineered pulp fibre reinforced concretes made with and without supplementary cementitious materials, Composites Part B, 172, 376–386.

[9] Booya E., Ghaednia H., Das S., Pande H. (2018) Durability of cementitious materials reinforced with various Kraft pulp fibers, Construction and Building Materials, 19.

[10] CSA A3001, (2013). Cementitious materials used in concrete, Canadian Standards Associations. Mississuiga, Ontario.

[11] Li C., Wang V.C., Wu S. (2001). Tensile strain-hardening behaviour of Polyvinyl Alcohol Engineered Cementitious Composites (PVA-ECC). ACI Materials Journal, 483–492.

[12] Li V.C. (2003). On engineered cementitious composites (ECC): A review of the material and its applications, Journal of Advanced Concrete Technology, 1, 215–230.

[13] Lu C., Li V.C., Leung C.K.Y. (2018). Flaw characterization and correlation with cracking strength in Engineered Cementitious Composites (ECC). Cement and Concrete Research, 107, 64–74.

[14] National Precast Concrete Association. https://www.precast.org. (Accessed June 2020)

[15] Şahmaran M., Li V.C. (2009). Durability properties of micro-cracked ECC containing high volumes fly ash, Cement and Concrete Research, 39, 1033–1043.

[16] Slag Cement Association. https://www.slagcement.org. (Accessed June 2020)

[17] Zhu Y., Yang Y., Yao Y. (2012). Use of slag to improve mechanical properties of engineered cementitious composites (ECCs) with high volumes of fly ash, Construction and Building Materials.

Upper Prosthetic Design based on EMG: A Systematic Review

Noof T. Mahmood[1], *, Mahmuod Hamza Al-Muifraje[2], Thamir R. Saeed[3], and Assel H. Kaittan4

[1]Collage of Electrical Engineering, University of Technology, Iraq
E-mail: noofthabit@gmail.com
[2]Collage of Electrical Engineering, University of Technology, Iraq
E-mail: drmahmood6@gmail.com
[3]Collage of Electrical Engineering, University of Technology, Iraq
E-mail: 50257@uotechnology.edu.iq
[4]Collage of Electrical Engineering, University of Technology, Iraq
E-mail: Moonlighta1992@yahoo.com
Noof T. Mahmood*: Collage of Electrical Engineering, University of Technology, Baghdad, Iraq, Tel: +9647833055317.

Abstract. In the past few years, physical therapy plays a very important role during rehabilitation. Numerous efforts have been made to demonstrate the effectiveness of medical/ clinical and human-machine interface (HMI) applications. The prevalent control methods are using electromyography (EMG) signals generated by muscle contractions to implement the prosthetic human body parts. This paper aims to provide and summarize ideas about recent researches in the field of Pattern Recognition (PR) based on EMG signals to save time and efforts for the readers working in this field. The first step starts by demonstrating a general overview of the various techniques to collect the database by taking into consideration the factors that affect the accuracy of the collected data. Hence, different types of filters are presented to process the signals and reduce the noise of the raw EMG signals. This research clarifies the features extraction methods using time-domain (TD), frequency domain (FD), and time-frequency domain (TFD) and which of these methods will be suitable to use for EMG signals. Finally, a group of studies is reviewed based on three classification methods i.e. artificial neural network (ANN), machine learning (ML), and deep learning (DL). Depending on these methods, the accuracy range can be specified for each classifier, also the factors which affect the accuracy percentage. Therefore, the researchers can avoid these issues that reduce accuracy.

Keywords: *EMG; signal processing; HMI; physical therapy and feature extraction.*

1. Introduction

Recently, the robotic has the effectiveness to increase the independence of the individuals living lifestyle with their disabilities. The objective is to improve the life quality that happens by empowering people to achieve a wide range of daily responsibilities within a few times. Humanoid robots can be prepared robotic limbs to make the physical actions of individuals as well as in the intelligent robotics industry. Nowadays, the most widely used is the robotic hands and arms, the robotic hands would have the ability to achieve the main skills like the transfer from one place to another and grasp of objects in the same nonamputees individuals do. Therefore, depending on the object shapes, the human hand actions must be trained and applied to the designed robotics hand. The intuitive approach is creating an interface that brings the activity of the muscle and records it by using electromyography (EMG) sensors (Oskoei et al., 2007).

EMG signals are a way for estimating the electrical signal that relatives to skeletal muscle, the electrical signal consists of some motor unit action potentials (MUAPs). The MUAP contains muscle fibers stimulated by harmonizes and solo alpha motor (Rissanen et al., 2007). Currently, the prosthetic technique is used as an artificial part to resolve the problem of amputee's parts. The appearance of the consumer-level 3D printers supports the prosthetics implementation. Thus, the develop prosthetics technique has gained popularity over traditional prosthetics which are expensive and unaffordable to many users (Amamcherla et al., 2018). The main challenge in the mechanical standpoint is how to mix many degrees of freedom (DOF). As a result, the increase in the DOF caused by decreasing in the gripping force that leads to an unstable system (Carrozza et al., 2002). Also, increasing in DOFs will lead to an increase in actuators number and high failure probability, so the prosthetic hand will need high maintenance and difficult manufacture (Choi et al., 2017).

This paper presents several problems that can happen in Pattern Recognition (PR), these problems start from the data collection passing through signal processing the feature extraction to end up with the required classification methods. These problems can be summarized as; myo armband position, number of electrodes that used to collect the data, the additive noise to the raw EMG signals, selecting the suitable features and how to reduce these features, and choosing the classifier depend on the a mounted of data. Each step in PR aims to solve these problems depending on the selected studies which used a different technique to reach the optimum solutions.

2. Electromyography Signal Processing

This part paper presents some essential information relevant to EMG signals. In addition, describing the methods that used for signal processing of raw EMG data, also presents the effectiveness and limitation of the environments and how it can be limited by understanding signals fundamental step and how to extract features that are related to pattern recognition (PR) to obtain a strong recognition system. Figure (1) shows a general overview of the EMG control signal.

2.1 Signal Overview

The innovation of electromyography signals has led to a countless application in clinical researches for motor unit employment, identification of neuromuscular disease, kinesiology, and motor unit disorders (C. J. J. I. T. o. B. E. De Luca, 1979). The EMG signals produce by the humanoid body in motor movement. The summing of the action potentials for the nerve cells will produce the EMG signals. The amplitude of these signals is very low and required suitable processing to give meaningful information (Prahm et al., 2017). The EMG signal is susceptible to the contamination that leads to the noise in the signals. For this reason, the authors proposed different methods to come over the noisy signals by using different types of filters (C. J. De Luca et al., 2010), other authors found a significant way to reduce the environmental noise by friction the surface of the skin with a rough conductive stick. Meanwhile, some authors work on the empirical mode decomposition (EMD) to cancel the noise of the signal (Huang et al., 1998), (X. Zhang et al., 2013).

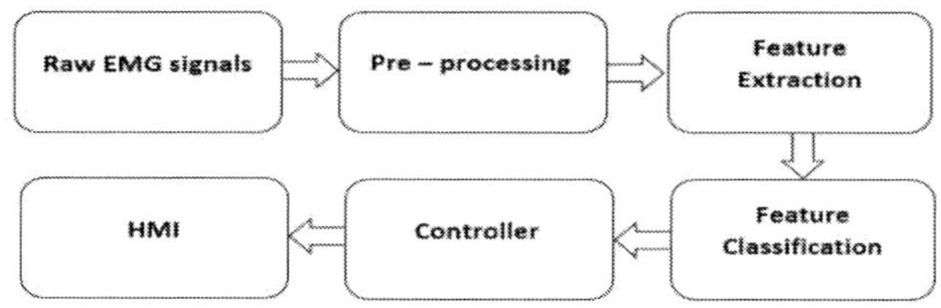

Figure (1). general overview of EMG control signal

2.2 Data Collection

The EMG (electromyography) electrode records the muscle movement. The principle of electrode based on muscle contraction that plays the role of electrical sparks to spread through the bone, tissue, and adjacent skin areas. This work depends on the data that have been collected from different types of sensors over the last ten years. There are different ways to recode EMG signal, Castiblanco, C., Parra, et al used Myo armband with 8 electrodes to detect the right forearm signals for 10 healthy persons (7males, 3 females) (Castiblanco et al., 2016b).

Park, Ki-Hee, et al used three separate electrodes to record the data of prismatic four fingers grasp, tip pinch grasp, parallel extension grasp, power grasp, opening a bottle, and lateral grasp for 27 subjects (Park & Lee, 2016). Ibrahim and Al-Jumaily used MyoScanTM T9503M Sensors to collect the signals of ten different finger movements, each movement recoded in 5 sec (M. F. I. Ibrahim & A. A. Al-Jumaily, 2016). In (Tortora et al., 2019) the author used two Myo armband to record 5 degrees of freedom (DOF) for non-amputees subjects. Other authors used the UCI database that recorded to detect 10 normal actions and 10 aggressive actions for 3 males and 1 female (J. Zhang et al., 2019). M. U. Khan, et al used the insertion level of the needle to record EMG signals for 6 normal males and 4 females, also to record 6 males and 4 females have lateral sclerosis (ALS) (Khan et al., 2019).

If electrode liftoff or shift, the desired muscle action has to stay the same topographical map but that will cause shifting in the direction. Therefore, the detection and correction of the shifted electrode add more complexity to the recognition system. The increment in electrodes number leads to an increment in the probability of pattern recognition. Therefore, increasing the number of electrodes leads to add redundancy that requires more computational power to post-process the generated signals also will effect on regression/classification features (Simão et al., 2019).

2.3 Signal Filtering

First of all, the quality of digitalizing of raw EMG signals must be improved by amplification the signals usually between 500-2000 fold. The raw EMG signals use different approaches to preprocessing. Usually, the band-pass filters (BPF) within the approximate range 5-500 Hz, low pass filter (LPF) of around 500Hz, and high pass filter (HPF) started with 5Hz (sometimes higher) these types of filter has been used in EMG sensors. Also, to cancel the power line interference (PLI) the notch filter is used at frequencies 50/60 Hz. After that, the filtered signals convert into a digital form by using an analog to digital conversion (ADC) (L. Xu & Adler, 2004).

One of the selective studies on EMG signals for pattern recognition the author amplified the signal 1000 time and uses BPF of 10 – 500Hz band with a sampling frequency of 1000 Hz (Park, S,1998). Some studies decreased the EMG signals noise by using HPF with a 500 Hz cutoff frequency and an LPF of 20 Hz cutoff frequency to decrease gesture remains (Balbinot & Favieiro, 2013). Al-Angari, Haitham M., et al used the BPF with a range of 5-500 Hz (Al-Angari et al., 2016). The authors presented an effective control using 4 sensors placed upper limb to measure the following EMG signals; triceps-long head, extensor carpi radialis, biceps-short head, and flexor carp radialis. The sampling frequency was 1000 Hz for the EMG signals, and this signal filtered using the BPF with a cut-off frequency of 90-450 (Riillo et al., 2014).

This study presents four raw signals with their filtered signal for three males and one female subjects ages from 25 to 30. The data signal is selected form the UCI Machine Learning Repository website. The data name is EMG Physical Action DataSet, the data was collected for ten normal actions i.e. waving, bowing, hand shaking, clapping, running, jumping, hugging, seating, standing, and walking. In this work the waving action for the right arm using bicep muscle is choosing for filtering process. This raw EMG signal has 10000 samples for each subject, the raw EMG signal is filtered to reduce the noise using first-order Butterworth LPF. Figure (2) shows the difference between the raw EMG signal and the filtered signal.

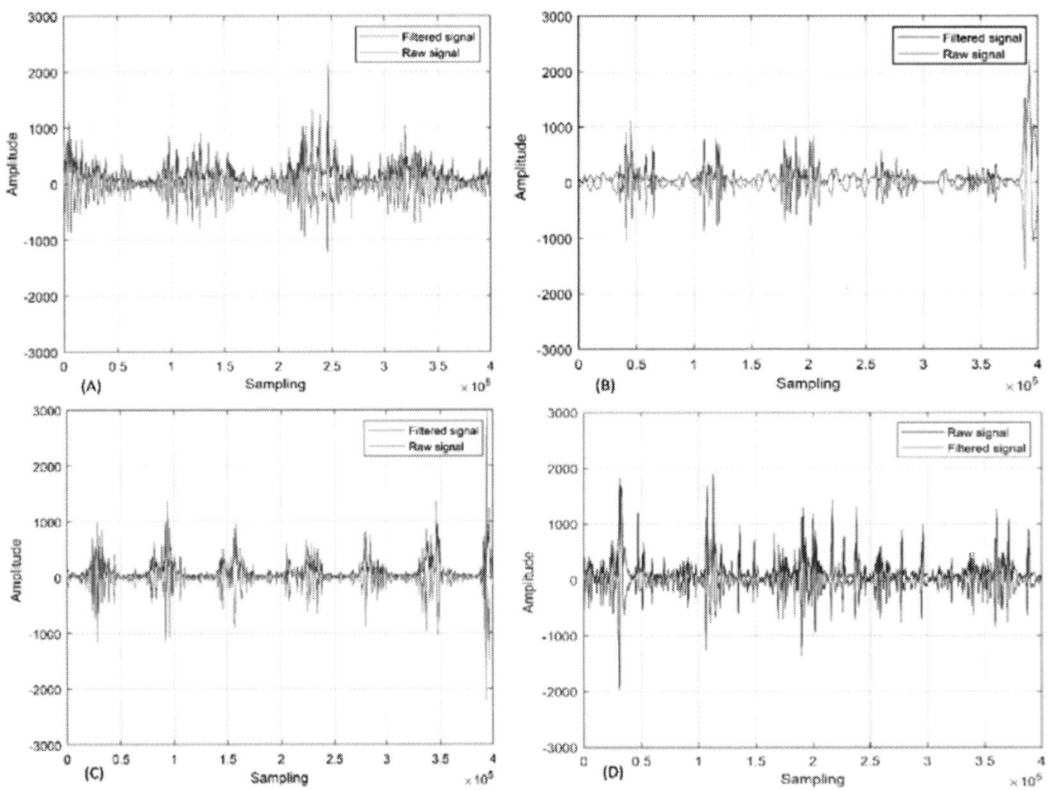

Figure (2). The raw EMG signal and the filtered signal: (A)subject1,(B)subject2, (C)subject3, and (D)subject4

Table 1. Summary on filtering and suppression of PLI of EMG signal.

Authors	Filter type	Filter Range (Hz)	Sampling Frequency(kHz)	Comments
(Jamal et al., 2019)	adaptive filter	40- 250 Hz	1000 Hz	Used 6th order BPF, learning rate μ as 0.005, and 100 coefficients.
(Wang et al., 2006)	amplifier filters	5- 1000 Hz.	2400 Hz	Used AR coefficients
(Y Paul et al., 2017)	BPF + butterworth filter	15- 500 Hz	500 Hz	Used notch filter at 50Hz.
(Maurya et al., 2019)	BPF + butterworth filter	5- 500 Hz	2000 Hz,	Used Butterworth filter to remove PLI.
(Li et al., 2017)	BPF	10-500 Hz	1024Hz	No comment
(Yang et al., 2019)	Butterworth bandpass filter	20–500 Hz	2000 Hz	Used 4th order Butterworth and notch filter at 50 Hz.
(Qi et al., 2019)	- LPF -HPF	Cutoff of 950 Hz Cutoff of 750 Hz	2000 Hz	Used 3rd order Butterworth filter. Used 3rd order Butterworth filter.
(Benalcázar et al., 2017)	LPF+ Butterworth filter		200 Hz	The IIR is preferred on the FIR digital because it is less expensive.
(A. Zhang et al., 2018)	BPF	20-50 Hz	1500 Hz.	Used notch filter at 50Hz.
(A. A. Adewuyi et al., 2015)	BPF	30-350Hz	1000 Hz	Used 8th order Chebyshev filter.
(Geng et al., 2018)	HPF	60 Hz.	1024 Hz	Used 5th order Butterworth.
(Amamcherla et al., 2018)	BPF	20 - 450Hz		Used 125ms as rectangular window of length and 25ms overlap.
(P. Xu et al., 2018)	BPF	20-500Hz	2000 Hz	Used notch filter at 50Hz.
(Young et al., 2012)	HPF	20 Hz	1000 Hz	Used 3rd Butterworth filter.
(Roy et al., 2018)	BPF	2-16 Hz		Used 2nd order to minimize DC components
(Bao, Zaidi, et al., 2019)	HPF	20 Hz	1024 Hz	Used 3rd order Butterworth filter.

A challenge of using electrode-based sensor systems that uses to record the EMG signal is how to reduce or avoid the external noise that affects signals detection. Reducing the noise will effect on the output signal, so rules must be applied to obtain good performance in the per-processing system:

- Skin preparation (shaving) to decrease the variance between the poles of the electrode (Piervirgili et al., 2014);
- cables arrangement for minimizing the coupling. (Merletti & Farina, 2016);
- Using differential amplifiers for common-mode rejection (Riillo et al., 2014).

This section pays little attention to the filtering stage, due to the importance of selecting the suitable pre-processing stage. Table 1 reviews some studies on filtering and suppression of PLI of EMG signal.

3. Feature Extraction of EMG

The traditional methods for extraction EMG features are frequency domain (FD) and time-domain (TD) analysis, which are commonly employed to analyze the data. Graupe, D. et al. were the first who employed the TD for feature extraction. furthermore, Autoregression (AR) was applied to recognize several arm actions, in this work the accuracy reached 85% (Graupe et al., 1982). Han and other researchers had a classification accuracy reached 92% by extracting the mean absolute values (MAV) as a main the features of the hand actions, then the Bayesian model was applied as a classifier (Han et al., 2013). The RMS appeared to be the most important feature if it compared with Waveform Length (WL), ZC, MAV, SSC, and MAX that gave good information for the selection of the electrode (Kendell et al., 2012). Moreover, the integrated EMG (IEMG) feature used TD for the fixed exterior force to calculate the increasing in the signal power, the period, and the amplitude that produces a higher muscle fiber employment (Al-Mulla et al., 2011).

Only a few research had employed FD features in PR, FD depends on the spectrum analysis such as Fourier transform is employed for analyzing the data of EMG. Bilodeau found a relationship between the change caused by the thickness of the skin volume conductor and the increase in force, this relationship came from the analysis of the power spectrum density (PSD) of EMG signals (Bilodeau et al., 1995). Therefore, the PSD, median power frequency (MNP) and mean frequency (MNF) of the power spectrum is commonly applied to indicate the characterize EMG signals (Merletti et al., 1997). researchers improved the median and mean in FD by taking the amplitude and calculate the median and mean for this amplitude instead of PSD. As a result, a new modified feature was obtained; modified median frequency (MMDF) and modified mean frequency (MMNF). After that, the MNP, MNF, bandwidth (BW), and normalized spectral moments (NSM) were extracted to identify muscle exhaustion for upper limbs (Rogers et al., 2013).

The researchers found a novel method that combined the characteristics of both TD and FD to present the wavelet transform. Thus, the time-frequency domain (TFD) analysis such as WT and short-time FT could be used to analyze the EMG signals. In (Patel & Gamit, 2016) the EMG features such as the RMS, MNF, and MNP are extracted with high accuracy by using continuous wavelet transform (CWT). In 2017 Fan extracted the EMG features using discrete wavelet transform (DWT) that have the strongest points of both TD and FD. After that, the DWT combined with a wavelet neural network (WNN), this work satisfies a high average ratio (Fan et al., 2017). The classifier performance affected by the EMG signal variations, to avoid this variation Abdullah et al, used the Wavelet Packet Decomposition (WPD) for feature extraction, the combination of Random Forest and WPD achieves better performance with high classification accuracy (Abdullah et al., 2017).

Surprisingly, the TFD is suffered from a high level of the resolution also complexity in the dimensionality of the feature vectors (Chowdhury, 2013). Therefore, efforts work to reduce the dimensionality and complexity of the vector that happened in TFD and preserving the discrimination capability (Rechy & Hu, 2015). Ibrahim and Al-Jumaily implemented a feature learning method that selects a subgroup from a large group of features, this method known as principal component analysis (PCA). The results of this study limiting the number of features gave a very closed accuracy result if it compared with the accuracy result without features reduction (M. Ibrahim & A. Al-Jumaily, 2016).

4. Classification Methods of EMG

The classifier takes the information extracted from the EMG signal to map several patterns and make matching between them. Therefore, the classifier is employed to recognized and extracted several features. After that, the gained result will be applied as control commands to the next stage of the controller. Many techniques are applied to classify the signals such as machine learning (ML) that contain the artificial neural networks (ANN), multilayer perceptron (MLP), support vector machines (SVM), linear discriminant analysis (LDA), K-nearest neighbor (KNN), and hidden Markov models (HMM).

Alkan and Ahmet made a comparison between the accuracy of LDA methods against an SVM, using the MAV as to classify upper arm movement, the results showed the accuracy of SVM is 99% better than the LDA model that achieved accuracy between 96% and 98% (Alkan & Günay, 2012). In 2015, (Gokgoz & Subasi, 2015) used the MAV, Standard deviation, Ratio of mean values and Average power to extract features from EMG signals. Then. The Random forest and the decision tree used to diagnose the neuromuscular disorder, the accuracy of the applied algorithm was 96.67%.

The fuzzy logic (FL) system is another technique that applied to classify the EMG signals, FL has the ability to control bio-signal (Ahmad, 2009). Therefore, the nonlinear dynamics methods were used on the EMG signals. The authors in (Chen et al., 2007) proposed a method that combining the multi-scale analysis and fuzzy entropy, the features extracted by the multi-scale fuzzy and the data classified by SVM the recognition rate was 97%.

In the past few years, the deep learning (DL) has emerged as a new technique, the main idea behind the DL is constructed from ML had numerous hidden layers that depend on taking a huge data such as a convolutional neural network (CNN) (LeCun, et al., 2015). In 2017 Cote-Allard presented a CNN using transfer learning (TL) techniques, the classifier was precise and robust for performing 6 DOF of the robotic arm. The CNN accuracy reached to 97.81% (Côté Allard et al., 2017). In 2019 the same authors presented a CNN classifier to train the raw EMG using CWT and Spectrograms, also the authors used the SampEn Pipeline, TD, Enhanced TD, and Nina Pro to extract features. For 17 participants the TL increased, and ConvNet achieved an accuracy of 98.31%. Moreover, when the NinaPro DB5 dataset had been used for 10 participants on a single Myo armband the average accuracy was 68.98% (Côté Allard et al., 2018). Table 2 reviews a list of studies that clarify several methods of feature extractions and classifications.

Table 2. Summary of some studies on feature extractions and classifications for EMG signal.

Author	Features	Classification	Accuracy
(L S et al., 2018)	RMS and Integrated Absolute Value (IAV)	KNN and NBP	92-94%.
(Kurniawan & Pamungkas, 2018)	RMS	ANN	92.68%.
(Y Paul et al. 2017)	(MAV), (RMS), Log Detector (LOG), Simple Square Integral (SSI), (WL), Difference Absolute Standard Deviation Value (DASDV), and Average Amplitude Change (AAC),	SVM and KNN	Best accuracy achieved of 63% using DASDV feature And SVM
(Caesarendra et al., 2016)	MAV, IEMG, MAV2, MAV1, RMS, VAR, WL, DASDV, SSI, Hjorth (Activity), AR , Hjorth (Complexity), and Hjorth (Mobility)	ANN	85.70%
(Maurya et al., 2019)	Approximate Entropy, Average Amplitude Change, IEMG, DASDV, Kurtosis, MAV, Log detector, RMS, Variance , Simple Square, Skewness, Total power, Integral, Mean Power, and Mean frequency	KNN and SVM	KNN of 99.7%, SVM of 99.6%.
(McCool et al., 2014)	ECG interference, PLI, ; motion artifact, additive white Gaussian noise (AWGN), and amplifier saturation;	SVM	95.75%
(Chung & Benalcázar, 2019)	RMS	ANN	85.08 ± 15.21%
(Yang et al., 2019)	R^2	CNN	E-F was 0.888 ± 0.054, U-R was 0.803 ± 0.087, and S-P was 0.810 ± 0.048.
(Qi et al., 2019)		CNN and TL	CNN 95%, TL 96%
(J. Zhang et al., 2019)	MAV, MAD, ZC, and SCS	Deep Belief Network (DBN)	99.79±0.29%
(Park & Lee, 2016)		CNN	Above 90%
(A. Zhang et al., 2018)	AR model with fifth order	LDA, BPNN and SVM	The LDA had the highest accuracy if it compared to BPNN with difference of 8.21%, p <103, and SVM with difference 4.55% $p < 10^3$.
(He et al., 2018)	LSTM network and MLP.	CNN	66+6.04%
(A. Adewuyi et al., 2015)	RMS, TD2, TD4, and TDAR	sparse representation based classification (SRC)	The TD4-SVM of RMS-SRC accuracy range from 12.60% to 16.96%. when the training set was contaminated the accuracy range from 18.24% to 34.79%.

(Castiblanco et al., 2016a)	IEMG, simple square, MAV, WL, variance, DASDV, average amplitude change, Willison amplitude , Myopulse percentage rate, MAVS, slope sign change, Cepstral coefficients, and AR.	SVM and k-means	SVM of 92%, K-mean1 of 55%, and K-mean2 of 75%.
(P. Xu et al., 2018)	RMS, maximum-slope and MF	BP and LVQ	BP of 88.3% LVQ of 94% BP and LVQ of 96.0%.
(Young et al., 2012)	MAV,ZC, number of slope sign changes, and WL	LDA	single classification-95%
(Khan et al., 2019)	mean, Kurtosis, Peak to Peak, Shape Factor, energy, jitter, ZC, Spectral Kurtosis, Spectral Skewness, Spectral Flux, Spectral Centroid, Spectral Crest, Spectral Roll Off, Mel-Frequency Cepstral Coefficients, and Spectral Spread.	LR and SVM	SVM of 94.1%, LR of 95.1%
(M. F. I. Ibrahim & A. A. Al-Jumaily, 2016)	Used Manhattan index 50 features are selected from PCA feature learning algorithm.	Quad SVM, Linear SVM, Fine Gauss SVM, Cubic SVM, Coarse Gauss SVM, and Medium Gauss SVM	Wavelet higher than 87%, the spectrogram of 82%.
(Demir et al., 2019)	AlexNet and VGG16	CNN	99.04%
(Roy et al., 2018)	VGG16	CNN	93%
(Bao, Zhang, et al., 2019)	MAV, RMS, VAR and 4th AR	CNN	90% for DOF1, 85.5% for DOF2, and 85.2% for DOF3
(Narayan et al., 2019)	20 subsets of best features	35 base classifiers	100%
(Jafarzadeh et al., 2019)	Features reduction use PCA	CNN	91.26%.

5. Discussion

This study covers an important element that is lacked in the previous studies is the data sets. The published data is difficult to compare among different methodologies. The main challenges for the researchers are to get the identical set up for the data acquisition, due to the restricted existence of the measuring equipment's and the EMG sensor's responsiveness, also taking into consideration the suitable place of the electrode (Jamal et al.,2019; Li et al., 2017; Maurya et al., 2019).

The current survey presents the most common PR approaches of EMG signals, that are involved in several applications, for instant the orthoses, the detection for neuromuscular disorders the rehabilitation devices, and the prostheses. The raw EMG signal contains unwanted signal sources. Therefore, filtering techniques are proposed to reduce the noise. Nevertheless, filtering processing can be reduced the noise level but that will affect the EMG signal quality. As a result, the researchers are focused on signal processing techniques to gain a simple, reliable, and accurately detects the signal. Hence, if the numbers of the electrode increase the numbers of control commands will increase too, so that will affect the computational time that classifier will take (Limem et al.,2016; Roy et al., 2018; Zhou et al.,2017).

The classification accuracy depends on the feature classifier combination. Therefore, the feature extraction of the time-domain and time-frequency domain is typically used. As noticed, from the previous studies frequency domain analysis is not promising, especially in isometric conditions. Therefore, the performance of time-frequency domain analysis using both ML and DL algorithms more than time-domain analysis to obtain high accuracies (Nazmi et al, 2016). However, the higher accuracies were gained using TD features with SVM and K-NN, correspondingly using the CNN with TL. Increasing the classification accuracies depend on a mixture of signal processing techniques and PR methods using the same contractions of the muscle (Demir et al., 2019; Qi et al., 2019). Hence, the combination of the method will increase the classification accuracies without using several muscle positions.

6. Conclusion

In this paper, the EMG signals are used for physical action recognition. To maintain an accurate pattern recognition some issues, have to be taken into consideration. First, to collect data correctly the position to the EMG sensors have to be fixed, and the number of sensors is preferred to cover the muscles that give the required action. Second, overcome the noises interference that appeared with the raw signals using a band-pass filter (BPF). Third, using the time-domain analysis to extract features and avoid using frequency domain analysis. Lastly, the classifiers such as support vector machine and k- nearest neighbor show the highest accuracy in machine learning algorithm, and if the amount of data is large would be better to use deep learning. Therefore, obtaining high classification accuracy is required to train the prosthetic parts accurately.

References

[1] Abdullah, A., Subasi, A., & Qaisar, S. (2017). Surface EMG Signal Classification by Using WPD and Ensemble Tree Classifiers. In (pp. 475-481).

[2] Adewuyi, A., Hargrove, L., & Kuiken, T. (2016). An Analysis of Intrinsic and Extrinsic HandMuscle EMG for Improved Pattern Recognition Control. IEEE transactions on neural systems and rehabilitation engineering: a publication of the IEEE Engineering in Medicine and Biology Society, 24. doi:10.1109/TNSRE.2015.2424371

[3] Ahmad, S. A. (2009). Moving approximate entropy and its application to the electromyographic control of an artificial hand (Doctoral dissertation, University of Southampton).

[4] Al-Angari, H. M., Kanitz, G., Tarantino, S., Cipriani, C. J. B. S. P., & Control. (2016). Distance and mutual information methods for EMG feature and channel subset selection for classification of hand movements. 27, 24-31.

[5] Alkan, A., & Günay, M. (2012). Identification of EMG Signals Using Discriminant Analysis and SVM Classifier. Expert Syst. Appl., 39, 44-47. doi:10.1016/j.eswa.2011.06.043

[6] Al-Mulla, M. R., Sepulveda, F., & Colley, M. J. S. (2011). A review of non-invasive techniques to detect and predict localised muscle fatigue. 11(4), 3545-3594.

[7] Amamcherla, N., Turlapaty, A., & Gokaraju, B. (2018). A Machine Learning System for Classification of EMG Signals to Assist Exoskeleton Performance. Paper presented at the 2018 IEEE Applied Imagery Pattern Recognition Workshop (AIPR).

[8] Balbinot, A., & Favieiro, G. J. S. (2013). A neuro-fuzzy system for characterization of arm movements. 13(2), 2613-2630.

[9] Bao, T., Zaidi, A., Xie, S., & Zhang, Z. (2019). Surface-EMG based Wrist Kinematics Estimation using Convolutional Neural Network. Paper presented at the 2019 IEEE 16th International Conference on Wearable and Implantable Body Sensor Networks (BSN).

[10] Benalcázar, M. E., Jaramillo, A. G., Zea, A., Páez, A., & Andaluz, V. H. (2017). Hand gesture recognition using machine learning and the Myo armband. Paper presented at the 2017 25th European Signal Processing Conference (EUSIPCO).

[11] Bilodeau, M., Cincera, M., Gervais, S., Arsenault, A. B., Gravel, D., Lepage, Y., . . . physiology, o. (1995). Changes in the electromyographic spectrum power distribution caused by a progressive increase in the force level. 71(2-3), 113-123.

[12] Caesarendra, W., Lekson, S. U., Mustaqim, K. A., Winoto, A. R., & Widyotriatmo, A. (2016, 29-31 Aug. 2016). A classification method of hand EMG signals based on principal component analysis and artificial neural network. Paper presented at the 2016 International Conference on Instrumentation, Control and Automation (ICA).

[13] Carrozza, M., Massa, B., Dario, P., Zecca, M., Micera, S., Pastacaldi, P. J. T., & Care, H. (2002). A two DoF finger for a biomechatronic artificial hand. 10(2), 77-89.

[14] Castiblanco, C., Parra, C., & Colorado, J. (2016a). Individual hand motion classification through EMG pattern recognition: Supervise and unsupervised methods.

[15] Castiblanco, C., Parra, C., & Colorado, J. (2016b). Individual hand motion classification through EMG pattern recognition: Supervise and unsupervised methods. Paper presented at the 2016 XXI Symposium on Signal Processing, Images and Artificial Vision (STSIVA).

[16] Chen, W., Wang, Z., Xie, H., & Yu, W. (2007). Characterization of Surface EMG Signal Based on Fuzzy Entropy. IEEE transactions on neural systems and rehabilitation engineering: a publication of the IEEE Engineering in Medicine and Biology Society, 15, 266-272. doi:10.1109/TNSRE.2007.897025

[17] Choi, K. Y., Akhtar, A., & Bretl, T. (2017). A compliant four-bar linkage mechanism that makes the fingers of a prosthetic hand more impact resistant. Paper presented at the 2017 IEEE International Conference on Robotics and Automation (ICRA).

[18] Chowdhury, R. H., Reaz, M. B., Ali, M. A. B. M., Bakar, A. A., Chellappan, K., & Chang, T. G. (2013). Surface electromyography signal processing and classification techniques. Sensors, 13(9), 12431-12466.

[19] Chung, E., & Benalcázar, M. (2019). Real-Time Hand Gesture Recognition Model Using Deep Learning Techniques and EMG Signals.

[20] Côté Allard, U., Fall, C. L., Campeau-Lecours, A., Gosselin, C., Laviolette, F., & Gosselin, B. (2017). Transfer Learning for sEMG Hand Gesture Recognition Using Convolutional Neural Networks.

[21] Côté Allard, U., Fall, C. L., Drouin, A., Campeau-Lecours, A., Gosselin, C., Glette, K., . . . Gosselin, B. (2018). Deep Learning for Electromyographic Hand Gesture Signal Classification Using Transfer Learning. IEEE Transactions on Neural Systems and Rehabilitation Engineering, PP. doi:10.1109/TNSRE.2019.2896269

[22] De Luca, C. J. J. I. T. o. B. E. (1979). Physiology and mathematics of myoelectric signals. (6), 313-325.

[23] De Luca, C. J., Gilmore, L. D., Kuznetsov, M., & Roy, S. H. J. J. o. b. (2010). Filtering the surface EMG signal: Movement artifact and baseline noise contamination. 43(8), 1573-1579.

[24] Demir, F., Bajaj, V., Ince, M., Taran, S., & Sengur, A. (2019). Surface EMG signals and deep transfer learning-based physical action classification. Neural Computing and Applications, 31. doi:10.1007/s00521-019-04553-7

[25] Fan, X., Dai, L., Chang, W., Ren, X., Sheng, S., & Duan, F. (2017). Verifying the gesture identification effect of WNN and DWT for an interference driven prosthetic hand.

[30] Geng, Y., Ouyang, Y., Samuel, O. W., Chen, S., Lu, X., Lin, C., & Li, G. J. I. A. (2018). A robust sparse representation based pattern recognition approach for myoelectric control. 6, 38326-38335.

[31] Gokgoz, E., & Subasi, A. (2015). Comparison of decision tree algorithms for EMG signal classification using DWT. Biomedical Signal Processing and Control, 18. doi:10.1016/j.bspc.2014.12.005

[32] Graupe, D., Salahi, J., & Kohn, K. H. J. J. o. B. E. (1982). Multifunctional prosthesis and orthosis control via microcomputer identification of temporal pattern differences in single-site myoelectric signals. 4(1), 17-22.

[33] Han, H., Jo, S. J. I. j. o. b., & informatics, h. (2013). Supervised hierarchical Bayesian model-based electromyographic control and analysis. 18(4), 1214-1224.

[34] He, Y., Fukuda, O., bu, N., Okumura, H., & Yamaguchi, N. (2018). Surface EMG Pattern Recognition Using Long Short-Term Memory Combined with Multilayer Perceptron (Vol. 2018).

[35] Huang, N. E., Shen, Z., Long, S. R., Wu, M. C., Shih, H. H., Zheng, Q., . . . sciences, e. (1998). The empirical mode decomposition and the Hilbert spectrum for nonlinear and non-stationary time series analysis. 454(1971), 903-995.

[36] Ibrahim, M. F. I., & Al-Jumaily, A. A. (2016). PCA indexing based feature learning and feature selection. Paper presented at the 2016 8th Cairo International Biomedical Engineering Conference (CIBEC).

[37] Ibrahim, M., & Al-Jumaily, A. (2016). PCA indexing based feature learning and feature selection.

[38] Jafarzadeh, M., Hussey, D., & Tadesse, Y. (2019). Deep learning approach to control of prosthetic hands with electromyography signals.

[39] Jamal, M. Z., Lee, D.-H., & Hyun, D. J. (2019). Real Time Adaptive Filter based EMG Signal Processing and Instrumentation Scheme for Robust Signal Acquisition Using Dry EMG Electrodes. Paper presented at the 2019 16th International Conference on Ubiquitous Robots (UR).

[40] Kendell, C., Lemaire, E. D., Losier, Y., Wilson, A., Chan, A., Hudgins, B. J. J. o. n., & rehabilitation. (2012). A novel approach to surface electromyography: an exploratory study of electrode-pair selection based on signal characteristics. 9(1), 24.

[41] Khan, M. U., Aziz, S., Bilal, M., & Aamir, M. B. (2019). Classification of EMG Signals for Assessment of Neuromuscular Disorder using Empirical Mode Decomposition and Logistic Regression. Paper presented at the 2019 International Conference on Applied and Engineering Mathematics (ICAEM).

[42] Kurniawan, S., & Pamungkas, D. (2018). MYO Armband sensors and Neural Network Algorithm for Controlling Hand Robot.

[43] L S, P., Nagananda, S., & Shankapal, P. (2018). Design and Development of Real Time Bionic Hand Control Using EMG Signal.

[44] LeCun, Y., Bengio, Y., & Hinton, G. (2015). Deep learning. nature, 521(7553), 436-444.

[45] Li, X., Fang, P., Tian, L., & Li, G. (2017). Increasing the robustness against force variation in EMG motion classification by common spatial patterns. Paper presented at the 2017 39th Annual International Conference of the IEEE Engineering in Medicine and Biology Society (EMBC).

[46] Limem, M., Hamdi, M. A., & Maaref, M. A. (2016, March). Denoising uterine EMG signals using LMS and RLS adaptive algorithms. In 2016 2nd International Conference on Advanced Technologies for Signal and Image Processing (ATSIP) (pp. 273-276). IEEE.

[47] Maurya, V. P., Kumar, P., & Halder, S. (2019). Optimisation and Classification of EMG signal using PSO-ANN. Paper presented at the 2019 Devices for Integrated Circuit (DevIC).

[48] McCool, P., Fraser, G., Chan, A., Petropoulakis, L., & Soraghan, J. (2014). Identification of Contaminant Type in Surface Electromyography (EMG) Signals. IEEE transactions on neural systems and rehabilitation engineering : a publication of the IEEE Engineering in Medicine and Biology Society, 22. doi:10.1109/TNSRE.2014.2299573

[49] Merletti, R., & Farina, D. (2016). Surface electromyography: physiology, engineering, and applications: John Wiley & Sons.

[50] Merletti, R., Conte, L. R. L. J. J. o. E., & Kinesiology. (1997). Surface EMG signal processing during isometric contractions. 7(4), 241-250.

[51] Narayan, Y., Singh, R., Mathew, L., & Chatterji, S. (2019). Surface EMG Signal Classification Using Ensemble Algorithm, PCA and DWT for Robot Control: Second International Conference, ICAICR 2018, Shimla, India, July 14–15, 2018, Revised Selected Papers, Part I. In (pp. 218-230).

[52] Nazmi, N., Abdul Rahman, M. A., Yamamoto, S. I., Ahmad, S. A., Zamzuri, H., & Mazlan, S. A. (2016). A review of classification techniques of EMG signals during isotonic and isometric contractions. Sensors, 16(8), 1304.

[53] Oskoei, M. A., Hu, H. J. B. s. p., & control. (2007). Myoelectric control systems—A survey. 2(4), 275-294.

[54] Park, K.-H., & Lee, S.-W. (2016). Movement intention decoding based on deep learning for multiuser myoelectric interfaces. Paper presented at the 2016 4th International Winter Conference on Brain-Computer Interface (BCI).

[55] Park, S. H., & Lee, S. P. (1998). EMG pattern recognition based on artificial intelligence techniques. IEEE transactions on Rehabilitation Engineering, 6(4), 400-405.

[56] Patel, J. M., & Gamit, N. C. (2016, 23-25 March 2016). A review on feature extraction techniques in Content Based Image Retrieval. Paper presented at the 2016 International Conference on Wireless Communications, Signal Processing and Networking (WiSPNET).

[57] Piervirgili, G., Petracca, F., & Merletti, R. J. P. m. (2014). A new method to assess skin treatments for lowering the impedance and noise of individual gelled Ag–AgCl electrodes. 35(10), 2101.

[58] Prahm, C., Kayali, F., Vujaklija, I., Sturma, A., & Aszmann, O. (2017). Increasing motivation, effort and performance through game-based rehabilitation for upper limb myoelectric prosthesis control. Paper presented at the 2017 International Conference on Virtual Rehabilitation (ICVR).

[59] Qi, S., Wu, X., Wang, J., & Zhang, J. (2019). Recognition of Composite Motions based on sEMG via Deep Learning. Paper presented at the 2019 14th IEEE Conference on Industrial Electronics and Applications (ICIEA).

[60] Rechy, E., & Hu, H. (2015). Bio-signal based control in assistive Robots: A survey. Digital Communications and Networks, 2. doi:10.1016/j.dcan.2015.02.004

[61] Riillo, F., Quitadamo, L. R., Cavrini, F., Gruppioni, E., Pinto, C. A., Pastò, N. C., . . . Control. (2014). Optimization of EMG-based hand gesture recognition: Supervised vs. unsupervised data preprocessing on healthy subjects and transradial amputees. 14, 117-125.

[62] Rissanen, S. M., Kankaanpää, M., Tarvainen, M. P., Nuutinen, J., Tarkka, I., Meigal, A., . . . Karjalainen, P. A. (2007). Extraction of typical features from surface EMG signals in Parkinson's disease. Paper presented at the 11th International Congress of Parkinson's Disease and Movement Disorders, Istanbul, Turkey.

[63] Rogers, D. R., MacIsaac, D. T. J. J. o. E., & Kinesiology. (2013). A comparison of EMG-based muscle fatigue assessments during dynamic contractions. 23(5), 1004-1011.

[64] Roy, R., Kumar, A., Mahadevappa, M., & Kumar, C. S. (2018). Deep Learning Based Object Shape Identification from EOG Controlled Vision System. Paper presented at the 2018 IEEE SENSORS.

[65] Simão, M., Mendes, N., Gibaru, O., & Neto, P. J. I. A. (2019). A review on electromyography decoding and pattern recognition for human-machine interaction. 7, 39564-39582.

[66] Tortora, S., Moro, M., & Menegatti, E. (2019). Dual-myo real-time control of a humanoid arm for teleoperation. Paper presented at the 2019 14th ACM/IEEE International Conference on Human-Robot Interaction (HRI).

[67] Wang, J., Wang, R., Li, F., Jiang, M., & Jin, D. (2006). EMG signal classification for myoelectric teleoperating a dexterous robot hand. Paper presented at the 2005 IEEE Engineering in Medicine and Biology 27th Annual Conference.

[68] Xu, L., & Adler, A. (2004). An improved method for muscle activation detection during gait. Paper presented at the Canadian Conference on Electrical and Computer Engineering 2004 (IEEE Cat. No. 04CH37513).

[69] Xu, P., Liu, Y., Gu, G., & Shen, X. (2018). Research on the Method of EMG Pattern Recognition Based on Neural Networks. Paper presented at the 2018 2nd IEEE Advanced Information Management, Communicates, Electronic and Automation Control Conference (IMCEC).

[70] Y Paul, Goyal, V., & Jaswal, R. A. (2017). Comparative analysis between SVM & KNN classifier for EMG signal classification on elementary time domain features. Paper presented at the 2017 4th International Conference on Signal Processing, Computing and Control (ISPCC).

[71] Yang, W., Yang, D., Liu, Y., & Liu, H. J. I. T. o. H.-M. S. (2019). Decoding Simultaneous Multi-DOF Wrist Movements From Raw EMG Signals Using a Convolutional Neural Network. 49(5), 411-420.

[72] Young, A. J., Smith, L. H., Rouse, E. J., & Hargrove, L. J. J. I. T. o. B. E. (2012). Classification of simultaneous movements using surface EMG pattern recognition. 60(5), 1250-1258.

[73] Zhang, A., Gao, N., Wang, L., & Li, Q. (2018). Combined Influence of Classifiers, Window Lengths and Number of Channels on EMG Pattern Recognition for Upper Limb Movement Classification. Paper presented at the 2018 11th International Congress on Image and Signal Processing, BioMedical Engineering and Informatics (CISP-BMEI).

[74] Zhang, J., Ling, C., & Li, S. (2019). Human Movements Classification Using Multi-channel Surface EMG Signals and Deep Learning Technique. Paper presented at the 2019 International Conference on Cyberworlds (CW).

[75] Zhang, X., Zhou, P. J. M. e., & physics. (2013). Filtering of surface EMG using ensemble empirical mode decomposition. 35(4), 537-542.

[76] Zhou, R., Luo, Q., Feng, X., & Li, C. (2017, December). Design of a wireless multi-channel surface EMG signal acquisition system. In 2017 3rd IEEE International Conference on Computer and Communications (ICCC) (pp. 279-283). IEEE.

IOP Conf. Series: Materials Science and Engineering **978** (2020) 012026 doi:10.1088/1757-899X/978/1/012026

Investigation the Properties of Waste Plastic Fiber Concrete Modified with HP-570 Super-Plasticizer

Osama M. Al-Hussaini[1], * Waseem H. AL-Baghdadi[2], Salama N. Al-Labban[3] and Qusay A. Jabal[4]

[1]Faculty of Engineering, University of Kufa, Iraq
E-mail: osama.alhussaini@uokufa.edu.iq
[2]Faculty of Engineering, University of Kufa, Iraq
E-mail: waseem.albaghdadi@uokufa.edu.iq
[3]Faculty of Engineering, University of Kufa, Iraq
E-mail: sallama.allaban@uokufa.edu.iq
[4]Faculty of Engineering, University of Kufa, Iraq - E-mail: qusay.alatiya@uokufa.edu.iq
* Corresponding author: Faculty of Engineering, University of Kufa, Najaf, Iraq. Tel: +9647814332941.

Abstract. Recycling plastic waste has always been vital for sustainability, besides exploiting that waste to improve the mechanical properties of concrete mixtures which contribute effectively in the field of civil engineering applications. In this research, the enhancement of compressive, tensile and flexural strength of concrete mixtures was explored using both adding plastic fibers and HP-570 super-plasticizer. Plastic waste, as fibers, was mixed with an Ordinary Portland cement for a total volume of concrete of 0.5, 1, 1.5 and 2%, then the HP-570 super-plasticizer was added and the procedure was re-done. At the age of 28 days of curing, the compression strength was increased by (13.2%) while both tensile and flexural strengths were increased significantly by (117.4%) and (207.5%) respectively when using the plastic as fiber with 2%. Further improvement was gained by adding the HP-570 super-plasticizer for the concrete mixture.

Keywords: *Plastic waste, Fiber reinforced concrete, concrete plasticizer, Concrete strength.*

1.Introduction

In modren world, one of the most concern is how to get sustainability by recycling waste materials. Numerous waste materials are generated from manufacturing processes, service industries and municipal solid wastes (Siddique, Khatib, and Kaur 2008), Plastics have been used in many industrial and healthcare applications besides packaging, automotive, medical delivery systems, artificial implants, water desalination, land/soil conservation, flood prevention, preservation and distribution of food, housing, communication materials, security systems, and other uses. With these large and diverse applications, plastics contribute significantly to increasing solid waste volume (Subramanian 2000). The consumption of

IOP Conf. Series: Materials Science and Engineering **978** (2020) 012026 doi:10.1088/1757-899X/978/1/012026

various forms of plastics is a challenging environment protection issue, all forms of consumed plastic become waste and require large areas of land for storage (Sharma and Bansal 2016). The researchers must find solutions also original, imaginative and brilliant to the reuse of the waste. With the lack of spaces for landfilling and due to continuous increasing cost, the attention is towards the recycle of waste as an attentive to disposal. Research is always more interested in the use of such products in the concrete mix. This makes the concrete more economic and, at the same time, there is a reduction in the problem of waste (Foti 2013).

On the other hand, In recent years, an emerging technology named ''Fiber-Reinforced Concrete (HPFRC)'' has become popular in the construction industry. The materials used in FRC depend on the desired characteristics and the availability of suitable local economic altetive materials (Afroughsabet, Biolzi, and Ozbakkaloglu 2016). Fibers –in general- increase the compressive, tensile and flexural strength of concrete and it used in large varieties of constructions like highly loaded members, bridges, rigid pavement. Also, fiber concrete used in the footing and precast piles (Fam et al. 2003). Using fibers can make concrete resist to deformations without cracks or very fewer cracks compared with normal concrete (Muwashee, Al-Jameel, and Jabal 2018), also,adding waste plastic in reinforced concrete beams decreases the ultimate deflection at failure and increase the ductility behavior at the same time (Al-Hadithi, Abdulrahman, and Al-Rawi 2020). Furthermore, the waste plastic fiber concrete will be slightly light weight (R. N. Nibudey; et al. 2013) and reduces 17.5% of concrete quantity and reduces 8.5% of the concrete cost (Singh and Kumar 2019).

One siginificant side effects of additons in concrete is the reduction in workability, the addition of waste plastics affects the free water available in concrete mixture and, consequently, the workability of the concrete (Sharma and Bansal 2016). The workability is inversely proportional to the ratio of waste addition and aspect ratio (Khatab, Mohammed, and Hameed 2019). For example, it decreases when the addition of waste plastics increases by up to 25% (Bhogayata and Arora 2017). As a result, using concrete plasticizer will be vaital to improve the workability when the plastic additions were used.

This research aims to investigate the variation of some mechanical properties in concrete when using plastic fibers made from waste bottles and produce low-cost mixtures at the same time, then using both HP-570 super-plasticizer with plastic fibers. Using the plastic fiber as strips has not been covered adequately besides using the HP-570 super-plasticizer to overcome the decrease in the concrete workability.

2.Experimental Work

2.1 Materials used in the study:

330 kg/m^3 of ordinary Portland cement was added to normal fine and coarse aggregate to produce a conventional concrete using drinkable water, the amounts of all other mix components were listed in table 1. All in-use material was tested to make sure that the concrete mix meets the Iraqi standard specifications. The grain-size distribution curves for the aggregates were shown in tables 2 and 3 and figure 1 which already meet Iraqi Standard - 45 (Specification 1884). The physical and chemical tests of Portland cement were shown in table 4 which already meet Iraqi Standard – 5 (Specification 1984).

Plastic fibers used in this study were obtained by slicing off used plastic bottles of water which are widely available in local markets. Along with using plastic fibers, HP-570 super-plasticizer was also used as an additive for the concrete mixture to enhance its compression, tensile and flexural strengths, the essential related properties of plastic and the admixtures were listed in tables 5 and 6.

2.2 Specimens and testing procedure:

Three types of specimens were used to determine the mechanical properties of the newly tested concrete as explained in table 7. While the compression test was gained by direct test, the tensile strength was gain from the approximate method of splitting test. Moreover, the flexural strength was calculated according to British standards B.S 1881(Standard 1881) using the following formula:

$$\sigma_{flex} = \frac{F.L}{b.d^2} \qquad (1)$$

where:
σ_{flex}: the flexural strength of concrete beam (modulus of rupture) due to bending stress.
F: the maximum load applied by the machine.
L: the distance between supports.
b: the width of the beam, and
d: the depth of the beam

All concrete members were categorized in three major groups, the first is the reference group (I) which means neither plastic fibers nor admixture is being used, the concrete with the plastic fiber was the second group(II), eventually, the final group (III) was the concrete specimens when using the plastic fiber coupled with the HP-570 super-plasticizer as shown in figure 2.

Each test was performed for three specimens to adopt the average value of each case. All specimens were tested after 28 days of curing by putting them in water as shown in figures 3 and 4.

Table 1. Mix proportion used in the study (for each 1m^3 of concrete)

Reference Mix.	Cement	Sand	Gravel	Water	HP-570 super-plasticizer	w/c
kg for each 1 cubic meter	330	620	1200	160	1.2 liter per 100 kg cement	0.3 for mixes with HP-super-plasticizer

Table 2. Grading of coarse aggregates used in the study

Sieve size	Percentage of passing (%)	IQS -85,limitations for coarse aggregate (%)
40 mm	100	100
20 mm	96.2	95-100
10 mm	43.8	30-60
4.75 mm	1.9	0-10

Table 3. Grading of fine aggregates used in the study

Sieve Designation	Percentage passing (%)	IQS -85,limitations for zone 1 fine aggregate (%)
10 mm	100	100
4.75 mm	93.6	90 - 100
2.36 mm	80.1	60- 95
1.18 mm	68.8	30-70
600 mic	37.3	15-34
300 mic	10.1	5-20
150 mic	3.9	0 -10

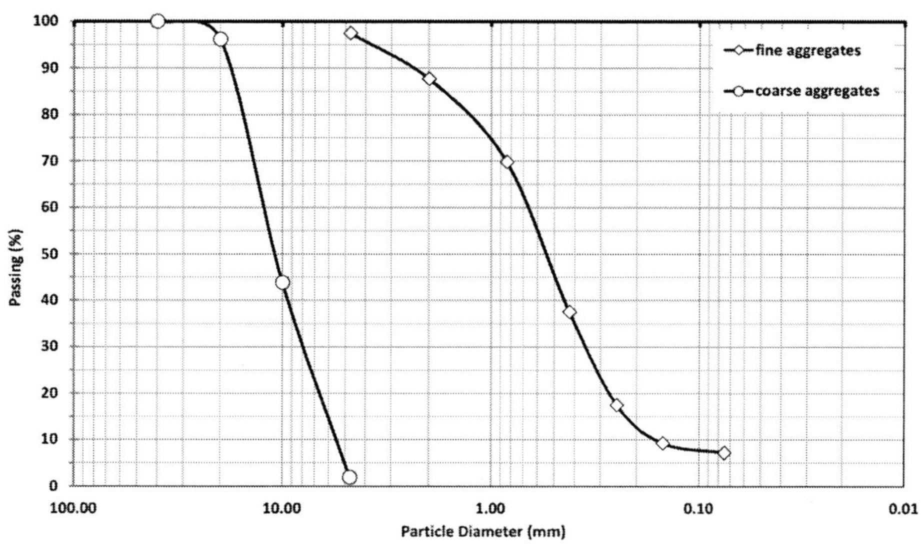

Figure 1. Grain size distribution of fine and course aggregate

Table 4. The physical and chemical tests of Portland cement

Groups	Test	Test results	Iraqi specification No5 /1984
Physical tests	Fineness, Blaine method (m^2/kg)	306	>250
	Initial setting time (hour:min)	1:37	>00:45
	Initial setting time (hour:min)	5:10	<10:00
	Compressive strength (MPa) 3 days 7 days	 16.6 24.9	 >15 >23
Chemical tests	CaO%	61.76	----
	Slicon Dioxide (SiO_2)%	20.6	----
	Aluminuum Oxide (Al_2O_3)%	3.50	----
	Iron Oxide (Fe_2O_3)%	4.72	----
	Lime Saturation Factor (LSF)	0.9	0.66-1.02
	Magnesium Oxide (MgO)%	2.22	<5%
	Tricalcium Aluminates (C_3A)%	0.38	>3.5%
	Sulfur Trioxide (SO_3)%	2.07	< 2.5 if C_3A less than 5% ≤ 2.8 if C_3A more than 5%
	Loss in Ignition (LOI)%	1.26	<4%
	Insoluble Residue (IR)%	1.12	<1.5%

Table 5. Properties of the plastic water bottles

Plastic fiber from water bottles	Chemical composition	Tensile strength	density	Melting point	Approximate Dimensions	Aspect ratio
DATA	[C2H4]n chains	48 -72 MPa	0.88 – 0.96 gr/cm^3	115 – 135 centigrade	0.5×1.5cm	75

Table 6. HP-570 super-plasticizer admixture properties

HP-570super-plasticizer type	Color	Specific Gravity	Dosage	Chloride content
Liquid super-plasticizer	light brown	1.1	500-220 ml for 100 kg cement	NIL

Table 7. Types of concrete specimens with its related test

Specimen shape	Dimensions (mm)	Related Mechanical Property	Type of Test
Cubes	100×100×100	Compressive strength	Direct Test
Cylinders	100×200	Tensile strength	The approximate method of splitting test
Prism	100×100×400	Flexural strength	Third point loading

Figure 2. Plastic fibers (Mixed with concrete) used in the study

Figure 3. Compression, splitting tensile strength testing for member used in the study

IOP Conf. Series: Materials Science and Engineering **978** (2020) 012026 doi:10.1088/1757-899X/978/1/012026

Figure 4. Flexural strength testing for member used in the study

3.Results and Discussion

The results of compression strength, splitting tensile strength and flexural strength for all types of concrete members were summarized in table 8, the first group of specimens was tested with 0% plastic fiber with no admixture also in order to be reference values for comparing other results with.

The plastic fibers were added gradually starting from 0.5, 1, 1.5 reaching up to 2% without any use of admixtures for the second group, then the procedure was repeated for the third group by adding the HP-570 super-plasticizer, these samples were tested to find the combined effect of adding plastic fibers together with admixture.

The compressive, tensile and flexural strengths were increased ultimately by 13.2%, 117.4% and 207.5% respectively when comparing the results of the second group with the reference one. The significant increase in strengths -especially in tensile and flexural- may be attributed to two major reasons; this first one is that the existence of plastic fibers in concrete will reduce the propagation of cracks under loaded specimens, this reduction gives more time of receiving loads which will increase the strengths. The second reason backs to the properties of the plastic fibers itself, which has high tensile strength as mentioned in table 5.

Additional enhancement was recorded in all previously mentioned properties when adding the HP-570 super-plasticizer for the last group of concrete, the increase of the values was rising by 14.6%, 151.7% and 369.8% respectively. This behavior was illustrated in figures 5, 6 and 7.

Table 8. Tests results

Groups	Mix Type	Compressive Strength (MPa)	Tensile Strength (MPa)	Flexural Strength (MPa)
I	Reference Mix: 0% Plastic Fiber with No HP-570 super-plasticizer	25.43	1.72	2.25
II	0.5% Plastic Fiber with No HP-570 super-plasticizer	26.15	2.80	3.86
	1% Plastic Fiber with No HP-570 super-plasticizer	27.40	3.54	6.24
	1.5% Plastic Fiber with No HP-570 super-plasticizer	28.43	3.70	6.87
	2% Plastic Fiber with No HP-570 super-plasticizer	28.80	3.74	6.92
III	0.5% Plastic Fiber with HP-570 super-plasticizer	26.73	2.97	5.12
	1% Plastic Fiber with HP-570 super-plasticizer	27.95	3.88	7.45
	1.5% Plastic Fiber with HP-570 super-plasticizer	28.70	4.17	8.76
	2% Plastic Fiber with HP-570 super-plasticizer	29.14	4.33	10.56

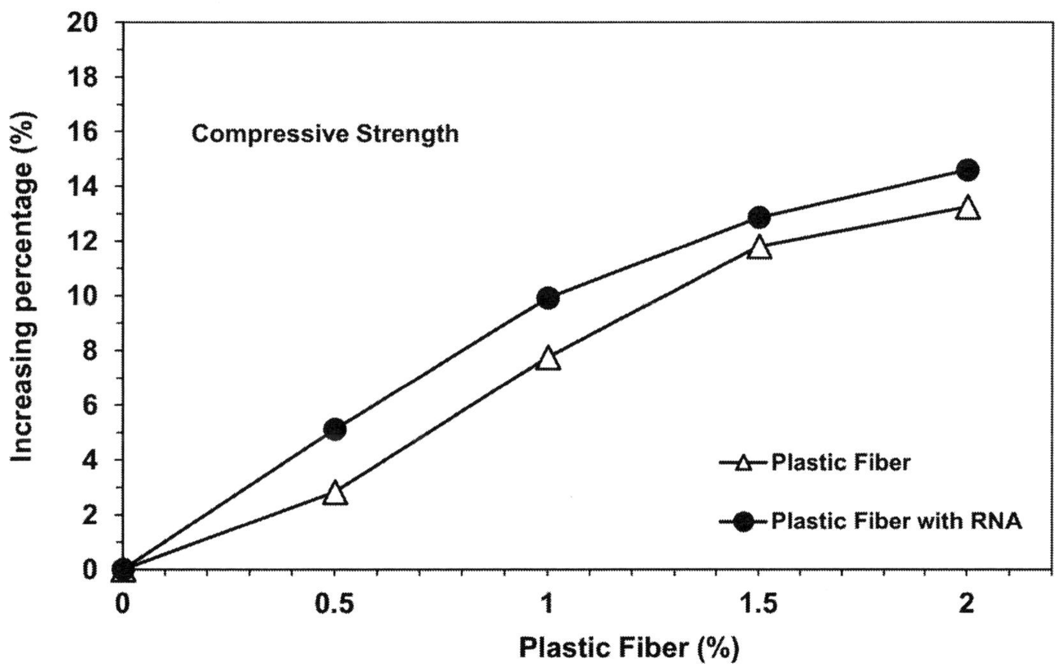

Figure 5. Increasing percentage in compressive strength

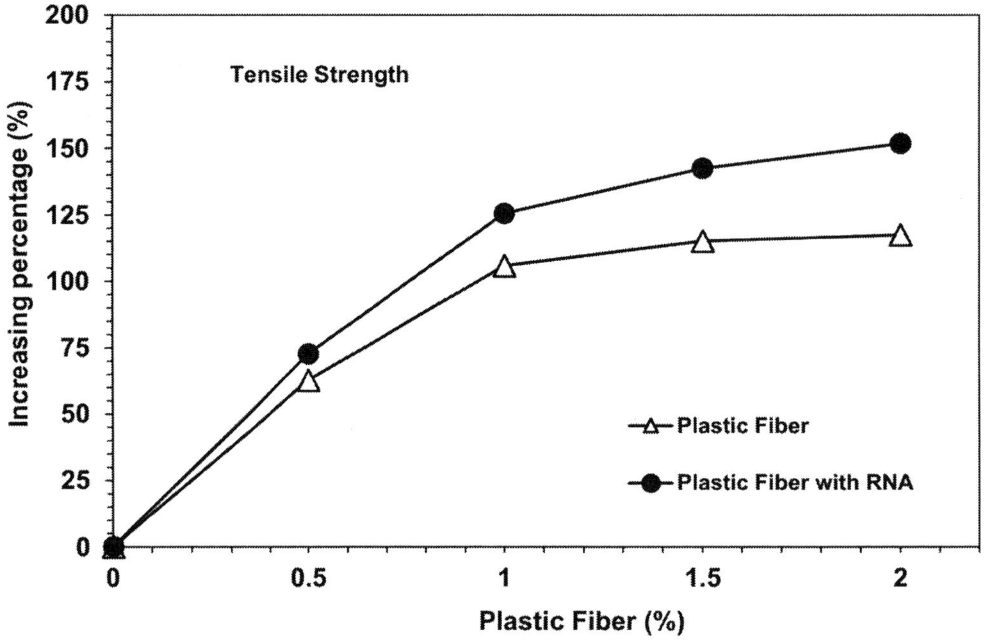

Figure 6. Increasing percentage in tensile strength

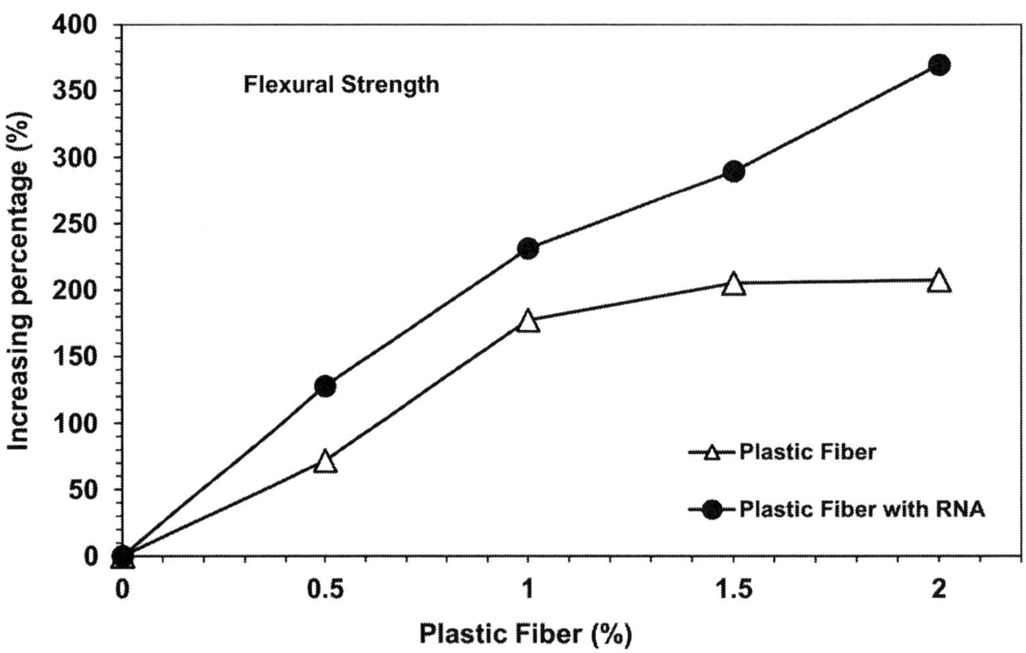

Figure 7. Increasing percentage in flexural strength

From figures 8 and 9 it can be seen that high increment of increase in tensile and flexural strength by using plastic fibers only, this increase could be attributed to the action of fibers which arrest the propagation of cracks under loading before failure (R. N. Nibudey; et al. 2013). This action gives long time to hold loads before failure, as a result, higher tensile and flexural strength. Moreover, fibers (in general) support concrete (which is already a brittle material) to resist cracks generation under loading (Mahadik, Kamane, and Lande 2014). The second reason is due to the high tensile strength of plastic fiber itself as mentioned in table 5. Compressive strength slightly increased by adding plastic fibers, as shown in figure 8. This is agree with the results of some studies such as Ramadevi and Manju (2012), Maqbool et al. (2019) and Prahallada and Prakash (2013).

By using both HP-570 super-plasticizer and fibers as shown in figure 8, the tensile and flexural strength increased significantly, that could be explained by the action of HP-570 super-plasticizer which reduce w/c ratio used in concrete and give higher strength comparing with ordinary concrete that have no super-plasticizer admixture (Salem, Alsadey, and Johari 2016).

Unlike the behavior of normal concrete during failure which happens all in a sudden, the concrete with fibers was taken longer time before complete the failure, this case was observed especially when testing the prism, this different behavior indicates that the ductility of concrete was increased by adding plastic to it as can be seen from figure 4.

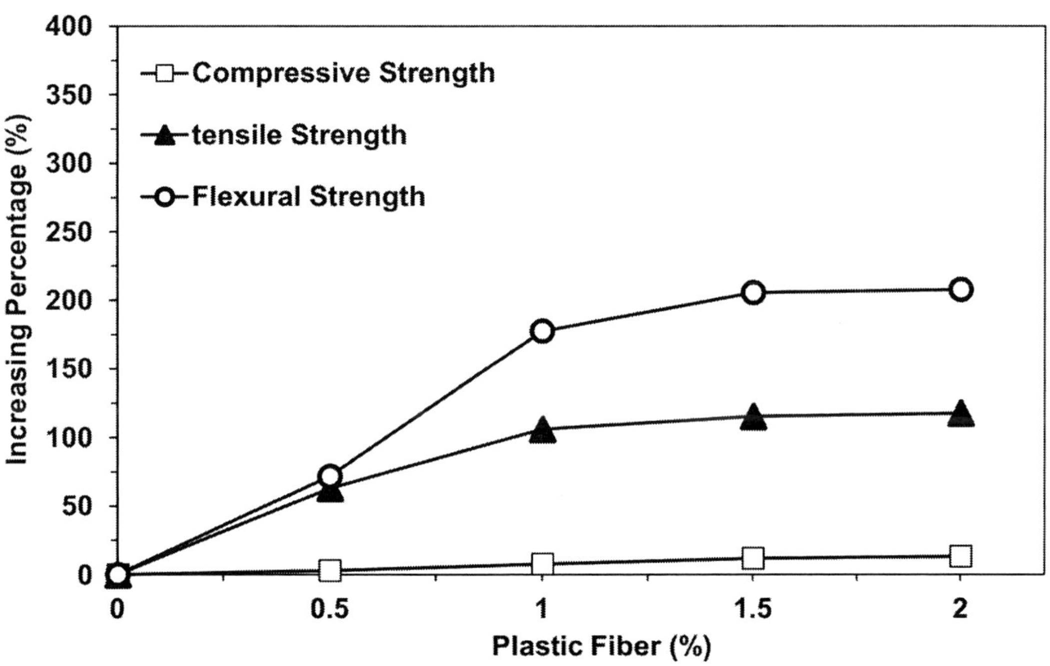

Figure 8. Increasing percentage in strengths as a result of adding plastic fiber only

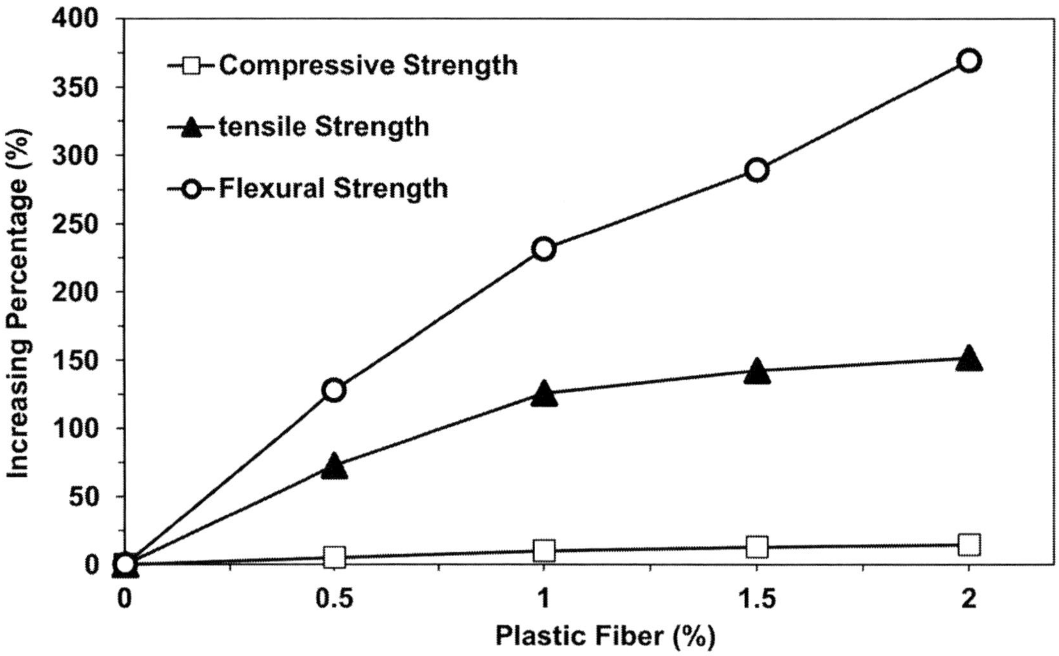

Figure 9. Increasing percentage in strengths as a result of adding plastic fiber and HP-570 super-plasticizer

4.Conclusions

- Using waste plastic as fibers improve the mechanical properties of concrete such as compressive, tensile and flexural strength
- A small increase in strength noticed when the percentage of plastic fibers exceeds 1.5%
- The combination effect of the plastic fiber and HP-570 super-plasticizer produce a significant increase in flexural strength.
- Reinforcing concrete by plastic fibers give ductile failure in flexure test comparing with ordinary concrete.

References:

[1] Afroughsabet, Vahid, Luigi Biolzi, and Togay Ozbakkaloglu. 2016. 51 Journal of Materials Science *High-Performance Fiber-Reinforced Concrete: A Review*. Springer US.

[2] Al-Hadithi, A. I., M. B. Abdulrahman, and M. I. Al-Rawi. 2020. "Flexural Behaviour of Reinforced Concrete Beams Containing Waste Plastic Fibers." *IOP Conference Series: Materials Science and Engineering* 737(1).

[3] Bhogayata, Ankur C., and Narendra K. Arora. 2017. "Fresh and Strength Properties of Concrete Reinforced with Metalized Plastic Waste Fibers." *Construction and Building Materials* 146: 455–63. http://dx.doi.org/10.1016/j.conbuildmat.2017.04.095.

[4] Fam, Amir, Miguel Pando, George Filz, and Sami Rizkalla. 2003. "Precast Piles for Route 40 Bridge in Virginia Using Concrete Filled FRP Tubes." *PCI journal* 48(3): 32–45.

[5] Foti, Dora. 2013. "Use of Recycled Waste Pet Bottles Fibers for the Reinforcement of Concrete." *Composite Structures* 96: 396–404. http://dx.doi.org/10.1016/j.compstruct.2012.09.019.

[6] Khatab, Hadeel R., Samaher J. Mohammed, and Laith A. Hameed. 2019. "Mechanical Properties of Concrete Contain Waste Fibers of Plastic Straps." *IOP Conference Series: Materials Science and Engineering* 557(1).

[7] Mahadik, S A, S K Kamane, and A C Lande. 2014. "Effect of Steel Fibers on Compressive and Flexural Strength of Concrete." *International Journal of Advanced Structures and Geotechnical Engineering* 3(4): 388–92.

[8] Maqbool, Younus et al. 2019. "Experimental Investigation on the Properties of Concrete with Plastic Pet Bottle Fibers as Partial Replacement of Fine Aggregates." *International Journal of Advance Research, Ideas and Innovations in Technology* 5(1): 524–29.

[9] Muwashee, Rawa Shakir, Hamid Athab Al-Jameel, and Qusay Abdulhameed Jabal. 2018. "Investigating the Behavior of Concrete and Mortar Reinforced with Aluminum Waste Strips." *International Journal of Engineering and Technology(UAE)* 7(4): 211–13.

[10] Prahallada, M.C, and K. B Prakash. 2013. "Effect of Different Aspect Ratio of Waste Plastic Fibres on the Properties of Fibre Reinforced Concrete-an Experimental Invistigation." *Journal, International* 2(2): 1–13.

[11] R. N. Nibudey;, Dr. P. B. Nagarnaik;, Dr. D. K. Parbat;, and Dr. A. M. Pande. 2013. "Strengths Prediction of Plastic Fiber Reinforced Concrete (M30)." *International journal of engineering research and applications* 3(1): 1818–25.

[12] Ramadevi, K K, and R Manju. 2012. "Experimental Investigation on the Properties of Concrete With Plastic PET (Bottle) Fibres as Fine Aggregates." *Journal of Emerging Technology and Advanced Engineering* 2(6): 42–46.

[13] Salem, Muhsen, Salahaldein Alsadey, and Megat Johari. 2016. "Effect of Superplasticizer Dosage on Workability and Strength Characteristics of Concrete." *IOSR Journal of Mechanical and Civil Engineering* 13(04): 153–58.

[14] Sharma, Raju, and Prem Pal Bansal. 2016. "Use of Different Forms of Waste Plastic in Concrete - A Review." *Journal of Cleaner Production* 112: 473–82. http://dx.doi.org/10.1016/j.jclepro.2015.08.042.

[15] Siddique, Rafat, Jamal Khatib, and Indepreet Kaur. 2008. "Use of Recycled Plastic in Concrete: A Review." *Waste Management* 28: 1835–52.

[16] Singh, Gursharan, and Haresh Kumar. 2019. "Effect on Compressive Strength of Concrete by Addition

of Polypropylene Fiber in M20 Grade of Concrete." *International Journal of Scientific Research in Science, Engineering and Technology* 1: 156–59.

[17] Specification, Iraqi Standard. 1984. "No. 45/1984, Aggregates from Natural Sources for Concrete and Construction." *Central Organization for Standardization & Quality Control (COSQC), Baghdad, Iraq.*

[18] Specification, Iraqi Standard. 1984. "No. 5/1984, Portland Cement." *Central Organization for Standardization & Quality Control (COSQC), Baghdad, Iraq.*

[19] Standard, British. 1881. "Part 118 (1983) Method for Determination of Flexural Strength." *British Standards Institution Publication, London.*

[20] Subramanian, P. M. 2000. "Plastics Recycling and Waste Management in the US." *Resources, Conservation and Recycling* 28(3–4): 253–63.

Statistical Model for Re-Refining of Used Lubricating Oil by Solvent Extraction and Bentonite Clay Adsorption Method

Ali Ghufran Khudhur[1,*] and Zakariya Ibrahim Mohammed[2]

[1]North Gas Company, Ministry of Oil, Iraq- E-mail: ali.ghufran1992@gmail.com
[2]North Gas Company, Ministry of Oil, Iraq- E-mail: zakariatazaly@yahoo.com
* Corresponding author: North Gas Company, Ministry of Oil, Iraq. Tel: +9647701303352.

Abstract. Lubricating oils are widely using for reducing friction and wear by interposing a film of material between rubbing surfaces, but these oils lose their properties after a certain period of use. For protection from environmental pollution and the economic side, used oil can be recycling and purifying by various methods. Solvent extraction+ clay adsorption technique is one of the cheapest, competitive, and most efficient processes experienced in the recycling of used lubricating oils. After vacuum filtration of the oil is distilled by vacuum distillation and extracted by Methyl Ethyl ketone as a solvent. Bentonite clay (400 gm) is activated by mixing (100 ml) diluted sulfuric acid under 100 °C for 4 hours. The extracted oil is treated by activated clay (Clay/Extracted oil =%15 wt) at (120°C) with mixing speed (500 rpm) for 90 minutes. However, the capability of this process largely depends on the optimum condition of each parameter in the recycling process. For this purpose, the experiments of three variables (Temperature, mixing speed, extracted solvent/ oil ratio) at three levels for the extraction process are designed by Design-Expert software. The optimum condition for the extracted process is (Solvent /used oil=3, Temperature =58 °C, mixing speed =710 rpm) confirmed by ANOVA. The recovered oil with 72% yield is analyzed in the laboratory, found it very close to base oil specifications.

Keywords: *Response surface methodology (RSM)- Re-refining used oil- Solvent extraction- Clay adsorption -Central composite design*

1. Introduction

Lubricating oil representing a mixture of hydrocarbons containing 18-40 carbon atoms with molecular weight 300-750 in boiling point 300-565 °C [1]. Its composition of two main parts: base oil and additives. After production of virgin base oil, additives are blended to the base oil for production of various types of lubricating oil for achieving the necessary protection of equipment and internal parts of machines from wearing which is specifically designed for the specified condition [2,3]. The main purpose of lubrication is besides reducing the friction, lube acts as an insulator layer between the moving part of the machine, which is elongated the age of the machine and lowering the required energy also of cleaning parts in bearing housing [4].

There are several additives inside the lubricant for increasing efficiency and performance features. Typically, a lubricant containing 10-20% of additives, and the rest is base oil [1,4]. For steam turbines that formed from high-quality base oil contain antifoaming, antioxidant, and anti-rust additives. [5]

After a specific period of operation of lubricating oil, as recommended by the manufactured company, lubricating oil should be changed to ensure equipment working efficiently [6,7]. That because the quality degraded by the decomposition of atoms and additives or contamination by metallic particle, dirt, or water after a period of operation [4,7]. These metals and dirty will deposited from warned up surfaces into the lubricating oil and contaminated it. At elevated temperature and after a long time of use lead to a loss in lube properties such as viscosity, viscosity index, and some oil molecules will oxidize and form corrosive organic acids [3]. These deposited metals at high-temperature work as an assistant factor for corrosion reaction.

This waste oil containing a large amount of Magnesium, lead, iron, etc.; which is mostly being environmentally hazardous [8]. The disposal of waste oil to the environment has large damage to human health and plants [9]. The underwater, live organism, and soil are all contaminated by the release of waste oil, because of the presence of PAHs (Polycyclic aromatic hydrocarbon) and metallic in additive which turns them to carcinogenic and toxic [10,11]. There are ways to get rid of this waste oil like burning as fuel after blending it with other constituents for furnaces. Environmental pollution is caused by the incineration of waste oil due to the diffusion of the PM (particulate matter) and gases and impact on the greenhouse effect [12]. It is good to know that the base oil is never spoilt and after removing the impurities and water by re-refining processes it is re-useable [13,4]. Re-refining of used oil shows its benefit not only as environmental preservation but also in the economic side 65-75% could be recovered [14].

Solvent extraction and clay treatment one of the effective methods for re-refining of the used oil [7]. Concentrated H_2SO_4 is used in this process for removing asphaltic material followed by the clay treatment to remove undesirable compounds such as the high polar compounds such as nitrogen and oxygen and aromatics [1,13]. Achievement of this process with high accuracy at optimum condition for saving energy and recovering material and reduce the cost to a minimum, statistical Response surface method by design expert software is applied [15].

RSM (response surface methodology) is a significant branch of design experiments, it is a technology for improving the formulation of a specific product, optimizing its performance, and developing new processes. One of the most advantageous of the RSM method illustrates graphically the relation between the response and independent variables, this graphical perspective led to the term Response Surface Methodology. Thus, the effect path of the input variable on the process is clear and optimization for the process optimal conditions are very straightforward.

One of the highest efficient ways in RSM to design and estimate second-order polynomial [16] is CCD (Central composite design), which is used in this study because it is higher efficient from other methods [17].CCD is known as experimental design in RSM for building a second order model without needing to use complete three level factorial experiment.

Experiments that are designed by the CCD method exist to three levels or more for each variable to cover more space of possible effects of input factors, therefore it gives enough information on the variable's behavior in the process.

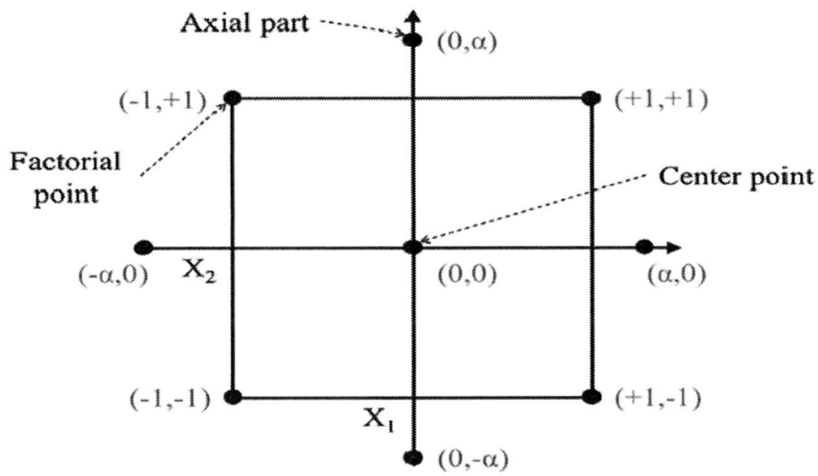

Figure 1. Central Composite Design point

The three distributed levels of points are: axial points, central points, and factorial points as it is obvious is Fig 1. Factorial points are consisting of two levels of points +1 and -1 of factors. Axial points put on the coordinate axes symmetrically with respect to the center point on distance α. Central point set on codded level (0,0) and are usually repeating to gain a good evaluation of error in experimental design.

2- Material and Methods

2.1-Material Used

Used lubricating oil was provided from the North Gas company after using in steam turbines 90 days by the recommendation of the manufactured Shell Company. The specification of the lubricant oil was illustrated in Table 1.

Table 1. The specifications of the used and fresh lubricating oil

Properties	Units	Used Oil	Fresh oil
Viscosity@ 40 °C	cSt	71.8	61.50
Viscosity @ 100 °C	cSt	8.6	8
Viscosity Index	-	89	95
Flash point	°C	230	218
Pour point	°C	-24	-15
Density	kg/m^3	882	890
TAN (Total acid number)	mg kOH/gm oil	0.12	0.115
H_2O content	ppm	75	Nil
Color value	-	Black	5

Removing suspended impurities in the oil done by filtration (filter paper Whatman, ϕ = 150mm) followed by the vacuum distillation for removing water and light hydrocarbons to 760 mmHg and 75 °C. Three kinds of solvents are used in this study:1-Butanol, Hexane, and MEK (methyl ethyl ketone). The solvent MEK is shown the highest PSR% (percentage sludge removal) between other solvents and it is more suitable for refining the steam turbine lubricating oil. Table 2 illustrates the highest PSR% by the MEK.

Table 2. Used Solvents

Solvent	Activity	PSR%	Type of solvent	Purity	Company
1-Butanol	Weak	0.36	Alcohol	99%	Sigma Aldrich
Hexane	Good	4.3	Alkane	99%	SRLChem.
MEK	Excellent	6.11	Ketone	99.5%	Chem-Lab NV

2.2 Exprimental Set-up and Methods

2.2.1. Extraction Process

Extraction by the solvent is the best process for removing the aromatic rings which affect adversely the viscosity of the lube oil. The solvents are blended with the 20 ml of the used sample according to the conditions: quantity of the Solvent /used oil Ratio variated from (2.5-3.5) while the temperature from (40 °C – 70 °C) with the mixing speed (400 rpm – 900 rpm) using Magnetic hot plate Stirring (Cimarec Company). For the study, the effects of each factor in each solvents Design expert software are used for the formation of the experiments as its obvious in Table 3. Each experiment is done in 10 minutes to obtain suitable homogeneity between the solvent and the used oil. Centrifuge (Petro-test) 1000 rpm used to separate and settle the sludge as a raffinate phase while the mixture of used oil-solvent is separated as an extracted phase. The remaining base oil in the sludge is washed by using the same solvent followed by drying in the Furnace (Heraeus-vacutherm: VT6025) half-hour in 130 °C for evaporating the solvent in sludge. PSR% can be calculated by the equation:

$$(P.\ S.\ R) = (W_{dry}/W_{oil}) \times 100\% \ \ldots.\ (1)$$

Wdry = weight of dry Sludge (gm)
Woil = weight of used oil (gm)

Table 3. Experiments of Extraction process for three Solvents

Runs	Solvent / Oil Ratio	Mixing Speed (RPM)	Temperature(°C)	PSR%(MEK)
1	3.7	650	55	5.2
2	3	650	55	3.8
3	2.5	400	40	0.84
4	3	650	55	6.11
5	3	650	76.21	0.4
6	2.5	900	70	0.56
7	3	650	55	4.15

8	3	296.4	55	0.302
9	3	650	33.78	0.14
10	3.5	900	40	0.65
11	3	650	55	3.9
12	2.3	650	55	0.86
13	3.5	400	70	0.47
14	3	650	55	4
15	3	1003.55	55	0.67

2.2.2. Experimental Design

RSM is a group of statistical and mathematical techniques suitable for improving, optimizing, developing process [23]. It has many extensive applications especially in manners where some of the input factors potentially impact quality features or performance parameters of a product or process.

The scope of RSM is consisting of experimental tactics discovering the space of independent variables (Solvent/Oil ratio, Temperature, mixing speed) empirical statistical model to develop suitable approximating relation by an equation between the input variable and output product, and various optimization approaches for finding that values or levels of independent variables (inputs) that leading to produce in-demand responses (output)[24]. The experiments are designed according to 15 trails of three variables by RSM methodology for building a second-order model. Design expert software version 10.1 is used for analyzing parameters of extraction.

2.2.3. Adsorption Process

Activated bentonite clay was used as an adsorbent for refining used oil. The clay first washed and dried in a furnace at a temperature of 110 °C for 4 hours. Subsequently, for activation 100 gm of the bentonite clay is blended with the 400 ml diluted H_2SO_4 for 4 hours in 100 °C. Separation of activated clay by Whatman filter paper from the sulfuric acid and washing it distilled water again in the last step.

3. Results

3.1. Analysis of Variance

The data obtained from Table 3 is analyzed by the RSM method. The accuracy of the result's experiments can ensure to be correct from Fig 2, results of actual and predicted are very close to each other.

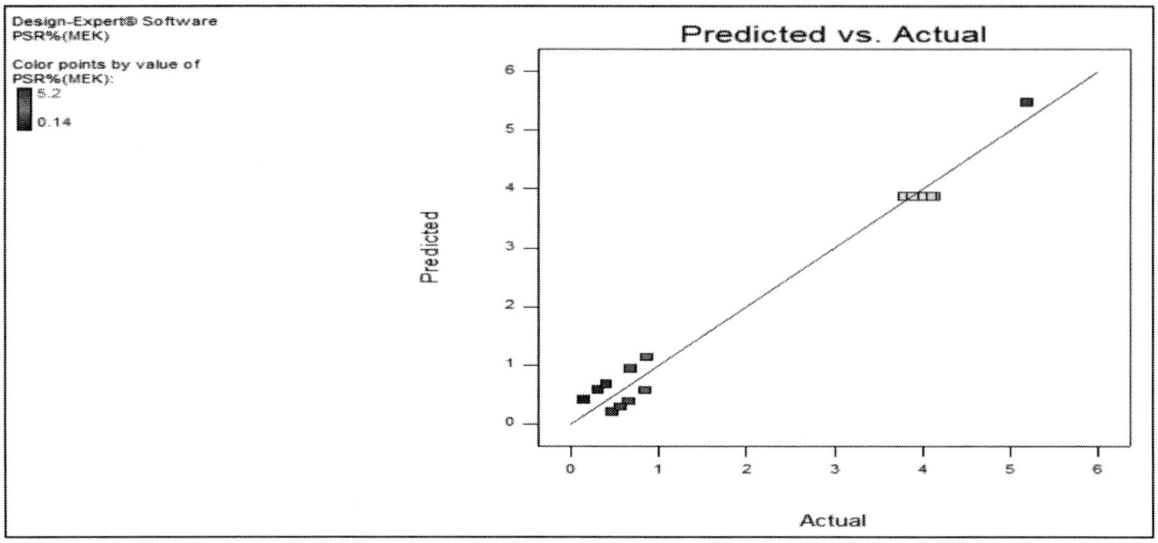

Figure 2. Actual and Predicted Results

The model has F value 30.41 that indicates the model of this process is significant. The F-statistic is simply a ratio of two variances. Variances are a measure of dispersion, or how far the data are scattered from the mean. Larger values represent greater dispersion, F value equal to (variation between sample means/variation within the samples).

P value is the probability of obtaining results as extreme as the observed results of a statistical hypothesis test, assuming that the null hypothesis is correct. In case the value of the P value is less than 0.05 this means that that parameter or model had a significant impact on the process. In this study the P value of the Model = 0.0008 less than 0.05, this means that the model is accepted.

The model has a coefficient of determination 98.21% which in agreement with the adjusted determination of coefficient adj-R^2 94.98%, both values are approximate 1.0. these high values support the relationship between the expected and the determined values. R^2 is a proportion of the variance in the dependent variable (the response) from the independent variables (the predictors) and can be used for model performance assessments, where adj-R^2 also indicates how well terms fit a curve or line, but adjusts for the number of terms in a model.

This point that the ANOVA (Analysis of Variance) shown in table 4, explained perfectly the relations of input variables and in addition to the response, where ANOVA is a collection of statistical models and their associated estimation procedures used to analyze the differences among group means in a sample.

Table 4. ANOVA for Response Surface Quadratic Model

Source	Squares Sum	df	Mean Square	F value	P value	
Model	48.83	9	5.43	30.41	0.0008	significant
A-Solvent/Oil (Ratio)	9.42	1	9.42	52.79	0.0008	
B-rpm	0.067	1	0.067	0.38	0.5656	
C-Temperature(C)	0.033	1	0.033	0.19	0.6849	
AB	0.085	1	0.085	0.48	0.5213	
AC	0.048	1	0.048	0.27	0.6263	
BC	5.20	1	5.20	29.13	0.0029	
A2	0.64	1	0.64	3.59	0.1166	
B2	18.81	1	18.81	105.47	0.0002	
C2	21.51	1	21.51	120.56	0.0001	
Residual	0.89	5	0.18			
Lack of Fit	0.81	1	0.81	38.33	0.0035	
Pure Error	0.084	4	0.021			
Cor Total	49.72	14				

* Note that, R^2=98.21%, Adj-R^2=94.98%

3.2. Optimum condition

The terms (A, BC, B^2 and C^2) had significant effect on the process. The P value of (B, C, AB and AC) was higher than 0.1, this means these terms had no significant effect on the process and can remove from the equation of model. The model equation of the extraction process by RSM method is:

$$PSR\% = -27.56 + 7.92 * A + 0.00429 * B + 0.479 * C - 1.170 * A2 - 0.000025\ B2 - 0.007419 * C2 + 0.00165\ A*B + 0.0206\ A*C + 0.000432\ B*C \quad \ldots\ldots (3)$$

3.2.1 Temperature

The temperature had its effect on the extraction process between used oil and MEK solvent due to an increase from solvent efficiency by increasing molecular movement and kinetic fluency which help more hydrocarbons will solve in the MEK solvent. However, the dissolving of hydrocarbons will decrease with increasing temperature due to the volatility of solvents in which the amount of solvent in the mixture will decrease at high temperature, so the temperature should be done in an acceptable range.

In this research the effect of the temperatures ranges from (40 °C – 70 °C) was studied, as MEK volatile solvent its volatility was noticed during experiments above 60 °C. The temperature effect begins adversary in opposite direction increasing it above 60 °C in experiments and it is shown in Fig 3. Many previous researcher's results from studies illustrated that the temperature participation in the extraction process specifically when it is above 45 °C. [18,19].

3.2.2 Mixing speed

Mixing speed is another factor impacted on extraction by MEK solvent and its effect obvious as it is shown in Fig 3. This increase in speed leads to a homogeneous distribution of solvent in used lube oil where the surface contact area of both two liquids used oil and solvent are increased that enhancing the extraction process. The P value of factors BC is less than 0.05, which means strong interaction of both factors towards the extraction process as it is depicted in Fig.4c. The mixing speed of the process effectively had a positive impact on the extraction process and 710.68 rpm is the optimum value as it is obvious in Fig.3. it is effluent begins to change and reducing the PSR% after further in increasing mixing speed, this is due to the reduction of the overall mass transfer coefficient [20].

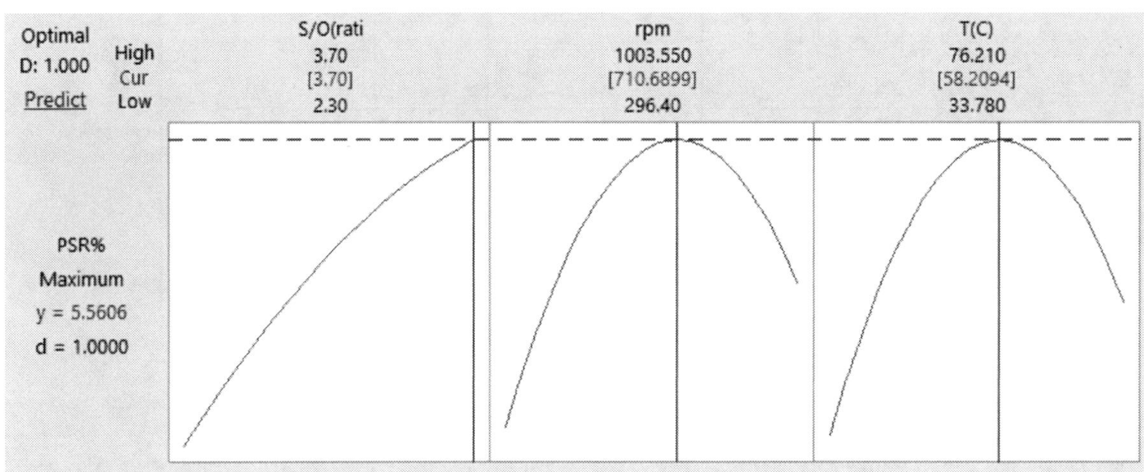

Figure 3. Optimum Recycling Condition of used oil.

3.2.3 Solvent to oil ratio

Equation 3 shows S/O ration coefficient is 7.92 and the positive towered extraction process has higher effects among other factors. The increasing ratio of solvent causes a reduction in oil viscosity and this is because the solvent extracted the sludge in used oil which contains aromatic compounds that had high molecular weight. Decreasing the viscosity enhanced from the velocity of the mixture molecule during the extraction better homogeneity obtain which leading more transfer rate and contact area. Paraffinic compounds remain in lube with lower viscosity in comparison with its aromatic counterpart's removal leading to lower viscosity [21]. The value of viscosity of used oil in table 71.8 cSt at 40 °C compared to extracted oil by MEK shows its viscosity reduced to 57.2 cSt at 40 °C and this is in agreement with (Temitayo et al .2018) [22]. PSR% was increased with the increasing solvent as it is shown in Fig.6.

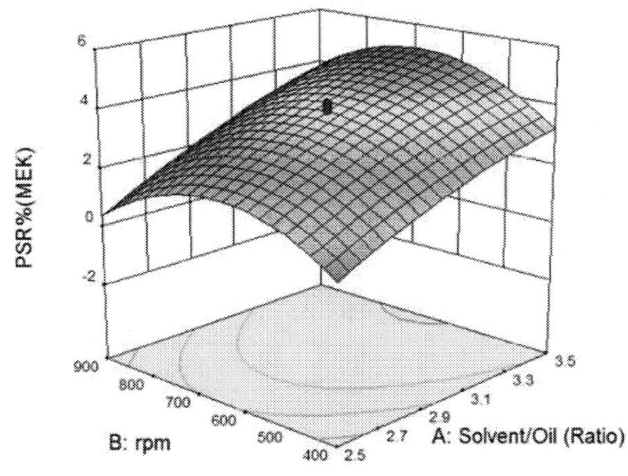

Figure 4. a.3D interaction (AB)

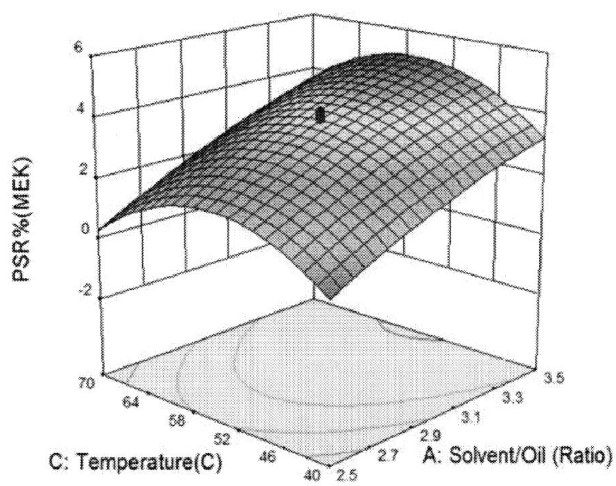

Figure 4.b.3D interaction (AC)

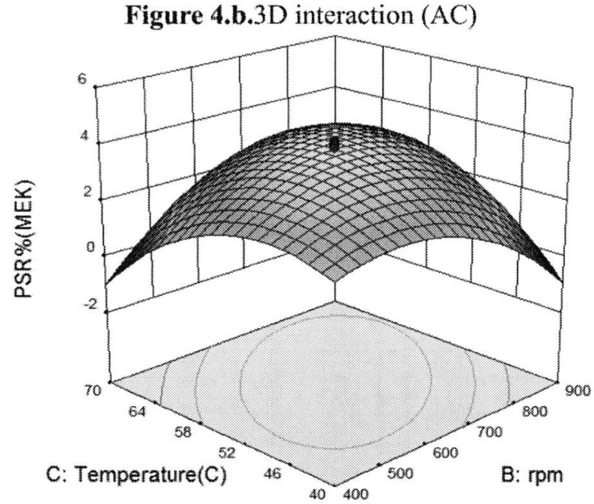

Figure 4. c.3D interaction (BC)

3.3. Effect of Process Variables

According to the Eq.3, the coefficient of Solvent/oil ratio(A) is +7.92, this value recorded the highest impact of these parameters. Increasing the MEK solvent to used oil ratio increasing from the sludge removal (PSR) as it is obvious is Fig.3.

The sludge removal percent is low at a low ratio of Solvent/oil ratio, that because at a low quantity of solvent doesn't dissolve all the base oil, the higher ratio of MEK solvent more sludge removal could obtain at optimum ratio 3.7:1.

Higher mixing speed contributes effectively to the removal of the sludge and improvement from the extraction process specifically a speed about 710 rpm, as shown in Fig.3.

The reason behind that the high speed helps to get a better distribution of both solvent and used oil with more contact area of mass transfer which provides a suitable environment for the extraction process.

Exposure to the mixture to higher mixing speed led to lower sludge removal, thus appropriate speed saves energy addition to the best results.

PSR increases by increasing the temperature of the blend to a certain limit because the heat provides higher kinetic energy of the molecules and increases the motion of the blend, but after the 58.2 °C shows an adverse influence on the process as obvious in Fig.3.

The interaction coefficient S/O% with the mixing speed (AB) +0.00165shows that both parameters interact positively towards the sludge removal from used oil.

Fig 4.a illustrates the interaction of the (A and B), although this relation has a positive effect on the sludge removal with the increasing of the mixing speed (B) the (PSR) are reduced at more high speed.

According to Equation3 Temperature and S/O% had the highest interaction over other recording coefficients 0.0206 for sludge removal shown in Fig.4. b.

Fig.4. c. illustrates the interaction by the Temperature and the mixing speed, the two variables are preferred to be used in the restricted region to continue the process path at the lowest cost and in a better manner.

4. Discussion of recovered base oil properties

Lubricating oil quality tests are performed after each step during the purification and recycling process. Importance of each extraction by solvent and activated bentonite are illustrated in Table 5. (Code No.4008, ISO Grade:68)

Table 5. Recovered base oil characterization

Properties	Unit	Extraction (MEK)	Recovered (MEK+Acticated clay)
Viscosity@ 40 °C	cSt	57.2	66.20
Viscosity @ 100 °C	cSt	7.6	8.4
Viscosity Index	-	94	95.5
Flash point	°C	215	220
Pour point	°C	-19	-17
Density	kg/m3	860	900
TAN	mg kOH/gm oil	0.011	0.020
H_2O content	ppm	0	Nil
Color value	-	Dark	4

Viscosity: denotes one of the most important of the performance test for lubricating oil. Lubricating oil is designing for providing viscosity to be an adequate minimum for starting engine lubricating in cold weather and high enough for supplying a thick layer in hot weather.

Figure 5. Recovered Base Oil

The increase of viscosity of used oil is an indication of the presence deposits, oxidizing products, deterioration of some additive that added to oil for enhancement its properties [25]. The presence of contaminants is reducing the viscosity; in our study, the viscosity is changed from 61.5 cSt to 71.8 cSt at 40 °C indicates the presence of impurities in the used oil.

Treatment by the MEK extraction can remove the aromatic hydrocarbon and waxes, so its viscosity reduced to 57.2 cSt. Adsorption by activated clay includes removing polar compounds that improve the chemical, thermal and color recovery led to viscosity 66.2 cSt.

VI: for fresh oil is 95 and for used oil is 89, because the increase of viscosity at 40°C but this value is reduced to 94 after MEK extraction. Bentonite Clay adsorption increased viscosity at 40°C which is led to increased VI to 95.5 as shown in Table 5.

The Density of the oil before using is 890 kg/m^3 changed to 882 kg/m^3 and this changeset depends on the nature and kinds of contaminants deposited in lubricating oil, the recovered base oil is 900 kg/m^{3}which is acceptable.

The Flashpoint: of the fresh oil 195 °C is increased to 230 °C after using because of saturation of the oil by contaminants. Wax removal by Extraction helps to enhance flashpoint to 215°C.

TAN: for fresh oil is 0.115 mg kOH/gm oil while for used oil is 0.12 mg kOH/gm, this increase because of the formation of carboxylic acid and other acids in the used oil through the oxidization process during using. The color and the all recovered properties are in the permitted range which is appropriate to use as in Fig.5.

5. Conclusions

1. MEK solvent is more active for removing sludge from used oil Turbine oil 4008.
2. RSM with R^2=98.21% is enhanced effectively and recovered base oil suitable for re-use.
3. The optimum condition of the process: Solvent to oil ratio 3.7, mixing speed 710.68 rpm, and the temperature is 58.2 °C, shown in Fig.3.
4. mixing speed and temperature had adversely affected when they are crossing optimum value as it shown in
Fig 6.

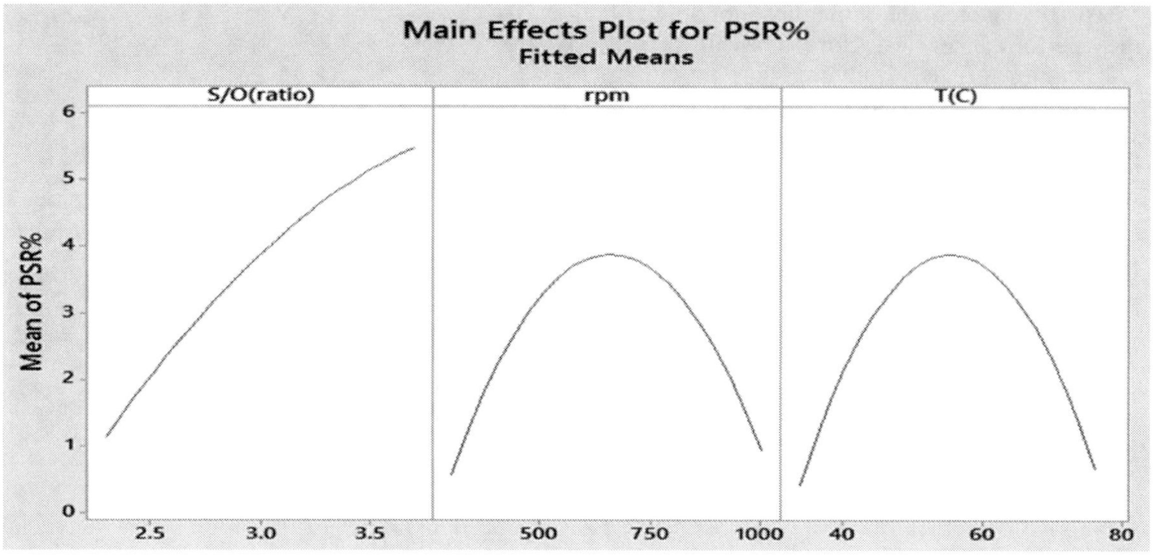

Figure 6. Influence of Extraction factors.

References

[1] James, G., & Douglas, I. (2014). *refining used lubricating oil.* (1st ed). New York, (Chapter 1). International Standard Book Number,13: 978-1-4665-5150-.

[2] Mohammad, A. , & Lina, A. (2007). Virgin and recycled engine oil differentiation: A spectroscopic study. *Journal of Environmental Management 90(1)*:187-95 , DOI: 10.1016/j.jenvman.2007.08.018

[3] Ihsan, H., Talal, Y., & Sardasht, R. (2013). Recycling of Waste Engine Oils Using a New Washing Agent. *Energies, 6(2)*, 1023-1049, DOI: 10.3390/en6021023

[4] Merai, P. (2015). Re-refining of used lubricating oil. *International Journal of Scientific & Engineering Research, 6 (3)*, 1211-1218. ISSN:2229-5518

[5] Marketing specification guide of Iraqi petroleum products, (2012), Ministry of oil, Iraq

[6] Eman, A., Abeer, M. (2013). RE-REFINING OF USED LUBE OIL, I- BY SOLVENT EXTRACTION AND VACUUM DISTILLATION FOLLOWED BY HYDROTREATING. *Petroleum & Coal, 55 (3)* 179-187.

[7] Ghassan, R., Adnan, A., & Adel, H. (2017). Re-refining of used lubricant oil by solvent extraction using central composite design method. *Korean J. Chem. Eng., 34(9)*, 2435-2444, DOI: 10.1007/s11814-017-0139-5

[8] Henry, M. (2015). RE-REFINING AND RECYCLING OF USED LUBRICATING OIL: AN OPTION FOR FOREIGN EXCHANGE AND NATURAL RESOURCE CONSERVATION IN GHANA. *ARPN Journal of Engineering and Applied Sciences, 10(2)*, 797-801. ISSN 1819-6608.

[9] Jhanani, S., & Kurian, J. (2011). Used oil generation and management in the automotive industries. *International journal of applied engineering research* 2(1), 135-146. ISSN :0976-4259.

[10] Lima, F., Araujo, M., & Borges, L. (2003). Determination of the carcinogenic potential of lubricant base oil using near infrared spectroscopy and chemometrics. *Tribology International*, *36(9)*, 691-696. https://doi.org/10.1016/S0301-679X(03)00028-8

[11] Rayckyte,T., Hargreaves. D., & Pawlak. Z, (2006). Determination of heavy metals and volatile aromatic compounds in used engine oils and sludges. *Fuel*, *85(4)*, 481-485. DOI: 10.1016/j.fuel.2005.08.004

[12] Michael, R., & Jan, K. (2001). Method for obtaining bade oil and removing impurities from used oil products, Patent No.: US 6,174,431 B1

[13] Kamal, A., & Khan, F. (2009). Effect of Extraction and Adsorption on Re-refining of Used Lubricating Oil, *Oil & Gas Science and Technology – Rev. IFP,* 64(2), 191-197, DOI: 10.2516/ogst/2008048

[14] Yasser, I., & Shaymaa, M. (2016). Recovery of Base Oil from Spent Automobile Oil Using Elementary and Binary Solvent Extraction, *College of Engineering Journal (NUCEJ) 91 (2)*, 6192,370 – 384.

[15] Montgomery, D. (1996). *Design and analysis of experiments*. (4th Ed). New York:Wiley,(Chapter one).

[16] Ponnala W., & Murthy, K. (2012). Modelling and optimization of end milling machining process, *International Journal of Research in Engineering and Technology*, 1 (3), 430-447.

[17] Chauhan, V., & Bhardwat, N. (2013). Application of response surface methodology and central composite design for the optimization of talc filler and retention aid in paper making, *Indian. J. Chem. Technol ,20* ,121-127.

[18] M. Hussein, A.A. Amer, used lubricating oils re-refining by solvent extraction, *Am. J. Eng. Sci.* 1 (2014) 44–50.

[19] F. Danane, A. Ahmia, A. Bakiri, N. Lalaoui, Experimental regeneration process of used motor oils (2014), *Rev. Des. Energies Renouvables* 17 (2) 345–351.

[20] M. Torab-Mostaedi,S. J. Safdari, M. A. Moosavian and M. Ghannadi Maragheh ,(2008) " MASS TRANSFER COEFFICIENTS IN A HANSON MIXER-SETTLER EXTRACTION COLUMN" *Brazilian Journal of Chemical Engineering* , Vol. 25, No. 03, pp. 473 – 481.

[21] J.D. Udonne, O.A. Bakare, Recycling of used lubricating oil using three samples of acids and clay as a method of treatment, *Int. Arch. Appl. Sci. Technol.* 4 (2) (2013) 08–14.

[22] Temitayo E. Oladimeji, Jacob A. Sonibare, James A. Omoleye, Abiola A. Adegbola, Hilary I. Okagbue (2018)" Data on the treatment of used lubricating oil from two different sources using solvent extraction and adsorption" Data in Brief 19 (2018) 2240–2252 , https://doi.org/10.1016/j.dib.2018.07.003

[23] A.M. Hatami; M.R. Sabour; A. Amiri (2018),"Recycling process of spent bleaching clay: Optimization by response surface methodology" *Global J. Environ. Sci. Manage.*,4(1):9-18,

DOI:10.22034/gjesm.2018.04.01.002

[24] Farzad Nezhdbahadori, Mohamad Ali Abdoli , Majid Baghdadi and Fereydoun Ghazban (2018), " A comparative study on the efficiency of polar and non-polar solvents in oil sludge recovery using solvent extraction" *Environ Monit Assess* (2018) 190: 389 , https://doi.org/10.1007/s10661-018-6748-6

[25] MangasM.A.SogorbE.Vilanova (2014) , "Lubricating Oils" *Encyclopedia of Toxicology* (Third Edition). Pages 670-676 , https://doi.org/10.1016/B978-0-12-386454-3.00525-X

Nonlinear Finite Element Analysis of Shear Strength for Steel Fiber Reinforced Concrete I-Section Beams

Nadiya Sadeek Al Saffar[1, *] and Ayad Amjad Abdul-Razzak Aghwan [2]

[1] Engineering College, Mosul University, Iraq -E-mail: nadiyasadeek@gmail.com
[2]College of Engineering, University of Mosul, Iraq
E mail: ayadaghwan960@gmail.com
* Corresponding author: Engineering College, Mosul University, Iraq. Tel: +9647736977287.

Abstract. This research presents a numerical model of I- section fiber reinforced concrete beams to evaluate the shear strength and complete behavior of the beams using the numerical formulation, which compared with the results of (twenty-three) tested I-section beams taken from previous investigations which are varying in shear span to effective depth ratio and volume fraction of steel fibers. Models using finite element method having the same dimensions and properties as the experimental test specimens. It was shown that the behavior of SFRC I-section beams could be modeled well using the proposed material constitutive relationships and there was a good agreement between the experimental and numerical failure load, crack pattern load deflection curves obtained from this study. The finite element model shows slightly more stiffness results than the test data ranges. The average ratio of the experimental to the calculated results is (1.027).

Keywords: *Finite element modeling, I-section beams, Shear strength, Steel Fiber Reinforced Concrete (SFRC).*

1. Introduction

The steel fibers addition in concrete develops the mechanical properties of the plain matrix as they decelerate the growth of crack and creates compressive forces at the top of cracks thus, the energy absorption for SFRC materials increases, moreover it would be the excellent choice for enhancing the seismic-resistance and blast-resistance for structures [1]

It has been established that steel fibers can offer significant improvement in the properties of plain concrete, mainly directed towards flexure strength and much less is known about shear strength of RC beams, especially I-beams [2].

Awoyera [3] used ADINA software to model a SFRC beams with dimensions of 100mm×100mm×500mm. The steel fiber was added in three different percentages (0, 0.75, and 1) % by volume. It was concluded that the existence of steel fibers delays the expansion of micro cracks due to the bond between the fibers and matrix and the failure occurs due to the fiber cutoff.

Min Sun et al. [4] analyzed two T-section beams, one of them made of SFRC, and the other one was made of NC. The effect of various volume fractions of steel fibers on ultimate shear capacity and crack spreading using ANSYS software was studied. The main conclusions were that the steel fiber improved the rigidity and ductility and the finite element results indicated a perfect conformity with test results

IOP Conf. Series: Materials Science and Engineering **978** (2020) 012028 doi:10.1088/1757-899X/978/1/012028

and a perfect emulation of T- section beams employing smeared crack model.

Aisyahira [5] made a nonlinear analysis using LUSUS software to model three beams with three types of concrete, Normal Concrete (NC), Steel Fiber Self Compacting Concrete (SFSCC) and Self-Compacting Concrete (SCC).

The SFSCC beams with 50% of traditional shear reinforcement were analyzed to investigate the steel fibers contribution to the shear reinforcement with percentage of (1%) by volume. It was established that the ultimate shear strength improved by 37.1% as compared as normal concrete, the deflection also improved between (15.6-35) percent. The difference between the theoretical using finite element method and experimental results was (8-18) percentage, which indicates that LUSAS software gives a good representation.

Jing et. al. [6] used ANSYS workbench 16.1 to model three categories of beams. All of them contain the same percentage of steel fiber of (0, 1, 2, 3, 4) % by volume, in the second category of beams the basalt fiber reinforced polymer (BFRP) bars were used, while the third one contained carbon fiber reinforced polymer (CFRP) bars. It was determined that when the percentage of fiber increases from 0% to 4% the failure load increases and the deflection decreases.

Studies by Mansur, Ong, and Paramasivam [7] suggest that the ultimate shear capacity predictions for SFRC beams can be made based on the post cracking tensile strength of fiber along with some empirical formulas for reinforced concrete beams, while Narayanan and Kareem Palanjian [8] and Sharma [9] recommend using tensile strength obtained using split cylinder tests. Also, there are studies by Lim, Paramasivam and Lee [10] indicating that, for ultimate load predictions for SFRC beams, the effect of steel fibers can be expressed in terms of an equivalent shear reinforcement.

The goals of this work are to study the efficiency of steel fiber on the shear strength of SFRC I-sections. Two–dimensional nonlinear finite element models are used to study the tested I-beams performance.

2. Finite Element Formulations

Plane stress state were used to represent the I-section beams, and eight nodded isoperimetric elements were used to exemplify the concrete. Cook [11] displays the formulation of this element and the appropriate matrices.

The embedded bar model was used to exemplify the conventional reinforcing bars. The concrete and the reinforcement are simulated so that a perfect bond between them; which indicates that the strain is the same for concrete and steel at any point.

2.1 Constitutive Relationships for Materials Used in Fibrous Concrete:

2.1.1 Compressive Stress-Strain Model

In the current investigation a strain–hardening approach was implemented, which consists of a linear part that begins from zero to 30 % of the compressive strength trailed by stress–strain curve with a parabolic shape, that sustains until reaching the compressive strength of fibrous concrete, then the behavior is assumed to be a perfect plastic till crushing happens as shown, Figure (1). Soroushian and Lee [12] provide the strain ε_{pf} at peak stress f'_{cf}, which is adopted in this study:

$$\varepsilon_{pf} = \frac{2f'_c}{E_c} + 0.0007\frac{V_f l_f}{d_f} \qquad (1)$$

where $f_c^{'}$ = compressive strength of standard cylinder for normal concrete, V_f = steel fiber percentage by volume and d_f and l_f = diameter and length of the steel fibers correspondingly.

To estimate the compressive strength for fibrous concrete, the equation suggested by Soroushian and Lee [12] and adopted by Abdul Razzak [13], can be used:

$$f_{cf}^{'} = f_c^{'} + 3.60 v_f . l_f / d_f \qquad (2)$$

The result of ultimate compressive strain obtained by Abdul-Razzak [13] is adopted in this analysis, which was obtained using the experimental test results by shah and Rangan [14]

$$\varepsilon_{cuf} = 3011.0 + 2295.0 v_f \ (\times 10^{-6}) \qquad (3)$$

2.1.2 *Tensile Stress–Strain Model*

Carreira and Chu [15] proposed a continuous function for the ascending and descending portions of the tensile stress-strain curve of plain concrete which is adopted in references [16and 17] for steel fiber concrete, see Figure (1), in Equation 4 that is adopted in this study:

$$\frac{f_t}{f_{tf}^{'}} = \frac{\beta.(\varepsilon / \varepsilon_{tf})}{\beta - 1.0 + (\varepsilon / \varepsilon_{tf})^{\beta}} \qquad (4)$$

where ε = the tensile strain at a stress equal to f_t, $f_{tf}^{'}$ = tensile strength of steel fiber concrete, ε_{tf} = strain corresponding to peak stress and β is a fibers parameter predicted in references [18, 19] as represented below:

$$\beta = 1.093 + 0.7132 R.I.^{-0.926}$$ Fiber type is hooked (Eq. 5a)

$$\beta = 1.093 + 7.4848 R.I^{-1.387}$$ Fiber type is smooth (Eq.5b)

$$\beta = 0.5811 + 1.93 R.I^{-0.7406}$$ Fiber type is crimped (Eq.5c)

$R.I$ is an index of reinforcing = $w_f . l_f / d_f$ and w_f is the ratio of the weight of the fibers to that of concrete. If there was no available experimental data, the strain ε_{tf} at ultimate stress $f_{tf}^{'}$ for fibrous concrete can be represented as suggested by Soroushian and Lee [12] as follow:

$$\varepsilon_{tf} = \varepsilon_t (1.0 + 0.35 N_f d_f . l_f) \qquad (6)$$

$$f_{tf}' = f_t'.\left(1.0 + 0.016 N_f^{(1/3)} + 0.05.\pi.d_f.l_f.N_f\right) \quad (7)$$

where, ε_t =the strain at which cracking is occurred in plain concrete $= f_t' / E_c$, f_t' =tensile strength of normal concrete, E_c = modulus of elasticity of normal concrete and N_f = Number of fibers for each unit area and it is equal to $\eta_o(4.0V_f / \pi.d_f^2)$, η_o is the factor of orientation, which is equal to 0.41 as proposed by Ezeldin [19] or 0.5 as suggested by Hannant [21] for three dimensional distribution.

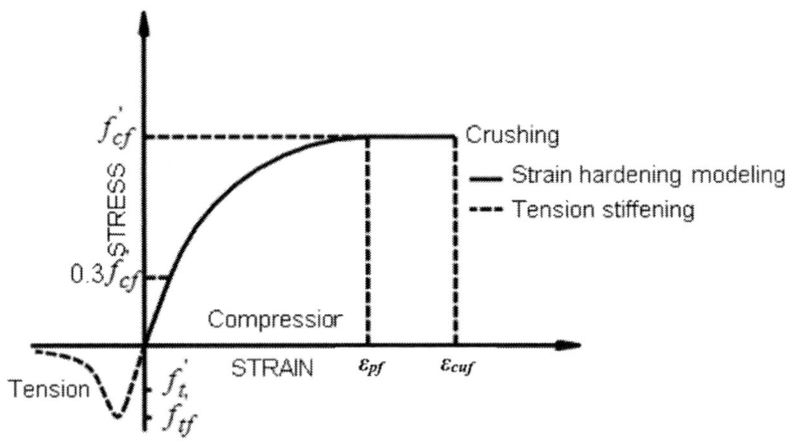

Figure 1 Uniaxial Stress-Strain in Compression Relationships for steel Fiber Concrete.

2.2 Biaxial Performance of Steel Fiber Concrete

In the state of compression-compression, plastic behavior of concrete supposed to begin when the effective stress transcends $0.3f_{cf}'$ and perfect plasticity supposed to take place when the effective stress arrives at f_{cf}'. A yield criterion was proposed to follow the plastic performance:

$$f(I_1, J_2) = \left[\beta_f(3J_2) + \alpha_f I_1\right]^{0.5} = f_{cf}' \quad (8)$$

where I_1 and J_2 are the 1st stress invariant and 2nd stress invariant and α_f, β_f are materials factors attained in reference [13] using fibrous concrete mixes obtained from previous experimental work:

$$\alpha_f = \frac{1-\omega^2}{\omega^2 - 2\omega}.f_{cf}', \quad \beta_f = \frac{1-2\omega}{\omega^2 - 2\omega} \quad (9)$$

and $\omega = e^x$ where:

$$x = \frac{1}{3.339 - 0.9772\ln(v_f l_f / d_f)} \quad (10)$$

An isotropic rule of hardening is assumed to express the uniform expansion for the yield surface.

In state of the tension-compression, the parabolic curve described below that has been proposed by Mahmood [22] and adopted by Reference [13] is adopted in this study:

$$\sigma_2 / f'_{cf} + S^2.[\sigma_1 / f'_{cf}]^2 = 1 \qquad (11)$$

where $S = f'_{cf} / f'_{tf}$.

In the state of tension-tension, the assumption is there was no reciprocal interaction, i.e., the tensile strength of concrete in the two orthogonal directions is the same as that in the uniaxial state.

2.3 Shear Modulus of Cracked Concrete

A stress principle is used for cracks beginning and when the concrete crack occurs in direction 1; the formula proposed by Al-Mahaidi [23] for predicting the shear modulus of the cracked plain concrete and was used soon after by Abdul-Razzak and AL-Hasan [13and16] for fibrous concrete is adopted in the present study:

$$\overline{G} = 0.40G/(\varepsilon_1 / \varepsilon_{tf}) \qquad (12)$$

If the cracking is occurred in direction 2, $\overline{G}$ turn into Eq. 13:

$$\overline{G} = o.40G/(\varepsilon_2 / \varepsilon_{tf}) \qquad (13)$$

In this region, the relationship proposed by Vecchio and Collins [23] treated the negative impact of cracks on the strength of concrete in the normal direction and was used by Abdul-Razzak [13] for steel fiber concrete.

$$\frac{f_{cf\,\max}}{f'_{cf}} = \frac{1.0}{0.80 + 0.34.\left(\frac{\varepsilon_1}{\varepsilon_o}\right)} \leq 1.0 \qquad (14)$$

Where:

ε_1 : Principal tensile strain in concrete.
ε_0: concrete cylinder strain at peak stress.

2.4 Reinforcement Stress-Strain Relationship

A bilinear relationship for tension and compression was proposed from zero stress up to the yielding strength with a slope E_s=200GPa trailed by a straight line with a suggested hardening modulus E'_s .

3. Numerical applications:

An interesting series of researches by Muhidin and Tan [24 and 25] have encompassed a wide-ranging study on fiber concrete I- section beams failing in shear and flexure, discussing the influence of several parameters including percentage of fibers, concrete compressive strength, a/d ratio with the same boundary restraint. Twenty-three SFRC I-section beams have been analyzed by using computer program described previously. Finite element formulation was employed to obtain the behavior of SFRC I-section beams without stirrups as shear reinforcement. The predictions obtained using the numerical model has been computed with the test results.

Table 1. Beams details with the experimental and calculated results

Beam No.	Beam Designation	Ref.	V_f %	l_f mm	d_f mm	f_{cu} N/mm^2	f_{sp} MPa	b_w mm	Experimental Ultimate Load (kN)	Theoretical Ultimate Load (kN)	Exp./Theo. Ratio
1	A1	24	2.13	60	0.64	45.7	—	50	71.3	65.0	1.09
2	A2	24	2.25	25	0.4	49.8	4.7	50	81.0	78.0	1.037
3	A3	24	0.75	25	0.4	66.1	4.9	50	64.5	64.8	0.99
4	A4	24	1.5	25	0.4	54.1	5.1	50	61.1	60.7	1.00
5	A5	24	3.0	25	0.4	52.5	4.9	50	78.9	76.3	1.034
6	A6	24	2.25	40	0.5	42.5	4.7	50	59.9	59.7	1.00
7	A7	24	2.25	60	0.64	42.3	5.3	50	76.4	73.6	1.037
8	A8	24	1.5	40	0.5	52.2	4.7	50	65.0	65.5	0.99
9	A9	24	1.5	40	0.5	24.3	3.2	50	54.9	49.5	1.11
10	A10	24	1.5	40	0.5	76.3	6.5	50	69.5	83.7	0.83
11	A11	24	1.5	40	0.5	55.0	5.2	70	93.8	96.0	0.97
12	A12	24	1.5	40	0.5	55.0	5.1	30	55.4	51.4	1.07
13	B1*	24	1.5	40	0.5	61.1	6.3	50	87.4	85.4	1.02
14	B2	24	1.5	40	0.5	49.0	4.5	50	81.4	80.0	1.01
15	B3	24	0.3	40	0.5	49.1	3.6	50	39.4	39.1	1.00
16	B4	24	0.75	40	0.5	42.1	4.2	50	44.9	45.3	0.99
17	B5	24	1.13	40	0.5	52.8	5.3	50	57.4	57.7	0.99
18	B6	24	2.25	40	0.5	54.1	7.1	50	60.0	60.8	0.98
19	C1**	25	0.5	30	0.5	42	3.32	60	218.0	160.0	1.36
20	C2**	25	0.75	30	0.5	39.6	3.3	60	180.9	182.0	0.99
21	C3**	25	1.0	30	0.5	43.2	4.03	60	210.3	210.0	1.00
22	C4***	25	1.0	30	0.5	43.2	4.03	60	154.2	184.0	0.83
23	C5****	25	1.0	30	0.5	43.2	4.03	60	307.0	252.0	1.21
											1.02

* hf =100 mm, d= 410mm
** a/d=2
*** a/d=2.5
**** a/d=1.5

3.1 Type (A) Beams:

Due to symmetry of the beam, only half of the beam is analyzed. The properties of the material for the tested beams are abbreviated in Table 1, dimensions; reinforcement's details and loading are shown in Figure 2. The beam is modeled by eight-nodded isoperimetric elements are adopted in the present study. For all beams the concrete mix proportions was 1: 2: 2 /0.5 except for beams (A9, A10) the concrete mix proportion were 1:3:4 /0.8 and 1:1:1.5/0.35 respectively and the steel fiber type is duoform.

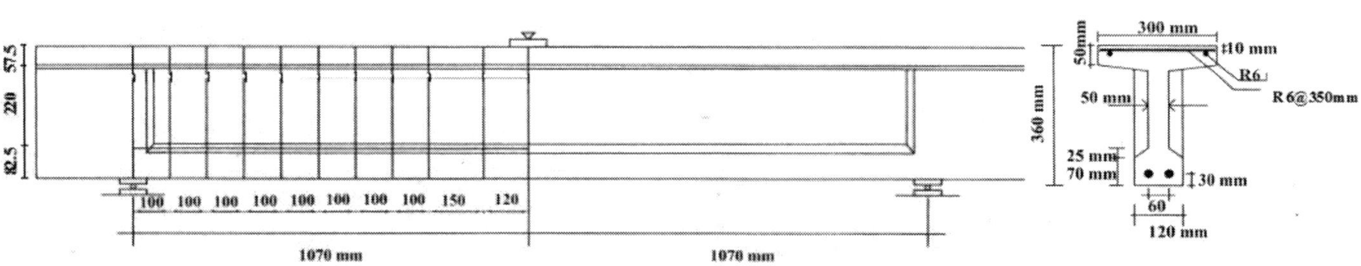

Figure 2. Beam details and Finite Element Mesh of Type (A) Beams.

Figure 3 represents the experimental and numerical load versus compressive strain curves for beam A5 near mid span. It can be noticed from this figure that the numerical results have an excellent approval with the experimental results at strain. Figure 4 shows the analytical load deflection curve for beam A5.

Figure 5 represents the crack pattern for beam A7 from the current analysis with the experimental one. Initial damage in the beam happened in the form of tension cracking in the vertical direction. With increasing load, crack spread widely through the depth and in vertical direction and in the inclined direction in region close to the support, such distribution of cracks forms a mechanism of shear failure. The experimental pattern of failure of the beam is shown in Figure 5. A good agreement was obtained between the numerical and experimental crack pattern.

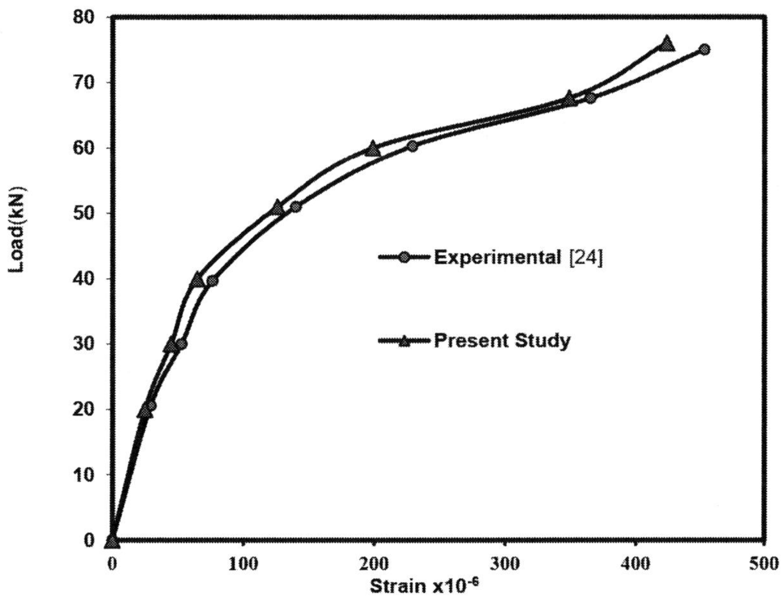

Figure 3. Experimental and Theoretical Load- Strain Curve for Beam A5.

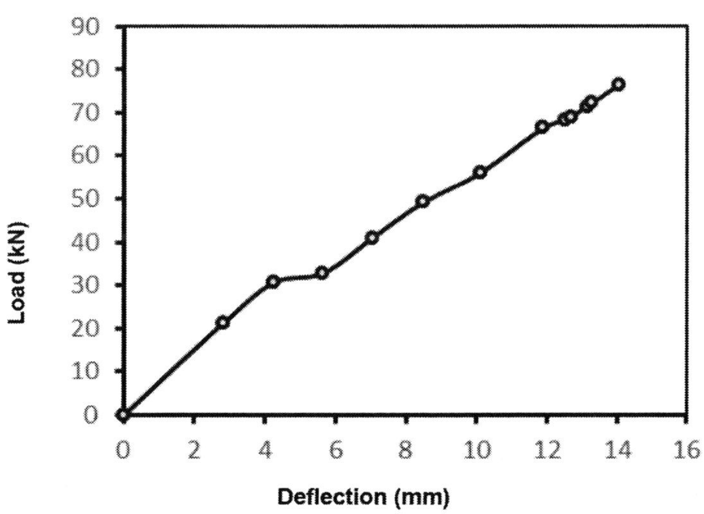

Figure 4 Analytical Load Deflection Curve for Beam A5.

IOP Conf. Series: Materials Science and Engineering **978** (2020) 012028 doi:10.1088/1757-899X/978/1/012028

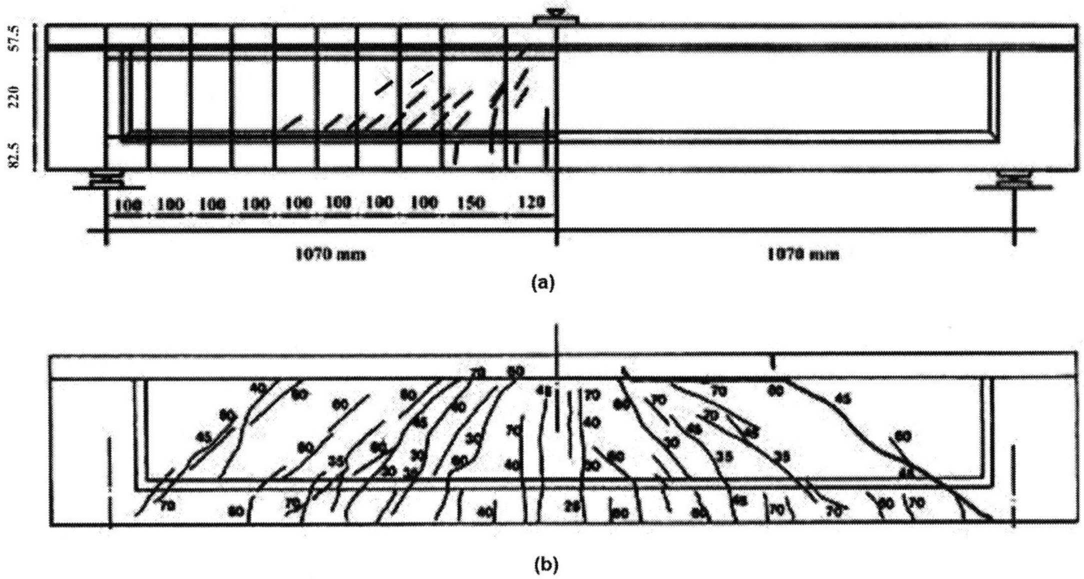

Figure 5 Theoretical and Experimental Crack Pattern for Beam (A7) at failure Load Respectively.

The effect of web-width of the beams on the response of the load versus deflection curve is shown in Figure 6. From this figure, it can be observed predictable influences of the web breath; there was a remarkable growth of ultimate load with the increase of beam web width.

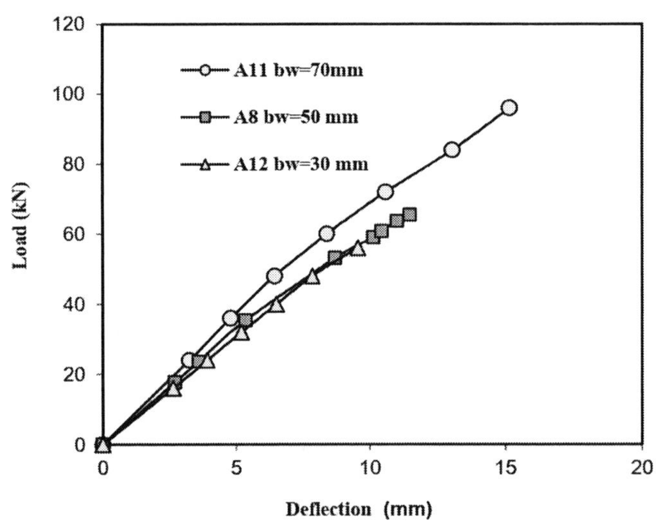

Figure 6 Load versus Deflection Relationship for beams A8, A10 and A11.

In Figure 7, there was a noticeable increase in load-deflection curve with the growth of concrete compressive strength. There was reduction in web deflection with the increase of compressive strength are employed. In addition, as the compressive strength increases the failure of the beam close to the brittle failure. Figure 8 displays the relationship between the load and deflection of beam A5 at different distance from the support.

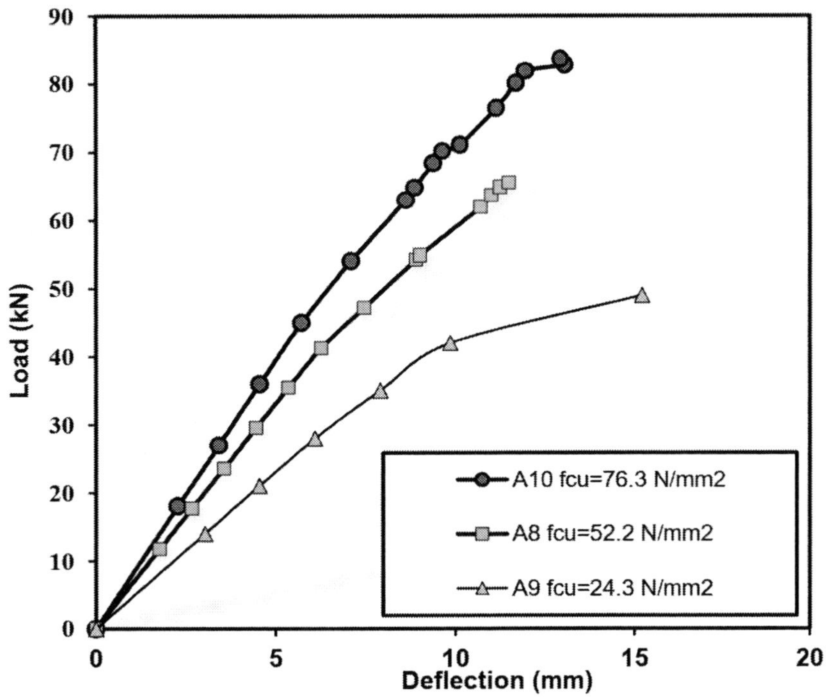

Figure 7. The Effect of Compressive Strength on the Load Deflection Curve.

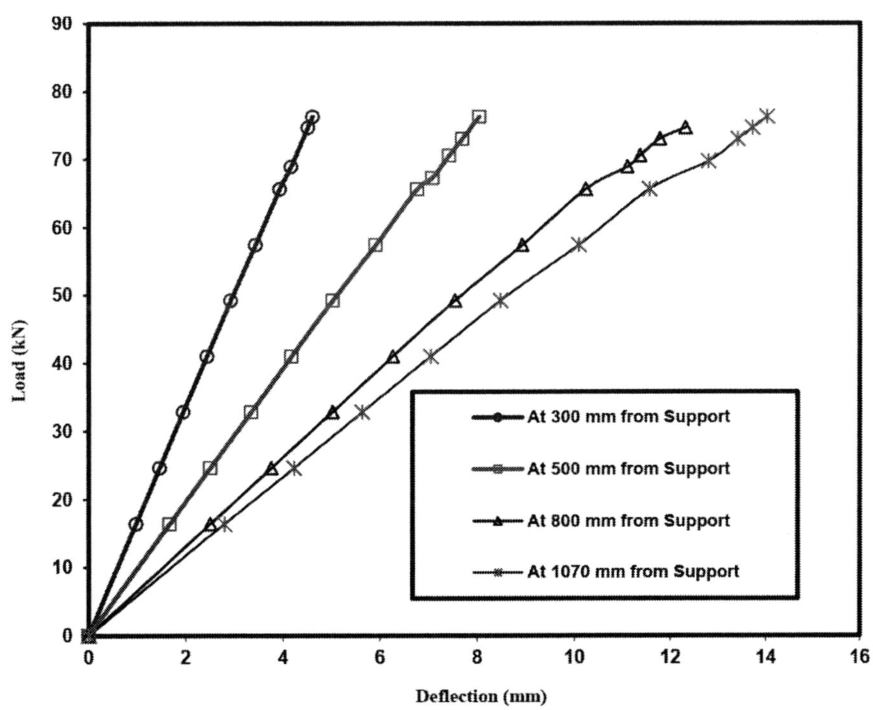

Figure 8. Load versus Deflection Curves at Different Distance from Support for Beam (A5)

3.2 Type (B) Beams

For this type of beams the geometry and beams cross section were described in Figure (9) with finite element mesh. Six beams were tested by Muhidin [24] were analyzed in the present study. The experimental results and beams properties are summarized in Table [1]. The concrete mix components ratios by weight for type (B) beams were 1:2:2 /0.5 and the steel fiber type is duoform.

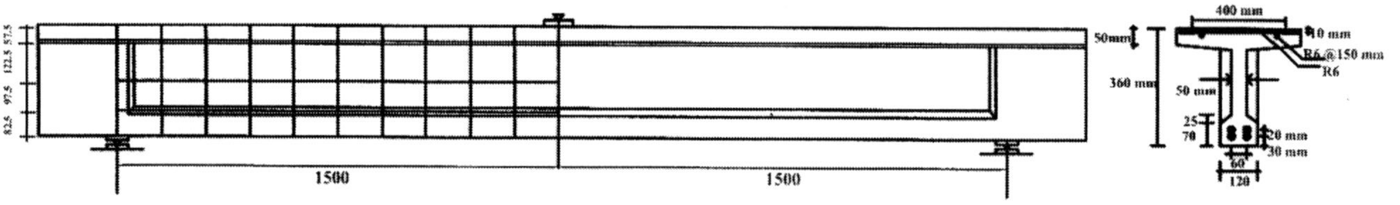

Figure 9. Beam Details and Finite Element Mesh of Type(B) Beams.

3.3 Type (C) Beams

Tan and Murugappan [25] tested five simply supported I-section beams under two points loading. Details of cross section are shown if Figure 10. The experimental failure load and the predicted values are given in Table [1].

The steel reinforcement bars had yield strength of 460 MPa. Hook-ended steel fibers 30mm long 0.5mm diameter were used, ordinary Portland cement, crushed granite and natural sand in the ratio 1:1.2:2.1 by weight, the water to cement ratio was (0.5). All beams are simply supported with a constant moment zone of 550 mm length.

In this study each beam is modeled by (8-node) element, Figure 9 shows the finite element mesh and the cross-section details. The effect of fiber volume fraction on the load versus deflection relationship was described in Figure 11. It is remarked that when the V_f increase the results become stiff and higher in ultimate load. The experimental versus the calculated loads for the 23 beams are shown in Figure 12.

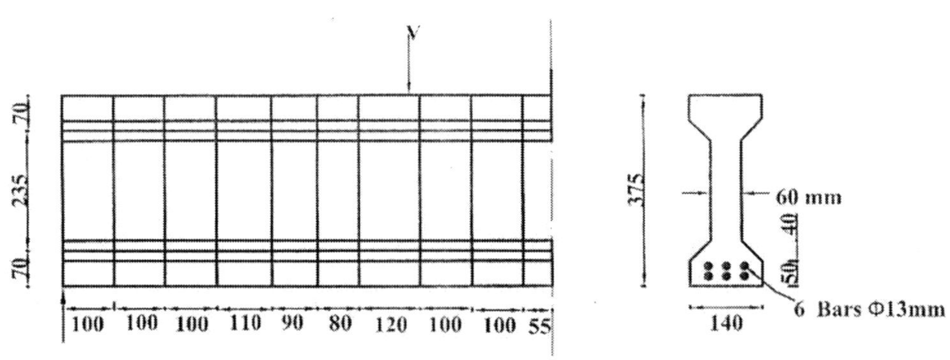

Figure 10. Beam Details and Finite Element Mesh of Type (C) Beams.

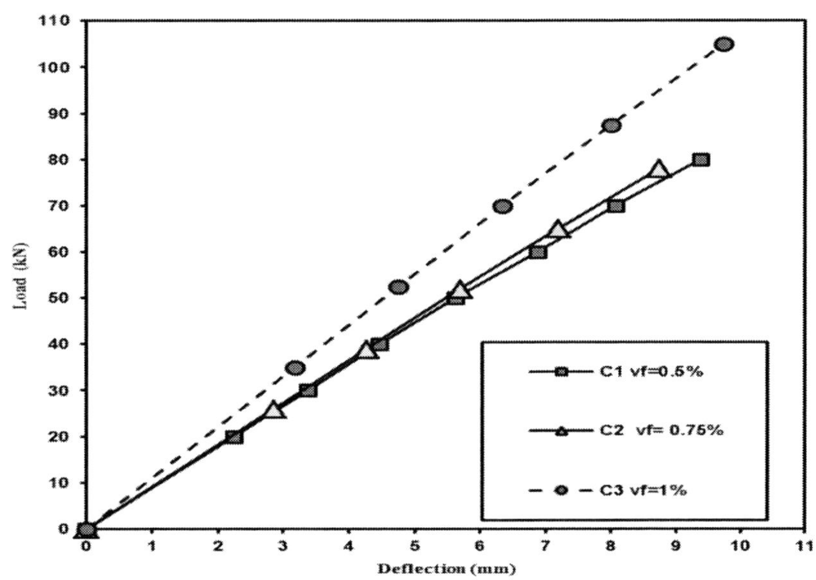

Figure 11. The Effect of Fiber Percentage by Volume on the Load Deflection

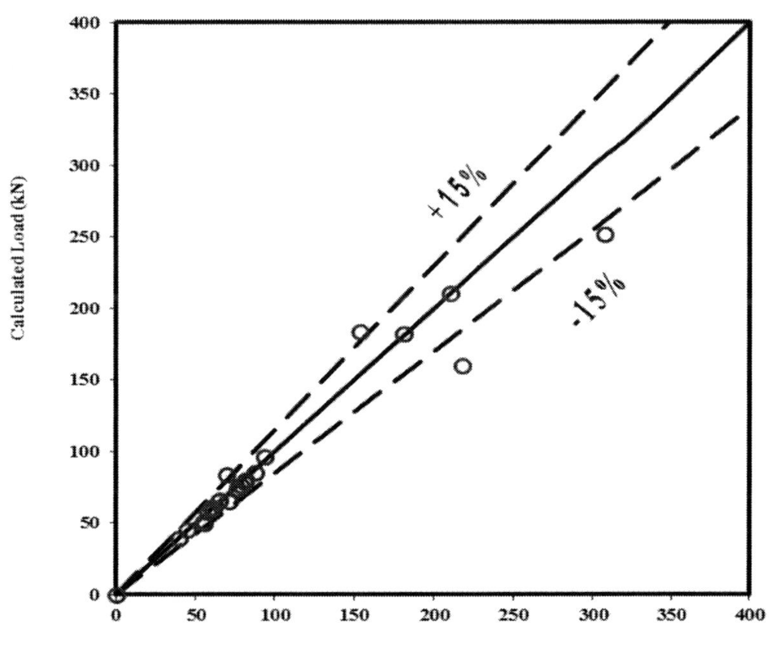

Figure 12. Experimental Versus the Calculated Values of the Beams Tested in the Present Study.

4. Conclusions and Recommendations:

The ultimate load for the SFRC I-section beams are well predicted using the present solution, algorithms and materials constitutive relationships for steel fibers concrete, which provided a secure and representative estimate of the behavior and strength of the studied beams.

The finite element model shows slightly more stiffness results than the test data ranges. Excluding the effects of bond slip in the finite element models leads to increasing the stiffness of the finite element models.

The current finite element model can be used to perform a parametric study rather than carrying out expensive experimental test programs. Executed relationships can be replaced in order to obtain better results.

The material model of the fiber reinforced concrete ascertained to give acceptable results for the analysis of SFRC I section beams exposed to progressive loading till reaching the failure load. The numerical compressive strain results using the present model, show excellent agreement with experimental results. From the theoretical results obtained in this study, the addition of steel fiber increased the ultimate shear strength by about 17.7% for beams A3 and A5 when the steel fiber volume fraction increased from 0.75 to 3% respectively. The increase in strength reaches 55.5 % for beams B3 and B6 when the fiber volume fraction increased from 0.3 to 2.25%.

References:

[1] Gunneswara Rao, T. D. and Rama Seshu, D. (2003). Torsion of Steel Fiber Reinforced Concrete Members, Cement and Concrete Research 33, 1783-1788.

[2] Hashemi, Rahgozar, R.and Maghsoudi A.A. (2007). Finite Element and Experimental Serviceability Analysis of HSC Beams Strengthened with FRP Sheats, American Journal of Applied Science 4(9), pp. 725-735.

[3] Awoyera, P.and Oluwaseun, P. (2016). Nonlinear Finite Element Analysis of Steel Fiber Reinforced Concrete Beam Under Static Loading, Journal of Engineering Science and Technology, V. 11, No. 12, pp. 1669 – 1677.

[4] Min Sun, Jiapeng Zhu, Ning Li, and Fu, C. C. (2017). Experimental Research and Finite Element Analysis on Mechanical Property of SFRC T-Beam, Hindawi, Advances in Civil Engineering, Article ID 2721356, 8 pages.

[5] Aisyahira Banta Melan (2018). Non Linear Finite Element Analysis of Steel Fiber Self Compacting Reinforced Concrete Beams, M.Sc. Thesis submitted to the School of Civil Engineering, University of Technology, Malaysia, August.

[6] Jigna Jagadish and Lekshmi L (2019). Non-Linear Finite Element Analysis of Steel Fiber Reinforced Concrete Beams, International Journal of Applied Engineering Research ISSN 0973-4562 V14, No. 12, pp. 108-115.

[7] Mansur, M. A., Ong, K. C. G., and Paramasivam, P. (1986). Shear Strength of Fibrous Concrete Beams without Stirrups, Journal of Structural Engineering, ASCE, V. 112, No. 9, Sept., pp. 2066-2079.

[8] Narayanan, R., and Kareem-Palanjian, A. S. (1986). Torsion, Bending, and Shear in Prestressed Concrete Beams Containing Steel Fibers, ACI Journal, Proceeding V. 83, No. 3, May –June, pp. 423-431.

[9] Sharma, A. K. (1986). Shear Strength of Steel Fiber Reinforced Concrete Beams, ACI Journal, Proceeding V.83, No. 4, July- Aug., pp. 624-628.

[10] Lim, T. Y., Paramasivam, P. and Lee, S. L. (1987). Shear and Moment Capacity of Reinforced

Steel Fiber Concrete Beams, Magazine of Concrete Research, V. 39, No. 140, Sept., pp. 148-160.

[11] COOK, R. D. (1981). Concepts and Applications of Finite Element Analysis, 2nd Edition, John Wiley and sons, Canada.

[12] Soroushian, P. and Lee, C. D (1989). Constitutive Modeling of Steel Fiber Reinforced Concrete under Direct Tension and Compression. Int. Conf. on Recent Developments in Fiber Reinforced Cement Concrete, Cardiff (UK), pp.363-377.

[13] Abdul-Razzak, A. A. (1996). Nonlinear finite element analysis of fibrous reinforced concrete structural members. Ph. D. Thesis, University of Mosul, IRAQ, August 1996.

[14] Shah, S. P. and Rangan, B. V. (1970). Ductility of concrete reinforced with stirrups, fibers and compression reinforcement. ASCE, June 1970, Vol. 96, No. ST-6, pp. 1167 –1184.

[15] Carreira, D. and Chu, K. H (1986). Stress-strain relationship for reinforced concrete in tension. ACI Journal, Proc., Vol.83, No. 1, Jan. 1986, pp. 21-28.

[16] AL-Hasan, B. J. (2004). Nonlinear finite element analysis of partially prestressed fibrous concrete beams. M.Sc. Thesis, Mosul University, IRAQ, Jan. 2004, 127pp.

[17] AL-Taan, S. A. and Shammas, S. S. Y. (2006). "Nonlinear time-dependent finite element analysis of fibrous reinforced concrete beams. Proc. of the 4th Jordanian Conference on Civil Engineering, Amman, April 2006.

[18] Nataraja, M. C., Dhang, N. and Gupta, A. P. (1999). Stress-strain curves for steel- fiber reinforced concrete under compression, Cement and concrete composites, 1999, Vol. 21, pp.383-390.

[19] Ezeldin N., A. S. and Balagura, P. N. (1992). Normal and high strength fiber reinforced concrete under compression, Jour. of Materials and Engineering, 1992, ASCE, Vol.4, No. 4, pp. 415-427.

[20] Hannant, D. J. (1978). Fiber Cement and Fiber Concrete, John Wiley and Sons Ltd., N. Y., 1978, 219pp.

[21] Mahmood, M. N. (1986). Nonlinear finite element analysis of reinforced deep beams, M.Sc. Thesis, Mosul University, IRAQ, Jan. 1986, 191pp.

[22] AL-Mahaidi, R. S. H. (1978). Nonlinear finite element analysis of reinforced concrete deep members, Ph.D., Thesis, Dept. of Structural Engineering, Cornell University, Ithaca, New York, May 1978, 374pp.

[23] Vecchio, F. and Collins, N. P. (1986). The modified compressive-field theory for reinforced concrete element subjected to shear, ACI Jour., Vol.83, 1986, No.2, pp.219-231.

[24] Muhidin N. A. and Regan P. E. (1977). Chopped steel fibers as shear reinforcement in concrete beams, paper 17 of Fiber Reinforced Material Institution of civil Engineers, London, 1977.

[25] Tan, K. H., Murugappan, K and Paramasivam P. (1993). Shear Behavior of Steel Fiber Reinforced Concrete Beams, ACI Structural Journal, January-February 1993, pp. 3-11.

The Nano Fluid Effect on the Plethora of Thermal Efficiency in Evacuated Tube Solar Collector

Khalid F. Sultan[1] Hosham S.Anead[2], Sura Abdul Jabber[3,*]

[1]College of engineering, University of Technology, Iraq
E-mail: 50084@uotechnology.edu.iq
[2]College of engineering, University of Technology, Iraq
E-mail: 50043@uotechnology.edu.iq
[3]College of engineering, University of Technology, Iraq
E-mail: suraabedalgabbar91@gmail.com
*College of Engineering, University of technology, Baghdad, Iraq. Tel: +9647812753142.

Abstract. Nano fluids are the most attractive means to boost heat transfer system efficiency, since they have high thermal conductivity. This research aims to investigate the effects of utilizing Nano fluids (Cu (30 nm) + DW) as the working fluid on the thermal effectiveness of the evacuated solar collector. Experiments for statistical analysis of techniques were Performed using nanoparticles in three volume concentrations (1%, 3% and 5% vol.), solar collector's thermal efficiency was tested at various volume flow rates (30, 60 and 90 Lit / min). This research indicated that percentages improvement of (Cu (30nm) + DW) with working fluid (distilled water) at 5 % volume concentration and different volume flow rates are (15.326% , 16.96% and 19.98%) while these percentages of improvement when used only distilled water at the same volume concentration and various volume flow rates are (2.41% , 4.36% , 5.43%) , The high thermal conductivity of the working fluids could enhance solar collector efficiency compared to that of the distilled water. The area under the curve (the thermal efficiency curve) was used as indicator to assess the results of the volume flow rate and nanoparticle concentrations on total collector efficiency. Nano fluids (Cu + DW), at concentrations ratio (1, and 5%vol) and volume flow rates (30, and 90 L/min) , Values of thermal solar characteristic value of (FR ($\tau\alpha$), – FRUL) are 0.407, 1.156 W/m^2.k, 0.522 and 0.945 W/m^2.k, Whereas for distilled water at volume flow rates 30 L/min and 90 Lit/min are 0.252, 0.945 W/m^2.k, 0.321 and 1.051 W/m^2.k respectively, where FR ($\tau\alpha$), – FRUL are two parameters used to describe how collector works , where FR ($\tau\alpha$) is an indication of how energy absorb and FRUL is an indication of energy is lost, the two parameters are constitute the simplest practical collector model , the values indicates that the solar collector efficiency when using Nano fluids (Cu(30)+ DW) is higher than efficiency when using distilled water due to high thermal conductivity for copper . The type of Nano fluid is a key factor for heat transfer enhancement, and improve performance of evacuated tube solar collector.

Keywords: *Evacuated solar collector tubes; Performance; Nano fluids; Thermal Performance.*

1. Introduction

Solar energy is converted into useful heat energy. Solar energy is amongst the cleanest, inexhaustible energy and cheap forms of renewable energy, Solar energy is now becoming widely used because it is safe and can also be accessed without constraints. In order to capture heat from solar energy, Solar collectors are the systems in use today this can be used for much of the buildings' energy usage, like air conditioning, water heating, swimming pool heating and so on. [1]. There are numbers of techniques to absorb solar radiation, but the evacuated tube method has a particular interest to the researchers because ETSC has higher efficiency than the traditional FPC, especially at low temperatures and insulation presence. An ETSC consists of parallel glass tubes (evacuated-pipes) that has been evacuated, it consists of two pipes one inside casing with the larger, and vacuum is produced between the two tubes that allows the solar radiation to pass through but does not allow the energy to be transferred thus the energy stays inside the internal pipes and absorbs radiation efficiently, making the ETSC the most efficient solar thermal energy collector [2]. Many ways to increase heat transfer by heat exchangers can be split into two groups: active and passive strategies, Passive strategies don't require exterior force. Using Nano fluids as a medium for heat transfer is a passive strategy to boost heat transfer [3]. Nano fluids are the most attractive mean to boost the heat transfer system efficiency, are created by incorporating Nano-particles just under 100 nm of metal and metal oxides to the base fluid. For solar collectors, renewable power makes special use of these Nano fluids [4]. Das et al. (2007) [5]; expressed that the Nano fluids could be utilized to enhance heat transfer from solar collectors to storage tanks and to increase the energy density. Otanicar and Golden. (2009) [6]; Reported solar collector experimental results depending on Nano fluids made up of Various Nano-particles such as carbon, graphite and silver; In solar energy systems, when using Nano fluid as an absorption mechanism, performance improvements were up to 5%. Experimental and mathematical findings an initial significant improvement in performance accompanied by an increase in productivity as the volume fraction keeps on increasing. Yong Yang Gan et.al (2018) [7]; Assesses thermal performance and performs an entropy (ETSC) analysis in which Nano fluid TiO_2 have been used as working fluids. Titanium dioxide (TiO_2 Nano fluid formed by scattering small a amounts of TiO_2 nanoparticles into distilled water). Higher thermal conductivity of the TiO_2 Nano fluids may enhance efficiency of solar thermal collector (ETSC) evacuated tube. And conclude the rise in nanoparticles volumetric concentration increases the TiO_2 Nano fluids thermal conductivity. Kamble D.P.1 et. al. (2014). [8]: Studied the Nano fluid's behavior to boost a circular heat pipe's thermal performance. The experiment, made of copper tubing and using hybrid Nano fluid (Al_2O_3 + CuO) with a water base as working fluids, explores the thermal dissipation of the heat pipe under different working conditions. Total thermal performance was predicted for pure water and water based on Nano fluid. The study discusses the impact of the working fluids charged volume ratio, the effect on thermal resistance of the filling volume ratio and of the volume of nanoparticles in the base fluid is also investigated. Contrasted to pure water, thermal effectiveness of the heat pipe increases with the growing (Al_2O_3 + CuO) water dependent Nano fluid. Tiwari et al. (2013) [9] Examine the influence of Al_2O_3 Nano fluid's theoretical use as an absorbent medium in a flat solar collector. They also investigate the impact on the collector's efficiency, volume flow rates and particle volume fraction. Their finding revealed that utilizing optimum fraction of the particle volume of 1.5% of Al_2O_3 Nano fluid increases solar collector's thermal efficiency by 31.64% contrasted to water as working fluids. Luie et al. (2011) [10]. Tested CuO Nano Fluid and it was found that the performance of the evacuated solar collector is 30 % greater than of the distilled water. Natarajan and Sathish (2011) [11] also believed to be new method for enhancing solar water heaters efficiency by adding Nano fluids instead of conventional heat transfer fluids.

2. Preparation of Nano fluids (Cu (30nm) +DW)

The Nano fluid studied is formed by nanoparticles of copper (cu (30 nm)) and two – step process applies to the base liquid by dispersing pre-weighted amounts of dry nanoparticles. The acidity (PH) of Nano fluid has been calculated in a standard procedure. The mixtures were then ultrasonically mixed [100 kHz, 300 W at 25 – 30°C , Toshiba, England] for two hour to break up any particle aggregates. The acidic pH is much less than the isoelectric point [iep] of these particles, thus ensuring positive surface charges on the particles. The surface enhanced repulsion between the particles, which resulted in uniform dispersions for the duration of the experiments, as shown in Figure (1), Figure (2) shows Nano fluids which containing copper (Cu (30 nm)).

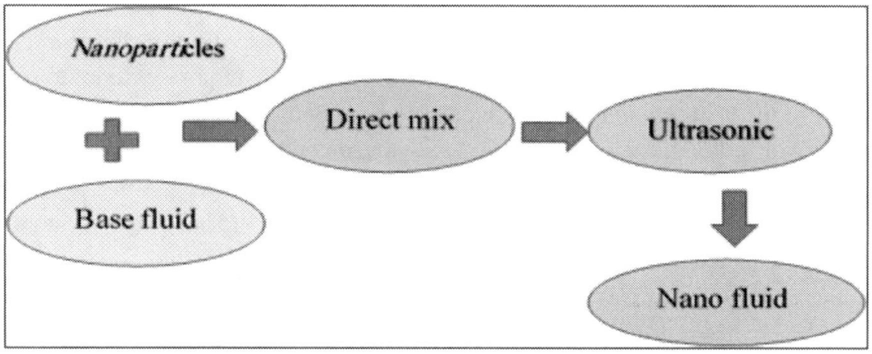

Figure 1: The preparation process of Nano fluids

Figure 2: The Nano fluid (Cu (30nm) +DW)

3. Experimental setup

Figure (3) displays the setups schematic view. An ETSC system consisting of 10 hollow tubes was added as shown in Figure (4), while heat exchanger was added to another hand which is used for heating process (transfer the heat from heating fluid to cooling fluid in the storage vessel), its consists of helical copper coil with 13 mm inner diameter, an outer diameter of 16 mm, 18 turns and total length of 10 m,. Coil immersed in 60 L storage vessel insulated by a reflective fiberglass sleeve the storage has an inside diameter of 370 mm, outside diameter of 385 mm and a length of 1000 mm. Circulation water pump used to transfer Nano fluids at different speeds. In addition, several thermocouples were also connected to the heat exchanger and on solar collector inlet and outlet.

Figure 3a: The test rig

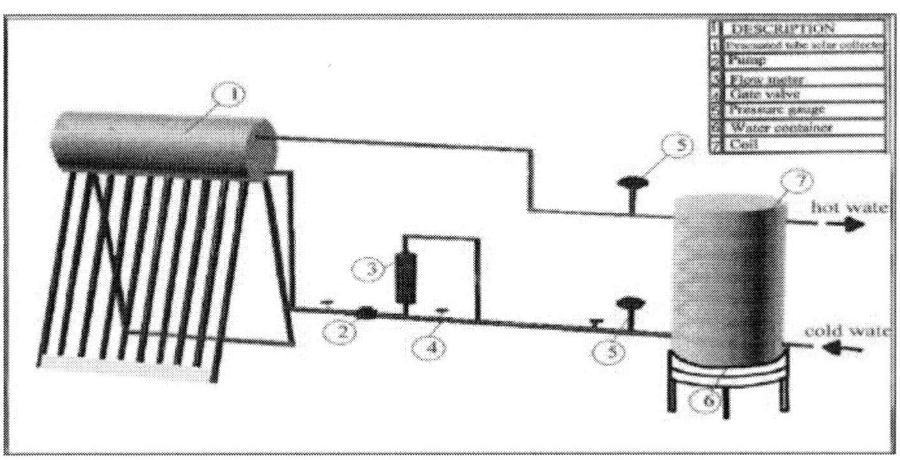

Figure 3b: Display Solar heating unit block diagram

IOP Conf. Series: Materials Science and Engineering **978** (2020) 012029 doi:10.1088/1757-899X/978/1/012029

Figure 4: Evacuated tube solar collector (ETSC)

Figure 5: Heat exchanger from inside and outside

Figure 6: circulation water pump

4. Measuring Apparatus

4.1 pressure gauge

Two pressure gauges with (0 — 150), (0 — 400) mbar ranges are mounted in the inlet and outlet of the exchanger to measure the difference at these stages, as shown in Figure (7). The readings of these two gauges should be registered in order to calculate the pressure drop and other relevant parameters.

Figure 7: pressure gauges

4.2 Temperature measurement

Real-time data logger SD card, 12 channel temperature recorder model: [BTM — 4208SD]. Thermocouple type K was used according to most common type of thermocouple, thermocouple grade wire, –454 to 2,300F (–270 to 1260°C) Accuracy of thermocouple type K (± 2.2°C or ± 75%), special Limits of error: (± 1.1°C) or 0.4%. Also it's cheap, precise, and reliable and has a wide range of temperature.

Figure 8: data logger

4.3 Solar power meter

The solar radiation was measured by digital solar meter type Make – Lutron, Made in Taiwan. Model No.: [SPM – 1116 SD] as shown in Figure (9), the solar power meter with accuracy ± 10 W/m^2. The solar meter measures the total solar radiation which falls on the inclined surface of the solar collector.

IOP Conf. Series: Materials Science and Engineering **978** (2020) 012029 doi:10.1088/1757-899X/978/1/012029

Figure 9: solar power meter

4.4 Flow Meter

Water variable-area flow meter (Rotr meter) with a range of (20 – 70) LPM to measure the volumetric flow rate of the system shown in Figure (10). Was installed in vertical position adjacent to the pump outlet. The transparent flow meter casing is fabricated from a Perspex glass. The range of flow meters is suitable for the pump operation range. The flow meter with full scale accuracy (± 4 percent).

Figure (10): flow meter

5. Analysis of data and validation

Useful thermal energy, temperature distribution, and efficiency of the collector was determined using equations (1– 7) which were derived from [12], [13] and [14]. The equation for useful thermal energy (Qu), as:

$$Q_U = A_C F_R\left[I\,\alpha\,\tau - \left(T_{nf,in} - T_a\right)\right] \quad (1)$$

The heat energy is converted into thermal energy of water in the pipes as:

$$Q = \dot{m}C_p\left(T_{nf,o} - T_{nf,i}\right) \quad (2)$$

Then

$$\dot{m}C_p\left(T_{nf,o} - T_{nf,i}\right) = A_C F_R\left[I\,\alpha\,\tau - \left(T_{nf,in} - T_a\right)\right] \quad (3)$$

Therefore,

$$\left(T_{nf,o} - T_{nf,i}\right) = \left\{\frac{A_C F_R}{\dot{m}C_p}\right\}\left[I\,\alpha\,\tau - U_L\left(T_{nf,in} - T_a\right)\right] \quad (4)$$

F_R may be obtained from

$$F_R = \frac{\dot{m}C_p}{A_C\, U_L}\left[1 - \exp\left(\frac{A_C\, U_L \dot{F}}{\dot{m}C_p}\right)\right] \quad (5)$$

Then the collector effectiveness was got by utilize the relation,

$$\eta_{s,c} = \frac{Q_U}{A_C I} \quad (6)$$

Substitute Equation (2) and Equation (3) in the Equation (5) yields,

$$\eta_{s,c} = F_R\left[\alpha\,\tau - \frac{U_L(T_{nf,in} - T_a)}{I}\right] \quad (7)$$

Since F_R, $\alpha\,\tau$ & U_L are constant,

$$\eta_{s,c}\ \alpha\ \left[\frac{(T_{nf,in} - T_a)}{I}\right] \quad (8)$$

6. Nano fluid thermal measurement-physical properties

The empirical relationship used in research to compare Nano fluid properties with practical readings. The previous equations were used to calculate thermal transfer-physical properties at average Nano fluid temperatures. Volume fraction (Φ) of the nanoparticles specified by [15].

$$\Phi = \frac{V_p}{V_p + v_f} = m\frac{\pi}{6}d_p^{-3} \quad (9)$$

Density [16]

$$\rho_{nf} = \rho_{nf}\,\Phi + (1-\Phi)\rho_{DW} \quad (10)$$

Viscosity [16]

$$\mu_{nf} = (1-\Phi)\mu_{DW} + \Phi\mu_{DW} \quad (11)$$

Specific heat [17]

$$cp_{nf}\rho_{nf} = \Phi(\rho_s cp_s) + (1-\Phi)(\rho_{DW} cp_{DW}) \quad (12)$$

an efficient model for conductivity in Equations (13) [17]

$$\frac{K_{nf}}{K_{DW}} = \left[\frac{cp_{nf}}{cp_{DW}}\right]^{-0.023}\left[\frac{\rho_{nf}}{\rho_{DW}}\right]^{1.358}\left[\frac{\mu_{nf}}{\mu_{DW}}\right]^{0.126} \quad (13)$$

Table 1. Thermal –physical properties of the Nano fluids used [18].

Particles of Nano scale	density (Kg/m^3)	Specific heat (J/kg k)	thermal transmittance (W/ m^2 k)	β *10^5 (k^{-1})	α *10^5 (m^2/s)
Copper (cu)	8933	385	401	1.67	11.7
Distilled water	997.1	4179	0.613	21	

7. Measurement of Nano fluid Thermal Properties

Many of the physical characteristics of Nano fluids (Cu + DW) and distilled water required for evaluation useful energy, thermal energy, collector effectiveness and the convective heat transfer. Dynamic viscosity (μ) measured with a digital viscometer. model (DV-II+pro) as shown in Figure (11), Thermal conductivity measured by using (HOT DISK Tps 500 of KIJTALEY) as shown in Figure (12), Specific heat are measured by using apparatus of specific heat (ESD –201) as shown in Figure (13), as well as measuring the density by weighing sample and the volume it as shown in Figure (14). All physical characteristics are evaluated at various volume concentrations (1%, 3%, and 5 % vol.).

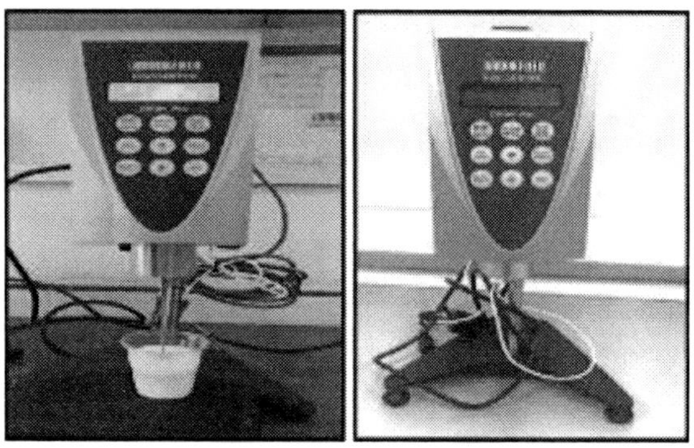

Figure 11: Digital viscometer model (DV-II+pro)

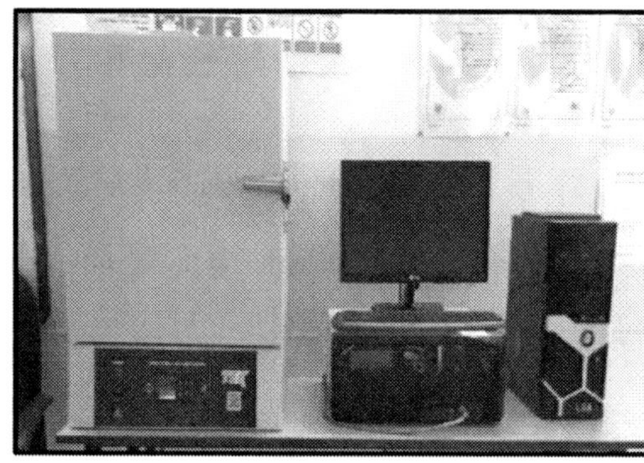

Figure. 12: Hot disk thermal constants analyzer apparatus

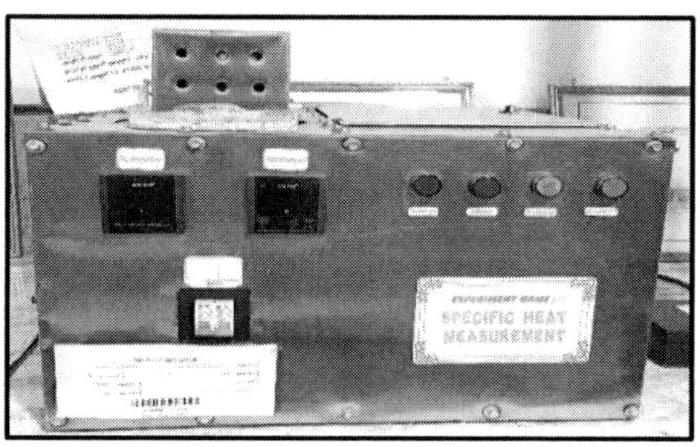

Figure. 13: Specific heat apparatus (ESD – 201)

Figure 14: Measurements of the density

8. Results and Discussion

The primary goal of the job is to make better solar collector's thermal efficiency. Performance of the evacuated solar was discussed in details when the nanoparticles were applied to the collector at concentration (1%, 3% and 5%) and volume flow rates (30, 60, 90) L/min. Useful heat gain is related to how collectors can accumulate a quantity of energy it can be calculated by equations (1) and (2) respectively. Useful heat is gained by collectors at different inlet temperatures, volume flow rate (30, 60 and 90 L/min) and volume fraction (1, 3 and 5%vol) are shown in **Figs (15)**, heat gained when using pure water are (40.058W, 42.6212W, 47.308W, 58.805W, 68.252W, 73.3055W, 74.6237W, 81.434W) for the volume flow rate (30 L/min). By adding the Nano fluids (Cu + DW) to the distilled water and at concentration (3 %) and at volume flow rate (30 L/min) noted the heat gained increased and became (54.118W, 55.876W, 62.247W, 69.053W, 75.063W, 78.8721W, 79.3106W, 84.43W) respectively,

Nano fluids (Cu + DW) at 5 %vol., has shown better results in comparison to distilled water about same flow rate, this means that, by increasing the concentration rate, the amount of heat gain is increased. The efficiency improvement values have been increased gradually as show in Figs (16), Notice that the efficiency enhancement increases as volume concentration and volume flow rates values increased, also it was noticed the effectiveness of collector when utilizing Nano fluids (Cu (30nm) + DW) better than that for distilled water due to high thermal conductivity of copper relative to distilled water. From another hand Table. (2) Show the levels of active enhancement of ETSC when using (Cu (30nm)) with distilled water as working fluids. The Nano fluids (Cu + DW), at concentrations (1% vol., and 5%vol) and volume flow rates (30, and 90 L/min), values of thermal solar properties (FR (τα), – FRUL) are 0.407, 1.156 W/m^2.k, 0.522 and 0.945 W/m^2.k, Whereas for distilled water at volume flow rates 30 L/min and 90 L/min are 0.252, 0.945 W/m^2.k, 0.321 and 1.051 W/m^2.k respectively. This meant that using of Nano fluids (Cu + DW) as a working fluid can enhance the productivity of collectors, other parameter for collector effectiveness comparison is 'A' (Area under the curve×100) which is better methods of recording effectiveness of the collector which implies the entire domain of the collector efficiency (X0 − X max), Area under the curve×100 (A) at various volume rates and volume concentration shown in Table 3. Figs (17) show the difference temperature between inlet and outlet collector and the flow rate for the Nano fluids (Cu (30nm) +DW) at volume concentrations (1, 3 and 5 % vol.), it could be shown tha comparison increased. The error rate was calculated in practical data by root mean square and it did not exceed ± 4%. t the difference in temperature between inlet and outlet is being smaller in with distilled water when the Nano fluid concentration was being increased. The error rate was calculated in practical data by root mean square and it did not exceed ± 4%.

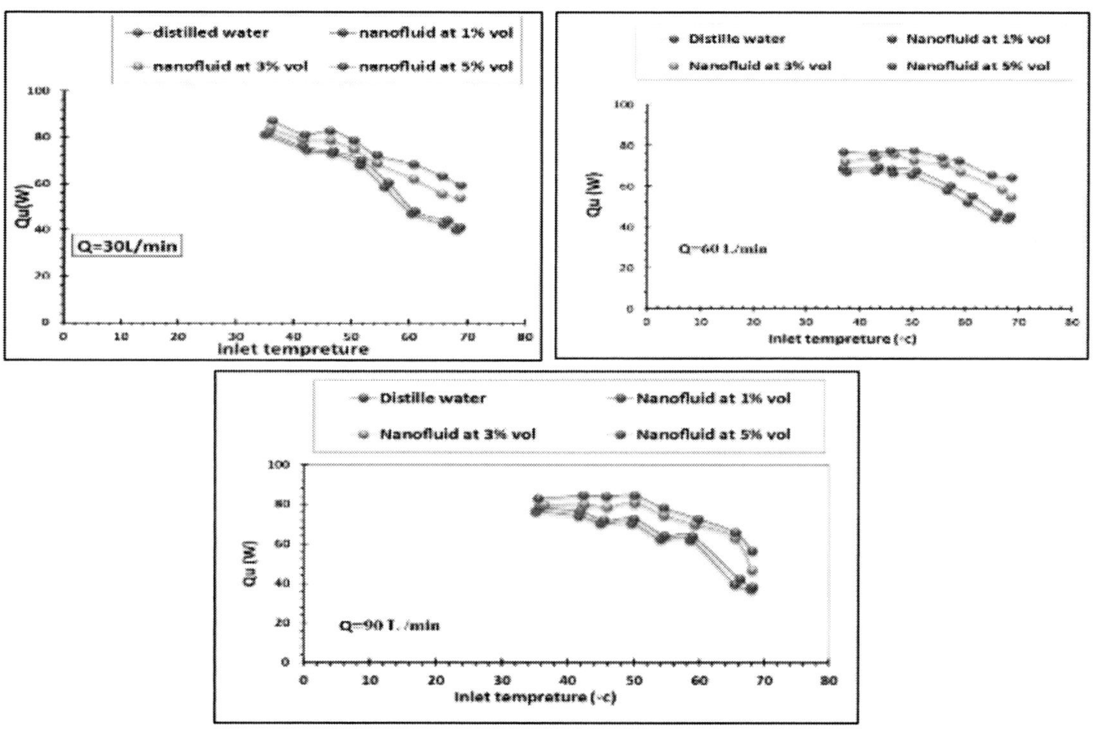

Figure 15: Gain beneficial heat of solar Collector at several Φ for Nano particles (Cu) with DW at different flow rates

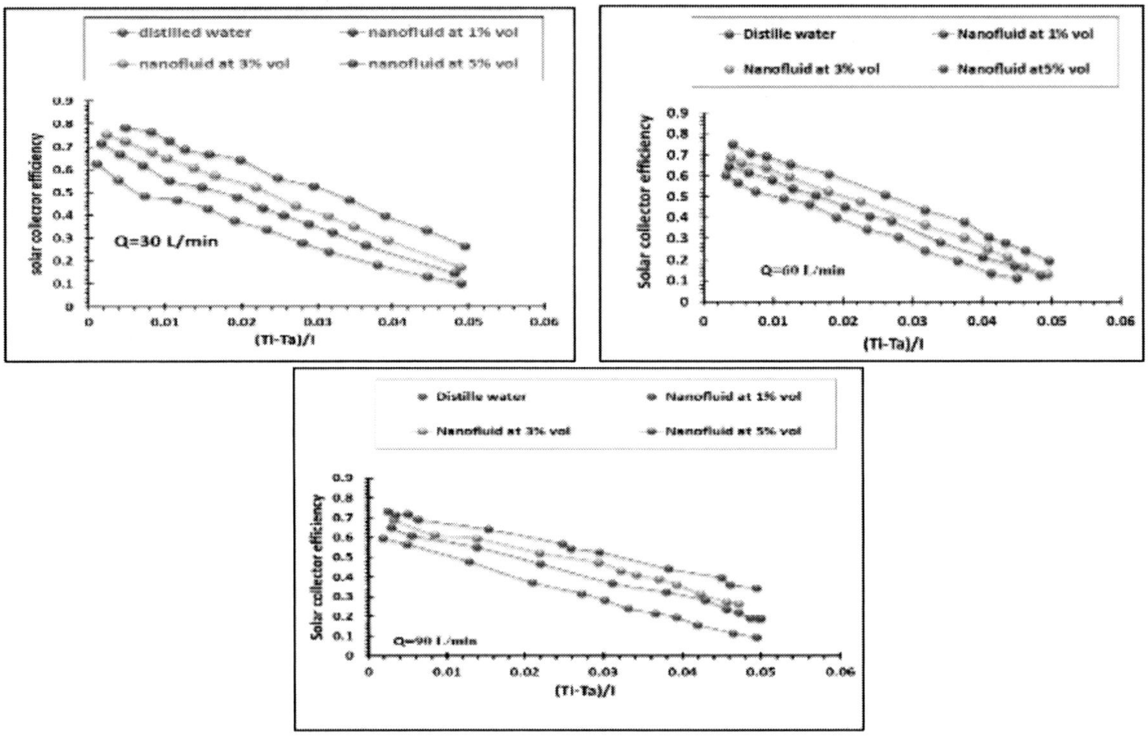

Figure 16: Efficiency Collector of Nano particles (Cu) with DW at various flow rates

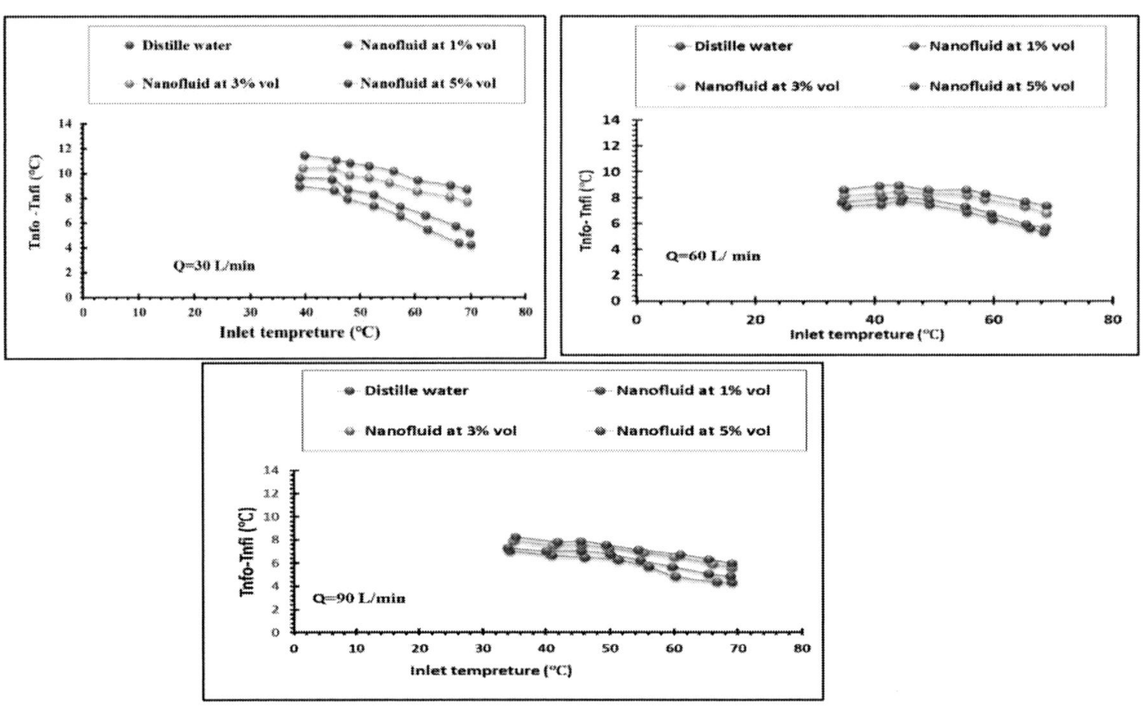

Figure 17: temperature changing between input and output solar collector at several Φ for nanoparticles (Cu) and distilled water

Table 2: the enhancements results

Components	Volume friction (Vol %)	volume flow rate L/min	Enhancement %
Distilled water	0%	30	2.41%
	0%	60	4.36%
	0%	90	5.43%
Nano fluids (Cu(30nm)+Dw)	1%	30	13.44%
	3%	30	14.28%
	5%	30	15.32%
	1%	60	12.096%
	3%	60	13.709%
	5%	60	16.96%
	1%	90	13.38%
	3%	90	14.17%
	5%	90	19.68%

Table 3: the experimental result

	Concentration (% Vol)	Q (LPM)	Model	Area under curve X100 (A)	R^2
Nano fluid (Cu (30nm) –DW)	1	30	$\eta = -1.156\,X + 0.407$	1.35	0.975
	3	30	$\eta = -1.163\,X + 0.421$	1.36	0.970
	5	30	$\eta = -1.192\,X + 0.444$	1.38	0.982
	1	60	$\eta = -1.201\,X + 0.465$	1.39	0.984
	3	60	$\eta = -1.215\,X + 0.484$	1.41	0.971
	5	60	$\eta = -1.225\,X + 0.506$	1.43	0.986
	1	90	$\eta = -1.233\,X + 0.513$	1.44	0.978
	3	90	$\eta = -1.245\,X + 0.534$	1.45	0.982
	5	90	$\eta = -1.257\,X + 0.552$	1.52	0.995
Distilled water (Dw)	0	30	$\eta = -0.945\,X + 0.252$	1.19	0.985
	0	60	$\eta = -0.974\,X + 0.274$	1.24	0.984
	0	90	$\eta = -1.050\,X + 0.312$	1.27	0.978

9. Conclusions

The thermal enhancement of solar collector performance was investigated with Nano fluids (Cu (30nm) +DW) as working fluid. The nanoparticles used in investigate with particles concentration ratios (1, 3 and 5 % vol.) and the based working fluid was distilled water. The summary results are as follows:

1. The Nano fluids (Cu + DW), at concentrations (1%vol., and 5%vol) and volume flow rates (30, and 90 L/min), the values of thermal solar properties (FR (τα), – FRUL) are 0.407, 1.156 W/m^2.k, 0.522 and 0.945 W/m^2.k, Whereas in the case of distilled water at volume flow rates 30 L/min and 90 L/min are 0.252, 0.945 W/m^2.k, 0.321 and 1.051 W/m^2.k respectively.

2. Use of Nano fluids (Cu (30nm)) as a working fluid could improve thermal performance of evacuated tube solar collector compared with distilled water, especially at high inlet temperature.
3. The increase in the rate of concentration result an increase in heat gain.
4. The nature and size of nanoparticles play a vital role in optimizing the rate of heat transfer and improving the efficiency of the (ETSC).

Nomenclature:

V_p Volume of nanoparticle (m^3).
V_f Volume of fluid (m^3).
Cp Specific heat at constant pressure (J/kg K).
Cp_{Dw} Heat capacity of distilled water (J/kg K).
Cp_s Heat capacity of nanoparticles (J/kg K).
K_{nf} Thermal conductivity of Nano fluid (W/m K).
K_{DW} Thermal conductivity of distilled water (W/m K).
Qu Rate of useful energy gained (W).
Q coll Heat rate from solar collector (W).
Ac Surface area of solar collector (m^2).
F_R Heat removal factor.
I solar radiation (W/m^2).
U_L Overall heat loss (W/m^2 °C).
DW distilled water.
Ui Overall loss coefficient of solar collector (W/m^2 K).
hfi Heat transfer coefficient inside the tube (W/m^2 K).
ETSC evacuated tube solar collector.
FLC flat plate collector.

Greek symbols
μ_{DW} The Viscosity of distilled water (m^2/s)
ρ_{nf} The density of Nano fluid (kg/m^3)
ρ_{DW} The density of distilled water (kg/m^3)
μ_{nf} The Viscosity of Nano fluid (m^2/s)
$\alpha\tau$ The transmittance - absorptance product

References
[1] H.A. Kazem, J.H. Yousif. " Comparison of prediction methods of photovoltaic power system production using a measured dataset. " Energy Conversion and Management, vol. 148, pp: 1070-1081, (2017).
[2] M.A. Sabihaa, R. Saidurb,n, Saad Mekhilefc, Omid Mahian "Progress and latest developments of evacuated tube solar collectors", Renewable and Sustainable Energy Reviews , vol. 51, pp:1038-1054, (2015).
[3] Kumar, R., & Rosen, M. A. "Thermal performance of integrated collector-storage solar water heater with corrugated absorber surface", Applied Thermal Engineering, vol. 30, pp. 1764-1768, (2010).
[4] M.A. Sharafeldin, Gyula Grof, "Evacuated tube solar collector performance using CeO_2/water Nano fluid", Journal of Cleaner Production, vol. 185, pp: 347-356, (2018).

[5] Das, S. K., Choi, S. U. S., Yu, W., & Pradeep, (2007). Nano fluid Science and Technology. John Wiley & Sons, Inc., Publication.
[6] Otanicar, T., & Golden, J., (2009) "Comparative environmental and economic analysis of conventional and Nano fluid solar hot water technologies". Environ. Sci. Technol., 43, 6082e7.
[7] Y. Yang Gan, W. Gao, "Comparative studies on thermal performance of water – in – glass evacuated tube solar water heaters with different collector tilt – angles", Journal of Solar Energy, vol. 85, pp:1381 – 1389, (2018).
[8] Kamble D.P., Gadhave P.S., M.A.Anwar ," Enhancement of Thermal Performance of Heat Pipe Using Hybrid Nano fluid", International Journal of Engineering Trends and Technology (IJETT) ,Vol.17, No.9 , (2014) .
[9] Tiwari, A., K. Pradyumna Ghosh, P., Sarkar, J "Solar Water Heating Using Nano fluids –A Comprehensive Overview and Environmental Impact Analysis ", International Journal of Emerging Technology and Advanced Engineering, vol. 3, pp: 221 – 224, (2013).
[10] Lu, L., Liu, Z.-H., Xiao, H.-S. "Performance of an open thermos syphon using Nano fluids for high-temperature evacuated tubular solar collectors," Solar Energy, vol. 85, pp: 379 – 387, (2011).
[11] Hashim A. Hussain, Qusay Jawad and Khalid F. Sultan "Experimental Analysis on Thermal Efficiency of Evacuated Tube Solar Collector by Using Nano fluids" Vol. 4, pp: 19 - 28, (2015).
[12] Koffi, P. M. E., Andoh, H. Y., Gbaha, P., Toure, S., & Ado, G. "Theoretical and experimental study of solar water heater with internal exchanger using thermos siphon system" Energy Conversion and Management, vol. 49, pp: 2279 – 2290, (2008).
[13] Jaisankar, S., Radhakrishnan, T. K., & Sheeba, K. N. "Experimental studies on heat transfer and friction factor characteristics of thermosiphon solar water heater system fitted spacer at the trailing edge of twisted tapes", Applied Thermal Engineering, vol. 29, pp: 1224 – 1231, (2009) a.
[14] Jaisankar, S., Radhakrishnan, T. K., & Sheeba, K. N. "Experimental studies on heat transfer and friction factor characteristics of forced circulation solar water heater system fitted with helical twisted tapes". Solar Energy, vol.83, pp: 1943 – 1952.
[15] Khanafer K, Vafai KA," Critical synthesis of thermophysical Characteristics of Nano fluids," International Journal of Heat and Mass transfer, (2011).
[16] Kumar, R., & Rosen, M. A Thermal performance of integrated collector – storage solar water heater with corrugated absorber surface. Applied Thermal Engineering, vol. 30, pp: 1764 – 1768, (2010)
[17] Suresh S., Chandrasekar M., Chandra Bose" Experimental investigations and theoretical determination of thermal conductivity and viscosity of Al_2O_3/ water Nano fluids", Experimental thermal and Fluid Science, vol. 34, pp: 210-216, (2010).
[18] J.P.Holman, Heat transfer, 8th ed. (2008).

Effect of Radiation on Natural Convection Heat Transfer from Heated Horizontal Cylinder in Vented Enclosure

Omar Mohammed Ali[1, *], Raid Ahmed Mahmood[2,3]

[1]Collage of Mechanical Engineering, University of Zakho, Kurdistan Region, Iraq
E-mail: omar.ali@uoz.edu.krd
[2]Collage of Mechanical Engineering, University of Zakho, Kurdistan Region, Iraq
E-mail: raid.ahmed@uoz.edu.krd
[3]School of Mechanical and Electrical Engineering, University of Southern Queensland, Queensland, Australia
* Corresponding author: Collage of Mechanical Engineering, University of Zakho, Kurdistan Region, Iraq. Tel: +9647504056592.

Abstract. The natural convection heat transfer from a heated circular cylinder placed in a square vented enclosure was investigated considering the effect of the thermal radiation from cylinder using finite volume method with simple algorithm that used pressure-velocity coupling to solve Navier-Stockes and energy equations using Computational Fluid Dynamics CFD (Fluent software). The flow domain is filled with air. The enclosure width is 12.5 cm, two symmetrical openings at its lower and upper walls with size O = 2.5 cm and the cylinder diameter is 5 cm. Different rang of cylinder's surface emissivity with Rayleigh number range between 10^3 to 10^6 was applied in this study as operating condition. The effect of the Ra, cylinder radiation for vented square enclosure on the convection, radiation and total Nusselt numbers Nu and flow and thermal patterns were investigated. The numerical results for pure natural convection were compared with available experimental results, good agreement was obtained. The results revealed that the natural convection heat transfer from the cylinder in the enclosure is increased as compared with that of the unbounded cylinder without radiation effect. The results also revealed that there is a linear relationship between the heat transfer enhancement and the emissivity, there is a significant effect of the cylinder surface radiation which gives high augmentation compared with that of the free cylinder and natural convection heat transfer without radiation effect. A maximum total Nusselt number is obtained at high emissivity compared with that of natural convection heat transfer, the maximum Nu enhancement ranged between 70 to 350 % for low Ra and 25 to 125% for high Ra.

Keywords: *natural convection, radiation, emissivity, heat transfer, cylinder, enclosure.*

1. Introduction

The increase of the free convection heat transfer from a horizontal cylinder received much attention to improve the performance of many applications such as cylindrical components for computer components, heat exchangers and solar collectors. There are numerous methods to improve the heat transfer that have been used recently.

Placement of the cylinder in the vented enclosure is one of the effective technique that can be used to improve the heat transfer (Ali, 2008). In this technique, the fluid moves through the channel and reaches the bottom of the cylinder. The thickness of the generated boundary layer around the cylinder is thinner than that of the free cylinder, which increase the natural convection heat transfer (Kahwaji et al., 2012).

There are many studies which investigated the enhancement of the natural convection from a heated cylinder using different methods and different fluid mediums. The placement of the cylinder in an open enclosure is one of the effective methods. (Sparrow & Pfiel, 1984) presented an experimental investigation to estimate the augmentation of the natural convection from a heated horizontal cylinder placed between two confiding walls. The study used 15 different heights of the channel H and wall spacings S to cylinder diameter D with different range of Rayleigh number (Ra). The results showed that the heat transfer increased by up to 40 percent for the confined cylinder compared with the unconfined cylinder. The results also revealed that there is a significant augmentation in heat transfer when the wall spacing S is decreased and the wall height is increased, maximum enhancement was achieved when $S/D = 1.5$ and $H/D = 20$.

(Kahwaji et al., 2013) provided an experimental study to investigate the heat transfer by natural convection from a cylinder has square cross section placed in a symmetrical vented square enclosure. The Nusselt number (Nu) was determined based on the temperature of the cylinder's surface and the temperature of air inside the enclosure during transient state for the free and bounded cylinder. Different range of Rayleigh number (Ra) was applied with different ratio of enclosure width to cylinder diameter and opening size to cylinder diameter. The results showed that the Nu is increased when the Ra is increased for all cases. The Nu is proportionally increased with the size of the opening for low enclosure widths. The results also revealed that the Nu is inversely proportional with opening size for large enclosure widths. The maximum percentage of enhancement was reached at 20% for enclosed square cylinder compared with free cylinder.

The radiation heat transfer has many applications in thermal engineering such as heat exchangers, solar energy, radiators, which combined with convection heat transfer to indicate the rate of heat transfer. However, the heat transfer coefficients for the natural convection heat transfer was very low values compared with that of forced convection. The radiation heat transfer was neglected and not considered in the forced convection heat transfer case, but it was considered in the natural convection heat transfer case.

In the natural convection heat transfer, the emissivity of the solid surface is close to zero and the heat transfer from the solid surface with a valuable emissivity is the combination of the convection and radiation components (Gengel & Ghajar, 2015). There are many studies about the effect of the surface radiation of the cylinder on the enhancement of heat transfer such as (Wei & Tao, 1996), (Saleh et al., 2015) and (Bello-Ochende, 2019). (Wang et al., 2011) investigated the combined effect of the natural convection-conduction and surface radiation in an open cavity with constant heat flux in a laminar region. The results showed that the radiation and conduction can increase the average total of the Nu. The average Nu was increased with the increasing of the emissivity for more than 0.2. The results also revealed that the conduction heat transfer for the wall can improve the total cooling effect and the enhancement of the heat transfer under surface radiation effect was between 54.1% to 64.0%.

(Ngo et al., 2014) provided a numerical study to assess the laminar natural convection with the effect of surface radiation in a hollow receiver using plate extended surface. Three-dimensional assessment simulation was applied to determine the effect of convection and radiation on the heat losses. The assessment considered the prediction of the temperature, emissivity, orientation and parameters of the geometry on the total heat loss. The results showed that the convective heat loss from the cavity receiver was affected by the inclination, the presence of fins and the number of fins. While the radiation heat loss was considerably affected by surface properties of the receiver. (Pandey & Kumar, 2017) investigated the radiation and convection influence of Cu-water nanofluid flow over an extending cylinder in a porous layer using slip and viscous dissipation boundary conditions. The dimensionless velocity and temperature distribution were displayed with skin friction and heat transfer coefficients. The results showed that the Nusselt number is decreased when the radiation parameter, thermal slip parameter and Eckert number increased. The Nu also increased when the slip velocity and natural convection increased.

(Nadjib et al., 2018) provided a numerical 2-dimensional study to determine the free convection and radiation of surface in a domain includes air connected with hot extended surface. The boundary conditions of the square enclosure were constant temperature and adiabatic boundary conditions. The position of the heating element was varied from a horizontal position (HPFU, HPFD) to a vertical position (VPFL). The effects of the Rayleigh number, fin length, position of the fin position and emissivity of the wall were considered in investigated. The results showed that the improvement in the heat transfer within the cavity and the optimum thermal performance was obtained for a vertical position. An empirical correlation was developed for Nusselt numbers with a maximum deviation less than 4%.

(El Moutaouakil et al., 2020) applied computational fluid dynamics (CFD) to assess the interaction between natural convection and radiation for Inner Wavy Body in a square body. Different Ra and emissivity were used with variable undulations number as operating conditions. The results revealed that the average radius and the amplitude influenced the total heat transfer and there was no effect on the undulations number of the wavy wall.

From the open literature, there are many studies considered the enhancement of natural convection from a heated horizontal cylinder using different methods to transfer the heat. However, the radiation effect from the cylinder's surface on the natural convection still needs more investigation using different range of enclosure size and dimension. This paper investigates the effect of the natural convection mixed with radiation on the heat transfer from a horizontal cylinder in a symmetrical vented square enclosure using finite volume numerical method that based on simple algorithm with employment of Fluent software. Wide range of Rayleigh number, outer cylinder emissivity and constant enclosure width and opening size were applied. The study predicts the thermal behavior for Nusselt number, fluid temperature inside the enclosure, streamlines and isotherms.

2. Mathematical formulation and numerical method model

Figure 1 presents the configuration of the investigated geometry. The circular cylinder diameter (D) is 50 mm and 500 mm length using a square enclosure with length (W) of 125 mm. The circular cylinder is placed in the middle of the enclosure. The opening size O is 25 mm. The temperature of the circular cylinder (Ts) and the enclosure surface temperature are assumed to be constant in this study. In numerical method, some assumptions have been applied to investigate the natural convection heat transfer from heated horizontal circular cylinder: 2-dimension, steady state, and constant properties for the laminar flow.

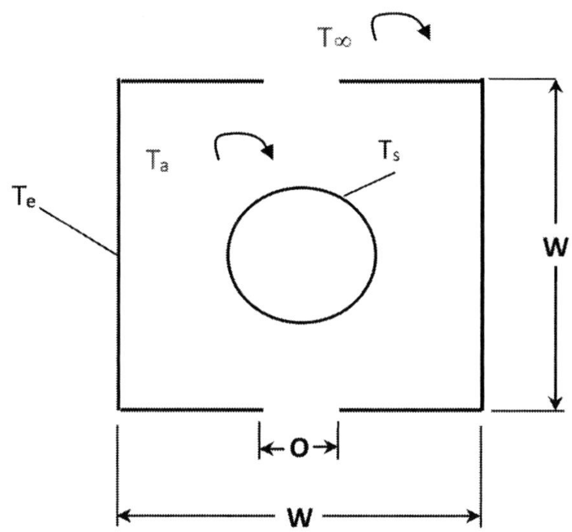

Figure 1. Combination of cylinder and enclosure.

2.1 Governing equations

The flow and thermal behavior of the fluid domain between the cylinder and enclosure walls can be described using the Navier-Stokes equations and energy equation. The flow in the computational domain is assumed as steady, laminar, two dimensional with constant property (incompressible flow). The Navier-Stokes equations (Nada et al., 2007) in the general coordinate system are:

$$\frac{\partial u_i}{\partial x_j} = 0 \quad (1)$$

$$\frac{\partial}{\partial x_j}(u_i u_j) = -\frac{\partial P}{\partial x_i} + Pr\frac{\partial^2 u_i}{\partial x_j \partial x_j} + RaPr\theta \quad (2)$$

Where xi, x_j indicates dimensionless Cartesian coordinates, ui, x_j indicates the corresponding dimensionless velocity components, p is the dimensionless pressure, θ is the dimensionless temperature, and Pr is the Prandtl number.

The energy equation can be presented as follows,

$$u_j\frac{\partial \theta}{\partial x_j} = \frac{\partial^2 \theta}{\partial x_j \partial x_j} \quad (3)$$

Buoyancy force is the dominant force in the natural convection when the density varies due to temperature change. The strength of the buoyancy can be measured by the Rayleigh number as follows,

$$Ra = \frac{g\beta \Delta T D^3 \rho}{\mu\alpha} \quad (4)$$

Where ρ is the fluid density, ΔT is the temperature difference between cylinder surface temperature and mean fluid temperature inside the enclosure, D is the diameter of the cylinder, μ is the fluid dynamic viscosity, α is the fluid thermal diffusivity and β represent the thermal expansion coefficient and can be estimated as follows,

$$\beta = -\frac{1}{\rho}\left(\frac{\partial \rho}{\partial T}\right)_p \quad (6)$$

$$\alpha = \frac{k}{\rho C_p} \quad (7)$$

Where k is the thermal conductivity of the air and C_p represents the specific heat at constant pressure. The average Nusselt number of the cylinder surface is calculated from the integration of the following equation (Wang et al., 2011):

$$Nu_{c,av} = \frac{h_{avg}D}{k} = -\frac{1}{2\pi}\int_0^{2\pi}\left(\frac{\partial \theta}{\partial n}\right)_{r=ro} rd\emptyset \quad (8)$$

Where

h_{avg} : is the average cylinder heat transfer coefficient, W/m^2.°K.
n : is the unit normal coordinate to the cylinder surface.
θ : is the dimensionless temperature.
ϕ : is the angle from the centerline of the cylinder.

For the surface radiation effect, the net heat fluxes of the radiation over the surfaces can be alienated into subdivisions based on the employed mesh. The equation formulations of the radiosity-irradiance can also be applied. The radiosity estimation of the radiative sub surfaces can be written as follow (Wang et al., 2011),

$$J_i = \varepsilon_i \sigma T_i^4 + (1-\varepsilon_i)\sum_1^{NF} J_i F_{ij} \quad (i = 1,2, \ldots, NF) \quad (9)$$

Where ε_i is the sub surface emissivity, Fij is the shape factor from the ith subsurface to the jth subsurface of the enclosure, and NF is the total number of sub surfaces along the enclosure surface.

The net radiation heat flux (q_r) for the ith subsurface of the enclosure's walls and can be determined by

$$q_{ri} = \sum_{j=1}^{NF}(J_i - J_j)F_{ij} \qquad (i = 1,2, \ldots, NF) \quad (10)$$

The average radiative Nusselt number can be presented as follows,

$$Nu_{r,av} = -\frac{1}{2\pi}\int_0^{2\pi}(Q_r)_{r=ro} rd\emptyset \quad (11)$$

The total Nusselt number is:

$$Nu_t = Nu_{c,av} + Nu_{r,av} \quad (12)$$

2.2 Boundary Conditions

The boundary conditions to predict the flow and thermal behaviors for the natural convection were applied. Geometry configuration presented in Figure (1) was used. The boundary conditions can be summarized as in the following table:

Boundary name	**Boundary conditions**
Cylinder surface	T = const., u = v = 0
Enclosure surface	T = const., u = v = 0
Lower opening	$T = T_{amb.}$, u = 0, v = const.
Upper opening	$dT/dy = 0$, m = const.

Where u is the x-component of velocity, v is the y-component of velocity and m is the mass flowrate from upper opening.

2.3 Numerical Method

The governing equations are solved using FLUENT 16.2. The method is assumed the Boussinesq approximation during the solution of the momentum equation The properties of the working fluid were based on the average temperature of the receiver surface and the ambient air. The surface-to-surface S2S radiation model is used in the FLUENT software to account the effect of the radiation exchange in an internal surface between cylinder surface and enclosure walls (ANSYS, 2018). The cylinder surface is assumed as gray and diffuse, while the enclosure surfaces are assumed as black. The combined natural convection and surface radiation from the heated horizontal cylinder in a cold square enclosure is investigated using a numerical model. For pressure velocity coupling, SIMPLE algorithm has been used, with second order upwind scheme for the discretization of momentum and energy equations. A convergence criterion of 10^{-6} was imposed on the residuals of the continuity equation, momentum equation and energy components as displayed in figure 2..

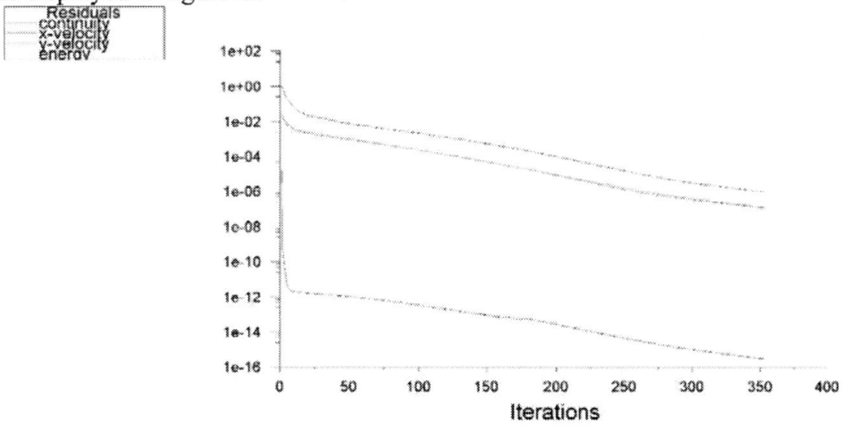

Figure 2: Convergence of the x and y velocities and temperature to the solution in the numerical method.

IOP Conf. Series: Materials Science and Engineering **978** (2020) 012030 doi:10.1088/1757-899X/978/1/012030

2.4 Grid independence study

Grid was generated in the 2D flow domain as displayed in Figure 3. Three different grid sizes were used to investigate the optimum grid number and size: 6417, 8618, and 15582. The different grid sizes are used to simulate the pure natural convection heat transfer from cylinder in a vented enclosure with W/D= 2.5 and O = 2.5 cm to keep the consistency of the numerical simulation using same operating condition at Ra = 5×10^5. The Nusselt number for the three cases was obtained and compared with the experimental results from previous study. The numerical results for the Nu for the three different grid numbers are 14.62, 14.57 and 14.48, respectively. The difference between the three values of Nu are not significant. However, the minimum percentage error was achieved at grid number of 15582 and grid size of 2 mm compared with the experimental value of the Nu, therefore, this grid size and number was selected for numerical simulation.

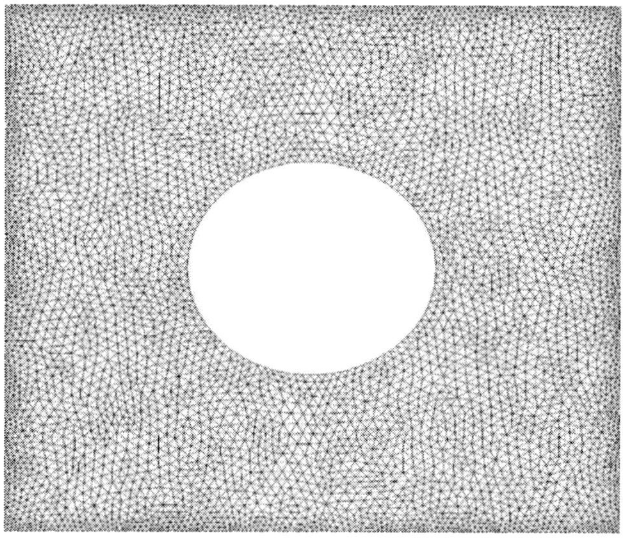

Figure 3. Grid distribution in the computational domain between the circular cylinder and the enclosure

2.5 Validation of numerical results

The experimental results from pervious study were used to validate the numerical results of this study because the same cylinder-enclosure with openings configuration and boundary conditions are used for both works. Figure 4. presents a comparison between the present results with the experimental results, presented by (Ali, 2008), for the heated horizontal cylinder with W = 12.5cm, D =5cm and O = 2.5 cm. Excellent agreement is obtained, the maximum percentage error between both results is about 1 %.

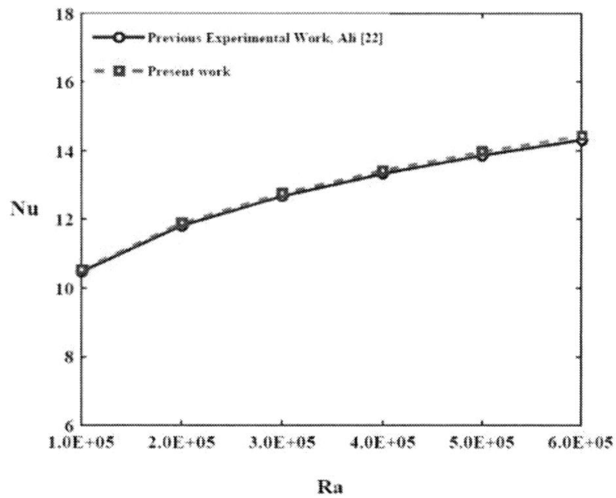

Figure 4: Comparisons between the present work and previous experimental work for W = 12.5 cm, O = 2.5 cm

3. Results and Discussions

This section will present and discuss the numerical results. The results of the present work for the natural convection from heated cylinder placed in a cold square enclosure with lower and upper openings are presenting in this section. Air was a medium in the enclosure and surrounded the cylinder.

Figure 5. presents the Nusselt number (Nu) for the convection, radiation, and total effect for the different emissivity of the surface of cylinder ($0 < \varepsilon \leq 1$) when W/D =2.5 and O = 2.5 cm. Figure 4a is the total Nu from the cylinder for the natural convection mode is presented at $\varepsilon = 0$ that it is increased with the increase of Ra in the presence of the cold square enclosure which its value is larger than those of unbounded cylinder due to the effect of confiding walls. The convection Nu is same for all cylinder emissivity values. The radiation heat transfer from the cylinder occurred when the cylinder surface emissivity increased. Figures from 5b to 5f present the Nu for the convection, radiation, and total effect when the emissivity increases from 0.2 to 1. The radiation heat transfer from horizontal cylinder to the enclosure surfaces influences the Nu as a result of the cylinder emissivity increasing from 0.2 to 1. The radiation Nu is approximately constant with the increase of Ra because the natural convection flow has no effect on the radiation Nu value for low emissivity values (0.2-0.4). The Nu from radiation is increased with the increase of Ra for high cylinder emissivity values due to the effect of high radiation heat transfer on the flow. The enhancement of the total heat transfer is obtained between 70% to 300% for the range of Ra from 10^3 to 5×10^4 and emissivity range from 0.2 to 1. The enhancement of the total Nusselt number is also obtained between 25 % to 125 % for Ra values between 5×10^4 to 10^6 and emissivity range from 0.2 to 1.

IOP Conf. Series: Materials Science and Engineering **978** (2020) 012030 doi:10.1088/1757-899X/978/1/012030

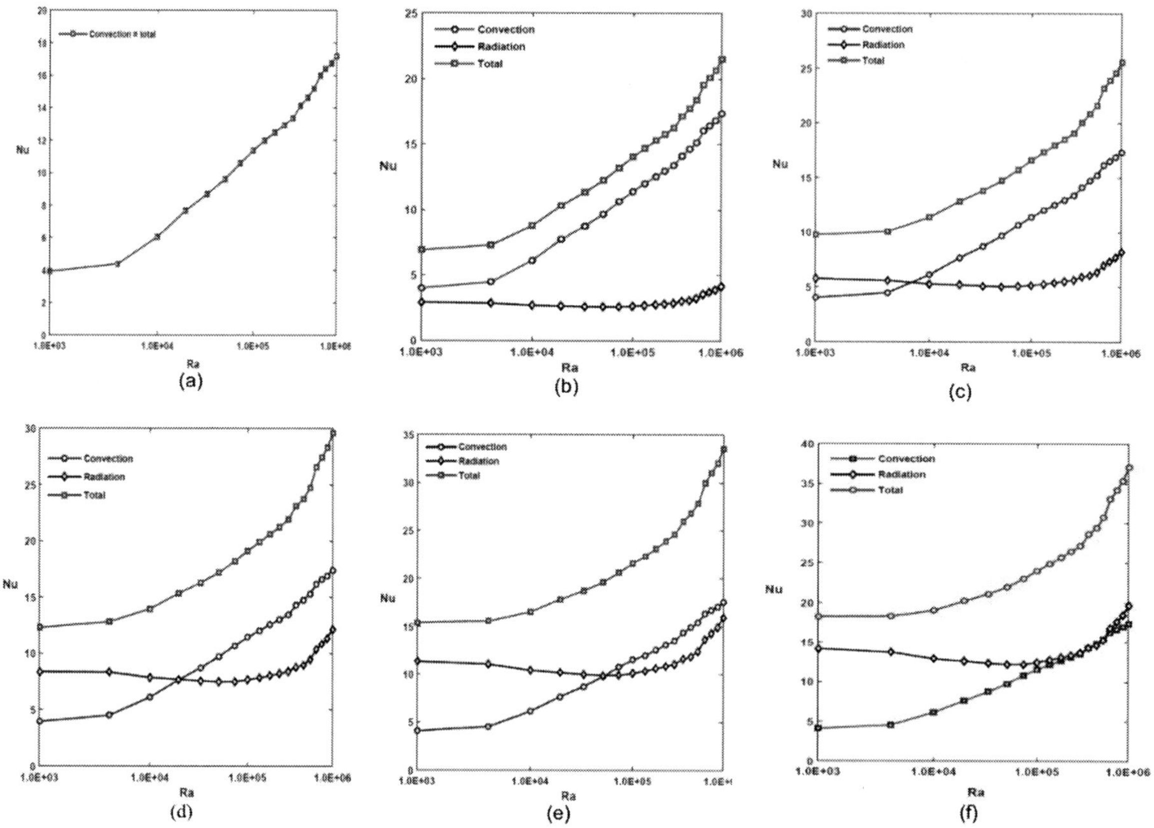

Figure 5. Convection, radiation and total Nusselt number for different Rayleigh numbers due to cylinder surface emissivity variation (a) ε =0, (b) ε =0.2, (c) ε =0.4, (d) ε =0.6, (e) ε =0.8, (f) ε =1.

Figure 6. presents the enhancement of the heat transfer as a result of the radiation heat transfer effect for the range of Ra from 10^3 to 10^6. Figures from 5a to 5e show that the emissivity variation of the cylinder's surface do not influence the Nu for the natural convection heat transfer for each Ra with the variation of the cylinder emissivity. For Ra = 10^3 and 10^4, the convection Nu is low because the conduction heat transfer is the dominant mode and the convection contribution is low. For Ra > 10^4, the convection heat transfer become more and it is the dominant mode due to the large temperature difference between cylinder surface and fluid temperature inside the enclosure that lead to the density variation causing high buoyancy force and Ra. There is a linear relationship between the total Nu and the emissivity because the radiation Nu is increased with the high emissivity cylinder surface that transfer to the enclosure surfaces. The enhancement for the Nu is very large when the Ra is in the low range, then it decreases when the Ra increase.

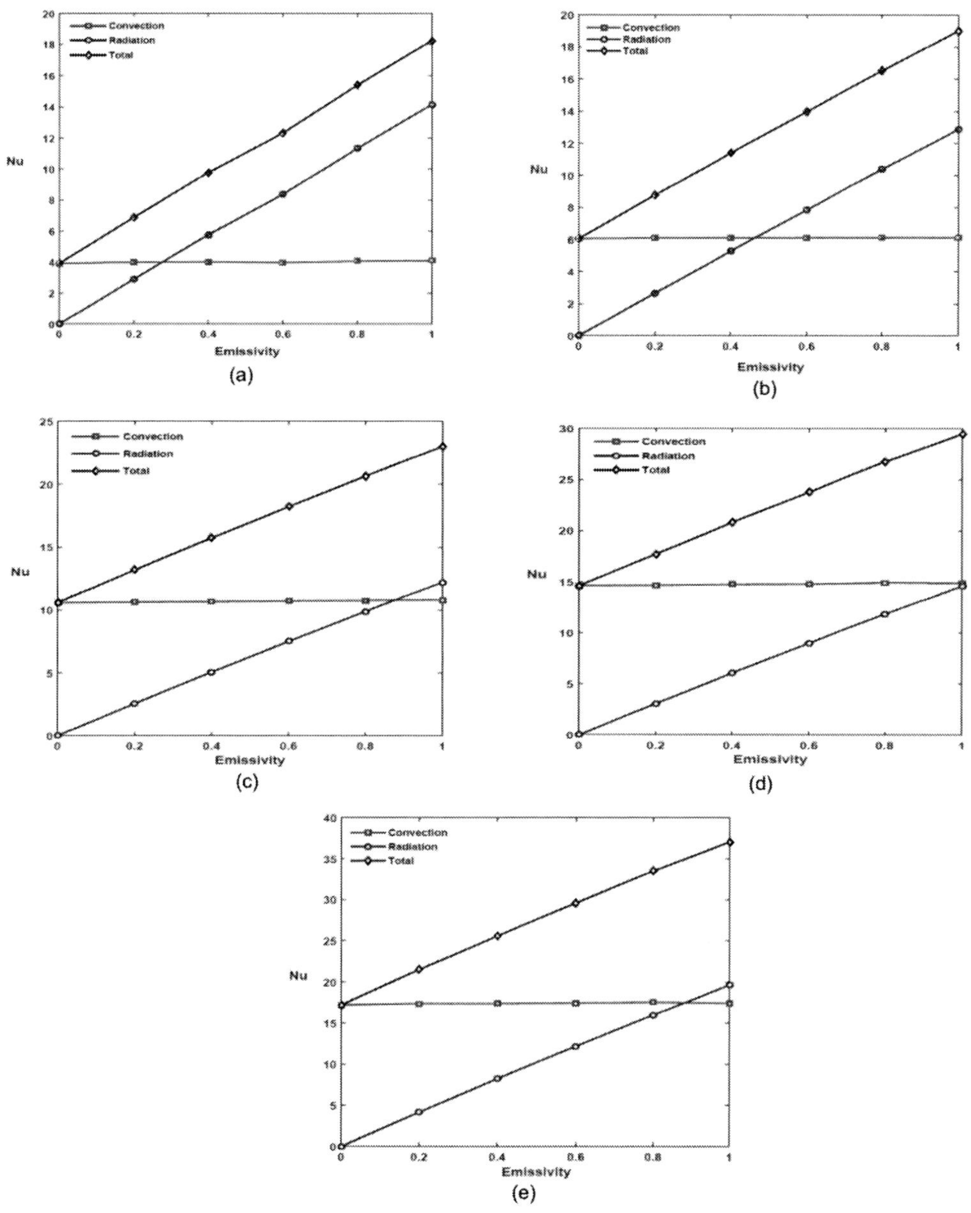

Figure 6. Enhancement of the Nusselt number for different surface emissivity's for Rayleigh numbers (a) Ra = 10^3, (b) Ra = 10^4, (c) Ra = 10^5, (d) Ra = 5×10^5, (e) Ra = 10^6.

Figure 7. presents the mean temperature of the fluid in the domain between the heated circular cylinder and vented enclosure for different Ra and emissivity. The mean fluid temperature (Tm) is increased when the emissivity (ε) increases for all Ra, but the increasing become sharp for high range of Ra. The fluid temperature is increased with high Ra because the flow movement become more that causes from the density variation due to high temperature difference. The increase of the emissivity lead to enhance in the heat transfer by radiation, therefore the fluid temperature is increased.

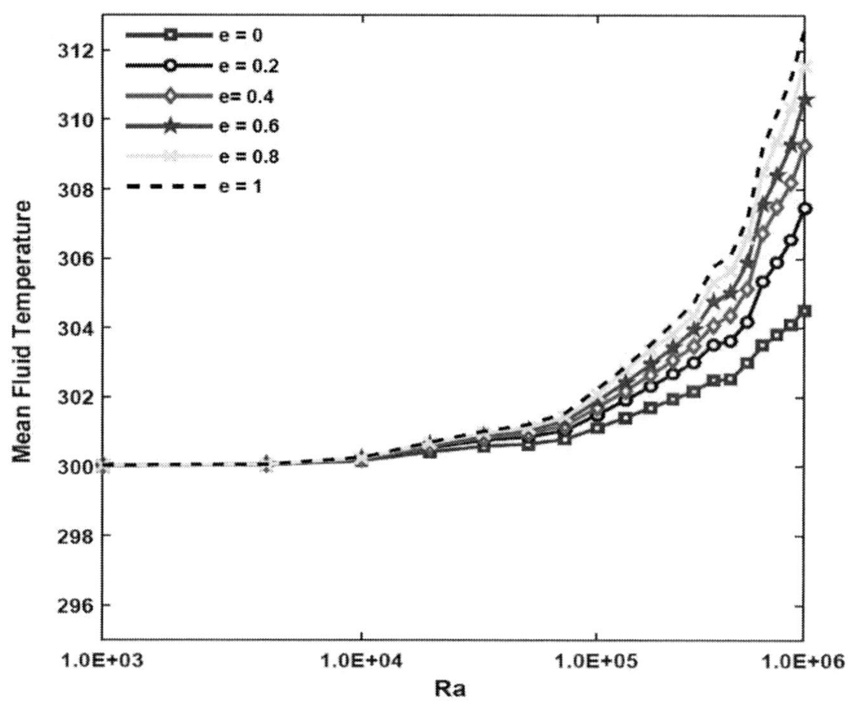

Figure 7: The mean temperatures of the fluid for different Ra and different emissivity.

Figure 8. shows the stream function and temperature distributions of natural convection heat transfer from the heated circular cylinder in a symmetrical enclosure with different Rayleigh number (Ra). The figure presents the flow and thermal behavior when the Ra increased from 10^3 to 10^6, there is a significant change in the behavior with increasing of Ra. For Ra = 10^3, the streamlines distributed in the whole fluid domain, the flow is symmetrical around y-axis, the stream function values are weak and the thickness of the hydrodynamic boundary layer with the cylinder is large. The temperature distribution shape appeared as circular rings around the cylinder which ensures that the conduction heat transfer is the dominant mode and thermal boundary layer is very thick. For Ra = 10^4, the flow around y- axis remains symmetrical but the stream function values are increased and the hydrodynamic boundary layer thickness is decreased. The isotherms shape is changed to become closer to the cylinder surface, a thermal plume is appeared above the cylinder and the thermal boundary layer thickness is decreased that ensures the convection heat transfer mode become more, but the conduction mode still effective. For high Ra, the hydrodynamic and thermal boundary layer thicknesses around the cylinder are decreased and become thinner and the strength of stream function is also enhanced when the Ra increased which ensures that the convection heat transfer is the governed mode. The streamlines become closer to the cylinder surface and two eddies are appeared along both sides of the enclosure with inner circular vortices at the upper corners of the enclosure due to the high flow movement. The temperature contours for high Ra displayed that the thermal plume become narrower as Ra is increased. The effect of conduction heat transfer decrease when the Ra is increased and the convection heat transfer become the dominant mode due to the thinner boundary layers.

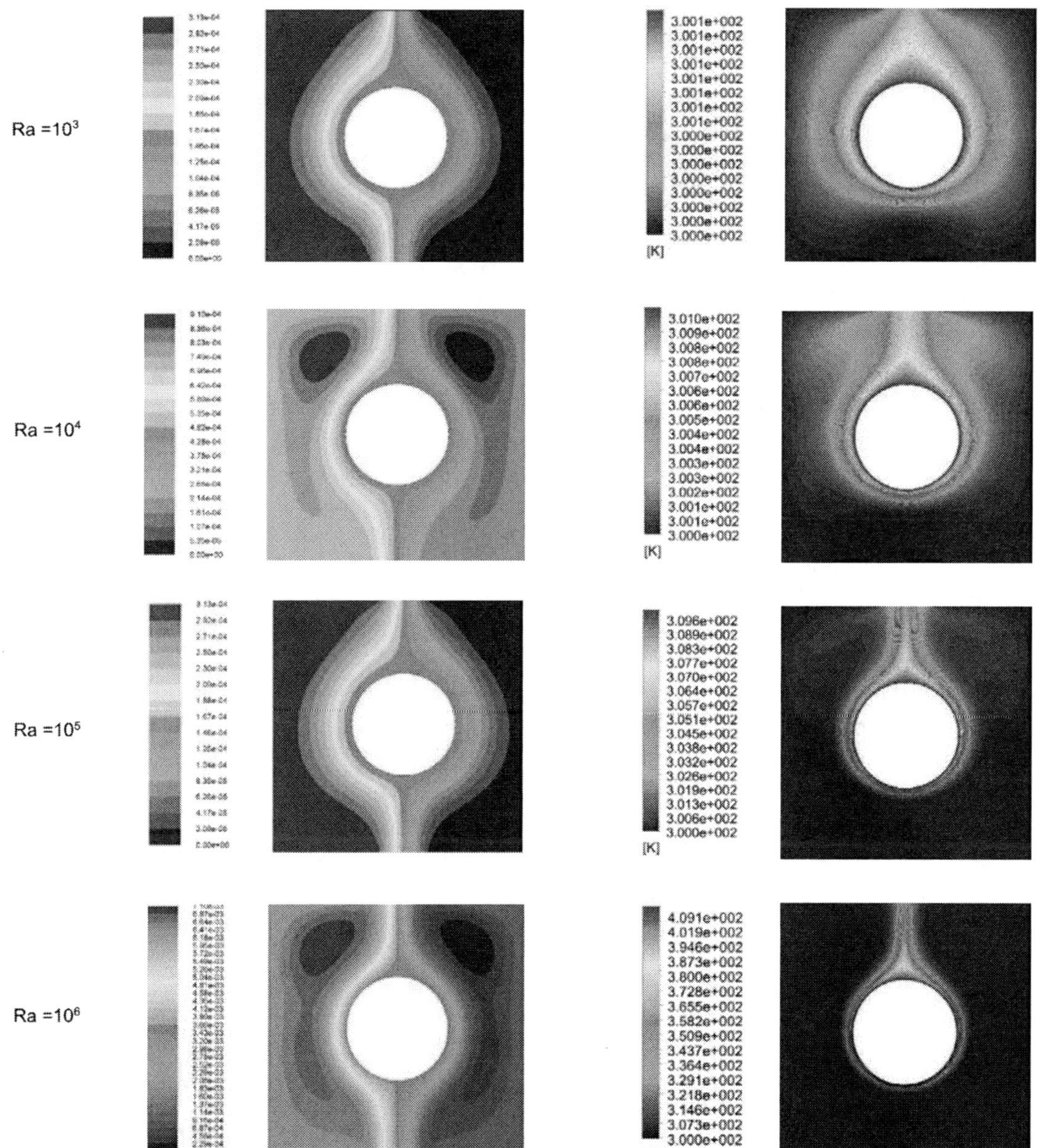

Figure 8. Streamlines and isotherms for pure natural convection heat transfer

Figure 9. shows the streamlines of the natural convection heat transfer from the circular cylinder for different range of Ra considering the effect of the surface radiation effect from the cylinder. The stream function values are slightly decreased as cylinder emissivity is increased due to the effect of radiation heat transfer on the flow movement. For Ra = 10^3, the arrangement of the streamline does not change with the emissivity of the cylinder's surface because the fluid flow is very slow and it distributed to cover most of the fluid domain. When Ra increased to 10^4, the streamlines with radiation heat transfer starts to be developed and changed compared with that of the pure natural convection. Two eddies are appeared above the cylinder surface and covered the sides of the enclosure, each eddy have a circular vortex at the upper portion of it. The arrangement of the streamlines starts to be changed when the emissivity is increased as a result of the radiation heat transfer effect. The eddies along enclosure sides are changed and move to the upper portion of the enclosure corners as cylinder emissivity is increased. When the Ra increased more to be equal to 10^5, a significant change and arrangement clearly appear for each cylinder's emissivity which have different arrangement compared with that of the natural convection flow. The shapes of the streamlines along y-axis still unchanged, but some eddies and vortices are appeared in the remaining flow domain near enclosure surfaces and corners. The eddies and vortices in the flow become more and disorder in the same region for Ra = 10^6 and $\varepsilon = 0.2$. For $\varepsilon = 0.6$ and 1, the eddies and vortices concentrated at enclosure corners below cylinder surface and plumes are displayed at the sides above the cylinder surface. These changes due to the combined effect of convection and radiation heat transfer on the flow that arises the fluid temperature and affected on the flow behavior inside the fluid domain.

Figure 10. presents the temperature distribution of the natural convection heat transfer from the circular heated cylinder for different range of Ra considering the effect of the radiation. For the low Ra from 10^3 to 10^4, the isotherm behavior are same as pure natural convection at the upper part of the enclosure, the temperature distribution are slightly changed as that at the upper part of the enclosure for natural convection heat transfer under different radiation effect, the temperature is increased and its layers are moved toward the enclosure surfaces. However, the isotherm behavior and temperature distribution at the lower opening vent are completely have different behavior when the emissivity change that a thermal plume is appeared from the lower vent due to an increase in the temperature layers under circular cylinder. The effect of radiation heat transfer is appeared apparently for Ra = 10^4 that the temperature of the layers near the enclosure surfaces is raised as cylinder emissivity is increased. When Ra increased more between 10^5 to 10^6, the temperature distribution and the thermal plume for all emissivities above the cylinder surface approximately are similar as that of the pure natural convection. The distribution of the isotherm and its arrangement away from the cylinder surface have different behavior compared with that of the natural convection as a result of the effect of radiation heat transfer from the cylinder, which the temperature of the layer are increased and displaced to the enclosure surfaces as the cylinder emissivity is raised.

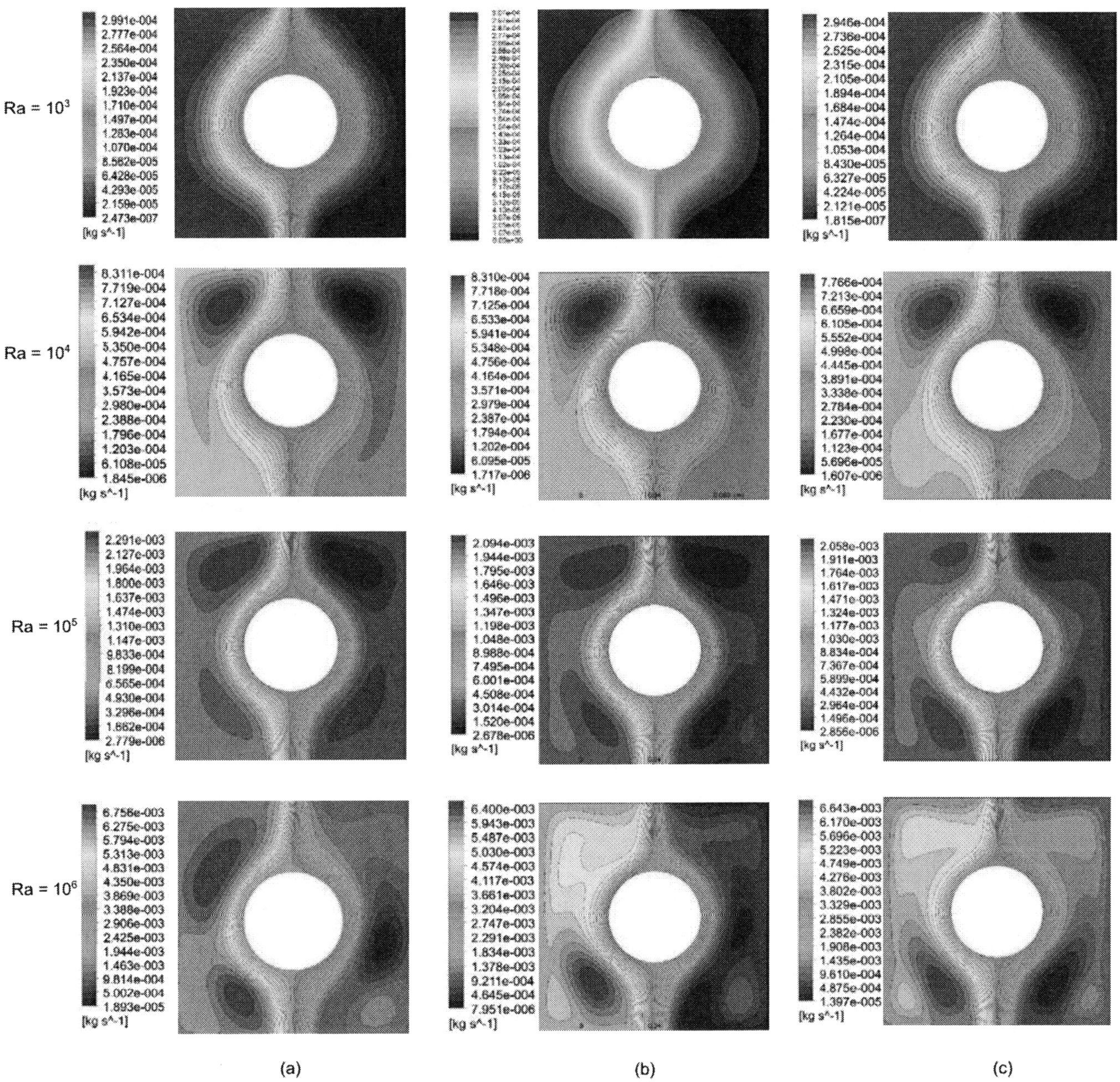

Figure 9: Streamline distribution around the cylinder at different Rayleigh numbers and different cylinder surface emissivity: (a) ε =0.2, (b) ε =0.6, (c) ε =1.

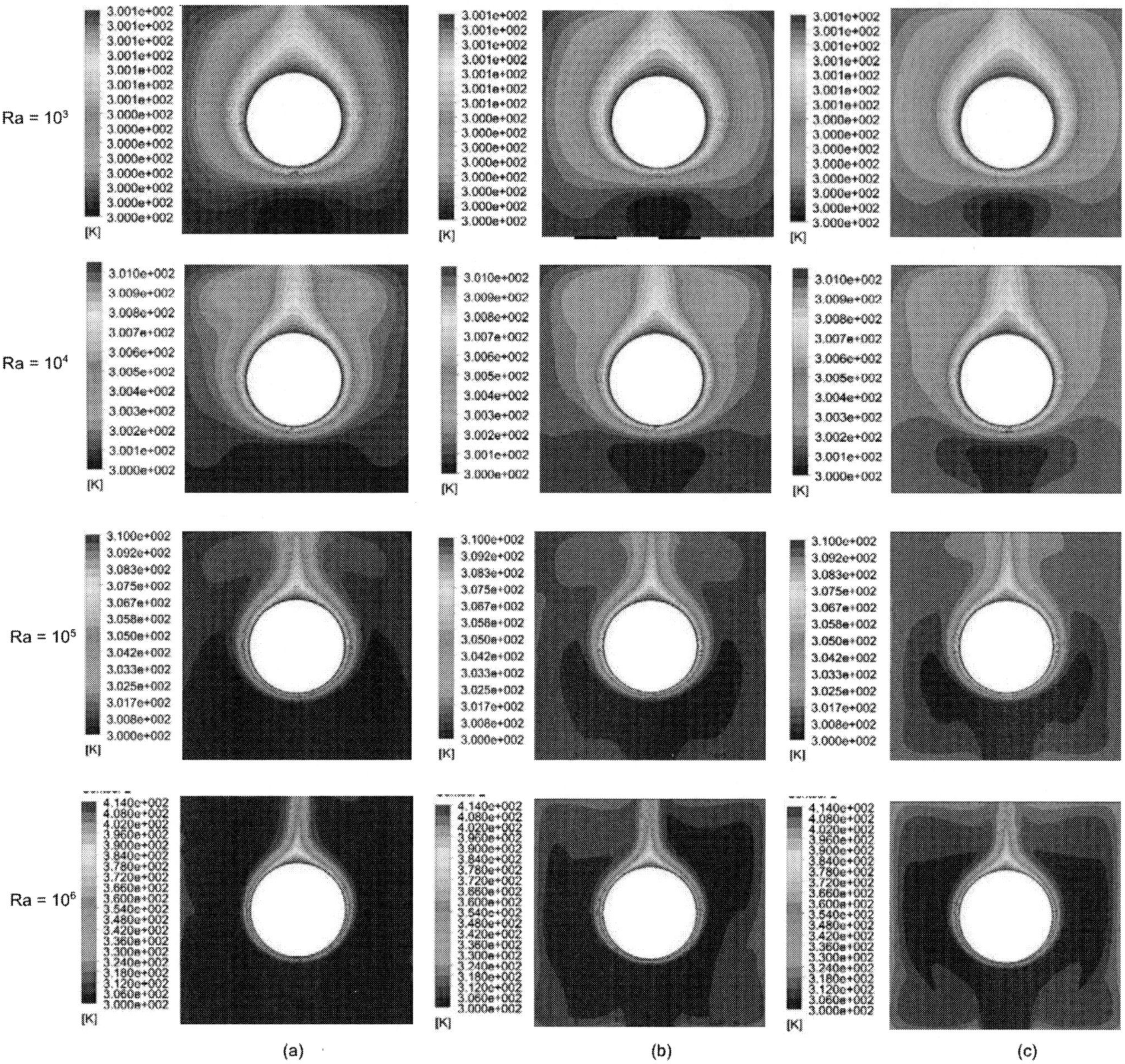

Figure 10: Temperature distribution inside the enclosure at different Rayleigh numbers and different cylinder surface emissivity: (a) ε =0.2, (b) ε =0.6, (c) ε =1.

4. Conclusions

The effect of surface radiation on the natural convection heat transfer from the heated cylinder in a symmetrical vented square cylinder has been investigated numerically. Experimental results from previous experimental work have been used to compare and asses the numerical results. The conclusion of this work can be summarized as follows,

1. The Nu of the natural convection is increased with Ra.
2. For low Ra, the dominant mode is the conduction heat transfer without and with surface radiation effect.
3. The strength of the natural convection heat transfer is increased with the increase of the Ra, and the convection heat transfer is the dominant mode for large Ra without and with surface radiation effect.
4. There is a linear relationship between the cylinder emissivity values and the radiation heat transfer, consequently; the total heat transfer is increased.
5. The Nu can be enhanced from 70 % to 350 % under the radiation effect at low rage of the Ra.
6. The heat transfer augmentation is decreased for medium and large Ra with the influence of the radiation heat transfer that reduced from 25% to 125% for total Nu as compared with those for natural convection heat transfer.
7. The streamlines and isotherms behavior displayed the conduction mode at low Ra.
8. The flow and temperature distribution can be used to identify the convection heat transfer mode when the Ra increases.
9. The radiation heat transfer has slight effect on the streamlines and isotherms behavior and arrangement above the cylinder surface and large effect beneath the cylinder surface at low Ra.
10. The is a significant effect on the thermal behavior when the radiation effect increased at high range of Ra, which the temperature layers are changed away from the cylinder surface and y-axis.

References

[1] Ali, Omar. 2008. "Experimental and Numerical Investigation of Natural Convection Heat Transfer from Different Cross Section Cylinders in a Vented Enclosure." Ph.D. Thesis, Mosul University.
[2] Ali, Omar. 2014. "Numerical Investigation of Prandtl Number Effects on The Natural Convection Heat Transfer from Circular Cylinder in an Enclosed Enclosure." Science Journal of University of Zakho 2(2): 358–74.
[3] Ali, Omar, and Ghalib Y. Kahwaji. 2014. "Numerical Investigation of Natural Convection Heat Transfer from Circular Cylinder Inside an Enclosure Using Different Types of Nanofluids." International Journal of Mechanical Engineering and Technology (IJMET) 5(5): 214–36. http://www.iaeme.com/MasterAdmin/Journal_uploads/IJMET/VOLUME_5_ISSUE_5/IJMET_05_0 5_022.pdf.
[4] Bello-Ochende, T. 2019. "Suppression of Natural Convection and Radiation Heat Losses in Solar Cavity Receivers: A Novel Approach." Journal of Physics: Conference Series 1378(4).
[5] Gengel, Younus A., and Afshin J. Ghajar. 2015. Heat and Mass Transfer Fundamentals and Applications. fifth. McGraw-Hill Series in Mechanical Engineering.

[6] Kahwaji, Ghalib Y., Abbas S. Hussien, and Omar Ali. 2013. "Experimental Investigation of Natural Convection Heat Transfer from Square Cross Section Cylinder in A Vented Enclosure." Journal of University of DUHOK Pure and Engineering Science 16(1).
[7] Kahwaji, Ghalib Y., Abbas S. Hussien, and Omar M. Ali. 2012. "Numerical Investigation of Natural Convection Heat Transfer from Square Cylinder in a Vented Enclosure." Applied Mechanics and Materials 110–116: 4451–64.
[8] El Moutaouakil, L., M. Boukendil, Z. Zrikem, and A. Abdelbaki. 2020. "Natural Convection and Surface Radiation Heat Transfer in a Square Cavity with an Inner Wavy Body." International Journal of Thermophysics 41. https://link.springer.com/article/10.1007/s10765-020-02688-7#citeas.
[9] Nada, S. A., H. El-Batsh, and M. Moawed. 2007. "Heat Transfer and Fluid Flow around Semi-Circular Tube in Cross Flow at Different Orientations." Heat and Mass Transfer/Waerme- und Stoffuebertragung 43(11): 1157–69.
[10] Nadjib, Hamici, Sahi Adel, Sadaoui Djamel, and Abderrahmane. 2018. "Numerical Investigation of Combined Surface Radiation and Free Convection in a Square Enclosure with an Inside Finned Heater." Fluid Dynamics and Manufacturing Process 12(3): 175–95. https://www.researchgate.net/publication/328598013_Numerical_investigation_of_combined_surface_radiation_and_free_convection_in_a_square_enclosure_with_an_inside_finned_heater.
[11] Ngo, Lloyd C., Tunde Bello-Ochende, and Josua P. Meyer. 2014. "Numerical Modelling of Combined Natural Convection and Surface Radiation Heat Transfer in Cavity Receiver with Plate Fins." In Proceedings of the 15th International Heat Transfer Conference, IHTC-15 August 10-15, 2014, Kyoto, Japan.
[12] Pandey, Alok Kumar, and Manoj Kumar. 2017. "Natural Convection and Thermal Radiation Influence on Nanofluid Flow over a Stretching Cylinder in a Porous Medium with Viscous Dissipation." Alexandria Engineering Journal 56(1): 55–62. http://dx.doi.org/10.1016/j.aej.2016.08.035.
[13] Saleh, Habibis, Ammar I. Alsabery, and Ishak Hashim. 2015. "Natural Convection in Polygonal Enclosures with Inner Circular Cylinder." Advances in Mechanical Engineering 7(12): 1–10.
[14] Sparrow, E. M., and D. M. Pfiel. 1984. "Enhancement of Natural Convection Heat Transfer from a Horizontal Cylinder Due to Vertical Shrouding Surfaces." ASME J. of Heat Transfer 106(1): 124–30. https://asmedigitalcollection.asme.org/heattransfer/article-abstract/106/1/124/414385/Enhancement-of-Natural-Convection-Heat-Transfer?redirectedFrom=fulltext.
[15] Wang, Zhiyun, Mo Yang, Ling Li, and Yuwen Zhang. 2011. "Combined Heat Transfer by Natural Convection - Conduction and Surface Radiation in an Open Cavity under Constant Heat Flux Heating." Numerical Heat Transfer; Part A: Applications 60(4): 289–304.
[16] Wei, J. G., and W. Q. Tao. 1996. "A Numerical Study of Natural Convection in a Vertical Annulus with Constant Heat Flux on the Inner Wall." International Journal of Numerical Methods for Heat and Fluid Flow 6(7): 31–46.

IOP Conf. Series: Materials Science and Engineering **978** (2020) 012031 doi:10.1088/1757-899X/978/1/012031

Preparation and Characterization of Nano silica Prepared by Different Precipitation Methods

Sabah M. Thahab Al-Abboodi[1,*], Eman Jabber Abed Al-Shaibani [2] and Enass. A. Alrubai [3]

[1]Nanotechnology and Advanced Materials Research Unit (NAMRU), Faculty of Engineering, University of Kufa, Najaf, Iraq
[1]Electronics and Communications Departement, Faculty of Engineering, University of Kufa, Najaf, Iraq: Email: sabah.alabboodi@uokufa.edu.iq
[2]Materials Engineering, Faculty of Engineering, University of Kufa, Najaf, Iraq
Email: iman.alshaibani@uokufa.edu.iq
[3]Dept. of Materials Engineering, University of Kufa, Najaf, Iraq
Email: eng.enaas1993@gmail.com
*Corresponding Author: Tel: +9647703996492

Abstract. Agricultural waste Rice husk ash (RHA) is large and massive by the output of rice milling manufacture and contains a high amount of silica. Taking into account this verdant technology is utilized to get SiO_2 Nanoparticles from rice shell ash. In this research, studies on the extraction methods of Nano-silica by combustion rice shell for 2 h at 700 °C were carried out and investigated. However, the Rice shell ash was undergoing to precipitation methods and sol gel method to manufacture Nano silica. precipitation methods involve refluxing silica from Rice shell ash in boiled (1.0,2.0 N NaOH). While the sol-gel method involves calcination of RHA was treated with HCl at 700 °C for 2 h followed by chemical treatment. Results of The Infrared spectral (IR) information display the presence of hydrogen-bonded silanol group and the siloxane groups in silica. The Optical properties through UV-visible spectroscopy of the Nano silica samples show a wide bandgap value and absorption peaks located in the UV region. As well as, the particles size analysis with values of (13.02, 16.83 and 23.96 nm) for the samples ($1NSiO_2$, $2NSiO_2$, and acid leaching) were obtained respectively.

***Keyword**: Rice husk waste; Nano-Silica; NaOH; FTIR; UV. Visible*

1. Introduction

Nanotechnology attracted much attention of scientists and producers due to the unique physicochemical properties of nanoparticles as compared to their bulk form [Patil et al., 2018]. Nano-sized materials have tremendous potential and used in kinds of industries, like electronics, optical, chemicals, environment, and so on [Nguyen et al., 2012]. Between them, nanosilica is an excessively utilized nanomaterial with applications in adhesives, food additive, sealants, paints, cosmetics, inks, fiber optic strands, polymer, coatings, and in cement-based building materials [Premaratne et al., 2013].

Silica nanoparticles (SNPs) have gained greater attention because of its highly reactive surface area to volume ratio, physical and chemical steadiness, low toxicity, and straight forward surface chemistry [Ghorbani et al.,2015]. there are two forms of Silicate materials amorphous and crystalline. Amorphous silica has large industrial applications. It is utilized in plastics, rubber, electronics, cosmetics, optics, coating, refractories, and also use as an abrasive agent. The crystalline form of Silica is widely utilized in the glass and ceramic industry [Singh et al., 2014]. The properties of Nano-Silica are shown in Table.1.

Table 1: Properties of Nano-Silica [Rao et al., 2016]

S. No	Characterization	Actual Analysis
1	Active Nano Content (%Wt. /Wt.)	30-32
2	pH(20C)	9-10
3	Specific Gravity	1.20-1.22
4	Texture	White Milky Liquid
5	Dispersion	Water

Tetraethoxysilane (TEOS) and tetramethoxysilane (TMOS) have been usually used as the silica source to produce Nano silica [Shen et al.,2014]. However, these chemical sources are rather expensive and toxic. RHA between the family of agricultural-wastes is the excessive silica-rich, containing silica 90% to 98% and some amount of metallic impurities. [Janasree et al.,2018] The Rice husk is an agro-waste residue that was chosen due to its local very low cost of availability, chemical stability, and its granular structure [Zhang et al.,2010]. The maximum length of rice husk is about 10mm and the width is 1 to 4mm. Due to high calorific value of Rice husk, its utilized to generate electricity [Vinukumar et al.,2016].

In the world, 500 million tons of paddies produce 20%of the Rice husk. Rice husks contain 25% volatile organic matter, with 75% of them turning ash during combustion, called Rice Husk Ash (RHA) [Kim et al., 2014]. Rice husks have been used in many applications, including as a source of energy, where they burn and lead to waste generation. Amorphous silica can be extracted from rice husks at low temperatures [Patel et al., 2017] . White-colored ash is produced If the rice husk is completely combusted. Approximately, every 100 g of rice husk gives 20 g of RHA [Vasamsetti et al.,2018]. After combustion, the organic constituent is decomposed and RHA is gained. [Janasree et al., 2018]. These ashes contain 72.1% SiO_2; 0.3% Al_2O_3; 0.5% Na_2O; 0.43% CaO; 0.72% K_2O; 0.05% TiO_2; 0.15% MnO, 24.3%; 0.06% P2O5 and ~0.7% MgO weight loss on ignition [Dang et al 2017]. to obtain high purity silica (above 99%) Eliminating these metallic impurities is required by hydrochloric acid leaching of rice husk and then controlled combustion [Bakar et al., 2016]. Rice husk ash is further suitable than rice husk in most applications [Ananthi et al.,2016].

Structural transformation of silica present in the ash occurs depending on the conditions (time and temperature of combustion, metals content, and the chemical composition after combustion). Amorphous ash is formed at 400-1200°C and crystalline ash is formed at a temperature above this [Ananthi et al., 2016]. This product is not suitable for human consumption. The disposal of the ashes and partially boring husk creates environmental hazards. Silica nano production is an inexpensive and important application of rice husk [Kaviyarasu et al., 2016]. There are sundry methods of manufacture Nano silica such as the precipitation method [Premaratne et al., 2013], Sol-Gel method [Rahman et al., 2012], and in situ emulsion polymerization [Moosa et al., 2017]. The main methods of producing silica are [Patel et al.,2017]:

1. Thermal treatment

2. Chemical treatment.

In the present, works involves the extraction of Nano-silica from rice husk by two methods. The first methods involve after thermal treatment, alkaline extraction followed by acid neutralization. while The second method involves the pretreatment of Rice Husk with acid and then thermal treatment. It may be concluded that Nano-silica extracted from rice husk ash can be used for different novel applications.

2. Materials and Methods

Jasmine rice husk was provided by a local rice milling company in Al-Najaf, Iraq. NaOH, Hydrochloric acid (HCL) (35 %, Merck, India) was utilized to eject the minerals impurities from the rice husk. Ethanol and distilled water were utilized wherever needed.

2.1 Preparation of rice husk ash

Firstly, the RH was sieved to remove dust and strange material. Consequently, RH was washed by distilled water several times until its reaching neutral PH value. An electric oven was used to dry the washed rice husk for 7 h at 120°C and then grinding the dried Rice husk by grinder. Followed, the nano-silica was produced by two methods as below:

2.2 Precipitation method

2.2.1 Preparation of sodium silicate solution

Silica was extracted using a method adapting from previous studies with some modification [Patel et al.,2017], [Janasree et al.,2018]. As shown in figure 1 about 50 g of the washed and dried rice husk (RH) was burned in muffles furnace at 700 ^{0}C for 2h. The goal of this step is to reduce the material to increase the silica content in the samples, as well as to burn out other undesirable components, and then the solid RHA was blended with 1N Na OH solution in 500 ml Erlenmeyer flask and is allowed to seethe for 2h at 100 ^{0}C with constant stirring to dissolve silica. The final mixture was allowed to cool down to room temperature and filtered to separate the filtrate which contains sodium silicate, while the residue was discharged.

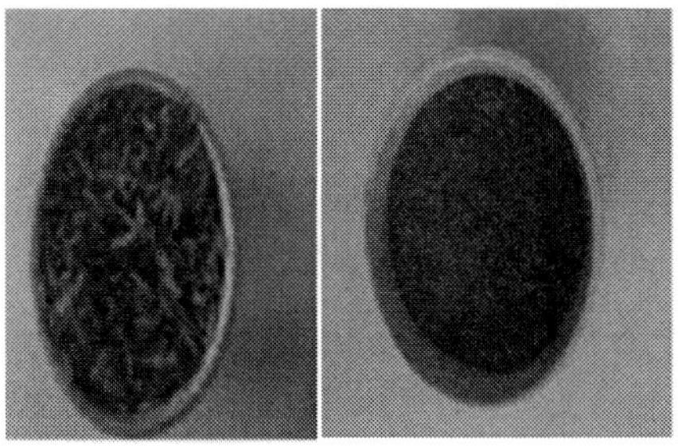

(a) RH

(b)RHA

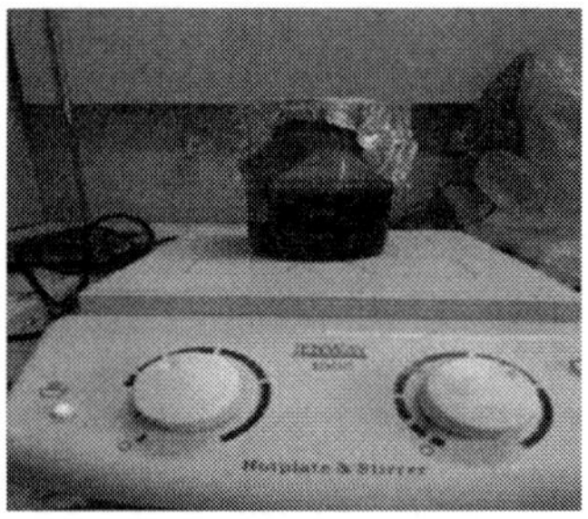

(c) RHA added to NaOH solution stirring for 2h at 100 °C

(d)Filtering

(e) Sodium silicate solution

Figure 1. Preparation of sodium silicate solution from rice husk

2.2.2 Precipitation of Nano silica

Figure 2 shows the synthesis of Nano silica by precipitation method.Concentrated Hydrochloric acid with regular addition (dropwise) was utilized to reduce the pH of the filter sodium silicate solution while being stirred magnetically. After the stirring period gained pure silica in gel form. An electric furnace at 200°C for 8 h was used to dry the gel. The solids were crushed to powder using the mill, to get the Nano silica particles.

In which the process of synthesis Nano silica by precipitation method display in figure 3 and by the following Eqs. (1 and 2) respectively.

$$SiO_2 + 2NaOH \longrightarrow Na_2SiO_3 + H_2O \qquad (1)$$

$$Na_2SiO_3 + HCl \longrightarrow SiO_2 + NaCl + H_2O \qquad (2)$$

In this work, RHA was pretreatment by two concentrations of Na OH (1, 2 N Na OH). The silica nanoparticles were left in a vacuum desiccator for more characterization.

IOP Conf. Series: Materials Science and Engineering **978** (2020) 012031 doi:10.1088/1757-899X/978/1/012031

(a)Silicate precipitate

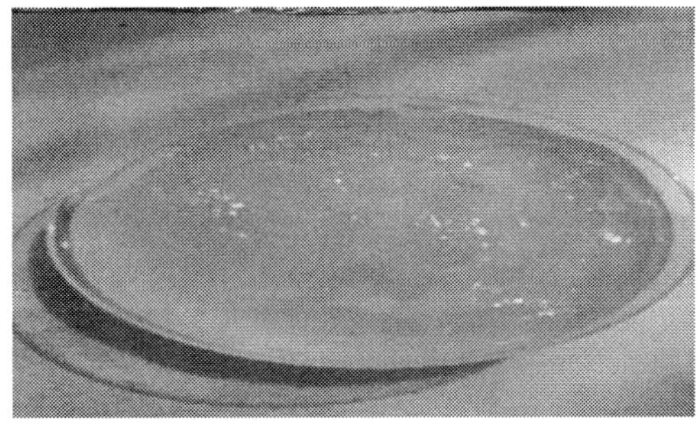

(b) Silica in gel form

(c) Drying at 200 ^{0}C for 8h

(d)Nano silica powder

Figure 2. Synthesis of Nano Silica by Precipitation Method

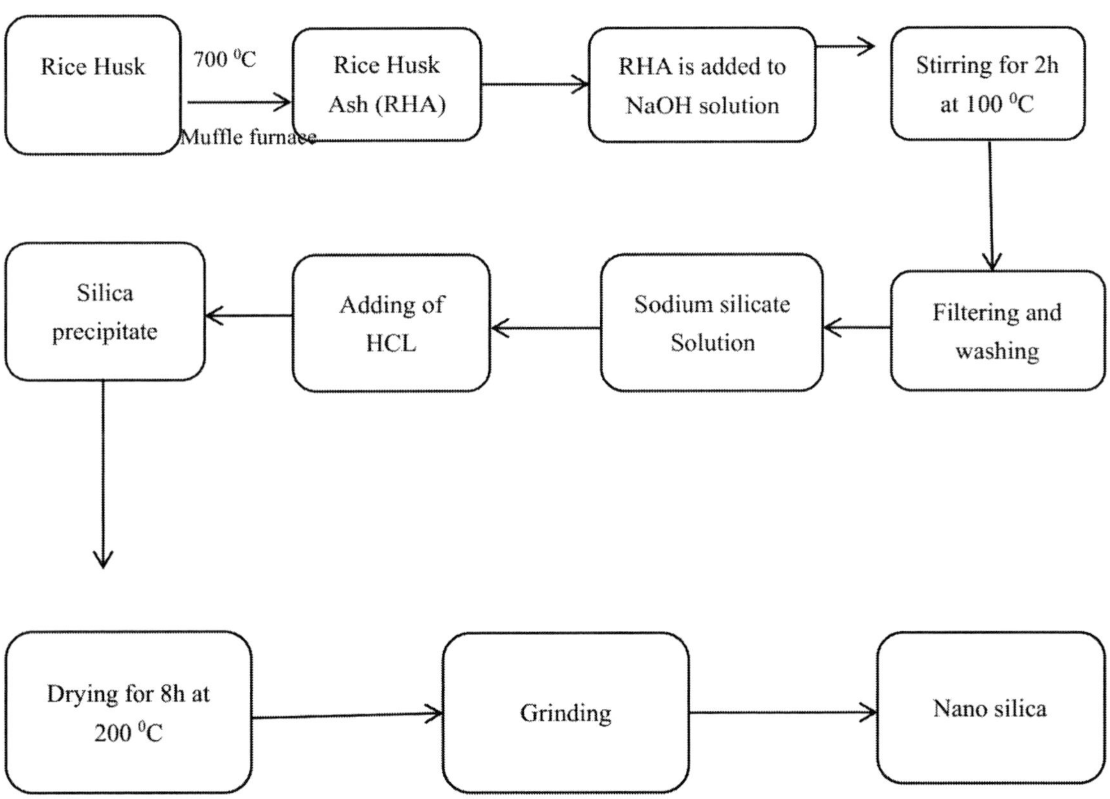

Figure 3. Flowchart for preparation of Nano silica by precipitation method [Janasree et al.,2018]

IOP Conf. Series: Materials Science and Engineering **978** (2020) 012031 doi:10.1088/1757-899X/978/1/012031

2.3 Sol gel method

Silica was extracted using a method adapting from previous studies with some modification [Dung et al., 2017], [Rafiee et al., 2012]. As shown in Fig. 4 a 5.0 g of RH was mixed with 1000 mL of 1N HCl for 15 minutes stirring time. and then left in the water bath for 2h at 100 ^{0}C to eject mineral ions in RHA. Then, acident-treated RH was purified, washed with filtered water, and then dehydrated at 110°C in the electric furnace. The acident-treated RH was burned in a muffle furnace at 700 ^{o}C for 2 h. The gained powder was specified as RHA. in which Fig. 4. show the synthesis process of Nano silica. RHA sample (41.835 g) was mixed with 1000 ml NaOH (2.5 N). The sodium silicate solution was heated in a closed beaker for 2h at 100 with constant whickering. It was then filtered and washed with filtered water. The gained transparent, viscous, and colorless solution was let to cold down to room temperature. Nano silica was synthesis by refluxing of taken away silica with 1NHCl fixed movement at controlled conditions until pH = 7. After the stirring period gained pure silica in gel form. an electric furnace at 200 ^{0}C for 8 h was used to dried The gel. The solids were crushed to powder using the mill, to get the Nano silica particles.

(a) Acid leaching RH

(b) Filtering add washing

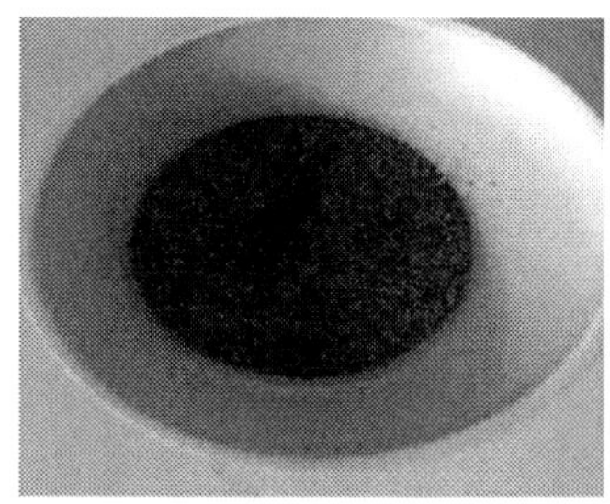

(c) RHA

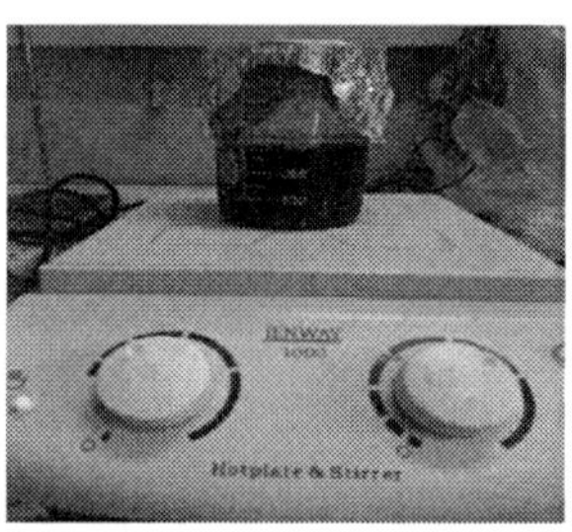

(d) RHA added to 2.5N NaOH
Stirring at 100 ^{0}C for 2h

(e) Filtering and washing

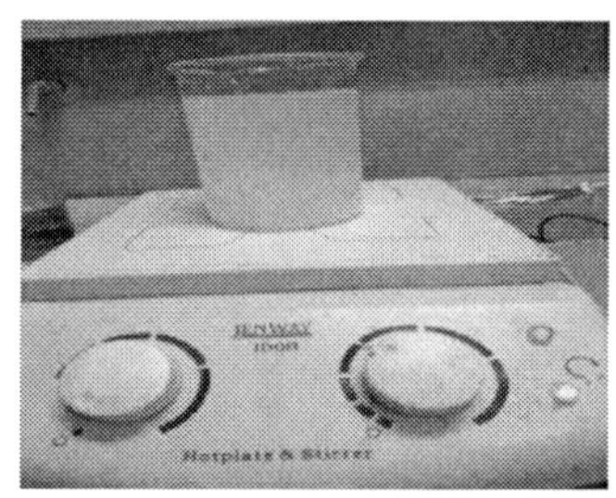

(f) Sodium silicate precipitate

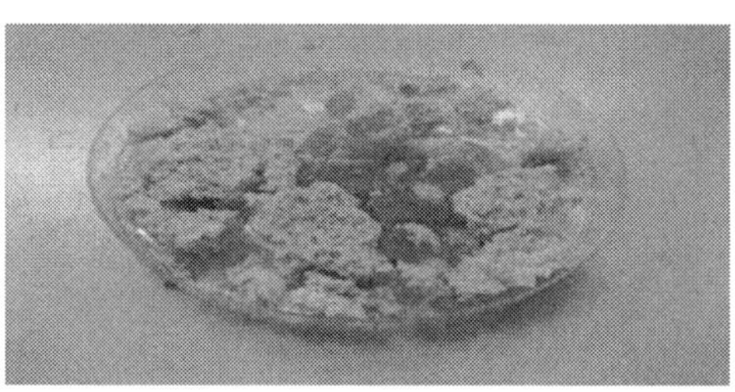

(g) Drying at 200 ^{0}C for 8h

(h) Nano silica powder

Figure 4: Schematic illustration of the synthesis mechanism for Nano silica powder from rice husk ash

3. Characterization Instruments

The prepared silica powder was inspected by FTIR, UV-Visible spectrometer, and particle size analysis. The functional groups of the SiO2 nanoparticles were examined by the FT-IR mechanism.in which Nano silica blended with Spectral-grade KBr powder at a weight ratio of 2 mg SiO2: 200 mg KBr in a deep mortar. and then pressed The blended powder into particles with a thickness and a diameter of (0.5 mm 13 mm), respectively. The infrared (IR) spectrum of Nano silica was measured by using FTIR spectroscopy (FTIR-SHIMADSO, Japan made) through the wavenumber domain from 4000 to 400 cm -1. The optical property was characterized using (T90 Double Beam Spectrophotometer by PG Instruments Ltd) UV-Visible spectrometer for optical absorption at wavelength 200-800 nm. After the powder milled by the mill, the particle size distribution and particle size were examined by using laser particle size analysis (Bettersize 2000 laser particle size analyzer) at the Department of Ceramics and Building Materials, Faculty of Materials Engineering, University of Babylon, IRAQ.

4. Results and Discussion

The FTIR spectrum in transmittance mode for (1N SiO_2) in Figure (5 a) shows peaks at 3456.44, 1641.42, 1425.40, 1103.28, 804.32, and 472.56 cm^{-1} respectively. The broadband between 1641.42 and 3456.44 cm^{-1} was because of OH groups and adsorbed water. The preponderant absorbance peak at 1425.40 cm^{-1} was because of bonds (Si ± O ± Si). The peaks between 1103.28 and 472.56 cm^{-1} are because of the shaking style of the gel [Patil et al., 2014].

FTIR spectroscopy was utilized to discover the presence of binding groups in the (2N SiO_2) SNPs (Fig. 6b). The shaking indication concerning 1101.35, 800.46 and 470.63 cm^{-1} are typical of Si–O–Si bands refer to the asymmetric extension, symmetric extension, and bending, respectively [Ghorbani et al.,2013]. These three crests are the essential index of the SiO_2, which explain the effective manufacture of SNPs. The absorption band for H–O–H bending shaking in water is at about 1635.64 cm^{-1} [An et al.,2010]. The shade shows broadband about 3450.65cm-1, which is because of adsorbed H_2O molecules [Rafiee etal., 2012]. There is no peak at about 1825_2800, which means after burning, there are no organic substances. The inorganic metallic ions impurities in RHA were separated into the liquid by HCl leaching RHA suspension. After acid treatment, the organic constituents and silica remained as solid ingredients. The chemical reaction along with the high-temperature calcination for the preparation of Nano silica powder can be described as follows Eqs. (3 and 4) respectively.

$$HCL \xrightarrow[RHA +]{100\ ^0C\ \ (2h),\ soaked\ in\ water\ bath} SiO_2(Ash) + MCL + H_2O \qquad (3)$$

Where M: Na; K; Ca; Fe; Al, Mg…

$$SiO2\ (Ash) \xrightarrow{700\ ^0C\ (2\ h)} SiO2\ (Nano\ silica\ powder) \qquad (4)$$

During the high-temperature calcination process, all of the organic constituents were burned out. Finally, white powder Nano silica was obtained (Fig. 5c). FTIR for acid leaching sample the peak at 1101.35 and 800.46 cm^{-1} because of Si–O–Si asymmetric and symmetric extension style, respectively. The band fixed at 466.77 cm^{-1} is because of the bending frequency of Si–O–Si. Great broadband about 3452.58 cm^{-1} refers to the presence of the O – H, extension hesitation to the

Solano series, and the residual adsorbed water. A squad about 1635.64 cm^{-1} is specific to the bending shaking of H2O molecules limited to the silica form. There is no crest at about 2900–3100 cm^{-1}, whose means after burning, there are no organic substances [Rafiee et al.,2012].

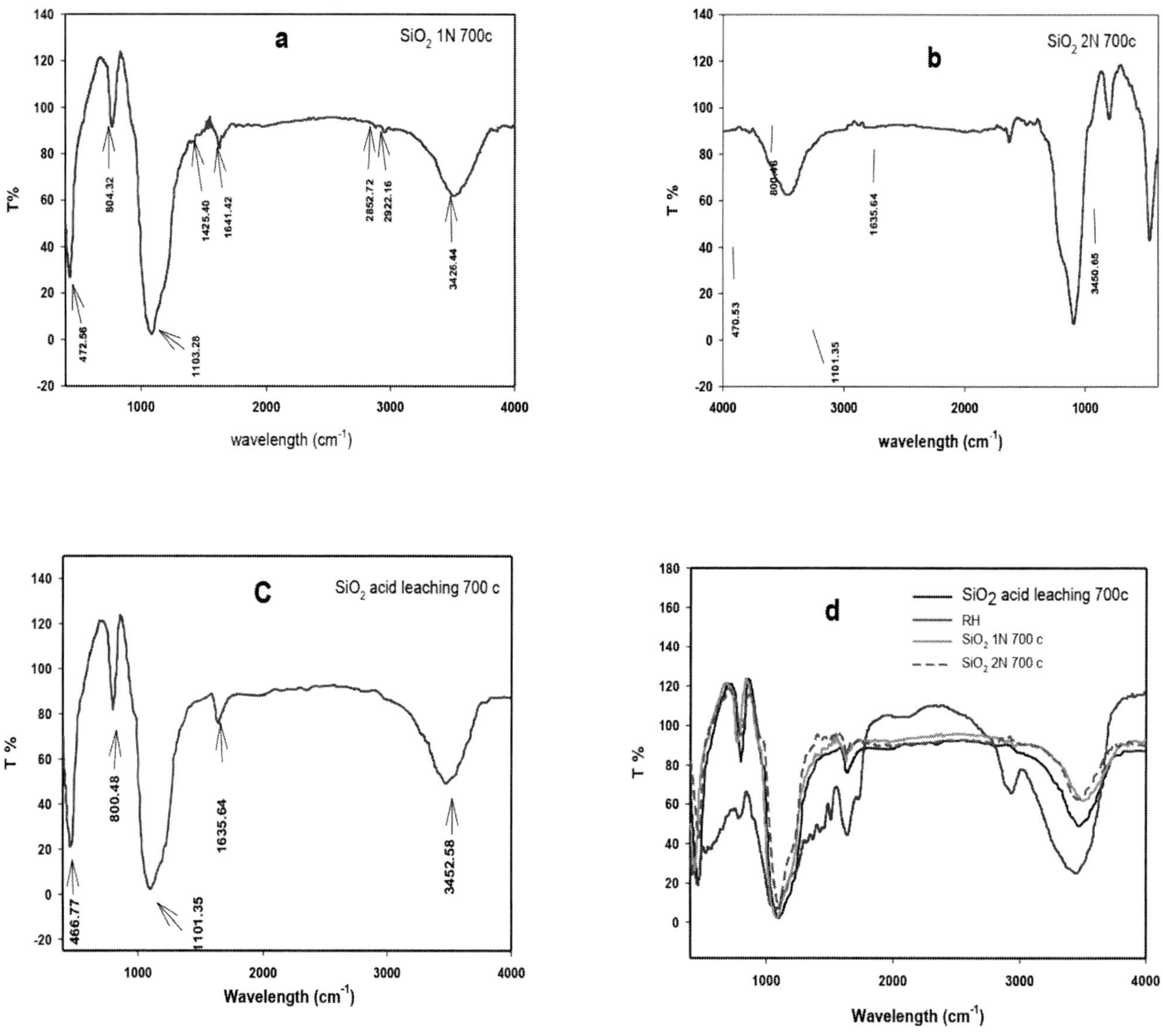

Figure 5. FTIR spectra of (a) 1N SiO2 (b) 2N SiO2 (c) SiO2 acid leaching (d) All samples within.

visual properties likewise FT-IR and Raman spectra, UV-Visible are generally susceptible to any nancrystal surface. Figure 6 shows the UV-Vis spectrum of amorphous silica nanoparticles with Different methods. The strong absorption edge is observed at 200 nm wavelength within all samples. However, another peak in 204 nm presented in $1NSiO_2$ and $2NSiO_2$ samples were observed [Engku Ali et al.,2017]. The absorption spectra of the prepared samples show a stronger absorption in the UV ray, mention that these samples could be a favorable process for increasing the catalytic activity. As well as, the presence of some other crest which may be because of pollution or because of bi-products this formed during synthesis. The results of UV-visible spectroscopy display absorbance of synthesized materials in the UV region and shown very low absorbance is in the visible region.

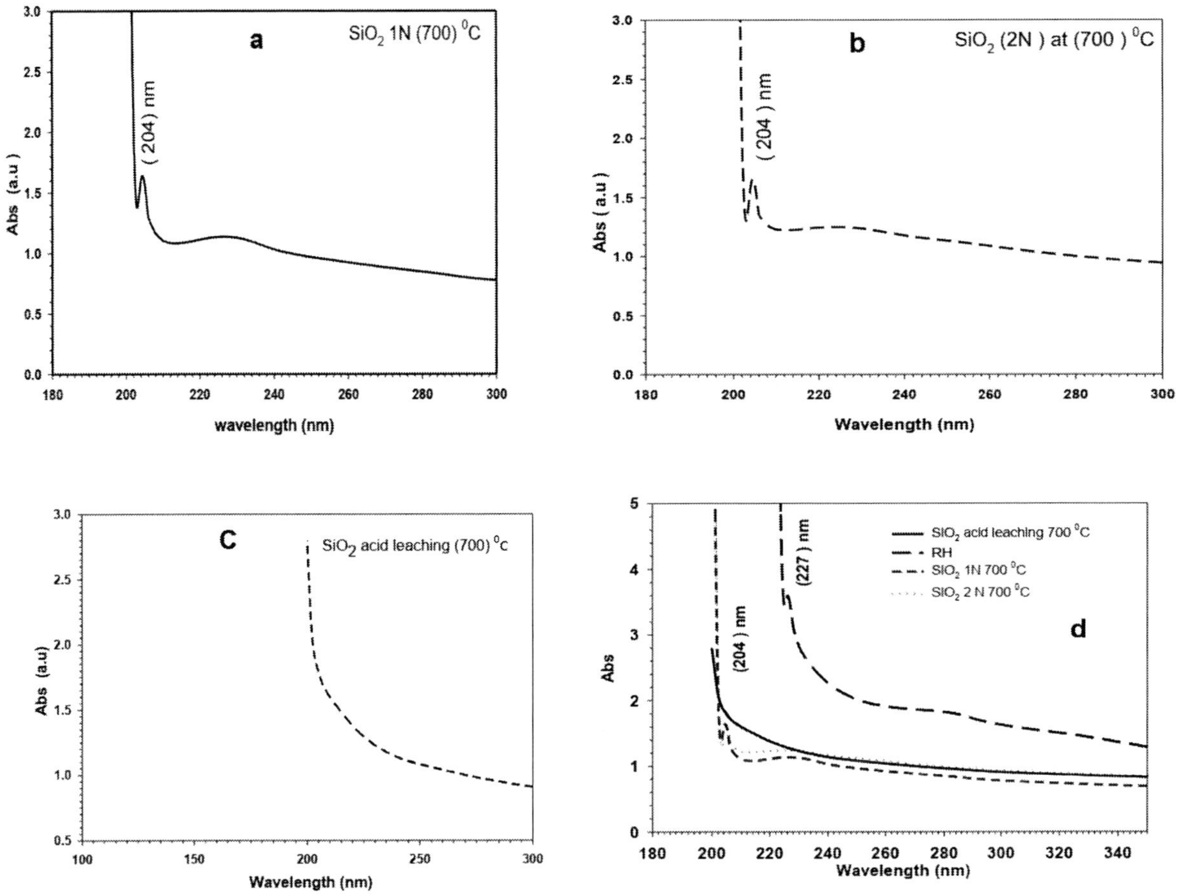

Figurer 6. UV-Visible absorbance spectra for (a) 1N SiO2 (b) 2NSiO2 (c) SiO2 acid leaching

Figure 7. Reveals the particle size analysis of Nano silica with purity (>95%) [Moosa et al., 2017],prepared by different methods. As can be seen from the chart, the medium size of the particles (1N SiO_2) (D50) was 13.02 nm, the medium size of the particles (2N SiO_2) (D50) was 16.83 nm and (D50) was 23.96 nm for particles (acid leaching).

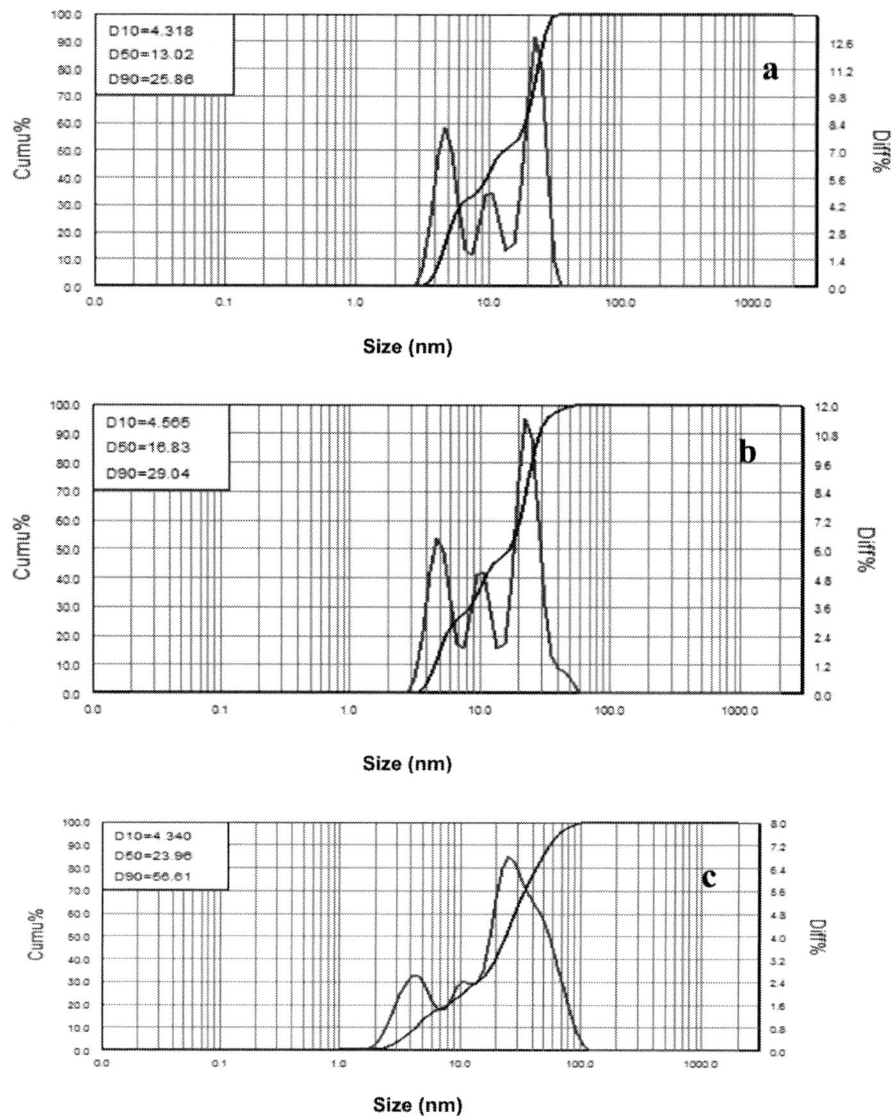

Figure 7. The particle size analysis of SiO_2: (a) 1N SiO_2 (b) 2NSiO_2 (3) SiO_2 acid leaching

5. Conclusion

In this research, an experiment work has been made to utilize cheap RHA to production Nano silica by two methods, where found precipitation method give very fine silica particles by using special solvents [Noushad et al., 2012]. And found precipitation method itis very easy and efficient methods for Nano silica synthesize and less time-time consuming, as compared with the sol-gel method [Janasree et al.,2018]. in the methods of extraction silica from RHA, in order to gain high purity and amorphous silica which was more favorable in various applications. The acid concentration must be not minimal than 1 M and The combustion temperature should not more than 700°C[Twej & Shihab, 2018]. FTIR spectra showed the formation of Si-O bond. The got Nano silica could be utilized in a variety of fields and helps to reduce the concerns of RHA disposal[Anh Tuan et al., 2017] .

Acknowledgments

The authors wish to admit the support given by Nanotechnology and Advanced Materials Research Unit (NAMRU) and Materials Engineering Department in Faculty of Engineering, University of Kufa, Najaf, IRAQ are highly appreciated.

References

[1] Patil, N. B., Sharanagouda, H., Doddagoudar, S. R., Ramachandra, C. T., & Ramappa, K. T. (2018). Effect of Rice Husk Silica Nanoparticles on Rice (Oryza sativa L.) Seed Quality. Int J Curr Microbiol App Sci, 7(12), 3232-3244

[2] Nguyen, T. C., Tran, N. H., & Hoang, V. D. (2012). Studying of Nano SiO 2 Preparation from Rice Husk Ash by Using High Gravity Reaction Precipitation Technology, Center for Material Technology, ITRRE, Vietnam Atomic Energy Institute 48 Lang Ha, Dong Da, Hanoi, Vietnam,319-324, (No. VINATOM-AR—12- 42).

[3] Premaratne, W. A. P. J., Priyadarshana, W. M. G. I., Gunawardena, S. H. P., & De Alwis, A. A. P. (2013), Synthesis of nanosilica from paddy husk ash and their surface functionalization, J. Sci. Univ. Kelaniya, 8, 33-48.

[4] Ghorbani, F., Sanati, A. M., & Maleki, M. (2015). Production of silica nanoparticles from rice husk as agricultural waste by environmental friendly technique. Environmental Studies of Persian Gulf, 2(1), 56-65.

[5] Singh, L. P., Bhattacharyya, S. K., Kumar, R., Mishra, G., Sharma, U., Singh, G., & Ahalawat, S. (2014). Sol-Gel processing of silica nanoparticles and their applications. Advances in colloid and interface science, 214, 17-37.

[6] Rao, D. P., & Navaneethamma, V. (2016). Influence of nano–silica on strength properties of concrete containing rice husk ash. International Journal of Advanced Research, 3(1), 39-43

[7] Shen, Y., Zhao, P., & Shao, Q. (2014). Porous silica and carbon derived materials from rice husk pyrolysis char, Microporous and Mesoporous Materials, 188, 46-76.

[8] Janasree, P., & Kumar, M. K. (2018). Synthesis and characterization of nanosilica from rice husk ash by precipitation method for chemically synthesized nanocement, Journal of Emerging Technologies and Innovative Research ,5(8), 314-319.

[9] Zhang, H., Zhao, X., Ding, X., Lei, H., Chen, X., An, D., & Wang, Z. (2010). A study on the consecutive preparation of d-xylose and pure superfine silica from rice husk. Bioresource technology, 101(4), 1263-1267

[10] Vinukumar, K., Azhagurajan, A., & Stephen, V. K. (2016). Synthesis of rice husk nano-particles for biodiesel production. J. Chem. Pharm. Sci., 9(4), 2671-2673.

[11] Kim, Y. Y., Lee, B. J., Saraswathy, V., & Kwon, S. J. (2014). Strength and durability performance of alkali-activated rice husk ash geopolymer mortar. The Scientific World Journal.

[12] Vasamsetti, S., Dumpala, L., & Subbarao, V. V. (2018). Synthesis, characterization and hardness studies of nano rice husk ash reinforced Al6061 nanocomposites. Journal of Engineering Science and Technology, 13(9), 2916-2929.

[13] Dang, N. T. T., Nguyen, T. T. A., Phan, T. D., Tran, H., Van Dang, P., & Nguyen, H. Q. (2017). Synthesis of silica nanoparticles from rice husk ash. Science and Technology Development Journal, 20(K7), 50-54.

[14] Bakar, R. A., Yahya, R., & Gan, S. N. (2016). Production of high purity amorphous silica from rice husk. Procedia Chemistry, 19, 189-195.

[15] Ananthi, A., Geetha, D., & Ramesh, P. S. (2016). Preparation and characterization of silica material from rice husk ash–an economically viable method. Chemistry and Materials Research,

8(6), 1-7.
[16] Kaviyarasu, K., Manikandan, E., Kennedy, J., Jayachandran, M., & Maaza, M. (2016), Rice husks as a sustainable source of high quality nanostructured silica for high performance Li-ion battery requital by sol-gel method–a review. Adv. Mater. Lett, 7(9), 684-696.
[17] Rahman, I. A., & Padavettan, V. (2012). Synthesis of silica nanoparticles by sol-gel: size-dependent properties, surface modification, and applications in silica-polymer nanocomposites—a review. Journal of Nanomaterial
[18] Premaratne, W. A. P. J., Priyadarshana, W. M. G. I., Gunawardena, S. H. P., & De Alwis, A. A. P. (2013). Synthesis of nanosilica from paddy husk ash and their surface functionalization, J. Sci. Univ. Kelaniya, 8, 33-48,
[19] Moosa, A. A., & Saddam, B. F. (2017). Synthesis and characterization of nanosilica from rice husk with applications to polymer composites. Am. J. Mater. Sci, 7(6), 223-231.
[20] Patel, K. G., Shettigar, R. R., & Misra, N. M. (2017). Recent Advance in Silica Production Technologies from Agricultural Waste Stream–. Journal of Advanced Agricultural Technologies Vol, 4(3).
[21] Dung, L. T. K., Hien, N. Q., Van Phu, D., & Du, B. D. (2017). Preparation and characterization of nanosilica from rice husk ash by chemical treatment combined with calcination. Vietnam Journal of Chemistry, 55(4), 455-459
[22] Rafiee, E., & Shahebrahimi, S. (2012). Nano silica with high surface area from rice husk as a support for 12-tungstophosphoric acid: an efficient nano catalyst in some organic reactions. Chinese Journal of Catalysis, 33(7-8), 1326-1333.
[23] Patil, R., Dongre, R., & Meshram, J. (2014). Preparation of silica powder from rice husk. J. Appl. Chem, 27, 26-29.
[24] Ghorbani, F., Younesi, H., Mehraban, Z., Çelik, M. S., Ghoreyshi, A. A., & Anbia, M. (2013). Preparation and characterization of highly pure silica from sedge as agricultural waste and its utilization in the synthesis of mesoporous silica MCM-41. Journal of the Taiwan Institute of Chemical Engineers, 44(5), 821-828.
[25] An, D., Guo, Y., Zhu, Y., & Wang, Z. (2010). A green route to preparation of silica powders with rice husk ash and waste gas. Chemical Engineering Journal, 162(2), 509-514.
[26] Engku Ali, E. A. G., Matori, K. A., Saion, E., Sidek, H. A. A., Zaid, M. H. M., & Alibe, I. M. (2017). Effect of reaction time on structural and optical properties of porous SiO 2 nanoparticles. Dig J Nanomater Biostruct, 12, 441-447.
[27] Anh Tuan, L. N., Kim Dung, L. T., Thanh Ha, L. D., Hien, N. Q., Phu, D. Van, & Du, B. D. (2017). Preparation and characterization of nanosilica from rice husk ash by chemical treatment combined with calcination. Vietnam Journal of Chemistry, 55(4), 455–459. https://doi.org/10.15625/2525-2321.2017-00490
[28] Moosa, A. A., & Saddam, F. (2017). Synthesis and Characterization of Nanosilica from Rice Husk with Applications to Polymer Composites. American Journal of Materials Science, 7(6), 223–231.
[29] Noushad, M., Rahman, I. A., Husein, A., Mohamad, D., & Ismail, A. R. (2012). A Simple Method of Obtaining Spherical Nanosilica from Rice Husk. International Journal on Advanced Science, Engineering and Information Technology, 2(2), 28–30.
[30] Patel, K. G., Shettigar, R. R., & Misra, N. M. (2017). Recent Advance in Silica Production Technologies from Agricultural Waste Stream– Review. Journal of Advanced Agricultural Technologies, 4(3), 274–279. https://doi.org/10.18178/joaat.4.3.274-279
[31] Rafiee, E., & Shahebrahimi, S. (2012). Nano silica with high surface area from rice husk as a support for 12-tungstophosphoric acid: An efficient nano catalyst in some organic reactions. Chinese Journal of Catalysis, 33(8), 1326–1333.
[32] Twej, W. A. A., & Shihab, B. F. (2018). Enhancement in extraction silica percentage from Iraqi Jasmine Rice Husk through controlling leaching process. International Journal of AdvancedResearch in Science, Engineering and Technology, 5(7), 6406–6410.

Reinforced Concrete Slabs Containing Recycled Concrete as Coarse Aggregate

Mazin Burhan Abdulrahman[1, *], Ammar Saleem Khazaal[2], Alyaa Abbas Al-Attar[3] and Saif Dawood Hamad[4]

[1]College of Engineering, Tikrit University, Iraq- E-mail: dr.mazinburhan@tu.edu.iq
[2]College of Engineering, Tikrit University, Iraq - E-mail: dr.ammarkhazaal@tu.edu.iq
[3]Northen Technical University, Iraq- E-mail: alattaralyaa@gmail.com
[4]College of Engineering, Tikrit University, Iraq- E-mail: saifcivileng@yahoo.com
*Corresponding author: College of Engineering, University of Tikrit, Iraq. Tel.+9647702009977.

Abstract. This research presents an experimental study on the performance of reinforced recycled aggregate concrete (RAC) that contains one of two types of recycled coarse aggregate, one from normal concrete and another from self – compacting concrete (SCC). The replacement percentages (by weight) of natural coarse aggregate with the recycled coarse aggregate are (0, 33.3, 66.7 & 100) % by weight. Moreover, Silica Fume was used as an admixture for all (RAC) mixtures. Concrete compressive strength, split tensile strength, and flexural strengths were studied for all of the reference and (RAC) mixtures at (28) days age, by testing concrete cubes, cylinders and reinforced concrete slabs specimens respectively. The results showed that the using of recycled waste concrete as concrete as aggregate will largely decreasing the slabs flexure strength while using of recycled (SCC) mixtures have relatively small effect in reducing the slabs flexure strength. The adding of silica fume to the concrete mix will increasing the slabs flexural strength and that increase is larger for recycled (SCC) aggregate mixtures than the recycled concrete aggregate mixtures. The results also indicate that the compressive strength and tensile strength are decreased by different ratio when using of recycled waste concrete as aggregate, while using of recycled (SCC) aggregate will also reduce the tensile strength and the modulus of rupture while the compressive strength increased when using this type of recycled aggregate. On the other hand, the mixtures with the Silica Fume showed a significant enhancement in the mechanical properties compared to those without the Silica Fume.

***Keywords**: Recycled Aggregate Concrete (RAC); Recycled Concrete Aggregate (RCA); Slabs; Strength; Flexure.*

1. Introduction

Since the coarse aggregate forms the largest portion in concrete, the strength and durability of the concrete is directly affected by the strength and durability of its coarse aggregate. As a practical fact, using recycled concrete aggregate (RCA) to produce a new concrete involves sequent stages starting with breaking, removing and then crushing the original concrete into a new material with specified properties such cleanness, size and quality.

Rao et al. (2011) defined recycled concrete aggregate (RCA) as the aggregate which is produced by the crushing of sound and clean demolition contains at least (95) % by weight concrete, and has a contamination level lower than (1) % of the bulk mass. Recycled concrete aggregate obtained from construction and demolition (C&D) waste has received an increasing attention recently, due its promising potential use in environmentally friendly concrete structures.

The basic method of recycling concrete is the crushing of (C&D) waste materials into a granular nature materials of a selected particle size distribution. The level of contamination of the recycled waste determines the quality of the (RCA). For the crushing process, many types of crushers can be used, but the main crushers are: jaw crusher, cone crusher and impact crusher. Most recycling plants use the jaw crusher as a primary crusher since it has the capability to process the large pieces of concrete as well as the residual reinforcement if existed. This crusher generally reduces the materials down to the size of (60 - 80) mm. In addition, some plants also use the impact crusher as a secondary crusher to produce more of aggregate without adhered mortar. The products of the secondary crusher are then screened to separate the aggregate into different sizes.

Anderson et al. (2009) stated that recycled aggregate – as compared to conventional or natural aggregate – has higher water absorption and lower specific gravity. Also, its density is lower than normal aggregate. For porosity, recycled aggregate has much higher porosity. It is a general fact that recycled aggregate's strength is lower than its corresponding value of natural aggregate, the grading curves for recycled aggregate are generally having a continuous form with a similar fineness modulus for the equivalent fracture.

For the fresh state of concrete, Obla et al. (2007) found that with the increase in (RCA) content in concrete mixtures, their workability decreases. They also stated that in order to produce a similar workability as that of natural aggregate concrete, (5) % more of mixing water was required when only the coarse (RCA) was used, and up to (15) % when both of fine and coarse (RCA) were used.

In the field of structural behaviour of reinforced (RAC), few investigations have been conducted. Maruyama et al. (2004) examined the flexural behaviour of (RRC) beams and reported that the crack width in (RRC) was wider than that of the conventional reinforced concrete. The deflection of the (RRC) beams was greater than the case of conventional concrete beams. Finally, they concluded that the use of (RAC) did not significantly decrease the flexural capacity, providing that the reinforcing steel was sufficiently anchored to yield before the concrete fails. Xiao et al. (2006) studied the seismic performance of (RAC) frame structures of (1/2) scale specimens made of (0,30, 50 &100) % replacement of (RCA). The specimens were subjected to a constant gravity load and a low - frequency cyclic lateral load. They reported that the failure pattern was similar for all frames, but the yield and ultimate load capacities of (RAC) frames were smaller. Their final conclusion was that a properly mixed (RAC) frame behaves well enough to stand an earthquake.

2. Aim of the study

This study aims to Using the produced self – compacting and waste concrete as recycled coarse aggregate to produce recycled aggregate concrete specimens containing different percentages of recycled aggregates.

Also, this research aims to studying the flexural behavior of the two-way reinforced concrete slabs which contain the recycled concrete aggregate.

3.The Experimental Program

The aim of this research is to compare the flexural strength of the reinforced reference concrete slab which is made with natural aggregate only, and the flexural strengths of reinforced concrete slabs made with different types and contents of recycled concrete as coarse aggregate. Five concrete types were tested in this research; mixture proportions of the tested concretes were determined as follows:

- Constant cement content,
- Constant maximum aggregate size of (14) mm,

- Constant type but variable content of fine aggregate,
- Constant workability about (75) mm,
- Variable type and content of coarse aggregate,
- Variable content of silica fume.

3.1. Concrete Mixtures

The reference concrete mixture was proportioned to obtain the slump of (75) mm, and the contents of this concrete are shown in Table (1) below.

Table 1. Contents of the reference concrete mixture (per cubic meter)

Cement (kg)	Water (kg)	w/cm	Coarse aggregate (kg)	Fine aggregate (kg)
471	194	0.412	1019	598

The recycled aggregate concrete mixtures were twelve mixtures, where the (RCA) was used with different type and percentage. In addition, silica fume was added with a constant amount of (10% of the cement's weight to enhance the strength of the reinforced (RAC) slabs (Abdul-Rahman, M. et al., 2018, Abdul-Rahman, M. et al. 2020, and Al-Attar et al., 2020). The details of the variables of each mixes are shown in Table (2), while Table (3) presents the materials quantities information of these mixtures. The replacement ratio of recycled coarse aggregate (for normal or self-compacted concrete aggregate) is considered by weight of the normal used aggregate.

Table 2. Variable details for the used recycled aggregate concrete mixtures

Mix Designation	Mix details
N33	Concrete with 33.3 % of recycled normal concrete as coarse aggregate
Ref	Reference concrete (no recycled aggregate or silica fume)
N66	Concrete with 66.7 % of recycled normal concrete as coarse aggregate
N100	Concrete with 100 % of recycled normal concrete as coarse aggregate
N33S	Concrete with 33.3 % of recycled normal concrete as coarse aggregate and added silica fume
N66S	Concrete with 66.7 % of recycled normal concrete as coarse aggregate and added silica fume
N100S	Concrete with 100 % of recycled normal concrete as coarse aggregate and added silica fume
S33	Concrete with 33.3 % of recycled Self-compacting concrete as coarse aggregate
S66	Concrete with 66.7 % of recycled Self-compacting concrete as coarse aggregate
S100	Concrete with 100 % of recycled Self-compacting concrete as coarse aggregate
S33S	Concrete with 33.3 % of recycled Self-compacting concrete as coarse aggregate and added silica fume
S66S	Concrete with 66.7 % of recycled Self-compacting concrete as Coarse aggregate and added silica fume
S100S	Concrete with 100 % of recycled Self-compacting concrete as coarse aggregate and added silica fume

IOP Conf. Series: Materials Science and Engineering **978** (2020) 012032 doi:10.1088/1757-899X/978/1/012032

Table 3. Contents of recycled aggregate concrete mixtures for one cube meter

Mixture Designation	Contents for a cubic meter (kg)					
	Cement	Water *	Fine Aggregate.	Natural coarse Aggregate	Recycled coarse Agg. (type-weight)	Silica Fume
N33	471.26	206.41	612.7	631.74	NC-332.86	-
N66	471.26	205.51	643.4	292.35	NC-618.0	-
N100	471.26	195.90	690.7	-	NC-853.24	-
S33	471.26	206.4	610.9	629.13	SCC-331.49	-
S66	471.26	194.28	650.4	289.75	SCC-620.45	-
S100	471.26	192.83	689.3	-	SCC-852.40	-
N33S	471.26	217.14	567.4	631.74	NC-332.86	47.1
N66S	471.26	215.24	598.8	292.35	NC-618.0	47.1
N100S	471.26	214.26	636.6	-	NC-853.72	47.1
S33S	471.26	217.17	565.3	629.13	SCC-335.84	47.1
S66S	471.26	215.73	596.2	289.75	SCC-620.54	47.1
S100S	471.26	214.29	635.1	-	SCC-852.40	47.1

* the w/c ratio is adjusted to achieving the requirements of workability limits.

3.2. Slabs Flexural Strength Test

This test was performed using a digital flexural testing machine of (200) kN capacity. Reinforced concrete slabs of dimensions (500 × 500 × 50)mm were subjected to a concentrated load by steel disc of (40) mm diameter. These slabs were simply supported in all directions by a steel frame, then a load- deflection curve was plotted with a precision of (1) kN. Details of the slab's supporting system, load and reinforcement are shown in Figure 1.

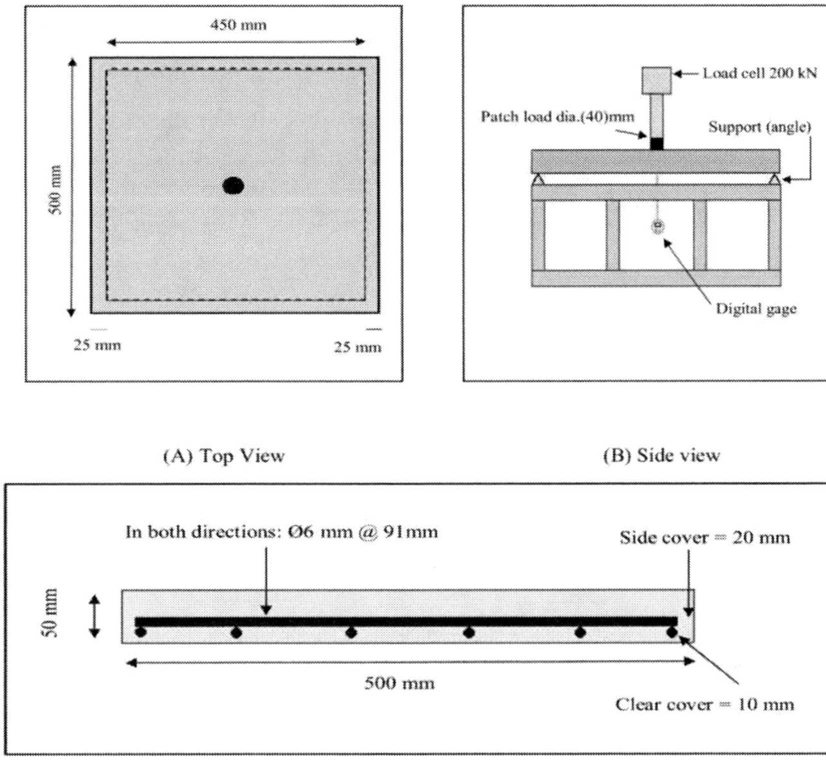

Figure 1. Details of the reinforced concrete slabs' flexural strength test

4.Results and Discussion

For an inclusive presentation of the results, they are discussed as follows:

4.1 Mechanical Properties of Concrete

4.1.1 Compressive Strength

The tests result of the recycled coarse aggregate concrete along with the types, percentages of the recycled aggregates with the presence of the silica fume are presented in Table (4).

It can be seen that the waste aggregate concretes with (33.3, 66.7 and 100) % replacement of coarse aggregate exhibited a decrease in the compressive strength (compared to the reference concrete) by the amount of (3.18, 5.05 and 6.12) % respectively. These results may be attributed to a number of reasons. First, the crushed waste (RCA) possesses a higher porosity and a lower density of about (20) % than the natural coarse aggregate. Therefore, lower density of the coarse aggregate (which has the largest contribution to the concrete mass) leads to a lower strength of coarse aggregate and consequently, lower strength of the concrete.

On the other hand, the replacement of recycled (SCC) coarse aggregate led to a significant increase in the compressive strength by (2.85, 6.45 and 9.95) % for the aggregate replacement percentages of (33.3, 66.7 and 100) % respectively when compared to the reference concrete. The above results can be interpreted as follows; the (SCC) has much more resistance to mechanical action than the ordinary concrete, hence a better recycled aggregate is gained. Also, the recycled (SCC) coarse aggregate consists of strong angular particles which tend to improve the interlocking action of the coarse aggregate and increase the strength. In addition, it is important to mention that the (SCC) has stronger mortar attached to coarse aggregate, due to the increased fine aggregate content and the use of powders and super plasticizers that produce denser and stronger mortar than other types of concrete.

Table 4. Compressive strength values for different concrete mixes

Symbol Designation	Compressive strength (fc') at (28) days (MPa)	Difference from Ref. (%)	Splitting tensile strength (*ft*) (MPa)	Difference from Ref. (%)
Ref	39.336	-	3.442	-
N33	38.085	- 3.18	3.43	-0.348
N66	37.349	- 5.05	3.396	-1.34
N100	36.93	- 6.12	3.323	-3.46
S33	40.458	2.85	3.559	3.4
S66	41.874	6.45	3.522	2.33
S100	43.25	9.95	3.405	-1.07
N33S	40.937	4.07	3.647	9.56
N66S	40.536	3.05	3.491	1.42
N100S	39.87	1.36	3.358	-2.44
S33S	41.464	5.41	3.811	10.7
S66S	42.159	7.17	3.694	7.32
S100S	43.73	11.17	3.438	-0.11

4.1.2 Splitting Tensile Strength

The splitting tensile strength test results of all concrete mixtures are shown in Table (4). These results show a slight decrease in strength when waste concrete aggregate is used with (33.3) % of replacement, this decrease is of (0.35) % compared to the reference concrete. This decrease may be attributed to that the recycled aggregate contains a high percentage of natural aggregate particles with a less percentage of mortar attached on the aggregate particles forming angular particles that limitedly enhanced the interlocking and gave a similar strength (Hansen, 1986).

When the recycled (SCC) aggregate was used, the splitting tensile strength witnessed an increase of (3.4 and 2.33) % for (33.3 and 66.7) replacement percentages respectively. While for (100) % replacement, the tensile strength decreased for about (1.07) %. These results can be explained when the nature of (SCC) is considered.

Nevertheless, compared to the mixtures without the silica fume, the use of this admixture resulted in an increase in the splitting tensile strength of (5.1, 2.8 and 1.05) % for (33.3, 66.7 and 100) % replacement of waste (RCA) in the same order. These results show that the use of the silica fume could result the same tensile strength of the reference concrete when the replacement of waste (RCA) is raised up to (66.7) %.

4.2 Ultimate Loading

The first crack and ultimate (failure) loads results of the (RAC) slabs are presented in Table (5). Also, the (load-deflection) relationships for the all tested concrete slabs are shown in Figures (2, 3, 4 and5). The test results showed a reduction in flexural strength of (2.19, 4.96 and 13.97) % for the replacement of normal (RCA) of (33.3, 66.7 and100) % respectively. The results confirmed that the flexural strength of concrete undergoes a reduction of less than (10) % and maximum of (20) %when the concrete is made of recycled coarse aggregate and natural fine aggregate as stated by ACI committee 555R-01 report (2001). The low reduction took place because the reinforced concrete is a composite material. Hence, weaker aggregate has low effect on its ultimate strength. Another reason for this reduction in strength is that the recycled coarse aggregate has a lower resistance to the mechanical action than the natural aggregate, so a weaker concrete is produced (Ravindrajah and Tam, 1985). In addition, the recycled aggregate concrete (RAC) has significantly weaker bond areas than the corresponding normal concrete since it has bonding areas between the round gravel and old or new mortar in addition to the bonding areas between the old and the new mortar (AlHussainy, 2011).

Oppositely, the recycled (SCC) aggregate showed an enhancement of the flexural strength of (6.23 and 4.58) % for (33.3 and 66.7) % of replacement, but for (100) %of replacement, a reduction of (2.09) %compared to the reference concrete occurred. When silica fume was added to the normal (RAC) mixtures, the flexural strength increased even beyond that of the reference concrete for (33.3 and 66.7) % of replacement with an increase of (7.94 and 8.9) %respectively, but for (100) % of replacement, the strength was still less than that of the reference concrete but with (10.2) % greater than that value of the concrete without the silica fume. Also, all concretes containing recycled (SCC) aggregate and silica fume gave flexural strength values more than the reference value, and the enhancements were (4.06, 1.05 and 4.72) % for the replacements of (33.3, 66.7 and100) % respectively.

The first crack loads for all normal (RAC) slabs were lower than that of the reference concrete slab, where the maximum reduction was (29.62) %, while for the recycled (SCC) slabs, the maximum reduction was (18.66) %. The addition of silica fume showed an increase in the first crack loads compared to those without it, and the differences were (20.52 &15.34) % for the normal (RAC) and recycled (SCC) slabs respectively, compared to the reference slab.

Table 5. Ultimate loading results of the reinforced concrete slabs

Concrete Symbol	Load at first crack (kN)	Difference from Ref (%)	Failure load (kN)	Difference from Ref. (%)
Ref	10.82	-	29.21	-
N33	8.6	-20.52	28.57	-2.19
N66	8.2	-24.21	27.76	-4.96
N100	7.615	-29.62	25.13	-13.97
S33	9.4	-13.12	31.03	6.23
S66	9	-16.82	30.55	4.59
S100	8.8	-18.66	28.6	-2.09
N33S	11.6	7.21	30.84	5.58
N66S	10.41	-3.79	30.23	3.49
N100S	8.6	-20.52	27.69	-5.20
S33S	9.57	-11.55	32.92	12.70
S66S	9.3	-14.05	30.87	5.68
S100S	9.16	-15.34	29.95	2.53

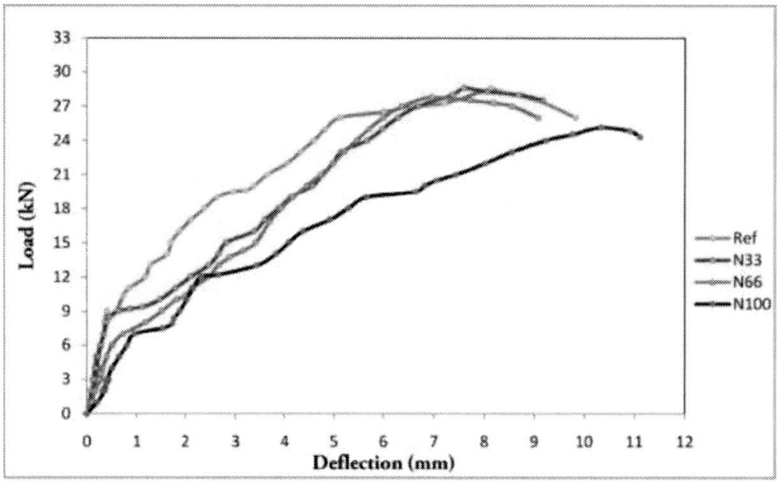

Figure 2. Load-deflection curves of the normal (RAC) slabs

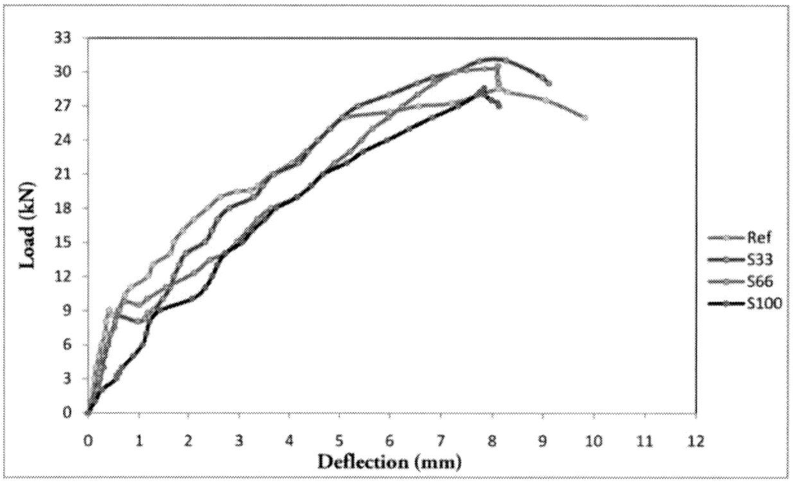

Figure3. Load-deflection curves the of self-compacting (RAC) slabs

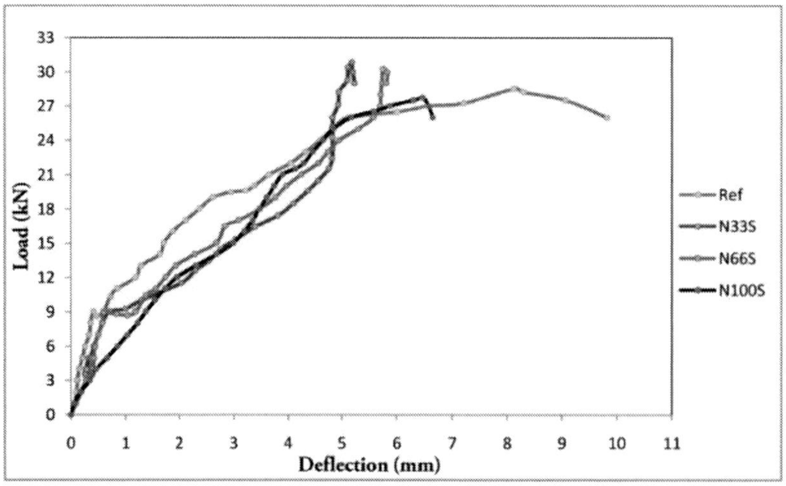

Figure 4. Load-deflection curves of the normal (RAC) slabs with SF

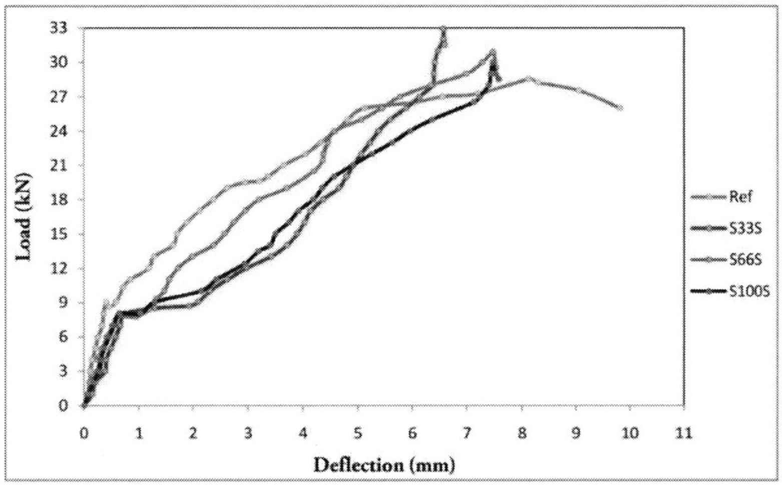

Figure 5. Load-deflection curves the of self-compacting (RAC) slabs with SF

4.3. Modes of Failure

In order to classify the failure, the (load-deflection) curves were used. According to Marzouk and Hussein (1991), modes of failure can be classified into three types: pure flexural failure, ductile shear failure and pure punching failure. The pure flexural failure occurs in the slabs in which most of their reinforcement yields before the punching takes place, hence; these slabs exhibit large deflection values prior to the failure. This type can be seen in Figure 6 for (Ref) slab and Figure 8 for (S33) slab, where the main wide cracks are located at the centre of the slab. As the cracks spread towards the corners, they gradually decrease in width until they reach the edges as hair line cracks.

Figure 6. Failure pattern of the reference concrete slab

Pure punching failure takes place when the slab shows small deflection values with the yielding of the reinforcement steel is being very localized at the loaded central area. The (N33S) and (N66S) slabs in Figure 7are examples of this failure. The cracks first occurred at the center, followed by radial extending from the center. As the load was increased, the final failure developed by the punching of the loading column through the slab.

Finally, the ductile shear failure is a transition between the previous cases of pure flexural and pure punching failures. The (N100S) and (S100S) slabs showed the ductile shear failure where they failed before the flexural strength was exceeded. See Figures 7 and 8.

IOP Conf. Series: Materials Science and Engineering **978** (2020) 012032 doi:10.1088/1757-899X/978/1/012032

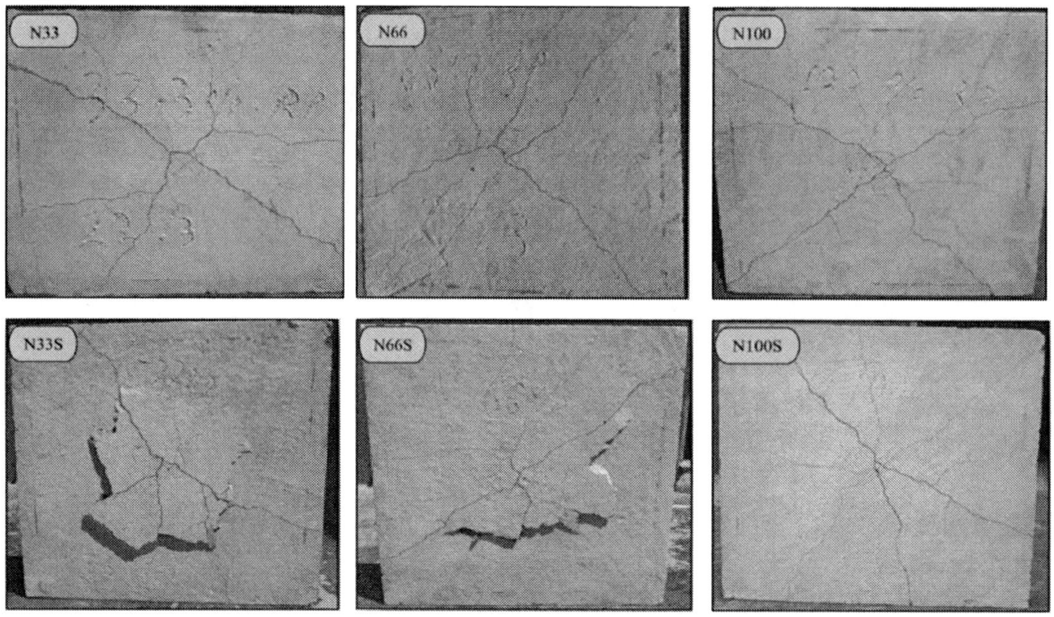

Figure 7. Failure pattern of the normal (RAC) slabs

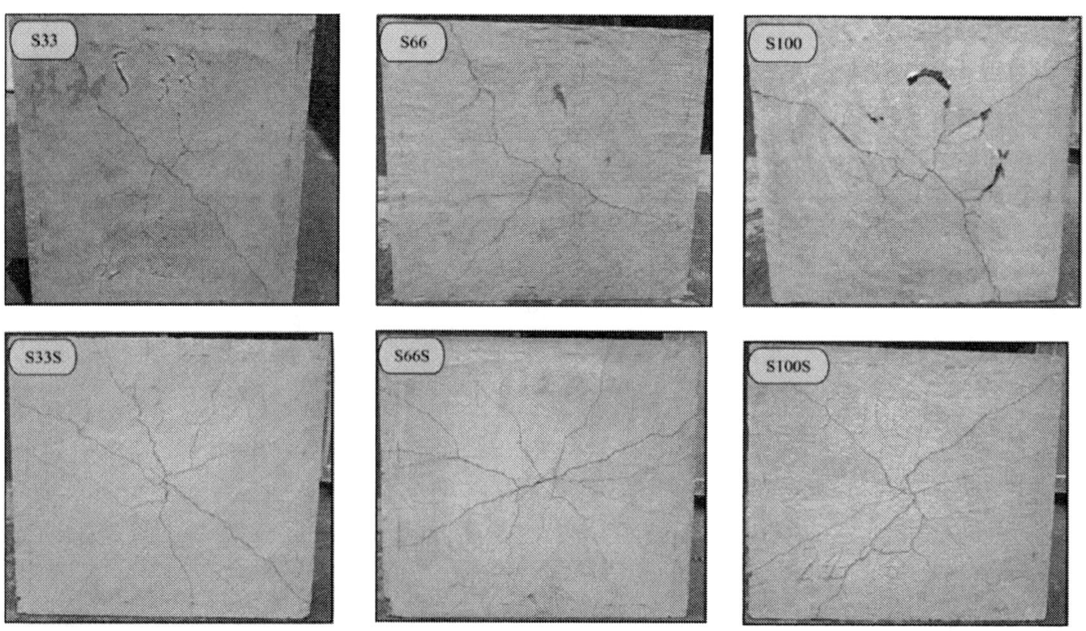

Figure 8. Failure pattern of the self-compacting (RAC) slabs

5. Conclusions

Based on the adopted experimental program, the following conclusions can be made:

1. Recycled aggregate concrete (RAC) made from waste coarse (RCA) showed a gradual decrease in the compressive strength for (33.3, 66.7 and 100) % of replacement. While (RAC) made from recycled (SCC) aggregate exhibited a gradual increase in the compressive strength compared to the reference concrete.
2. The splitting tensile strength decreased when waste (RCA) was used. The use of recycled (SCC) aggregate presented an increase and a slight reduction for (100) % of replacement, compared to the reference concrete. When the silica fume was used, the
3. The flexural strength of the reinforced concrete slabs decreased when only waste (RCA) was used. While the flexural strength increased for (33.3 & 66.7) % replacement of recycled (SCC) where only the (100) % of recycled (SCC) showed a slight decrease in the flexural strength.
4. The addition of Silica Fume improved the mechanical properties as well as the flexural strength of all mixtures for recycled (SCC) mixtures, and for (33.3 & 66.7) % of waste (RCA) mixtures, except for the (100) % replacement of waste (RCA) mixture where a reduction occurred compared to the reference concrete.
5. In general, reinforced (RAC) slabs showed less deflections at the ultimate load than that of the reference slab.
6. The addition of Silica Fume resulted even less deflections thanthe cases of its absence.

References

[1] Abdul-Rahman, M., Ali, A., &Younus, A. (2018) Effecting of Steel Fibers and Fly Ash on the Properties of Concrete, *Tikrit Journal of Engineering Sciences*; 25(4):30-36.

[2] Abdul-Rahman, M., Al-Attar, A., Hamada, H., &Tayeh, B. (2020) Microstructure and structural analysis of polypropylene fibre reinforced reactive powder concrete beams exposed to elevated temperature, Journal of Building Engineering, 29(2020).

[3] ACI Committee 555. (2001). Removaland Reuse of Hardened Concrete. ACI Manual of Concrete Practice, Part 1.

[4] Al Hussainy, F.A. (2011). Using Recycled Concrete as Coarse Aggregate to Produce Self-Compacting Concrete. [M.Sc. Thesis]. College of Engineering, University of Babylon.

[5] Al-Attar, A., Abdul-Rahman, M., Hamada, H., and Tayeh, B. (2020). Investigating the behaviour of hybrid fibre-reinforced reactive powder concrete beamsafter exposure to elevated temperatures, 9(2):1966–1977.

[6] Anderson, K.W., Uhlmeyer, J., & Russell, M. (2009). Use of Recycled Concrete Aggregate in PCCP. WSDOT research report, Washington State, Department of Transportation, WA-RD 726.1.

[7] Hansen, T. (1986). The Second RILEM State of the Art Report on Recycled Aggregates and Recycled Aggregate Concrete. Materials and Structures. 1(111), 201-246.

[8] Maruyama, M., Sogo, T., Sogabe, R., Sato, K., & Kawai. (2004, November). Flexural Properties ofReinforced Recycled Concrete Beams. Conference on the Use of Recycled Materials in Building and Structures, Barcelona, Spain.

[9] Marzouk, H., & Hussein, A. (1991). Experimental Investigation on the Behavior of High-Strength Concrete Slabs. ACI Structural Journal, Technical Paper,88(6), 701-711.

[10] Obla, K.., Kim, H., & Lobo, K. (2007). Crushed Returned Concrete as Aggregates for New Concrete. Final Report to the RMC Research & Education Foundation. Project 05-13,1-44.

[11] Rao, M., Bhattcharyya, S., and Barai, S. (2011). Influence of Recycled Aggregate on Mechanical Properties Materials and Structures, 44, 205-220.

[12] Ravindrajah, R., and Tam, T., (1985). Properties of Concrete Made with Crushed Concrete as Coarse Aggregate. Magazine of Concrete Research.37(130), 29-38.

[13] Xiao, J., Sun, Y., &Fulkner, H. (2006). Structural Performance of Frame Structures with Recycled Aggregate Concrete. Engineering Structures, 28(1), 1-8.

CFD Analysis for The Effect of Personal Ventilation Combined with Mixing Ventilation on Performance Index (ADPI) and Thermal Human Comfort

Hussien Aziz Saheb1[*] Ala'a Abbas Mahdi[2] and Qusay Rasheed Al-amir[3]

[1] College of Engineering, Babylon University, Iraq
E-mail: hassan1245konh@gmail.com
[2] College of Engineering, Babylon University, Iraq
E-mail: alaa.mahdi1959@gmail.com
[3] College of Engineering, Babylon University, Iraq - E-mail: qusay1972@gmail.com
* Corresponding author: Mechanical Engineering Department, College of Engineering, Babylon University, Iraq. Tel: +7815005345.

Abstract. Good indoor air quality and ideal thermal comfort are very important indicators for the comfort of occupant's space. The Iraqi climate is hot dry in summer, so the use of air conditioners is indispensable. This work focuses on solving the IAQ problems in office room environments, and it evaluated a new principle of ventilation which is combined personal and mixing ventilation systems. Personal and mixing ventilation system can provide individual control of indoor climate. The airflow motion and distribution of temperature investigated numerically with the best flow rate from combined system. CFD study is used for simulation the indoor airflow and temperature distribution by using (AIRPAK3.0.16) for solving the turbulence equations, Naiver-Stocks, energy equations, and using (FVM). The Renormalization Group RNG K-Ɛ turbulence model was used for simulation this study cases, the airflow rate from ATD was 5 l/s and 10 l/s. For mixing ventilation supply temperature of 17°C and The temperature of the supplied air from the personal ventilation system ranges from (17 to 23) °C, it was found that if using personal ventilation system at airflow rate of 10 l / s it would improve the quality of the inhaled air present in the breathing zone. The personal and mixing ventilation system provides good human thermal comfort based on the values of effectiveness temperature (ε_t) and air distribution performance index (ADPI) which give (1.241) and (63.226%) respectively. The increased airflow rate from (ATD) leads to a decrease in the temperature in the breathing zone, this increases the human thermal comfort of the occupants inside the room.

Keywords: Thermal comfort, AIRPAK, Computational fluid- dynamics (CFD), Mixing ventilation (MV), Indoor air quality (IAQ)., Personalized ventilation (PV).

1. Introduction

Nowadays a lot of people spend about 80% of their time inside an industrial environment, i.e. office, factory, public building, home, transport vehicle, etc. Therefore, it is essential to provide comfortable and healthy indoor environments for their occupants. Both thermal comfort and indoor air quality are important factors affecting the quality of the indoor environment. Recent research has improved the understanding of the influence of the thermal environment on human comfort and performance.

The range of temperature of room should be between (20 to 26) °C with relative humidity from (30 to 60%).For the purpose of avoiding drafts, the average air speed is from (0.2 to 0.6) m/s, depending on temperature of room and turbulence intensity [1]. MV system provides air to the room from the diffuser at a relatively high speed, usually located at or near the roof over the occupied zone, jet of air is provided in The top of the room (roof or wall at a high level) at high speed usually greater than 2 m.s-1 to ensure air circulation in the room, the resulting distribution of temperature and concentration of pollutants in the occupied zone must be uniform [2]. The air conditioning provided from outside the room mixes with the polluted air inside the room before reaching the occupant, which leads to a decrease in the quality of the air inhalation while the PV system provides fresh air directly to the occupant breathing zone, PV systems have the ability to improve inhaled air quality and thermal comfort, and also provide individual control of the temperature of air, air direction, and possibly speed of air as well [3].

PV system aims to provide fresh, cool air directly to occupant's breathing zone. The positive effects of personal ventilation, that is, clean air breathing, are documented for health, comfort and performance [4]. PV system has the ability to protect occupant from infectious agents causing disease, each PV user can control the appropriate local cooling for him and also can achieve a good environment based on individual preferences, [5, 6].

Additionally, since personal ventilation systems target the occupant breathing zone, the interaction between the supplied air and the flow in the delicate environment surrounding the occupant is the determining factor for personal ventilation systems performance [7]. The thermal comfort of the human was greatly improved as well as the perceived indoor air quality was improved and the PAQ were documented, and the symptoms of SBS were reduced when using PV, which was performed in field laboratories. [8, 9, 10]. PV system has the ability to protect space occupants from diseases transmitted through the exhaled air of sick people [11]. Many simulation and experimental works such as elmaghraby, (2014) [12]. Hussein et al. [13] have been done to study airflow pattern, temperature, and concentration distribution in mixing and Personal ventilation systems. Chen (1995) [14], studied five k-ε models, conclude The RNG k-ε model is best than the standard k-ε model and is therefore recommended for simulations. The RNG k model is widely used for indoor airflow for different configurations as it gives accurate results and is closer to experimental results, [15].

Kaczmarczyk et al.,2002 [16] studied the impact of PV system on the perceived air quality, the experiment was conducted on an office containing six occupants, through a personal ventilation system, the occupant is allowed to control the amount and direction of air. Four experiments were performed: (I) PV system provides outdoor air at 20 °C; (II) PV system supplies external air at 23 °C; (III) PV system provides recycled room air; And (IV) MV. Temperature of room was kept constant at 23 °C and relative humidity at 30%. The results showed that the best case with PAQ was when PV system introduced outdoor air at 20 °C. PAQ was much better in the Case-I.

Hooff & Blocken, 2019 [17] introduced MV driven by two oppositely located supply jets with a time-periodic supply speed, theoretical analysis using (CFD) program. Temperature of air in the small office, with higher cooling load, so it was chosen the change of air rate carefully to avoid a higher temperature gradient. This study showed that the use of the turbulence model (RNG k-ε) gave very good and accurate results as well as the ability to predict well on jet spreading rate and the behavior of recirculating air.

Al-ssaad et al., 2017 [18] investigated improvement of ventilation by combining personal ventilation with mixing ventilation, the study was carried out numerically and experimentally, at ideal conditions, a thermal comfort about 0.95 and also ventilation effectiveness 77% was obtained, the cost of energy for PV system was reduced by 21.34% compared to a constant personal ventilation system at equal thermal comfort. This study demonstrated that RNG k-ε model with enhanced wall treatment is used to model turbulence it is characterized by accuracy and very good predictability of jet spreading rate and the behavior of re-circulated air.

Al-ssaad et al., 2018 [19] studied the best performance of a MV system Compound with PV, Simulated were performed using the CFD technique to find out the airflow the most suitable supply frequency, Fresh air is emitted from (PV) towards the upper part of the person. The purpose of this study was to achieve good thermal comfort and good IAQ in breathing zone. In this study, the test

IOP Conf. Series: Materials Science and Engineering **978** (2020) 012033 doi:10.1088/1757-899X/978/1/012033

model was simulated numerically using CFD technology, with a transient 3D test under laboratory conditions at a temperature of 25 ° Cand PV jet temperature of 22°C. This study showed that when the frequency increases at a constant flow rate, thermal comfort increased by 15.2%. However, when the medium flow rate increased at a fixed frequency, thermal comfort decreased at a low frequency of 0.3 Hz but remained acceptable at a higher frequency of 0.5 Hz.

Yakoob et al., 2019 [20] studied personal ventilation with displacement ventilation (DV) for the purpose evaluated the efficiency of air exchange in rooms, the study was practical and theoretical, this study using RNG, K-ε turbulence model for simulating temperature profile. This study concluded, that the arrangements at an office rooms for the air supply diffuser personal and DV are combined gives Acceptable human thermal comfort depending on the value of ($\mathcal{E}_t$) and (ADPI) which gives (71%) and (1.8) respectively which improved the air quality and thermal comfort.

(Katramiz et al., 2020) [21] studied the interaction between breathing of human and the (IPV) jet and investigated the effectiveness of IPV in supplying fresh and cold air of breathable. Developed a transient three-dimensional CFD mock-up of a breathing occupant in a chamber equipped with an IPV coupled with an MV system. Two modes of breathing were considered: breathing of nose and mouth. To evaluate air quality of the breathing zone, using the tracer gas method. The same model was practically conducted, and the same experiments were conducted to validate numerical results and compare them with experimental results. The results of this study gave that maximum effectiveness of ventilation ε_v during inhalation was obtained at an IPV frequency of 0.5 Hz ranging from 36.75% to 86.6 % for breathing through the nose, 49.6% to 87.3% for breathing through the mouth. The use of intermittent personal ventilation system improves the quality of air in the breathing zone.

Most researchers discussed in their studies, whether experimental or theoretical (the location, Geometric shapes for diffusers, number of diffusers, comparing DV with MV system, the position of the inlet and outlet of air and effect of an air change per hour on thermal comfort).

The main objectives of this study are:

- This study aims to investigate how air quality in the breathing zone is affected when PV system is added.
- Studying the effect of adding PV on the thermal comfort of the occupant in office environment.
- Calculation the actual values of the input airflow rate and temperature of air supply needed by the testing room depending on the cooling loads of room.
- To investigate Air Diffusion Performance Index indoor air quality and comfort issues by CFD modeling by ventilation the room under the Iraqi climate.
- Study the possibility used PMV system type under the Iraqi climate.

In this work an office room was simulated using a program AIRPAK3.0.16 For the purpose of conducting the numerical study for airflow rates, distribution of temperature around occupants due to multiple sources of heat by used the CFD method to get a better level of IAQ and improved human thermal comfort.

2. Description room configuration and assumptions

An office room was simulated to numerical study for the airflow and distribution of temperature with the MV and PMV system. Steady state three-dimensional computational (CFD) mode is used to simulate the speed and temperature fields in an office room. Dimensions of the simulation room3.0 m (length) × 2.5 m (height) × 2.5m (width). Includes different ranges of speeds for personal ventilation with a constant flow rate for (MV) device to evaluate the performance of the various configurations facing local and general thermal comfort. The figures 1 & (2) show the schematic diagram of the ventilation systems (MV) & (PMV), respectively. MV device using to supply air into the office room located at the middle of the room's south wall is below the ceiling. Simulated two persons, two computers and one lump were placed as a heat source. Description of various details in the simulated office room numerically, with the ventilation system, is only MV system(case I) and PMV system (cases II&III) are given in tables (1,2) respectively, it shows us all the contents of the office room

With the two types of ventilation systems used, a set person is 1.1m height, heat gain(only Sensible Heat) for person is (75W/person), computer (60W/computer) and lump (100W), and ΔT_{hf} is 2 °C. Table (3) shows the required Iraqi conditions inside the office buildings during hot and dry climates. It was used [Cooling Load Temperature Difference Method (CLTD)] to calculate the cooling loads as shown in the equations listed below [22].

$$Q_{person} = \dot{NO}.\times qs/person \times CLF \qquad (1)$$

Where, Q_{person} =Sensible Cooling Load, NO=number of persons, q_s sensible heat for one person, CLF= person cooling load factor [22]. While The luminescent load is calculated by the following equation

$$Q_L = \dot{NO}.\times W \times Fb \qquad (2)$$

Where, Q_L = Sensible Cooling Load, NO=number of lump, W= lump power, F_b= factor equal to1.2, [22].

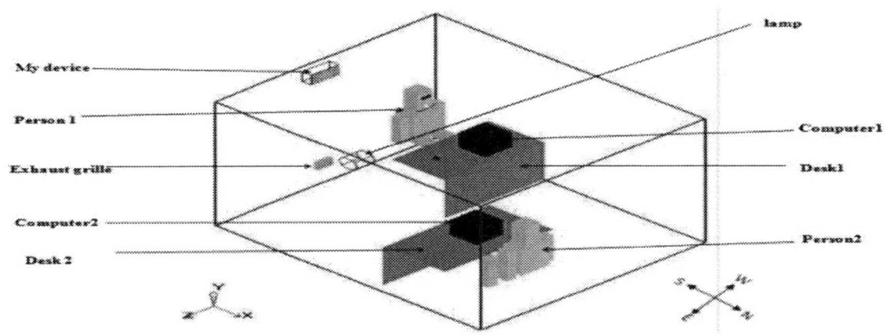

Figure1. Model of the room in Airpak for MV system.

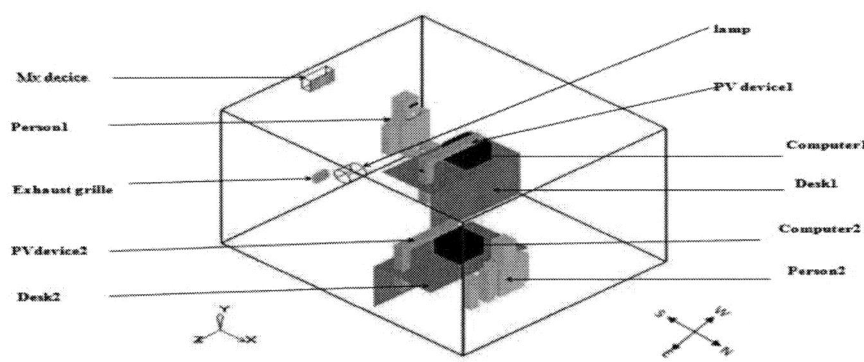

Figure2. Model of the room in Airpak for PMV system

Air is the supposed working fluid and the flow assumed to be steady state, 3D flow and incompressible, [22]. The test room at indoor temperature (25°C). It was measured before the start of the test. The tested room was thermally insulated, so the values of (q_l& q_{ex}) are equal to zero in this work. (PV) system at different flow rate 5 l/s and 10 l/s. The air is supplied with different speeds and flow rates from (PV) is (0.6 m/s) at flow rate 5 l/s and (1.2 m/s) at flow rate 10 l/s. The following two

steps are used to estimate the supply ventilation rate for mixing ventilation and total heat applications [23]. The supplied air temperature from mixing air terminal is 17 °C [23]. The indoor room temperature before air is supplied is 25 °C.
The necessary air volume is calculated (airflow rate), The heat transfer due to the ventilation process is calculated by the following equations which are illustrated as,

$$Q_t = \dot{m} \bullet \times cp \times \Delta T \quad (3)$$

$Q_t = q_{oe} + q_l + q_{ex}$ =2×person+2×computer+lump=2×75+2×60+100=370W ,
$370 = \dot{m} \bullet \times 1005 \times (25 - 17)$ =0.046kg/s
From the value of the mass flow rate $\dot{m}$ in above equations can be calculated design airflow rate (l/s)

$$Q_s = \frac{m\bullet}{\rho} = 0.046/1.2 = 0.04 m^3/s \quad (4)$$

The effective draft temperature, is given as, [24]:

$$EDT = (Tx - Tr) - (8Vx - 1.2) \quad (5)$$

Where: T_x : local temperature(K), T_r : mean temperature of room(K),and V_x : local speed of air $m.s^{-1}$
Because it represents local cooling not suitable for the human body due to movement of air and low temperature of air in space [25]. The draft temperature limit was taken as -1.7 ° C < EDT <1.1 ° C and the higher value for speed of air was taken as 0.35 $m.s^{-1}$ [25&26]. The measured values of temperature and speed of air at uniformly spaced points throughout the occupied zone or through a vertical, centerline plane through the air supply [24].
(ADPI) used to satisfy the comfort limits, a percentage of the points number of draft temperature measured in the occupied zone in which (-1.7 °C < EDT <1.1 °C) to the total number of draft temperature points measured in the occupied zone, [12]. Kansas test data showed that ADPI depends on the type of ATD, load of room, airflow rate, and geometry of room. ADPI is useful for turning on the cooling mode; a value of 80% is usually considered an acceptable minimum, [24].

$$ADPI = \frac{N_\theta}{N} \times 100 \quad (6)$$

Where N_θ in which -1.7 ° C < EDT < 1.1 ° C
Air change per hour (ACH) It can be calculated by the following equation. (7), [24].

$$ACH = \frac{Q_s}{Vroom} \times 3600 \quad (7)$$

In MV system there is a set of goals that must be achieved represented by clam operation, thermal comfort, suitable speed, low energy consumption, and low noise diffuser. In this present work rectangular diffusers were used [27]. by assumed the air supply speed 2.5m/s (According to the recommendations of ASHRAE), the Q_s It was calculated by the equation (4) and found its value (0.04) m^3/s. Eqn. (8) used to calculate the area of the diffuser and found the dimensions of air supply unit are (0.016) m^2.

$Q_s = Ux \times A_s$ where U_x =2.5 [23]. (8)
Then A_S= 0.04/2.5=0.016 m^2
Where: U_x: local air speed (m/s), As: Surface area of supply air diffuser (m^2), the area of exhaust grills is about 75% of the air supply unit area, [28].

3. Computational Fluid Modeling

3.1 Governing Equations:

The field of fluid flow can be described by a set of equations including continuity equation, momentum equation and energy equation in addition to other equations, [28]

$$\frac{\partial}{\partial x}(\rho u) + \frac{\partial}{\partial y}(\rho v) + \frac{\partial}{\partial z}(\rho w) = 0 \quad (9)$$

$$\frac{\partial}{\partial x}(\rho uu) + \frac{\partial}{\partial y}(\rho uv) + \frac{\partial}{\partial z}(\rho uw) = -\frac{\partial p}{\partial x} + \frac{\partial}{\partial x}\left(\mu\frac{\partial u}{\partial x}\right) + \frac{\partial}{\partial y}\left(\mu\frac{\partial u}{\partial y}\right) + \frac{\partial}{\partial z}$$

$$\left(\mu\frac{\partial u}{\partial z}\right) + \frac{1}{3}\frac{\partial}{\partial x}\left[\mu\left(\frac{\partial u}{\partial x} + \frac{\partial v}{\partial y} + \frac{\partial w}{\partial z}\right)\right] + \frac{\partial}{\partial x}\left(-\rho\overline{u'u'}\right) + \frac{\partial}{\partial y}\left(-\rho\overline{u'v'}\right) + \frac{\partial}{\partial z}\left(-\rho\overline{u'w'}\right) + \rho g_x \quad (10)$$

$$\frac{\partial}{\partial x}(\rho uv)+\frac{\partial}{\partial y}(\rho vv)+\frac{\partial}{\partial z}(\rho vw)=-\frac{\partial p}{\partial y}+\frac{\partial}{\partial x}\left(\mu\frac{\partial v}{\partial x}\right)+\frac{\partial}{\partial y}\left(\mu\frac{\partial v}{\partial y}\right)+\frac{\partial}{\partial z}$$

$$\left(\mu\frac{\partial v}{\partial z}\right)+\frac{1}{3}\frac{\partial}{\partial y}\left[\mu\left(\frac{\partial u}{\partial x}+\frac{\partial v}{\partial y}+\frac{\partial w}{\partial z}\right)\right]+\frac{\partial}{\partial x}\left(-\rho\overline{u'v'}\right)+\frac{\partial}{\partial y}\left(-\rho\overline{v'v'}\right)+\frac{\partial}{\partial z}\left(-\rho\overline{v'w'}\right)+\rho g_y \quad (11)$$

$$\frac{\partial}{\partial x}(\rho uw)+\frac{\partial}{\partial y}(\rho vw)+\frac{\partial}{\partial z}(\rho ww)=-\frac{\partial p}{\partial z}+\frac{\partial}{\partial x}\left(\mu\frac{\partial w}{\partial x}\right)+\frac{\partial}{\partial y}\left(\mu\frac{\partial w}{\partial y}\right)+\frac{\partial}{\partial z}\left(\mu\frac{\partial w}{\partial z}\right)+\frac{1}{3}\frac{\partial}{\partial z}\left[\mu\left(\frac{\partial u}{\partial x}+\frac{\partial v}{\partial y}+\frac{\partial w}{\partial z}\right)\right]+$$
$$\frac{\partial}{\partial x}\left(-\rho\overline{u'w'}\right)+\frac{\partial}{\partial y}\left(-\rho\overline{v'w'}\right)+\frac{\partial}{\partial z}\left(-\rho\overline{w'w'}\right)+\rho g_z \quad (12)$$

$$\frac{\partial}{\partial x}(\rho ut)+\frac{\partial}{\partial y}(\rho vt)+\frac{\partial}{\partial z}(\rho wt)=\frac{\partial}{\partial x}\left(\Gamma\frac{\partial t}{\partial x}\right)+\frac{\partial}{\partial y}\left(\Gamma\frac{\partial t}{\partial y}\right)+\frac{\partial}{\partial z}\left(\Gamma\frac{\partial t}{\partial z}\right)+\frac{\partial}{\partial x}\left(-\rho\overline{u't'}\right)+\frac{\partial}{\partial y}\left(-\rho\overline{v't'}\right)+$$
$$\frac{\partial}{\partial z}\left(-\rho\overline{w't'}\right)+S_t \quad (13)$$

Where:

(Γ) is the coefficient of diffusion ,which is given by: $\Gamma=\frac{\mu}{\sigma}$, ($\sigma=\mu\times C_p/_Y$) is the Prandtl or Schmidt number for the fluid. The terms $-\rho\overline{u't'}$, $-\rho\overline{v't'}$ and $-\rho\overline{w't'}$ are the turbulent heat fluxes, S_t is a source term allowing for the rate of thermal energy production.

3.2 Models of Turbulent:

In this numerical work, the turbulent model RNG K-Ɛ was chosen for use Chen, [29] investigated eight K-Ɛ models for mixed convection flow and found that the Renormalization Group (RNG K-Ɛ) model was the best, closest to reality and acceptable results. Yakhot [30]. It is the best type of Turbulence models has tested. Yuan, [31], Studied of predicting the distribution of indoor pollutants in the displacement ventilation chamber using the model (RNG K-Ɛ), the equation for model of turbulent used are [32]:

$$(\rho U_i)\frac{\partial k}{\partial x_i}=\mu_t S^2+\frac{\partial}{\partial x_i}\left[\alpha_i\mu_{eff}\frac{\partial k}{\partial x_i}\right]-\rho\varepsilon \quad (14)$$

Where $S\equiv\sqrt{2S_{ij}S_{ij}}$ $S_{ij}=\frac{1}{2}\left(\frac{\partial U_j}{\partial x_i}+\frac{\partial U_i}{\partial x_j}\right)$ (15)

$$\rho u_i\frac{\partial\varepsilon}{\partial x_i}=C_{1\varepsilon}\frac{\varepsilon}{k}\mu_t S^2+\frac{\partial}{\partial x_j}\left[\alpha_\varepsilon\mu_{eff}\frac{\partial\varepsilon}{\partial x_i}\right]-C_{2\varepsilon}\rho\frac{\varepsilon^2}{k}-R \quad (16)$$

The model constants are the following values: $C1\varepsilon=1.42$ and $C2\varepsilon=1.68$

3.3 Numerical solution and mesh generation:

In addition to the measurement of the Heat removal effectiveness (ε_t), a simulation was performed to calculate the ADPI of the ventilation of office room While using the software system (CFD), ANSYS Airpak 3.0.16. Airpak simulation software has been widely applied to numerical simulation based on the limited size method, Airpak uses the FLUENT CFD engine to solve thermal calculations and flow of fluid to solve equation to conserve mass, energy, and air momentum. RNG, K-ε turbulence model was chosen to solve flow equations. In this study, the cells number was found (440145) for MV and (482985) for PMV, while using Hexa unstructured geometry to discretize. For this function, all types of element were used to fit the grid to the geometry. Simulations were repeated to a level of approximation 10^{-3} until the solutions were stable as shown in the figure. (3). In addition, a network improvement study was conducted to identify and reduce the error due to the estimate. Four different mesh systems were coarser type, course type, medium type, and fine were generated, to perform the test. The AIRPAK3.0.16 software was used to create the model and meshed study cases based on several testing meshes as shown in Figures 4&5, the dimensions of room edge in (x, y, z) meshed as interval size of (0.04), mesh parameter normal, with Max side ratio (2), The selection was made after several attempts. After applying the meshing strategy.

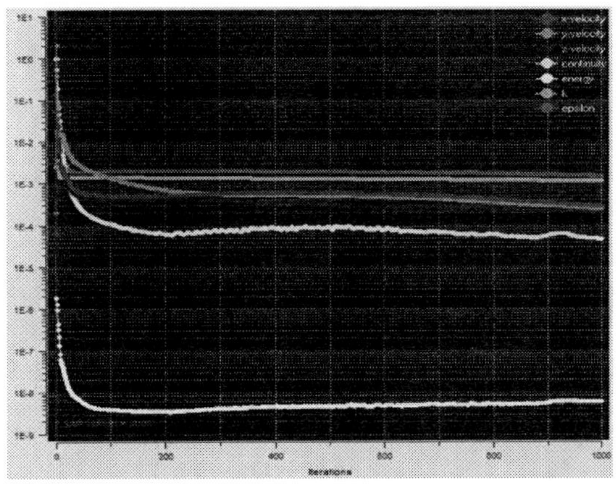

Figure 3. Residuals Plot

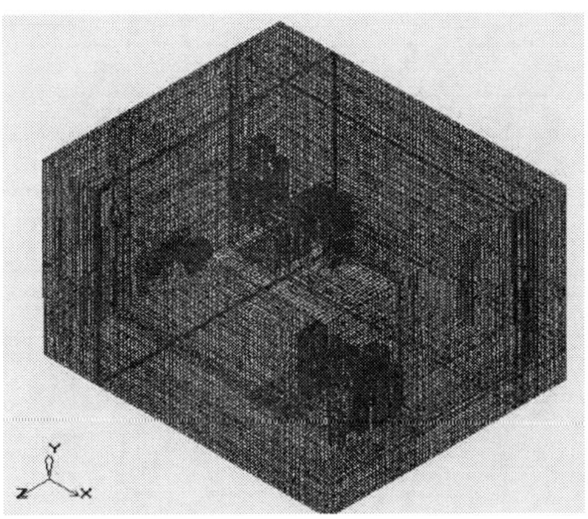

Figure 4. A part from meshed model for only MV

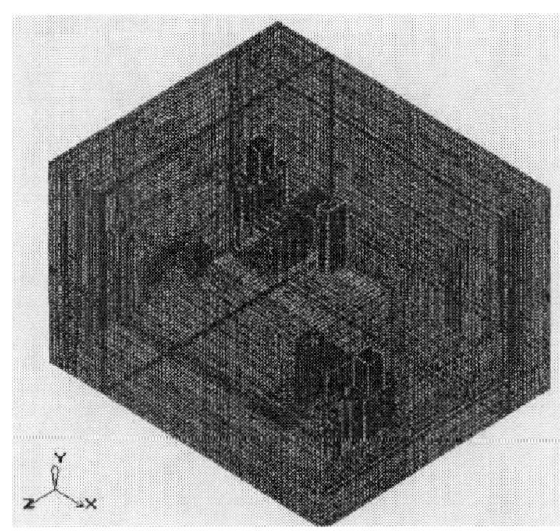

Figure 5. A part from meshed model for PMV

3.4. Boundary conditions

Boundary conditions have a major impact on the success of a CFD simulation in giving a reliable result of a problem solved. In this work, boundary conditions are assumed in a CFD simulation consisting of one type speed. inlet and outflow as tabulated in the table (4) wall boundary included isothermal walls, When the problem is solving numerically it is impossible to get very precise solution, therefore accepted scaled residuals of error must be specified for different terms such as continuity, components of speed and energy.

Finally, it is necessary to verify the validity of the use of the turbulent model RNG K-with another experimental study, The comparison is based on the vertical air temperature, which is measured by five points in the vertical pole as shown in Figure 6.The comparison provides a good fit between the numerical data for this study and experimental data [33], The temperature of air was simulated using the turbulent model RNG K-epsilon .The error rate between experimental and numerical data was calculated and found1.5%.

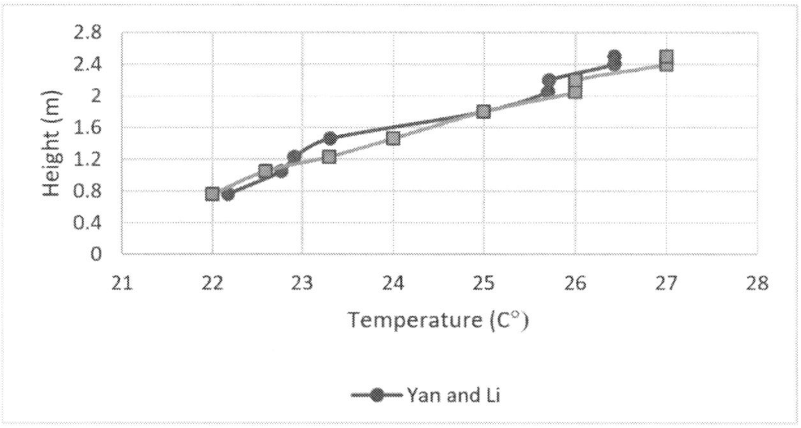

Figure 6 . Validation between the numerical data of this study and an experimental data. [33].

Table 1: room configuration for MV system only (Case-I)

Item	Location (m)			Size (m)			Heat(W)
	X-start	Y-start	Z-start	X-end	Y-end	Z-end	
Office Room	0	0	0	3	2.5	2.5	
Mixing device	0	2.15	1.17	0	2.25	1.33	
Exhaust grille	0	0.3	1.19	0	0.4	1.31	
Person no.1	0.2	0	0.4	0.4	1.1	0.6	75
Person no.2	2.7	0	1.7	2.8	1.1	2	75
Computer_no.1	0.65	0.76	0.3	0.95	1.06	0.6	60
Computer_no.2	2.1	0.76	1.6	2.4	1.9	2.2	60
Light	1.45	2.4	2.3	1.55	2.45	2.5	100
Computer table1	0.6	0.76	0.1	1	0	1.1	
Computer table2	2	0.76	1.4	2.5	0	2.4	

Table 2: room configuration for PMV system (Case-I &Case-II)

Item	Location (m)			Size (m)			Heat(W)
	X-start	Y-start	Z-start	X-end	Y-end	Z-end	
Office room	0	0	0	3	2.5	2.5	
Mixing device	0	2.15	1.17	0	2.25	1.33	
Exhaust grille	0	0.3	1.19	0	0.4	1.31	
Person no.1	0.2	0	0.4	0.4	1.1	0.6	75
Person no.2	2.7	0	1.7	2.8	1.1	2	75
Computer_no.1	0.65	0.76	0.3	0.95	1.06	0.6	60
Computer_no.2	2.1	0.76	1.6	2.4	1.9	2.2	60
Light	1.45	2.4	2.3	1.55	2.45	2.5	100
Computer table1	0.6	0.76	0.1	1	0	1.1	
Computer table2	2	0.76	1.4	2.5	0	2.4	
Personal device1	0.95	1.15	0.3	1.05	1.2	0.54	
Personal device 2	2.1	1.15	1.6	2.2	1.2	1.84	

Table 3: The required conditions inside the office buildings in the Iraqi climate during the summer season, [21].

Item \ Property	DBT	RH	Recommended air velocity inside the office room
The conditions required within the office buildings	23 °C – 26 °C	40% – 50%	0.25m/s_0.13m/s

Table 4: Boundary Conditions

part	types	Condition of momentum	
		Motion of wall	Shear condition
Person	wall	Stationary	Not Slipping
computer	wall	Stationary	Not Slipping
side walls, floor ceil,	wall	Stationary	Not Slipping
tables and lights	wall	Stationary	Not Slipping
Supply air diffuser	Speed inlet	magnitude, normal to boundary	
Extract Grill	Pressure outlet	-Gauge Pressure = (0 Pa), [constant]. -Backflow direction specification Method: (Normal to Boundary).	

Table 5: Different temperature values for Cases (I, II and III)

Types of Ventilation system type	No-case	Setup temperature °C	Average temperature °C	Different in temperature °C
Mixing ventilation system (MV) only	I	25	24.1327	0.8673
MV and PV combined system(5L/s per ATD)	II	25	23.8516	1.1484
MV and PV combined system(10L/s per ATD)	III	25	22.803	2.197

Table 6: numerical values of ADPI and effectiveness ($\mathcal{E}_t$) temp. for each ventilation type

Ventilation devices used	Only mixing	MV combined with PV	
CASE	Case II	Case II	Case III
Air flow rate from PV device	------	5 l/s	10 l/s
ADPI%	53.0277%	62.125%	63.226
Effectiveness	0.9224	1.0955	1.241

4. Results and discussion

This section of the current work is dedicated to presenting and discussing the numerical results obtained from the room tested under the Iraqi climate. CFD technique was used to study the temperature behavior and speed fields in addition to other parameters in the simulated test room. The numerical details that are obtained are more than those obtained from experimental studies. Also, to check the appropriateness of the numerical method and to see whether the data obtained from the calculations can be used in developing the specific model.

- *Air speed distribution (only MV) (Case. I)*

The contour of air speed magnitude in (m/s) in the (z-plane through center=1.25m), In the middle of the room wall, toward the z axis, there is an opening to supply air to the room below the ceiling in addition to the exhaust grille on the same side which is slightly above the floor of the room, is shown in Figure 7. The average speed of air at these locations was about (0.153769 m/s) The air supply speed is 2.5m.s^{-1}, the maximum speed was (4.5217 m/s) at the center of exhaust grille because of it less area than the opening supply air, Therefore, noted the speed of the air leaving the office room higher than the speed entering the room.

figure. 8 shows the location in the x-plane through center x = 1.5m of the office room. The average air speed about 0.153769 m/s and the maximum speed was about 1.47523 m/s in front of the air supply hatch. figure. 9 shows the location in the z-plane, z = 0.4m pass-through person no.1 and computer

IOP Conf. Series: Materials Science and Engineering **978** (2020) 012033 doi:10.1088/1757-899X/978/1/012033

no.1. The average air speed was about 0.153769 m/s and the maximum speed was about 0.516450 m/s (above the person's head and shoulders, as well as the computer) located above the occupant's head and shoulders and in a thin layer near to the room's floor. figure 10 shows the location in the z-plane, z=1.75m pass-through person no. 2 and computer no. 2. The average air speed was about 0.153769 m/s and the maximum speed was about 0.494293 m/s (above the person's head and shoulders, as well as the computer) located above the occupant's head and shoulders. Figure. 11 shows the location at the breathing zone at 1.1 m in the y- plane through 1.1m. The average speed near the person's nose is close to 0.368 and is suitable for breathing and does not cause heat discomfort.

figures. 7&11 show the air speed field around the simulated occupant. A buoyancy-driven thermal plume develops in the vicinity of the body because of gradient of temperature between ambient air and the body surface. Vertical air movement between layers is caused by strong convection forces associated with heat sources (occupant & computer). The movement of a fluid element noted that it consists of a translation, a rotation, and rate of deformation inside the room, then passes vertically to a high level in the office room where it is exhausted from the exhaust grills during the first airflow cycle the speed of air will be high, but as the quantity of room air that becomes mixed increases, the speed of air will decrease.

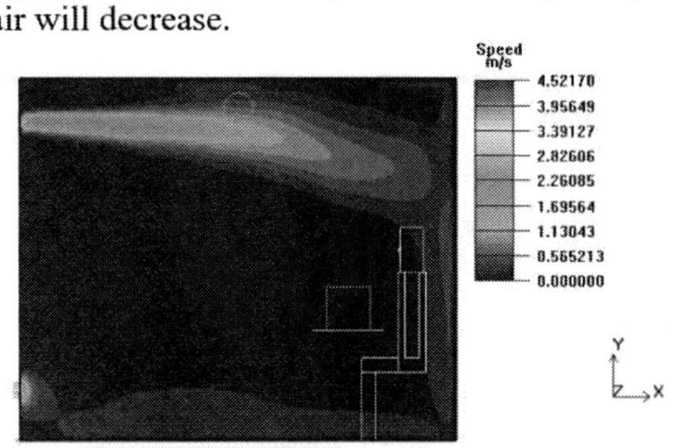

Figure 7 . Contour of air speed magnitude (m/s) at the centers of the supply vents in case 1

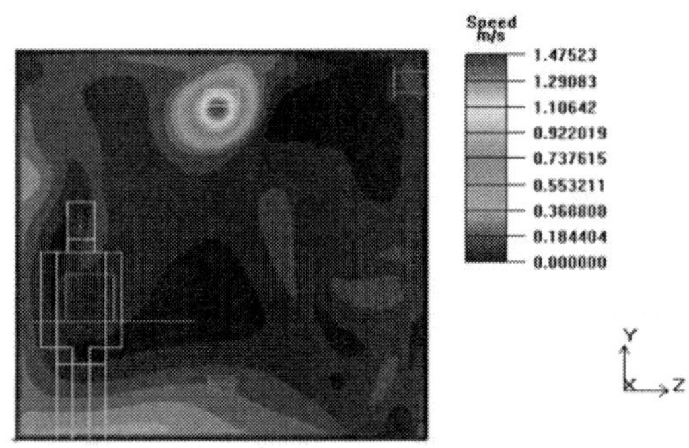

Figure 8 . Contour of air speed magnitude (m/s) in plane(x=1.5) within the office room in case

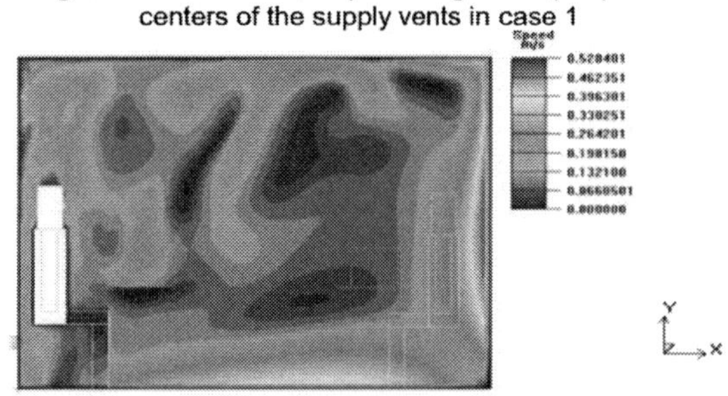

Figure 9 . Contour of air speed magnitude (m/s) in Z-plane(pass through person1 and computer1)within the office room (only mixing ventilation

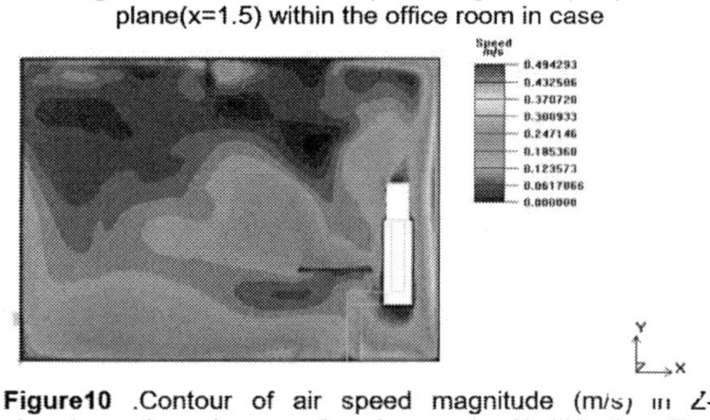

Figure10 .Contour of air speed magnitude (m/s) in Z-plane(pass through person2 and computer 2)within the office room (only mixing ventilation)Temperature distribution contours

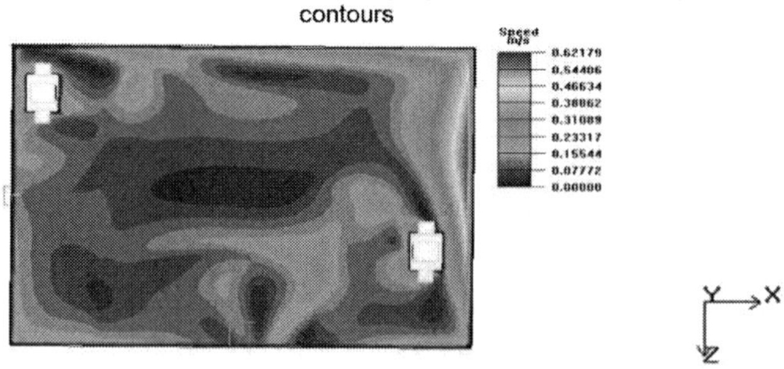

Figure 11. Contour of air speed magnitude (m/s) in (y- plane through 1.1m)within the office room (only mixing ventilation)Temperature distribution contours

IOP Conf. Series: Materials Science and Engineering **978** (2020) 012033 doi:10.1088/1757-899X/978/1/012033

- *Air temperature distribution (case only mixing ventilation) (Case I)*

The temperature contours in the z-plane at z = 0.4m through person1 and computer of the room is shown in figure 12. The average temperature of air around the occupant was 24.235 °C while the minimum air temperature in the room was 22.9475 °C in most parts of it. The mean temperature in the room (except for the (34to 36) °C exhalation temperature out of the occupant's mouth and also near the simulated human body) was about 31.1 °C found closely around the computer, vertically above his head and above the computer's monitor.

The temperature contour in the x-plane through the person are shown in figure 13. The average temperature of air around the occupant was 24.134°C with minimum temperature of 17.42 °C and maximum temperature of 36.6 °C above and around the occupant's head and shoulders.

The temperature contour in (z-plane, z=1.75m) ((z plane through person2 and computer)of the room is shown in figure 14. The average temperature of air around the occupant was (24.236) °C while the minimum temperature of air in the room was (23.4124) °C in most parts of it. The mean temperature in the room (except for the (34to 36) °C exhalation temperature out of the occupant's mouth and also near the simulated human body was about 33.134 °C found closely around the computer, vertically above his head and above the computer's monitor.

The temperature contour in the x-plane through the person is shown in figure 15. The average temperature of air around the occupant was 24.134°C with a minimum temperature of 22.54°C and a maximum temperature of 36.6°C above and around the occupant's head and shoulders.

figures. 12 to 15 display numerically the contours of distribution of air temperature at different planes in the tested room for MV only, an increase in temperature is observed from 17 ° C closely at the supply air terminal and reaches about 36 ° C near the person and (PC) simulator.

MV is distinguished by the fact that the air is transported at high speed outside the occupied zone, the high speed of air in the air jet will generate extra pressure, resulting in airflow from the room and inserting it in the air direction. The air is supplied to the office room at high speed and high momentum from the (air diffuser)(MV device) located at below the ceiling, which leads to mixing the room air with fresh air, and thus the temperature of room decreases and reduces the percentage of pollutants inside the room , an air stream is formed, The air becomes hot due to the heat exchange between the air and the heat sources inside the room, then it vertically rises upward to mix again with the air coming from MV device ,thus carries with it the heat produced by (occupants & computers) and put it out of the room with air outside (exhaust grille). Thermal plumes are generated by convection due to the differences in temperature between the heat sources and the surrounding air. It found that the zone below the ceiling region near the supplied diffuser zone shows the lowest temperature values, due to the cooling effect of the supply entering air. Numerical results showed a decrease in temperature in and near the floor of the room, a gradual increase in temperature was observed, with an increase in the height above the floor of the room, and then it started to decrease again due to the cold air coming from MV device Installed to the top of the wall just below the ceiling. The increase in the indoor temperature of room can be observed near the persons, computers and lighting, this increase in temperature leads to excessive kinetic energy in the stagnation region.

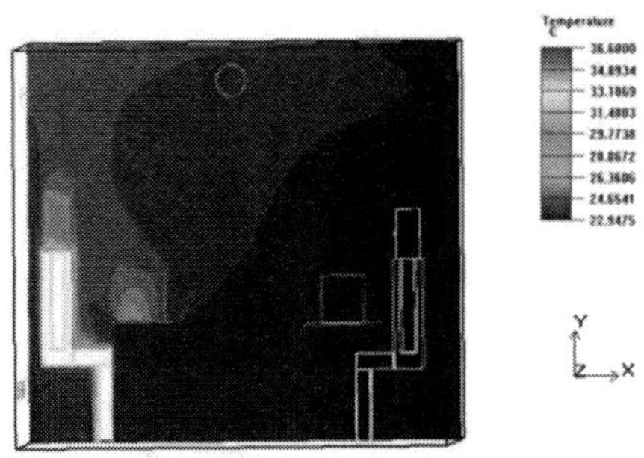

Figure 12 .Contour of air temperature in z-plane at z=0.4m through person no.1 within the office room in case I

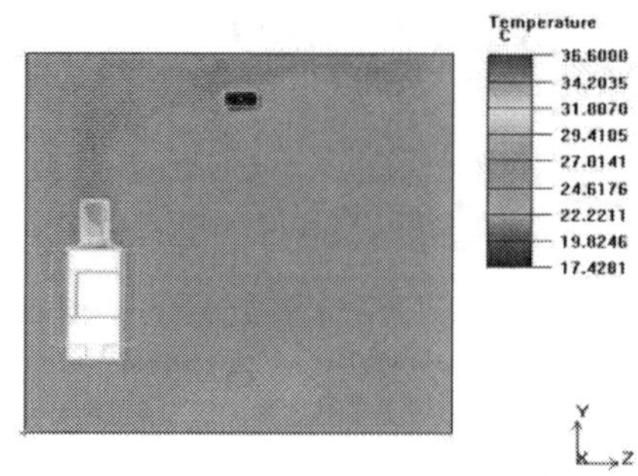

Figure 13 . Contour of air temperature in x plane through person no.1 at x=0.25m within the office room in case I

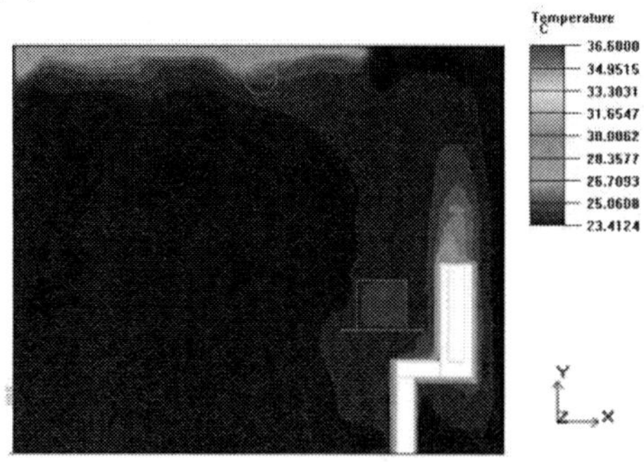

Figure 14. Contour of air temperature in z-plane at z=1.75m through person no. 2 in case 1

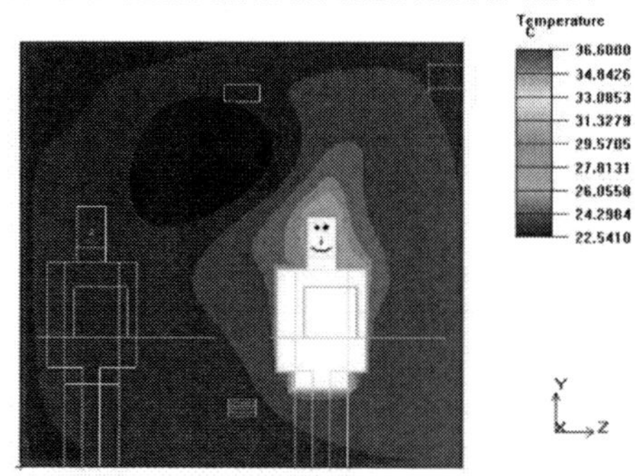

Figure 15. Contour of air temperature in at x plane through person no. 2 at x=2.75m in case II

- *The Effect of Personalized Ventilation Flow Rate (L/s)*

The effect of (PV) system conditioned airflow rate in (L/s) on the thermal comfort conditions and indices (air speed, air temperature) is investigated through the case studies (5&10) L/s. The same fresh air supply temperature, the same speed of air supplied from (MV) device. figures 16 &17 shows path lines of air coming from the PV ATD set on the desk and directed towards the occupant's face at 5 L/s per ATD and 10L/s per ATD colored by speed magnitude

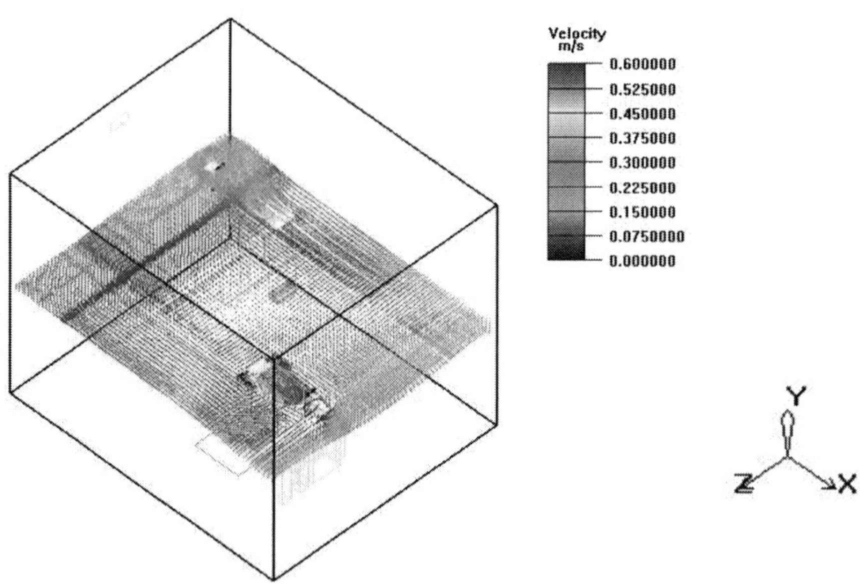

Figure 16. the PV ATDs set on the desk and directed towards the occupant's face at 5 L/s per ATD(CASEII)

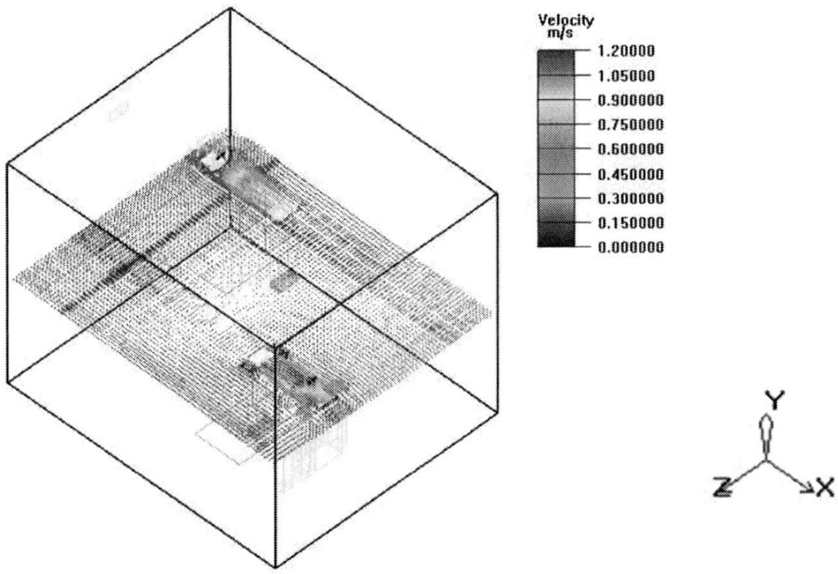

Figure 17. the PV ATDs set on the desk and directed towards the occupant's face at 10 L/s per ATD(CASEIII)

- Air speed distribution, Case II (5L/s per ATD)

The contour of air speed magnitude in (m/s) in the Z- plane through Z=0.4m of the office room (through person1) are shown in figure 18. The average air speed around the occupant's body at this plane was 0.45 m/s and the maximum speed was 1.19542 m/s exactly at the lower part of the occupant's face and decreases gradually over his head.

The contour of air speed magnitude in m/s in the Z- plane through z=1.75m (through person2) of the office room are shown in figure 19. The average speed of air around the occupant's body at this plane was 0.3125 m/s, the maximum speed was 1.311 m/s exactly at the lower part of the occupant's face and decreases gradually over his head.

The contour of air speed magnitude in (m/s) in the (x-plane through 0.25m) are shown in Figure20. The average speed of air around the occupant's body was 0.435 m/s around the occupant's head then speed decreases gradually above that level in an air vortex up to the ceiling, while the maximum speed at that plane was 2.3 m/s found it is located in front of the air supply opening.

The contour of air speed magnitude in (m/s) in the (x-plane through 2.75m) are shown in figure 21. The average speed of air around the occupant's body was 0.213m/s around the occupant's head then speed decreases gradually above that level in an air vortex up to the ceiling, while the maximum speed at that plane was 0.376 m/s.

The contour of air speed magnitude in (m/s), in the y- plane through y=1.1m of the office room is shown in Figure22. The speed near to the simulated human body about 0.1243 and max speed 0.241m/s.

Figures 18 to 22 display the calculated air speed distribution of the cases studied through airflow rate (5 l/s per ATD) for (PV) system. The flow will be studied in two zones. Firstly, high values of air speed obtained from supplied air terminal at the below the ceiling of the room, due to location of MV device in this zone, Secondly, when the air flow pulls by PV system at level 0.4 m (Air intake) to level 1.1m, as shown in Figures.18&19, then it spreads over a large zone lead to the jet momentum will reduce, but still has a sufficient force to reach long distances. Generally, the air speed was increased at level 1.1 m (Figure 22). The maximum speed region where MV exists, remains at a value of 2.5 m/s for MV as stipulated in ASHRAE standard. Then it gradually decreases as the air blowing away from the mixing device.

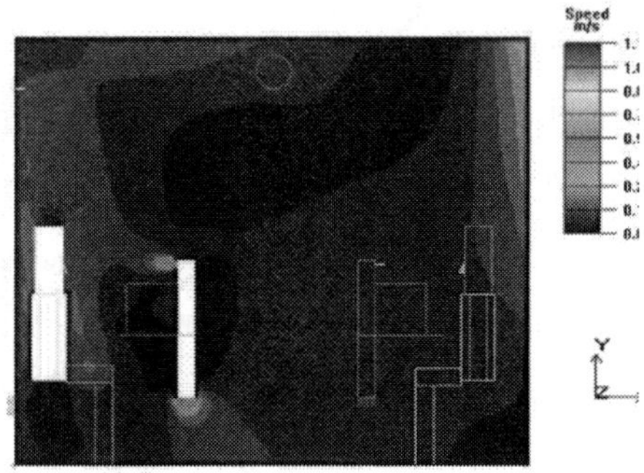

Figure 18 .Contour of air speed in the Z- plane through z=0.4m of the office room through person no.1 at 5l/s per ATD

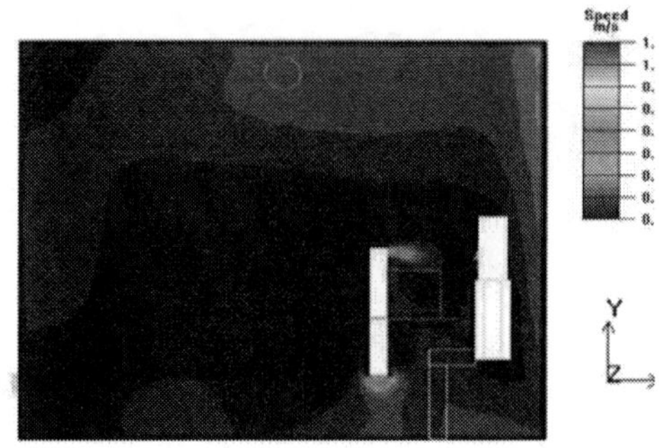

Figure 19 .Contour of air speed in the Z- plane through z=1.75m of the office room through person no.2 at 5l/s per ATD.

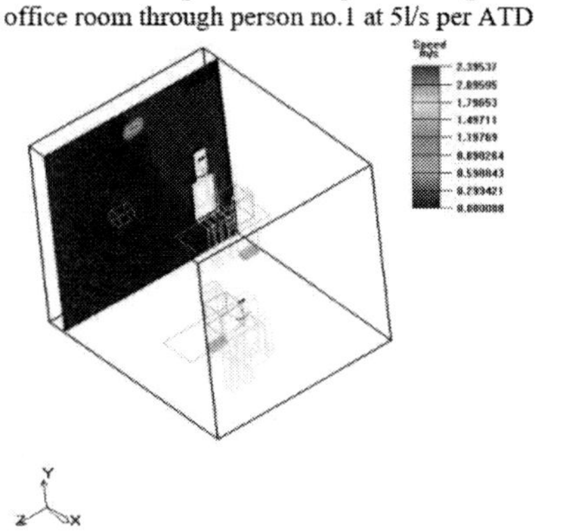

Figure 20 . Contour of air speed in the X- plane through X=0.25m of the office room through person1 at 5l/s per ATD.

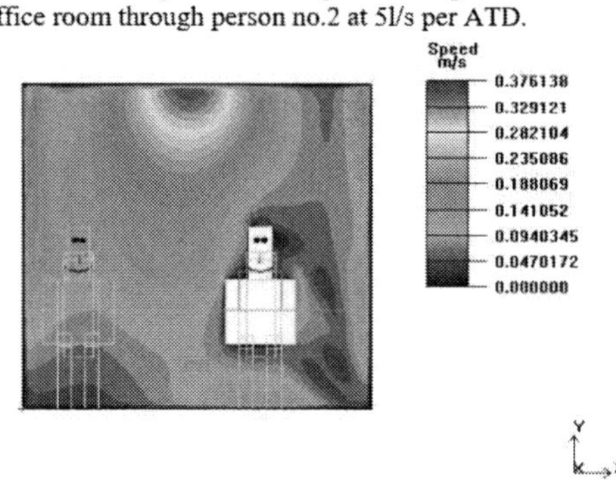

Figure 21. Contour of air speed in the X- plane through X=2.75m of the office room through person no.2 at 5l/s per ATD.

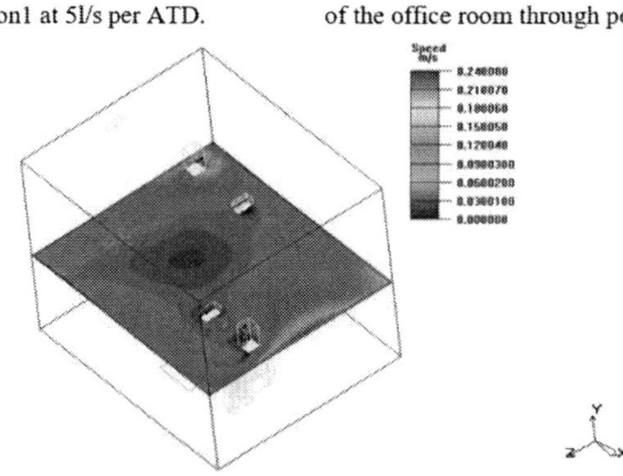

Figure 22. Contour of air speed magnitude in (m/s) in the Y- plane through Y=1.1m of the office room (5L/s per ATD).

- *Air temperature distribution, Case II (5L/s per ATD)*

The temperature contour in the (x-plane through x=0.25m) sectional plane of the room is shown in figure 23. While the minimum temperature of air in the room was 24.145 °C in most parts it due to the heating effect caused by the walls and exactly at the occupant's face front. The maximum air temperature was 34.3315 °C found very closely around the occupant's body and above his head in the thermal plume generated from the occupant's body and blown away by the PV air to a further distance than that of the cases.

The temperature Contour in the (x –plane through) (x=2.75m) sectional plane is shown in figure24. The average temperature of air around the occupant was 24.368 °C with a minimum temperature of 22.541 °C and a maximum temperature of 35.8426 °C above the occupant's head and shoulders. The temperature Contour in the (z –plane through 1.75m) sectional plane is shown in figure 25. The minimum temperature of 23.2142 °C and maximum temperature of 35.3033 °C above the occupant's head and shoulders and continue upwards in a thermal plume with about 18°C air vortex up to the ceiling. The temperature Contour in the (z –plane through 0.4m) sectional plane is shown in figure 26. The minimum temperature of 17.2837 °C and maximum temperature of 34.1980 °C above the occupant's head and shoulders and continue upwards in a thermal plume with about 18°C air vortex up to the ceiling. The temperature Contour in the (y–plane through 1.1) shown in figure27 the temperature in the breathing area is approximately 23 ° C or slightly higher the lowest temperature is approximately 22.4136 ° C.

These figure 23 to27 also display numerically the contours of air temperature distribution at planes in the tested room where the temperature of air increases from 17 ° C closely to MV device and reaches about 34° C near the person and PC-simulator. It was observed in this case (II) that the temperature of air surrounding the simulated person is lower than the air temperature in the case (I). A decrease in temperature of air was observed near the ceiling of the room due to the presence (MV device) near the ceiling of the room on the side wall of the room there a gradual increase in temperature obtained as elevation increase inside the room, then the air temperature decreases due to the cold air coming from(MV)device, Also, an increase in indoor temperature can be noted near the person, pc, lighting, Thermal plumes are generated by natural convection due to the temperature differences between the sources of heat and the surrounding air. Also, the increase in indoor temperature can be noted near the person, PC-simulator and lighting, and noted the existence of personal ventilation in level 1.1m as shown [see figure 27] reduce the heat generated from person and PC-simulator, besides that refreshing air breathing at this zone due to the continuous pumping of air from the layers of cold air in the bottom coming and produce excessive kinetic energy at the stagnation region and distribute approximately in the longitudinal and lateral directions of room, that found from (ADPI) and effectiveness temperature the supply air device of combined (MV and PV) ventilation gives good ventilation.

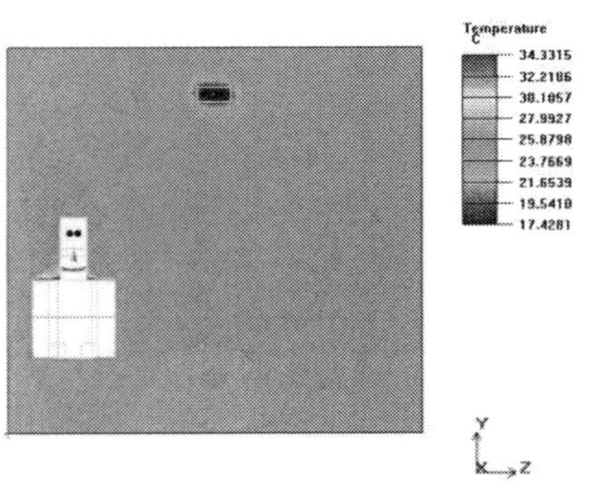

Figure 23 . Contour of air temperature (°C) in(x plane through person1)(x=0.25) within the office room in case (5L/s per ATD)

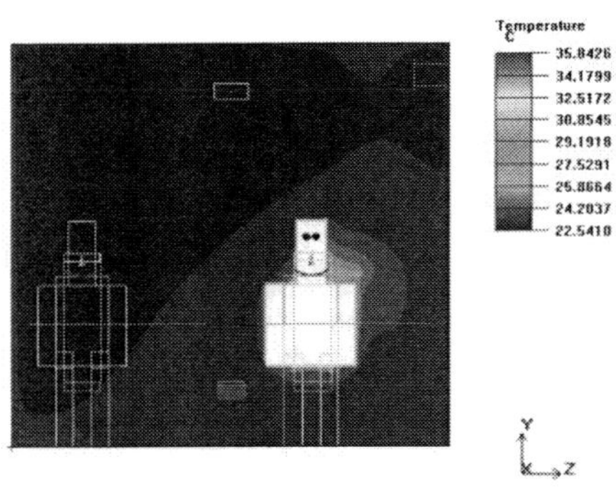

Figure 24 . Contour of air temperature (°C) in(x plane through person2)(x=2.75m) within the office room in case (5L/s per ATD)

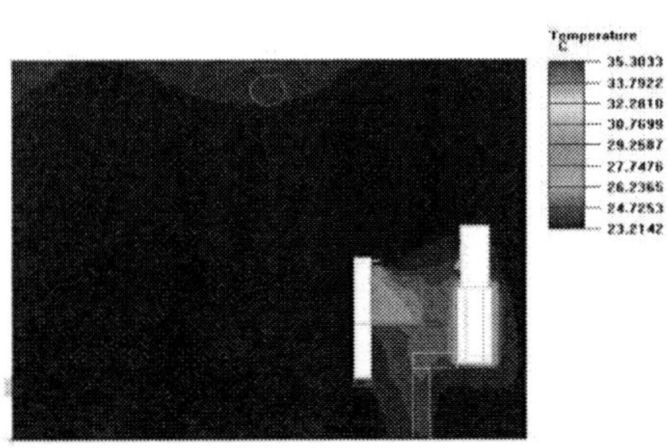

Figure 25: Contour of air temperature (°C) in plane(z=1.75m through person2) within the office room in case (5L/s per ATD)

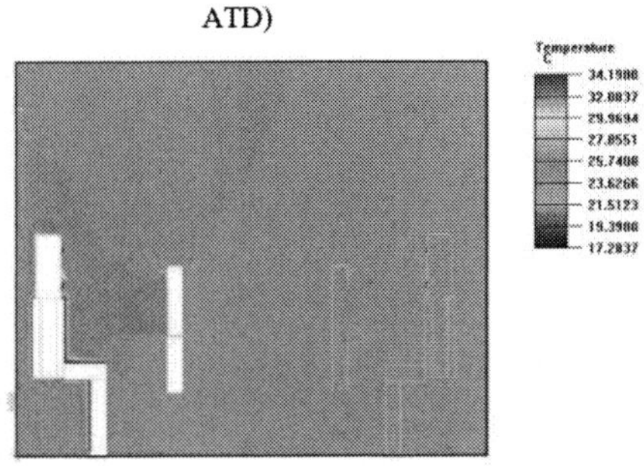

Figure 26: Contour of air temperature (°C) in plane(z=0.4m through person1) within the office room in case (5L/s per ATD)

IOP Conf. Series: Materials Science and Engineering **978** (2020) 012033 doi:10.1088/1757-899X/978/1/012033

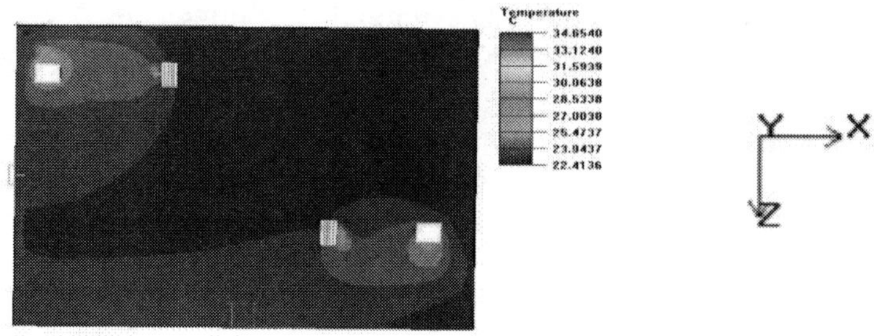

Figure 27 .Contour of air temperature (°C) in plane(y=1.1) within the office room in case (5L/s per ATD)

- *Air speed distribution, Case III (10L/s per ATD)*

The contour of air speed magnitude in (m/s) in the Z- plane through Z=0.4m of the office room is shown in figure.28 The average speed of air around the occupant's body at this plane was 0.45 m/s and the maximum speed was 1.4 m/s (at intake air of personal device) exactly at the lower part of the occupant's face and decreases gradually over his head.
The contour of air speed magnitude in (m/s) in the Z- plane through Z=1.75m of the office room is shown in Figure.29 The average speed of air around the occupant's body at this plane was 0.45 m/s and the maximum speed was 1.5m/s (at intake air personal device) exactly at the lower part of the occupant's face and decreases gradually over his head.
The contour of air speed magnitude in (m/s) in the (x-plane through 0.25m) is shown in figure .30, The average speed of air around the occupant's body was 0.435 m/s around the occupant's head then speed decreases gradually above that level in an air vortex up to the ceiling, while the maximum speed at that plane was 2.3 m/s found It is located in front of the air supply opening.
The contour of air speed magnitude in (m/s) in the (x-plane through 2.75m) is shown in figure.31,The average speed of air around the occupant's body was 0.213m/s around the occupant's head then speed decreases gradually above that level in an air vortex up to the ceiling, while the maximum speed at that plane was 0.376 m/s.
Contour of air speed magnitude in m/s, in the y- plane through y=1.1m of the office room are shown in figure 32 the speed near to person about 0.48m/s and max speed 1.1m/s.
Figures.28 to 32 display the calculated air speed distribution through mass flow rate 10 l/s and for two situations. The flow can be considered in two zones. Firstly, high values of air speed obtained from supplied air terminal from the air diffuser located at below the ceiling of the room due to location of supply air for MV system in this zone, Secondly, when the air flow is pulled by PV system at level 0.4 m (Air intake level) as shown (28&29) to level 1.1m (level of breathing zone for The person sitting) as shown in figure. 32 and the resulting temperature mix should be also higher than the surrounding air temperature. The effect of the PV airflow will be only on the plume air temperature and flow rate, then it spreads over a large area causing the jet momentum to recede but still has a sufficient force to reach long distances. Generally, the air speed was increased at level 1.1 m. Air speed distribution at a mass flow rate of 10 l/s is more acceptable than mass flow rate 5 l/s. The PV airflow speed is assumed to be low enough when reaching the thermal plume so it will not destroy its rising profile.

IOP Conf. Series: Materials Science and Engineering **978** (2020) 012033 doi:10.1088/1757-899X/978/1/012033

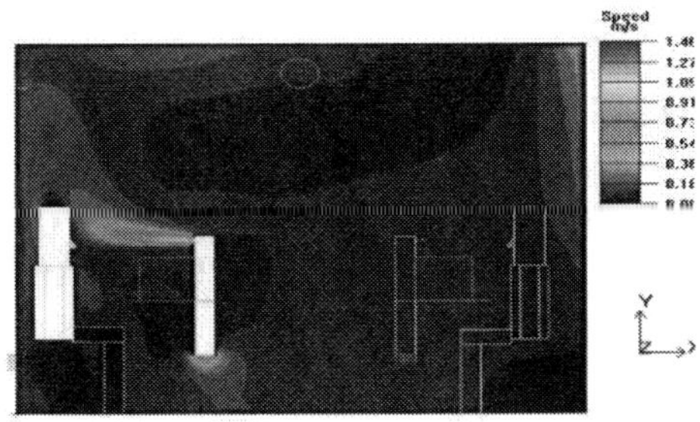

Figure 28. Contour of air speed magnitude in m/s in the Z- plane through z=0.4m(through person1) of the office room, case(10L/s per ATD)

Figure 29. Contour of air speed magnitude in m/s in the Z- plane through z=1.75m(through person2) of the office room 10L/s per ATD

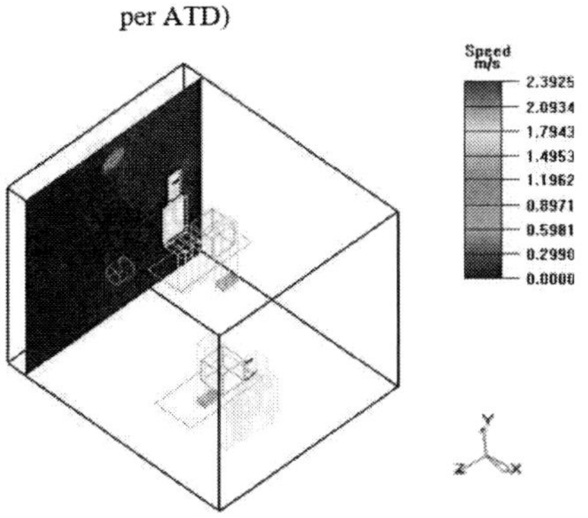

Figure 30. Contour of air speed magnitude in m/s in the X- plane through x=0.25m(through person1) of the office room, (10L/s per ATD)

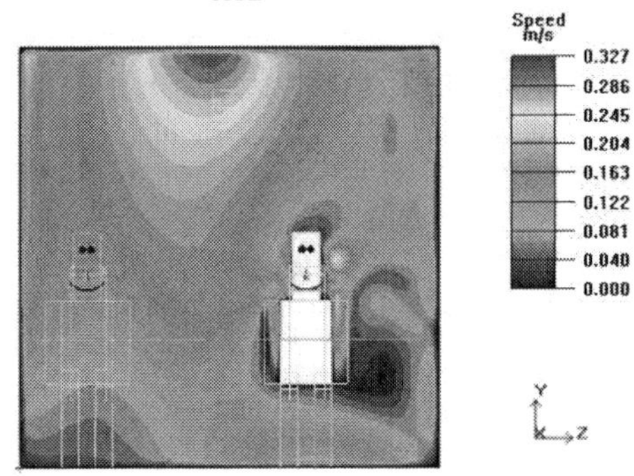

Figure 31. Contour Contours of air speed magnitude in m/s in the X- plane through x=2.75m of the office room (10L/s per ATD)

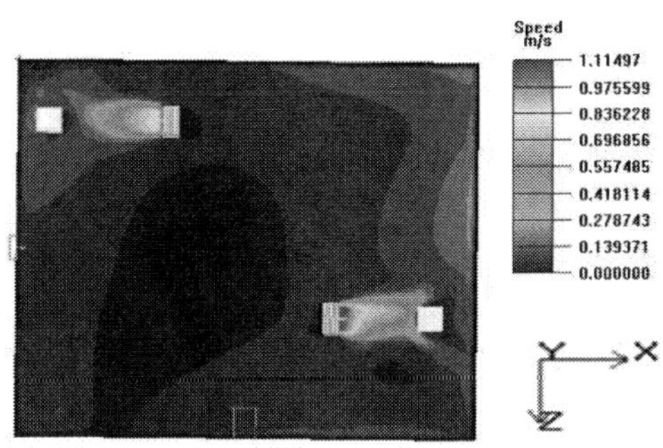

Figure 32. Contour of air speed magnitude in m/s in the Y- plane through Y=1.1m of the office room case II

- *Air temperature distribution, Case III (10L/s per ATD)*

The temperature contour in the (x-plane through x=0.25) sectional plane of the room is shown in figure.30 the minimum temperature of air in the room was 17.8 °C in most parts due to the heating effect caused by the walls and exactly at the occupant's face front. The maximum temperature of air was 32.21 °C found very closely around the occupant's body and above his head in the thermal plume generated from the occupant's body and blown away by the PV air to a further distance than that of the cases.

The temperature contour in the (x –plane through x=2.75) sectional plane is shown in Figure 34. The average temperature of air around the occupant was 24.368 °C with minimum temperature of 22.75 °C and a maximum temperature of 34.8688 °C above the occupant's head and shoulders. The temperature contour in the (z –plane through 0.4m) sectional plane is shown in figure 35. Minimum temperature of 22.45 °C and a maximum temperature of 31.1C above the occupant's head and shoulders.

The temperature contour in the (z –plane through 1.75m) sectional plane is shown in Figure 36. The minimum temperature of 22.45 °C and a maximum temperature of 31.24°C above the occupant's head and shoulders. The temperature contour in the (y–plane through 1.1) shown in Figure 37 The temperature in the breathing area is approximately 24°C or slightly higher The lowest temperature is approximately 22.456 ° C. the maximum temperature 31.246° C.Figures. 33 to 37 show the air temperature distribution contours for the vertical and the horizontal planes in the tested room for the turbulence model used for the mass flow rate of PV at 10 l/s. These figures also display numerically the contours of air temperature distribution at planes in the tested room, where the air temperature increases from 17 ° C closely to the supply terminal and reaches about 31° C near the person and PC-simulator. It found that PV can improve the inhaled air quality in the tested room and each occupant has delegated the authority to optimize and control temperature, flow rate (local air speed) and direction of the locally supplied personalized air according to his/her preference, and thus to improve his/her thermal comfort conditions.

In each layer, the temperature inside the rising plumes is different than the homogenous surrounding air temperature zone and from the wall plume air temperature. The airflow of the PV could disturb the buoyant plume airflow, and it might create some mixing with the surrounding air zone, especially if the PV air is injected at high speed of air or if the resulting mix temperature between the thermal plumes and the surrounding air is lower than the surrounding air temperature forcing the rising air plume to descend. Also, the increase in indoor temperature can be noted near the person, PC-simulator and lighting, and noted the existence of personal ventilation in level 1.1m as shown [see figure. 37] reduce the heat generated from the person and PC-simulator, besides that refreshing air-breathing at this zone due to the continuous pumping of air from the layers of in the bottom and produce excessive kinetic energy at the stagnation region and distribute approximately in the longitudinal and lateral directions over the floor, that found from (ADPI) and $\mathcal{E}_t$ the supply air device of combined (MV & PV) ventilation gives good ventilation, this is consistent with the research results Elmaghraby, (2014) [12].

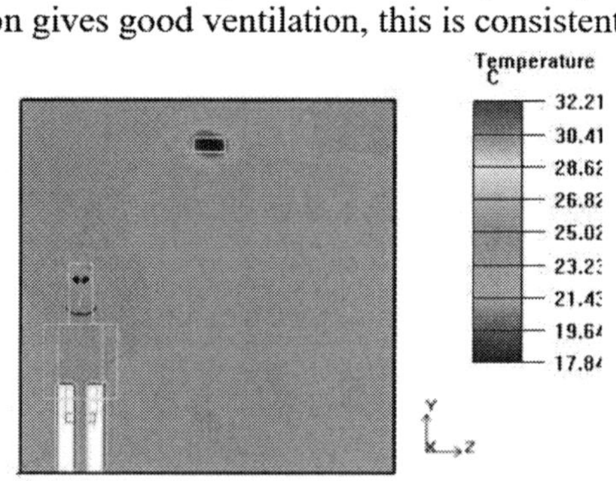

Figure 33. Contour of air temperature (°C) in(x plane through person1)(x=0.25m) within the office room in case (10L/s per ATD)

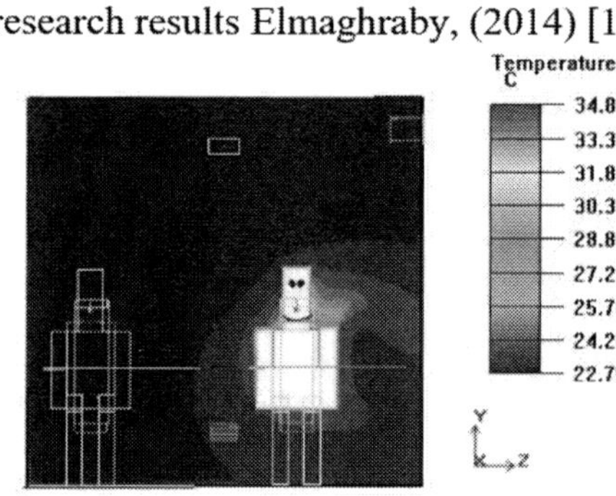

Figure 34. Contour of air temperature (°C) in(x plane through person2)(x=2.75m) within the office room in case (10L/s per ATD)

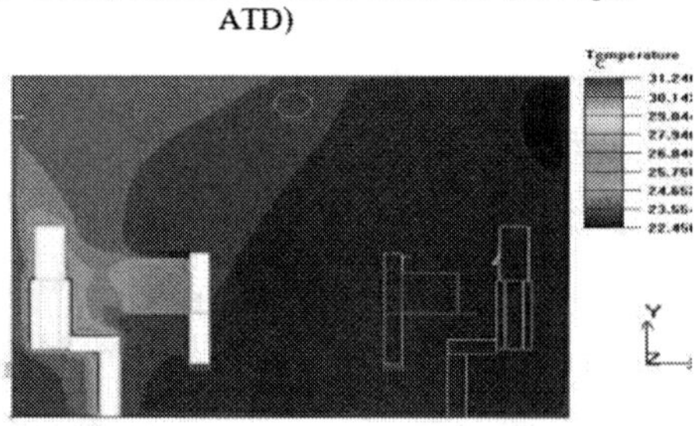

Figure35. Contour of air temperature (°C) in plane(z=1.75m,through person2) within the office room in case 10L/s per ATD

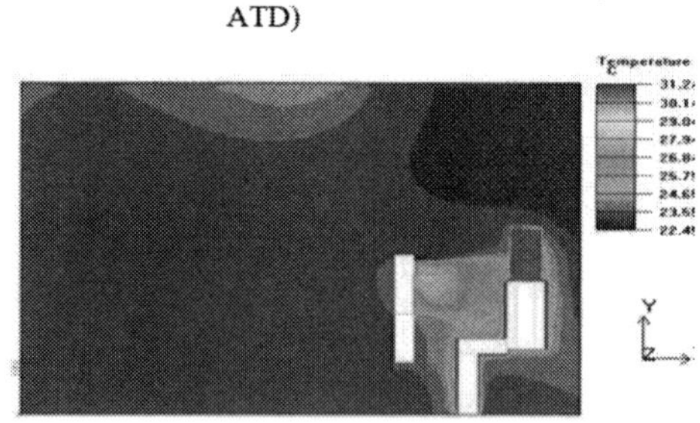

Figure36. Contour of air temperature (°C) in plane(z=0.4m,through person1) within the office room in case 10L/s per ATD

IOP Conf. Series: Materials Science and Engineering **978** (2020) 012033 doi:10.1088/1757-899X/978/1/012033

Figure 37. Contour of air temperature (°C) in plane(y=1.1) within the office room in case (10L/s per ATD)

The effect of changing the PV system air flow rate in (L/s) on the thermal comfort conditions and indices was investigated through case studies (II and III). As a predictable result, increasing the PV ATDs flow rate raised the overall average speed of air in the room and around the occupant's body and decrease the temperature near the person in the breathing zone, these results were compared with the experimental results of the experimental study (Li et al., 2010) [34], and it was noticed that there is a great deal of accuracy and convergence in the results. The effect of altering the MV and desktop PV systems air flow rates concurrently on the thermal comfort conditions and indices was investigated through case studies (II and III). From one case to the other, increasing the PV ATDs flow rate simultaneously made the overall average speed of air in the room and around the occupant's body increasing and maximum speed changing with no specific trend.

The increase in the speed and the amount of air flowing from the PV ATDs also led to a decrease in the temperature near the computers and near the head and shoulders of the simulated human body.

The speed of the air flowing from the PV ATDs should not increase beyond the permissible limit in order not to cause discomfort, as well as cause dry eyes and dry nose and mouth.

- *Relative humidity*

In all the cases studied(I)(II) and (III), it was found that the relative humidity was within the range from 40% to 50% and the effect of personal ventilation systems on it was very little. And as shown in Figures 38, 39 and 40. Where it is noticed that with an increase in the rate of airflow from (ATD), the relative humidity will increase slightly, which does not clearly affect the thermal comfort of the human.

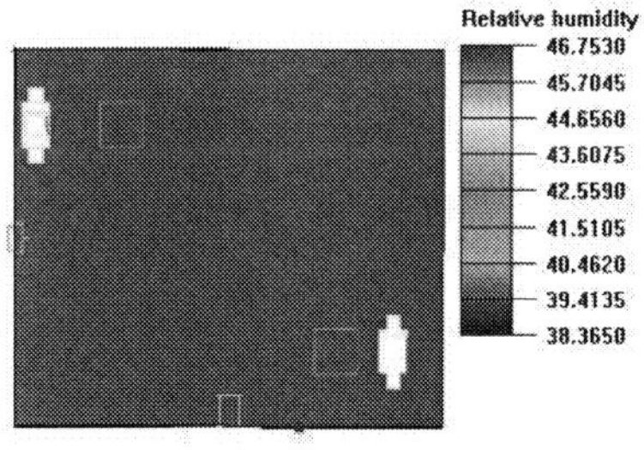

Figure 38. Contour of relative humidity in plane(y=1.1)case I

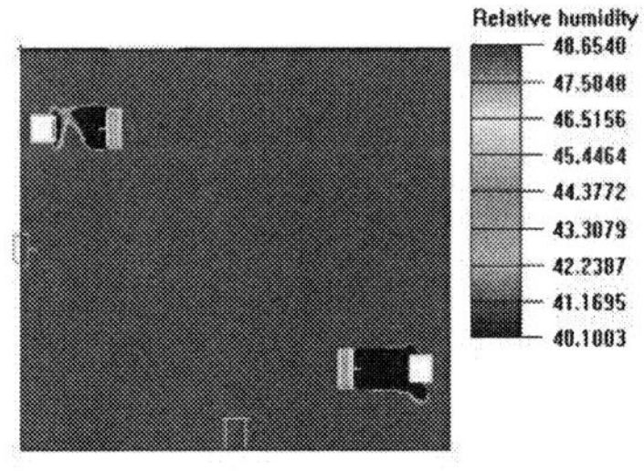

Figure 39. Contour of relative humidity in plane(y=1.1)case II

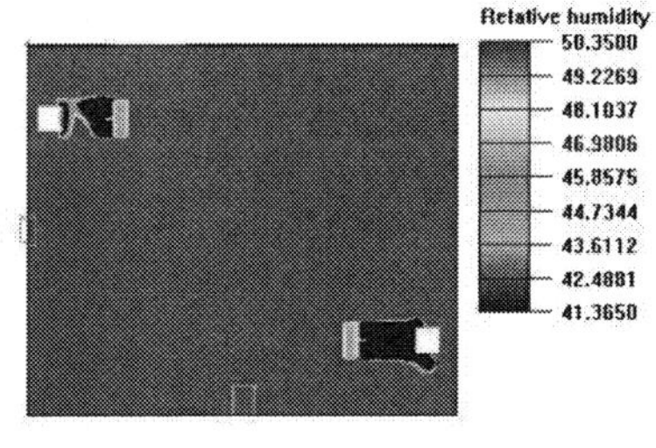

Figure 40 contour of relative humidity in plane (y=1.1)case III

5. Conclusions

The present work focuses on the numerical analysis of air quality and human thermal comfort in the office room by adopting MV and PV system under the Iraqi climate. Where conclusions of this study can be summarized:

1- Using the mixing ventilation combined with personal ventilation systems are more efficient than used mixing ventilation only.
2- Numerical results at combined MV with PV devices give acceptable result at the air volume flow rate of 10 l/s.
3- It has been proven that the supply of fresh cold and clean air to the occupants breathing zone by using PMV system will improve greatly human thermal comfort, PAQ, and inhaled air quality.
4- The turbulence model (RNG k-ε) gives the best compatibility between experimental and numerical results in verification cases by predicting thermal behavior and air flow patterns for MV and PV devices in simulated office rooms.
5- Through numerical results comparisons, ADPI values were calculated for all cases. It was found that the highest value (63.226%) and with respect to (ε_t), the maximum value (1.241) was obtained from-case III.

Nomenclature	Description	Unit
A	Surface Area for Wall	(m2)
C_p	Specific heat of the air at pressure constant	(kJ/kg.K)
C1ε, C2ε	Coefficient in the specific dissipation rate	
D	Diameter	(m)
dx dy dz	Control volume	(m)
G	Gravitational acceleration	(m/s^2)
N	Total number of draft temperature points measured in occupied zone	
N_θ	Number of points of draft temperature measured in occupied zone	
P	Pressure	(N/ m²)
Q	Heat transfer through the wall.	(W)
St	Source term for the rate of thermal energy Production	(J/kg)
T	Temperature	(°C)
u,v,w	component of speed in x,y and z-directions	(m/s)
ui,j,k	Speed at cell (i,j,k)	(m/s)
x,y,z	Cartesian coordinates	(m)
Vx	Local air speed	(m/ s)
Vroom	Volume of room (m3)	
Q_s	design flow rate of the supply air	(l/s)
q_{ex}	Cooling load for the heat conduction through the walls and transmitted solar radiation	(W)
q_l	Cooling load for the overhead lighting	(W)
q_{oe}	Cooling load for occupant, desk lamp and equipments	(W)
Q_t	Total heat transfer	(W)
T_{av}	average room temperature	(°C)
Te	Exhaust Air Temperature	(°C)
T_m	outlet design temperature	(°C)
T_o	outlet temperature	(°C)
T_x	local temperature	(°C)
T_{sp}	Setup (design) temperature.	(°C)
Greek Symbols		
ρ	Air density	(kg/m3)
ε	Turbulent energy dissipation rate.	(J/kg .s)
σ	Prandtl or Schmidt number	-
Σk	Model constant	-
Σε	Model constant	-
Ω	Specific dissipation rate	(1/s)
Γ	Diffusion coefficient	(m2/s)
μ_t	Turbulent viscosity	(N.s/m2)
Ω	Rotation speed	(rad/s)
ΔThf	The difference in temperature from head to foot level.	(°C)
αk, αε	Coefficient in the specific dissipation rate	-
ε_t	Effectiveness temperature	-
τ_t	Turbulent Reynolds stress	-

Sub-Scripts	
av.	Average
c	Correct
e	Exhaust Air
ex	External
f	Floor
hf	Head to foot level.
H	Hydraulic
i	Inside
i,j,k	Location of point in a Cartesian grid
l	Overhead light
Abbreviations	
ACH	Change of air per hour
ADPI	Air Distribution Performance Index
ASHRAE	American Society for Heating, Refrigeration, and Air Conditioning Engineers
CFD	Computational Fluid Dynamics
PV	Personalized ventilation
PMV	Personal-mixing Ventilation
FVM	Finite Volume Method
EDT	Effective Draft Temperature
IAQ	Indoor air quality
MV	Mixing Ventilation
RNG	Re-Normalization Group
SBS	Sick Building Syndrome
ATD	Air terminal device
PAQ	Perceived air quality
DV	Displacement Ventilation

References

[1] Zeng, Qingfan, and Rongyi Zhao. "Prediction of perceived air quality for personalized ventilation systems." Tsinghua Science and Technology 10.2, 227-232, 2005.

[2] Awbi, Hazim B. "Ventilation for good indoor air quality and energy efficiency." Energy Procedia 112 (2017): 277-286.

[3] Melikov, Arsen Krikor. "Personalized ventilation." Indoor air 14, 157-167, 2004.

[4] Kaczmarczyk, Jan, A. Melikov, and Povl Ole Fanger. "Abstract." Indoor Air 14, 17-29, 2004.

[5] Cermak, Radim, et al. "Performance of personalized ventilation in conjunction with mixing and displacement ventilation." Hvac&R Research 12.2, 295-311, 2006.

[6] Nielsen, Peter V., et al. "Chair with integrated personalized ventilation for minimizing cross infection." The International Conference on Air Distribution in Rooms, Roomvent. FINVAC ry, 2007.

[7] Melikov, Arsen K. "Human body micro-environment: The benefits of controlling airflow interaction." Building and Environment 91, 70-77, 2015.

[8] Kroner, Walter M. "Environmentally responsive workstations and office-worker productivity." ASHRAE Transactions 100.2, 750-755, 1994.

[9] Chen, Huijuan. Experimental and numerical investigations of a ventilation strategy–impinging jet ventilation for an office environment. Diss. Linköping University Electronic Press, 2014.

[10] Olesen, Bjarne W. "The philosophy behind EN15251: Indoor environmental criteria for design and calculation of energy performance of buildings." Energy and Buildings 39.7, 740-749, 2007.

[11] Wouters, Peter, and Christophe Delmotte. "Ventilation, good indoor air quality and rational use of energy." Pollution atmosphérique 1, 65, 2005.

[12] ElMaghraby, Hossam A., "Numerical Simulation for Thermal Comfort using Conditioned Air through Mixing and Personalized Ventilation Systems in Field Environmental Chamber (FEC)."2014.

[13] Mohammed, Hyder. A theoretical study of a cold air distribution system with different supply patterns. Diss. M. Sc Thesis, University of Technology, Iraq, 2013.

[14] Chen, Q. "Comparison of different k-ε models for indoor air flow computations." Numerical Heat Transfer, Part B Fundamentals 28.3, 353-369, 1995.
[15] Bunn, R. "Finding the right mix." Building Services Journal 18.8, 43-4, 1996.
[16] Kaczmarczyk, Jan, et al. "The effect of a personalized ventilation system on perceived air quality and SBS symptoms." 9th International Conference on Indoor Air Quality and Climate. 2002.
[17] Van Hooff, T., and B. Blocken. "Mixing ventilation driven by two oppositely located supply jets with a time-periodic supply velocity: A numerical analysis using computational fluid dynamics." Indoor and Built Environment: 1420326X19884667, 2019.
[18] Al Assaad, Douaa, et al. "Mixing ventilation coupled with personalized sinusoidal ventilation: Optimal frequency and flow rate for acceptable air quality." Energy and Buildings 154, 569-580, 2017.
[19] Al-Assaad, Douaa, Nesreen Ghaddar, and Kamel Ghali. "Performance of Mixing Ventilation System Coupled with Dynamic Personalized Ventilator for Thermal Comfort." ASME 2017 Heat Transfer Summer Conference. American Society of Mechanical Engineers Digital Collection, 2017.
[20] Yakoob, Mahdi, Alamir, "Evaluation of Air Exchange Efficiency in Rooms with Personal Ventilation in Conjunction with Displacement Ventilation Systems", International Journal of Mechanical & Mechatronics Engineering IJMME-IJENS Vol:19 No:06, 2020.
[21] Iraqi cooling code, 2012.
[22] Katramiz, E., Al Assaad, D., Ghaddar, N., & Ghali, K. "The effect of human breathing on the effectiveness of intermittent personalized ventilation coupled with mixing ventilation". Building and Environment, 106755. 2020.
[23] Awbi, Hazim B. Ventilation systems: design and performance. Routledge, 2007.
[24] Awbi, Hazim B. Ventilation of buildings. Taylor & Francis, 2003.
[25] Cho, Soolyeon, Piljae Im, and Jeff S. Haberl. Literature review of displacement ventilation. Energy Systems Laboratory, Texas A&M University, 2005.
[26] John, David A. "Designing air-distribution systems to maximize comfort." ASHRAE Journal 54.9, 20, 2012.
[27] Al-Shemmeri, Tarik. Engineering fluid mechanics. Bookboon, 2012.
[28] Halton Co. Care for Indoor Air, AHD - Exhaust Grille guide, 2012.
[29] Melikov, Arsen Krikor, et al. "Impact of airflow interaction on inhaled air quality and transport of contaminants in rooms with personalized and total volume ventilation." 7th International Conference on Healthy Buildings 2003. Natiol University of Singapore, Department of Buildings, 2003.
[30] Versteeg, H. K., and W. Malalasekera. "An introduction to computational fluid dynamics." Finite Volume Method, Essex, Longman Scientific & Technical, 1995.
[31] Yuan, Xiaoxiong, et al. "Measurements and computations of room airflow with displacement ventilation." Ashrae Transactions 105, 340, 1999.
[32] Yakhot, V. S. A. S. T. B. C. G., et al. "Development of turbulence models for shear flows by a double expansion technique." Physics of Fluids A: Fluid Dynamics 4.7, 1510-1520, 1992.
[33] YAN, Shuai; LI, Xianting. Analytical expression of indoor temperature distribution in generally ventilated room with arbitrary boundary conditions. Energy and Buildings, 208: 109640, 2020.

[34] Li, R., Sekhar, S. C., & Melikov, A. K. "Thermal comfort and IAQ assessment of under-floor air distribution system integrated with personalized ventilation in hot and humid climate. Building and Environment", 45(9), 1906-1913. 2010.

Mechanical Properties of Polyester Matrix Composites Reinforced with Waste Marble Particles

Niveen Jamal Abdulkader[1,*], Aseel Mahmood Abdula[2] and Sura Salim Ahmed[3]

[1]Department of Materials Engineering, University of Technology, Baghdad-Iraq
E-mail: 130012@uotechnology.edu.iq
[2]Department of Materials Engineering, University of Technology, Baghdad-Iraq
E-mail: 130027@uotechnology.edu.iq
[3]Department of Materials Engineering, University of Technology, Baghdad-Iraq
E-mail: 130016@uotechnology.edu.iq
* Corresponding author: Department of Materials Engineering, University of Technology, Baghdad-Iraq. Tel: +96407700669754.

Abstract. In this study a preparation of composite materials with a polyester base reinforced by low cost industrial waste and weight fraction from of powder (1,3,5,7,9) % wt of marble particles prepared by manually technique. Mechanical properties have been studied and discussed. The results showed that the new composite materials supported by marble particle gave high results in resistance to hardness, bending and tensile strength compared to the based matrix. This is due to the presence of marble particle, which reduces the concentration of stresses of the base material, and as a result reduces the growth of cracks on the base material. Therefore, the mechanical properties increase with adding of the reinforced particles.

Keywords: *Polyester resin, marble particle, mechanical properties, composite.*

1. Introduction

Composite material is formed by combining two or more materials together which were combined physical interaction and do not react chemically [1]. The constituents are combined in such a way that they keep their individual physical phases and do not form a new chemical compound [1]. That is why a composite is considered to be any multiphase material systems that exhibit greater properties of composite than both of constituent phase [2]. This criterion has motivated the researchers to develop the composite materials over all the world. Composites involve of two phases, which is called reinforcement, embedded in continuous phase, which is called matrix. The boundaries between the two phases are called interface. There are three major types of composite materials which are metallic, polymeric or can even be ceramic [3,4]. Polymers are considered one of the most important challenges

IOP Conf. Series: Materials Science and Engineering **978** (2020) 012034 doi:10.1088/1757-899X/978/1/012034

by adding a quantity of filler with different weights. Adding particles of filling material improves the performance of the overlapping material in industrial and structural applications and gives good properties. The most important goal of the new is to recycle waste, so in this research, marble powders were used as a filling material in polyester [5,6]. Borsellino et.al. [7] concluded the effect of composite materials reinforced with marble powder by various weight fractions on the many properties for the composite with different matrix materials (epoxy and polyester resins). It was concluded that adding marble powder by 60% to epoxy gives superior properties of the material. Raffi Mohammed et al. [8] studied the effect of adding marble powder as a filler material to the epoxy on mechanical properties such as tensile strength, young modulus, compression strength and impact strength of the composites. All these properties of the composite with 5% of marble powder were the best among of all of composites. Husaini et al, [9] determined the tensile and hardness strength which consisting of resin and marble powder in order to obtain the highest values of the compositions of the composite material. High mechanical properties were obtained from yield strength, modulus of elasticity and hardness. The best compositions values for the composites were 60/40 and 70/30. Subhrajit Ray et al., [10] investigated the materials with superior properties of importance were used in the industrial use of superimposed polymeric based materials reinforced by waste of industrial materials in its place of ceramic materials which were highly cost. Where used glass fiber epoxy with different weight content of marble powder. It was observed that the mechanical properties of hybrid composites have been enhanced compared using the non-reinforced material.

The aim of the current study to estimate the using of recycled materials which are waste marble particles as a filler in the polyester.
This is for the purpose of improving mechanical properties and obtaining the best percentage of marble particles addition to the composite material with different weight fraction.

2. Materials and procedures

2.1 Material and Equipment:

2.1.1 Materials

Unsaturated polyester resin (UPR) used as the matrix. The resin is a liquid transparently viscosity which is hardened in thermally at room temperature called (Thermoset) polymer. In table 1. illustrations the specifications of polyester resin.

Table 1. the specifications of polyester resin.

Density (gm/cm3)	Tensile strength (MPa)	Thermal conductivity (w/m.c)
1.225	70.3-103	0.17

Marble powders are added to polyester resin as reinforced materials. The waste marble powders were collected from different processing such as (shaping and polishing), then the powders sieving by a fine sieve. Fig. (1) shows the marble powders which are using.

Figure 1. Marble powder

2.1.2 SEM and EDS of Marble Powder

Figure 2) shows the SEM marble wastes were grinded in the mill and then sieving. polyester matrix was combined with marble wastes at different weight fraction.

Figure 2. SEM marble powder

The EDS analysis was used for detecting the element composition of the marble waste powder, which X-ray spectra are existing by energy in keV on the x-axis and the number of counts on the y-axis as observed in figure (3).

Spectrum: Acquisition 838

El	AN	Series	unn. C [wt.%]	norm. C [wt.%]	Atom. C [at.%]	Error (1 Sigma) [wt.%]
O	8	K-series	49.10	54.54	61.16	10.61
Ca	20	K-series	24.67	27.40	12.27	0.80
C	6	K-series	15.82	17.58	26.25	4.24
Al	13	K-series	0.25	0.28	0.19	0.06
Si	14	K-series	0.18	0.20	0.13	0.05
		Total:	90.02	100.00	100.00	

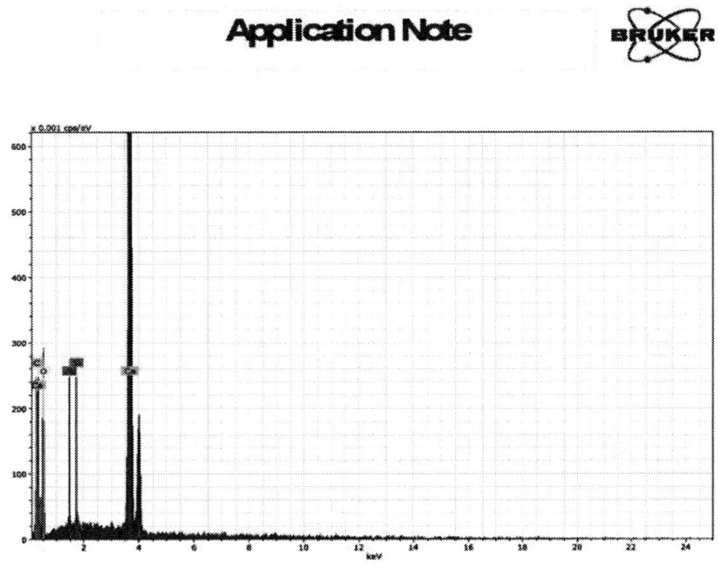

Figure 3. EDS analysis of marble powder

2.1.3 Composite preparation

The test specimen preparation by using unsaturated polyester resin reinforced with different weight percentages (1,3,5,7,9) % of the marble powder unsaturated polyester resin. The materials were mixed at room temperature gradually for voiding bubbly in mixture with adding the hardener material. Then all the mixture was pouring in the mold.

The mold used in the research is made of cast iron, as shown in the figure (4). The mixture was lifting in the mold for (24) hours at room temperature to cure.

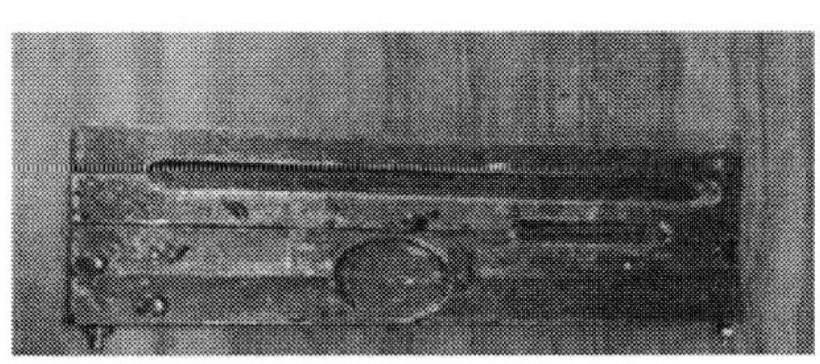

Figure 4. Cast iron mold

2.2 Impact Test:

For polymeric impact testing were used Izod test. This test is performed according to (ISO-180) and the standard sample is shown in (Fig. 5) at room temperature.

Figure 5. Samples of Impact test

2.3 Hardness Test

Shor hardness test was done agreeing to (ASTM D 2240) at room temperature. Figure (6) shows a standard sample for hardness test. Seven measurements of hardness were made at different positions the specimens to determine the average value.

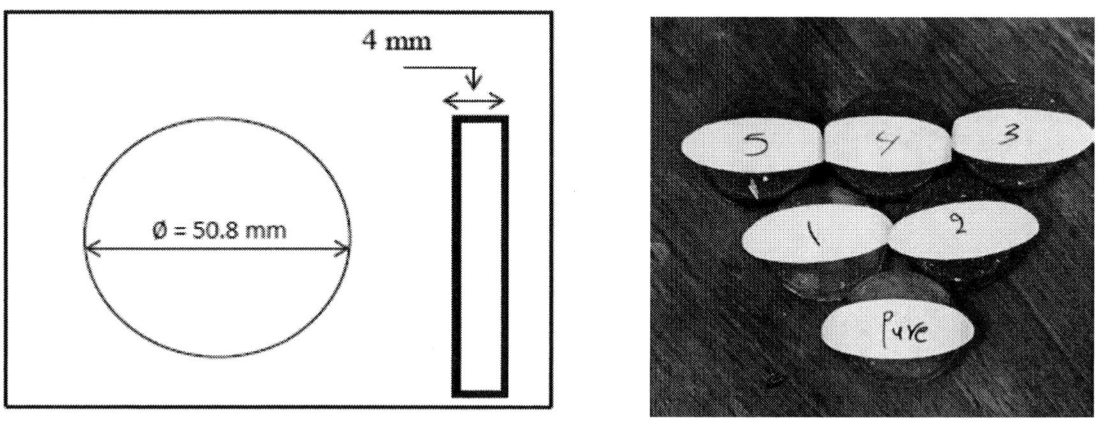

a) Schematic b) Experimental

Figure 6. Hardness test specimen.

2.4 Flexural (Bending) Strength Test:

The bending strength performs for characterization of a fundamental constituent which exposed a perpendicular loading for the axis section.
Bending strength signifies the fracture and were used the three-point test (ASTM D790). In (Fig. 7) demonstrations bending strength test.

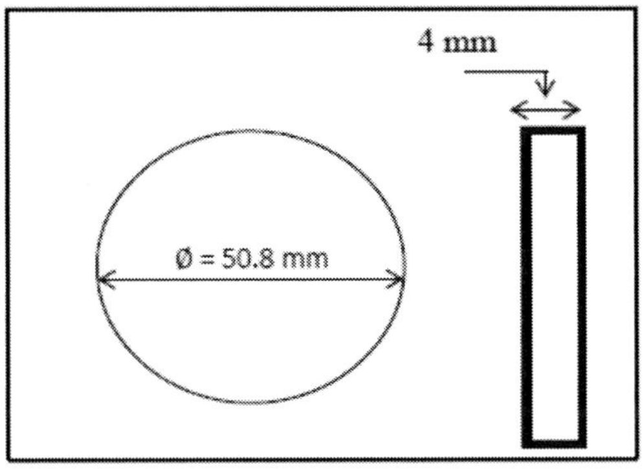

a) Schematic b) Experimental

Figure 7. Bending test specimen.

3. Results and Discussion

3.1 Impact Strength

Impact strength were increased by increasing weight faction of marble powders as shown in figure (8). This is due to the fact that the quantity of powder was good distribution within the matrix, which that lead good interface and reinforcement [8].

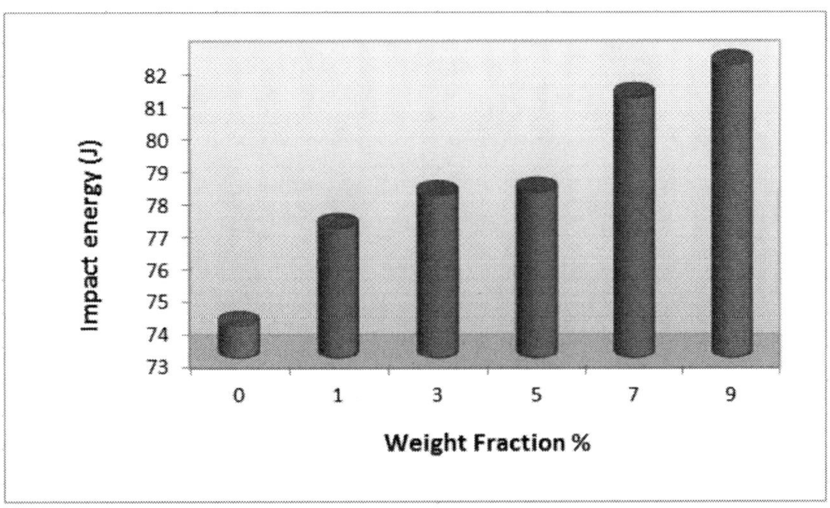

Figure 8. Relation between weight fraction and impact energy

3.2 Micro Hardness

From figure 9 it can be notice from the hardness examination the effect of adding marble powder to the polyester. The addition of 9% of marble powder had been shown a maximum hardness of 0.38 HV among the all of the composites. This is due to the absence of voids and pores which were increased the hardness value [11].

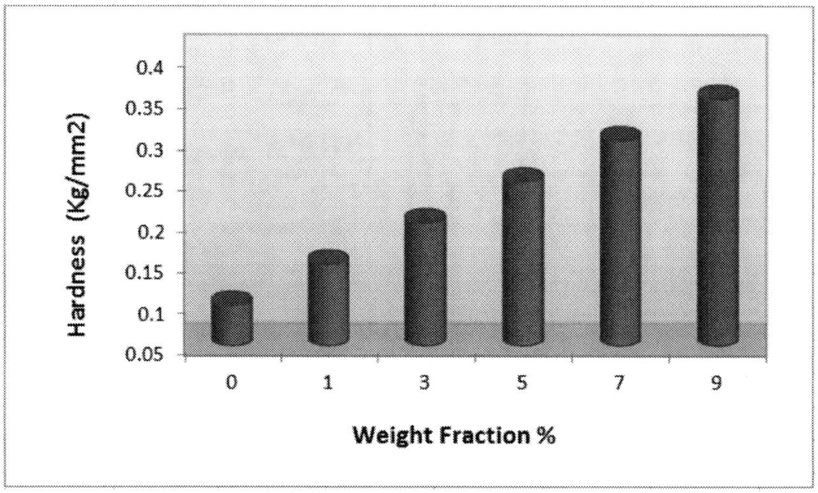

Figure 9. Relation between weight fraction and hardness

3.3 Tensile strength

The results of the tensile test shown in Figure 10. The marble powder additives that were increased the tensile strength value compared to the polyester matrix.

As a result, resistance to interfacial bond increased between marble powder and polyester matrix. As the marble powders were reduced of the stress concentration

the base material during the load tensile test. As a result, the tensile strength increased by increasing the weight fraction of the marble powders due to good interface between marble and matrix. So indicates lower porosity in composite structures.

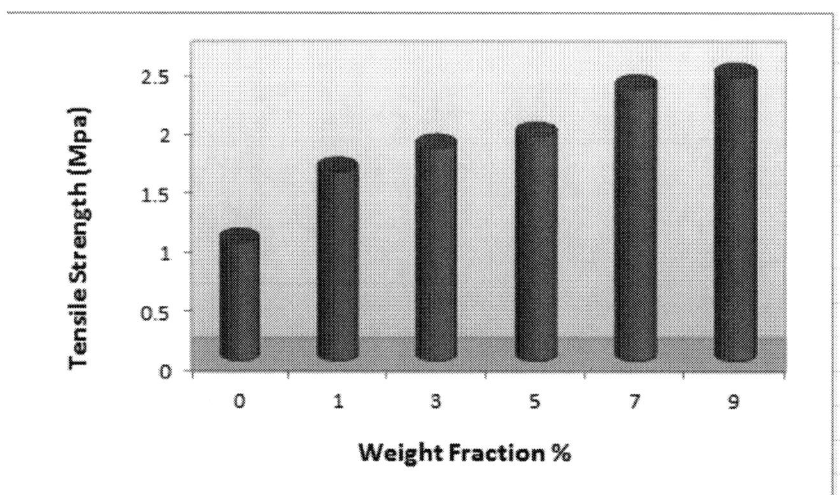

Figure 10. Relation between weight fraction and hardness

4. Conclusions

1- The impact energy for the prepared composite materials were increased by increasing the weight fraction of marble powders. The optimum value of marble powders was at 9% wt.
2- The hardness of composite material increases with increase in weight fraction percentage of Marble (hardness and impact).
3- The prepared composite was improving mechanical properties by using (9% wt.) of marble powder.
4- The use of recycled marble powder has been assignation effect in improving the mechanical properties of the polyester.

References

[1] Callister Willium D, (2007). "Material Science & Engineering – An Introduction", USA, John Wiley & Sons, Inc.
[2] Hull D. and Clyne T.W. (1995). "An Introduction to Composite Materials", Second Edition, Cambridge Solid State Science Series.
[3] Bahar Demirel, (2010). "The effect of the using waste marble dust as fine sand on the mechanical properties of the concrete", International Journal of the Physical Sciences Vol. 5(9), pp. 1372-1380, 18 August.
[4] Imran Oral, (2014). "Ultrasonic Properties of Epoxy Resin/Marble Waste Powder Composites", Polymer Composites, DOI 10.1002/pc.
[5] Aruniit A., Kers, J., Tall, K., Goljandin, D., Saarna, M., Majak, J., Herranen, H. (2011). "Particulate Filled Composite Plastic Materials from Recycled Glass Fibre Reinforced Plastics", Materials Science (Medziagotyra) 17 (3).
[6] Guru M, Tekeli S, Akın E., (2007). "Manufacturing of polymer matrix composite material using natural stone dust and fly ash", Key Eng Mater, 36–338:1353–6.
[7] C. Borsellino, L. Calabrese, G. Di Bella, (2009). "Effects of powder concentration and type of resin on the performance of marble composite structures", Construction and Building Materials 23.
[8] Raffi Mohammed, Raja Kumar, Yvenkatesh, Rahul Kumar, Shak Abdull Azeez, (2018). "Characterization of Marble Powder Filled Epoxy Composites", International Journal of Scientific Research and Review Volume 7, Issue 4, ISSN NO: 2279-543X.
[9] Husaini, Ihsan, M. Nizar Machmud, Sandi Yudha BZ and M P Anhar, (2019). "Mechanical Properties Characterization of Marble and Resin Composite Materials", Materials Science and Engineering 506.
[10] Subhrajit Ray, Arun Ku. Rout, Ashok Ku. Sahoo, (2019). "Effect of Marble Particle on Physico-Mechanical Behaviour of Glass Fiber Reinforced Epoxy Composites", International Journal of Integrated Engineering, Vol.11, No.6.
[11] Binici H, Kaplan H, Yılmaz S., (2007) Influence of marble and limestone dusts as additives on some mechanical properties of concrete, Sci Res Essay, 9:372–9.

Compressibility Characteristics of Soft Clays Treated by Graphene

Mohammed Y. Fattah[1,*], Sinan A. Al-Haddad[2] and Kumail R. Al-Khafaji[2]

[1]Civil Engineering Department, University of Technology, Iraq
E-mail: myf_1968@yahoo.com
[2]Civil Engineering Department, University of Technology, Iraq
E-mail: sinanalhaddad1@gmail.com
[3]Civil Engineering Department, University of Technology, Iraq
E-mail: kumailrasheed@gmail.com
* Corresponding author: Civil Engineering Department, University of Technology, Iraq. Tel: +9647901678315.

Abstract. It is extremely difficult to obtain a change in soil properties by incorporating graphene directly into the soft soil. Therefore, the stability of the soil or cementitious materials can be improved by treating them with graphene. Graphene is easily manufactured from pure graphite powder. Based on this aggregation, it may alter the final properties of graphene, because it completely depends on the raw chemicals and the method of building this material. The mechanical behavior of a new compound graphene was studied through consistency and compression. The soft clay used in this study is characterized with shear strength of 33.0 kPa. It is stabilized with 0.1 % of graphene with curing periods of 1 day and 7 days. It was observed that the void ratio decreased with addition of graphene under a given consolidation pressure. In addition, each of the swelling potential, the coefficient of volume compressibility and the coefficient of consolidation decrease with increasing the period of curing of graphene.

Keywords: *Compressibility; soft clays; improvement; graphene; consolidation.*

1. Introduction

Soils containing a high percentage of clay fraction are a type of problematic soil. Before erecting any structure on this type of soil, cautious through design analysis must be carried out to characterize the softening and high compressibility of the soil. These unfavorable issues merit a thorough investigation of the basic characteristics and shear strength of the soil. Apart from clay, sand and silt make up the other two principal types of soil. A certain percentage of clay is in soil since it holds other particles together and is resistant to water.

Soil settlement is a big problem when relating to structures such as roads, buildings and construction of dams on soft soils. These structures might be unstable and undergo bearing capacity and/or tilting failures due to the high compression and strength deficiency properties of soft soil.

Several methods have been used to treat the problem of low shear strength and high compressibility of soft clays. Different materials were added to soft clays such as lime (Alrubaye et al., 2016), silica fume (Fattah et al., 2014; Alrubaye et al., 2018), graphene (Al-Saadi et al., 2017), graphene oxide (Zhou et al., 2017) and geo-foam beads and bypass cement dust (El-Kady and Badrawi, 2018).

Due to the contrast in interaction, composition, and characteristics between different constituents in nano-composites, a number of controlling factors influence the action that nano-particles play as reinforcement fillers in a polymer-matrix. As an example: (i) fillers must exhibit special mechanical characteristics like strength and modulus of elasticity; (ii) they must include high aspect-ratio and high surface-area to maintain interaction with polymers; and (iii) they must be well-dispersed and away from agglomeration. The dispersion of nano-particle in the composition to the point at which individual particles become coated with the polymer is extremely critical. Increasing dispersion assist achieving good transfer of load to the nano-filler network, causing more uniform distribution of stress. The mismatch between the characteristics of nano-particle and polymer is mitigated by the increasing of interfacial area between the filler and composition, resulting in improved strength. Alignment of the filler in the composition, which is essential, is not critical.

Graphene is recently used in various applications in geotechnical engineering, and the effects of this nanomaterial showed great importance. Due to its high solubility in water, and its presence in soil significantly changes the physical and engineering properties of the soil.

Zhu et al. (2013) made a review on the studies related to the characteristic analysis of graphene and graphene nano-composites (GNCs) and their demonstrated superior actions in storage of energy and conversion, electrochemical- and bio-sensing, environmental remediation and flame retardant application and atomic thickness membrane separation. The applications of graphene and GNCs are highly dependent on their chemical composition, synthetic method, nano-scale morphology, and structure of the hybrids.

Naseri et al. (2016) demonstrated that the use of graphene oxide (GO) nanotubes affect the properties of cemented soils. Various concentrations of GO (0.02, 0.05 and 0.1% by weight of cement) were mixed with the soil to assess the effect of GO on soil compression properties, consistency limits, and shear strength parameters. The results demonstrate that the addition of GO as a stabilizing agent reduced the plasticity and improved shear strength parameters of the treated cement soil mixture.

Bhattachary (2016) concluded that nano-filled polymeric compositions have achieved noticeable electrical, mechanical and thermal characteristics. The treatment of carbon nano-tube, clay montmorillonite platelets and graphene was evaluated as suggested nano-fillers to make nano-composites. The different functionalization procedures of developing the nano-fillers to improve the interaction with polymers were reported. The essentials of filler dispersion in the polymeric composition was presented. In addition, the difficulties and future vision for nano-filled polymeric products were highlighted.

Graphene nano-materials (GMs), such as graphene-oxide (GO) and "reduced graphene oxide" (GO), were extensively used in different applications. For this purpose, He et al. (2017) showed the effect of the highly relevant environmental indices on the aggregation, stability and transformation of GMs in aquatic environments with concerned risks. In addition, the transport of GMs and micro-bial communities developments were also presented depending on the present conclusions.

Pateriya et al. (2019), investigated the use of graphene oxide (GO) as a nano material in low-cement fly ash "Class F" to improve the properties of available local alluvial soils. Several tests such as unconfined compression test, box shear test, and Atterberg's limits tests on the newly formed soil mixture were performed to verify their behavior to stimulate the site condition on the specified soil mixture in the laboratory. A microscopic scanning (SEM) analysis was also used to study the newly formed soil structure mixture. Adding a small amount of GO to the soil-fly ash mixture reduces the plasticity index, and increase maximum dry density, unrestricted compressive strength and cohesive value of the new soil mixture.

The aim of this study is to evaluate the effectiveness of using of graphene as a stabilizing agent in order to decrease the compressibility of soft clays. Laboratory tests were carried out on natural clayey soil and clayey soil–graphene mixtures with different curing periods.

2. Experimental Work

2.1 Soil

The soil was brought from an earth embankment site south of Baghdad. Specific tests were conducted on the remolded soil to find physical properties of the soil. The test results are shown in Table 1. The distribution curve of grain size particles of the soil used is shown in Fig. 1. The soil is classified as "CL" according to the Unified Soil Classification System (USCS).

2.2 Graphene

Graphene used is pure graphite powder. Before the first stage of preparation, experimental tests were conducted to specify the effectiveness of the preparation method. Quality control tests were performed to verify two main points for the homogeneous bottom floor preparation. The first is to determine the shear strength variation at different water contents with time (20%, 22%, 28%, 32% and 36%). The tests determine the time required to restore the treated soil after the mixing process. To achieve this test, five soil samples were prepared individually and laid out in three layers in the Proctor compaction mold. Each layer is gently tamped with a special tamper to extract any trapped air. The soil samples were then covered with a waterproof sheet and left for seven days. The shear strength was measured every day, using a portable shear device. The results show that the shear strength of the soil decreases with an increase in the liquidity index value. It was found also, that the time required to recover the remodeled soil is about 5 days. The second point is to determine the shear strength variance after 5 days against different liquidity indices. Figure 2 illustrates the results of the shear strength variance unrestricted by different liquidity indices.

Table 1. Physical properties of the soil used

The Property		Value
Consistency limits	Liquid limit, LL (%)	32
	Plastic limit, PL (%)	17
	Plasticity index, PI (%)	15
	Activity	0.30
	Liquidity Index, LI	0.266
Specific gravity, G_s		2.70
Grain size distribution	Gravel % (larger than 4.75 mm)	0.0
	Sand % (0.075 to 4.75 mm)	7.0
	Silt % (0.005 to 0.075 mm)	44
	Clay % (less than 0.005 mm)	49
Standard compaction	Maximum dry unit weight, $\gamma_{d,max}$ (kN/m^3)	16.7
	Optimum moisture content, ω_{opt} (%)	21
Unconfined compression strength, q_u (kN/m^2)		67
Classification of the soil (USCS)		CL

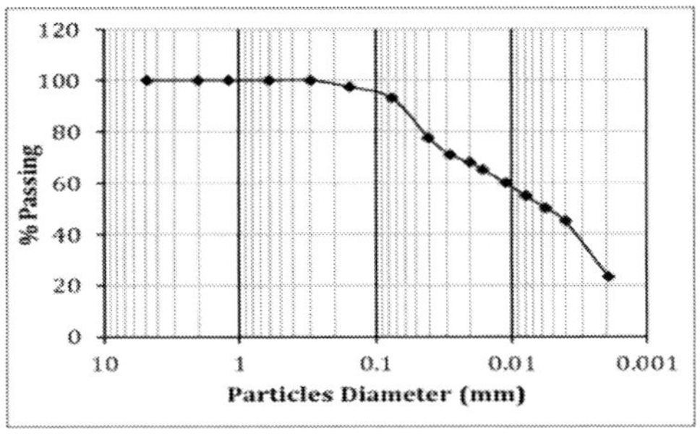

Figure 1: Particle grain size distribution of the soil used

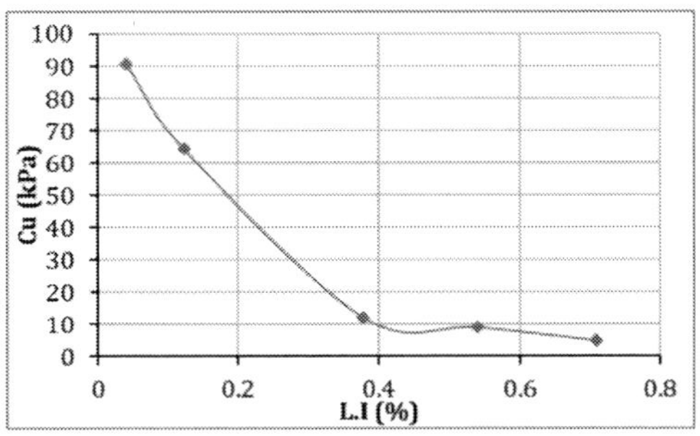

Figure 2: Undrained shear strength with liquidity index after 5 day-curing at different water contents

The soil water content was controlled at 21%. Aqueous solutions of different concentrations of graphene were performed by soaking, before being mixed well with water and the soil to obtain graphene soil-mixture percent by weight. The soil particles in the graphene solution were gradually dropped under stirring. Since the average size of soil particles was approximately 0.2 mm, which is the order size of orders higher than graphene, they were easily dispersed in a graphene solution.

Before gridding, the clayey soil samples were air dried at 105°C, and then mixed with the specific amount of stabilizer (graphene aqueous solution). The mixed soil was tested without and after 1 and 7-day curing.

To understand the rate and volume of soil consolidation upon lateral restriction, loading and depletion, a standardization test was performed (AS1289.6.6.1, 2001). The increase in excess pressure on a layer of increasingly compact cohesive soil leads to a reduction in void ratio with view of long-term strength which or the stability of the soil layer. The settlement process can be predicted based on the value and duration of settlement in the soil layer.

3. Results and Discussion

Table 2 presents the values of Atterberg limits before and after mixing the soil with 0.1% of graphene.

Table 2: Change of Atterberg limits of soft clay treated with graphene.

Graphene (%)	Curing age	LL (%)	PL (%)	PI (%)
0.0	Natural soil	32	17	15
0.1	No curing	34	17	17

Soil plasticity depends on water content associated with the type of soil particles. Graphene nanosheets contain a large set of functional groups on the surface, giving graphene high water-properties in attracting a large amount of pore water. This film of coating soil particles with graphene may be expected to increase the quantity of bound water, and lead to an increase in the soil boundary liquid. This behavior may cause the observed effect of graphene on the plasticity index and liquid soil boundaries. It shows that adding graphene in the clayey soil gives a little effect on the plastic boundary of the soil (Zhou et al., 2017).

Figures 3 and 4 show the void ratio with applied pressure plotted in logarithmic and normal scale, for e-log p and e-p relationships, respectively of natural and treated soils for 1-day and 7-day curing periods.

The most important parameter related to the amount of anticipated consolidation settlement is the compression index (Cc), that a soil layer subject to loads that are greater than experienced in the past (Fattah et al., 2014). It is simply defined as the slope of the linear compression segment of the e-log p curve. The results of compression and swelling indices which are obtained from consolidation tests of the samples with different curing periods are presented in Table 3.

Graphene apparently acts as a binder between soil particles and fill the fine inner fracturing within the soil that reduces the voids and this in turn improve the strength and durability of the treated soil mixture. This improvement in soil strength is referred to the formation of a denser soil mixture as a result of graphene addition. The same trends have been reported by former researchers Zhou et al., (2017) and Pateriya et al., (2019). Researchers concluded that the higher the graphene ratio, the recent the soil matrix density increases with a decrease in voids. Therefore, treated soil responds to more solid behavior compared to untreated soil.

Figures 3 and 4 show that the percentage of voids in the soil has decreased continuously with the uniformity pressure as a result of graphene addition and that the "free" pores within the soil block, is reduced through the application of external pressure.

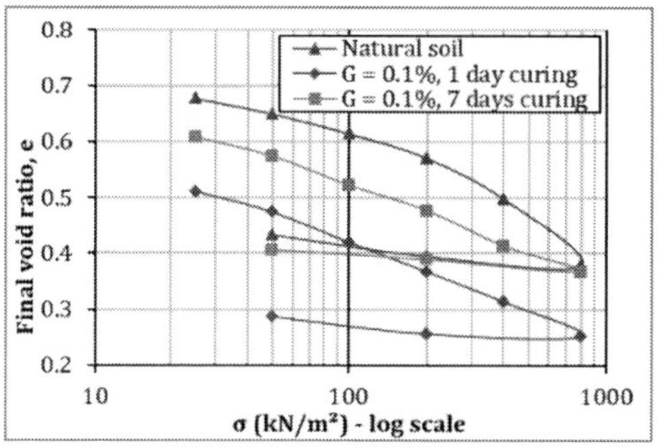

Figure 3: Void ratio – logarithm of pressure relationships of treated and untreated soil

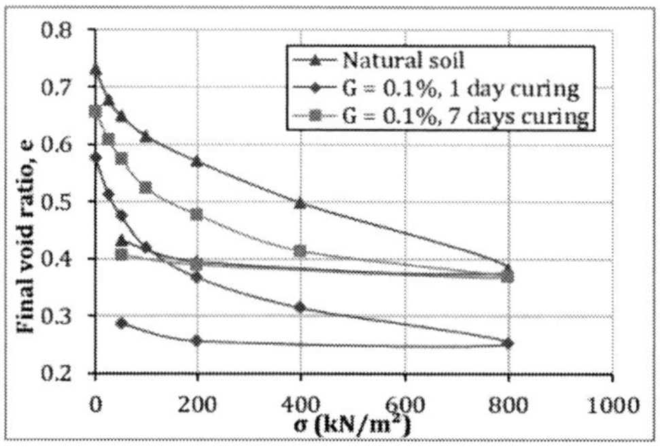

Figure 4: Void ratio – pressure relationships of treated and untreated soil

Table 3. Values of (Cc) and (Cs) of treated and untreated soils with different curing periods

					Decrease in indices (%)
Untreated soil	Loading	Cc		$= \frac{e_1 - e_2}{\log\left(\frac{\sigma_1}{\sigma_2}\right)} = 0.2300$	
			Empirical equations	$\approx 0.009(LL - 10) = 0.1980$	
				$\approx 0.2343\left(\frac{LL(\%)}{100}\right)G_s = 0.2024$	
	Unloading	Cs		$= \frac{e_1 - e_2}{\log\left(\frac{\sigma_1}{\sigma_2}\right)} = 0.0440$	
			Empirical equations	$\approx \frac{1}{5} \text{to} \frac{1}{10}\ C_c = 0.0396$	
				$\approx 0.0463\left(\frac{LL(\%)}{100}\right)G_s = 0.0400$	
Treated soil with graphene, 1-day curing	Loading	Cc		0.176	23
	Unloading	Cs		0.029	34.1
Treated soil with graphene, 7-day curing	Loading	Cc		0.172	25
	Unloading	Cs		0.030	31.8

The nanosheet can actually fill the pores within the soil and coating soil particles. It is difficult to expel the water trapped in the pores between the soil particles and the nanosheets graphene from the soil samples, for reducing the percentage of voids obtained from compression tests, but as the curing period increases, the graphene absorbs from the treated soil the free confined water and the emptiness ratio will increase as shown in Fig. 3. Table 4 presents the values of the coefficient of compressibility (a_v), and coefficient of volume change (m_v) of treated and untreated soils under different applied pressures and curing periods.

Another important parameter in consolidation tests is the coefficient of consolidation (Cv) which relates to how long a time will take to occur an amount of consolidation.

Figures 5 and 6 show the variation of the coefficient of consolidation versus pressure of treated and untreated soils under different curing time periods using Cassagrande logarithm of time method (based on t_{50}; 50% consolidation) and Taylor square root of time method (based on t_{90}; 90% consolidation), respectively.

Table 4. Values of (a_v) and (m_v) of treated and untreated soils with different pressures and curing periods

	σ (kN/m²)	(50 – 100) x 10⁻⁴	(100 – 200) x 10⁻⁴	(200 – 400) x 10⁻⁴
Untreated soil	a_v (m²/kN) $= -\frac{e_1 - e_2}{\sigma_1 - \sigma_2}$	7.000	4.400	3.600
	m_v (m²/kN) $= \frac{a_v}{1 + e_o}$	4.046	2.543	2.081
Treated soil	a_v (m²/kN)	1.100	5.200	2.650

with graphene, 1-day curing	m_v (m^2/kN)	6.9841	3.302	1.683
Treated soil with graphene, 7-day curing	a_v (m^2/kN)	10.200	4.600	3.150
	m_v (m^2/kN)	6.1560	2.776	1.901

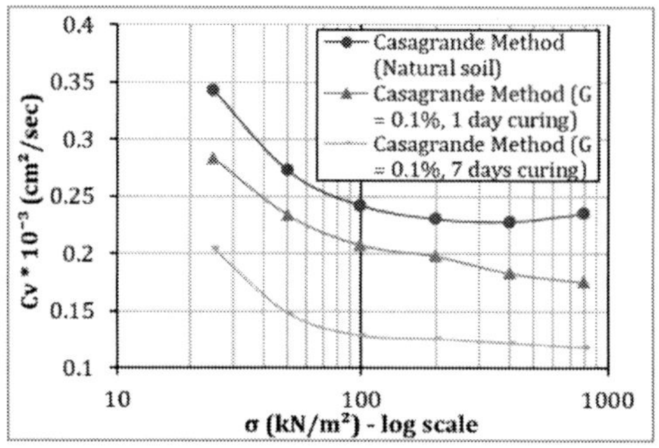

Figure 5. Variation of coefficient of consolidation versus pressure of treated and untreated soil under different curing time periods using Casagrande method.

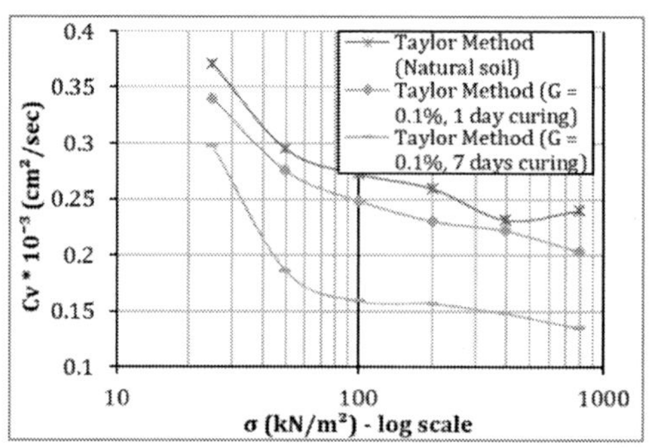

Figure 6. Variation of coefficient of consolidation versus pressure for treated and untreated soil under different curing time periods using Taylor method.

Figures 5 and 6 show that (C_v) decreases with increasing curing time as a result of the change in the soil texture due to interaction within the soil. So, the required time to achieve a specific degree of consolidation and specific drainage path increases for the treated soil with graphene.

The decreases in Cv due to increase in percentage of curing time is a result of change in the soil texture due to reaction taking place within the soil. Thus, the time required to achieve primary consolidation increase for graphene treated soil for a given degree of consolidation and a given drainage path.

4. Conclusions

According to the laboratory results of this study, the following conclusions can be drawn:

1- Adding graphene sheets leads to coating of particles and filling the pores around soil particles and causing to increase the absorbed confined water, and leads to a decrease soil compressibility.

2- The void ratio decreased with addition of graphene under a given consolidation pressure. No swelling potential was indicated.

3- With increasing the curing time, the void ratio of the soil treated with graphene will increase due to absorbing the confined free water.
4- The coefficient of volume compressibility (mv) and the coefficient of consolidation (Cv) decrease with increasing of curing time period of graphene.

Acknowledgements

The tests were performed is Soil Mechanics Laboratory of Civil Engineering Department, University of Technology, Iraq. The authors would like to thank the Laboratory stuff for their help in preparing the setup and machine operations.

References

[1] Alrubaye, A. J., Hasan, M., Fattah, M. Y., (2016), "Engineering Properties of Clayey Soil Stabilized with Lime", ARPN Journal of Engineering and Applied Sciences, Vol. 11, No. 4, February, pp. 2434-2441, www.arpnjournals.com.

[2] Alrubaye, A. J., Hasan, M., Fattah, M. Y. (2018), "Effects of using silica fume and lime in the treatment of kaolin soft clay", Geomechanics and Engineering, Vol. 14, No. 3, pp. 247-255 DOI: https://doi.org/10.12989/gae.2018.14.3.247, Techno press, Korea.

[3] Al-Saadi, T. M. and Jihad, M. A. K., (2016), "Preparation of graphene flakes and studying its structural properties", Iraqi Journal of Science, Vol. 57, No., pp. 145 – 153 (in Arabic).

[4] Bhattachary, M. (2016), "Polymer Nanocomposites—A Comparison between Carbon Nanotubes, Graphene, and Clay as Nanofillers", Materials, 9, 262; pp. 2-35, doi:10.3390/ma9040262.

[5] El-Kady, M. S., and Badrawi, E. F., (2018), "Soft clay treatment using Geo-foam beads and bypass cement dust", Civil Engineering Research Journal, 7(1): CERJ.MS.ID.555701, DOI: 10.19080/CERJ.2018.07.555701 Civil Eng Res J Copyright.

[6] Fattah, M. Y., Al-Saidi, A. A., Jaber, M. M., (2014), "Consolidation Properties of Compacted Soft Soil Stabilized with Lime-Silica Fume Mix", International Journal of Scientific & Engineering Research, Volume 5, Issue 7, July-2014, pp. 1675-1682.

[7] He, K., Chen, G., Zeng, G. Peng, M., Huang, Z., Shia, J. and Huanga, T., (2017), "Stability, transport and ecosystem effects of graphene in water and soil environments", Nanoscale, 9, pp. 5370–5388.

[8] Naseri, F., Irani, M., Dehkhodarajabi, M. (2016), "Effect of graphene oxide nanosheets on the geotechnical properties of cemented silty soil", Archives of Civil and Mechanical Engineering, Vol. 16, Issue 4, pp. 695-701.

[9] Pateriya, A., Kishan, D. and Kushwaha, S. S. (2019), "Evolution of the Geotechnical Properties of Fly Ash Stabilized Silty Soil Activated by Graphene Oxide (GO)", International Journal of Recent Technology and Engineering (IJRTE) ISSN: 2277-3878, Volume-8 Issue-2, pp. 4732 – 4737.

[10] Zhou, G. X., Zhong, J., Zhang, H. Hu, X., Wu, J., Koratkard, N. and Shi, X., (2017), "Influence of releasing graphene oxide into a clayey sand: physical and mechanical properties", RSC Advances, 7, pp. 18060 – 18067, The Royal Society of Chemistry.

[11] Zhu, J. Chen, M. He, Q., Shao, L., Wei, S. and Guo, Z. (2013), "An overview of the engineered graphene nanostructures and nanocomposites", RSC Advances, 3, pp. 22790–22824.

Reinforced Concrete Beams Capacity with Various Concrete Compressive Strengths

Ahmed Abdullah Mansor[1,*], Ahlam Sader Mohammed[2] and Maysoon Adullah Mansor[3]

[1] College of Engineering, University of Diyala, Iraq
E-mail: aamansor@engineering.uodiyala.edu.iq
[2]Civil Engineering Department, University of Technology, Iraq
E-mail: 40159@uotechnology.edu.iq
[3] College of Engineering, University of Tikrit, Iraq
E-mail: dr.maysoonabdullah@tu.edu.iq
*Corresponding author: College of Engineering, University of Diyala, Iraq, Tel: +9647703916610

Abstract. The use of high-strength concrete (HSC) started about four decades ago in world. Using (HSC) supports the applications in civil engineering such as reduce the size of buildings columns, increase the girders length, increase the buildings stores as well as the economics benefits. The paper aims to investigate the effect of using various compressive strength of concrete mixes on the behavior and load capacity of reinforced concrete beams in terms of cracking, yield, failure loads, deflection and ductility. The experimental program included the cast of six reinforced concrete beams with 1.75 m length and cross section of 200 × 300 mm. The beams specimens were tested under two points loads. Several concrete mixes were tried to obtain the concrete compressive strengths 30.9, 40.3, 51.2, 60.1, 71.6 and 89 MPa that used for every specimen beam. The experimental results showed that the increase of concrete compressive strength developed the load capacity strength and ductility. By increasing the compressive strength ratio with (30%, 67%, 95%, 132%, and 188%) caused increase in cracking load by 25%, 66%, 94%, 116% and 152% respectively, and increased the yield and failure loads by (18%, 31%, 41%, 51%, and 69%), (14%, 32%, 47%, 59% and 77%) respectively. As well as the deflection in yield was decreased by (13%, 22%, 33%, 26% and 13%), while it was increased by 10%, 16%, 23%, 26%, and 39% respectively in failure stage as a result of increased of load failure. Adding to that, ductility is increased by 27%, 50%, 84%, 70% and 61%, respectively with the of increasing of compressive concrete strength in ratios mentioned above.

Keywords: *High strength Concrete, Beams, Strength, Deflection, Ductility.*

1. Introduction,

High strength concrete (HSC) could be defined as a concrete with a specified cylinder compressive strength of 41 MPa or more, considering shall not include concrete made using exotic material or techniques [1,2]. Current concrete code of practice, specification and quality control procedures are not necessarily applicable to HSC. Also, the batching, mixing, transporting and placing procedures for HSC are almost similar to the principle from procedure utilized for normal strength concrete. Changes, refinement and increased emphasis on some certain parts of right now utilized concrete quality control procedures may be needed to insure successful concreting operations [1.3].

As results of late advancements in material innovation and requirements for HSC, it is now employed over a much broader range of member types including columns; beams and long- span bridge girders [1, 3, 4]. HSC life time cost may be less than NSC in several cases of construction, since the former requires less repair and renovations [5.6] in addition to that, structural performance and even more economical than the less expensive NSC [5,7].

Using HSC in tall buildings allows for smaller column sizes, thus increasing rentable floor space; also, deflections of tall slender buildings can be reduced because of the increased modulus of elasticity of HSC [1, 2]. As in the bridge girders, utilizing high strength concrete would take into account the utilization of longer spans, which in turn would decrease the number or size of piers required for support then reduce the cost of instruction [3,8].

For standard precast prestressed bridge girders, increasing concrete compressive strength from (34.5 to 48.3 MPa) increases span capabilities of AASHTO girders by about 15% [9,10]. Using HSC increases the available number of stores that can be constructed in buildings subject to a given maximum permissible height [10,11,12].

The present paper points to investigate experimentally the impact of using different compressive concrete strength on the capacity of reinforced concrete beams in terms of cracking, yield, failure loads, deflection and ductility. Six different values of compressive strength were used (30.9, 40.3, 51.2, 60.1, 71.6 and 89 MPa) for this purpose. As well as to predict the benefits of using the HSC in reinforced concrete beams in construction of multi-story buildings and girders.

2. Experimental Program

In this article, the details of the experimental program of the present work are presented, these include materials used, beams specimen's preparation and test procedures.

2.1 Description of Beams Specimens and details

The six beams are identical simply supported reinforced concrete beams were reinforced, molded, cast and tested to assess and evaluate the viability of utilization of a different compressive concrete strength on the load capacity of reinforced concrete beams. All of the beams specimens were of 1.75m length and cross area of 200 x 300 mm. As well as, the specimens have the same ratio of shear span to effective depth which was equal to (a/d= 2.25). 2Φ12 mm deformed steel bars were used in the main longitudinal tension reinforcement. Additionally, the beams were strengthened with 2Φ8 mm bars in compression zone and Φ10@150mm used as stirrups which were given along the shear span to avoid shear failure. Figure 1 illustrates geometry and reinforcement detailing of the beam specimens; this figure shows likewise the configuration of the test. The tensile steel ratio was kept constant as 0.46%. Table 1 sum up the properties of the used steel reinforcement, while Table 2 show the details of concrete compressive strength and steel reinforcement of specimens. All specimens were tested under the effect of two-point loads. All the beams were tested up to failure and the beams central deflections were measured using mechanical strain gauges with a precision of 0.01 mm which was situated at the beam middle-span.

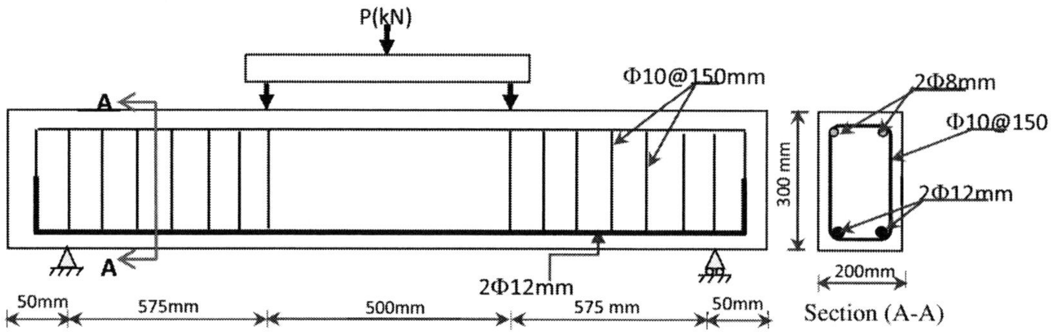

Figure 1. Beam specimen's details

Table 1. Steel reinforcing proprieties

Reinforcing bar	Nominal bar Diam. (mm)	Measured bar Diam. (mm)	Steel Yield stress (fy)MPa	Ultimate steel strength (fu) MPa	*Elastic Modulus of elasticity Es (GPa)*
Main reinforcement	12	12.35	460	648	207
Shear reinforcement	8	8.01	460	663	202

The outcomes are the average of testing three samples from each size of bars with 1000mm long.

Table 2. Properties of beam specimens

Beam	f'c (MPa)	Ten. Reinforcement	Comp. Reinforcement
G1	30.9	2Φ12	2Φ8
G2	40.3	2Φ12	2Φ8
G3	51.2	2Φ12	2Φ8
G4	60.1	2Φ12	2Φ8
G5	71.6	2Φ12	2Φ8
G6	89	2Φ12	2Φ8

2.2 Concrete Ingredients

The most important step in production (HSC) is the selection of the proper types of materials. Also, the proportioning of HSC mix design is more critical than the NSC design mixtures. In the present work, efforts were made to choose acceptable materials to get good quality of concrete mix or HSC.

Ordinary Portland cement is usually used in production of HSC because of the high C3S content percentage (about 60%). Kubaysa cement was chosen to be used in this work. This cement was found to satisfy the ASTM C150-16 [13]. It should be mentioned that, prolonging the curing period was adopted to provide enough time for C2S to hydrate in adequate amount sum to acquire the necessary compressive quality. Tables 3 and 4 show respectively the chemical and physical properties of the used cement.

Table 3. Cement Chemical Results

Compositions	*symbol*	*(%),by weight*	*ASTM, C150-16 [8]*
Lime	CaO	61.0	-
Silicon dioxide	SiO_2	21.0	-
Aluminum oxide	Al_2O_3	5.40	-
Ferric oxide	Fe_2O_3	3.0	-
Magnesium oxide	MgO	2.82	5 (max.)
Sulfur trioxide	SO_3	2.37	2.5 (max.)**
Loss on ignition	L.O.I	2.5	4 (max.)
Insoluble residue	I.R	1.22	1.5 (max.)
Lime saturation factor	L.S.F	0.83	0.66-1.02
Bogie's equation: (ASTM C150-04)			
Tricalcium aluminate	C_3A	9.25	-
Tricalcium silicate	C_3S	41.33	-
Dicalcium silicate	C_2S	29.1	-
Tetracalciumaluminoferrite	C_4AF	9.25	-

Table 4. Cement Physical Results

property	*Test outcomes*	*ASTM,, C150-16 [8].*
Fineness, ((Blain air permeability apparatus)) (m^2/kg)	280	230.0 (min.)
Soundness (by autoclave method)	0.41%	0.8%
Setting Time		
Initial Setting Time, (min)	190	00: 45 (min.)
Final Setting Time, (hr)	4.5	10 (max.)
Compressive Strength (MPa)		
3 Days,	19.35	15.00 (min.)
7 Days,	27.23	23.00 (min.)
28 Days,	41.20	Not applicable
56 days,	56.87	Not applicable

Regular sand taken from Al-Rahalia region(zone) is utilized as a fine aggregate, while crushed gravel from Al-Nibaey region(zone) was utilized to be the coarse aggregate with (14 mm) maximum size. Both fine and coarse aggregate were fulfilling the ASTM C33-11 specification limits [14]. Table 5 shows grading of the used natural sand. Table 6 illustrates the grading of the used crushed gravel aggregate after being sieved on (14 mm) sieve to evacuate particles with size more than 14 mm.

Table 5. fine aggregate grading

size of Sieve (mm)	*Percent passing (%)*	
	Present work results	the ASTM C33-11 specification limits [9]
5.0	98	89-100
2.36	93	65-100
1.18	68.1	45-100
0.60	57	25-80
0.30	14.5	5-48
0.15	6	0-15

Table 6. course aggregate Grading

Sieve size (mm)	*Percent passing(%)*	
	Present work results	the ASTM C33-11 specification limits [9]
20.0	100	100
14.0	100	90-100
10.0	73	50-85
5.0	4	0-10
2.36	0	-

The commercially known Melment L-10 (Melamine formaldehyde condensate) was utilized as a super plasticizing admixture in the production HSC in present work, in view of the facility of more than 12% water reduction. This admixture is classified as type F according to ASTM, C 494[15]. The optimal dosage was found to be 4% by the weight of cement and the reduction in water for this dosage was about 25%. A special consideration was taken to governor the water- cement ratio in the HSC mix.

2.3 Mix Design

Several trial mixes had been made in order to get the desired compressive strength. British Standard BS 5328-83[16] mix design method was used because it yields mixes with strength range higher than the compressive strength ACI 211 method [17]. The concrete compressive strength was 30.9, 40.3, 51.2, 60.1, 71.6 and 89 MPa. The experimental program had been done in the structural laboratory of civil engineering department// university of Diyala. The mix proportions details are shown in Table 7.

Table 7. Mix details

Spe. Name	Experimental f'c(N/mm^2)	Mix Properties (Kg/m^3)					
		W/c-ratio	Water	Cement	Sand	Gravel	S.p%
G1	30.9	0.55	192	350	750	1128	-----
G2	40.3	0.50	200	400	720	1100	----
G3	51.2	0.42	200	475	700	1045	----
G4	60.1	0.37	200	555	670	1000	----
G5	71.6	0.313	180	575	675	990	2%
G6	89	0.267	160	600	670	990	4%

2.4 Preparation of Beam specimens

The main steel reinforcing bars were cuts to the chosen length, and formed at the end of each bars with a 90-degree hook according to ACI 318-14 code [18]. No stirrups were used in middle third zone (pure bending zone). After the reinforcement placement in the molded form, it was fixed carefully to prevent its movement during casting and vibration of concrete. In adding to the beam specimens, standard cylinders control specimens were cast from each batch of concrete. The control specimens and the beam were vibrated after casting in their molds by using table vibrator. The concrete was cast in two layers with a compaction time of about 1-2 minute for each layer. The beams and control specimens were covered with moist burlaps and left in the laboratory for twenty-four hours. After that, the molds removed and all specimens placed in water paths for curing with 24C^{O} temperature degree. All the casted specimens were tested at the age of 32 days after their curing time and drying for about 4 days. The test of compressive strength had been done according to ASTM C39 [19] utilizing (150*300mm) standard cylindrical specimens which was cast with every concrete beam specimen. A computerized testing machine of 2000kN was utilized for concrete compressive strength.

3. Results and Discussion…

3.1 Beams General behavior

The six beam specimens were tested with two-point loads. The load was applied using a hydraulic testing machine with increments of 10kN. Figure 2 shows the shape of the failed G6 specimen. All tested beams were showed an elastic behavior and they were free of cracks in early stages loading. After increasing the applied load, the flexural cracks in mid span were appeared gradually. Then the shear cracks were developed and expanded in the shear span zone (range zone) with the loads further increasing. In general, the planes of failure were formed alongside the diagonal line joining the supports and the loading points. This was also noticed by Amer et al [20]. The first crack was recorded by microcrack equipment and the yield loads are recorded according the load deflection curves.

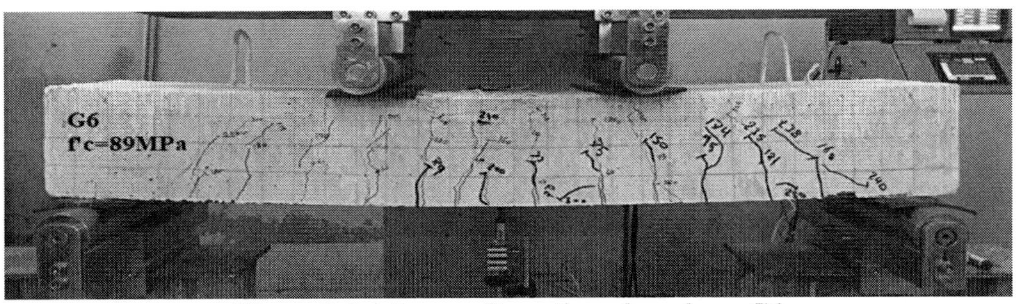

Figure 2: loading configuration of specimen G6

3.2 First cracking, Yielding and Ultimate Loads Capacity

Table 8 presents the experimental results concerning the first cracks, yield and ultimate loads capacity of the six beam specimens adopted in the present work. The results of beam (G1) were served to be as a control one and the results of the other specimens of beams were compared to G1. As well as concrete compressive strength (**f'**$_c$) values are presented in Table 8. It can be seen that the initial cracks developed at a higher load with the increase of compressive strength, the increase in compressive strength provide a good restraint to the crack's growth [21]. Also, test results demonstrate that the beam with higher compressive strength values gave more load capacity due to reduce in block stress in compression zone. For example, increasing the compressive strength form 30.9MPa for (G1) to 60.1MPa for (G4) causes improving in the ultimate load capacity by 47%. It could be notified from Table 8 that the highest value for the ultimate load capacity was recorded for beam (G6) which was about 77% higher than that of the control one.

Table 8. Experimental results

Beam name	f'_c	% Different	Cracking loads f_{cr}(kN)	% Different	Yield load P_y(kN)	% Different	Ultimate load P_u (kN)	% Different
G1	30.9	---	32	---	90	---	133	---
G2	40.3	30	40	25	106	18	152	14
G3	51.2	67	53	66	118	31	176	32
G4	60.1	95	62	94	127	41	196	47
G5	71.6	132	69	116	136	51	212	59
G6	89.0	188	76	138	152	69	236	77

3.3 Load- Deflection Responses

The performance of the beams under different phases of loads for different compressive strength could be described through Load-deflection relation. The central deflection of all tested beams specimens at yield and ultimate loads is presented in Table 9. As expected, the values of deflection depend mainly on the values of compressive strength for the tested beams. At the ultimate load, the central deflection was increased with increasing the compressive strength in companied with capacity strength [22]. The highest value for mid deflection at ultimate load was recorded for beam (G6) which was about 39% higher than that of the control one (G1) companied with 77% increasing in load capacity as mentioned in paragraph 4.2.

Table 9. Load – central deflection results

Beam name	f'c	% Difference	Deflection at yield load (mm) Δ_y	% Difference	Deflection at ultimate load (mm) Δ_u	% Difference	Ductility μ_d	% Difference
G1	30.9	---	5.5	---	31	---	5.6	----
G2	40.3	30	4.8	-13%	34	10%	7.1	27%
G3	51.2	67	4.3	-22%	36	16%	8.4	50%
G4	60.1	95	3.7	-33%	38	23%	10.3	84%
G5	71.6	132	4.1	-26%	39	26%	9.5	70%
G6	89.0	188	4.8	-13%	43	39%	9.0	61%

The load-central deflection curves for all values of the adopted compressive strength are given in Figure 3. These curves seem to be identical in profile to their reference one (G1) with diverse extremes. It could be seen from Figure 3 that in the pre-cracking stage, the deflection is linearly increased with applied load increasing for all the tested beams because at this stage, the strains in the concrete as well as in steel reinforcements are comparatively small and both materials are in elastic portion of their respective responses. Gradually, the slope of the load- deflection curves is changed as results for concrete cracks with increasing the load in the post-cracking stage. The cracks caused a reduction in the effective moment of inertia value for the cross sections of beams.

The deflection again increases almost linearly with the further increase of the applied load up to the point at which the tensile steel starts to yield. After steel yielding stage (post-yielding stage), the deflections immediately increased non-linearly due to the reduction of neutral axis depth. Each beam showed a different post-yield load-deflection response.

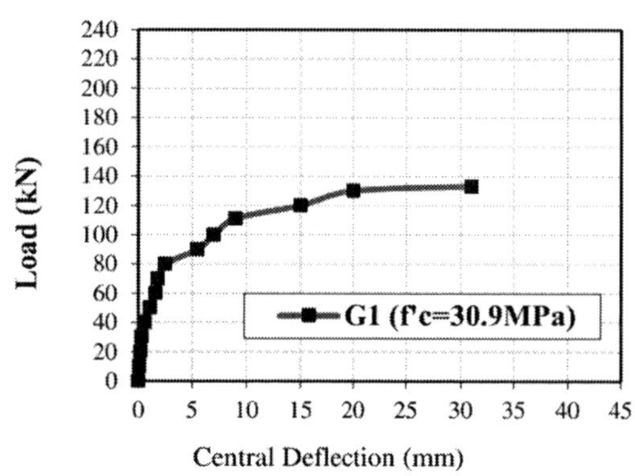

(a): Load- central deflection curve for beam G1

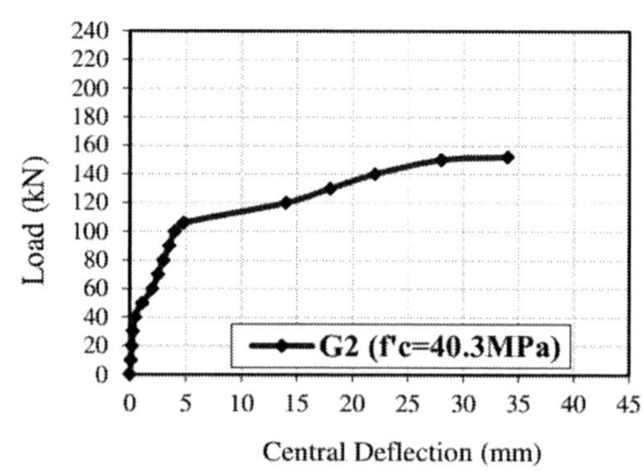

(b): Load- central deflection curve for beam G2

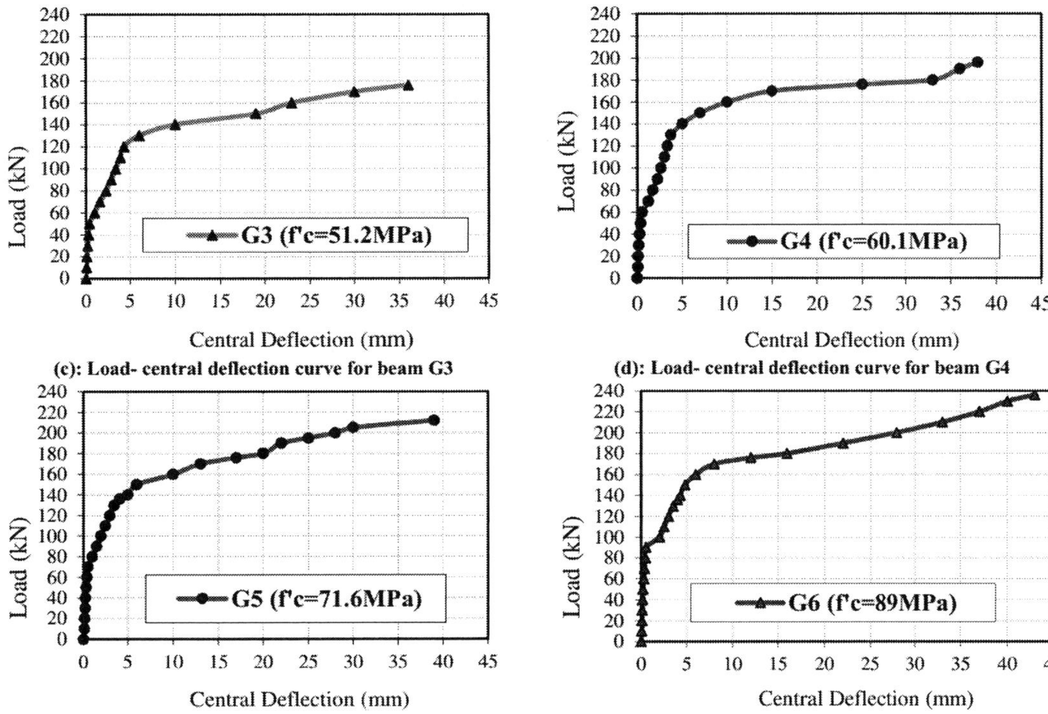

(c): Load- central deflection curve for beam G3
(d): Load- central deflection curve for beam G4
(e): Load- central deflection curve for beam G5
(f): Load- central deflection curve for beam G6

Figure 3. load-central deflection curves for the tested beams

For comparison, the curves of load versus central deflection are exhibited in Figure 3 for all the tested beams. This comparison curves shows that the maximum compressive strength value (89MPa) gave a higher deflection value companied with more quality limit, more strength capacity and more protection from cracks developed and cracks propagations.

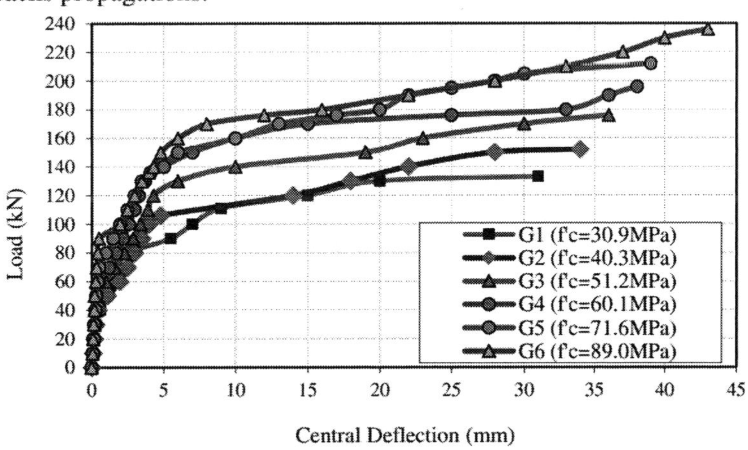

Figure 4. load-central deflection curves of all tested beams specimens

3.4. Ductility

Ductility could be defined as a measure of a material's ability to undertake plastic deformation before breaking or rupture. Ductility could be expressed as percent from the ultimate load to the yield load for the concrete members or the deflection value at ultimate stage to the deflection at yield stage [23,24].

Member ductility is best expressed in terms of deflections. In this work, the deflection ductility index is taken as in Eq. (1) below:

$$\mu_d = \frac{\Delta_u}{\Delta_y} \qquad (1)$$

Where: Δ_u is the member deflection at ultimate load and $\Delta_y$is member deflection at yielding of the tension steel bars. Ultimate is definite as the stage beyond which the specimen would not be able to sustain further deformations at the same load intensity. For the present work, the ductility index for the six tested beams is given in Table 8. This table includes also a summary of the test results for deflection at yield and ultimate stages as mentioned previously.

It could be seen by Table 8 and Figure 3 that the ductility index increases with increasing the concrete compressive strengths for the equal amount of longitudinal reinforcements [25,26 and 27]. This may due to increasing in ρ_b, thus at a constant value of $\rho$the value of ρ/ρ_b ratio decrease and then the ductility increase. Knowing that ρ_b could be calculated according to ACI 318M-2014[18] as in eq.(2) below

$$\rho_b = \beta_1 \frac{f'c}{fy}\frac{\varepsilon_u}{\varepsilon_u+\varepsilon_y} = \rho_b = \alpha \frac{f'c}{fy}\frac{0.003}{0.003+\frac{fy}{E_s}} \qquad (2)$$

Here, (f'c): concrete compressive strength;

(fy): reinforcing steel's yield strength

β_1=0.85 for f'c ≤ (17- 28) MPa[13]

β_1=0.65 for f'c > 56 MPa [13]

$\beta_1=0.85-(\frac{0.05}{7})$ (f'c -28) for 28 < f'c < 56 MPa[13]

4. Conclusion

Based on the present work experimental investigation, the following conclusions are made: -

1- It could be noticed from the experimental results and the load deflection curves that the increasing of concrete compressive strength has an apparent influence on the beams load capacities and deflections.

2- The initial cracks developed at a higher applied load with increasing of compressive strength, so that the increase in compressive strength provide a better restraint to growth of cracks.

3- Although high strength concrete (f'c≥60MPa) show a sudden failure in compression test, high strength reinforced concrete beams exhibit a higher ductility when compared with normal strength concrete beams for same amount of steel reinforcement.

4- The flexural strength capacity of the tested beams increases as the increasing in concrete compressive strength due to reduction in compression part of the beam and this means increase in their curvature ductility that lead to increase in moment capacity or in another word increase in load capacity of the concrete section.

5- High strength reinforced concrete beams generally have no softening part. Therefore, softening may not contribute to additional ductility in high strength concrete beams.

References

[1] Revanth J.; and Kumar, G. V. (2017). High Strength Concrete. *International Journal of Engineering Sciences & Research Technology*, 6(2), 394-407.

[2] Preethi, M.; Ashveen, P. K.; and Hamraj, M., (2019). Strength Characteristics of High Strength Concrete (HSC) with and without Coarse Aggregat.*Journal of International Journal of Engineering and Advanced Technology (IJEAT)*, 9(2), 4701-4704.

[3] Carrasquillo, P. M., & Carrasquillo, R. L. (1988). Evaluation of the use of current concrete practice in the production of high strength concrete. *Materials Journal*, *85*(1), 49-54.

[4] Paulson, K. A., Nilson, A. H., & Hover, K. C. (1991). Long-term deflection of high-strength concrete beams. *Materials Journal*, *88*(2), 197-206

[5] Ahlam, S. Mohammed, (2018). Flexural Behavior of Spliced Ultra High Strength Concrete Beams using Finite Element Analysis.*ARPN Journal of Engineering and applied sciences*, 13(16), 4598-4609.

[6] Al-Quraishy, Q. A., (2018). Behavior of Normal, High and Ultrahigh Strength Concrete in Direct shear. *International Journal of Civil Engineering and Technology (IJCIET)*, 99(2), pp.349-359.

[7] Chong, W.; Changhui, Y.; Fang, L., Chaojun, W.; and Xincheng, P, (2012). Preparation of Ultra-High-Performance Concrete with common technology and materials. *Cement and Concrete Composites*, 34(4),538–544.

[8] Qais F. Hasan, Maan A. Al-Bayati and Dlear A. Al-Amamany. (2019). Flexural Behavior of High Strength Reinforced Concrete Beams Strengthened with Hybrid Fibers. *International Journal of Civil Engineering and Technology* 10(1):1147-1158

[9] Fiorato, A. E. (1994) PCA Research on High- Strength Concrete. *ACI Compilation*, High-Strength Concrete, Compilation 17, 4-10.

[10] Mansor A. Ahmed, (2001). Flexural cracks in high strength reinforced concreted concrete beams. *M.Sc. thesis, University of AL -Mustansriya*,1-3.

[11] Cook, J. E. (1989). 10000 psi Concrete. *Concrete International: Design & Construction*.V.11, No.10, 67-75.

[12] ACI Committee 363. (1984) State-of-the-Art Report on High-Strength Concrete. *ACI Journal*, V. 81, No. 4, July-August, 1984, 364 - 411.

[13] ASTM C150/C150M-16e1. (2016). Standard Specifications for Portland Cement. Developed by ASTM Subcommittee C01.10 on *Concrete and Concrete Aggregates*, Vol. 04.01, West Conshohocken, PA, USA, 10pp.

[14] ASTM C33/C33M-11. (2011). Standard Specifications for Concrete Aggregate. *Developed by ASTM Subcommittee C09-20 on Concrete and Concrete Aggregates*, Vol.04.02, West Conshohocken, PA, USA, 11pp.

[15] ASTM C494/C494M-15a. (2015). Standard Specifications for Chemical Admixtures for Concrete. *Developed by ASTM Subcommittee C09*.23, Vol. 04.02, West Conshohocken, PA, USA, 10.

[16] BS 5328 Part 2. (1991). Method for Specifying Concrete Mixes.

[17] ACI 211, Part 1. (1991). Standard Practice for Selecting Proportions for Normal, Heavyweight, and Mass Concrete. American *Concrete Institute.* Detroit, 38pp. Also, *Structural Journal*, Vol.93, No.1, January – February 1996, 30 – 35.

[18] ACI Committee 318. (2019). Building Code Requirements for Structural Concrete (ACI 318M-19) and Commentary on Building Code Requirements for Structural Concrete (ACI 318R-19). *American Concrete Institute*, Farmington Hill, MI, 2019, 628.

[19] ASTM C 39/C 39M-16a. (2016) Standard Test Method for Compressive Strength of Cylindrical Concrete specimens. *Developed by ASTM Subcommittee C09.61 on Concrete and Concrete Aggregates*, Vol. 04.02, West Conshohocken,PA, USA, 7pp.

[20] Amer M. Ibrahim., Mansor A. Ahmed, and Muthafer Hameed (2017). Structural Behavior of Strengthened RC Beams in Shear using CFRP Strips.*The Open Civil Engineering Journal*, 11, pp.205-215. DOI: 10.2174/1874149501711010205

[21] Mansor A. Ahmed, Salman, W.D., Abbas, A.L., and Khalifa, H.J. (2018). Additional Steel Reinforcement Surrounded Opening as Strengthening in Reinforced Concrete One-Way Slabs. *International Journal of Civil Engineering and Technology (IJCIET)*, 9(11), 1662-1671.

[22] Salman, W. D., Mansor A. Ahmed, and Mohammed, M. (2018). Behavior of Reinforced Concrete One-Way Slabs Strengthened by CFRP Sheets in Flexural Zone. *International Journal of Civil Engineering and Technology (IJCIET)*,9, (10), 1872–1881.

[23] Amer M. Ibrahim, Mansor A. Ahmed, Salman, W.D., and Hamood, M.J. (2018). Strength and Ductility of Bubbled Wide Reinforced Concrete Beams with Diverse Types of Shear Steel Plates. *International Journal of Engineering & Technology*, 7(4.20), 502-506.

[24] Shin S.W., Ghosh S. K., and Mereno J. (1989). Flexural Ductility of Ultra High Strength Concrete Members. *ACI Structural Journal,* 86(4), 340-400.

[25] Budynas, R. G. (2015). Shigley's Mechanical Engineering Design. 10th ed. *McGraw Hill.* p. 233, ISBN 978-0-07-339820-4.

[26] Salman, W. D., Safie Mahdi Oleiwi, and Mansor A. Ahmed, (2019). Flexural Behavior and Mechanical Properties of Steel Fibers Ferro geopolymer One Way Slabs Mortar. *Journal of Engineering and Applied Sciences*, 14(15), 1267-1281.

[27] Hussain N. Laith, Mohammed S. Ahlam and Mansor A. Ahmed, (2020). Finite Element Analysis of Large-Scale Reinforced Concrete Shell of Domes. *Journal of Engineering Science and Technology (JESTEC)*, 15(4), 782-791.

Automated Approach for Extraction of Retinal Blood Vessels

Dheyaa M. Abdulsahib[1, *] and Hussain F. Jaafar[2]

[1]College of Engineering, university of Babylon, Iraq - E-mail: diyaaalkaby@gmail.com
[2]College of Engineering, university of Babylon, Iraq E-mail: hussain_engi@yahoo.com
* Corresponding author: College of Engineering, university of Babylon, Iraq. Tel: +9647706050150.

Abstract. In the field of ophthalmology, retinal image analysis is crucial to extract many details that help doctors identifying retinal diseases at early stages, such details are the blood vessels, optic disk, and fovea. The exponential increase in number of diabetic retinopathy patients necessitates the use of computers to help doctors to diagnose and treat retinal diseases of the human eyes. Computer vision effectively helps doctors by analyzing and treating human retinas. In the field of ophthalmology, a color image (RGB image) of the eye fundus is captured by ophthalmoscopy. In this work, an automatic technique for extraction of human retinal blood vessels is proposed. The proposed method is based on three main stages, namely image pre-processing, initial segmentation of blood vessels and image post-processing. In the first stage, the contrast-limited adaptive histogram equalization is applied to the green component image to improve its contrast, and then it is remerged again with the red and blue components. In the second stage, the blood vessels are extracted using mean-C thresholding. Finally, in the third stage, many morphological operations are used to refine the segmented blood vessels image. The proposed method is validated using the expert ground truths with the DRIVE dataset in terms of pixels, and experimental results show sensitivity, specificity, accuracy and positive predictive value of 0.770816, 0.977575, 0.959993 and 0.7611835, respectively. The performance measures are compared with many recent related works and found to outperform most of them. The superior performance of this method proves that it is promising for mass screening of human fundus images.

Keywords: *Biomedical imaging; Human retinal blood Vessels; Image Enhancement; Mean-C thresholding; Mathematical morphology.*

1. Introduction

The exponential growth in the medical field necessitates the use of computers to help doctors diagnose and treat retinal diseases. Computer vision effectively helps doctors by analyzing and treating human retinas. In the field of ophthalmology, a color image (RGB image) of the eye fundus is captured by ophthalmoscopy. This method is called fundus imaging, and it shows all the details of the retina including the blood vessels (BVs), the fovea and the optic disk. Fundus imaging is used to diagnose numerous eye diseases, like hypertension, cardiovascular, diabetic retinopathy (DR), and glaucoma. This method uses a specialized camera which works with low-power microscope Saine and Tyler (2002). Figure 1 shows the main structures of a retinal fundus image.

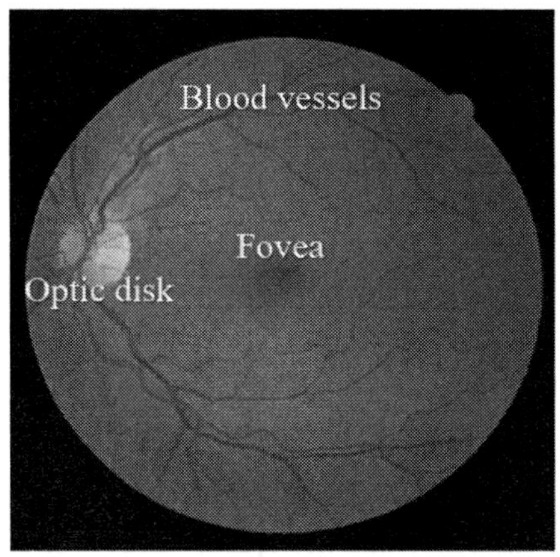

Figure 1. Fundus image with the main structures, (Niemeijer et al.).

Retinal BVs are spread throughout the whole retina, and their details can provide significant information for the detection of many eye diseases including DR, hypertension, glaucoma, and arteriosclerosis. Moreover, extraction of the BVs can help locate other major retinal structures, namely the fovea and also help the doctors diagnose many measurable features of different eye diseases.

Many BVs segmentation methods are proposed in the literature. An automatic approach for extraction of BVs was proposed by Fraz et al. (2011) using combination of morphological operations and differential filtering. To extract the centerlines, the Gaussian first-order derivative is applied in four orientations. BVs shapes and orientations map are obtained by applying a multidirectional morphological operator, and a bit plane slicing of a grayscale image with enhanced BVs. The Gaussian first-order derivative is a very efficient tool in segmentation of BVs. This tool can detect BVs perfectly by Gaussian-shaped profile, but it is not suitable when the arterioles of the retinal image show prominent light reflexes, such as those of younger patients.

Fraz et al. (2013) proposed a new method using combination of retinal BVs skeleton extraction and multidirectional morphological operations. Directional differential operators are used to detect the skeleton of the main vessels, and mathematical morphology is used to obtain the retinal vasculature in RGB fundus image. Zhao et al. (2014) used contrast-limited adaptive histogram equalization (CLAHE) and 2D Gabor wavelet to improve image quality as pre-processing stage, and the image is then filtered by an anisotropic filter. The technique of region growing and models of region-based active contour are used to detect the BVs. This method could extract all wide and thin vessels precisely and because it is unsupervised method, it does not require manual segmentation in the training stage. The performance of this approach can be further improved by removing the active contour which moves to the pathological region in several abnormal retinal images.

Sreejini and Govindan (2004) used particle optimization to find the suitable filter among the Gaussian-matched filters and then to improve the accuracy of the retinal BVs detection. The green component image is filtered using median filter followed by applying optimized matched filter, and then the filtered image is thresholded using global thresholding. The experimental results showed that the multi-scale matched filter can achieve better performance than the single-scale matched filter in the BVs extraction.

Dash and Bhoi (2017) proposed a method which works on the basis of local adaptive thresholding by mean-C thresholding using three stages. Firstly, the image was enhanced by the CLAHE and median filter. Secondly, the mean-C thresholding was used to detect initial BVs, and in third stage, morphological operations were used to refine the initial results from the artifacts. Dash and Bhoi (2019) proposed morphological approach to extract the BVs by enhancing and smoothing the image using CLAHE. Kirsch's method was used to extract the BVs, and morphological operations were used to refine the results. The main advantage of this approach is in its ability to identify the BVs without small artifacts.

Although the presented related methods could extract BVs with good performance measures but there are noticeable limitations and drawbacks to be overcome. These limitations are in the presence of high undesired false positives (FPs) in the BVs results, which affect the positive predictive value (PPV) dramatically. In this work, a novel method to extract retinal BVs in efficient and fast approach is proposed. This method is based on three stages: image pre-processing to enhance image quality, extraction of course BVs as preliminary BVs image and then refining the BVs image by filtering and morphological operations. In the image pre-processing, the green component image is used because it has better contrast than those in the other components, so it is pre-processed to improve its contrast and then the pre-processed image is merged again with the red and blue channels to achieve additional details that are helpful in the segmentation stage. In the second stage, mean-C thresholding is used to detect the BVs, and the C value is set by many experimental processes to achieve optimal outcomes. In the third stage, the segmented BVs image is refined to obtain the final result with superior performance measures.

The organization of this paper is as as follows: section2 includes pre-processing operations, segmentation of initial BVs, and refining of the segmented image. Section 3 presents the databases, used in this work, experimental results and discussions. Sec. 4 presents the conclusions and future works.

2. Methodology

The proposed method consssists of three stages: the first stage is the image pre-prossing to achieve image with better quality, the secound stage is the segmentation stage to extract BV candidates, and the final stage is the post-prossing stage to refine the image from non-BVs. In the Figure 2 a block diagram for the stages of the proposed method is presented.

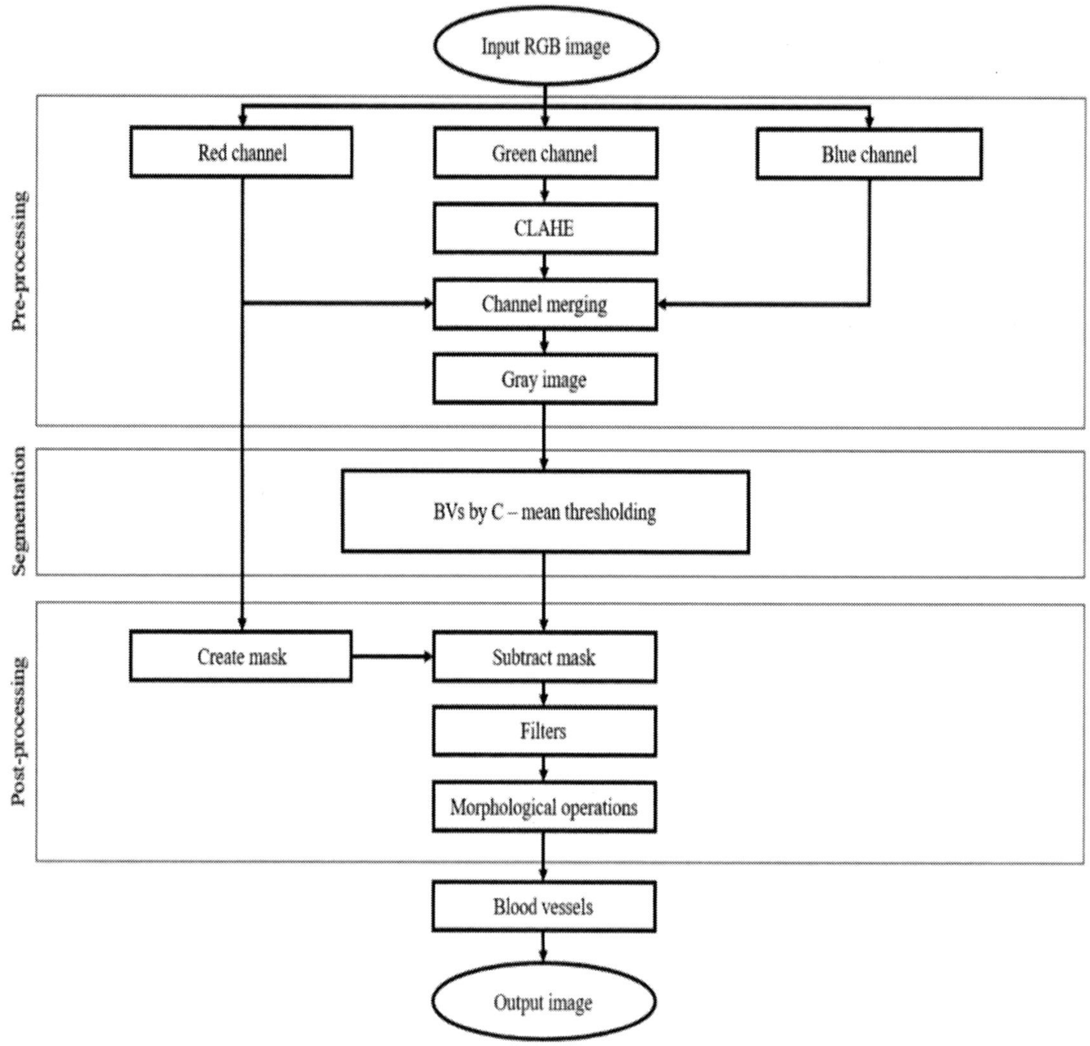

Figure 2. A block diagram indicating the proposed method steps for detection of BVs.

2.1 Image pre-processing

Because of variability in the acquisition of fundus images and physical features of patient retinas, like iris, skin and color, most of retinal images are different in the illumination, and they are sometimes poorly contrasted. The BVs are not easily extracted from the background when the retinal image is uneven illuminated and/or poorly contrasted. Then, contrast enhancement is essential to improve the contrast of the image and to facilitate the extraction of the BVs accurately. In this work, image pre-processing for image improvement is proposed, where the green component image is obtained from the RGB color image and it is used for extraction of the BVs because it appears more contrasted against the background than the other components. The features of the green component are clearly obvious by its histogram as shown in Figure 3.

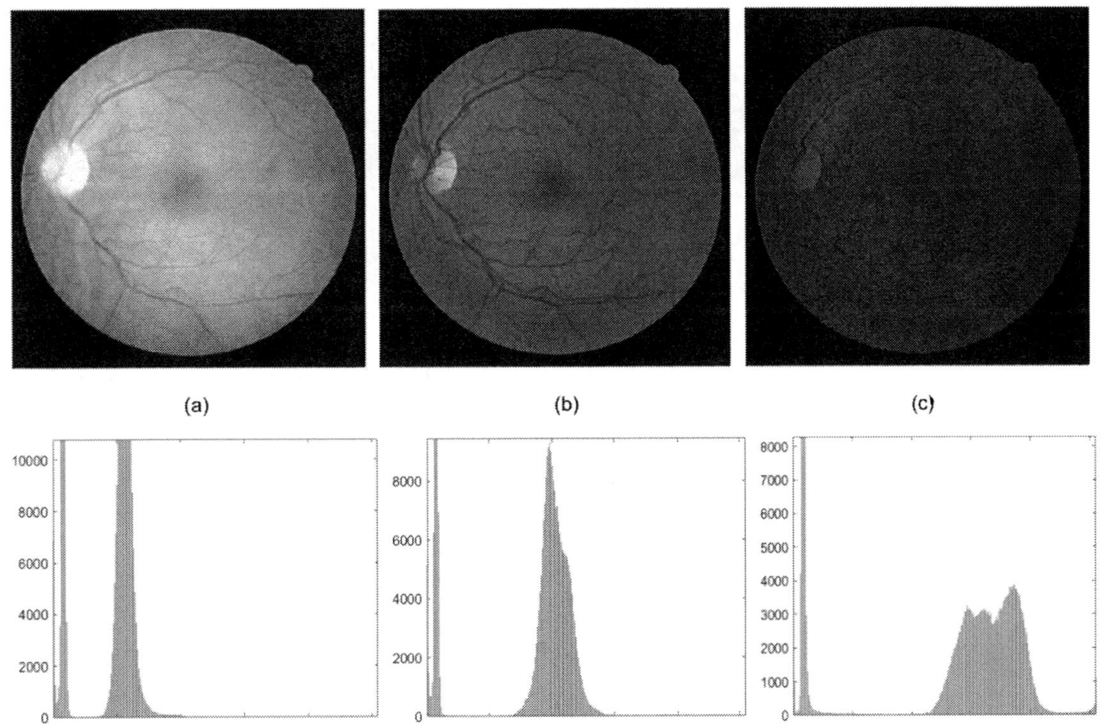

Figure 3. Image components and their histograms. (a) Red component. (b) Green component. (c) Blue component.

In this stage, the green image is enhanced using the CLAHE. The CLAHE is designed to operate on small tiles in the image, and not on the whole image. For each particular tile, the contrast transform function is calculated with adaptive histogram equalization. The contrast of each tile is improved in such a way that the histogram of the output region will match that specified region by the distribution value. The neighboring tiles are then combined using bilinear interpolation to eliminate artificially induced boundaries. The contrast, especially in homogeneous areas, can be limited to avoid amplifying any noise that may be presented in the image. Thus, the contrast of the image can be increased, and the BVs appear more contrasted. The last pre-processing image is achieved by merging and concatenating the pre-processed green component with the original red and blue components. Figure 4 illustrates visual comparison between original image and the result of pre-processing image for both color image and green component image.

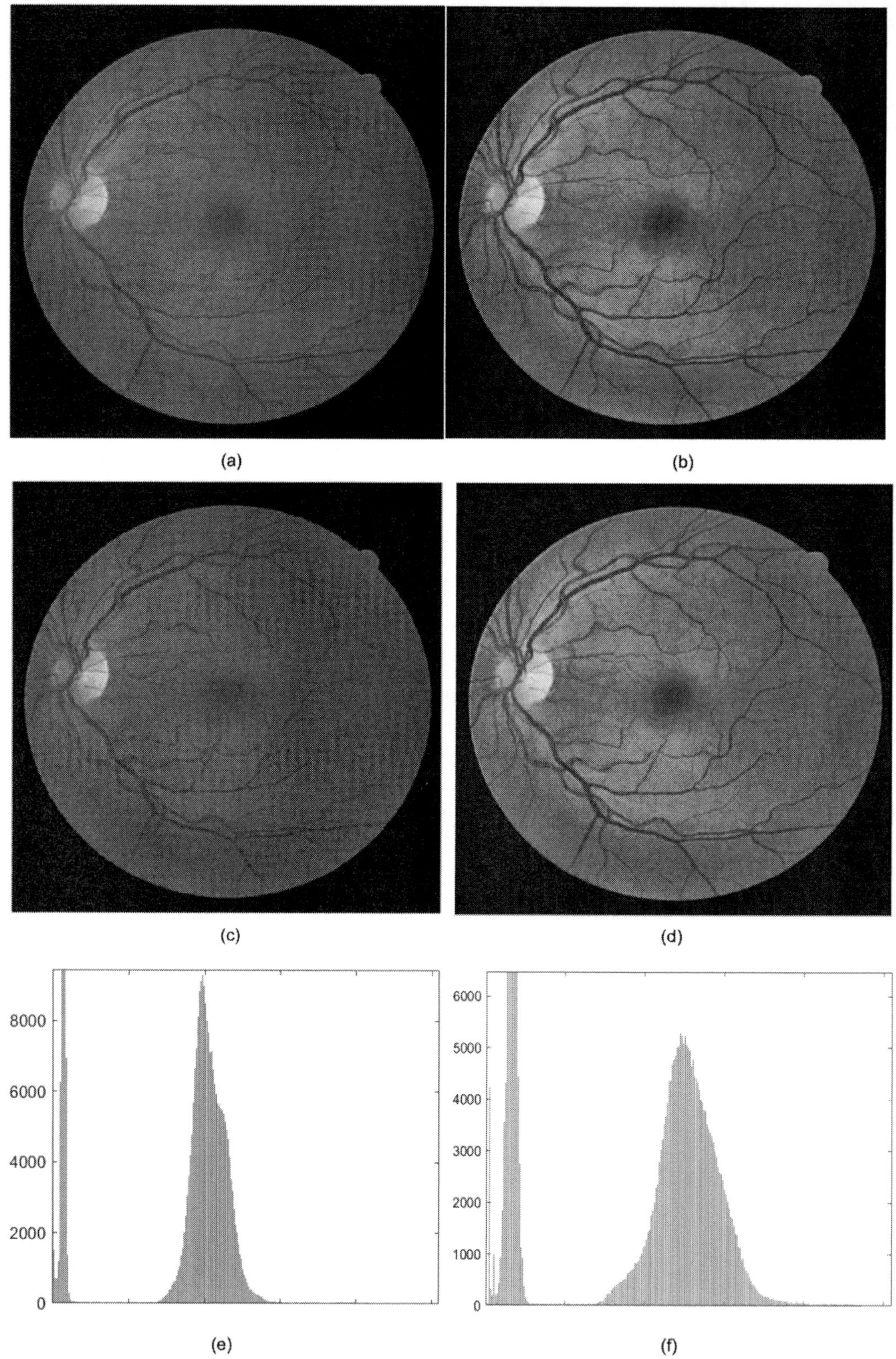

Figure 4. Visual comparison for the pre-processing results (a) Original color image. (b) Color image after the pre-processing (c) Green component image (d) Pre-processed green component image. (e) Histogram of the green component image. (f) Histogram of the pre-processed green component image.

2.2 Segmentation of blood vessels

Image segmentation is a process of subdividing an image into multiple objects, it is also known as image objects (the objects represent a set of pixels with particular features). The purpose of image segmentation is to make it increasingly meaningful and easy to analyze. The greyscale image (known as black-and-white image) could be generated by setting image pixels to black for the values which are beyond a certain threshold, while the other pixels to white. In this work the mean-C thresholding has been used to extract the BVs from the pre-processed image. The advantage of using this method is that it can be used even for bad quality and uneven illuminated images. The mean-C thresholding (where C is a threshold value) is used to extract the BVs. The mean-C segmentation could be used by applying the following steps:

1. Select N as window of the mean filter.
2. Select value of C threshold.
3. Convolve the pre-processed image with the mean filter corresponding to the window size.
4. Subtract the convolved image from the pre-processed image.
5. The subtracted image is segmented with the threshold C.
6. Obtain the complement of the segmented image.

For selecting optimal value of the threshold C, a number of values between 0.02 and 0.06, with a step of 0.002 were examined. As shown in Figure 5, the PPV and sensitivity (SE) values are inversely correlated. When the value of C is increased, the PPV is increased, but the SE value will be decreased. Dash and Bhoi (2017) proposed that the optimal value of C is 0.042 to achieve reasonable balance between SE and PPV. This value is similar to the results that we obtained to achieve the best compromise between the PPV and SE. Dash and Bhoi proposed that the optimal value of window size is 13 × 13 which is the same we found as the optimal window size.

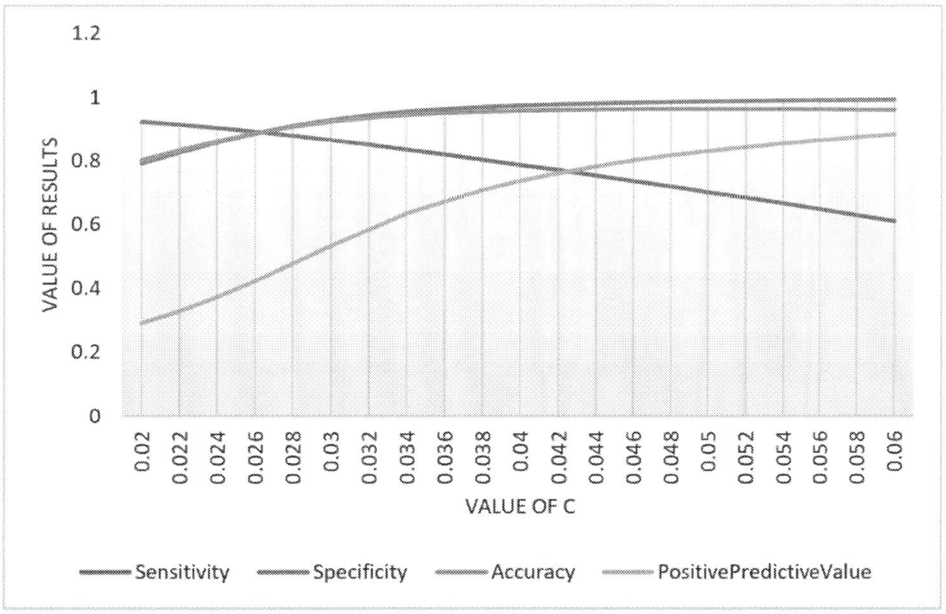

Figure 5. The response of performance measures due to variation in the C.

The two-dimensional convolution is used to convolve the pre-processed image with the mean filter. In spatial form, implementation of the two-dimensional convolution could be mathematically expressed as follows:

$$c(n_1,n_2)=\sum_{k_1=-\infty}^{\infty}\sum_{k_2=-\infty}^{\infty}a(k_1,k_2)b(n_1-k_1,n_2-k_2) \quad (1)$$

where c refers to the convolution operation, a and b represent the functions of the discrete variables n_1 and n_2.

The resulting image from the segmentation phase contains a lot of noises as shown in Figure 6. Thus, post-processing is crucial to remove the noises and then to achieve better results.

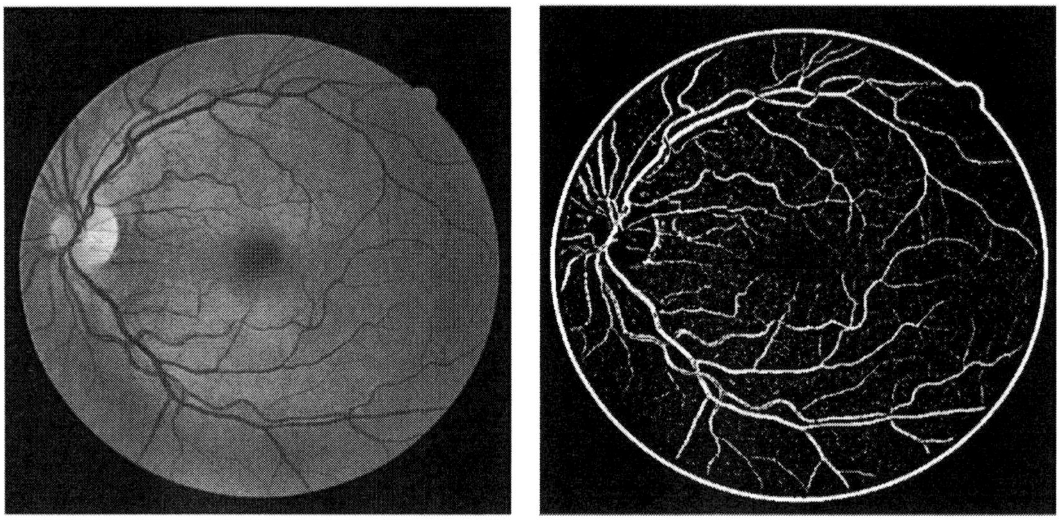

(a) (b)

Figure 6. Segmentation result. (a) Original green component image. (b) Blood vessels image using the mean-C thresholding.

2.3 Post-processing

In the segmentation stage we have achieved a coarser BVs image with some artifacts (non-BVS). In this stage we try to refine the coarser BVs image to increase the accuracy of the results. As shown in Figure 6, a circle structure around the segmented vessels which is not BV can be observed. This circle structure should be removed to enhance the BVs image. Mask of fundus image refers to the parts of the fundus image which contain the eye details and area out of the mask is the background. Therefore, it is essential to generate the mask and then to be used in removing the circle structure around the detected BVs.

The retinal image consists of two main regions, region of interest (ROI) and dark surround region (DSR). The main ocular details of the retina are concentrated in the ROI, which is circular or semicircular-shaped and it occupies about 75% of the retinal fundus image, while the DSR is just a dark background region that surround the ROI and occupies about 25% of the fundus image Jaafar (2012). To detect the FROI of the image, the red channel of the main RGB image has been used because it is the highest brightness compared to the green and blue channels. To obtain the mask of the fundus retinal image (as a binary image) Otsu's method is used with red component image. The Otsu's method is based on choosing the global threshold value to minimize the interclass variance of the thresholded black and white pixels (see Figure 7a). To remove the circle structure, the mask image is subtracted from the coarser BVs as illustrated in Figure 7b. The final step in the refining phase is based on using some filters and morphological operations, such as closing, dilatation and filling operations. The aim of filtering and morphological operations is to join the gaps in the image together by filling between them and by smoothing their outer edges to remove noises and non-BVs.

The morphological closing is based on dilation followed by erosion; hence it is effective to be used for smoothing the edges of detected vessels. To compensate the reduction in the detected vessels, due to smoothing operation, a morphological dilation is used to retrieve the original size of the BVs, and the suitable structuring element is found to be disk-shaped of radius 3 for both closing and dilation operations. The morphological filling is used to fill some holes in the thick vessels, created during the course BV segmentation. Figure 8 illustrates the final BVs image and the ground truth image (hand labeled by experts).

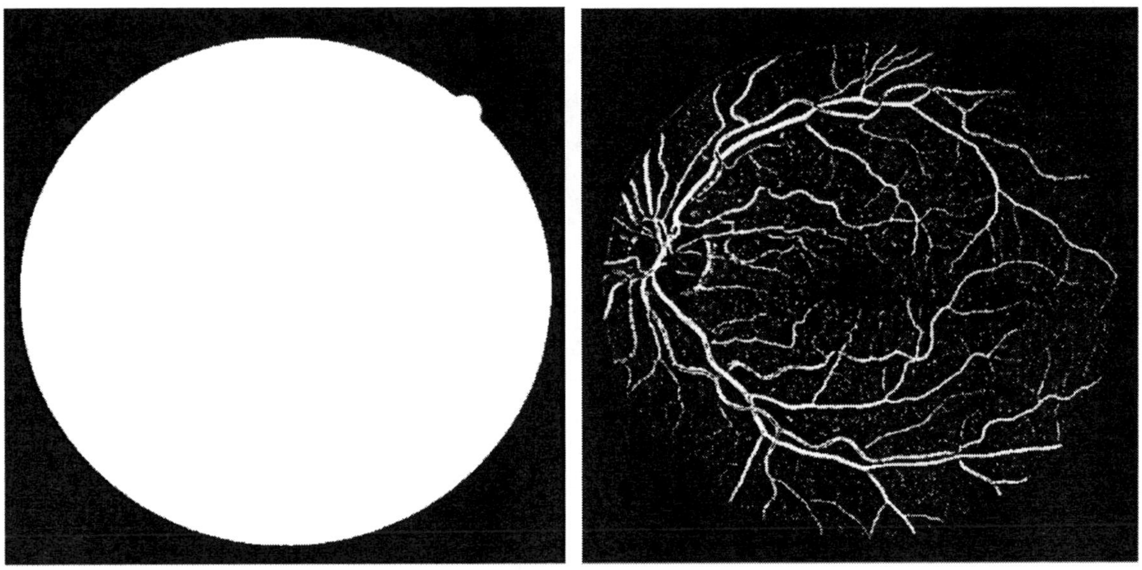

(a) (b)

Figure 7. Circle removal. (a) The mask of the image. (b) Result of the blood vessels image after removing the circle.

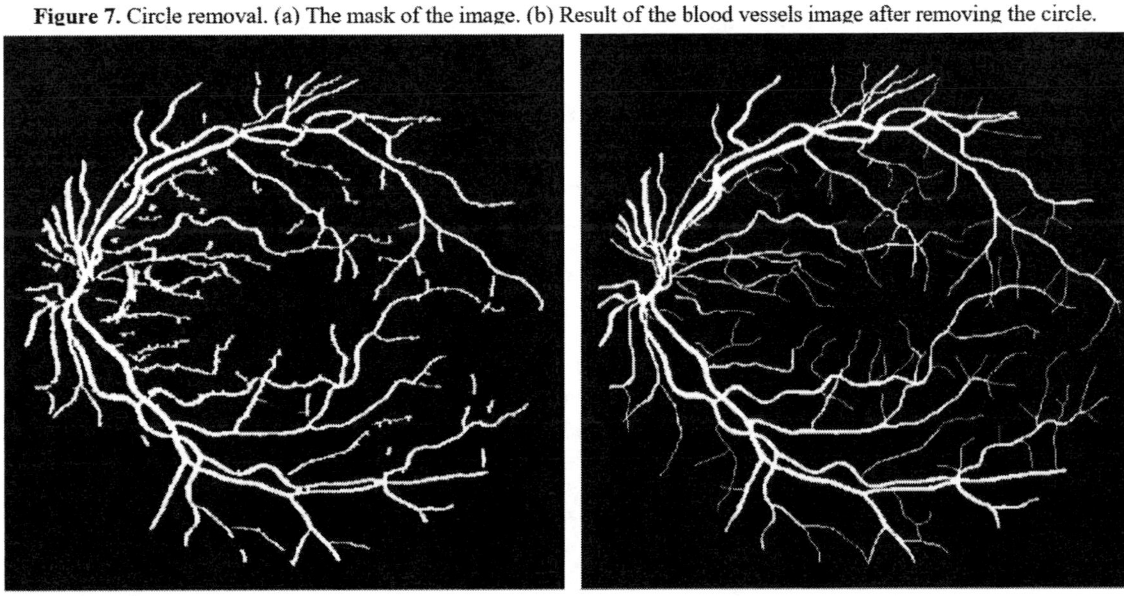

(a) (b)

Figure 8. (a) The blood vessels image produced by the proposed method, (b) The blood vessel image annotated by experts.

3. Results and Discussions

In this work, the proposed method was trained using 20 images from the DRIVE database* and then tested using other 20 images provided with their ground truth images from the DRIVE database as well. Performance measures of the proposed method are calculated using the Equations (2-5) in term of pixels. In these equations, the following four pixel types are used:

a. True positives (TPs): number of vessels pixels that are correctly detected as vessels.

b. True negatives (TNs): number of background pixels that are identified correctly as background.

c. False positives (FPs): number of background pixels that are wrongly detected as vessels.

d. False negatives (FNs): number of vessel pixels that are wrongly identified as background.

For each retinal image, the preceding pixel types are determined to calculate the performance measures individually. Then the average of each performance measure is calculated, from the whole image measures, to get the proposed method performance in terms of pixels.

$$Sensitivity = \frac{TP_S}{TP_S + FN_S} \quad (2)$$

$$Specificity = \frac{TN_S}{TN_S + FP_S} \quad (3)$$

$$Accuracy = \frac{TP_S + TN_S}{TP_S + FP_S + TN_S + FN_S} \quad (4)$$

$$Positive\,predictive\,value = \frac{TP_S}{TP_S + FP_S} \quad (5)$$

To evaluate the proposed method robustness, a comparison of the proposed method against the other previous related works, in terms of performance measures, are presented in Table 1. In the proposed method it appears that the average values of SE, specificity (SP), accuracy (ACC) and PPV are; 0.771, 0.978, 0.960 and 0.761 respectively. The superiority of SE and PPV indicates that the BVs are marked with more exactness. Although the approach proposed by Dash and Bhoi is similar to our proposed method in some parts, the SE and ACC of the proposed method outperform those in Dash and Bhoi. The reason behind that improvement in the performance is the efficient pre-processing step and the morphological operations which have very effective role in achieving the superior performance. The proposed method is also compared against other related works, which used different approaches, and showed superior performance as illustrated in Table 1.

In this work, the big challenge for obtaining competitive performance is in selecting the main factors, i.e. the C in the mean-C thresholding and the window size of the mean filter. Based on many experiments, it appears that the window of 13 × 13 can ensure optimal influence on the results in terms of average of the performance measures. The manipulation in the C of the mean-C threshold has a noticeable effect on the compromise between the TPs and the FPs and as a result on the compromise between the most informative measures, i.e. the SE and the PPV. Experiments on the influence of the threshold C showed that an increase in the C will cause decrease in the SE and increase in the PPV and vice versa. Experimental tests have showed that the best value of the C for reasonable balance between SE and PPV is 0.042. Influence of the proposed method parameters on the rates of the performance measure demonstrates that it is robust and reliable, and the intervention on these parameters values is based on diagnostic requirements which are taken by the doctor. Figure 9 shows three color images, their binary BVs images by the proposed method, and their binary BVs images, annotated by experts.

Table 1. Comparison between the proposed method and other related works using DRIVE database.

Author	SE	SP	ACC	PPV
Fraz et. al. (2011)	0.715	0.977	0.943	0.821
Fraz et. al. (2013)	0.730	0.974	0.942	0.811
Zhao et. al. (2014)	0.735	0.979	0.948	0.836
Sreejini & Govindan (2015)	0.713	0.987	0.963	Not reported
Dash & Bhoi (2017)	0.719	0.976	0.955	0.746
Dash & Bhoi (2019)	0.703	0.985	0.951	0.810
Proposed method	0.771	0.978	0.960	0.761

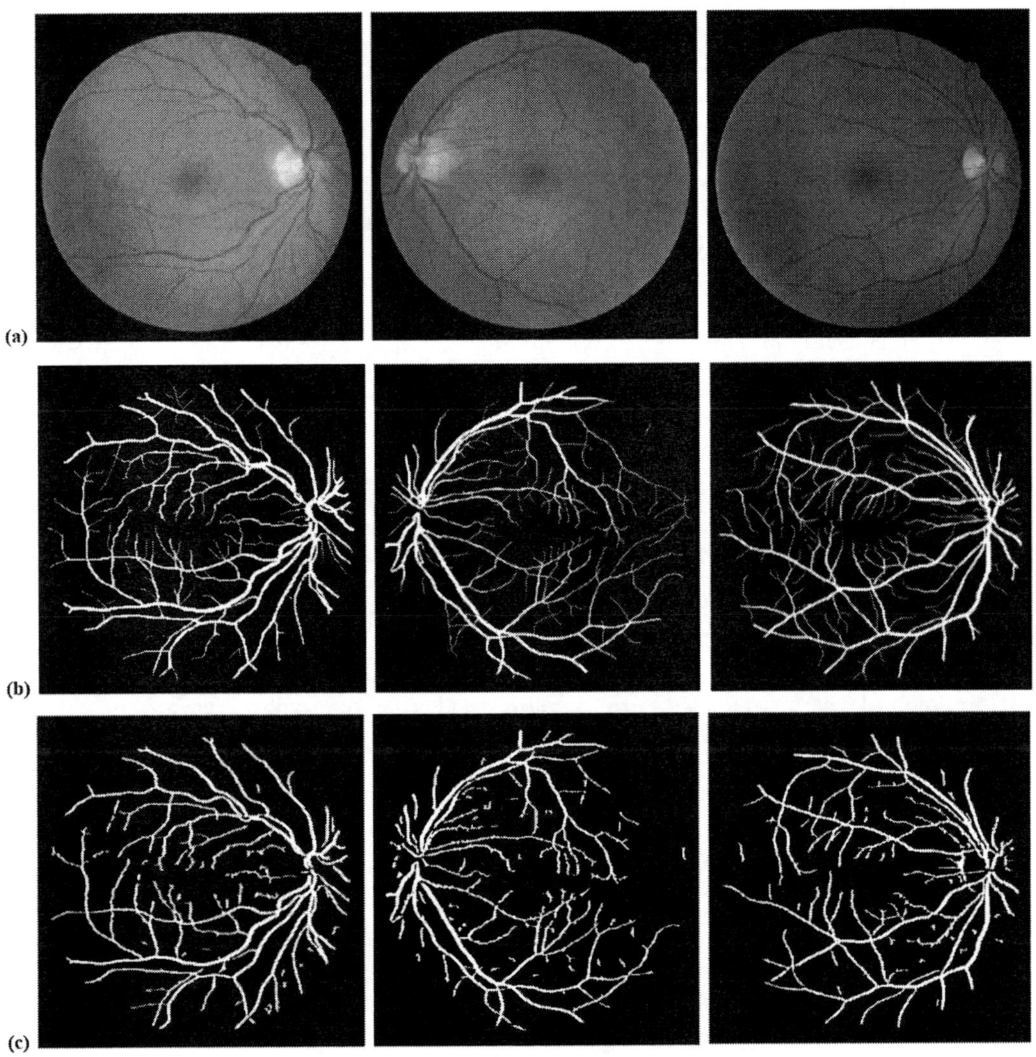

Figure 9. Inputs and results of blood vessels extraction for three images (a) Original color retinal images (b) Binary blood vessels images by the proposed method. (c) Binary blood vessels images, annotated by experts.

4. Conclusions

In this work, a novel approach for extraction of retinal BVs is proposed. The contribution of the proposed approach is in using the technique of remerging the pre-processed green component image with the other raw components, i.e. the red and blue components. The new color image after remerging operation will have additional details compared to the raw color image. Hence, it will be more informative to extract retinal information. In the green image enhancement, the CLAHE is used to increase image contrast. In the extraction of BVs, the mean-C thresholding is used to separate BVs structures from the background. The BVs image may contain some non-BVs due to existence of artifacts which may have similar features of true BVs. To remove detected artifacts and noises, morphological operations are used to refine the BVs image and hence the performance measure will be improved. The DRIVE dataset was used to test the proposed method and it is found to achieve superior performance compared to the state of the art methods. From many experimental results, we can conclude that the behavior of the proposed method with intervention to parameter values is robust. Intervention on the parameters is based on the diagnostic requirements and may be taken by the clinician to manipulate between the performance measure rates. The proposed approach has obtained SE, SP, ACC, and PPV of 0.771, 0.978, 0.960 and 0.761, respectively. In future, we will try to develop the current algorithm to detect the other retinal structures i.e. the fovea and the optic disk with competitive performance to recent related works.

Acknowledgements

The authors would like to thank the DRIVE Database Centre, (Staal et al.)* for the cooperation in providing retinal images.

References

[1] Dash, J., & Bhoi, N. (2017). A thresholding based technique to extract retinal blood vessels from fundus images. *Future Computing and Informatics Journal*, 2(2), 103-109.

[2] Dash, J., & Bhoi, N. (2019). Retinal Blood Vessel Extraction Using Morphological Operators and Kirsch's Template. *In: Soft Computing and Signal Processing*, Springer, Singapore, 603-611.

[3] Fraz, M. M., Remagnino, P., Hoppe, A., Uyyanonvara, B., Owen, C. G., Rudnicka, A. R., & Barman, S. A. (2011). Retinal vessel extraction using first-order derivative of Gaussian and morphological processing. *In: International Symposium on Visual Computing,* Springer, Berlin, Heidelberg, 410-420.

[4] Fraz, M. M., Basit, A., & Barman, S. A. (2013). Application of morphological bit planes in retinal blood vessel extraction. *Journal of digital imaging*, 26(2), 274-286.

[5] Hasby, M., & Khodra, M. L. (2013). Optimal path finding based on traffic information extraction from Twitter. *In: International Conference on ICT for Smart Society, IEEE*, 1-5.

[6] Jaafar, H. F., Nandi, A. K. and Al-Nuaimy, W. (2011). Decision support system for the detection and grading of hard exudates from color fundus photographs. *Journal of biomedical optics*, *16*(11), p.116001.

[7] Niemeijer, M., Staal, J. J., Ginneken, B. V., Loog, M., & Abramoff, M. D. (2017). DRIVE: digital retinal images for vessel extraction; 2004. WebLink: http://www. isi. uu. nl/Research/Databases/DRIVE

[8] Saine, P. J., & Tyler, M. E. (2002). Ophthalmic photography: retinal photography, angiography, and electronic imaging, Boston: Butterworth-Heinemann., 132.

[9] Sreejini, K., & Govindan, V. (2015). Improved multiscale matched filter for retina vessel segmentation using PSO algorithm. *Egypt. Inform. J.*, 16 (3), 253–260.

[10] Staal, J. J., Abramoff, M. D., Niemeijer, M., Viergever, M. A., van Ginneken, B. (2004). Ridge based vessel segmentation in color images of the retina, *IEEE Transactions on Medical Imaging*, 23, 501-509.

[11] Zhao, Y. Q., Wang, X. H., Wang, X. F., & Shih, F. Y. (2014). Retinal vessels segmentation based on level set and region growing. *Pattern Recognition*, 47(7), 2437-2446.

IOP Conf. Series: Materials Science and Engineering **978** (2020) 012038 doi:10.1088/1757-899X/978/1/012038

Dynamic Response of Machine Foundation Resting on End Bearing Piles

Mohammed Y. Fattah[1*], Mohammed J. Hamood[2], and Ibtihal A. M. Al-Nakdy[3]

[1]Civil Engineering Department, University of Technology, Iraq
E-mail: myf_1968@yahoo.com
[2] Civil Engineering Department, University of Technology, Iraq
E-mail: mjh_smm@yahoo.com
[3]Chief Engineer, Ministry of Education.
* Corresponding author: Civil Engineering Department, University of Technology, Iraq. Tel: +9647901678315.

Abstract. Herein, a finite element software (ANSYS.11) is adopted which provides an effective tool for dynamic analysis and structural design of machine-foundations. As a case study, machine foundation on piles in sandy soil is investigated. Machine foundations supported by end-bearing piles are introduced. Sinusoidal dynamic load is applied. A study of parameters is conducted to investigate the influence of a number of parameters that include: geometry of the piles below the foundation, the dynamic load amplitude, damping ratio and frequency of the load. Linear elastic model is adopted for modeling the piles and their cap for machine foundation using eight node isoparametric (solid 65) element, while elastic model is adopted to model the soil behavior and eight node isoparametric elements are used to model the soil through (solid 45) element. It is concluded that the frequency ratio decreases with increase of spacing between piles values due to increase in natural frequency, except in the case of 3 m spacing where it increases due to decrease in natural frequency as the mass was increased obviously. The increase in spacing caused the increase of natural frequency by about 6% in the model having 1.5 m spacing more than the 1.25 m case and by about 3% for the 1.75 m spacing more than the 1.5 m reference case. The increase in pile cap length caused the natural frequency to decrease by about 18% for the 3.75 m pile cap length as compared with the reference case and by 29% in the case of 5 m less than the reference case. In the case of increasing pile cap thickness, there was only a slight difference in the values of natural frequency.

Keywords: *Dynamic, finite elements, machine foundation, end bearing pile, displacement.*

IOP Conf. Series: Materials Science and Engineering **978** (2020) 012038 doi:10.1088/1757-899X/978/1/012038

1. Introduction

The foundations and structures design when subjected to vibration loads is considered to be a complicated step involving problems including structural and soil mechanics engineering and the mechanical vibrations. The structural shape of the foundation of a machine is usually selected by the properties supplied by the geotechnical engineer and the producer of the machine.

There are subjects that must be accounted for in the criteria for acceptable action of a machine foundation subjected to dynamic load: There should be no resonance. The frequency ratio (which the ratio of applied frequency to natural-frequency) should be either less than 0.5 or more than 2 for important machines (Rao, 2011). The motion amplitudes at the working frequencies should be less than the limiting amplitudes. The vibrations must not be annoying to the persons who work in the buildings or damping to the other machines (Prakash and Puri, 2006). Foundation mass must be structurally adequate to endure the loads. The combined gravity center of machine and foundation and the contact area center (with the soil) should be along the same vertical line if possible (Rao, 2011), in order to increase the system's bearing capacity. Where possible, the foundation should be designed in a manner as to allow a subsequent change in the natural frequency by varying the area of the base or the foundation mass as it is decided to be essential. The natural-water-level should be so low, and it should be at least deeper by 1/4 of the of the foundation width under the foundation plane. Foundations of machines must be separated from nearby structural elements through expansion-joints. Foundation of a chine must be constructed at a level below the foundation level of the nearby structures, in order that the transmitted wave through soil will not result in damage to the nearby foundation (Arora, 1997).

Deng et al. (2007) presented the results of a three dimensional seismic analysis of a battered pile group design for a heavy machine foundation located in a seismically active region. Results of the analysis showed that the battered piles tend to attract significantly larger seismic loads in terms of the axial loads when compared with the vertical piles and require a careful attention for design of the pile connection to the pile caps. This result is a quantitative confirmation of the commonly known field performance problem. Further study of the results indicated that the main cause of the increase in axial loads in battered piles is the kinematic interaction between the soil layers and that of the pile group.

Fleischer and Trombik, (2008) explained that for all kinds of foundation, the under-structure (piles, soil, columns and springs) could be modeled by spring-type elements, the dynamic behavior can be represented with accurate sense. Till today, sufficient computational estimating capacity is usually available in the design bureaus, it is proposed for foundations to induce the whole structure and substitute the soil or piles by spring-type elements. The foundation material stays usually through the linear region. All machine foundations must be designed to be able to carry full earthquake loads. This illustrates that the foundation ductility must be a low value, and the focus on conventional ways to increase the bearing capacity, if required, must be focused on pile heads and connections of columns.

A case study by the use of finite element program for the dynamic-analysis of a machine foundation supported by piles in homogeneous sand soil was carried out by Fattah et al. (2015). A study of parameters was conducted to study the influence of the geometry of foundation, the frequency and amplitude of the dynamic load, in addition to the damping ratio. It was found that when the pile-cap thickness is increased, the variation of displacement decreases because of material damping existing in the pile cap concrete. A limit exists of the size of pile cap at which its stiffness controls its dynamic behavior. Exceeding this size, the pile cap weight overrides the influence of its stiffness, and the overweight results in an increase in displacement of pile. When increasing the size of pile group, the frequency where maximum displacement takes place increases, and therefore, the system becomes more stable to resist resonance.

The response and behavior of machine foundations resting on dry and saturated sand was investigated experimentally by Fattah et al. (2017). The manufactured physical model could be used to simulate steady state harmonic load at different operating frequencies. Total of (84) physical models were performed. The parameters that were taken into considerations include loading frequency, size of footing and different soil conditions. The footing parameters were related to the size of the rectangular footing and depth of embedment. The response of the soil to dynamic loading includes measuring the stresses inside the soil using piezoelectric sensors as well as measuring the excess pore water pressure by using pore water pressure transducers. It was found that the rate of increase in excess pore water pressure ratio decreased remarkably at a depth of 0.5 B–1.5 B (B is the footing width) for medium and loose dense sand, respectively. In addition, the rate of variation of pore water pressure with eccentricity for loose sand is less than that for medium dense sand. The dynamic stress increments resulting from the dynamic load on the foundation reduce with depth. In addition, the dynamic stresses under the corner are slightly greater than the stresses at the center by a percentage of about 10.0%.

The present study consists of numerical solution by the finite element method to the machine foundation problem. The problem studies the influence of load frequency by considering a range of frequencies and amplitudes of dynamic load through studying the response of pile machine foundation in sandy soil. The pile machine foundation is taken as a continuum including soil; the soil is not modeled as springs as in previous studies. This study includes studying the effect of different parameters on the behavior of the pile foundation, these parameters include (pile diameter, spacing among piles, thickness and length of cap).

2. Machine Foundations Resting on Piles

Utilization of piling in machine foundations is shown to provide more flexibility to the designer and quite possibly result in more economical design for some cases. It can also provide a solution to some difficult foundation problems. The key for useful utilization is a complete analysis method which provides for a choice of practical models.

ANSYS software provides an effective procedure for dynamic analysis and structural design of the machine foundation. This paper covers the results of study on the pile machine foundation. Analyses are carried out by studying the effect of various factors of loading, vertical displacement and vertical stress in the pile machine foundation cap are presented. The chosen factors are related to the geometry of the pile machine foundation (cap thickness and length), the geometry of piles (piles diameter and spacing between piles), and the increase in the dynamic load. The analysis was performed using (ANSYS) program.

IOP Conf. Series: Materials Science and Engineering **978** (2020) 012038 doi:10.1088/1757-899X/978/1/012038

3. Description of the Piled Machine Foundation Problem

A circular cross-section of piles is depended in the finite element analysis; the soil is discretized with the 8-node hexagonal elements. Figure 1 displays the machine foundation on piles with the position of the load application, and Figure 2 presents the meshed model of machine foundation on piles with soil. The side boundaries of the domain are selected to be at an adequate distance from the influence zone under the vertical load (five x the pile cap width). End bearing piles are taken into account with specified damping ratio. A lower boundary which represents the soil layer depth, is also modeled at a sufficient distance from the pile end (0.25L, where L is the length of pile) since the piles are considered to be floating piles (i.e., the piles are not pressed to a rigid layer). A boundary condition describing the vertical and lateral displacement with a value of zero is depended, while the vertical is not for the pile tips. The results of analysis process are calculated at the top central point of the pile cap (node A) where the vertical dynamic load is supposed. Interface elements (contact-surfaces) were inserted between the pile and its cap and also the soil The surface to surface contact element (CONTA172) and (TARGE169) are used to simulate the boundary between the pile and its cap and the soil.

The pile machine foundation analysis is conducted for foundation supported by finite isotropic elastic homogenous soil since the machines can result in a limited vibration amplitude (Ottavini, 1975 and Liu and Novak, 1991). The concrete is considered linearly elastic also (Liu and Novak, 1991). The value of Poisson's ratio for most types of soils can be found in Bowles (1997), and the static stress-strain modulus values E_s, for selected soils could be also proposed depending on the guides of Bowles (1997).

The soil used in this study is sand. Hence, a typical values range is:
The sandy soil modulus of elasticity is considered: Es = (25-250) MPa.
The Poisson's ratio range for sandy soil is (0.2 - 0.3).
Three amplitude values were taken in the analysis; 80 kN, 100 kN and 120 kN.
The harmonic forces in any direction perpendicular to longitudinal rotation axis are:

$$F(t) = P \sin(\omega t) \quad (1)$$

where:

$P = m_i \, e \, \omega^2$

m_i = proportional part of rotating-mass,

e = mass eccentricity,

t = time,

$\omega = 2\pi f$ = circular operating frequency of the machine, and

f = operating frequency.

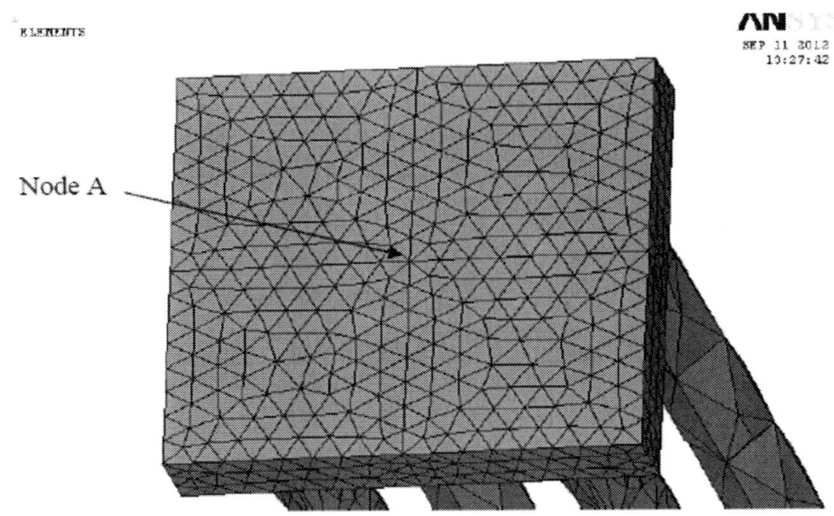

Figure 1. Finite element of pile machine foundation, which illustrates the location of the applied load.

Nine cases of pile machine foundation with dimensions shown in Table (1) are chosen for analysis.

Table 1. Parameters for the machine foundation resting on end-bearing pile.

case	B (m)	W (m)	h (m)	d (m)	L (m)	$S_{c/c}$ (m)
1	2.5	2.5	0.5	0.5	20	1.5
2	2.5	2.5	0.75	0.5	20	1.5
3	2.5	2.5	1.0	0.5	20	1.5
4	2.5	2.5	0.5	0.8	20	1.5
5	2.5	2.5	0.5	1.0	20	1.5
6	2.5	2.5	0.5	0.5	20	1.25
7	2.5	2.5	0.5	0.5	20	1.75
8	3.75	2.5	0.5	0.5	20	1.5
9	5	2.5	0.5	0.5	20	1.5

where:

B = pile cap length,
W = pile cap width,
h = pile cap thickness,
d = pile diameter,
L = pile length,
$S_{c/c}$ = spacing between piles, center to center, and
P = amplitude of harmonic dynamic load.

Figure 2 presents the details of pile machine foundation adopted in this study, and Figure 3 presents the top view of the pile cap with different pile cap lengths. The material properties are given in Table 2.

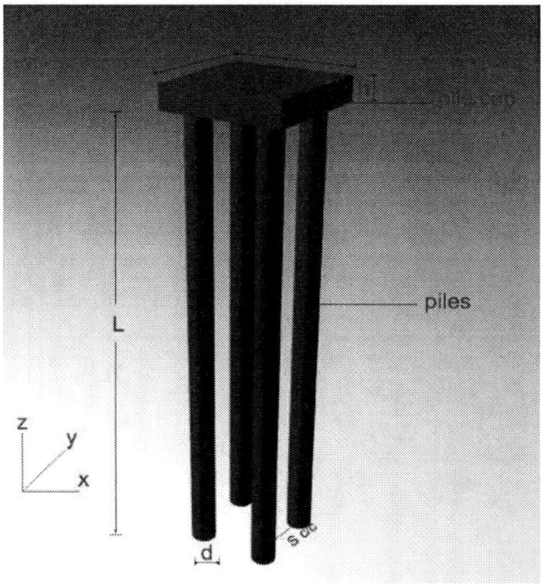

Figure 2. Details of pile machine foundation.

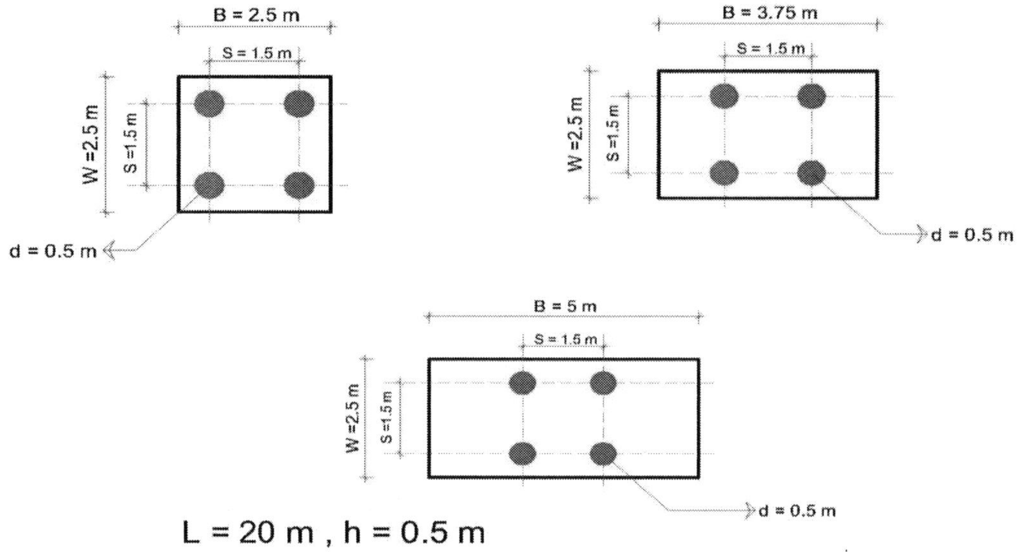

Figure 3. Top view of pile cap with different cap lengths.

4. Effect of Spacing on Vertical Displacement and Frequency Ratio of Pile Machine Foundation

In order to study the effect of spacing between piles on the response of pile machine foundation, different values of pile spacing were studied.

In Table 3, values for different spacings are considered, and the relation between the maximum displacements and spacing is shown in Figure 4. The figure reveals that the maximum displacement is found to be at minimum value at spacing of 1.5 m and increases in the case of 1.25 m spacing by about 62% but it increases about 317% for the 1.75 m spacing and also by about 625% for the 3 m spacing than the 1.5 m spacing.

IOP Conf. Series: Materials Science and Engineering **978** (2020) 012038 doi:10.1088/1757-899X/978/1/012038

5. Effect of Spacing on the Vertical Displacement and Frequency Ratio of Pile Machine Foundation

In order to study the effect of spacing between piles on the response of pile machine foundation, different values of pile spacing were studied.

In Table 3, values for different spacings are considered, and the relation between the maximum displacements and spacing is shown in Figure 4. The spacing 1.5 m is considered the reference case. The figure reveals that the maximum displacement is found to be at minimum value at spacing of 1.5 m and increases in the case of 1.25 m spacing by about 62% but it increases about 317% for the 1.75 m spacing and also by about 625% for the 3 m spacing than the 1.5 m spacing.

The frequency ratio (r) (is defined as the ratio of operating frequency to natural frequency) should be either less than 0.5 or more than 2 for important machines to avoid resonance condition (Rao, 2011). The natural frequency of the whole system can be obtained from the equation:

$$\omega_n = \sqrt{\frac{K}{m}} \qquad \text{..... (2)}$$

where:

K= total stiffness of piles plus pile cap,

M= total mass of piles, pile cap and soil included in the pile group.

$$r_f = \frac{\omega}{\omega_n} \qquad \text{..... (3)}$$

where:

r_f = frequency ratio,

ω = operating frequency, and

ω_n = natural frequency.

$$M = \sum V.\gamma/g \qquad \text{.... (4)}$$

$$\sum V * \gamma = V_1\gamma_1 + V_2\gamma_2 + V_3\gamma_3 \qquad \text{.... (5)}$$

where:

M = Total mass of piles, pile cap, and soil included by the pile group in (kg),

$V_1\gamma_1$ = Volume of piles multiplied by its unit weight in (kN).

$V_2\gamma_2$ = Volume of pile cap multiplied by its unit weight in (kN).

$V_3\gamma_3$ = Volume of soil multiplied by its unit weight in (kN).

g = acceleration of gravity in (m/sce^2).

Table 2. Material properties (Al-Nakdy et al., 2014).

	Name	Definition	Values	
Concrete	E	Young's modulus (MPa)	$4700\sqrt{f'c}$	25742.96
	γ_c	Density of concrete (kg/m^3)	2400	2400
	ν	Poisson's ratio	0.15*	0.15
	β_o	Open shear transfer coef	0.2*	0.2
	β_c	Closed shear transfer coef	0.7*	0.7
	f_t	Uniaxial Cracking stress (MPa)	$0.62\sqrt{f'c}$	3.3958
	f'_c	Uniaxial Crushing stress (MPa)	f'c*	30.000
	f_{cb}	Ultimate biaxial compressive strength (MPa)	1.2f'c	36.000
	σ_h^a	Hydrostatic stress	1.157f'c	34.710
	f_1	Ultimate compressive strength for a state of biaxial compression superimposed on (σ_h^a) (MPa)	1.45f'c	43.500
	f_2	Ultimate compressive strength for a state of uniaxial compression superimposed on (σ_h^a) (MPa)	1.725f'c	51.750
Interface	μ	Coefficient of friction	0.5*	0.5
Soil	Es	Young's modulus (MPa)	30000*	30.000
	ρ_s	Density of soil (kg/m^3)	1800*	1800
	ν	Poisson's ratio	0.3*	0.3
Steel	ρ_{st}	Steel ratio	0.002*	0.002

* Assumed values.

Table 3. Parameters for the pile machine foundations.

Case	B (m)	W (m)	h (m)	d (m)	S (m)
1	2.5	2.5	0.5	0.5	1.25
2	2.5	2.5	0.5	0.5	1.5
3	2.5	2.5	0.5	0.5	1.75
4	2.5	2.5	0.5	0.5	3

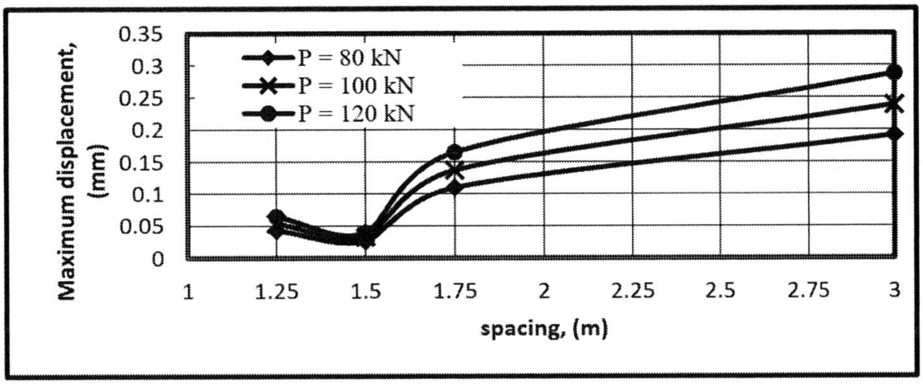

Figure 4. Relation between the maximum displacement and spacing between piles.

Figure 5 shows the relation between the frequency ratio and maximum displacement for the four different spacings stated in Table 3. The maximum displacement is found to be minimum at frequency ratio for 1.5 m spacing at operating frequency of 60 rad/sec and natural frequency of 52.7 rad/sec. The frequency ratio decreases with increase of spacing values due to increase in natural frequency, except in the case of 3 m spacing where it increases due to decrease in natural frequency as the mass was increased obviously.

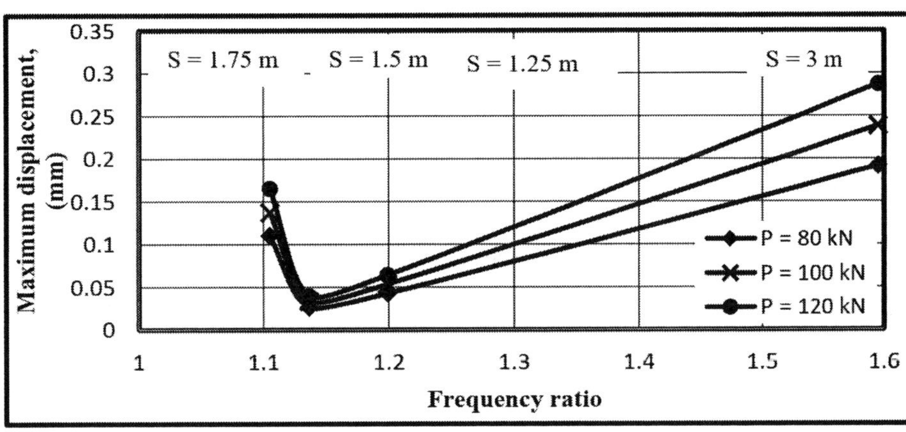

Figure 5. Relation between maximum displacements and frequency ratio for four different spacing.

Frequency ratio is found to increase by about 5% for 1.25 m spacing than that of 1.5 m spacing and to decrease by about 3% for the 1.75 m spacing than that of 1.5 m spacing and increase by 40% for the 3 m spacing than that for 1.5 m spacing due to increase in mass that reduces the natural frequency and consequently increases the frequency ratio. These results indicate that in the end bearing pile machine foundation studied, the pile spacing ratio (S = 1. 5 m) does not reveal the optimum spacing from a displacement point of view.

The present results are compatible with those obtained by Abdul Kaream et al. (2020) who found that the rate of increase in settlement decreased remarkably when increasing the frequency for both types of sand. While increasing the soil relative density under the same load and frequency resulted in a decrease of the settlement. Moreover, the amplitude displacement decreased when increasing frequency and relative density.

6. Comparison between Pile and Unpiled Machine Foundations

Introducing piles alters the displacement of pile cap obviously. With depth of soil 20 m below the (raft) cap without introducing piles with the other properties remain the same; a contact surface is created between the raft and the soil. Figures 6 and 7 show the response in the case of (2.5 m*2.5 m) cap raft with thickness of 0.5 m. It is found that the maximum displacement decreases with the use of piles by about 65% under the three amplitudes (80 kN, 100 kN, 120 kN). The maximum displacement for the unpile cap is found at frequency of (36 rad/sec). Figures 8 and 9 illustrate the relation between the maximum displacements with frequency for the case of (3.75 m*2.5 m) unpiled cap with 0.5 m thickness. The maximum displacement is found to decrease without introducing piles by about 91% for the three amplitudes of dynamic load. The maximum displacement for the unpiled raft is at frequency of 12 rad/sec.

For the case of (5 m × 2.5 m) unpiled cap with 0.5 m thickness, Figures 10 and 11 show that the maximum displacement decreases with the use of piles by about 65%. The maximum displacement for the unpiled raft is found at frequency of 24 rad/sec.

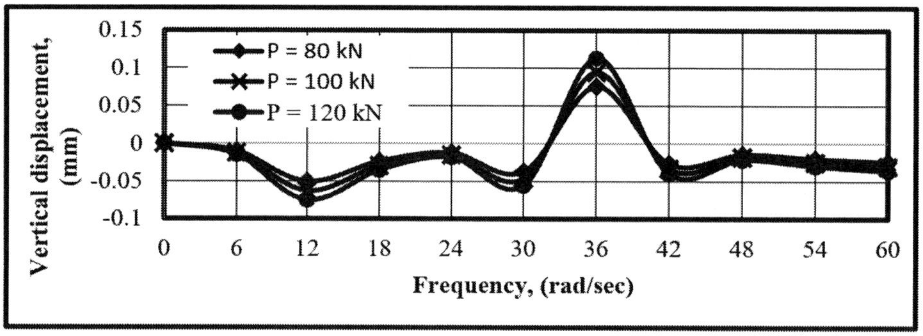

Figure 6. Change of vertical displacement with load frequency for the model pile machine foundation (h = 0.5 m, d = 0.5 m, B*W = 2.5 m*2.5 m, L = 20 m, S = 1.5 m) under vertical dynamic load.

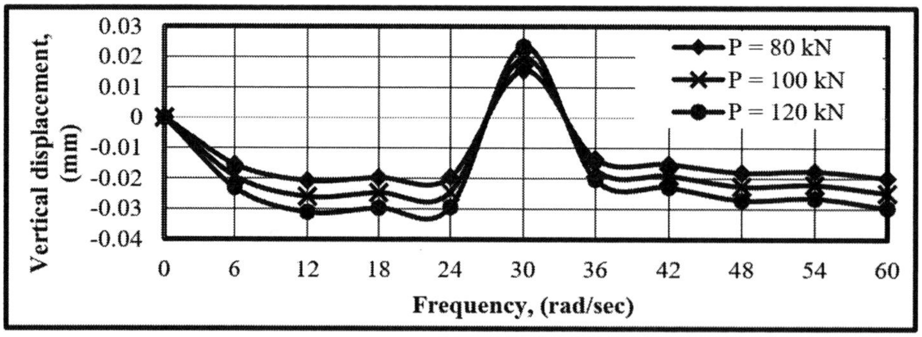

Figure 7. Change of vertical displacement with load frequency for the un-piled machine foundation (h = 0.5m, d = 0.5 m, B*W = 2.5 m*2.5 m, S = 1.5 m) under vertical dynamic load.

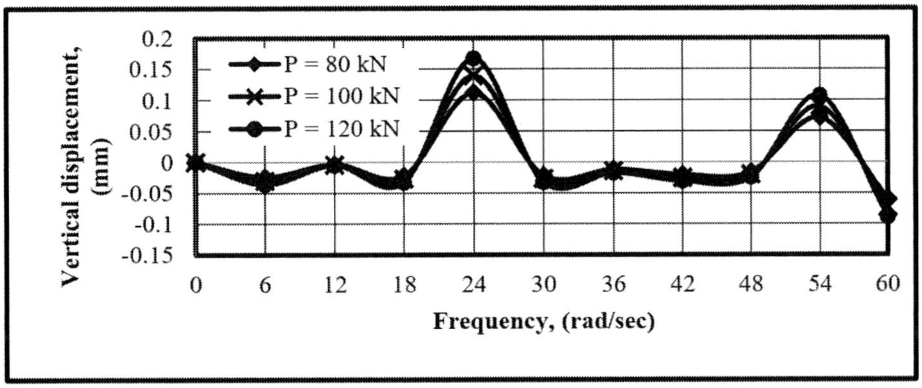

Figure 8. Change of vertical displacement with load frequency for the pile machine foundation (h = 0.5 m, d = 0.5 m, B*W = 3.75 m*2.5 m, L = 20 m, S = 1.5 m) under vertical dynamic load.

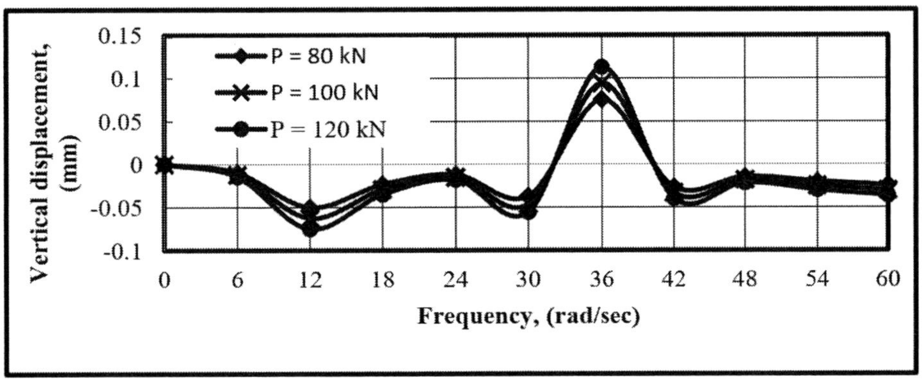

Figure 9. Change of vertical displacement with load frequency for the un-piled machine foundation (h = 0.5 m, d = 0.5 m, B*W = 3.75 m*2.5 m, S = 1.5 m) under vertical dynamic load.

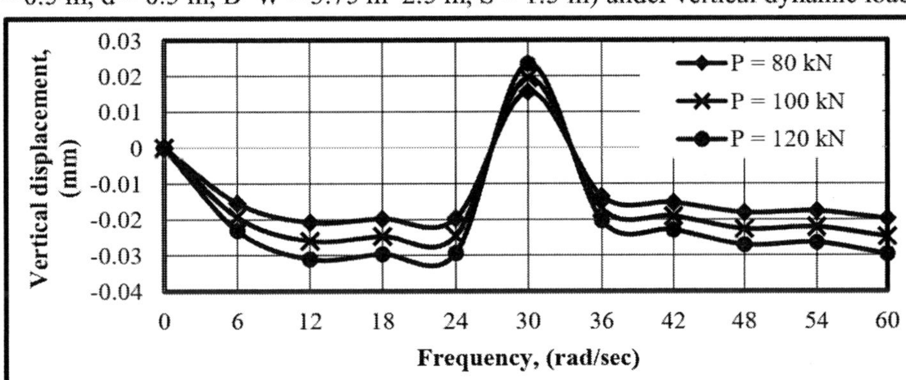

Figure 10. Change of vertical displacement with load frequency for the pile machine foundation (h = 0.5 m, d = 0.5 m, B*W = 5 m*2.5 m, L = 20 m, S = 1.5 m) under vertical dynamic load.

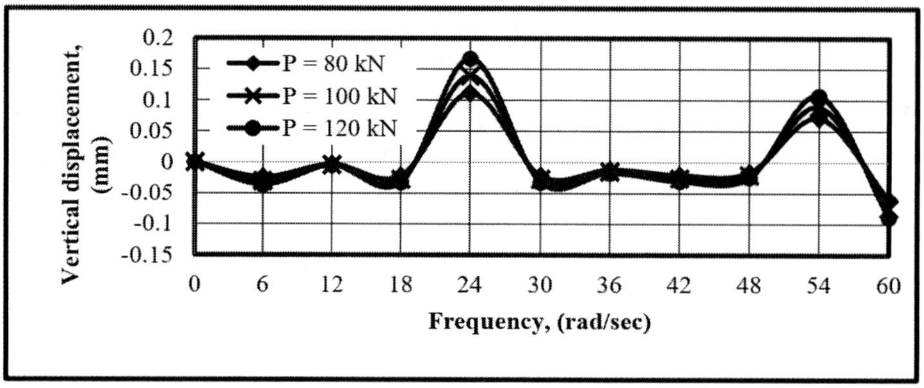

Figure 11. Change of vertical displacement with load frequency for the un-piled machine foundation (h = 0.5m, d = 0.5 m, B*W = 5 m*2.5 m, S = 1.5 m) under vertical dynamic load.

7. Conclusions

1. The frequency ratio decreases with increase of spacing values due to increase in natural frequency, except in the case of 3 m spacing where it increases due to decrease in natural frequency as the mass was increased obviously.
2. The increase in spacing caused the increase of natural frequency by about 6% in the case of 1.5 m spacing more than the 1.25 m case and by about 3% for the 1.75 m spacing more than the (1.5 m) reference case.
3. The increase in pile cap length caused the natural frequency to decrease by about 18% for the 3.75 m pile cap length as compared with the reference case and by 29% in the case of 5 m less than the reference case. In the case of increasing pile cap thickness, there was only a slight difference in the values of natural frequency.
4. The maximum displacement is found to be at minimum value at spacing of 1.5 m and increases in the case of 1.25 m spacing by about 62% but it increases about 317% for the 1.75 m spacing and also by about 625% for the 3 m spacing than the 1.5 m spacing.
5. Frequency ratio is found to increase by about 5% for 1.25 m spacing than that of 1.5 m spacing and to decrease by about 3% for the 1.75 m spacing than that of 1.5 m spacing and increase by 40% for the 3 m spacing than that for 1.5 m spacing due to increase in mass that reduces the natural frequency and consequently increases the frequency ratio.

References

[1] Abdul Kaream, K. W., Fattah, M. Y., Khaled, Z. S. M., (2020), Response of Circular Machine Foundation Resting on Sandy Soil to Harmonic Excitation, Journal of Engineering Science and Technology, Vol. 15, No. 2, pp. 831 – 845.

[2] Al-Nakdy, I. A. M., Fattah, M. Y., Hamood, M. J., (2014), Finite Element Analysis of Machine Foundation Resting on End Bearing Piles, Engineering and Technology Journal, University of Technology, Vol. 32 Part(A), No. 1, pp. 132-153.

[3] Arora, K. R., (1997), Soil Mechanics and Foundation Engineering, 4th ed., Delhi Standard Publishers Distributors, India.

[4] Bowles, J. E, (1997), Foundation Analysis and Design, Fifth Edition, Engineering Computer Software, Peoria, Illinois.

[5] Deng, N., Kulesza, R. and Ostadan, F., (2007), Seismic Soil- Pile Group Interaction Analysis of a Battered Pile Group, 4th International Conference on Earthquake, Geotechnical Engineering, June 25-28, Greece 2007, paper No. 1733.

[6] Fattah, M. Y., Al-Nakdy, I. A. M., Hamood, M. J., (2015), Finite-Element Analysis of a Piled Machine Foundation, Structures and Buildings, Proceedings of the Institution of Civil Engineers, Vol. 168, Issue 6, June 2015, pp. 421-432, http://dx.doi.org/10.1680/stbu.14.00053, Paper 1400053.

[7] Fattah, M. Y., Al-Mosawi, M. J., Al-Ameri, A. F. I., (2017), Stresses and Pore Water Pressure Induced by Machine Foundation on Saturated Sand, Ocean Engineering, Vol. 146, pp. 268–281, Elsevier.

[8] Fleischer, P.St. and Trombik P.G. (2008), Turbogenerator Machine Foundations Subjected to Earthquake Loading, The 14 th World Conference on Earthquake Engineering, Beijing, China.

[9] Liu, W. and Novak. M., (1991), Soil Pile Cap Static Interaction Analysis by Finite and Infinite Elements, Canadian Geotechnical Journal, Vol.28, p.p. 771-783.

[10] Ottaviani, M. (1975), Three-Dimensional Finite Element Analysis of Vertically Loaded Pile Groups, Geotechnique, Vol. 25, No.2, p.p.159-174.

[11] Prakash, S. and Puri, V.K., (2006), Foundation for Vibrating Machines, Special Issue, April-May 2006, Journal of Structural Engineering, SERC, Madras, India, http://yoga10.org/Documents/SPVKPSERCpaper.pdf.

[12] Rao, K., N. S. V., (2011), Foundation Design Theory and Practice, a book, university Malaysia Sabah, Malaysia.

Effect of TiO_2 Nanoparticles Addition on Contact Angle of Epoxy Resin

Ali. F. Nyief[1,*], Kadhum. M. Shabeeb [2] and Ayad. K. Hassan [1]

[1] Materials Engineering Department, University of Technology, Iraq
E-mail: engali2861981@gmail.com
[2] Materials Engineering Department, University of Technology, Iraq
E-mail: kadhum.muttar@gmail.com
[3] Materials Engineering Department, University of Technology, Iraq
E-mail: Ayad_kadhm@yahoo.com
* Corresponding author: Materials Engineering Department, University of Technology, Baghdad, Iraq. Tel: +9647717704522.

Abstract. The contact angle is very important parameter in improving the self-cleaning property as well as in increasing the wear resistance of engineering surfaces in addition to improving the thermal transfer properties of materials in general. In this study, titanium dioxide nanoparticles were added to the epoxy resin and sprayed on a substrate of the brass tube used in heat exchangers for the purpose of studying its effect on the angle of contact of the substrate where different proportions of the nanomaterial were added to the epoxy resin in (1%, 3%, 5%, and 7%)wt. The results showed that the contact angle increased with increasing the rate of titanium dioxide nanoparticles.

Keywords: *contact angle, epoxy resin, Nano composite coating, TiO2 Nanoparticles, self-cleaning.*

1. Introduction

Epoxy is one of the highly corrosion resistance materials, and this feature can be improved with some other properties by adding nanomaterials to the epoxy when used as a coating to protect surfaces from corrosion(Chaudhari, Shaikh, & Pandey, 2013). Fouling and corrosion relates to the main problem with heat exchangers. Such a problem causes the heat exchanger to stop getting it out of work to disassemble it and clean it can be very expensive(Rathish, Dorothy, R M, Pandiarajan, & Rajendran, 2013). Fouling and corrosion in chemical factories and power plants, leading to the need to use large quantities of water for the purpose of improving the efficiency of the heat exchanger(Zhao, Liu, Wang, Wang, & Müller-Steinhagen, 2005). The phenomenon of fouling is the main problem that causes the efficiency of heat exchanger to decrease and is defined as the accumulation of unwanted materials on the inner or outer surface of the heat exchanger tube, such as , scale, algae, suspended solids and insoluble salt(Al-Haj Ibrahim, 2012). Fouling and corrosion layer forms on the walls of tubes due to the low contact angle of the tube surface. It can lead to corroded of the tube metal, which increases Potential problems for conventional and nuclear power plants(Videla & Characklis, 1992). As this layer has a much lower thermal conductivity than the thermal conductivity of the tube metal and thus a deterioration in the heat transfer coefficient of the tube and the

IOP Conf. Series: Materials Science and Engineering **978** (2020) 012039 doi:10.1088/1757-899X/978/1/012039

heat exchanger in general, also high drop in pressure, this requires the use of pressure-lifting equipment(Karabelas, 2002). Due to the adhesion of the fouling layer on the surface of the pipes and the problems that lead to it in the cooling water systems and heat exchangers, it has taken a large portion of attention for the purpose of studying the means that limit or prevent this fouling and its growth(Hamza, Pham, Matsuura, & Santerre, 1997). The process of changing the surface characteristics is one of the most important methods by which the rate of fouling on the surface of tube of heat exchanger is reduced, it changes the nature of surface and make it less attractive to the fouling(Dexter, Sullivan, Williams, & Watson, 1975). Numerous attempts were made to increase the contact angle by coating, whether by metal or polymeric materials, for the purpose of reducing this problem. The results showed that the coating with nanomaterials increases the corrosion resistance and improves thermal transfer(Takata et al., 2005).

Sometimes composite coating consisting of polymeric material and nanomaterials are used for the purpose of improving surface properties and to take the advantage of the properties of both materials(Liu, Wang, Wang, & Müller-Steinhagen, 2005).

The aim of the work is enhancement of the contact angel of the brass tube surface grade (B-111) after coating by only epoxy and nanocomposite coating of epoxy with different weight ratio of TiO_2 nanoparticles.

2. Experimental work

2.1 Materials

Materials used in this experiment are epoxy resin type Quick mast 105 made in Focsroc Company in Jordan) that is in liquid state where it was used as basic coating material in preparing the nanocomposite coating materials. The hardener is used to harden the epoxy and add a ratio 2:1 with continuous mixing. The TiO_2 nanoparticles (Eprui Nanoparticles & Microphares Co. Ltd) is used and add at four ratios (1%, 3%, 5% and 7%).wt

Properties

Pot life:	60 minutes at 20°C 15 minutes at 40°C
Density:	Approx. 1060 kg/m³
Tensile strength:	30 N/mm² at 7 days
Flexural strength:	55 N/mm² at 7 days
Compressive strength:	55 N/mm² at 1 day 68 N/mm² at 3 days 93 N/mm² at 7 days
Modulus of elasticity: (ASTM C580 7 days)	16 KN/mm²
Viscosity:	250 Cps at 20°C 100 Cps at 35°C

Figure1. properties of epoxy resin

2.2 Coating Technique

Brass grade (B-111) samples with chemical composition shows in table No (1a) tested it by X-Ray Fluorescence and table (20 for ASTM stander it was taken from the same alloy of heat exchanger tube with dimensions 15×10 mm for purpose of its coating and then characterizing the coat and measuring the contact angle before and after the coating with the mentioned ratios. Before the coating process, firstly, in order to prepare the sample for etching, mechanical Polishing is used and then the sample is etching (Mixing 5 g

IOP Conf. Series: Materials Science and Engineering **978** (2020) 012039 doi:10.1088/1757-899X/978/1/012039

Ferric Chloride, and 10 ml Hydrochloric Acid with 100 ml water and the specimen is immersed), then the samples are washed and clean to remove impurities from the surface where they washed with distilled water and then dried.

Table 1: Chemical composition of brass tube

Cu	Zn	Pb	P	Sn	Mn	Ti	Fe
73.32	25.08	0.0084	0.0017	0.001	0.006	0.0042	0.09
Sb	Mg	Al	Si	Cd	V	Zr	Ni
0.001	0.032	0.057	0.0026	0.003	0.0021	0.02	0.097
Nb	Mo	W	Cr	Co	As	Ag	S
0.0021	0.175	0.021	0.0579	0.045	0.0005	0.0011	0.002

Table 2. ASTM B111 Tube Chemical Composition

Element	Ratio weight
Cu	76.0- 79.0
Zn	Reminder
Al	1.8-2.5
As	0.02-0.06
Pb	0.07
Fe	0.06 max

2.3 Coating Preparation

20 ml of epoxy mixed with 10 ml hardener, the weight of mixture 35 gm mixed by magnetic stirrer for 5 minuts and then using airbrush spray gun for coating the samples with epoxy alone. After that, the different proportions of titanium dioxide nanoparticles are added to the epoxy, where every time a percentage of the TiO_2 is added, it is mixed with the epoxy by using the magnetic stirrer for 1 hr and then placed in an ultrasonic homogenizer for five minutes, after which the hardener is added to the mixture and Place it on the magnetic stirrer for five minutes, after which it is sprayed on the samples for the purpose of coating them. At 5% and 7% wt of TiO_2 in epoxy resin a 3ml volatile diluent was used to dilute the coating and facilitate the spraying of samples.

3.Characterization

3.1 Atomic Force Microscopy

AFM was conducted for the purpose of characterizing the coating on the substrate of the disinfection. Microscopic examination of the AFM was carried out for the purpose of measuring surface roughness, as well as measuring the diameter of particles and surface morphology in addition to structures.

3.2 Scanning Electron Microscopy

A scanning electron microscopy SEM test was conducted for the purpose of characterizing surface topography after coating and knowledge of composition and information whereby the surface can be described as if it is woven, smooth or rough

3.3 Measurement of Contact Angle

This test was conducted to knowledge the nature of surface whereas hydrophobic or hydrophilic and determined the affected of TiO_2 on the contact angle of epoxy resin. The test carried out by using the

IOP Conf. Series: Materials Science and Engineering **978** (2020) 012039 doi:10.1088/1757-899X/978/1/012039

instrument by put 3 droops on the surface of sample before and after coating from distilled water with 9μL volume at different positions on the surface of the samples, by using the needle in the position very near to the surface of sample. The repeated tow times and the images in fig 3 shows the measurements of contact angle for samples before and after coating

4. Results and discussion

4.1Atomic Force Microscopy (AFM)

Atomic Force Microscopy (AFM) It is used to investigated the topography and roughness of the surface of the coated samples respectively. Where the results of the examination showed a rate of superficial roughness at a value 4.422, 7.912, 14.941 and 22.098 respectively for coated samples Epoxy with (1,2,3 and 7%) weight ratio of TiO_2 nanoparticles, as shown in the figure (1). Where we notice a decrease in the surface roughness for the samples that were coated with reinforced epoxy by 1% and 3%. This indicates that the titanium dioxide nanoparticles have a high dispersion within the epoxy resin and have a strong cross-linked. While the high surface roughness in the samples coated with epoxy containing 5% and 7% weight ratio of TiO_2 is due to the high ratio of TiO_2 nanoparticles to agglomerate, which leads to an increase in the tendency of nanoparticles, thus the collective size of the particles will increase(Goyat, Rana, Halder, & Ghosh, 2018).

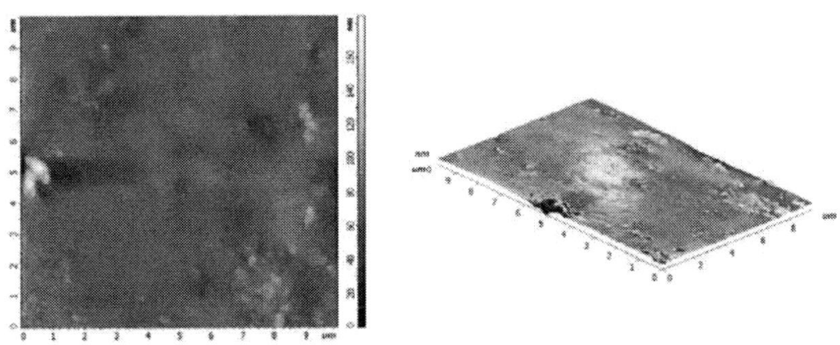

Figure (1-a): 2D and 3D AFM examination of sample No1 (epoxy with 1% TiO_2)

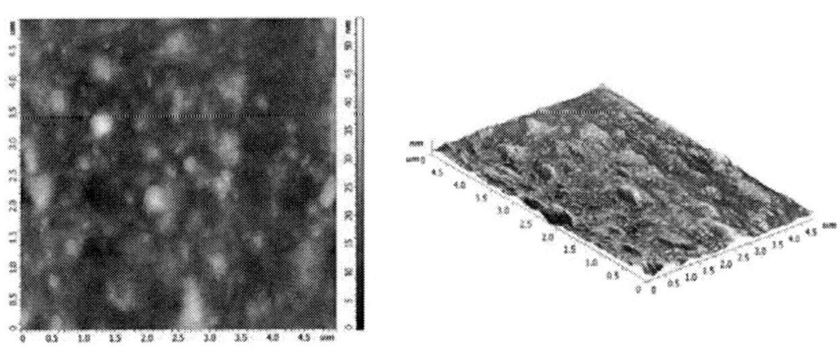

Figure (1-b): 2D and 3DAFM examination of samples No2 (epoxy with 2%TiO_2)

IOP Conf. Series: Materials Science and Engineering **978** (2020) 012039 doi:10.1088/1757-899X/978/1/012039

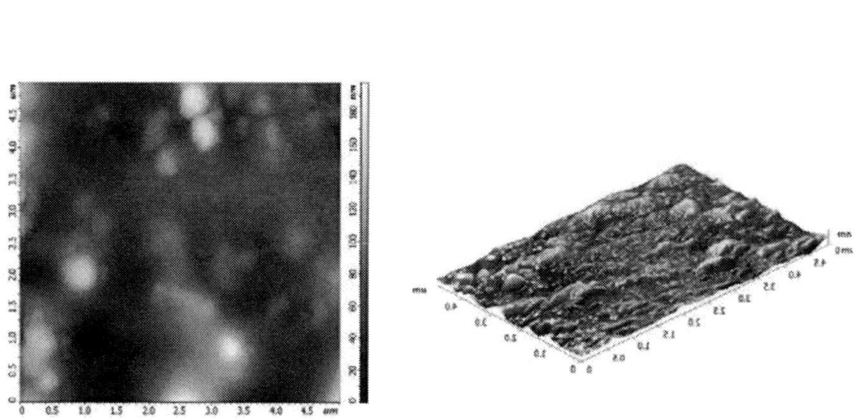

Figure (1-c): 2D and 3DAFM examination of samples No2 (epoxy with 3%TiO_2)

Figure (1-d) :2D and 3D AFM examination of samples No2 (epoxy with 4%TiO_2)

4.2 Scan Electron Microscopy

SEM examination of all samples showed that the presence of individual titanium dioxide nanoparticles in the epoxy resin is possible but very limited within the coating. It is also noticed that there is a agglomeration state of the embedded particles in the epoxy, but in spite of this, there is a good dispersion condition of the nanoparticles of titanium dioxide as show in fig (2) a and b.

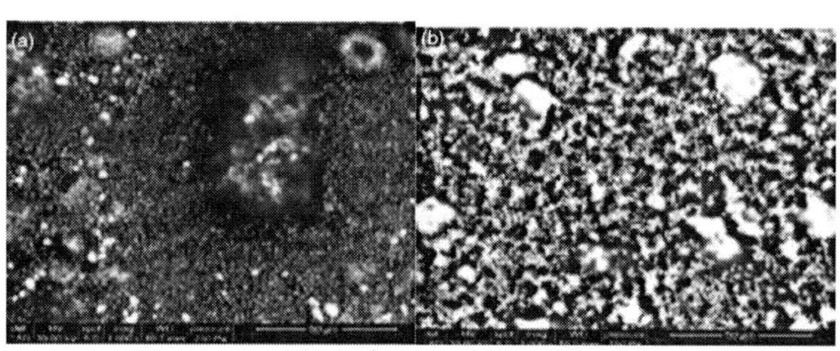

Figure 2a: SEM of 1% and 3% TiO_2 samples

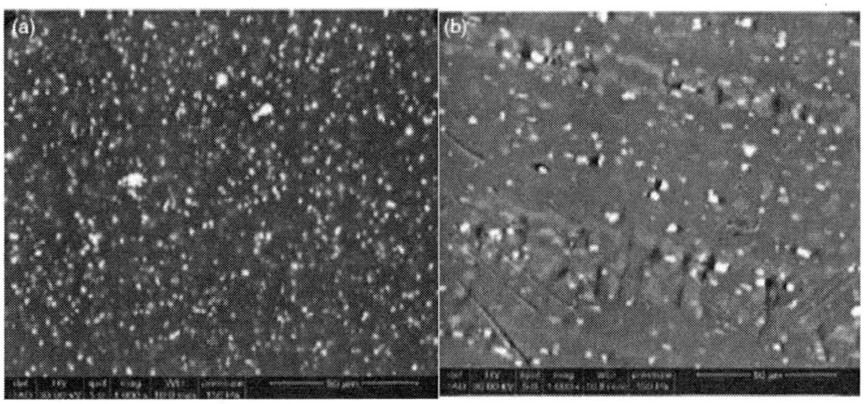

Figure 2b: SEM of 5% and 7% TiO_2 samples

4.3 Measurement of contact angle (CA)

A contact angle measurement test was performed in order to know the wettability of the surface to the samples before and after coating and to know the extent of the impact of the TiO_2 nanoparticle on that property by performing this examination as show in fig (5). The result showed that an increase in the ratio of TiO_2 nanoparticle on epoxy resin to some extent increases the contact angle as shown in the fig (4) the Max contact angle at 7%TiO_2 is 68.369^0.

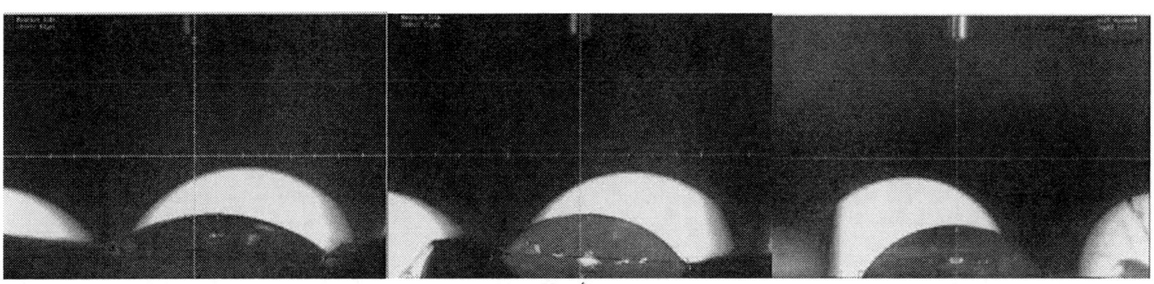

a b c

Figure 3. Samples a-Nano-coating (b-epoxy and c-5%)wt of TiO_2 nanoparticle (38.714^0, 51.715^0 and 57.391^0) respectively.

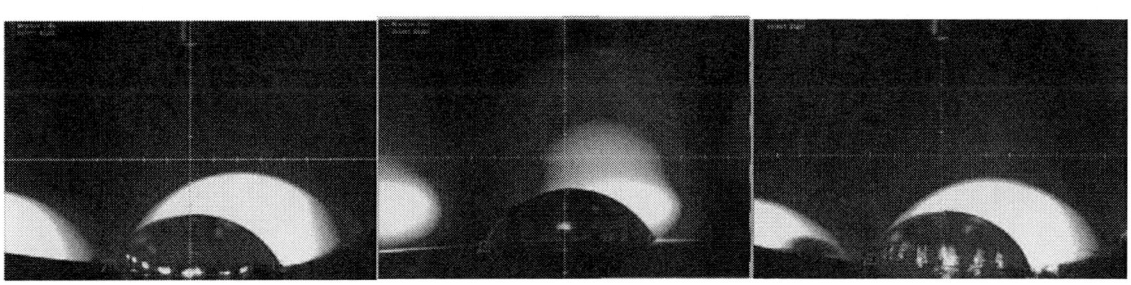

Figure 3. Samples with (d-3%, e-5% and f-7%**) wt of** TiO_2 nanoparticle (60.904^o, 65.585^o and 68.369^o) respectively.

5. Conclusion

The addition of different ratios of TiO_2 nanoparticle to the epoxy resin increases the contact angle of the nanocomposite coating, which leads to a decrease in the build- up of the fouling and improvement of the surface properties coated with the nanocomposite coating. The characterizations performed on the coating, AFM and SEM, showed a good cross-linking between nanoparticles and the epoxy.

References

[1] Al-Haj Ibrahim, H. (2012). Fouling in Heat Exchangers. In (Vol. 40, pp. 57-96).

]2] Chaudhari, D. S., Shaikh, T., & Pandey, P. (2013). A Review on Polymer Tio2 Nanocomposites. *Int. Journal of Engineering Research and Application, 3*, 1386-1391.

[3] Dexter, S. C., Sullivan, J. D., Williams, J., & Watson, S. W. (1975). Influence of substrate wettability on the attachment of marine bacteria to various surfaces. *Appl Microbiol, 30*(2), 298-308.

[4] Goyat, M. S., Rana, S., Halder, S., & Ghosh, P. K. (2018). Facile fabrication of epoxy-TiO2 nanocomposites: A critical analysis of TiO2 impact on mechanical properties and toughening mechanisms. *Ultrasonics Sonochemistry, 40*, 861-873. doi:https://doi.org/10.1016/j.ultsonch.2017.07.040

[5] Hamza, A., Pham, V. A., Matsuura, T., & Santerre, J. P. (1997). Development of Membranes with Low Surface Energy to Reduce the Fouling in Ultrafiltration Applications. *Journal of Membrane Science, 131*, 217-227. doi:10.1016/S0376-7388(97)00050-1

[6] Karabelas, A. (2002). Scale formation in tubular heat exchangers - Research priorities. *International Journal of Thermal Sciences - INT J THERM SCI, 41*, 682-692. doi:10.1016/S1290-0729(02)01363-7

[7] Liu, Y., Wang, C., Wang, S., & Müller-Steinhagen, H. (2005). Effect of surface free energy on the adhesion of biofouling and crystalline fouling. *Chemical Engineering Science, 60*, 4858-4865. doi:10.1016/j.ces.2005.04.006

[8] Rathish, J., Dorothy, R., R M, J., Pandiarajan, M., & Rajendran, S. (2013). CORROSION RESISTANCE OF NANOPARTICLE - INCORPORATED NANO COATINGS. *European Chemical Bulletin, 2*, 965-990.

[9] Takata, Y., Hidaka, S., Cao, J., Nakamura, T., Yamamoto, H., Masuda, M., & Ito, T. (2005). Effect of surface wettability on boiling and evaporation. *Energy, 30*, 209-220. doi:10.1016/j.energy.2004.05.004

[10] Videla, H. A., & Characklis, W. G. (1992). Biofouling and microbially influenced corrosion. *International Biodeterioration & Biodegradation, 29*(3), 195-212. doi:https://doi.org/10.1016/0964-8305(92)90044-O

[11] Zhao, Q., Liu, Y., Wang, C., Wang, S., & Müller-Steinhagen, H. (2005). Effect of surface free energy on the adhesion of biofouling and crystalline fouling. *Chemical Engineering Science, 60*(17), 4858-4865. doi:https://doi.org/10.1016/j.ces.2005.04.006

Preparation and Characterization of Chitosan/ZnO/Ag Nanocomposites as Antibacterial Hydrogel for Wound Dressing

Sabah M. Thahab Al-Abboodi[1,*], Nibras Yahya Abdulla[2] and Thamar A. Almussawy[3]

[1]Nanotechnology and Advanced Materials Research Unit (NAMRU), Faculty of Engineering, University of Kufa, Najaf, Iraq- Email: sabah.alabboodi@uokufa.edu.iq

[1]Electronics and Communications Departement, Faculty of Engineering, University of Kufa, Najaf, Iraq.

[2]Dept. of Biology, College of Science, University of Kufa, Najaf, Iraq
Email: nibras.alsallami@uokufa.edu.iq

[3]Dept. of Materials Engineering, University of Kufa, Najaf, Iraq
Email: engthamar241@gmail.com

*Corresponding Author: Tel: +9647703996492.

Abstract. The aim of this study is the preparation of chitosan/Ag/ZnO (CS/Ag/ZnO) composite to investigate the antibacterial activity of the prepared nanocomposites. Chitosan is a linear biodegradable natural polymer having excellent characteristics, such as biocompatibility. Hydrogel as a wound dressing could be made to act as antibacterial, antimicrobial, and anti-inflammatory agents in the presents of ZnO and Ag Nanoparticles. In this research, pure chitosan (Cs) and Cs/ZnO/Ag nanocomposites were papered to evaluate the antimicrobial effect against gram-positive and gram-negative microorganisms. The results show that the pure chitosan with (6mM) concentration has the best antibacterial effect against *K-pneumoniae* bacteria as compared to S.*aureus* and *E. coli* bacteria with an Inhibition Zone of (20.66-65.33) mm at pH value in the range of (5-6), while the prepared nanocomposite of Cs/Ag/ZnO (30/10) nano-composites shows that the antibacterial behavior against *S-aureus* bacteria is better than the antibacterial effect against *K-pneumonia*bacteria. The results illustrated that the bi-nanocomposites shows a higher antibacterial effect due to the combined effect of Cs and nanoparticles (NPs).

Keywords: *Antibacterial; Nanocomposites; Chitosan; Nanoparticles; Wound dressing*

1. Introduction.

Nanocomposite has been used for many applications like wound healing, tissue engineering, thermal therapy, and drug delivery. Nanocomposite hasvarious properties make it appropriate for wound healing process such as mechanical and thermal characteristics, antimicrobial properties, balance, and structure with the porosity to allow absorbance, and capability to provide perfect humidity in the wound bandage [Shabunin et al., 2019].

The most important properties of wound dressing are, biocompatible, biodegradable, infection protection, stable shape, good mechanical characteristics, non-toxic, and cost effectiveness [Noori et al., 2018]. Chitosan is a biocompatible polymer consist of polysaccharide which has antibacterial properties that have been added to wound dressings [Cremar et al., 2018].

Chitosan is a bio-copolymer including N-acetylglucosamine and glucosamine. It has many characteristics suchas non-toxic, biocompatiblity, biodegradability, moisture retentive and low cost [Lu et al., 2017]. In the wound healing process, chitosan work as a painkiller drug and anti- provocative factor, which can show a good effect when used to an open wound [Cremar et al., 2018]. In vivo examinations have revealed that the pain is decreased when chitosan fabric/membrane is used as wound dressings to the burns, skin abrasions, skin grafts and, ulcers [Ohshima et al., 1987]. The mechanism of chitosan as a wound has been studied by using a suspension of chitosan/acetic-acid composites on mice, which showed a decrease in the inflammatory pain because of an increase in the pH from protonation of $-NH2$ to $-NH3+$ [Okamoto et al., 2002]. Various shapes of chitosan employed for wound dressing have been studied, like hydrogels, films, scaffolds, and sponges [Ma et al., 2014, Zakhem et al., 2012].

Nano-sized ZnO exhibits varying morphologies and shows significant antibacterial activity over a wide spectrum of bacterial species explored by a large body of researchers [Buzea et al., 2007, Brayner et al., 2006, Raghupathi et al., 2011]. ZnO NPs in the micro and the nano scale are used as antibacterial material especially when their size is reduced into the nanometer scale, ZnO NPs can interact with the surface of the bacteria or enter into the cell wall of the bacteria, as a result shows its antibacterial mechanisms. The interactions between NPs and bacteria are typically toxic, which have been subjugated for antimicrobial applications like food industries and wound healing [Seil et al., 2012].

Metal NPs like Ag, ZnO, TiO, Cu, CuO, and Au have revealed wonderful antimicrobial activity to antibiotics, particularly versus antimicrobial resistant. These metal NPs create reactive oxygen species (ROS) which destroy microbial DNA and interrelate with proteins and organic constituents of the microbe cause electrolyte inequity leading to killing of microbes [Dizaj et al., 2014].

For instance, ZnO inhibits the adhesion and internalization of enterotoxigenic *E. coli* (ETEC) into enterocytes (Roselli et al., 2003). In addition, ZnO nanoparticles (ZnO-NPs) exhibit antibacterial activity and can reduce the attachment and viability of microbes on biomedical surfaces (Brayner et al., 2006; Yamamoto, 2001). Interestingly, several results suggest a selective toxicity of ZnO-NPs preferentially targeting prokaryotic systems, although killing of cancer cells has also been demonstrated (Hanley et al., 2008; Reddy et al., 2007; Taccola et al., 2011).

Furthermore, Ag^+ ions and Ag-based compounds are highly toxic to several microorganisms, which make them interesting candidates for multiple applications in the medical field (Furno et al., 2004; Prakash et al., 2013). Ag is generally used as nitrate salt, but in the form of Ag nanoparticles (Ag-NPs) the surface area is increased and thereby antimicrobial efficacy is greatly enhanced.

In this study, we produced an antibacterial wound dressing by loading ZnO/Ag nanocomposite to chitosan. The antibacterial activity was investigated against three types of microorganisms gram-positive (S-aureus and K-pneumonia) and gram-negative (E-coli).

2. Materials and methods

Chitosan powder (Molecular weight of 161.16 g/mol - Sigma Aldrich-UK), Acetic acid with chemical formula CH3COOH was made in India with purity equal to 99.5%. ZnO was made in USA with purity equal to 99.8% and the particle size is in the range of 10-30nm.AgNPs were made in India with purity equal to 99% and the particle size is in the range of 20nm. Three different types of bacteria *S-aureus, K-pneumonia and E-coli* were gently supplied from laboratories of biology department/college of science/ university of kufa.

2.1 Preparation of chitosan/ acetic acid solution

In order to prepare solutions of chitosan, the chitosan in powdered form was weighed using a sensitive electronic balance device (type HT series- Japan) Vibra company and added into (100 ml) of acetic acid to prepare a three different viscous polymeric solution as illustrated in table (1).

Table1. Illustration different concentration of chitosan.

Solution no.	Chitosan weight (g)	AA (ml)	Concentrations (mM)
1	0.1	100	6
2	2	100	80
3	1.62	100	100

2.2 Preparation of Cs/ZnO and Cs/Ag Nanocomposites

To prepare the nanocomposites, ZnO NPs and Silver nanoparticles were incorporated with two different weight percentages (0.05wt%, 0.10wt %) and mixed by sonicator for 10 minutes, see table (2).

The following equation was used to evaluate the weight of the NPs required to be added into the solution of chitosan to obtain the nanocomposites. (William. D., 2007):

$$V_f = \frac{\frac{W_m}{\rho_m}}{\frac{W_m}{\rho_m} + \frac{W_p}{\rho_p}} \quad \text{...Eq.1}$$

Where:

Wm: weight of matrix

Wp: weight of particles

ρm: density of matrix

ρp: density of particles

Vf: volume fraction

The NPs were added into chitosan and the nanocomposite solution was subjected to ultrasonic device (1200W ultrasonic processor) from MTI company in USA to homogenizing NPs in the polymeric solution.

Table 2. Weight percentages of Nano composite formulations

Sample No.	Concentrations
1	Cs/Ag=20ml/0.10wt%
2	Cs/ZnO=20ml/0.05wt%
3	Cs/Ag=60ml/0.10wt%
4	Cs/ZnO=60ml/0.05wt%
5	Cs/Ag/ZnO=10/10
6	Cs/Ag/ZnO=30/10
7	Cs/Ag/ZnO=10/30

Antibacterial activities of the prepared nanocomposite solutions were tested versus Gram-positive and Gram-negative germs with S. aureus and E. coli and K.pneumoniae correspondingly. The procedure for culturing the prepared nanocomposites and nanofibers can be explained as follows:

1. All the used Petri dishes were autoclaved by using autoclave before the experimental work.
2. To prepare the medium for culturing bacteria, five to six different colonies were allocated in petri dishes and incubated for 24 hr at 37°C.
3. Three different types of bacteria were sprayed on the prepared medium.
4. For culturing nanocomposites, many holes are created in the medium with (6mm) in diameter, after that the nanocomposites are added to the holes using micropipette (10μl).
5. For culturing nanofibers obtained from Electrospinning process using Electrospinning system that laboratory manufactured in Nanotechnology and Advanced Materials Research Unit (NAMRU) at Faculty of Engineering in University of Kufa/IRAQ, the prepared nanofibers were shaped into circular having diameter of (6mm), and attached to the surface of the bacteria.
6. All the prepared petri dishes were incubated for 24hr.

3. Results and Discussion

When the pH of Cs decreases, the antibacterial activity of Cs is increased. This is likely due to the fact that the amino groups of chitosan become positively ionized below pH 6[Erdem et al., 2016].

From table (3) can be observed that the maximum inhibition zone is for the concentration of (0.1M), as shown in fig.1. The mechanism which explain the antibacterial behavior of chitosan can be describe by interaction of positive charge in chitosan such as protonated NH_3 and the negative charge present in the cell wall of bacteria through electrostatic forces. These interactions destroying the function of microbial cell membrane such as escaping of intracellular compounds and also prevent transformation of nutrients which leading to kill the bacteria [Pasaribu et al., 2018, Liu et al., 2006]. Srinath et al., 2019 observed that the (IZ) of the Gram-negative bacteria (E. coli) is larger than the Gram-positive bacteria (S. aureus).

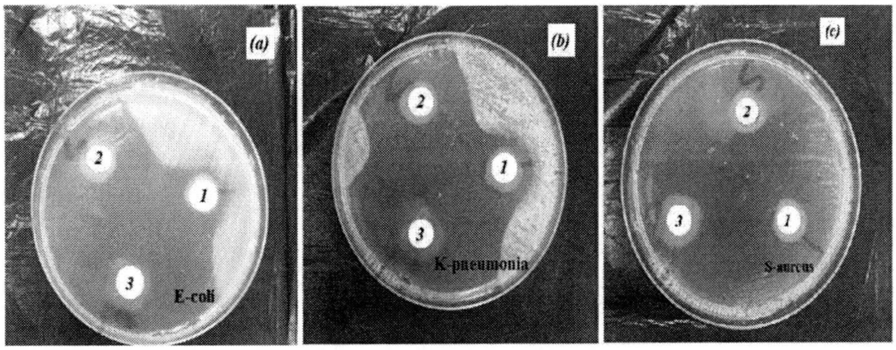

Figure 1. Shows how the inhibition zone effected by antibacterial activity of chitosan concentrations on the gram positive and gram negative bacteria.

Table 3. Illustrate the diameters of the Inhibition Zone(IZ) of chitosan.

Concentration (mM)	S-aureus IZ(mm)	E-coli IZ(mm)	K-pneumoniae IZ(mm)
100	38.33	49.33	65.33
80	25	38.33	64
6	20.66	44	61.33

The cell walls of the bacteria containing many layers of protein and lipids (give the hydrophobic properties) in specific structure known as bi- layer [Haider et al., 2015]. Table (4) shows the concentrations of prepared Cs/Ag/ZnO nanocomposites and the diameters of the inhibition zone of three types of bacteria.

Table 4. Illustrate the diameters of the Inhibition Zone (IZ) of chitosan nanocomposites.

Sample No.	**Concentrations**	**S-aureus IZ (mm)**	**E-coli IZ (mm)**	**K-pneumonia IZ (mm)**
1	Cs/Ag=20ml/0.10wt%	42.33	37.66	31
2	Cs /ZnO=20ml/0.05wt%	42	36	29.66
3	Cs/Ag=60ml/0.10wt%	45.33	39	32.66
4	Cs/ZnO=60ml/0.05wt%	40.33	32.33	34.33
5	Cs/Ag/ZnO=10/10	37.33	34.33	32
6	Cs/Ag/ZnO=30/10	45.66	36.66	40
7	Cs/Ag/ZnO=10/30	40.33	31.66	35.33

From table (4) the Cs/Ag/ZnO nanocomposites shows that the best antibacterial behavior was against S-aureus. The addition of ZnO NPs into chitosan/acetic acid aqueous solution leading to formation of different types of ROS like (•OH), (O2• −), and (H2O2). (•OH) and (O2• −), have negative charge so they cannot go through the cell wall due to the repulsion of negative charges, while (H2O2) has positive charge and due to attraction between the different charges, the (H2O2) enter the cell membrane. Li-Hua et al., 2010, reported that reported that the CS/ZnO nanocomposite showed antibacterial activity against Bacillus subtilis, Escherichia 317 coli, and Staphylococcus aureus. The formation of ROS relies on the surface area of the nanoparticles (higher surface area produces higher concentration of ROS) [Shi et al., 2014]. Cs has strong tendency towards metal ions since the Cs possess many amine and hydroxyl groups [Sanpui et al., 2008].

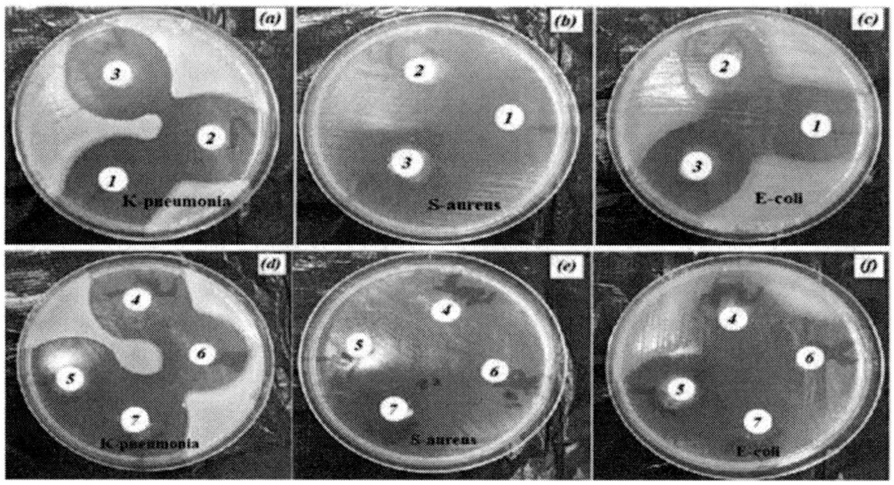

Figure 2. Shows the antibacterial behavior of Cs/Ag/ZnO nanocomposite against (a), (d) *K-pneumonia*, (b), (e) *S-aureus*, and (c), (f) *E-coli.*

Fig.2 shows that the Cs/Ag/ZnO nanocomposite has the best inhibition zone against *S-aureus* bacteria in all prepared compositions, and the lowest antibacterial effect was against *K-pneumonia.* The combined effect of chitosan and nanoparticles can be explained by various mechanisms such as changing the chemistry of the surface area of the particles, consequently, changing their biological characteristics. Li et al., 2010 reported that the increase of Ag and ZnO contents causing increase in the diameter of (IZ). Claudia et al., 2014 reported that the systems CS/ Ag/ZnO have higher antimicrobial activity that of the chitosan indicating that ZnO and Ag enhance the antimicrobial activities of chitosan. The most possible methods are (i) as the chitosan wrapper the surface of the NPs causing "steric stabilization" and reduced the agglomeration, that consequently rising the effectual concentration of NPs; (ii) improving the binding forces between the positive charges of chitosan and the negative charges from the bacteria [Potara et al., 2011].

4. Conclusions

To conclude, we have produced nanocomposites of chitosan loaded with different weight percentages of silver and zinc oxide nanoparticles to enhance the antimicrobial efficiency against both grams positive and gram negative pathogens. All the prepared compositions showed a range of bactericidal behavior and the bimetallic formulations exhibited a synergistic antibacterial effect due to the high antibacterial activity of ZnO and Ag nanoparticles which improve the antibacterial activity of the bi-nanocomposites (ZnO/Ag/Cs).

Acknowledgments

The authors would like to acknowledge the assistance offered by Nanotechnology and Advanced Materials Research Unit (NAMRU) at Faculty of Engineering in University of Kufa/IRAQ. The authors would like to express their appreciation to the University of Kufa/College of Science/ Department of biology for its support of this work.

References

[1] Brayner, R., Ferrari-Iliou, R., Brivois, N., Djediat, S., Benedetti, M. F., &Fiévet, F. (2006.) Toxicological impact studies based on Escherichia coli bacteria in ultrafine ZnO nanoparticles colloidal medium. Nano letters, 6(4): 866-870.

[2] Buzea, C., Pacheco, I. I., & Robbie, K. (2007). Nanomaterials and nanoparticles: sources and toxicity. Biointerphases, 2(4).

[3] Cremar, L., Gutierrez, J., Martinez, J., Materon, L., Gilkerson, R., Xu, F., & Lozano, K. (2018). Development of antimicrobial chitosan based nanofiber dressings for wound healing applications. Nanomedicine Journal, 5(1): 6-14.

[4] Dizaj, SM., Lotfipour, F., Barzegar-Jalali, M., Zarrintan, MH.,Adibkia, K. (2014). Antimicrobial activity of the metals and metal oxide nanoparticles. Materials Science and Engineering: C. :278–84.

[5] Erdem, B., Kariptas, E., Kaya, T., Tulumoglu, S., &Görgülü, Ö. (2016). Factors influencing antibacterial activity of chitosan againstAeromonashydrophila and Staphylococcus aureus. International Current Pharmaceutical Journal, 5(5), 45-48.

[6] Furno, F., Morley, K. S., Wong, B., Sharp, B. L., Arnold, P. L., Howdle, S. M., & Reid, H. J. (2004). Silver nanoparticles and polymeric medical devices: a new approach to prevention of infection? Journal of Antimicrobial Chemotherapy, 54(6): 1019-1024.

[7] Hanley, C., Layne, J., Punnoose, A., Reddy, K. Á., Coombs, I., Coombs, A., & Wingett, D. (2008). Preferential killing of cancer cells and activated human T cells using ZnO nanoparticles. Nanotechnology, 19(29): 295103.

[8] Haider, A., & Kang, I. K. (2015). Preparation of silver nanoparticles and their industrial and biomedical applications: a comprehensive review. Advances in materials science and engineering.

[9] Liu, N., Chen, X. G., Park, H. J., Liu, C. G., Liu, C. S., Meng, X. H., & Yu, L. J. (2006). Effect of MW and concentration of chitosan on antibacterial activity of Escherichia coli. Carbohydrate polymers, 64(1): 60-65.

[10] Ma, B., Qin, A., Li, X., Zhao, X., & He, C. (2014). Structure and properties of chitin whisker reinforced chitosan membranes. International journal of biological macromolecules, 64: 341-346.

[11] Noori, S., Kokabi, M., & Hassan, Z. M. (2018). Poly (vinyl alcohol)/chitosan/honey/clay responsive nanocomposite hydrogel wound dressing. Journal of Applied Polymer Science, 135(21):46311. 1-12.

[12] Ohshima, Y., Nishino, K., Yonekura, Y., Kishimoto, S., & Wakabayashi, S. (1987). Clinical application of chitin non-woven fabric as wound dressing. European Journal of Plastic Surgery, 10(2): 66-69.

[13] Okamoto, Y., Kawakami, K., Miyatake, K., Morimoto, M., Shigemasa, Y., & Minami, S. (2002). Analgesic effects of chitin and chitosan. Carbohydrate Polymers, 49(3), 249-252.

[14] Pasaribu, S. P., Kaban, J., Ginting, M., &Sinaga, K. R. (2018). Synthesis and evaluation antibacterial activity of phosphate buffer solution (pH 7.4)-Soluble acylated chitosan derivative. In AIP Conference Proceedings, 2049(1).

[15] Potara, M., Jakab, E., Damert, A., Popescu, O., Canpean, V., &Astilean, S. (2011). Synergistic antibacterial activity of chitosan–silvernanocomposites on Staphylococcus aureus. Nanotechnology, 22(13), 135101.

[16] Prakash, P., Gnanaprakasam, P., Emmanuel, R., Arokiyaraj, S., & Saravanan, M. (2013). Green synthesis of silver nanoparticles from leaf extract of Mimusops elengi, Linn. for enhanced antibacterial activity against multi drug resistant clinical isolates. Colloids and Surfaces B: Biointerfaces, 108, 255-259.

[17] Raghupathi, K. R., Koodali, R. T., & Manna, A. C. (2011). Size-dependent bacterial growth inhibition and mechanism of antibacterial activity of zinc oxide nanoparticles. Langmuir, 27(7): 4020-4028.

[18] Reddy, K. M., Feris, K., Bell, J., Wingett, D. G., Hanley, C., & Punnoose, A. (2007). Selective toxicity of zinc oxide nanoparticles to prokaryotic and eukaryotic systems. Applied physics letters, 90(21): 213902.

[19] Roselli, M., Finamore, A., Garaguso, I., Britti, M. S., & Mengheri, E. (2003). Zinc oxide protects

cultured enterocytes from the damage induced by Escherichia coli. The Journal of nutrition, 133(12): 4077-4082.

[20] Sanpui, P., Murugadoss, A., Prasad, P. D., Ghosh, S. S., &Chattopadhyay, A. (2008). The antibacterial properties of a novel chitosan–Ag-nanoparticle composite. International journal of food microbiology, 124(2): 142-146.

[21] Seil, J. T., & Webster, T. J. (2012). Antimicrobial applications of nanotechnology: methods and literature. International journal ofnanomedicine, 7:2767– 2781.

[22] Shabunin, A. S., Yudin, V. E., Dobrovolskaya, I. P., Zinovyev, E. V., Zubov, V., Ivan'kova, E. M., &Morganti, P. (2019). Composite wound dressing based on chitin/chitosan nanofibers: processing and biomedical applications. Cosmetics, 6(1): 1-10.

[23] Shi, L. E., Li, Z. H., Zheng, W., Zhao, Y. F., Jin, Y. F., & Tang, Z. X. (2014). Synthesis, antibacterial activity, antibacterial mechanism and food applications of ZnO nanoparticles: a review. Food Additives & Contaminants: Part A, 31(2):173-186.

[24] Taccola, L., Raffa, V., Riggio, C., Vittorio, O., Iorio, M. C., Vanacore, R., & Cuschieri, A. (2011). Zinc oxide nanoparticles as selective killers of proliferating cells. International journal of nanomedicine, 6, 1129.

[25] William, D., (2007). Materials science and enginnering. 7th edition. John wily & sons, Inc.

[26] Yamamoto, O. (2001). Influence of particle size on the antibacterial activity of zinc oxide. International Journal of Inorganic Materials, 3(7): 643-646.

[27] Zakhem, E., Raghavan, S., Gilmont, R. R., &Bitar, K. N. (2012). Chitosan-based scaffolds for the support of smooth muscle constructs in intestinal tissue engineering. Biomaterials, 33(19): 4810-4817.

[28] L.H. Li, J.C. Deng, H.R. Deng, Z. L. Liu, L. Xin, Synthesis and characterization of 504 chitosan/ZnO nanoparticle composite membranes, Carbohydrate Res. 345 (2010) 994.

Geometrical and Material Nonlinear Finite Analysis of Fiber Reinforced Concrete Slabs

Saad A. Al-Taan[1*], Ayad A. Abdul-Razzak[2]

[1]College of Engineering, University of Mosul, Iraq, E-mail: saad.altaan@yahoo.com
[2]College of Engineering, University of Mosul, Iraq
E-mail: drayad_aghwan@yahoo.co.uk, drayad.aghwan@uomosul.edu.iq
* Corresponding author: College of Engineering, University of Mosul, Iraq.
Tel: +964 770 879 7197.

Abstract. Geometrical nonlinearity arises from the continuous change of shape under increasing loads on thin or long span members, and the materials nonlinearity due to the nonlinear behavior of the materials composing the structural element. In this study the finite element method is used to analyze fibrous reinforced concrete slabs, material and geometric nonlinearities were considered. Concrete is represented by degenerate quadratic thick shell element. Concrete behavior in compression is modeled by elastic perfectly plastic or strain hardening plasticity. A tensile strength criterion is used to follow crack initiation and propagation. A parabolic stress degradation function, which is based on the fracture energy concept, is used to model the tension stiffening effect of steel fibers. The reinforcement is considered as smeared within the concrete layers, and assumed as either an elastic perfectly plastic material or as an elastic-plastic material with linear strain hardening. Von Karman assumptions which is based on the total Lagrangian approach is used for the geometric nonlinearity. The numerical results using for both normal and lightweight fiber reinforced concrete slabs were predicted satisfactorily. The average predicted failure load of the investigated cases to the experimental failure load was 0.899 with a standard deviation of 0.092.

Keywords: *Finite element; Geometrical nonlinearity; Reinforced concrete slabs; Steel fibers; Tension stiffening.*

1. Introduction

Reinforced concrete slabs are important structural member in any reinforced concrete building, since it is the only member that takes the live loads directly. The thickness of the flat plates or flat slabs is mostly controlled by the punching shear stresses created in the vicinity of the columns. To prevent the catastrophic and brittle punching shear failure, the ACI Code, (2019) recommend the increase of the slab thickness or use of shear reinforcement. The shear reinforcement may be in the form of single leg, multiple legs, bent-down bars, shear studs, or shear head reinforcement.

Steel, glass, polymer and other types of short discrete fibers have been added as reinforcement to concrete, thus producing a composite material with enhanced tensile strength, ductility, well defined post cracking behavior, and strain capacity, ACI Committee 544 (2002), and ACI Committee 544.4R (2018). The ACI Code, (2019) recommends the use of deformed steel fibers complying with ASTM Specifications, (2016) as shear reinforcement with length to diameter ratio ranging between 50 and 100. Steel fibers has been used to improve the behavior and strength of reinforced concrete slabs in shear and flexure, Swamy and Ali, (1982), Theodorakopoulus and Swamy, (1993), Yaseen, (2006), Min et al., (2011), Jalate and Kalurkar, (2013), Tan and Venateshwaran, (2017), and Abdel-Rahman et al., (2018).

The finite element (FE) is used extensively for the nonlinear analysis of fiber reinforced concrete (FRC) members, Al-Taan and Ezzadeen, (1995), and Xu et al., (2008). Geometrical and material nonlinearity due to tensile cracking, bond-slip of the bars embedded in concrete, nonlinear behavior of concrete in compression have been taken into account. The (FE) is used also by many researchers for the nonlinear analysis of fiber reinforced concrete slabs, Abdel-Rahman et al., (2018), Tazaly, (2012), Mpegetis, (2012), Irani and Abadi, (2013), Alvrez, (2013), Sayhood et al., (2014), Puddicome, (2018).

Mpegetis, (2012) used nonlinear finite element to analyze interior square fiber reinforced concrete pile supported slabs. Three types of steel fibers at a dosage of 35-50kg/m^3 were used, 35×0.75 mm hooked fibers, 35×0.55 mm hooked fibers, and 25×0.5 mm twisted wire fibers. Three and four nodded shell elements with reduced integration were used to model the slab. In the post cracking stage, the analysis adopts multilinear stress – displacement (σ-w) relationship derived from inverse analysis on notched beams, round determinate panels (supported on three equally spaced supports), and round indeterminate panels (supported on six equally spaced supports) rather than a stress –strain (σ-ε) one. Discrete and smeared cracking were used in the analysis and showed comparable results when displacement control is used.

The nonlinear finite element method is used by Irani and Abadi, (2013) for the analysis of two-way octagonal concrete slabs with a side length of 1.0 m using the program DIANA. Three different configurations of reinforcement were considered: Conventional reinforcement, steel fiber reinforcement (hooked ends, length = 60 mm, 0.9 mm diameter, and V_f =0.45%) and a combination of both. Concrete, and steel bars were represented by two-dimensional curved shell elements (with five degrees of freedom) and embedded bar elements respectively. Concrete behavior in compression is assumed as an elastic perfectly plastic and in tension six different models were used including multi-linear stress-strain diagram that is not dependent on fracture energy. Smeared cracking model was used to model cracking of concrete. Von Mises plasticity model was used for the reinforcement and to specify the hardening. Displacement control was used in static analyses, with Secant method for iterations. Similar numerical results were obtained for the used tensile models except the constant model. Slabs with steel fiber reinforcement only, higher capacity and stiffer behavior were obtained compared to experimental results. Slabs with combined steel fiber and conventional reinforcement, showed larger cracking load while the load from yielding until failure was lower in the analysis than in the experiments.

Abdel-Rahman et al., (2018) tested fourteen reinforced concrete slabs reinforced longitudinally with 1.2% in both directions under concentric and eccentric load. Steel fibers with dimensions (50×0.52×0.72 mm) with volume fractions of 0.5 to 1.5% were used to study the influence of steel fibers on the punching shear strength of the slabs. The test results showed an increase in the punching shear strength up 24%. A nonlinear (FE) analysis is conducted using the program ANSYS R14.5 to compare the numerical with the experimental results. Concrete is represented by eight node solid elements, solid65, and three-dimensional link elements 180 were used to represent the reinforcing bars. The average ratio of the numerical-to the experimental load is 0.98 with a standard deviation of 0.14, which indicates a good agreement.

Puddicome, (2018) presented a (FE) model using the program ABAQUS for the analysis of (FRC) slabs. The nonlinear behavior of concrete is simulated using the Concrete Damaged Plasticity constitutive model. An exponential decay expression with variables such as concrete compressive strength, flexural reinforcement ratio, steel fibers volume fraction, and the reinforcing steel yield strength is used to derive a tension – stiffening model. Eight-nodded hexahedral (brick) elements were used to model concrete with reduced integration, and the reinforcement is modeled as two-node linear truss elements The model is applied to published test results of nine slabs from the literature, and the ratio of the numerical to the experimental failure load varied from 0.78 to 1.281 with an average of ratio of 1.068 and a standard deviation of 0.195.

Al-Taan and Abdul-Razzak, (2020) used the computer program written by Hinton and Owen (1984) for the nonlinear (FE) analysis of fiber reinforced lightweight and normal weight square concrete slabs. Thick shell element was used to represent concrete and the thickness is discretized into layers. The steel reinforcement was smeared within the concrete layers. Concrete in compression assumed to follow an elastic-plastic or strain hardening behavior. In tension Tensile strength criterion is used to detect crack initiation and propagation. The descending post peak stress-strain relationship in tension is modeled using a parabolic stress degradation function that is based on the fracture energy concept of steel fiber concrete. The numerical results and the experimental results of square lightweight and normal weight fibrous reinforced concrete slabs showed good agreement.

In this paper the computer program written by Hinton and Owen, (1984) for the nonlinear analysis of reinforced concrete plates and shells is modified by including the constitutive material relationships of steel fiber concrete for the nonlinear geometrical and material (FE) analysis of (FRC) slabs under increasing monotonic loading. Modified version of the Newton-Raphson method has been used for the solution of the nonlinear problem.

2. Materials Constitutive Relationships

2.1 Compression

The first part of the curve in compression can be treated using the elasticity theory, Figure 1, and the second part using the theory of plasticity. The growth of yield surface during plastic loading is described using a hardening rule. A yield function depending on the mean normal stress, I_1 and the shear stress invariant, J_2 is used:

$$f(I_1, I_2) = \sqrt{\beta_f(3J_2) + \alpha_f I_1} = \sigma_o \qquad (1)$$

α_f and β_f are material parameters, and σ_o = equivalent effective stress, taken as the compressive strength f'_{cf} of fibrous concrete. For the biaxial compression state:

$$\sigma_1 = \sigma_2 = \omega f'_{cf} \qquad (2)$$

ω is the biaxial stress ratio $\omega = \sigma_1/\sigma_2$.

Abdul-Razzak, (1996) used the experimental results reported by Abdul-Ahad and Abbas, (1989), Leonard and Mansur, (1991), and Yin, et. al. (1989), in a regression analysis, and the following equation is derived for (ω) by including the influence of the stress ratio, fiber aspect ratio and volume fraction:

$$\omega = e^x \qquad (3)$$

$$x = \frac{1}{3.339 - 0.9772 \ln\left(\frac{V_f L_f}{D_f}\right)} \qquad (4)$$

α_f and β_f can be written as follow:

$$\alpha_f = \frac{1-\omega^2}{\omega^2 - 2\omega}\sigma_o \qquad (5)$$

$$\beta_f = \frac{1-2\omega}{\omega^2 - 2\omega} \qquad (6)$$

The stress components are used to express and the yield condition:

$$f(\sigma) = \left[\beta_f\{(\sigma_x^2 + \sigma_y^2 + \sigma_{xy}) + 3(\tau_{xy}^2 + \tau_{xz}^2 + \tau_{zy}^2)\} + \alpha_f(\sigma_x + \sigma_y)\right]^{1/2} \quad (7)$$

To predict the crushing state of concrete, the yield criterion is converted into strain instead of stress components. The experimental results reported by Abdul-Ahad and Abbas, (1989), Leonard and Mansur, (1991), and Yin, et. al. (1989), are used by Abdul-Razzak, (1996) to estimate the ultimate compression strain of steel (FRC):

$$\epsilon_{cfu} = 0.003011 + 0.002295V_f \quad (8)$$

2.2 Tension

As for plain concrete, the first ascending part of the stress-strain relationship in tension, up to first cracking, Figure 2 is linear. Due to the process of micro-cracks initiation and propagation, it starts deviates from linearity. After that a formation of an unstable micro-crack system will occur, which grows rapidly under the increasing tensile stress bringing the material to its post-peak region, Bentur and Mindess, (2007). The equations derived by Soroushian and Lee, (1989) for the tensile strength f'_{tf} and its corresponding strain ε_{tf} for (FRC) are used in this study:

$$f'_{tf} = f'_t(1 + 0.016N_f^{1/3} + 0.05\pi D_f L_f N_f) \quad (9)$$

$$\epsilon_{tf} = \epsilon_t\ (1 + 0.35 D_f L_f N_f) \quad (10)$$

where N_f is the number of fibers crossing a unit area [30]:

$$N_f = \frac{1.64V_f}{\pi D_f^2} \quad (11)$$

After the formation of a continuous crack, a stress drop occurs and the crack is arrested at a certain level by the pullout resistance of the fires. It is assumed that the fibers bridging the crack start resisting the applied stress immediately after composite cracking. This average tensile stress is defined here as f_u, and can be calculated as the product of the average bond stress by the interfacial area of all fibers crossing the cracks, Bentur and Mindess, (2007):

$$f_u = \eta_o \tau_u V_f \frac{L_f}{D_f} \quad (12)$$

where η_o is the orientation factor. Different values were proposed for this factor and the probabilistic value of 0.41 is used in this study, Hannant, (1978). τ_u is the average bond strength and is equal to Soroushian and Lee, (1989):

$$\tau_u = 2.62 - 0.0036N_f \quad (13)$$

After the sudden drop in stress from f_{tf}' to f_u a descending nonlinear stress-strain relationship is used in this investigation to simulate the post-peak constitutive model as shown in Figure 2. The derivation of the nonlinear curve is based on the fracture energy concept, Abdul-Razzak, (1996) using experimental load-deformation curve reported by Visalvanich and Naaman, (1983). The stress-strain relation can be expressed as follows:

$$\sigma_i = f_u\left(\frac{\varepsilon_i - \varepsilon_m}{\varepsilon_{tf} - \varepsilon_m}\right)^2 \quad (14)$$

where ε_m is the limiting tensile strain. This value is calculated by Abdul-Razzak, (1996) assuming that the descending stress-strain curve is similar to that of a load- displacement curve of a fiber exhibiting pull-out, and the area under this curve is equal to the fracture energy G_f:

$$\epsilon_m = \frac{3G_f}{hf_u} + \epsilon_{tf} \quad (15)$$

h = characteristic length of the Gauss point. The fracture energy G_f which is derived by Visalvanich and Naaman, (1983) is adopted in this study:

IOP Conf. Series: Materials Science and Engineering **978** (2020) 012041 doi:10.1088/1757-899X/978/1/012041

$$G_f = 0.171\beta\tau_u V_f\left(\frac{2L_f^2}{D_f}\right) \quad (16)$$

The tension stiffening effect is considered as indicated in Figure 1 by assuming a gradual release of the concrete stress component normal to the cracked plane. Unloading and reloading of cracked concrete is supposed to follow the linear behavior shown in Figure 2 with a fictitious secant modulus E_{ts}. In the biaxial tension-compression zone, the parabolic relationship derived by Al-Taan and Mahmood (1988) is used:

$$\sigma_{2p} = \left[\frac{\sqrt{1+4(\alpha_2^2 S^2)}-1}{2(\alpha_2^2 S^2)}\right] f'_{cf} \quad (17)$$

$S = f'_{cf} / f_{tf}$, and the peak tensile strength as:

$$\sigma_{1p} = \sigma_{2p}\alpha_2 \quad (18)$$

where σ_{1p} and σ_{2p} are the limiting principal tensile and compressive strengths respectively in this zone and α_2 is the principal stress ratio $= \sigma_2/\sigma_1$.

After cracking, the concrete is assumed to be in a state of uniaxial compression parallel to the cracks and exhibiting characteristics of the strain hardening elasto-plastic model, and the relationship proposed by Vecchio and Collins, (1986) is used in this investigation:

$$f'_{cfmax} = \frac{f'_{cf}}{0.8+0.34\frac{\varepsilon_1}{\varepsilon_2}} \leq f'_{cf} \quad (19)$$

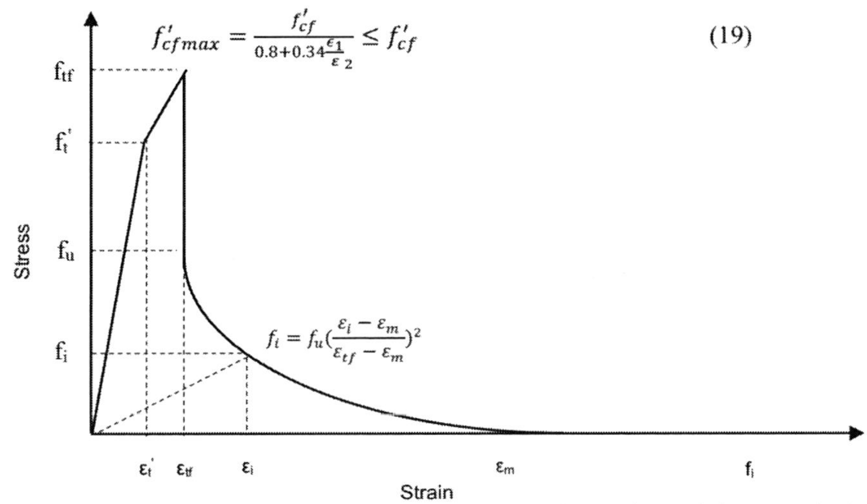

Figure 1. Stress - Strain curve for fibrous concrete in tension, Abdu-Razzak, (1996)

Perpendicular to the cracks, the elastic modulus and Poisson's ratio is reduced to zero and a reduced shear modulus is used to represent the aggregate interlock.

2.3. Shear Modulus of Cracked Concrete

Four different relationships are used for the reduced shear modulus of cracked concrete G,

(a) Approach G1, in this relationship the value of G assumed to be decreasing linearly with the current tensile strain as assumed by Cedolin and Poli, (1977), for concrete cracked in in direction 1,

$$\bar{G}_{12} = 0.25G\left(1-\frac{\varepsilon_1}{\varepsilon_m}\right) \quad \text{for } \varepsilon_1 < \varepsilon_m \quad (20a)$$

$$\bar{G}_{12} = 0 \quad \text{for } \varepsilon_1 > \varepsilon_m \quad (20b)$$

(b) Biaxial orthotropic relationship G2: This approach has been proposed by Chen, (2007), the cracked shear

IOP Conf. Series: Materials Science and Engineering **978** (2020) 012041 doi:10.1088/1757-899X/978/1/012041

modulus $\overline{G}$ can be calculated as follows:

$$\bar{G}_{12} = E_1E_2/(E_1 + E_2 + 2\upsilon E_2) \quad (21)$$

If Poisson's ratio (υ) equal to zero Eq. (21) becomes:

$$\bar{G}_{12} = E_1E_2/(E_1 + E_2) \quad (22)$$

E1 and E2 are the elastic moduli in directions 1and 2 respectively.

(c) Approach G3: In this expression the value of G is decreasing parabolically with the current tensile strain, using a function similar to that used in Eq. (14), Abdul-Razzak, (1996), for concrete cracked in direction 1,

$$\bar{G}_{12} = 0.25G[(\epsilon_1 - \epsilon_m)/(\epsilon_{tf} - \epsilon_m)]^2 \quad \text{for } \varepsilon_1 < \varepsilon_m \quad (23a)$$

$$\overline{G}_{12} = 0 \quad \text{for} \quad \varepsilon_1 \geq \varepsilon_m \quad (23b)$$

(d) Approach G4: This relationship was proposed initially by Al-Mahaidi, (1978) and modified in this study by substituting ε_{tf} for ε_t. The reduced shear modulus of cracked concrete G can be calculated as follows:

$$\bar{G}_{12} = \frac{0.4G}{\epsilon_1\epsilon_{tf}} \quad (24)$$

3. Finite Element Formulation

The eight-node serendipity thick shell element is used in the present study. At each node, three displacements and two independent rotations of the normal are considered. The formulation of this element is mentioned by Onate et al., (1978). This element is the most popular degenerated shell element in isoparametric formulation and it is the simplest element considered in the original work of Ahmad, Onate et al., (1978), and this element requires C^0 continuity. Shear locking occurs even in moderately thin situations when using full integration. A great improvement of the results was obtained when the reduced integration was used, Hinton and Owen, (1984). Smeared approach is implemented for the steel reinforcement which is incorporated with the layered approach. Across the thickness, each element is divided into a number of concrete layers. The stresses of each layer are computed at the Gauss points. Smeared approach is implemented for the steel reinforcement which is incorporated with the layered approach.

3.1 Geometrical Nonlinearity

The nonlinearity in structural members may be classified into two types; material and geometric. The material nonlinearity, arises from the nonlinear behavior of the material composing the structural element. While the geometrical nonlinearity arises from the changes of the geometry associated with the increased deformation of the monotonically loaded member. The problem is usually solved by following the geometrical change of both the structural member and its elements in a step by step manner. In this study a specific total Lagrangian formulation is adopted in which large deflections and moderate rotations (in the sense of the Von Karman hypotheses) are accounted for. The strain- displacement matrix is calculated once during the nonlinear process and its nonlinear part is updated using the current displacement by a simple matrix product. The total strain is divided into parts, the linear part $\{\varepsilon\}_O$ and the nonlinear part $\{\varepsilon\}_L$

$$\{\varepsilon\} = \{\varepsilon\}_O + \{\varepsilon\}_L \quad (25)$$

and

$$d\{\varepsilon\} = d\{\varepsilon\}_O + d\ \{\varepsilon\}_L \quad (26)$$

The strain-displacement matrix is also divided into two parts,

$$d\{B\} = d\{B\}_O + d\{B\}_L \quad (27)$$

Where $\{B\}_o$ is the linear part of the strain - displacement matrix, and $\{B\}_L$ is the nonlinear part of the strain- displacement matrix. The nonlinear part of the strain vector can be written as follows, Hinton and Owen, (1984):

$$\{\varepsilon\}_L = \begin{Bmatrix} \frac{1}{2}\left(\frac{\partial w}{\partial x}\right)^2 \\ \frac{1}{2}\left(\frac{\partial w}{\partial Y}\right)^2 \\ \frac{\partial w}{\partial x}\frac{\partial w}{\partial x} \\ 0 \\ 0 \end{Bmatrix} = \frac{1}{2}\begin{Bmatrix} \frac{\partial w}{\partial x} & 0 \\ 0 & \frac{\partial w}{\partial Y} \\ \frac{\partial w}{\partial Y} & \frac{\partial w}{\partial x} \\ 0 & 0 \\ 0 & 0 \end{Bmatrix} \begin{Bmatrix} \frac{\partial w}{\partial x} \\ \frac{\partial w}{\partial Y} \end{Bmatrix} \quad (28)$$

or

$$\{\varepsilon\}_L = \frac{1}{2}[S]\{R\} \quad (29)$$

where

$$[S]^T = \begin{bmatrix} \frac{\partial w}{\partial x} & 0 & \frac{\partial w}{\partial Y} & 0 & 0 \\ 0 & \frac{\partial w}{\partial Y} & \frac{\partial w}{\partial x} & 0 & 0 \end{bmatrix} \quad (30)$$

The vector $\{R\}$ can be defined as:

$$\{R\} = \begin{Bmatrix} \frac{\partial w}{\partial x} & \frac{\partial w}{\partial Y} \end{Bmatrix}^T = [G]\{a\} \quad (31)$$

$[G]$ is a rectangular matrix having two rows and columns equal to the total number of element nodal variables. The geometrical stiffness matrix may be written as,

$$[K]_\sigma = \int [G]^T\{\sigma\}[G]dv \quad (32)$$

$\{\sigma\}$ are the components of the Piola- Kirchhoff stress vector.

4. Numerical Examples

The first example is a normal weight reinforced concrete square slab-column connection with a side length of 1800 mm and overall thickness of 125 mm tested by Swamy and Ali, (1982), Figure 2. The slab was designed to fail in shear. The slab was simply supported along all four edges, with span of 1690 mm. The slab was loaded through the column stub over an area of 150×150 mm. The geometry was chosen to simulate a flat plate-column connection with a column spacing of 4.0 m center to center in both directions. One quarter of the slab is analyzed due to symmetry, Figure 2. Dimensions, reinforcement details and loading are shown in Figure 2a. the materials properties are summarized in Table 1. Two steel layers at the top and two at the bottom are used to represent the reinforcement and six layers to represent concrete are found to be enough for the analysis. This slab was analyzed by Al-Taan and Abdul-Razzak, (2020) using material nonlinearity, and in the present study it is analyzed using the effect of geometrical nonlinearity. The slab is modeled by three meshes, four, nine, and eleven elements. However, the loading, geometry of the slab and the obtained results showed that using eleven elements gives a reliable simulation and results.

IOP Conf. Series: Materials Science and Engineering **978** (2020) 012041 doi:10.1088/1757-899X/978/1/012041

Table 1. Materials properties of slab S-2, Swamy and Ali, (1982)

E_{cf} (GPa)	E_s' (GPa)	V_f^* %	f_{cuf}' (MPa)	f_t' (MPa)	f_y (MPa)	A_s' bars	A_s bars	ϵ_{cuf}^{**} ×10^{-3}	Experimental failure load (kN)
33.29	17	0.5	34.87	4.2	425	7- ϕ 8	12-ϕ 10	4.4	243.6

* Crimped steel fibers 0.5×50 mm, ** calculated from Equation (8), v_c = 0.15

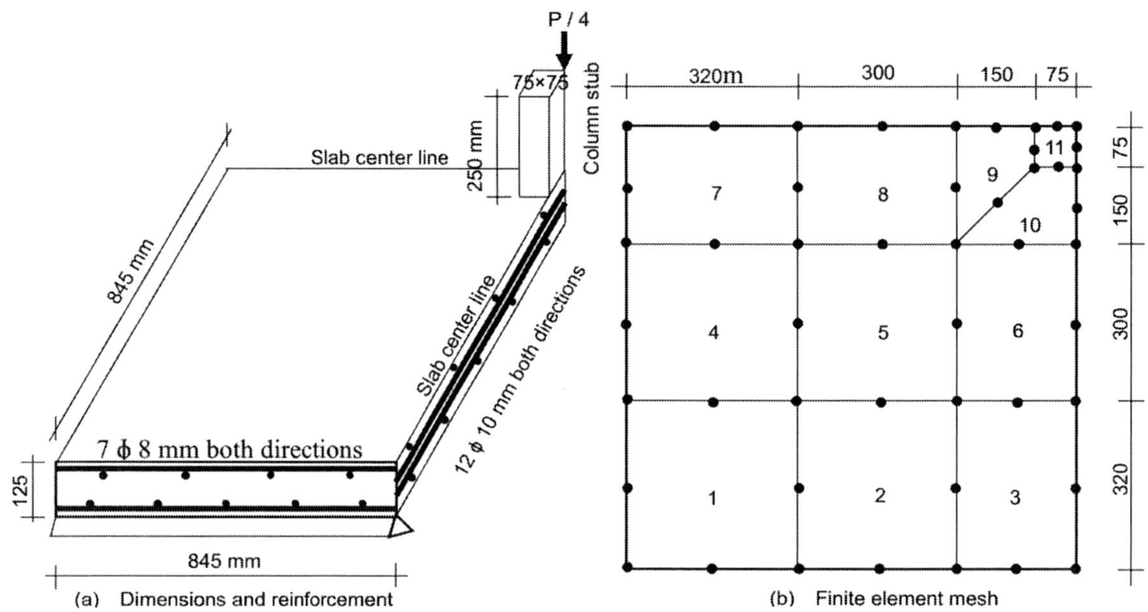

Figure 2. Details and finite element mesh for slab S-2, (a) Dimensions and reinforcement, (b) Finite element mesh

Figure 3 shows a comparison between the four shear moduli (G1-G4) models with elasto-perfectly plastic and strain hardening models. From this figure it is seen that G1 with elasto-perfect plastic model and G3 model with strain hardening model give very good results compared with the experimental results, while G4 gives good response but with slightly smaller failure load. The shear crack model G2 underestimates the strength and in some cases leads to premature failure.

Figure 4 below, show that there is an effect of the concrete compression model used in the analysis, if the compression model changed from elasto-perfectly plastic to stain hardening. Also, when the geometrically nonlinear analysis is used G1 with elasto-perfectly plastic model and G3 model with strain hardening model give very good results compared to experimental ultimate load.

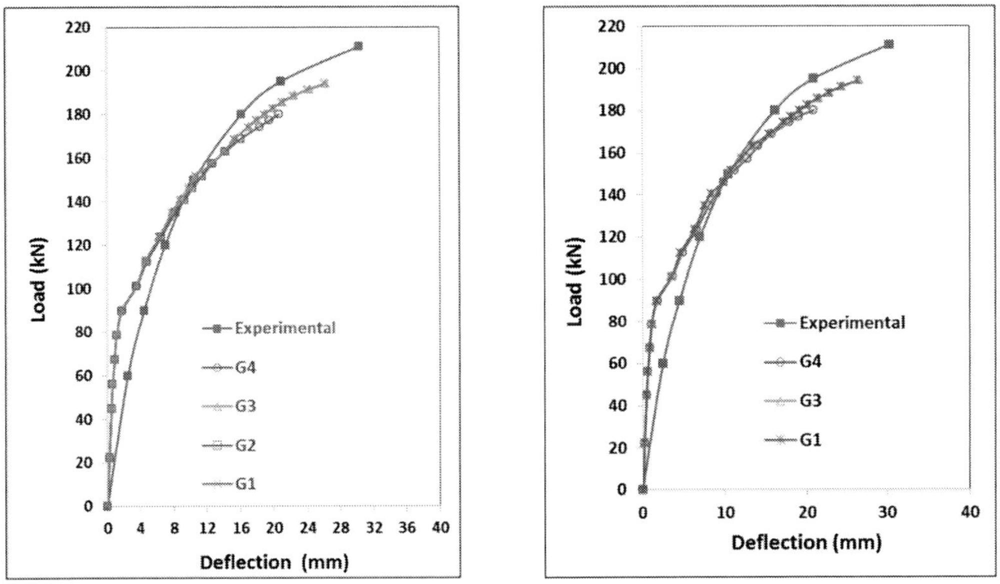

(a) Elasto-perfectly plastic Model (b) Strain-Hardening Model

Figure 3. Load-deflection curves for slab S-2, (a) elasto-perfectly plastic model, (b) Strain hardening model

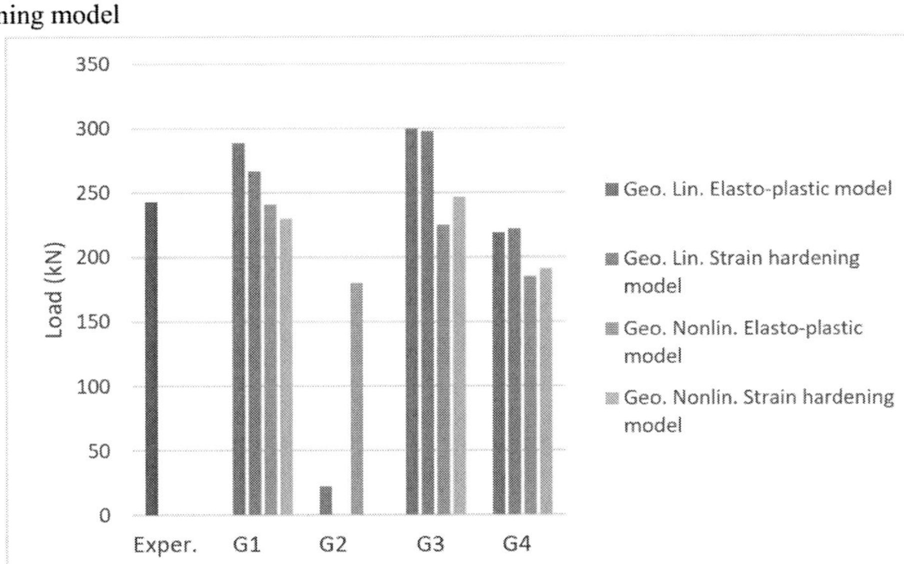

Figure 4. Variation of the ultimate load with the shear model, slab S-2

Initial damage of the slab occurs in the form of tension cracking in the radial direction, Figure 5. The damage zone adjacent to the column stub prior to failure was dominated by two way tension cracking action, high shear and compression stresses and strains leading to crushing of concrete under the load, Figure 5.

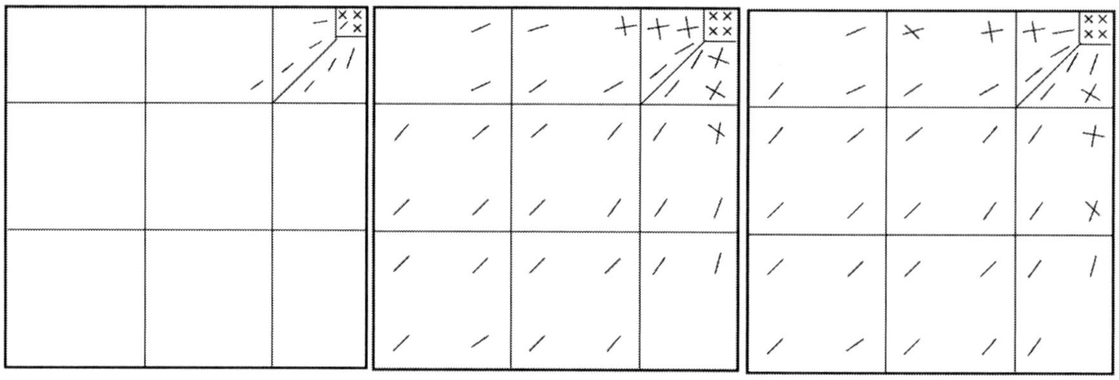

(a) Load 67.5 kN (b) cracking at 180 kN (c) Cracking at failure load

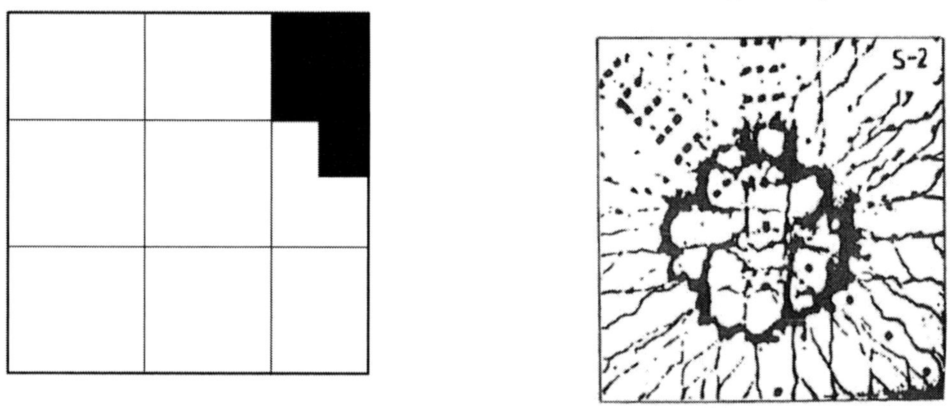

(d) Crushing at the compression zone (e) Experimental cracking at failure

Figure 5. Cracking and crushing patterns, (a) cracking at load 67.5 kN, (b) cracking at load 180 kN, (c) cracking at failure load, (d)crushing at failure load, (e) Experimental cracking at failure.

The second example is a lightweight concrete slab-column connection tested by Theodorakopoulos and Swamy, (1993) of the same geometry as the previous example. The steel reinforcement used in the tested slab was cold-worked high-tensile deformed bars with 8 bars 10 mm diameter placed in the tension zone in both directions. Due to symmetry of the slab, only quarter of the slab is analyzed. Table 2, shows the material properties of the tested slab. Dimensions detail and loading are as shown in Figure 2 above. As for the previous example the slab is modeled in the present study by many elements namely four, nine, and eleven elements. However, the obtained results showed that using eleven elements which is shown in Figure 2 above give a reliable simulation and results. Two steel layers at the bottom are used to represent the tension reinforcement and eight layers to represent concrete are found to be enough for the analysis. This slab was analyzed also by Al-Taan and Abdul-Razzak, (2020) using material nonlinearity, and in the present study the slab is analyzed using the effect of geometrical nonlinearity.

Table 2. Materials properties of slab FS-20, Theodorakopoulos and Swamy, (1993)

E_{cf} (GPa)	$E_s^{'}$ (GPa)	V_f^{*} %	$f_{cuf}^{'}$ (MPa)	$f_t^{'}$ (MPa)	f_y (MPa)	A_s bars	ϵ_{cuf}^{**} ×10^{-3}	Experimental failure load (kN)
19.4	17	1.0	46.3	4.38	425	8-ϕ 10	5.3	211

* Crimped steel fibers 0.5×50 (mm), ** calculated from Eq. (8), v_c = 0.15, E_s = 204 GPa.

It is shown in Figure 6 that the geometrically nonlinear analysis underestimates the ultimate load for all the investigated reduced shear moduli due to the relatively large deformations prior to failure (25% of the thickness) and probably due to the relatively low elastic modulus of the lightweight concrete. In Figure 6b the stain hardening model with the geometrically nonlinear analysis shows that the results are similar to those of Figure 6a, this means that changing the compression model has no influence on the results.

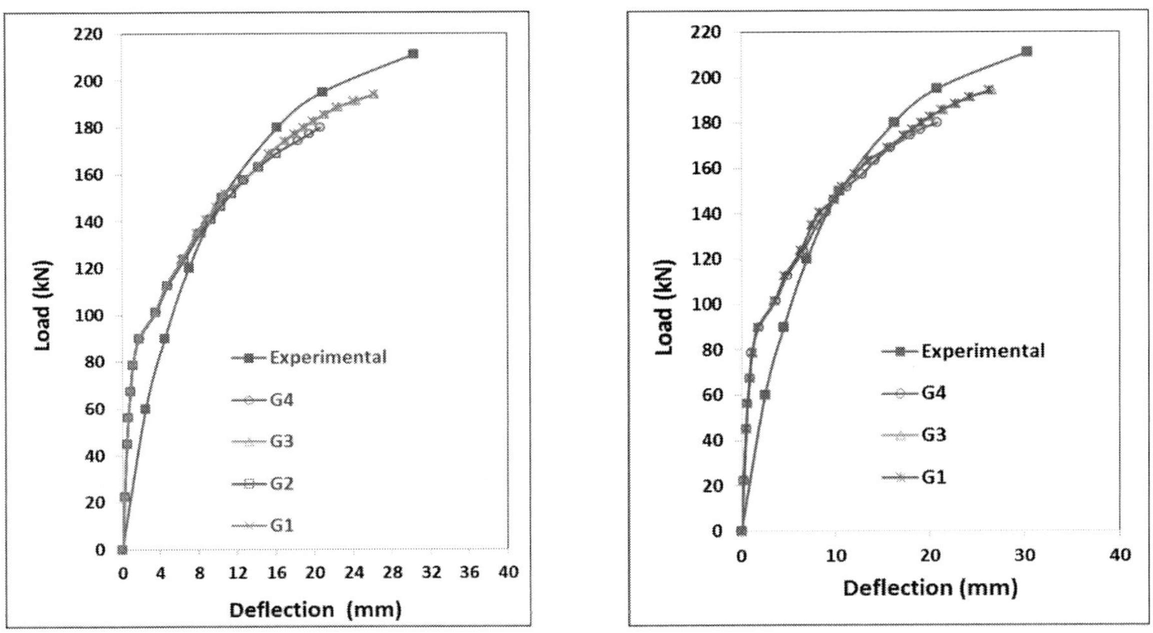

(b) Elasto-perf. plastic Model (c) Strain-Hardening Model

Figure 6. Load-deflection curves for slab FS-20, (a) elasto-perfectly plastic model, (b) Strain hardening model

Figure 7, shows that when the material nonlinearity (geometrical linear analysis) is used, G4 model gives very good results compared to the experimental results presented by Al-Taan and Abdul-Razzak, (2020). But when geometrical nonlinearity is used in the analysis G1 and G3 models give good results for both elasto-perfectly

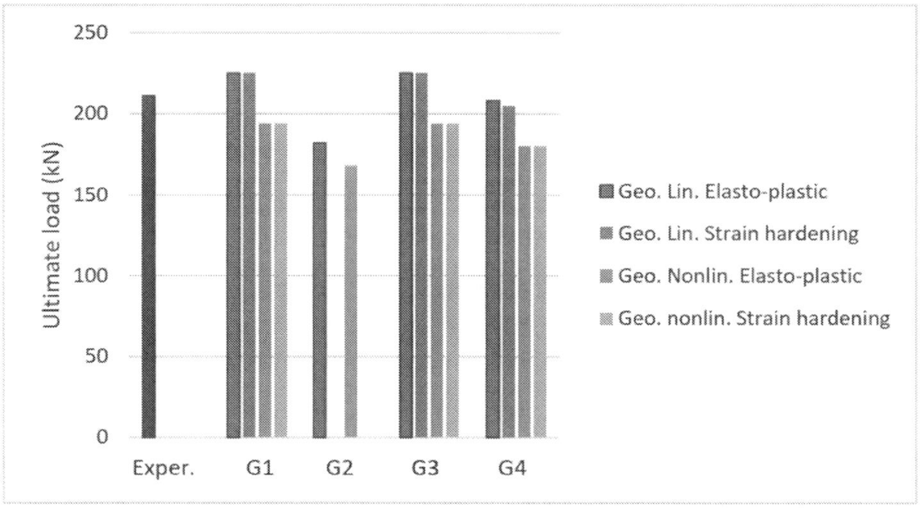

Figure 7. Variation of the ultimate load with the shear model

plastic and stain hardening models compared to experimental results. Also, it is seen that there is small effect of the compression models on the results of the ultimate load because the failure of the slab is punching shear failure and may be due to the relatively low elastic modulus of the lightweight concrete. The shear crack model G2 underestimates the strength and in some cases leads to premature failure.

5. Conclusions

The presented numerical together with the experimental results showed that the used eight nodes (Serendipity) general shear deformable shell elements with five degrees of freedom proved to be efficient for the nonlinear geometrical and material analysis of the studied fiber reinforced concrete slabs. The layered approach is suitable for modeling these types of slabs for both the structural response and ultimate load. The adopted constitutive models for fiber reinforced concrete concerning the strain hardening, tension stiffening, and reduced shear crack models proved to give satisfactory results (such as; load-deflection curves, cracks initiation and propagation, and ultimate loads) for the analysis of normal weight and lightweight fibrous reinforced concrete slabs subjected to monotonic loading up to failure.

Geometric nonlinearity incorporating the present models plays an important role in determining the response of fibrous reinforced concrete slabs. For the investigated fibrous reinforced concrete slabs, it is generally shown that, when geometrical nonlinearity is used G1 and G3 models give good results compared to the experimental results. The shear crack model G2 underestimate the strength for the examples investigated and, in some cases, leads to premature failure.

References

[1] Abdel-Rahman, A. M., Hassan, N. Z. & Soliman, A. M. (2018). Punching Shear Behavior of Reinforced Concrete Slabs using Steel Fibers in the Mix. Housing and Building National Research Center Jour. (Egypt), 14, 272-281.

[2] Abdul-Ahad, R.B.& Abbas, J. M. (1989). Behavior of Steel Fiber Reinforced Concrete under Biaxial Stresses. Proceedings of the International Conference on Recent Developments in Fiber Reinforced Cements and Concrete, 126-135, Cardiff School of Engineering (U.K.), Edited by Swamy, R.N. and Barr, B., Elsevier Applied Science.

[3] Abdul-Razzak, A. A. (1996). Nonlinear Finite Element Analysis of Fibrous Reinforced Concrete Structural Members," Ph.D. Thesis, University of Mosul, Iraq.

[4] ACI Committee 318-19 (2019). Building Code Requirements for Structural Concrete and Commentary. American Concrete Institute, Farmington Hills, MI 48331, USA.

[5] ACI Committee 544 (2002). "Report on Fiber Reinforced Concrete, Reported by ACI Committee

IOP Conf. Series: Materials Science and Engineering **978** (2020) 012041 doi:10.1088/1757-899X/978/1/012041

544, American Concrete Institute, Farmington Hills, MI, 48331, USA.

[6] ACI Committee 544.4R (2018). Guide to Design with Fiber-Reinforced Concrete. ACI Committee 544, American Concrete Institute, Farmington Hills, MI, 48331, USA.

[7] Al-Mahaidi, R. H. S. Nonlinear Finite Element Analysis of Reinforced Concrete Deep Members. (1978). Ph.D. Thesis, Dept. of Structural Engg., Cornell University, Ithaca, New York.

[8] Al-Taan, S. A. & Abdul-Razzak, A. A. (2020). Nonlinear Finite Element Analysis of Fiber Reinforced Concrete Slabs. Accepted for publication in the Jour. of Engineering and Technology, University of Technology (Baghdad), Iraq.

[9] Al-Taan, S. A. & Ezzadeen, N.A. (1995). Flexural Analysis of Reinforced Fibrous Concrete Members Using the Finite Element Method. Jour. of Computers and Structures, 56 (6), 1065-1072.

[10] Al-Taan, S. A. & Mahmood, M. N. (1988). Biaxial Stress-Strain Relationship for Concrete. Proceeding of the 2nd International Conference on Computer Applications in Concrete, C1-C14, Singapore.

[11] Álvarez, A. B. (2013). Characterization and Modelling of Steel Fibre Reinforced Concrete Elements. Ph.D. Thesis, Universitat Politècnica de Catalunya, Departament d'Enginyeria de la Construcció, 242 pp.

[12] ASTM A820M-16 (2016). Standard Specification for Steel Fibers for Fiber-Reinforced Concrete. ASTM International, West Conshohocken.

[13] Bentur, A. & Mindess, S. (2007). Fibre Reinforced Cementitious Composites. Tylor and Francis Group, 2nd Ed., New York.

[14] Cedolin, L. & Poli, S. D. (1977). Finite Element Studies of Shear. Critical R/C Beams. Jour. of Engineering Mech. Div., ASCE, Vol. 103 (3), 395-410.

[15] Chen, W. F. (2007). Plasticity in Reinforced Concrete. 1st Ed., McGraw-Hill Book Co., New York.

[16] Hannant, D. J. (1978). Fibre Cements and Fibre Concretes," Wiley-Interscience,, 1st Ed., New York, 1978.

[17] Hinton, E.& Owen, D. R. J. (1984). Finite Element Software for Plates and Shells. Pineridge Press, Swansea (U.K.).

[18] Irani, F. & Abadi, B. M. (2013). Finite Element Analysis of Conventional and Fiber Reinforced Concrete Slabs. M.Sc. Thesis, Chalmers University of Technology, Sweden.

[19] Jatale, V. B. & Kalurkar, L. G. (2013). Punching Shear High Strength Steel Fiber Reinforced Concrete Slabs. Inter. Jour. of Engineering Research and Applications, 3 (6), 269-275.

[20] Leonard, A. T. & Mansur, S. A. (1991). Biaxial Strength and Deformational Behavior of Plain and Steel Fiber Concrete. ACI Material Jour., 88 (4), 354-362.

[21] Minh, L. N., Rovnak, M., Quoc, T. T. & Kim, K. N. (2011). Punching Shear Resistance of Steel Fiber Reinforced Concrete Slabs. Procedia engineering, 1-8.

[22] Mpegetis, S. O. (2012). Behavior and Design of Steel Fibre Reinforced Concrete Slabs. Ph.D. Thesis, Imperial College of Science and Technology, Britain.

[23] Onate, E., Hinton, E. & Glover, N. (1978). Techniques for Improving the Performance of Ahmed Shell Elements. Report No. C/R/313/78, Civil Engg. Dept., University of Wales, Swansea (U.K.).

[24] Puddicome, t. (2018). Finite Element Analysis of Reinforced Concrete and Steel Fiber Reinforced Concrete Slabs in Punching Shear. M. Engg. Thesis, Memorial University of Newfoundland, Canada.

[25] Sayhood, E. K., Yaakoub, S. P. & Hussien,H. F. (2014). Nonlinear Finite Element Analysis for Punching Shear Resistance of Steel Fibers High Strength Reinforced Concrete Slabs. Jour. of Engineering and Technology (Iraq), 32 (6), 1411- 1432.

[26] Soroushian, P. & Lee, C. D. (1989). Constitutive Modelling of Steel Fiber Reinforced Concrete under Direct Tension and Compression. Proc. of the International Conference on Recent Developments in Fiber Reinforced Cements and Concrete, 363-377, Cardiff School of Engineering (U.K.), Edited by Swamy, R.N. and Barr, B., Elsevier Applied Science.

[27] Swamy, R. N. & Ali, S. A. R. (1982). Punching Shear Behavior of Reinforced Concrete Slab – Column Connections Made with Steel Fibers Concrete. ACI Jour., 79 (5), 392 – 406.

[25] Tan, T. K. & Venkateshwaran,A. (2017). Punching Shear in Steel Fibre Reinforced Concrete Slabs without Traditional Reinforcement," IOP Conference Series: Materials Science and Engineering, 246, 1-9.

[26] Tazaly, Z. (2012). Punching Shear Capacity of Fibre Reinforced Concrete Slabs with Conventional Reinforcement: Computational Analysis of Punching Models. M.Sc. Thesis, Royal Institute of Technology (KTH), Dept. of Civil and Architectural Engineering, Stockholm, Sweden.

[27] Theodorakopoulus, D. D. & Swamy, R.N. (1993). Contribution of Steel Fibers to the Strength Characteristics of Lightweight Concrete Slab Column Connections Failing in Punching Shear. ACI Structural Jour., 90 (4), 342 – 355.

[28] Vecchio, F. & Collins, M. P. (1986). The Modified Compressive - Field Theory for Reinforced Concrete Elements Subjected to Shear. ACI J., 83 (2), 219-231.

[29] Visalvanich, K. &. Naaman, A. E. (1983). Fracture Model for Fiber Reinforced Concrete. ACI Jour., 80 (2), 128-138.

[30] Xu, L., Chei, Y., Su, J., and Xia, D. (2008). Nonlinear Finite Element Analysis of Steel Fiber Reinforced Concrete Deep Beams. Wuhan University J. of Natural Sciences, 13 (8), 201-206.

[31] Yaseen, A. A. (2006). Punching Shear Strength of Steel Fiber High Strength Reinforced Concrete Slabs. M.Sc. Thesis, Dept. of Civil Engineering, University of Salahaddin, (Iraq), 111p.

[32] Yin, W. S., Eric, C. M., Mansur, M. A. & Hsu, T. T. C. (1989). Biaxial Tests of Plain and Fiber Concrete. ACI Material Jour., 86 (3), 236 -243.

Generation and Evaluation of a New Time-Dependent Dynamic S-Box Algorithm for AES Block Cipher Cryptosystems

Hayder T. Assafli [1,*] and Ivan A. Hashim [2]

[1] Department of Electrical Engineering, University of Technology, Iraq
E-mail: 140030@uotechnology.edu.iq
[2] Department of Electrical Engineering, University of Technology, Iraq
E-mail: 30095@uotechnology.edu.iq
* Corresponding author: Department of Electrical Engineering, University of Technology, Iraq. Tel: +9647901772796.

Abstract. Cryptographic communication systems depend extensively on the Advanced Encryption Standard (AES) block cipher algorithm. The encryption strength in the AES algorithm is derived from the nonlinearity introduced by the application of s-boxes inside the encryption process. Today, most s-boxes are either static in which the s-box content does not change during the encryption process or dynamic which continuously changes depending on the encryption key. Two major problems exist in such systems that increase the risk of unauthorized decryption. First, the encryption of the same data yields the same cipher key repeatedly. Seconds, the key-dependent s-boxes require sharing the encryption key each time a new s-box is generated. In this paper, a dynamic s-box is introduced that constantly changes independent of the encryption key. The new s-box depends on the epoch timestamp that exists in every digital system as well as transmitted with digital satellite communication systems. The generation of the nonlinear s-box occurs inside the transmitter and receiver identically. The main strength of the proposed approach is that the ciphertext changes while keeping the encryption key constant ensuring different encryption results for the same data. This paper also investigates and analyzes the strength and quality of the new s-box using the avalanche effect method. The results showed that different ciphertext is produced after every encryption process. Moreover, the modified dynamic s-box reached a high avalanche effect exceeding the Strict Avalanche Criterion (SAC).

Keywords: *Encryption; Advanced Encryption Standard; S-box, Avalanche effect; and Epoch time.*

1. Introduction

Block cipher cryptographic algorithms are becoming more susceptible to attack by hackers with the developments in computation power and speed. Therefore, high-security systems are required to protect important and valuable information from the access of unauthorized users (Park et al., 2018). Cryptographic algorithms are among the most effective methods in which the desired information is hidden inside cyphered text. Only users with the encryption key have access to the information by the decryption process. When the encryption and decryption keys are identical, the cryptographic algorithm is called symmetric as in Advanced Encryption Standard (AES) and Data Encryption Standard (DES). On the contrary, using a different key for decrypting data is called asymmetric method such as Rivest-Shamir-Adleman algorithm (RSA) (Kumar & Tiwari, 2012). The cryptographic algorithms are required to include nonlinearity in the encryption process. Several encryption methods depend on s-boxes for providing strength to the diffusion process such as Data Encryption Standard (DES), International Data Encryption Algorithm (IDEA) (Al-Dweik et al., 2019), and PRESENT algorithms (Jithendra & Shahana, 2019). Advanced Encryption Standard became a major focus for several studies in the previous decade. The AES is used as a major encryption process in several applications including media such as images (Alsaffar et al., 2020; Singh et al., 2019) and videos (Adiguna & Hendrawan, 2017). Researches have been going on to improve the performance of AES encryption from several criteria such as strength, speed, and power consumption (Ueno et al., 2020). This research mainly focuses on introducing a new dynamic key independent s-box for increasing the strength of the AES algorithm (Patil et al., 2016).

2. Related Work

Using dynamic s-box for substitution in AES or other forms of cryptographic algorithms is not new to the field of cryptography. Indrajit Das et al. (Das et al., 2012) suggested that the safety of the AES algorithm can be increased by generating random dynamic s-boxes depending on irreducible polynomial. However, the polynomial must be sent to the receiver with the encryption key. Moreover, the polynomial occupies extra bandwidth and is exposed to cyber-attacks.

Lamiaa A. Elazm et al. (Elazm et al., 2010) introduced a high speed with a low power consumption s-box for the AES algorithm. They depended on the composite Galois field for the substitution box architecture. The size and delay of the s-box circuit is reduced compared with the standard complementary metal-oxide-semiconductor (CMOS) based systems. On the other hand, Manjula G. and Mohan H. S. (Manjula & Mohan, 2018) investigated the hash function for generating dynamic s-boxes for AES encryption. Their experiment results showed that the time of the encryption process is increased with the AES encryption strength and the time of the decryption process is decreased. Their method requires additional time for the dynamic s-box generation.

E. Siva Ganesh et. al. (Ganesh et al., 2012) proposed a method for generating dynamic s-boxes depending on the biometric scheme. They showed that the nonlinearity of the suggested s-box is increased resulting in a higher strength AES system. The speed of their method qualifies their algorithm for encryption applications. However, the biometric scheme must be represented in both the encryption and decryption process. Kamsiah Mohamed et al. (Mohamed et al., 2014) and Hui Shi et al. (Shi et al., 2011) studied the nonlinearity effect of the s-box in the AES encryption process. Since the s-box is the only step that introduces nonlinearity effect to the process, the effect must be high enough not to be hacked by attackers. They both concluded the avalanche effect is the most important criterion that must be investigated to analyze the strength of the s-boxes.

Pham Anh Hung et al. (Hung et al., 2013) used two different chaotic maps for generating dynamic s-boxes for usage in image encryption. Their suggested method adds more shuffling to the s-box matrix elements increasing the overall strength of the AES encryption method. Do Quang Huy (Huy et al., 2019) used Galois field (GF) instead of the standard lookup tables for generating dynamic s-boxes. They proposed that the usage of s-box and inverse s-box on the same hardware chips makes it suitable for applications such as wireless local area networks. Julia Juremi et al. (Juremi et al., 2012) rotated the static s-box of the basic AES encryption structure depending on the encryption key. This method does not require sharing additional information between the sender and receiver. However, any change in the encryption key must be updated in the decryption key in order to continue the AES operation.

Kazys Kazlauskas and Jaunius Kazlauskas (Kazlauskas & Kazlauskas, 2009) suggested and analyzed a random s-box generator depending on the encryption key. They proposed that the usage of random operation in the generation process results in higher strength AES algorithm without prediction of the generated s-box. Lijuan Li and Shuguo Li (Li & Li, 2017) used a shared s-box in the encryption-decryption process instead of the inverse s-box. The sharing process resulted in higher throughput according to their hardware implementation on an FPGA device. On the other hand, M. Kermani and A. Masoleh introduced a fault detection method for s-boxes in AES algorithms. Their approach is based on the low-cost composite field implementations. Another research implemented the s-box on hardware is done by Shreenivas Pai N (Shreenivas Pai et al., 2017). They applied Rijndael s-box depending on the combinational logic circuit in AES for application-specific integrated circuits. The reduced the occupancy area and the consumption of power in their suggested s-box module.

Previous researches focused on generating dynamic s-boxes depending on resources that must be present in both the sender and receiver. Any modification on one side requires an update on the other side which represents a major source of risk of encryption key exposure. The solution for such risk is to provide a nonce number for both senders and receivers such that the digits mapping process is concealed inside the encryption algorithm. This process keeps the encryption key hidden and achieves all previous research goals.

3. Advanced Encryption Standard (AES)

In 2001, the National Institute of Standards and Technology (NIST) selected the AES algorithm to succeed the older Data Encryption Standard (DES) which became weak and unable to stand hacking attacks [ref]. The name Rijndael algorithm is also applied to the AES algorithm because of the developers J. Daemen and V. Rijmen (Daemen & Rijmen, 2002). AES encryption and decryption method is a block cipher algorithm. In this type, the stream of data is divided into 128 bits blocks, and then each block is encrypted with 128, 192, or 256-bit key depending on the desired application requirement. The data blacks are named states and are arranged as 4x4 matrices. The blocks go through serial data modification processes resulting in cyphered blocks that are completely not understandable to the public. The flow chart represents the complete encryption process is shown in Figure 1 (Huy et al., 2019). The four basic operations used in AES encryption are adding a round key, byte shifting, mixing columns, and substituting bytes.

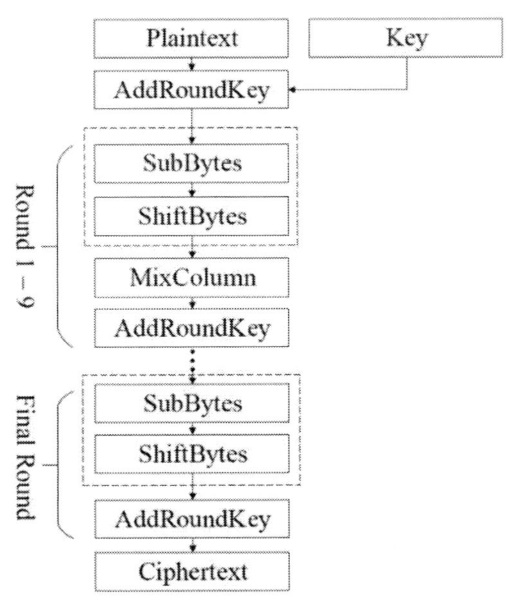

Figure 1. Flow chart of AES encryption

Substitute bytes:

A high nonlinearity is introduced to the encryption process by substituting bytes in the s-box. This step adds more confusion to the encryption operation.

Shift rows:

The rows of the 4x4 blocks are shifted depending on the position of each row. The first row is unaltered and the rest rows are shifted in cycles according to their positions. This process increases the scrambling of the encrypted data.

Mix columns

A fixed matrix is used to mix the data block columns by multiplication. The matrix must satisfy the Galois Field (8th power of 2).

Add round key:

The working data block and the round key of each cycle are XORed as a simple addition process (Chennam et al., 2017).

The addition process is the first step in the encryption operation in which the encryption key ix XORed with the first message block. Then, a loop start including all the four operations mentioned above. The number of cycles depends on the size of the encryption key used to protect data. 10, 12, or 14 cycles are used for 128, 192, or 256 respectively (Lee et al., 2010). The final step is adding the round key again to the encrypted block resulting in the final cyphered text (Vaidehi & Rabi, 2014). Inversing the encryption process exactly results in the decrypted text with the same encryption key. Advanced Encryption Standard can be operated in six different modes depending on the requirement of the application. All six modes share the four encryption process except slight differences in arrangement or technique. Electronic Code Book mode is the simplest operation mode in which a single key and a static s-box are used without initialization vector (Abdelrahman et al., 2018).

3.1 The S-Box

The substitution process in the AES encryption system performs nonlinear alterations on the encryption of data. It fulfills one of the main requirements for data security described by Shannon by applying the confusion property inside the algorithm (Shannon, 1998). Shannon suggested that repeating transposition with substitution leads to secure ciphers. Substitution permutation networks (SPN) were used in early cryptographic systems for combining permutation and substitution circuits. On the other hand, modern cryptographic systems must contain at least one nonlinear process for strengthening the encryption process (Hosseinkhani, 2012). As mentioned earlier, the s-box is the most essential step in the AES algorithm which requires the same length for both the message and the key. However, the highest delay in the AES algorithm is the s-box substitution process (Shivkumar & Umamaheswari, 2011). There are two types of s-boxes: static and dynamic. Static s-boxes have fixed contents during each round in the encryption process. Oppositely, the inverse s-box is fixed during the decryption process and it is the inverse transformation of the original s-box. Table 1 shows the static S-box used in the AES algorithm (Hosseinkhani, 2012). The first row and column represents the input to the s-box. When the two input numbers intersect, the addressed number is submitted as output. The static s-box is built as a simple logic circuit. The second type of the s-boxes is a dynamic s-box. In this type, the s-box is generated instantaneously with the key or depending on other sources of information. Generating s-box consumes time resulting in a slower encryption process. Nevertheless, the cipher key generated by dynamic s-boxes is stronger and have higher security than static s-boxes (Schneier & Sutherland, 1995). Special considerations must be taken for s-box when designing the AES algorithm. The most important consideration is the avalanche effect (Shi et al., 2011).

IOP Conf. Series: Materials Science and Engineering **978** (2020) 012042 doi:10.1088/1757-899X/978/1/012042

Table 1. The static s-box of AES encryption

	00	01	02	03	04	05	06	07	08	09	0a	0b	0c	0d	0e	0f
00	63	7c	77	7b	f2	6b	6f	c5	30	01	67	2b	fe	d7	ab	76
10	ca	82	c9	7d	fa	59	47	f0	ad	d4	a2	af	9c	a4	72	c0
20	b7	fd	93	26	36	3f	f7	cc	34	a5	e5	f1	71	d8	31	15
30	04	c7	23	c3	18	96	05	9a	07	12	80	e2	eb	27	b2	75
40	09	83	2c	1a	1b	6e	5a	a0	52	3b	d6	b3	29	e3	2f	84
50	53	d1	00	ed	20	fc	b1	5b	6a	cb	be	39	4a	4c	58	cf
60	d0	ef	aa	fb	43	4d	33	85	45	f9	02	7f	50	3c	9f	a8
70	51	a3	40	8f	92	9d	38	f5	bc	b6	da	21	10	ff	f3	d2
80	cd	0c	13	ec	5f	97	44	17	c4	a7	7e	3d	64	5d	19	73
90	60	81	4f	dc	22	2a	90	88	46	ee	b8	14	de	5e	0b	db
a0	e0	32	3a	0a	49	06	24	5c	c2	d3	ac	62	91	95	e4	79
b0	e7	c8	37	6d	8d	d5	4e	a9	6c	56	f4	ea	65	7a	ae	08
c0	ba	78	25	2e	1c	a6	b4	c6	e8	dd	74	1f	4b	bd	8b	8a
d0	70	3e	b5	66	48	03	f6	0e	61	35	57	b9	86	c1	1d	9e
e0	e1	f8	98	11	69	d9	8e	94	9b	1e	87	e9	ce	55	28	df
f0	8c	a1	89	0d	bf	e6	42	68	41	99	2d	0f	b0	54	bb	16

3.2 The Avalanche Effect

The avalanche effect is one of the important criteria that evaluates block cipher algorithms and describes the cryptographic function property. It is used to evaluate the effect of change in the output bits with respect to the input bits. In other words, it measures the flipping of output bits when one bit is flipped in the input. Strong cryptographic algorithms have a radical change in the output cyphered text due to a small change in the input data or encryption key. Failing to fulfill the requirement of the avalanche effect at a specific degree results in insignificant randomization and leads to the prediction of the encrypted information without the encryption key. The Strict Avalanche Criterion (SAC) was introduced by Webster and Tavares in 1985. This criterion states that at least 50% of the output bits are required to flip with the change of one bit in the input. The SAC became a standard for examining and evaluating cryptographic algorithms against cyber-attacks (Webster & Tavares, 1986).

4. Methodology

We propose a new method for modifying the AES-ECB encryption algorithm to increase the encryption strength. This method depends on the usage of a dynamic s-box instead of static s-box. Previously proposed algorithms generate the s-box dynamically with the encryption key. However, the proposed approach generates the s-box with each encryption process without any modification in the encryption key. The dynamic s-box generation depends on the Epoch time or Unix time on every encryption cycle. The Unix time has a unique one time value at every moment. This value is synchronized globally with all digital systems including mobile phones, computers, and global positioning system satellites. So, systems can generate their dynamic s-boxes independently without transmitting any information related to the encryption process. In other words, the encryption key is kept securely inside systems while the s-boxes are changed continuously. Ten digits are used to represent the UNIX time. These digits are fed into a dynamic s-box generator after being multiplied by a constant each called shift number. The proposed dynamic s-box generator accepts up to 16 double numbers. This gives the algorithm six spare digits that can be used for additional applications such as addressing. The

IOP Conf. Series: Materials Science and Engineering **978** (2020) 012042 doi:10.1088/1757-899X/978/1/012042

distribution of the epoch digits alongside with the shift number introduces extra strength to the AES-ECB encryption algorithm. Only the encryptor and the decryptor are supplied with the correct mechanism and information for the cryptographic process. Figure 2 demonstrates the suggested encryption method for higher security AES.

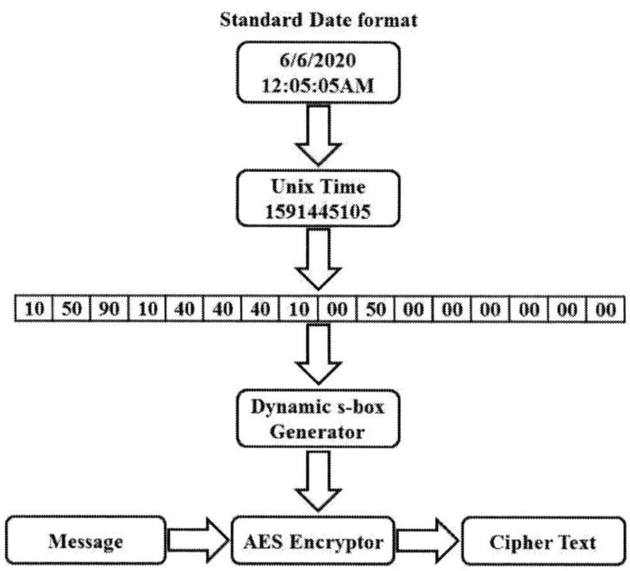

Figure 2. Dynamic s-box generation with Unix time.

Each digit of the Unix time is increasing continuously and the speed of change of each digit is related to a certain period. Minutes, hours, months, or years can be associated with a specific digit. This feature gives the encryption algorithm the ability of limited time operation. Thus, an expiration date can be embedded inside a system such that a system can decrypt data only for a specific limited time. Different encryption algorithms depend on the s-boxes for introducing the nonlinearity effect in the encryption process. The time dependent s-box can be modified to be used in other block cipher algorithms making the hacking process impossible even when the encryption key is known.

The simulation setup starts with a simple input text "1234567090abcde" which is converted into hexadecimal before encryption. Any combination of text sources can be used as a source for input text. Then, the hexadecimal result is entered to the AES encryption algorithm as a text source. Meanwhile, a dynamic s-box generator with internal epoch time generates an s-box at each request. The s-box is used in the substitution step. The rest of the processes are the same AES encryption steps. Using a new s-box each time ensures the generation of a new ciphertext after each execution of the encryption algorithm.

5. Results and Evaluation

To evaluate the dynamic s-box for AES encryption, several tests have been conducted using Matlab 2020a. There is no specific toolbox for algorithm testing in Matlab. Therefore, a software program is written, executed, tested and verified for the Avalanche effect testing. The AES algorithm is executed repeatedly to encrypt a simple text. The execution depended on different epoch time on each run. Different ciphertext is produced from the encryption process depending on different dynamic s-boxes. The resulted ciphertext is passed through the avalanche effect test. Results strongly suggested that encrypting with the suggested dynamic s-box leads to a completely different ciphertext each time. Moreover, the avalanche effect satisfies the SAC by maintaining higher than 0.5 most of the time. Table 2 shows the resulted ciphertext from the same message text for the first ten runs. Figure 3 illustrates the change in the avalanche effect during the first ten runs.

Table 2. AES encryption results with dynamic s-box

Message Type	Hexadecimal code	Avalanche Effect	Unix Time
Encoded Message	31323334353637303930616263646560	----	----
Encryption Key	87006B08B4C29AEAE0BB7EC871619DD6	----	----
Encryption 1	A7000F08B0AED68684B33AA03121DDBA	0.53906	1593355648781
Encryption 2	CF682B60B0A292EAE8D37EA83D6D91D6	0.52343	1593355648785
Encryption 3	E3046F4CB08EF6C284F31EA47565FDDE	0.51562	1593355648787
Encryption 4	C5606F08908EE292A4D77EA01101E9DA	0.53906	1593355648790
Encryption 5	CB4C6B24B8E29EE6E0BB7CE03961D1DA	0.46875	1593355648792
Encryption 6	C7002768B8AE9EE6E0BB7EE075619DD6	0.50781	1593355648795
Encryption 7	A7402F6CD48EF6C680975AA05521D99A	0.46875	1593355648797
Encryption 8	87244768B4C2B2C6AC9F12CC59619DF6	0.49218	1593355648799
Encryption 9	C32C6300BAE2D2E2ACB77ECC59259D96	0.52343	1593355648801
Encryption 10	87006B08B4C29AEAE0BB7EC871619DD6	0.50781	1593355648803

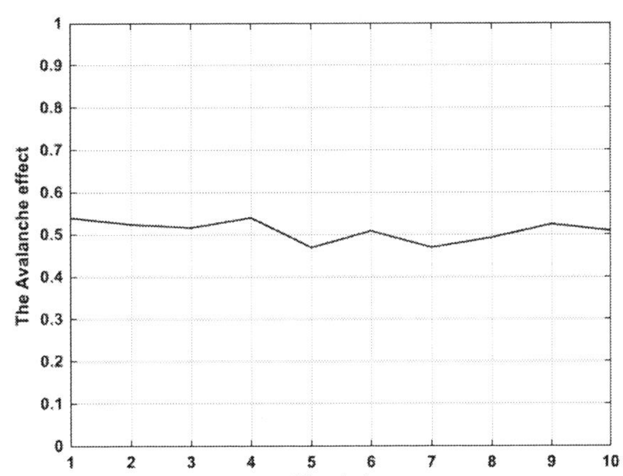

Figure 3. The Avalanche effect for a discrete time

Executing the algorithm continuously yields continuous fluctuations of the Avalanche effect as shown in Figure 4. The statistics for the observed fluctuations is shown in Table 3. It can be noted that the usage of the dynamic s-box satisfies the Strict Avalanche Criterion (SAC) for 75% of the time with a maximum of 0.63 exceeding the preferred SAC which is 0.5. It is worthwhile to mention that using constant s-box results in constant avalanche effect during the encryption process. The avalanche effect of static s-box depends on both the encryption key and input text. This raises the possibility of reducing the avalanche effect into undesired value (below 0.5). On the other hand, The suggested method does not require using different binary codes for mapping the input text to acquire such high avalanche effect as in (Dewangan et al., 2012). Moreover, the suggested approach stays higher than 0.5 most of the time exceeding the original static s-box which stays fairly at 0.5 (Shi et al., 2011). Furthermore, replacing the static s-box with a dynamic s-box in previous works did not achieve such high results and the avalanche effect remained close to 0.5 (Nejad et al., 2014; Tran et al., 2009; Bhoge & Chatur, 2014). Averaging the Avalanche effect during the runtime shows that it is also above 0.5 satisfying SAC. In other words, every encryption process uses the same but generates different outputs for the same input text. This increases the encryption security and does not require sharing any modification in the encryption key as in (Arrag et al., 2013; D'souza & Panchal, 2017; Hung et al., 2013) and (Nejad et al., 2014)

Table 3. Statistics of the Avalanche effect

Encryption Runs	Minimum	Maximum	Mean	Standard Deviation	Satisfies SAC	Dissatisfies SAC
10000	0.4063	0.6328	0.5129	0.0263	75.51%	24.49%

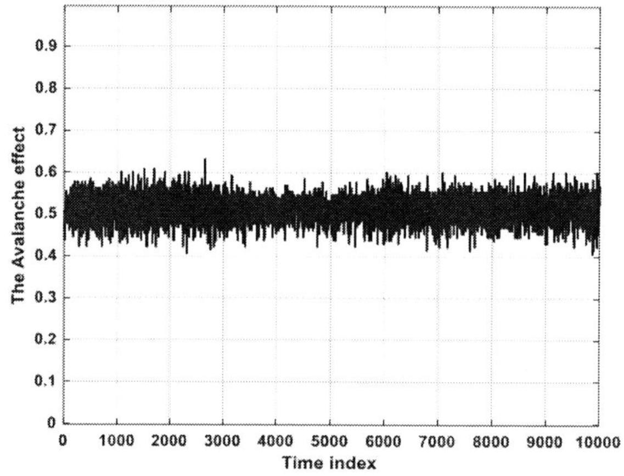

Figure 4. The Avalanche effect for continuous-time

6. Conclusion and Future Work

An enhanced mode of AES-ECB having high security is presented in this paper. The suggested algorithm is different from the original AES method such that it depends on the generation of dynamic s-boxes instead of a single static s-box. The creation of the dynamic s-box depends on the Unix time as a source of nonrepeatable number. The algorithm is tested and analyzed using the Avalanche effect in both discreet and continuous time. The results clearly showed that using the proposed algorithm generates a new ciphertext at each execution decreasing the risk of leaking the encryption key. Moreover, the achieved results satisfied the Strict Avalanche Criterion more than 70% of the execution time. Using the algorithm in future applications makes the communication process safer and more vulnerable to attacks. The future work is to focus on the implementation of the proposed algorithm in FPGAs for encrypting database systems.

References

[1] Abdelrahman, A. A., Fouad, M. M., Dahshan, H., & Mousa, A. M. (2018). High performance CUDA AES implementation: A quantitative performance analysis approach. *Proceedings of Computing Conference 2017, 2018-Janua*(July), 1077–1085. https://doi.org/10.1109/SAI.2017.8252225

[2] Adiguna, T., & Hendrawan. (2017). Secure H.264 Video Coding using AES/CFB/PKCS5 padding encryption on various video frames (I, P, B). *Proceeding of 2016 10th International Conference on Telecommunication Systems Services and Applications, TSSA 2016: Special Issue in Radar Technology*, 5–9. https://doi.org/10.1109/TSSA.2016.7871104

[3] Al-Dweik, A. Y., Hussain, I., Saleh, M. S., & Mustafa, M. T. (2019). *A Novel Method to Generate Key-Dependent S-Boxes with Identical Algebraic Properties*. 1–20. http://arxiv.org/abs/1908.09168

[4] Alsaffar, D. M., Sultan Almutiri, A., Alqahtani, B., Alamri, R. M., Fahhad Alqahtani, H., Alqahtani, N. N., Mohammed alshammari, G., & Ali, A. A. (2020). Image Encryption Based on AES and RSA Algorithms. *2020 3rd International Conference on Computer Applications & Information Security (ICCAIS)*, 1–5. https://doi.org/10.1109/ICCAIS48893.2020.9096809

[5] Arrag, S., Hamdoun, A., Tragha, A., & Khamlich Salah, E. (2013). Implementation of stronger AES by using dynamic S-box dependent of master key. *Journal of Theoretical and Applied Information Technology*, *53*(2), 196–204.

[6] Bhoge, J. P., & Chatur, P. N. (2014). Avalanche Effect of AES Algorithm. *International Journal of Computer Science and Information Technologies*, *5*(3), 3101–3103. http://ijcsit.com/docs/Volume 5/vol5issue03/ijcsit2014050394.pdf

[7] Chennam, K. K., Muddana, L., & Aluvalu, R. K. (2017). Performance analysis of various encryption algorithms for usage in multistage encryption for securing data in cloud. *RTEICT 2017 - 2nd IEEE International Conference on Recent Trends in Electronics, Information and Communication Technology, Proceedings*, *2018-Janua*, 2030–2033. https://doi.org/10.1109/RTEICT.2017.8256955

[8] D'souza, F. J., & Panchal, D. (2017). Advanced encryption standard (AES) security enhancement using hybrid approach. *Proceeding - IEEE International Conference on Computing, Communication and Automation, ICCCA 2017*, *2017-Janua*, 647–652. https://doi.org/10.1109/CCAA.2017.8229881

[9] Daemen, J., & Rijmen, V. (2002). The Design of Rijndael. In *New York*. Springer Berlin Heidelberg. https://doi.org/10.1007/978-3-662-04722-4

[10] Das, I., Nath, S., Roy, S., & Mondal, S. (2012). Random S-Box generation in AES by changing irreducible polynomial. *Proceedings of the 2012 International Conference on Communications, Devices and Intelligent Systems, CODIS 2012*, *2*, 556–559. https://doi.org/10.1109/CODIS.2012.6422263

[11] Dewangan, C. P., Agrawal, S., Mandal, A. K., & Tiwari, A. (2012). Study of avalanche effect in AES using binary codes. *Proceedings of 2012 IEEE International Conference on Advanced Communication Control and Computing Technologies, ICACCCT 2012*, *978*, 183–187. https://doi.org/10.1109/ICACCCT.2012.6320767

[12] Elazm, L. A., El-Moursy, M. A., Elsimary, H., Dessouky, M. I., & Shawki, F. (2010). High speed low power composite field SBOX. *IDT'10 - 2010 5th International Design and Test Workshop, Proceedings*, 24–27. https://doi.org/10.1109/IDT.2010.5724400

[13] Ganesh, E. S., Velayutham, R., & Manimegalai, D. (2012). A secure software implementation of nonlinear AES S-box with the enhancement of biometrics. *2012 International Conference on Computing, Electronics and Electrical Technologies, ICCEET 2012*, 927–932. https://doi.org/10.1109/ICCEET.2012.6203796

[14] Hosseinkhani, R. (2012). Using Cipher Key to Generate Dynamic S-Box in AES Cipher System. *International Journal of Computer Science and Security (IJCSS), Volume (6) : Issue (1) : 2012*, *6*, 19–28.

[15] Hung, P. A., Klomkarn, K., & Sooraksa, P. (2013). Image encryption based on chaotic map and

dynamic S-box. *ISPACS 2013 - 2013 International Symposium on Intelligent Signal Processing and Communication Systems*, 435–439. https://doi.org/10.1109/ISPACS.2013.6704589

[16] Huy, D. Q., Duc, N. M., Khai, L. D., & Lung, V. D. (2019). Hardware implementation of AES with S-Box using composite-field for WLAN systems. *RIVF 2019 - Proceedings: 2019 IEEE-RIVF International Conference on Computing and Communication Technologies*, 1–6. https://doi.org/10.1109/RIVF.2019.8713711

[17] Jithendra, K. B., & Shahana, T. K. (2019). A New Efficient Sbox for Strengthening PRESENT Like Block Ciphers Against Linear Cryptanalysis. *2019 2nd International Conference on Intelligent Computing, Instrumentation and Control Technologies, ICICICT 2019*, 507–511. https://doi.org/10.1109/ICICICT46008.2019.8993397

[18] Juremi, J., Mahmod, R., & Sulaiman, S. (2012). A proposal for improving AES S-box with rotation and key-dependent. *Proceedings 2012 International Conference on Cyber Security, Cyber Warfare and Digital Forensic, CyberSec 2012*, 38–42. https://doi.org/10.1109/CyberSec.2012.6246172

[19] Kazlauskas, K., & Kazlauskas, J. (2009). Key-Dependent S-Box Generation in AES Block Cipher System. *Informatica*, *20*(1), 23–34. https://doi.org/10.15388/informatica.2009.235

[20] Kumar, A., & Tiwari, N. (2012). Effective implementation and evaluation of AES in Matlab. *Advances in Intelligent Systems and Computing*, *176 AISC*(VOL. 1), 95–101. https://doi.org/10.1007/978-3-642-31513-8_11

[21] Lee, H., Lee, K., & Shin, Y. (2010). Implementation and performance analysis of AES-128 CBC algorithm in WSNs. *International Conference on Advanced Communication Technology, ICACT*, *1*, 243–248.

[22] Li, L., & Li, S. (2017). High throughput AES encryption/decryption with efficient reordering and merging techniques. *2017 27th International Conference on Field Programmable Logic and Applications, FPL 2017*, 12–15. https://doi.org/10.23919/FPL.2017.8056803

[23] Manjula, G., & Mohan, H. S. (2018). Improved Dynamic S-Box generation using Hash function for AES and its Performance Analysis. *Proceedings of the 2nd International Conference on Green Computing and Internet of Things, ICGCIoT 2018, November 2001*, 109–115. https://doi.org/10.1109/ICGCIoT.2018.8753074

[24] Mohamed, K., Mohammed Pauzi, M. N., Hj Mohd Ali, F. H., Ariffin, S., & Nik Zulkipli, N. H. (2014). Study of S-box properties in block cipher. *I4CT 2014 - 1st International Conference on Computer, Communications, and Control Technology, Proceedings*, *I4ct*, 362–366. https://doi.org/10.1109/I4CT.2014.6914206

[25] Nejad, F. H., Sabah, S., & Jam, A. J. (2014). Analysis of avalanche effect on advance encryption standard by using dynamic S-Box depends on rounds keys. *2014 International Conference on Computational Science and Technology, ICCST 2014*, *2014*(2005). https://doi.org/10.1109/ICCST.2014.7045184

[26] Park, T., Seo, H., Kim, J., Park, H., Kim, H., & Kim, C. H. (2018). Efficient Parallel Implementation of Matrix Multiplication for Lattice-Based Cryptography on Modern ARM Processor. *Security and Communication Networks*, *2018*. https://doi.org/10.1155/2018/7012056

[27] Patil, P., Narayankar, P., Narayan, D. G., & Meena, S. M. (2016). A Comprehensive Evaluation of Cryptographic Algorithms: DES, 3DES, AES, RSA and Blowfish. *Procedia Computer Science*, *78*(December 2015), 617–624. https://doi.org/10.1016/j.procs.2016.02.108

[28] Schneier, B., & Sutherland, P. (1995). *Applied Cryptography: Protocols, Algorithms, and Source Code in C* (2nd ed.). John Wiley & Sons, Inc.

[29] Shannon, C. E. (1998). Communication theory of secrecy systems. 1945. *M.D. Computing : Computers in Medical Practice*, *15*(1), 57–64.

[30] Shi, H., Deng, Y., & Yu, G. (2011). Analysis of the avalanche effect of the AES S box. *2011 2nd International Conference on Artificial Intelligence, Management Science and Electronic Commerce, AIMSEC 2011 - Proceedings*, 5425–5428. https://doi.org/10.1109/AIMSEC.2011.6009935

[31] Shivkumar, S., & Umamaheswari, G. (2011). Performance Comparison of Advanced Encryption Standard (AES) and AES key dependent S-box - Simulation using MATLAB. *Proceedings of 2011 International Conference on Process Automation, Control and Computing, PACC 2011*, 1–6. https://doi.org/10.1109/PACC.2011.5979007

[32] Shreenivas Pai, N., Raghuram, S., Chennakrishna, M., & Karthik, A. S. V. (2017). Logic optimization of AES S-Box. *International Conference on Automatic Control and Dynamic Optimization Techniques, ICACDOT 2016*, 1042–1046. https://doi.org/10.1109/ICACDOT.2016.7877745

[33] Singh, A., Agarwal, P., & Chand, M. (2019). Image Encryption and Analysis using Dynamic AES. *2019 International Conference on Optimization and Applications, ICOA 2019*, 1–6. https://doi.org/10.1109/ICOA.2019.8727711

[34] Tran, B. N., Nguyen, T. D., & Tran, T. D. (2009). A New S-Box structure to increase complexity of algebraic expression for block cipher cryptosystems. *ICCTD 2009 - 2009 International Conference on Computer Technology and Development*, *2*(2), 212–216. https://doi.org/10.1109/ICCTD.2009.235

[35] Ueno, R., Homma, N., Morioka, S., Miura, N., Matsuda, K., Nagata, M., Bhasin, S., Mathieu, Y., Graba, T., & Danger, J. L. (2020). High Throughput/Gate AES Hardware Architectures Based on Datapath Compression. *IEEE Transactions on Computers*, *69*(4), 534–548. https://doi.org/10.1109/TC.2019.2957355

[36] Vaidehi, M., & Rabi, B. J. (2014). Design and analysis of AES-CBC mode for high security aVaidehi, M., & Rabi, B. J. (2014). Design and analysis of AES-CBC mode for high security applications. 2nd International Conference on Current Trends in Engineering and Technology, ICCTET 2014, 499–50. *2nd International Conference on Current Trends in Engineering and Technology, ICCTET 2014*, 499–502. https://doi.org/10.1109/ICCTET.2014.6966347

[37] Webster, A. F., & Tavares, S. E. (1986). On the Design of S-Boxes. In H. C. Williams (Ed.), *Advances in Cryptology --- CRYPTO '85 Proceedings* (pp. 523–534). Springer Berlin Heidelberg.

IOP Conf. Series: Materials Science and Engineering **978** (2020) 012043 doi:10.1088/1757-899X/978/1/012043

Optimizing Study of Some Factors That Influence the Performance of Valve Trays in Crude Oil Distillation Towers

Ghufran J. Matrood [1,*], Ahmed M. Al-Gaban [2] and Hussain M. Yousif [3]

[1] Materials Engineering Department, University of Technology, Iraq
E-mail: ghufranjassim92@gmail.com
[2] Materials Engineering Department, University of Technology, Iraq
E-mail: a.ghabban@gmail.com
[3] Ministry of industry and minerals / state company for steel industries, Iraq
E-mail: hussainyousif68@yahoo.com
* Corresponding author: Materials Engineering Department, University of Technology, Baghdad, Iraq. Tel: +964735845499.

Abstract. Valve trays play an important role in the productivity of crude oil distillation towers. The performance efficiency of these valves depends on their erosion-corrosion and fouling corrosion. In this research, NiCrAlY powder with constituents of different ratios of Molybdenum (Mo) was used as a deposited coating layer on the surface of 446 stainless steel. The scanning electron microscope and XRD were used to study the microstructure of the coating layer. Microhardness test was carried out to evaluate coating layer hardness. Erosion – corrosion test was performed to evaluate the performance of the coating layer under an aggressive medium of glass sand and a mixture of sulphuric acid and hydrochloric acid. Contact angle test was carried out to evaluate the contact angle of coated and uncoated samples. Taguchi / ANOVA technique was used to tabulate data and to optimize the results. The results showed that the coating layer containing 4% Mo exhibit high erosion-corrosion resistance and good contact angle, which improving the fouling resistance of the samples under working conditions.

Keywords: *Valve trays, erosion-corrosion resistance, fouling corrosion, Mo addition, design of experiments Taguchi/ ANOVA technique.*

1. Introduction

Valve trays are popular because of their efficiency and low cost. The hole of the tray is covered by moveable valves. The vapour load controls the entrance area for the vapour, and thus the movement of the valve. At low vapour loads, the valves are closed, and the vapour enters the tray through the open hole of the valves. Increasing the vapour load caused a wider opening of the valves. The valves are in some way self-adjusting to the vapour load. This is the main advantage of valve trays(Gmehling, Kleiber, Kolbe, & Rarey, 2019). Erosion- corrosion is an important problem for valves trays because of the crude oil's containment (sand) and naphthenic acid. The valves are exposure to fouling corrosion as a result of the presence of sludge and sulfur, and thus decrease the efficiency of the tower.

Erosion – corrosion of metallic materials describes the process of material degradation in which mechanical wear by solid particles, liquid or the combination of both processes combines with corrosion induced by dissolution or oxidation of surfaces (Rajahram, Harvey, & Wood, 2009). The naphthenic acid erosion-corrosion (NAEC), and corrosion at high temperatures of naphthenic acid (NAC) is the major problems of corrosion of crude oil refineries. These two types of corrosion have become a prominent feature of transmission lines and valves in which flow velocity at high rates in addition to occurring in the other equipment such as heat exchangers, furnace tubes atmospheric and vacuum columns (Slavcheva, Shone, & Turnbull, 1999). The (NAC) and (NAEC), cause significant economic losses to the crude oil refineries and decrease productivity.

Fouling is the accumulation of unwanted material on solid surface of equipment and structures. Deposits, sludge, scaling, sediments, precipitates, unused residue are related to fouling (Coletti & Hewitt, 2014). Corrosion in the equipment of oil refining cause increases in the fouling rate, and vice versa (Kapusta, Ooms, Buijs, Fort, & Fan, 2001; Litschewski, 1996; Yanes, 2012). Corrosion products such as Iron oxide and sulfides and other typical scale materials can cause fouling. Also, clay or dirt from poor desalting/washing of brine salts and catalyst fines may all cause corrosion. (Speight, 2014).

Taguchi method is a standardized form of experimental design technique. It involves reducing the variation in a process through the robust design of experiments. The overall objective of the method is to produce a high-quality product at a low cost to the manufacturer (Anderson & Whitcomb, 2004)

2. Experimental procedure:

The technique of thermal spraying was used in this research. The automated spraying was used by linking the spray gun to an automated control system, that was locally manufactured. The amount of powder entering the spray gun was controlled using an electronic pulse technology system to ensure, uniform distribution of the deposited metal on the substrate surface, that was fixed on a table moving with a bi coordinate system (X, Y) as shown in figure (1).

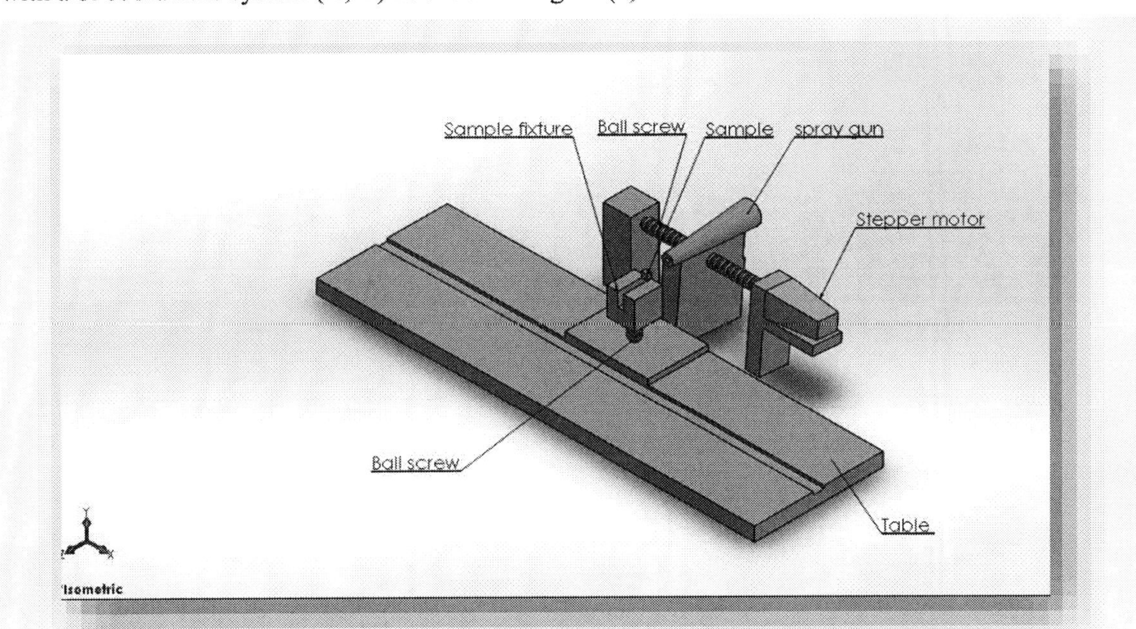

Figure1: Thermal Spray device

IOP Conf. Series: Materials Science and Engineering **978** (2020) 012043 doi:10.1088/1757-899X/978/1/012043

2.1 Sample and powder materials:

(AISI 446) stainless steel was used in this research with dimensions 22mm in diameter and 6mm in thickness. Sample pretreatment (degreasing and scale removing) were performed according to standard specifications. The roughening of the samples was carried out using sand shoot blast technique to obtain a rough surface (N12) according to (DIN) standard. NiCrAlY powder material was used with particle size range (25-53) micron. Molybdenum powder with particle size (less than 53 microns) was added with ratios (2,4, and 6%).

2.2 Thermal spray parameters:

In this work,thermal spray coating was performed by using an oxyacetylene thermal spray gun. The work sample was fixed in front of a spray gun on a flat table. The sample fixture comprised a ball screw with a stepper motor which makes the upward and downward movements possible. The distance between the spray gun and the sample was 100 mm (Fakhri, Al-Gaban, & Yousif, 2019)and other parameters that related to the thermal spray process are illustrated in table(1).

Table 1: thermal spray parameters

Oxygen pressure (bar)	Acetylene pressure (bar)	Powder feeding (g/min)	Traverse velocity (mm/min)
5	0.8	10	10

2.3 Erosion corrosion test:

To evaluate the performance of the coating layer, erosion-corrosion test was carried out by using the erosion-corrosion system as shown in figure (2).

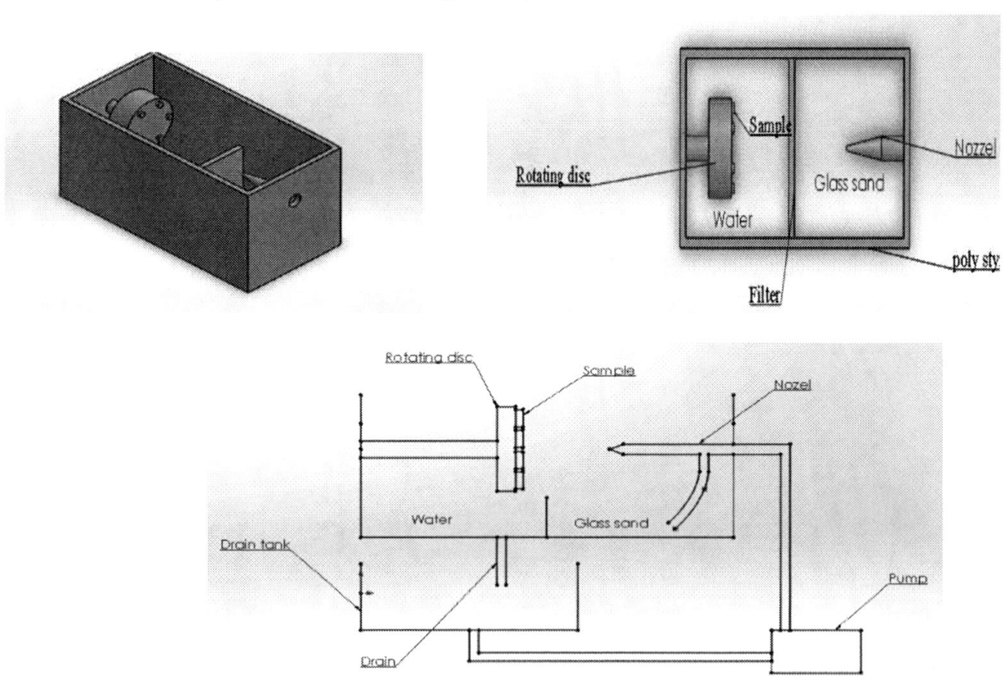

Figure 2: erosion-corrosion system

IOP Conf. Series: Materials Science and Engineering **978** (2020) 012043 doi:10.1088/1757-899X/978/1/012043

The system consists of polystyrene tank with two-zones separated by a filter, one of them filled with (a mixture of the hydrochloric acid solution and sulphuric acid solution) and the other filled with (250) gram glass sand with particle size(50 to 150) micron. The concentration of the hydrochloric acid and sulphuric acid were (5,10, and 15 M).

Taguchi/ ANOVA technique was used to tabulate data and to optimize the results. Four parameters with three levels for each parameter were studied as illustrated in the table(2).

Table 2: Factors and their levels

Factor	Level 1	Level 2	Level 3
Mo %	2	4	6
H_2SO_4 (M)	5	10	15
HCl (M)	5	10	15
Exposure time (hr.)	12	24	36

3. Results and discussion:

Taguchi orthogonal array was used to determine the design of the experiment. A total of nine experiments were executed, and the results are shown in table(3).

Table 3: Taguchi orthogonal array results

	Mo%	Hcl conc. (M)	H2SO4 conc.(M)	Expos. time (hr.)	Trial 1	Trial 2	Trial 3	Mean	S/N ratio
1	2	5	5	12	0.126	0.119	0.113	0.119333	18.4672
2	2	10	10	24	0.139	0.140	0.139	0.139333	17.1190
3	2	15	15	36	0.144	0.143	0.144	0.143667	16.8530
4	4	5	10	36	0.133	0.132	0.132	0.132333	17.5666
5	4	10	15	12	0.136	0.134	0.135	0.135000	17.3933
6	4	15	5	24	0.132	0.134	0.131	0.132333	17.5666
7	6	5	15	24	0.219	0.221	0.223	0.221000	13.1121
8	6	10	5	36	0.236	0.231	0.233	0.233333	12.6404
9	6	15	10	12	0.233	0.237	0.236	0.235333	12.5663

Where Trial 1, 2, and 3 are weight loss (g) due to erosion-corrosion action.

To determine which factor has a strong significant effect on minimizing weight loss, the response table for means was created. From which it shown that the percentage addition of molybdenum has a high delta value (0.0967) with rank1. While hydrochloric acid concentration, sulphuric acid concentration and exposure time have a weak significant effect on minimizing weight loss, as shown in table (4).

Table 4: response table for means

Level	Mo%	HCl (M)	H2So4 (M)	Exposure time (hr.)
1	0.1341	0.1576	0.1617	0.1632
2	0.1332	0.1690	0.1690	0.1642
3	0.2299	0.1704	0.1666	0.1698
Delta	0.0967	0.0129	0.0073	0.0066
Rank	1	2	3	4

Data listed in table (4) are used to plot the main effects plot of the means. From which it has been shown that the lowest value for weight loss can be achieved at (4% Mo), (5 M HCl), and (5 M. H_2SO_4), as shown in figure (3).

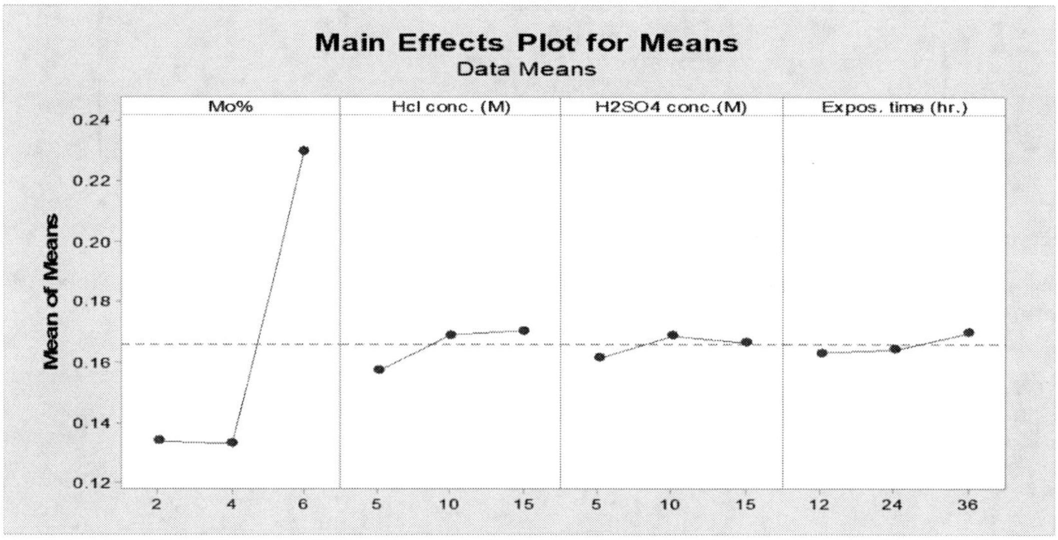

Figure 3: Main Effect Plot for Means

In addition, examining figure (3), indicates that the values of the weight loss of the factors HCl concentration, H_2So_4 concentration and exposure time, are somewhat close, which indicates the weakening of these factors' effect on the average weight loss.

3.1 Analysis of variance ANOVA:

Interaction plot is usually used in conjunction with an analysis of variance.

The effect of two parameters on the weight loss was investigated, as shown in figure (4). From which it shown that Molybdenum at a range of 4% in combination with HCl concentration, H_2So_4 concentration, and exposure time with all their ranges has a significant influence on minimizing the weight loss.

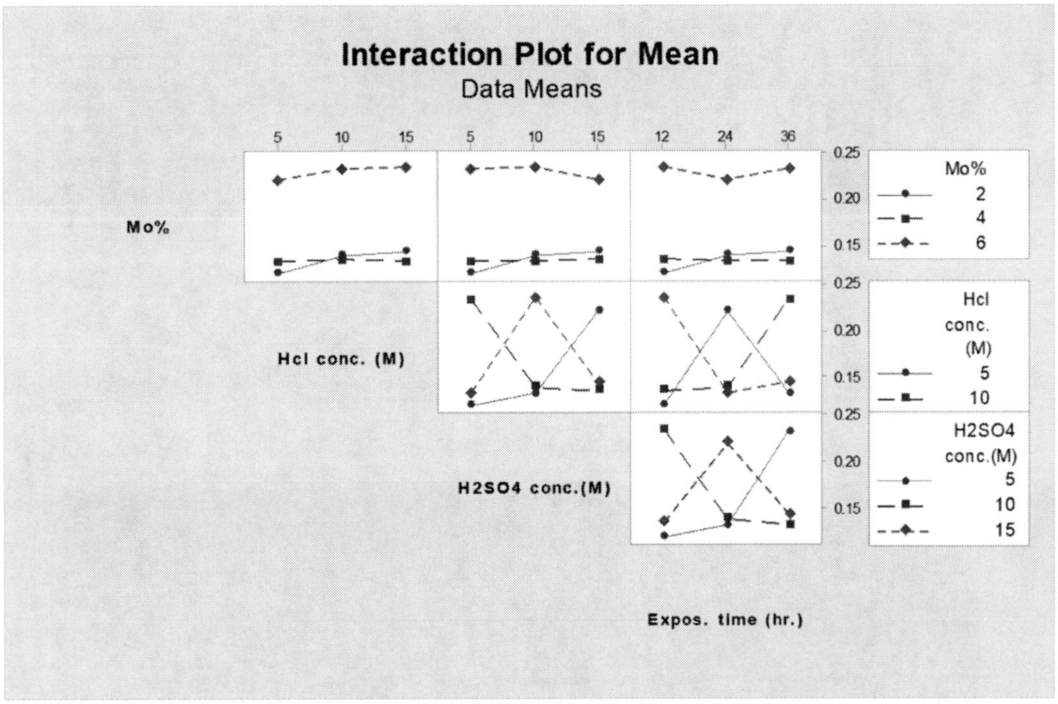

Figure 4: Interaction plot for means

The interaction plot also shows the effect of HCl concentration and H_2So_4 concentration on weight loss, it can be observed that the weight loss increases when the concentration of both acids at rang (5 M). When the coating layer exposed to HCl and H_2SO_4 acids, the coated layer start to dissolve and release ions. Then the beginning stage of the formation of a molybdenum-rich oxide stable film (MoO_3) starts, can acts as an effective barrier against the diffusion of species through the film and, therefore, decrease the dissolution rate. The effect of HCl concentration in combination with exposure time is the same as the effect of H_2So_4 in combination with exposure time on weight loss.

Based on the main effect plot for means and interaction plot, characterization and microstructure analysis were performed for samples with (2%, 4% and 6% Mo) using the scanning electron microscope. Figure (5) shows scanning electron microscope images for samples with 2,4, and 6% molybdenum. From which it shown that a sample with 2% Mo has a coarse structure, while the sample with 4% molybdenum has a more smooth and homogeneous structure which make it more resistant to erosion-corrosion. As for sample with 6% Mo, a crack can be notice in its structure along the grain boundary, making it weak to resist erosion-corrosion, because the electrolyte penetrates through these defects to reach the substrate. This gives the impression that the percentage of molybdenum addition plays an important role in increasing the performance efficiency of the coating layer in corrosion resistance and that the 6% addition ratio leads to initiate cracks in the coating layer structure. These cracks will enable the electrolyte solution to penetrate to the metal substrate.

IOP Conf. Series: Materials Science and Engineering **978** (2020) 012043 doi:10.1088/1757-899X/978/1/012043

a. sample with 2% Mo

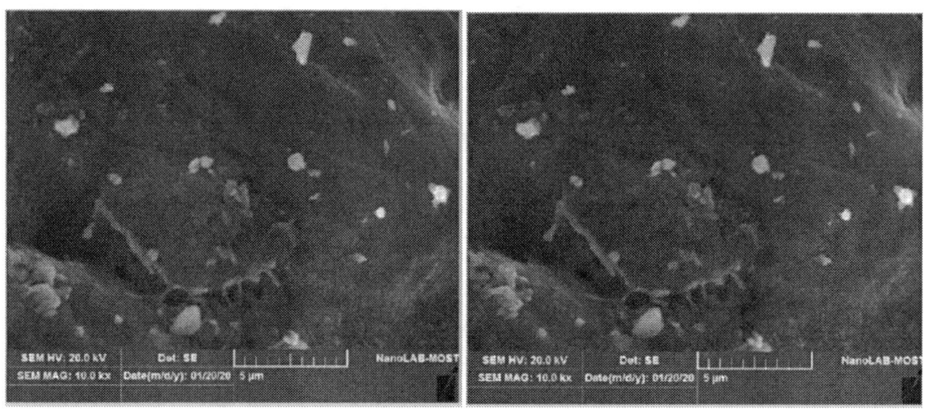

b. sample with 4% Mo

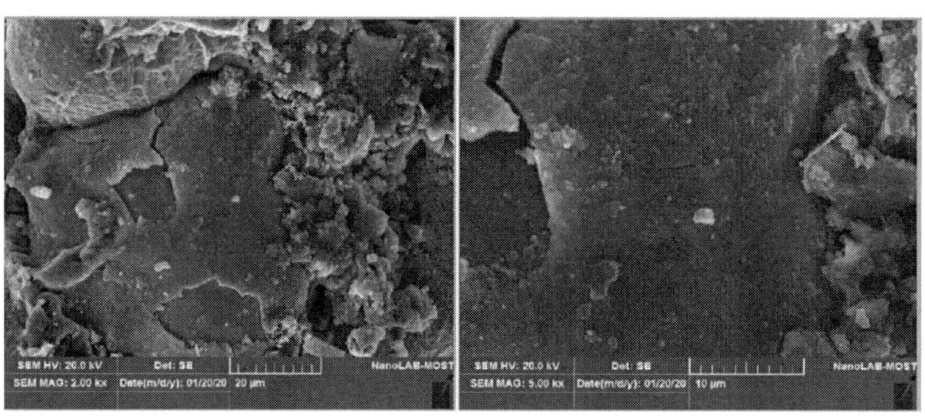

c. sample with 6% Mo

Figure 5: Scanning Electron Microscope images for testing samples

The XRD pattern of the coating layer shown β-Ni(NiAl) and γ-Ni(NiCr) phases, which behaves as aluminium reservoir and MoO_3 which can acts as an protective barrier against the diffusion of species that cause corrosion through the film and, therefore, decrease the corrosion rate .

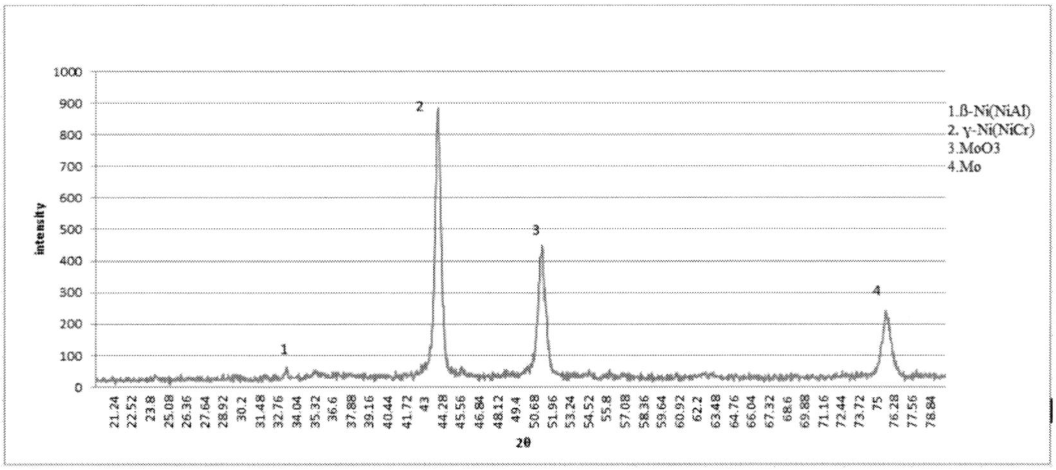

Figure 6 : XRD pattern of the surface of the coating layer

3.2 Confirmation experiments:

The above experimental analysis, main effect plot for means show that the maximum adhesion force can be obtained with the following conditions presented in Table 5.

Table 5: optimal design conditions

Factor	Best level
Mo (%)	Level 2
HCl (M)	Level 1
H2So4 (M)	Level 1
Exposure time (hr.)	Level 1

Three experiments of erosion-corrosion (weight loss) were performed under the optimal conditions in order to validate the optimization, from which it shown that the results of confirmation experiments and the predicted weight loss by optimization procedure are considered to be valid based upon the results of Taguchi experiments and within the 95% confidence interval.

3.3 contact angle

After the execution of the confirmation, the contact angle was inspected after and before the coating process .The results show that the surface of the uncoated sample was a hydrophilic and the coating changes the nature of the surface to hydrophobic. This means that the samples will have higher resistance to fouling corrosion Figure(6) represents a Photograph of water droplet of Stainless Steel (AISI 446) substrate with and without coating.

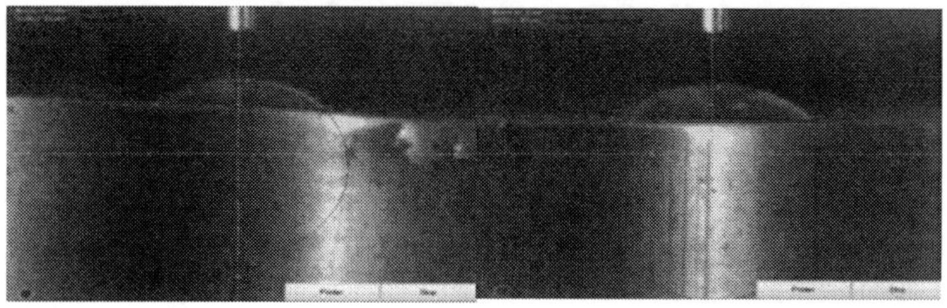

Ɵ=46.859 Ɵ=47.7222

a. the contact angle of uncoated samples

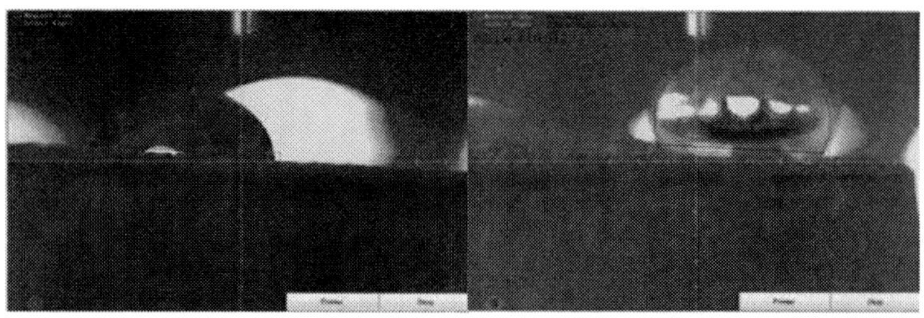

Ɵ=80.603 Ɵ=110.612

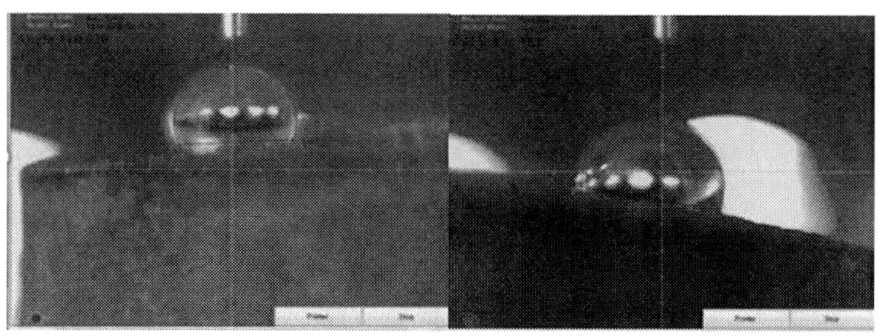

Ɵ=110.979 Ɵ=102.481

b. the contact angle of the coated sample

Figure 7: Photograph of water droplet of Stainless Steel (AISI 446) Substrate with and Without Coating

4. Conclusion

In this work, NiCrAlY with different ratio of Mo (2, 4, and6%) were applied on (AISI 446) stainless steel by using a flame spray technique. Taguchi / ANOVA technique was utilized in this research to optimize the results of erosion-corrosion and tabulate the data. The parameters investigated were Molybdenum percentage, HCl concentration, H_2So_4 concentration, and exposure time. From the results of Taguchi, it is observed that the percentage of Mo has a strong and significant effect on minimizing weight loss, which has high delta value (0.0967) with rank1. The other factors (HCl concentration, H_2SO_4 concentration, and exposure time) have an approximately close and weak effect on weight loss with delta value 0.0129, 0.0073, and 0.0066 respectively. The main effect plot for means showed that the low weight loss was achieved at level 2 for Mo% and level 1 for HCl concentration, H_2SO_4 consternation, and exposure time. SEM image showed that coating layer with 4% Mo makes the structure smooth and homogeneous, which as a result make it more resistant to erosion-corrosion. Contact angle test showed that the surface of uncoated samples have contact angle 46.859, and 47.772 ,while the contact angle for the surface of the coated samples were 80.603, 110.612, 110.979, and 102.481, which makes the coated samples more resistance to fouling corrosion.

References

[1] Anderson, M. J., & Whitcomb, P. J. (2004). *Screening process factors in the presence of interactions.* Paper presented at the Annual quality congress proceedings-American society for quality control.
[2] Coletti, F., & Hewitt, G. (2014). *Crude oil fouling: deposit characterization, measurements, and modeling*: Gulf Professional Publishing.
[3] Fakhri, A. S., Al-Gaban, A. M., & Yousif, H. M. (2019). *Oxidation behavior of NiAl coatings applied on 304 stainless steel by flame thermal spray*
[4] Gmehling, J., Kleiber, M., Kolbe, B., & Rarey, J. (2019). *Chemical thermodynamics for process simulation*: John Wiley & Sons.
[5] Kapusta, S. D., Ooms, A., Buijs, J. W., Fort, W. C., & Fan, D. (2001). *Systematic Approach to Controlling Fouling and Corrosion in Crude Unit Overheads and Hydrotreater Reactor Effluents.* Paper presented at the CORROSION 2001.
[6] Litschewski, M. J. (1996). *More Experiences With Corrosion and Fouling in Refinery Amine Systems.* Paper presented at the CORROSION 96.
[7] Rajahram, S., Harvey, T., & Wood, R. J. W. (2009). Erosion–corrosion resistance of engineering materials in various test conditions. *267*(1-4), 244-254.
[8] Slavcheva, E., Shone, B., & Turnbull, A. J. B. C. J. (1999). Review of naphthenic acid corrosion in oilrefining. *34*(2), 125-131.
[9] Speight, J. G. (2014). *Oil and gas corrosion prevention: From surface facilities to refineries*: Gulf Professional Publishing.
[10] Yanes, M. J. (2012). *Corrosion Problems Might Begin During Engineering Design.* Paper presented at the CORROSION 2012.

Variation of Erbil Municipal Wastewater Characteristics Throughout 26 Years (1994-2020) with Possible Treatments and Reusing: A Review

Shuokr Qarani Aziz

Department of Civil Engineering, College of Engineering, Salahaddin University-Erbil, Kurdistan Region, Iraq.
Emails: shuokr.aziz@su.edu.krd, shoker71@yahoo.com Tel: +964754625426.

Abstract. This research aimed to study characteristics variation of Erbil municipal wastewater (EMWW) during 26 years (1994-2020), appropriate treatment using different methods, and suitability of the treated wastewater (WW) for disposal to the natural environment or using for irrigation purpose. Forty-seven WW quality parameters were studied. A number of EMWW characteristics such as five-day biochemical oxygen demand (BOD_5), chemical oxygen demand (COD), ammonia nitrogen (NH_3-N), total suspended solids (TSS) were exceeded the WW discharge standards. Consequently, EMWW needs treatment process prior to disposal to the environment. Primary treatment units with lagoons or oxidation ditch or wetland were applied as first scenario for treatment of EMWW; While, using only lagoons or oxidation ditch or wetland directly was the second situation. EMWW normally regarded as weak/low to medium WW type and it is classified as good to injurious irrigation water kind. Commonly, time had not great effect on EMWW characteristics. Life style, climate, sewerage system (combined or separate systems), climate, and areas/zones had effect on the quality of the municipal WWs. Primary units plus wetland led to removal efficiency of 94.75 %, 93.07 %, 89.47 %, 96.72 %, and 57.68 % for BOD_5, COD, NH_3-N, TSS and PO4, respectively. Treatment of EMWW using both primary units and wetland resulted in achieving effluents agreed with the standards for disposal of WW. Generally, treated EMWW can be used for cooked vegetables and irrigating green areas.

Keywords: *Erbil; Lagoon; Oxidation ditch; Wastewater Characteristics; Wetland.*

1. Introduction

Municipal wastewater (MWW) denotes to the domestic wastewater (WW) in addition to those discharged from commercial, institutional, and similar services. MWW includes of WW produced by residences, businesses such as restaurants and shopping centres, institutions such as schools, universities, hospitals, prisons, and rest homes, recreational facilities, storm water, infiltration and industries in a definite community (Al-Zboon and Radaideh, 2012). MWW is the most abundant kind of WW that located into the category of low-strength WWs, characterized by low organic strength and high particulate organic matter content (Sikosana et al., 2019).

Sewerage system in Erbil City, Kurdistan Region-Iraq covers both storm water and grey water. Commonly, black water from toiles are treated using cesspools and septic tank with cesspools (in some cases). Therefore, neither full combined and nor full separate sewerage systems are available in Erbil City. In the recently constructed cities and villages in Erbil City and due to the investment laws, small scale WW treatment plant (WWTP) is compulsory for treatment of the produced WWs and in some areas it uses for irrigating green areas. Erbil MWW (EMWW) sometimes used for irrigation directly and in some cases it reaches to the Greater-Zab River water at Gwer area without treatment (Mustafa and Sabir, 2001; Amin and Aziz, 2005). To date, centralized WWTP is not available in Erbil City. Accordingly, treatment of EMWW is necessary for irrigation and prior disposal to the natural environment or water sources. In this work, two options were presented for treatment of EMWW. In the first option, EMWW is treated using primary treatment units plus aerated lagoons or oxidation ditch or wetland; while in the second scenario, EMWW is treated by lagoons, oxidation ditch or wetland. In literature, primary units, lagoons, oxidation ditch, and wetland were applied widely for treatment of MWWs (Asano and Tchobanoglous, 1987; USEPA, 2000; Teleman et al., 2004; Lian-Feng et al., 2011; Al-Zboon and Radaideh, 2012; Gikas and Tsihrintzis, 2014; Hadisoebroto et al., 2014; Metcalf and Eddy, 2014; INDITEX, 2015; Jasim, 2020).

Although, there are some published works on EMWW characteristics, impact, and treatment (Shekha, 1994; Aziz et al., 1998; Mustafa and Sabir, 2001; GDIWR, 2001; Ali, 2002; Aziz, 2004; Amin and Aziz, 2005; Ganjo *et al.*, 2006; Shekha, 2008; Bapeer, 2010; Shekha *et al.*, 2010; Shekha, 2013; Amine and Shekha, 2016; Shekha *et al.*, 2016; Aziz and Ali, 2017; Aziz and Ali, 2018). However, studying EMWW characteristics variation, treatment via different scenarios and reusing of treated water over a period of 26 years have not been published yet. Subsequently, the objectives of the current research were to study: 1) characteristics of EMWW during 26 years (1994-2020), 2) treatment of EMWW using different options, and 3) the suitability of reusing of treated EMWW for irrigation purposes.

2. Materials and Methods

2.1 EMWW (Erbil Municipal wastewater)

The main wastewater (WW) channel in Erbil City is located at the left side of Erbil-Mosul Main Road at Tooraq Q., Fig. 1. EMWW at Tooraq Q. commonly consists of WWs produces at residential areas, shops and super markets, restaurants, hotels and motels, car washing places, north industrial area, universities and schools, worship places, governmental and administration buildings, private sector houses and buildings, washings, infiltration, and losses of water supply system. Additionally, storm water is mixed with the MWWs during rainy seasons and it dilute the concentration of pollutants.

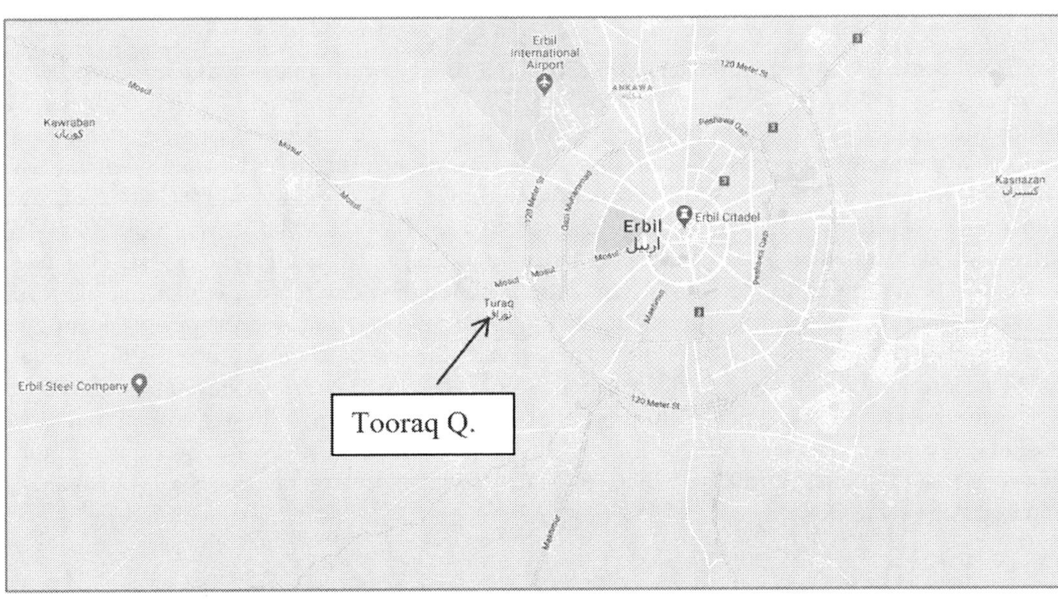

Fig. 1. Erbil Municipal Wastewater Main Channel

IOP Conf. Series: Materials Science and Engineering **978** (2020) 012044 doi:10.1088/1757-899X/978/1/012044

2.2 Data Collection and Analysis

Data from published works were collected. Forty-seven quality parameters for EMWW were collected, arranged, and compared with WW disposal standards. In the current study, EMWW characteristics data during 26 years (1994-2020) were collected and studied. First and last recognized published data on EMWW were published in 1994 and 2020 (Shekha, 1994; DEI, 2020); therefore, the period between 1994 and 2020 was selected. Data on pH, temperature, electrical conductivity (EC), total salts (T. salts), total solids (TS), total suspended solids (TSS), total dissolved solids (TDS), turbidity, chloride, total acidity (T. acidity), total hardness (T. hardness), five day biochemical oxygen demand (BOD_5), chemical oxygen demand (COD), ammonia (NH_3-N), nitrite (NO_2-N), nitrate (NO_3-N), sulphate (SO_4), dissolved oxygen (DO), phosphate (PO_4), sodium percentage (Na%), sodium adsorption ratio (SAR %), total coliform, sodium (Na), calcium (Ca), magnesium (Mg), cadmium (Cd), cupper (Cu), zinc (Zn), lead (Pb), total volatile solids (TVS), total non-volatile solids (TnVS), biodegradability ratio (BOD_5/COD), color, manganese (Mn), total organic carbon (TOC), phenols, oil and grease, oxidation-reduction potential (ORP), salinity, total alkalinity (T. alkalinity), Alkalinity %, TVC bacteria, Phytoplankton density, total bacteria count, total bacteria cell, total fungi cells, and discharge for EMWW were collected, arranged and studied.

3. Results and Discussions

3.1 Characteristics of EMWW

Collected data were arranged into three parts (Tables 1 to 3). Additionally, range and WW discharge standards are shown in the tables as well. pH, temperature, chloride, NO_3-N, and SO_4 found within the allowable limit of disposal of WW. While, TSS, BOD_5, COD, NH_3-N, NO_2-N, color, oil and grease, Mn, PO_4, Mg, Cd, Cu, Zn, Pb, and phenols were surpassed the WW disposal standards (EPA, 2003; Iraqi Environmental Standards, 2011). Consequently, EMWW requires treatment processes so as to be fit with WW disposal standards and reusing. EMWW observed been as weak/low to medium WW type according to (Henze and Comeau (2008) and Metcalf and Eddy (2014) and this found agreeing with Sikosana *et al.* (2019). BOD_5 values of 6.3 mg/L to 304 mg/L for EMWW are agree with published data by JSWA (2013). Lower COD values were reported in a number of South Korean WWTPs by Choi *et al.* (2017). Additionally, lower BOD_5, COD, TSS, PO_4, NO_3-N, and NO_2-N for two municipal WWPTs in Indonesia were reported (Wijaya and Soedjono, 2018). On the other hand, lower BOD_5, COD, NO_3-N, NO_2-N, and turbidity values for MWW in there WWTPs in Malysia were reported (Sabeen *et al.*, 2018). Teleman *et al.* (2004) reported lower values of COD, BOD_5, PO_4, TSS, and turbidity for MWW in Germany WWTP. EMWW parameters normally agree with the MWW characteristics in developing countries (Sperling, 2007). Life style, season, sewerage system (combined or separate systems), climate, and areas/zones had effect on the quality of the MWW. Fluctuations were noticed in the EMWW quality; Commonly, time had not great effect on EMWW characteristics. Mustafa and Sabir (2001) reported that EMWW discharge was 0.85 m^3/s. While, discharge for 2020 calculated as 5.56 m^3/s (Metcalf and Eddy, 2014; GDWS, 2020). The discharge increase may be attributed to increase in population, losses in water supply, expansion of Erbil City and sewerage system, and extra storm water. Mixing of storm water and water supply bad using leaded to dilution of EMWW, especially in rainy seasons (Aziz; 2004). Biological problems are available in EMWW; accordingly, treatment and disinfection is necessary for EMWW. Nutrients (organic matter, nitrogen compounds, and phosphate) are present in EMWW, it is useful for agriculture and irrigation purpose. Heavy metals such as Cd, Cu, Zn, and Pb were reported in EMWW and exceeded the WW discharge standard limits, normally effects on the biological treatment processes and needs extra treatments (Aziz *et al.*, 2011).

Table 1. Characteristics of Erbil MWW-Part 1

References	Parameters										
	pH	Temp. (°C)	EC (μs/cm)	T. Salts (mg/L)	TS (mg/L)	TSS (mg/L)	TDS (mg/L)	Turbidity (NTU)	Chloride (mg/L)	T. Acidity (mg/L)	T. Hardness (mg/L)
Shekha (1994)	6.8-8.4	10-23	430-946				441-972	6-128	30-101	0.18-5.62	207-432
Aziz *et al.* (1998)	6.75-7.6		531-574.9	340					4.97-20.4		
Mustafa and Sabir (2001)	7.4-8.85	13.5-18							40-80	5.2-6.8	268-280
GDIWR (2001)			284-873						13.5-43		
Ali (2002)	6.5-8.06		296-756					20-225		0.4-1.62	120-292
Aziz (2004); Amin and Aziz (2005)	6.82-8.2	11-28	491-835	314.2-534.4				0.41-1000	7.1-53.2		200-326
Ganjo *et al.* (2006)	6.1-6.6	15-18	1400-2300	996-1800					148-165		
Shekha (2008)	6.5-7.82	11.5-31.5	496-831								
Bapeer (2010)						40-60					
Shekha *et al.* (2010)	7.73				434.64	80.15	354.5				
Shekha (2013); Amine and Shekha (2016)	6.99-7.78		486.15-838.8	236.8-491.2					0.86-2.68		226-346
Shekha *et al.* (2016)	7.0-7.5	10-28	455-825								
Aziz and Ali (2017)	6.78-8.81	16.2-25.1	490-590	273-577	300-900	200-400	100-500	18.3-44.2	28-33.98	30.67-60	175.3-590
Aziz and Ali (2018)	6.78	21.86	582.6	372.9	10000	1800	8200	19.77	30.5	40	194

DEI (2020)	7.3		709	354.5				33.2	43		341
Range	6.1-8.85	10-31.5	284-2300	236.8-1800	300-10000	40-1800	100-8200	0.41-1000	0.86-165	0.18-60	120-590
Standards	6-9.5*	35*, 40**				60*, 35**			750 **		

*: Iraqi Environmental Standard (2011)
**: Environmental protection regulations (EPA) (2003).

Table 2. Characteristics of Erbil MWW-Part 2

References	Parameters										
	BOD_5 (mg/L)	COD (mg/L)	NH_3-N (mg/L)	NO_2-N (mg/L)	NO_3-N (mg/L)	SO_4 (mh/L)	DO (mg/L)	PO_4 (μg/L)	Na (%)	SAR (%)	Total coliform cell/100 ml X 10^5
Shekha (1994)	145		0.04-0.49 +	0.004-0.086			0.01-10.4	0.0015-0.113	15.17-16.4	0.81-0.97	0.34-42.5
Aziz *et al.* (1998)	30.6-304		0.006-0.10 +	0.001-0.027				0.0016-0.0382			4-180
Mustafa and Sabir (2001)	80-105						3.3-4				
Ali (2002)	17.5-219		0.016-0.13 +	0.0014-0.022			0.4-1.15	0.018-0.1004			
Aziz (2004); Amin and Aziz (2005)	13-110	20-188			7.6-22				6.1-38.0	0.19-1.96	
Ganjo *et al.* (2006)									69.8-73	14-16	200-380
Shekha (2008)	9.3-208	130-901	0.004.-0.19 +	0.029.-0.616	0.005-4.449		0-3.8	0.419-1.167			5.7-18
Hama Saeed et al. (2010)	110-170	132-195					0.0				160

Bapeer (2010)	8.2-60	12.2-126.1						1.55-6.97			2-240
Shekha *et al.* (2010)								0.0384			
Shekha (2013); Amine and Shekha (2016)						4.17-28.65			11.63-41.43	0.32-1.84	
Shekha *et al.* (2016)	6.3-104				0.003-4.2	170-1220	0.2-3.3	0.4-1.2			
Aziz and Ali (2017)	40-56	53-108	3.25-11.4	20-26	5.4-22.3	0.082-0.25					
Aziz and Ali (2018)	44										
DEI (2020)					47	62					16
Range	6.3-304	12.2-901	0.004-11.4	0.001-26	0.003-47	0.008-1220	0-10.4	0.0015-6.97	6.1-73	0.19-16	0.34-380
Standards	40*	100*	Nil*, 1 **	1 **	50*, 10**	1500 **		3*			

*: Iraqi Environmental Standard (2011)
**: Environmental Protection Regulations (EPA) (2003).
+ :NH_4-N (mg/L)

Table 3. Characteristics of Erbil MWW-Part 3

Parameter	Value	Range	Standards	References
Na (mg/L)	30.6-32.1	0.38-62		Ganjo *et al.* (2006)
	0.38 -2.3			Shekha (2013), Amine and Shekha (2016)
	62			DEI (2020)
Ca (mg/L)	5.4-6.4	1..8-85		Ganjo *et al.* (2006)
	1.8-4.8			Shekha (2013), Amine and Shekha, (2016)
	85			DEI (2020)
Mg (mg/L)	2.8-3.4	0.1-30.8	0.5*	Ganjo *et al.* (2006), Shekha (2013), Amine and Shekha (2016)
	0.1-0.42			
	30.8			DEI (2020)
Cd (mg/L)	0-46.73	0-46.73	0.01*	
Cu (mg/L)	0-18.69	0-18.69	0.2*	Shekha (2013), Amine and Shekha (2016)
Zn (mg/L)	0-76.92	0-76.92	0.2*	
Pb (mg/L)	0-61.76	0-61.76	0.1*	

TVS (mg/L)	206.8	100-300		Shekha *et al.* (2010)
	100-300			Aziz and Ali (2017)
TnVS (mg/L)	227.9	100-600		
	100-600			
BOD_5/COD	0.487-0.830	0.487-0.830		
Color (Pt.Co,)	186-379	186-379	Nil *	
Mn (mg/L)	1.3-4.6	1.3-4.6	0.2 *	
TOC (mg/L	19-180	19-180		
Phenols (mg/L)	0.044-0.102	0.044-0.102	0.01-0.05*	
Oil & grease (mg/L)	0.04-1.05	0.04-1.05	Nil*, 10 **	
ORP (Mv)	-107.4 - (-33.2)	-107.4 - (-33.2)		
Salinity	0.26-057	0.26-057		
T. Alkalinity (mg/L)	260-340	157.3-340		Mustafa and Sabir (2001)
	157.3-236			Aziz and Ali (2017)
	206			Aziz and Ali (2018)
	301			DEI (2020)
Alkalinity (%)	17.67-19.11	8.93-40.15		Shekha (1994)
	8.93-40.15			Aziz (2004), Amin and Aziz (2005)
TVC Bacteria (Cfu/mL)	$110*10^5$-$176*10^5$	$110*10^5$-$176*10^5$		Aziz and Ali (2017)
Phytoplankton density	21787.5	21787.5		Shkekha *et al.* (2010)
Total Bacteria Count ($X10^8$)	0.002-0.74	0.002-0.74		Shekha (2008)
Total bacteria cell/L x 10^5	0.047-77	0.047-193		Shekha (1994)
	0.87-193			Aziz *et al.* (1998)
Total fungi cells/L X 10^4	0.49-107.5	0.035-240		Shekha (1994)
	0.035-240			Aziz *et al.* (1998)
Discharge (m^3/s)	0.85-1.7	0.85-5.56		Mustafa and Sabir (2001)
	1.23-2.61			GDIWR (2001)
	1.696-3.582			DEI (2013)
	5.56			GDWS (2020)

*: Iraqi Environmental Standard (2011)
**: Environmental protection regulations (EPA) (2003).

3.2 Treatment of EMWW

As mentioned earlier, some EMWW parameters such as BOD_5, COD, TSS etc. were exceeded the effluent standards. Therefore, EMWW requires treatment. COD values for EMWW were ranged between 12.2. mg/L to 901 mg/L, Table 2; Sikosana *et al.* (2019) reported that aerobic processes are most suitable for MWWs with COD values of less than 1000 mg/L. Biodegradability ratio for EMWW was close to and higher than 50 %; this mean that biological processes are efficient (Aziz and Ali, 2017; 2018). Presence of toxic materials such as Cd, Cu, Zn, and Pb effects on the biological treatment processes. Consequently, further physical/chemical treatment technologies are required to remove/minimize the effect of toxic matters (Aziz *et al.*, 2011; Metcalf and Eddy, 2014). For treatment of EMWW, two scenarios were proposed: 1) treatment using primary treatment units plus aerated lagoon or oxidation ditch or wetland methods, and 2) Using only aerated lagoon or oxidation ditch or wetland. For both scenarios, data for parameters BOD_5, COD, NH_3-N, TSS, and PO_4 reported in Tables 1 and 2 were used.

3.2.1 First Scenario

Normal WWTP consists of primary, secondary/biological, advanced treatment units. Primary treatment units involve screens and comminution, grit removal, flow equalization, primary sedimentation tank units (Metcalf and Eddy, 2014; Jasim 2020). Removal efficiencies for some pollutants such as BOD_5, COD, NH_3-N, TSS, and PO_4 in the primary treatment units are shown in Table 4 (Asano and Tchobanoglous, 1987; Teleman *et al.*, 2004; Metcalf and Eddy, 2014). Effluent of primary unit become influent for the further treatment processes such as aerated lagoons, oxidation ditch and wetland. Removal efficiencies for BOD_5, COD, NH_3-N, TSS, and PO_4 using aerated lagoons, oxidation ditch and wetland are illustrated in Table 5. It can be noticed from the results shown in Table 5 that commonly wetland was efficient, when compared with the other methods. Treatment of EMWW using both primary units and wetland resulted in the effluent parameters of WW disposal to satisfy with the standards of EPA (2003) and Iraqi Environmental Standard (2011).

Table 4. Removal of pollutants by primary treatment units

Parameters	Removal Efficiency (%) **	Influent concentration (mg/L)	Average effluent concentration (mg/L)
BOD_5	25-50 (37.5*)	6.3-304 (155.15*)	96.97
COD	19.29-36.36 (27.82*)	12.2-901 (456.6*)	329.57
NH_3-N	26.4	0.004 -11.4 (5.7*)	4.19
TSS	50-70 (60*)	40-1800 (920*)	368
PO_4	15	0.0015-6.97 (3.45*)	2.93

* Average Value

** Sources: Asano and Tchobanoglous (1987); Teleman *et al.*, (2004); Metcalf and Eddy (2014)

Table 5. Performance of aerated lagoons, oxidation ditch and wetland methods after primary treatment units

Parameters	Treatment Methods			Effluent Standard
	Aerated lagoon	Oxidation ditch	Wetland	
Avg. influent BOD (mg/L)	96.97	96.97	96.97	40 mg/L $^+$
BOD removal (%)	70-95 (82.5*)	82-99 (90.5*)	90.8-92.3 (91.6*)	
Effluent BOD (mg/L)	16.97	9.21	8.15	
Avg. influent COD (mg/L)	329.57	329.57	329.57	100 mg/L$^+$

COD removal (%)	62.5	61.5	89-91.7 (90.4*)	
Effluent COD (mg/L)	123.59	126.88	31.64	
Avg. influent NH_3-N (mg/L)	4.19	4.19	4.19	1 mg/L ++
NH_3-N removal (%)	≈ 90	90-94 (92*)	83.8-87.5** (85.7*)	
Effluent NH_3-N (mg/L)	0.42	0.34	0.60	
Avg. influent TSS (mg/L)	368	368	368	35 mg/L ++
TSS removal (%)	60.3	69-97 (83*)	90.4-93.2 (91.8*)	
Effluent TSS (mg/L)	146.1	62.56	30.18	
Avg. inlet PO_4	2.93	2.93	2.93	3 mg/L+
PO_4 Removal	15-50 (37.5*)	80	38.8-61.3 (50.1*)	
Effluent PO_4	1.83	0.59	1.46	
References	INDITEX (2015)	USEPA (2000); Hadisoebroto *et al.* (2014) Lian-Feng *et al.* (2011)	Gikas and Tsihrintzis (2014)	+Iraqi Environmental Standard (2011) ++(EPA) (2003)

* Average Value
** Removal efficiency for NH_4-N.

3.2.2 Second Scenario

In this scenario, EMWW treated directly by aerated lagoons, oxidation ditch and wetland methods. Table 6 illustrated performance calculations and calculations of the mentioned methods. In general, wetland technology offered better removal efficiencies than aerated lagoons and oxidation ditch techniques. Treated EMWW remain within the standards for WW disposal, except TSS which was a slightly higher than the standards (35 mg/L) (EPA, 2003; WHO, 2006; Iraqi Environmental Standard, 2011). First scenario showed presenting better effluent quality than second scenario, Tables 4 to 6.

Table 6. Performance of aerated lagoons, oxidation ditch and wetland methods

Parameters	Treatment Methods			Effluent Standard
	Aerated lagoon	Oxidation ditch	Wetland	
Avg. influent BOD (mg/L)	6.3-304 (155.15*)	6.3-304 (155.15*)	6.3-304 (155.15*)	40 mg/L +
BOD removal (%)	70-95 (82.5*)	82-99 (90.5*)	90.8-92.3 (91.6*)	
Effluent BOD (mg/L)	27.15	14.74	13.03	

Avg. influent COD (mg/L)	12.2-901 (456.6*)	12.2-901 (456.6*)	12.2-901 (456.6*)	100 mg/L+
COD removal (%)	62.5	61.5	89-91.7 (90.4*)	
Effluent COD (mg/L)	171.22	175.79	43.83	
Avg. influent NH_3-N (mg/L)	0.004 -11.4 (5.7*)	0.004 -11.4 (5.7*)	0.004 -11.4 (5.7*)	1 mg/L ++
NH_3-N removal (%)	≈ 90	90-94 (92*)	83.8-87.5** (85.7*)	
Effluent NH_3-N (mg/L)	0.57	0.46	0.82	
Avg. Influent TSS (mg/L)	40-1800 (920*)	40-1800 (920*)	40-1800 (920*)	35 mg/L ++
TSS removal (%)	60.3	69-97 (83*)	90.4-93.2 (91.8*)	
Effluent TSS (mg/L)	365.24	156.4	75.44	
Avg. inlet PO_4 (mg/L)	0.0015-6.97 (3.45*)	0.0015-6.97 (3.45*)	0.0015-6.97 (3.45*)	3 mg/L+
PO_4 Removal (%)	15-50 (37.5*)	80	38.8-61.3 (50.1*)	
Effluent PO_4 (mg/L)	2.16	0.69	1.72	
References	INDITEX (2015)	USEPA (2000); Hadisoebroto *et al.* (2014) Lian-Feng *et al.* (2011)	Gikas and Tsihrintzis (2014)	+Iraqi Environmental Standard (2011) ++(EPA) (2003)

* Average Value
** Removal efficiency for NH_4-N.

3.3 Reusing

In the current study and based on the pH, EC, TDS, and SAR results, degree on restriction on use for EMWW is slight to moderate (WHO, 2006; Aziz *et al.*, 2019). According to EC, T. salts, and Na% values, EMWW is good to injurious type (Amin and Aziz, 2005). In other classification and regarding to T. salts, Cl, SAR, and alkalinity % figures, EMWW is considered as intermediate for certain crops (Amin and Aziz, 2005). Researchers reported that WW in Erbil City was not safe for all kinds of irrigation and they reported that EMWW is suitable for irrigating green areas and for cooked vegetables (Amin and Aziz, 2005). Authors stated that WW in Erbil cannot be used for irrigation directly (Aziz *et al.*, 2019). The achieved results are agreeing with the published data (Aziz and Amin, 2005; Aziz *et al.*, 2019). Of course, treatment of EMWW using various systems decreases contaminants such as organic matter, suspended solids, nitrogen compounds etc. in the EMWW (INDITEX, 2015; USEPA, 2000; Hadisoebroto *et al.*, 2014; Gikas and Tsihrintzis, 2014). As a result, treatment of WW enhancing the quality of EMWW for the irrigation purpose. Disinfection is necessary to overcome biological problems. Based on oil and grease, PO_4, Mg, and Ca values for EMWW, reclaimed EMWW can be used for irrigation (WHO, 2005). Regarding to BOD_5 and TSS values of treaded EMWW, it can be used for irrigation of ornamental fruit trees and fodder crops (WHO, 2006). According to the values of BOD_5, COD, TSS, DO, pH, turbidity, and NO_3-N for treated EMWW, it can be used for cooked vegetables, parking areas, playgrounds and side of roads inside cities (WHO, 2006).

4. Conclusions

EMWW generally considered as weak/low to medium WW type, it needs treatment prior to disposal to the natural environment, and it is classified as good to injurious type irrigation water. Although, time did not have a big effect on EMWW characteristics; life style, season, combined or separate sewerage system, climate condition, and areas/zones had effected the quality of the MWWs. Treatment of EMWW using both primary units and wetland led to removal efficiency of 94.75 %, 93.07 %, 89.47 %, 96.72 %, and 57.68 % for BOD_5, COD, NH_3-N, TSS and PO_4, respectively which, resulted the disposed effluents characteristics to be within WW standards. Hence, treated EMWW can be used for cooked vegetables and watering green areas.

References

[1] Ali, L. A. (2002). *Algal studies in sewage water within Erbil city*. M.Sc. Thesis, College of Education, Salahaddin University-Erbil.

[2] Al-Zboon, K., & Radaideh, J. (2012). Chapter Title: *MUNICIPAL WASTEWATER TREATMENT*. Edited by Hung, Y. –T, Wang, L. K., & Shammas, N., Book Title: *Handbook of Environmental and Waste Management*. World Scientific Publishing Co, Singapore.

[3] Amin, K. N., & Aziz, S. Q. (2005). Feasibility of Erbil wastewater reuse for irrigation. *Zanco Journal of Pure and Applied Sciences, 17*, 63–77.

[4] Amine, J. K. M., & Shekha, Y. A. (2016). Multivariate statistical characterization of water quality analysis for Erbil wastewater channel. *ZANCO Journal of Pure and Applied Sciences, 28 (3)*, 142-151.

[5] Asano T., & Tchobanoglous G. (1987). *Municipal wastewater treatment and effluent utilization for irrigation*. Paper prepared for the Land and Water Development Division. FAO, Rome. http://www.fao.org/3/t0551e/t0551e05.htm [Accessed: 30 May 2020].

[6] Aziz, F. H., Darokha, S. N., & Shekha,Y.A. (2001). Ecological and microbiological studies of Erbil city's sewerage. *Scientific conference of water- Erbil. Brayeti-Center. No. 18.*

[7] Aziz, S. Q. (2004). Seasonal variation of some physical and chemical properties of water and wastewater in Erbil City. *Journal of Dohuk University 7*, 76–88.

[8] Aziz, S. Q., Saleh, S. M., & Omar, I. A. (2019). Essential Treatment Processes for Industrial Wastewaters and Reusing for Irrigation. *ZANCO Journal of Pure and Applied Sciences*, 31(s3), 269-275.

[9] Aziz, S.Q., & Ali, S. M. (2017). Characterization of Municipal and Dairy Wastewaters with 30 Quality Parameters and Potential Wastewater Treatment by Biological Trickling Filters. *International Journal of Green Energy, 14(13)*, 1156-1162.

[10] Aziz, S.Q., & Ali, S. M. (2018). Characteristics and potential treatment technologies for different kinds of wastewaters. *Zanco Journal of Pure and Applied Sciences, 30 (S1)*, s122-s134.

[11] Aziz, S.Q., Aziz, H.A., Yusoff, M.S., & Mohajeri, S. (2011). Removal of phenols and other pollutants from different landfill leachates using powdered activated carbon supplemented SBR technology. *Environmental Monitoring and Assessment, 184(10)*, 6147- 6158.

[12] Bapeer, U. H. K. (2010). A study on the main sewage channel in Erbil city Destiny and its matching to irrigation purpose. *Journal of Kirkuk University–Scientific Studies, 5(2)*, 61-75.

[13] Choi, Y. –Y., Baek, S.-R, Kim, J.-I, Choi, J. -W, Hur, J., Lee, T. -U, Park, C. –J, & Lee, B. J. (2017). Characteristics and Biodegradability of Wastewater Organic Matter in Municipal

Wastewater Treatment Plants Collecting Domestic Wastewater and Industrial Discharge. *Water, 9,* 1-12.

[14] Directorate of Erbil Irrigation (DEI) (2013). *General Directorate of Dams and Water Resources, Ministry of Agricultural and Water Resources*, Kurdistan Region, Iraq.

[15] Directorate of Erbil Irrigation (DEI) (2020). *General Directorate of Dams and Water Resources, Ministry of Agricultural and Water Resources*, Kurdistan Region, Iraq.

[16] Environment Protection Agency (EPA) (2003). *Standards for effluent discharge, Regulations*. General Notice No.44. The Environmental Protection Act 2002. Regulations made by the Minster under section 39 and 96 of the Environmental Protection Act 2002.

[17] Ganjo, D. G. A, Aziz, F. H., & Shekha, Y. A. (2006). An attempt for reuse of the wastewater of Erbil City for irrigation purposes. Erbil, Kurdistan Region of Iraq.I. Typha angustifolia L. macrophyte as a biological waste removal from wastewater. *Zanco Journal of Pure and Applied Sciences, 18(2)*, 148- 175

[18] General Directorate of Irrigation and Water Resources (GDIWR) (2001). *Ministry of Agriculture and Water Resources*, Kurdistan Region-Iraq.

[19] General Directorate of Water and Sewerage (GDWS) (2020) *Ministry of Municipality and Tourism,* Kurdistan Region-Iraq.

[20] Gikas, G. D., & Tsihrintzis, V. A. (2014). Municipal wastewater treatment using constructed wetlands. *Water Utility Journal 8*: 57-65.

[21] Hadisoebroto, R., Kamil, I. M., Notodarmojo, S., & Bindar, Y. (2014). Improving Performance of Water Treatment on Oxidation Ditch Using Modification of Reactor Hydrodynamic. *Jurnal Teknologi*, DOI: 10.11113/jt. v69.3249.

[22] Hama Saeed, M. A., Lak, M. H. H., & Toma, J. J. (2010). Environmental and Biological study of Arab-Kand wastewater channel in Erbil Governorate Kurdistan Region–Iraq. *Tikrit Journal of Pure Science, 15 (3)*, Pp. 91-100.

[23] Henze, M. & Comeau, Y. (2008). Chapter Title: *Wastewater Characterization Biological*. Edited by Henze, M., M.C.M. van Loosdrecht, M. C. M., Ekama, G. A. & Brdjanovic, D., Book Title: *Wastewater Treatment: Principles, Modelling and Design*. ISBN: 9781843391883. Published by IWA Publishing, London, UK.

[24] INDITEX (2015). *Aerated Ponds (Fs-Bio-011) Series:* Secondary Treatments, Technology Fact Sheets for Effluent Treatment Plants on Textile Industry.

[25] Iraqi Environmental Standards (2011). Contract No.: W3QR-50-M074, Rev. No.: 03Oct. Morning Star for General Services, LLC Iraq, West Qurna I Project, EXHIBIT Eight.

[26] Japan Sewage Works Association (JSWA) (2013). *Design Standard for Municipal Wastewater Treatment Plants*. Second Edition, 2-10-12 Uchikanda, Tokyo, 1010047, Japan.

[27] Jasim, N. A. (2020). The design for wastewater treatment plant (WWTP) with GPS X modelling. *Cogent Engineering, 7*,1-33. https://doi.org/10.1080/23311916.2020.1723782.

[28] Lian-Feng, S., Rong-ping, C., Fa-wen, Z, Jian-tao, Y., & You, L. (2011). Discussion on Nitrogen and Phosphorus Removal Process Characteristics of Improved Oxidation Ditch. *Procedia Environmental Sciences, 11*, 1291-1295.

[29] Metcalf, E., & Eddy, E. (2014). *Wastewater engineering: treatment and reuse*. McGrawHill. Inc., London.

[30] Mustafa, B. Y. & Sabir, S. (2001). Reuse of Erbil City Sewage for Irrigation Purposes, *Brayeti Center Erbil -Iraq, 4th year, 13.*

[31] Sabeen, A. H., Ngadi, N., Noor, Z. Z., Raheem, A. B., Agouillal, F., Mohammed, A. A., & Abdulkarim, B. I. (2018). Characteristics of the Effluent Wastewater in Sewage Treatment Plants of Malaysian Urban Areas. *CHEMICAL ENGINEERING TRANSACTIONS, 63,* 691-696.

[32] Shekha, Y. A. (1994). *An Ecological Study on the Main Sewage Channel of Erbil city*. M. Sc. Thesis, College of Science, Salahaddin University-Erbil, Iraq.

[33] Shekha, Y. A. (2008). *The effect of Erbil city wastewater discharge on water quality of Greater Zab river, and the risks of irrigation*. PhD Thesis, Baghdad University, Iraq.

[34] Shekha, Y. A. (2013). Multivariate statistical characterization of water quality analysis for Erbil wastewater channel. IOSR Journal of Environmental Science. *Toxicology and Food Technology (IOSR-JESTFT), 5(6),* 18-26.

[35] Shekha, Y. A., Haydar, N. A., & Al-Barzingy, Y. O. M. (2010). The effect of wastewater disposal on the water quality and phytoplankton in Erbil wastewater channel. *Baghdad Science Journal, 7 (2),* 984-993.

[36] Shekha, Y. A., Toma, J. J., & Al- Barzingy, Y. O. M. (2016). Algal Survey in Wastewater Channel of Erbil City, Iraq. *Diyala Journal for Pure Sciences, 12(4),* 39-57.

[37] Sikosana, M. L., Sikhwivhilu, K., Moutloali, R., & Madyira, D. M. (2019). Municipal wastewater treatment technologies: A review. *2nd International Conference on Sustainable Materials Processing and Manufacturing (SMPM 2019), Procedia Manufacturing 35 (2019)*, 1018–1024.

[38] Sperling, M. V. (2007). *Wastewater Characteristics, Treatment and disposal.* Volume 1, IWA Publishing. Alliance House, 12 Caxton Street, London SW1H 0QS, UK.

[39] Teleman, D., Wilderer , P.A., Teodosiu, C., & Kötzle, T. (2004). Evaluation of Physical-Chemical Treatment Alternatives to Increase Removal Efficiency of Primary Wastewater Treatment. *Environmental Engineering and Management Journal, 3 (3)*, 265-274.

[40] Uinted States Environmental Protection Agency (USEPA) (2011). *Principles of Design and Operations of Wastewater Treatment Pond systems for Plant Operators, Engineers, and Managers.* August 2011 | www.epa.gov /nrmrlSystems.

[41] United States Environmental Protection Agency (USEPA) (2000). *Wastewater Technology Fact Sheet Oxidation Ditches.* Office of Water Washington, D.C., EPA 832-F-00-013.

[42] Wijaya, I. M. W., & Soedjono, E. S. (2018). Physicochemical Characteristic of Municipal Wastewater in Tropical Area: Case Study of Surabaya City, Indonesia. *IOP Conference Series: Earth and Environmental Science 135 (2018)*, doi :10.1088/1755-1315/135/1/012018.

[43] World Health Organization (WHO) (2005). *A regional overview of wastewater management and reuse in the Eastern Mediterranean Region.* Regional Office for the Eastern Mediterranean, Regional Centre for Environmental Health Activities, CEHA.

[44] World Health Organization (WHO) (2006). *A compendium of standards for wastewater reuse in the Eastern Mediterranean Region.* Regional Office for the Eastern Mediterranean, Regional Centre for Environmental Health Activities CEHA.

An Operational Study and Modeling of a Reverse Osmosis Desalination System

Marwa Asad Salih [1,*], Asmaa H. Dhiaa [2] and Zainab Hasan Abdulabbas [3]

[1]Department of Chemical Engineering, Faculty of Engineering, University of Kufa, Najaf , Iraq - E-mail: marwa.salih@uokufa.edu.iq
[2]Department of Chemical Engineering, Faculty of Engineering, University of Kufa, Najaf , Iraq - E-mail: asmaah.alhusseini@uokufa.edu.iq
[3]Department of Civil Engineering, Faculty of Engineering, University of Kufa, Najaf, Iraq - E-mail: zainab.alhasnawi@uokufa.edu.iq
* Corresponding author: Chemical Engineering, University of Kufa, Najaf, Iraq.
Tel: +9647709751899.

Abstract. The desalination of brackish water is one of the common issues in the world and spiral wound membrane is mostly used in desalination systems in home and industrial level, in this study a spiral wound brackish water membrane element were used to study the reverse osmosis system practically with the effect of multiple operational parameters and by creating Matlab Simulink modeling to analyze the RO system performance depending on concentration polarization theory then compare both results. The parameters that studied were flow rate range (0.6, 1, 1.3, 1.6, and 2 cm^3/s), concentration (100, 200, 300, 400 and 500 ppm), and temperature (15, 25 and 35 °C), of the feed and mass transfer coefficient for the permeate water, the salts that used ($CaCl_2$ and KCl) and by the concentration polarization (a phenomenon of accumulation of solute) the salts will tangle up at the surface of the membrane and affect membrane efficiency. The results indicated that the membrane operation is response sensitively to variations in feed temperature. The modeling results indicated that the major effect on the membrane parameters is the feed temperature and concentration of the salt.

Keywords: *Desalination; RO Simulation; Reverse Osmosis; mass transfer coefficient; Spiral Wound Membrane.*

1. Introduction

Water covers about 70% of earth's surface and it's a vital element for all living beings. Despite that the major part of earth is covered by water, still there is a huge shortage of drinking water across the world in many countries. Desalination of sea water considers as one of the solutions of this problem. Applying desalination methods to seawater and brackish water can be depended as successful alternatives in many situations, desalination is a technology where extra salts removed from brackish and seawater and it categorized into two processes thermal and membrane. (Thimmaraju et al., 2018). Water quality is usually improved with the use of reverse osmosis systems RO. This method was used to treat water on a large scale to convert salt water or sea water into drinking water, and is also used in cleaning sewage, and in industrial processes to recover salts and reuse them. These systems for water purification have become an essential part of many homes and are common in the market as well, where home owners always want to ensure that there are no chemicals to disturb the quality of drinking water (taste, color and smell) and pollutants that may affect their health. RO system is working to removes several inorganic impurities from drinking water, its performance depends on feed water, temperature, feed flow, and, pressure in addition to the type of the membrane it's differ in capacity and vary in their impurities rejection characteristics. (KNEEN et al., 2005). The polymer membrane affected in a sensitive way to any changes in the feed temperature. Its noticed that when feed temperature raised from 20^0C to 40^0C the permeate flux goes up to a 60 %, while it degreased when decreasing the feed flow rate from 17.8 to 9.0 L/min this happened under same temperature. Enhancing solute reinforcement at the surface of the membrane will increase the resistance to flux. There is a linear relationship between flux and pressure. A Spiegler-Kedem film theory model results match the experimental data that suggest the spiral wound type membrane is affected by smaller differences amount the feed flow rate and feed temperature. The modeling experiments showed that the salt concentration of the feed and the temperature are affect the mass transfer coefficient. (Goosen et al., 2004, cited by Santos et al., 2016).

RO model were built using Matlab and Thermolib to perform a numerical analysis for brackish water desalination production for several purposes found that all salts rejection could lowered by 9% to get high quantities water suitable for agricultural use, and physically by increasing the average pore size will reduce power consumption of the system (Sarai Atab el al., 2016). The comparison between model and practical results showed that these parameters matches with the practical data with comparatively high exactness. It found that at the temperature range from 0^0C to 60^0C the RO system operate in high stability (Jiang et al., 2014). A semi-rigorous mathematical model used to simulate an industrial brackish water RO plant, comparing the real data with model results showed a high accuracy in prediction the steady-state behavior of the RO were the general recovery value of water is 0.37 and rejection is 1.33%, a deterioration of the quality of the product noticed when increasing the pressure further than an assured value, water production could decreased if there a trans-membrane pressure difference reduced due to a drop in high frictional pressure. (Abbas, 2005, cited by Alsarayreh et al., 2020).

A drop in the permeate quality and a failing in product quality notice due to high flow feed rates and high operating pressures respectively (Abbas and Al-Bastaki, 2005, cited by Alsarayreh et al., 2020). Concentration polarization can be abandoned at higher feed flow rate. Increasing feed flow rate caused a decrease in the permeate concentration. the water flux with non-linear behavior increased by increasing operating pressure (Kaghazchi et al., 2010). It found that raising in the concentration level at the wall of the membrane caused by increasing module area, feed concentration, and feed pressure. Increasing feed flow rate and temperature would decrease concentration value of membrane wall (Kotb et al., 2015). The feed flow rate and feed pressure positively affected the salinity of permeate water (Al-Obaidi et al., 2018). It confirmed that the RO modeling showed a lower deviation after comparing its predicted data with the real data. In same subject, many simulation studies working to make assessment and prediction to the effect of variation in operating conditions of the RO system on the process indicators (Alsarayreh et al.,2020).

2. Material and methods

2.1 Experimental Process

The spiral wound membrane is made of flat membrane sheets connected from all sides except one to connect the Collector channel materials permeate to be like a sheet, typically the polymer spiral wound module containing between 26 and 28 layers with about 40 m^2 of active membrane area and it is 9 cm in diameter and 15 cm in length.

Brine feed solutions were prepared in the glass vessels and fed to the Ro system by gravity of 9 m height (88.2 Kpa) to gives uniform feed of flow, it passes through three filters (5 microns) to remove

IOP Conf. Series: Materials Science and Engineering **978** (2020) 012045 doi:10.1088/1757-899X/978/1/012045

macromolecules and suspended solids then to the membrane by a high-pressure pump. The experiment was carried out for (KCl and $CaCl_2$) individually with different concentrations (100, 200, 300, 400 and 500ppm) to get brine feed, flow rate was varied (0.6 - 2 cm^3/s) and for feed temperature (15 - 35 °C).

2.2 Modeling Process

Solution diffusion model were used to describe our reverse osmosis mathematically. It assumes that the driving force of permeation is the gradient in chemical potential of the solute (Wijmans and Baker, 1995). the terms of solvent flux (J) were used to express the transport equation as below:

$$\mathrm{Jv} = \mathrm{A}\,(\Delta \mathrm{P} - \sigma \Delta \pi) \qquad (1)$$

Aw: permeability of solvent (m s^{-1} Pa^{-1})

ΔP: pressure (Pa).

Δπ: difference in osmotic pressure (permeate - membrane surface) (Pa).

σ: reflection coefficient.

Most of recent reverse osmosis and Nano filtration separations adopt equation (1) at the beginning of the design, it describes the behavior of the membrane where a layer of local concentrated (CP) will be created by high salt concentration at or close to the surface of membrane, this elevated salt concentration is a result of the rejection of ionic species, this layer shortly reaches a steady state, and the transverse solute flux through the CP layer is constant (Eric et al., 2002, Jeffrey et al., 2007).

The solvent flux (J) will be calculated by the below equation with steady state mass balance through CP:

$$Jv.Cp = Jv.C - \langle D\frac{dc}{dx}\rangle \qquad (2)$$

Jv: permeate flux across the membrane L/m^2 h

Cp: concentration of permeate solute (kg/m^3).

C: concentration of solute at the boundary layer (kg/m^3).

D: coefficient of solute diffusion in water (m^2/s) (Dharmesh, 2002).

Mixing the finite mass boundary layer thickness (δ), out to the convection diffusion mass balance, produces the relationship between CP and permeate flux:

$$CP = \frac{Cm - Cp}{Cb - Cp} = \exp\langle\frac{Jv}{K}\rangle \qquad (3)$$

Cm: concentration at the membrane surface of the rejected salt (kg/m^3).

Cb: solute concentration (bulk) (kg/m^3).

K: Mass transfer coefficient (m s^{-1})

Jv: permeate water flux (L/m^2 h) (Kim, S. Eric M.V. Hoek, 2005).

2.3 Simulation procedure

The Matlab/Simulink blocks software version R2017b (9.3.0.731579) were used to build a numerical model to analyze the design and examine the performance of any reverse osmosis system, this model designed based on theory of solution diffusion and the results obtained from the experimental work.

The properties and mass transfer coefficient compared with the real data obtained from the experimental work, since the designed model affected by feed temperature and flow rate obviously.

IOP Conf. Series: Materials Science and Engineering **978** (2020) 012045 doi:10.1088/1757-899X/978/1/012045

3. Result and Discussion

3.1 Effect of feed flow rate on Mass transfer coefficient

Mass transfer coefficient increases with the increase in feed Flow Rate as shown in figures (1 and 2) for calcium chloride and potassium chloride respectively for comparison between both results experimental and modeling.

The figures show that increasing in feed Flow Rate leads to an increase in driving pressure for fluid flow and increase in turbulent flow respectively. This would prevent or decrease the concentration buildup in the solution at the surface membrane. In other words, the thickness of the concentration boundary solution will be less while the diffusivity of solute stays constant. Therefore, the mass transfer coefficient increases steadily relatively to the increase in the Feed Flow Rate.

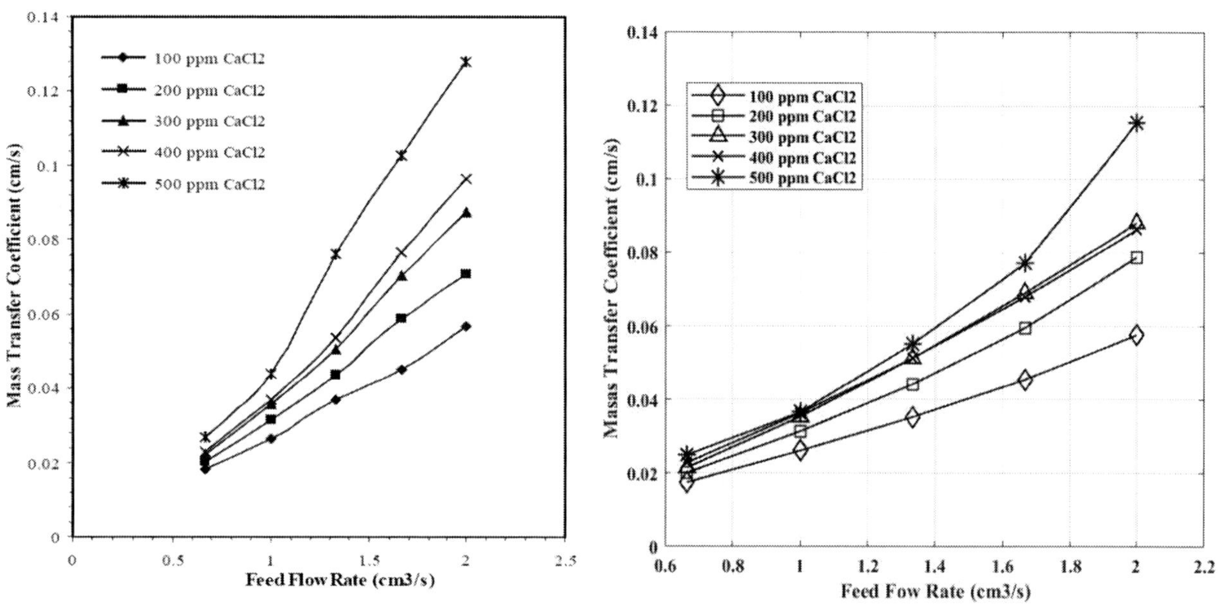

(a) Real data (b) Modeling data

Figure 1. Effect of feed Flow Rate on K for $CaCl_2$

IOP Conf. Series: Materials Science and Engineering **978** (2020) 012045 doi:10.1088/1757-899X/978/1/012045

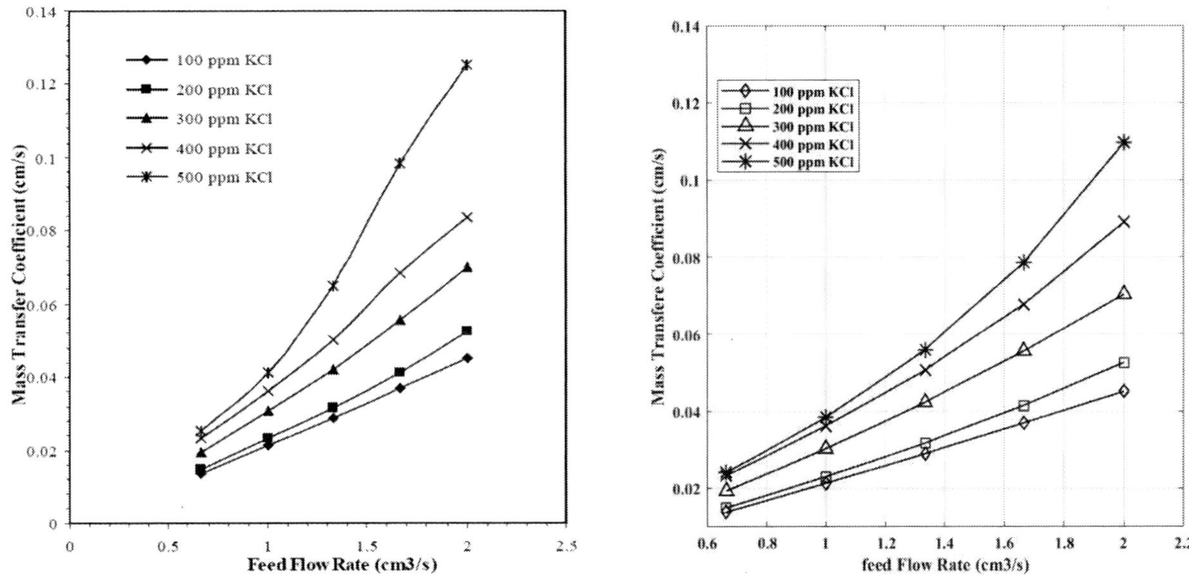

(a) Real data (b) Modeling data

Figure 2. Effect of feed Flow Rate on K for KCl

3.2 Effect of feed temperature on Mass transfer coefficient

By increasing the feed temperature, the rate of water permeation increase through the membrane and then we obtained higher diffusion rate of water through the membrane due to the reduction in solution viscosity.

The real data show that the spiral wound polymer membrane is reflecting high sensitivity to feed temperature. The membrane productivity is very sensitive to deviation in feed water temperature. As water temperature increases, water flux linearly increases, mainly this because high level of diffusion rate of water through the membrane. Therefore, (K) increases with the increase in feed temperature as shows in figures (3 and 4) for calcium chloride and potassium chloride real and model respectively, while figure (5) illustrate the difference between the effects of feed flow rate on different salts with same concentration.

IOP Conf. Series: Materials Science and Engineering **978** (2020) 012045 doi:10.1088/1757-899X/978/1/012045

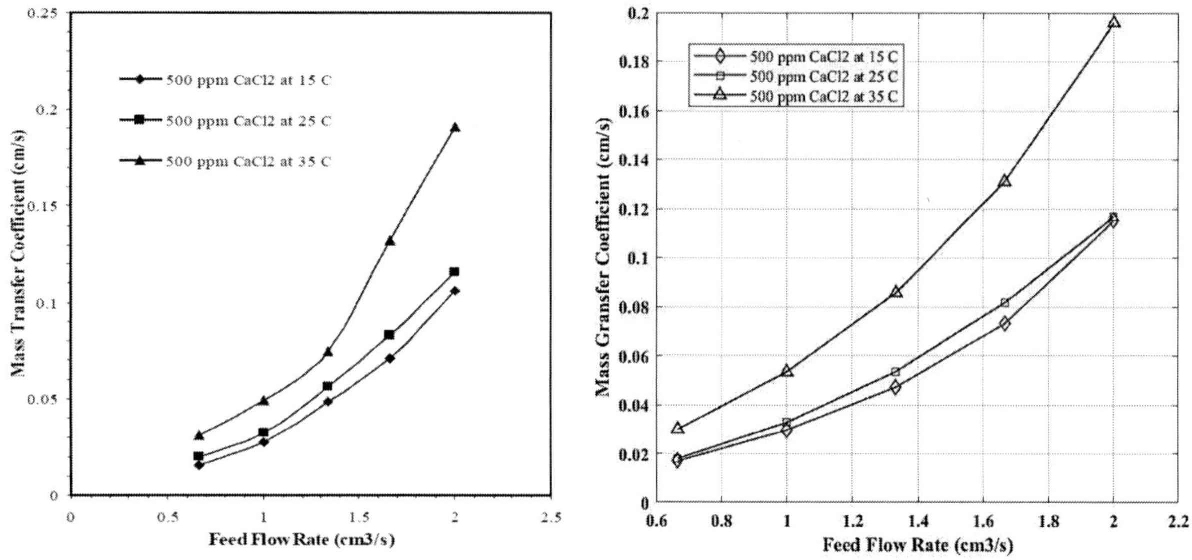

(a) Real data (b) Modeling data

Figure 3. Effect of feed Temperature on Mass Transfer Coefficient $CaCl_2$

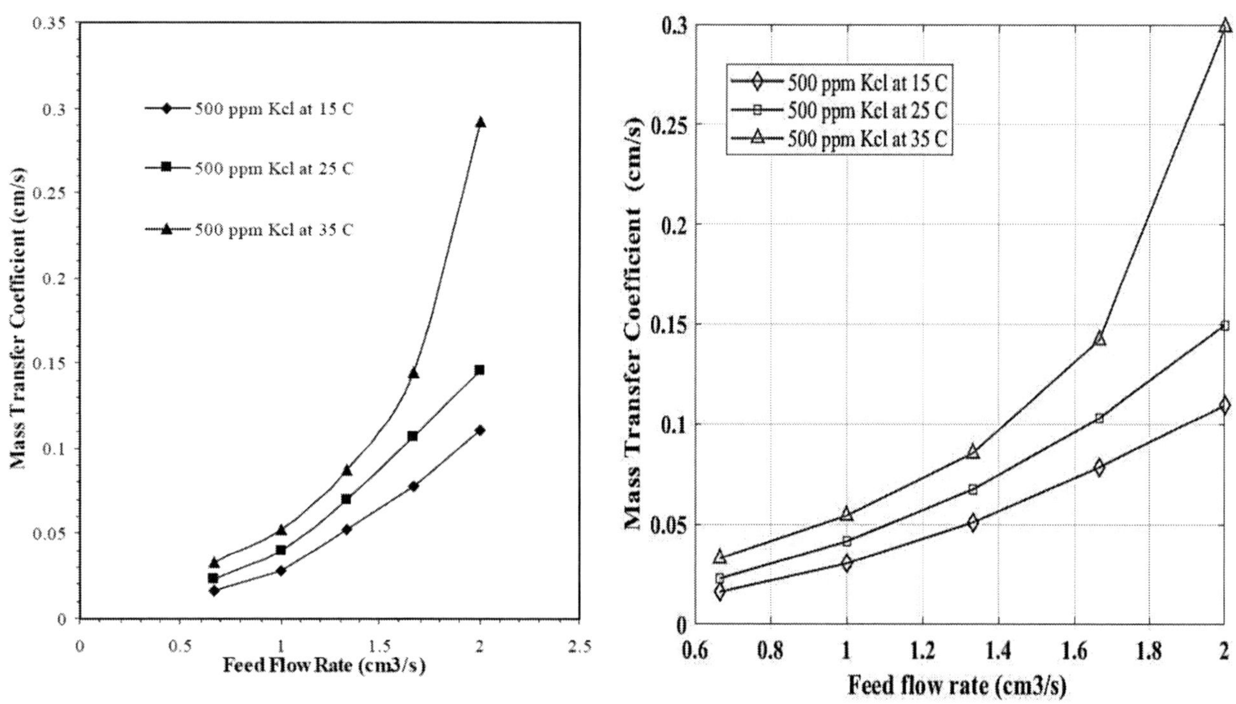

Real data (b) Modeling data

Figure 4. Effect of feed Temperature on Mass Transfer Coefficient KCl

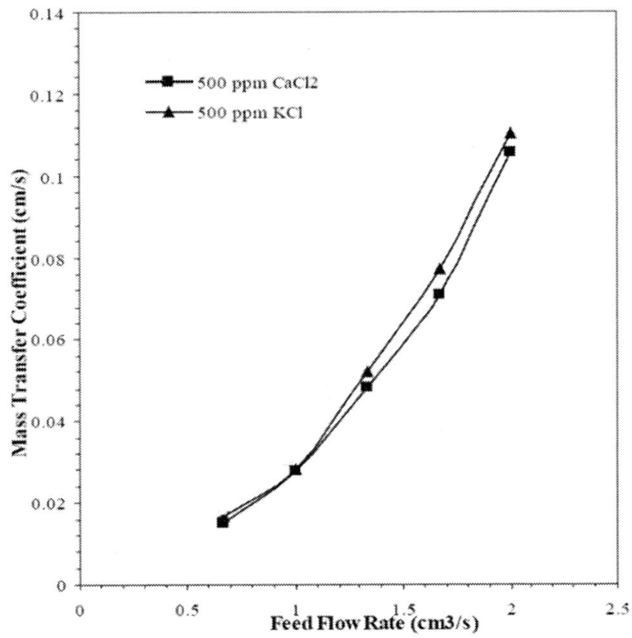

Figure 5. Effect of Feed Concentration on K for Different Salts.

4. Conclusions

After making this comparison between the results showed that, the calculated values were consistent with experimental data. Evaluation of the operational data with model results presented that the model predicts the steady-state performance of the plant with respectable accuracy. Increasing the operating feed flow rate will reduce concentration polarization (CP) value and decreases the permeate concentration, the experiment data shows that when temperature of feed water is increased from 15 to 35 °C, the permeate salinity ppm TDS increases from 70 - 115 ppm TDS. Mass transfer coefficient increase with the increase in feed Flow Rate and feed temperature. For different salt feed is recognized that the selectivity of reverse osmosis is not to be based on a separating mechanism.

References

[1] Thimmaraju,M., Sreepada,D., Babu,G.S., Dasari,B.K, Velpula,S.K, & Vallepu,N.(2018). Desalination of Water. 333-347.

[2] KNEEN, B., LEMLEY, A., & WAGENET, L. (2005). Reverse Osmosis treatment of Drinking Water. (November 2005). https://docplayer.net/1739733-Water-treatment-notes-cornell-cooperative-extension-college-of-human-ecology-reverse-osmosis-treatment-of-drinking-water.html

[3] Goosen, M.F.A., Sablani, S.S., Al-Maskari, S.S., Al-Belushi, R.H. (2004). Effect of feed temperature and flow rate on permeate flux in spiral wound reverse osmosis systems. Second LACCEI International Latin American and Caribbean Conference for Engineering and Technology (LACCEI'2004), Miami, USA).(B Santos, JG Crespo, MA Santos, S Velizarov, (2016). Oil refinery hazardous effluents minimization by membrane filtration: An on-site pilot plant study. Journal of Environmental Management).

[4] Sarai Atab, M., Smallbone, A.J., Roskilly, A.P. (2016). An operational and economic study of a reverse osmosis desalination
system for potable water and land irrigation. Desalination (174-184).

[5] Jiang A., Ding Q., Wang J., Jiangzhou S., Cheng W., Xing C., (2014). Mathematical modeling and simulation of SWRO process based on simultaneous method. Journal of Applied Mathematics.

[6] Abbas, A. (2005). Simulation and analysis of an industrial water desalination plant. Chemical Engineering and Processing. (44, 999–1004). (Alsarayreh A. A., Al-Obaidi M.A., Patel R., & Mujtaba I.M. (2020). Scope and Limitations of Modelling, Simulation, and Optimisation of a SpiralWound Reverse Osmosis Process-BasedWater Desalination. MDPI. AG.

[7] Abbas, A.; Al-Bastaki, N. (2005). Modeling of an RO water desalination unit using neural networks. Chemical Engineering and Processing. (114, 139–143). (Alsarayreh A. A., Al-Obaidi M.A., Patel R., & Mujtaba I.M. (2020). Scope and Limitations of Modelling, Simulation, and Optimisation of a SpiralWound Reverse Osmosis Process-BasedWater Desalination. MDPI. AG.)

[8] Kaghazchi, T., Mehri, M., Ravanchi, M.T., & Kargari, A. (2010) A mathematical modeling of two industrial seawater desalination plants in the Persian Gulf region. (Desalination 2010, 252, 135–142).

[9] Kotb, H., Amer, E., & Ibrahim, K. (2015). Effect of operating conditions on salt concentration at the wall of RO membrane. (Desalination 2015, 357, 246–258).

[10] Al-Obaidi, M.A., Alsarayreh, A.A., Al-Hroub, A.M., Alsadaie, S., & Mujtaba, I.M. (2018). Performance analysis of a medium-sized industrial reverse osmosis brackish water desalination plant. (Desalination 2018, 443, 272–284).

[11] Wijmans, J.G. & Baker, R.W. (1995). The Solution-Diffusion Model. (107(1-2) (1-21).

[12] Eric M.V. Hoek, Albert S. Kim, & Menachem Elimelech. (2002) Environmental Engineering Science 19(2002) 6.

[13] Jeffrey, R. McCutcheon, & Menachem Elimelech. (2007) American Institute of Chemical Engineers, 53, July (2007)7.

[14] Dharmesh S. Bhanushali. (2002). Solvent- Resistant Nanofiltaration Membranes Separation Studies and Modeling. Ph.D.Thesis, University of Kentucky, (2002).

[15] Kim, S. Eric M.V. Hoek. (2005). Modeling concentration polarization in reverse osmosis processes. Desalination 186 (2005) 111 128.

IOP Conf. Series: Materials Science and Engineering **978** (2020) 012046 doi:10.1088/1757-899X/978/1/012046

Human Facial Aggressive Detection System Based on Facial-Width-to-Height Ratio

Tabreer T. Hasan[1, *] and Abbas H. Issa [2]

[1] Department of Electrical Engineering, University of Technology, Iraq
E-mail: tabo.tariq@gmail.com
[2] Department of Electrical Engineering, University of Technology, Iraq
E-mail: 30050@uotechnology.edu.iq
* Corresponding author: Department of Electrical Engineering, University of Technology, Iraq. Tel: +9647715273477.

Abstract. Recent researches have shown closely related evidence between the human individual social behavior and precisely measurable facial features. The Facial-width-to-height ratio (FWHR) has become quite an interesting topic concerning human aggressive behavior. Recent studies presented evidence showing that the precise measurement of FWHR can be used to predict human aggressive behavior based on facial landmark extraction. In this paper, the Facial-width-to-height ratio is extracted and analyzed among men, women, and children using the recently presented Convolutional Experts Constrained Local Model (CE-CLM). Then, extracted features are used to train the Numeral Virtual Generalizing Random Access Memory (NVG-RAM) pattern recognition technique. The results show promising clues in depending on this feature extraction method for the Facial-width-to-height ratio, and depending on SVG-RAM classifier for aggressive behavior. Moreover, the proposed method is less susceptible to facial rotation error ensuring accurate FWHR extraction.

Keywords: *Facial expression; Aggression; Facial-width-to-height ratio (FWHR); facial landmark extraction; and Numeral Virtual Generalizing Random Access Memory (NVG-RAM).*

1. Introduction

Social interaction depends mostly on human facial expression. Human emotions, ideas, and personality can be recognized from the face and its features. The highest informative channel for social communication is represented by nonverbal signs and it possesses 7% of the communication message (Mehrabian, 1968). While vocal voice retains 34% of communication, facial expression occupies 55% of the communication channel. Thus, the facial expression became a main subject for studying and analyzing for different sciences such as behavior, psychology, computer, and medicine for more than a century (Darwin & Darwin, 2009). Face recognition, feature extraction, and tracking went through intense advances in the last two decades. Such techniques become the major focus in many applications such as sleepy driver detection (Pinto et al., 2019) marketing (Spivak et al., 2019) surveillance (Ya et al., 2017) and more. Race, age, sex, and other important information can be extracted from a human face (Oladipo et al., 2019). Additionally, human personality, emotions, and behavior can be obtained by analyzing the human face. One particular characteristic that draws attention recently is the tendency to aggression. Aggressive people are people who may arise physical threats to the surrounding social community. Predicting and preventing such behavior helps to protect and survive public life before becoming higher risk to other people (Sell et al., 2009). Although anger expression provides a good signal of a serious threat, the facial structure provides valuable evidence about the tendency of aggression (Carré & McCormick, 2008). The dominant definition of aggression is "aggression is any form of behavior directed toward the goal of harming or injuring another living being who is motivated to avoid such treatment" (Fischer et al., 2013).

Aggression behavior is found in both animals and humans. Any creature might use aggression for maintaining its position, control, dominate, or compete for resources. In other words, aggression is related to the survival of individuals. It is interesting to study the mechanism behind aggression and its effect on social interaction in life, because individuals may suffer from negative influence in their social communications. Social sciences and especially psychology focused intensely on the human aggression field of study. Many studies have been done for quantifying aggression. Some evaluation depended on self-assessment questionnaires such as the Buss–Perry Aggression Questionnaire (BPAQ) (Buss & Perry, 1992). The questionnaire consists of four scales: physical aggression, verbal aggression, and hostility. Other forms of evaluation depend on facial expression such as facial-width-to-height ratio (FWHR). Although there have been many studies, to our knowledge, the work proposed in this paper is the first to explore using the NVG-RAM classifier on aggressive behavior classification. Our research attempts to find a relationship between the human facial to width ratio (FWHR) and Aggressive behavior. The FWHR landmark is used as data source to feed the classifier network for both training and testing.

2. Previous work

Previous works have been done trying to investigate the relationship between the FWHR and aggression. The studies included different social group individuals under different circumstances. Other studies distinguished the aggressive behavior between men and women with different age groups.

Early investigations on aggression depended on stimuli such as questions. In such studies, individuals were asked to answer questions related to their personality. Then, the results were scored and analyzed to understand the relationship between human aggression behaviors. The oldest and the most known questionnaire is (Buss and Perry aggression questionnaire (Buss & Perry, 1992). They constructed a group of questions with different ascended answers. Their results yielded four scales: Verbal Aggression, Physical Aggression, Hostility, and Anger. They have found that anger is the interconnection between hostility and both verbal and physical aggression. Their experiments included both men and women. Men scored high in verbal, physical, and hostility while there was no sex difference in Anger scoring.

Piątkowska & Martyna (Piątkowska & Martyna, 2012) tried to extract aggression from facial expressions by analyzing spontaneous facial expression. They depended on tracking specific points on face trajectories calculated by following human face movement. Then, they used Local Binary Patterns (LBP) and Gabor filter for extracting and analyzing facial features. Finally, they used the support vector machine (SVM) classifier for classifying spontaneous face features and obtained 85% accuracy.

Other studies tried to link between the aggressive questionnaire scaling and FWHR such as in (Özener, 2012). This study depended on studying both questionnaire aggression and FWHR on Turkish people only. They depended on a sample of 470 Turkish students from both male and female. The study concluded that there is no relationship between self-reported aggressive questionnaire and facial width to height ratio for either sex.

Carré et al. (Carré et al., 2009) exposed photographs of both neutral and aggressive men into a group of observers. The observers had a limited time for estimating the propensity of the photographs. The study suggested that the facial width to height ratio contains evidence that can be used in predicting the aggression of others.

Another research conducted by Taylor and Jose (Taylor & Jose, 2014) explored the influence of physical aggression on the identification of facial expression among healthy participants. The expressions of different faces are evaluated by the participants in addition to the measure of self-report aggression. The mistake decision ratio was high among high physically aggressive participants in identifying non-angry facial expressions. The differences explanations did not depend on response time or gender. The results extremely supported the idea that aggressive people express confrontational biases when identifying facial expressions.

Arriaga and Aguiar (Arriaga & Aguiar, 2019) focused on studying the gender difference of aggression depending on facial expression signs. The aggression is induced by a fictional opponent. Their results presented that men are more aggressive than women when they receive emotional cues of anger and sadness from the sender. On the other hand, both genders evaluated equally when neutral expressions were showed. Moreover, men expressed the same stage of aggression in all conditions, whereas women expressed higher aggressive behavior in no feedback and neutral conditions than in the anger or sad conditions. This research confirms the gender difference in aggression between men and women without studying FWHR.

Both Krenn and Meier (Krenn & Meier, 2018) studied the FWHR and aggression in football players in sports only. The aggression is a fair part of the game and most players depend on aggressive behavior while playing the game. The complete details and terms of the investigation are included in their paper. The study ensured that there is no relationship between sport and the FWHR researches with aggressive behavior. In other words, the investigation of aggression on football players and normal individuals yielded close results with social position effects.

Goetz et al. (Goetz et al., 2013) tried to find an association between FWHR aggressive study and both subjective and objective social status. The study reported that there is a positive correlation between FWHR and aggression among men students with low social status. Additionally, there was a positive correlation among hockey players who received low salaries. So, the study provided convincing evidence for the role of social status in controlling the relationship between aggression and facial structure.

Haselhuhn et al. (Haselhuhn et al., 2015) focused on the argument of depending on the facial width-to-height ratio (FWHR) on men for predicting aggressive behavior. The research investigated the success and failure of researchers on finding a reliable cue between aggression and FWHR. Their results showed that there is enough evidence to ensure that a connection does exist between men's aggression and FWHR.

Lefevre and Lewis (Lefevre & Lewis, 2014) studied the robustness of the FWHR connection with the aggressive behavior in both men and women. Moreover, the study compared the results with self-reported aggressive behavior. Their results showed that the faces of both females and males with higher facial masculinity and higher FWHR masculinity were observed as more aggressive. However, they did not find any systematic clues for the effects of self-assessed control on the perception of other individuals' aggression. Furthermore, their results supported the robustness of FWHR and facial expression as evidence of aggressive behavior.

3. Facial Width-to-Height Ratio and Aggression

Facial width to height ratio is a face feature determined by calculating the ratio between the distance from the mid-brow to the upper lip and bizygomatic width. The latter is calculated by measuring the maximum horizontal distance between the right facial boundary to the left facial boundary (Hehman et al., 2015; Weston et al., 2007). An example of facial width to height calculation is shown in Figure 1. (Ma et al., 2015b). The calculated FWHR for the shown neutral face is 1.9521 and for the aggressive face 2.0594. The FWHR can be calculated either from face images directly or from landmarks that have been extracted from the face image. Longer faces have smaller FWHR and wider faces have higher FWHR (Haselhuhn & Wong, 2012).

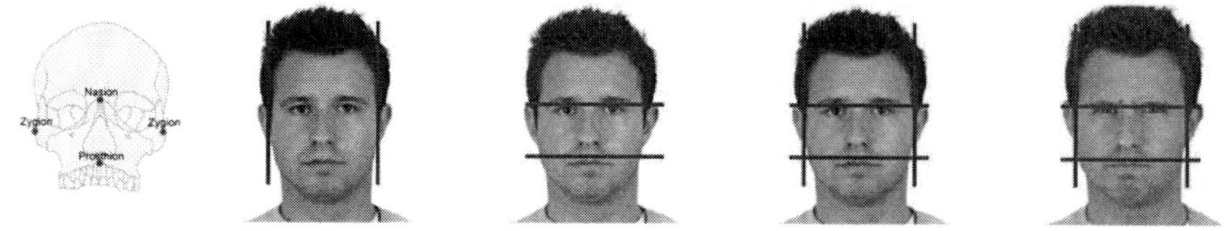

Figure 1. Facial Width to Height ratio calculation and face source conducted from (Ma et al., 2015a)

4. Methodology

This work depends on different hardware and software resources connected as shown in Figure 2. Most libraries become open source and can be implemented using different platforms or programming languages. Moreover, a slight modification of the currently available sources makes them possible to operate with different programming languages. This work consists of three phases: Landmark extraction phase, training phase, and classification phase which are executed and analyzed by Matlab R2020a software.

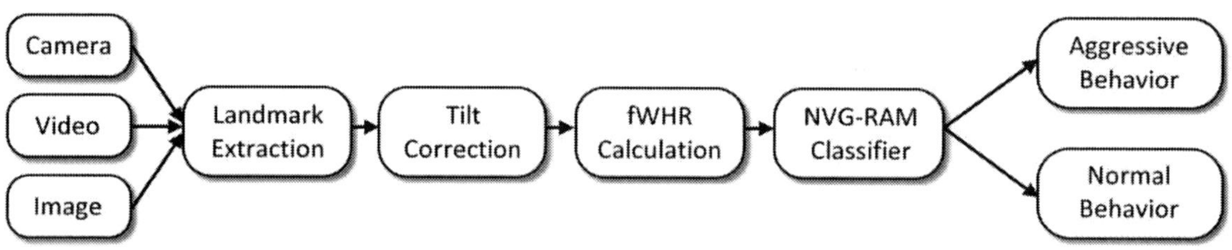

Figure 2. Proposed FWHR and Aggressive behavior estimation

4.1 Landmark Extraction

The OpenFace library is used for the extraction of face landmarks. The OpenFace tool is designed by developers for researchers who are interested in building applications for face recognition and analysis of facial behavior (Baltrusaitis et al., 2016). The OpenFace tool can extract landmarks from live video cameras, recorded videos, or captured images. During the first step, Matlab calls the OpenFace library and hands the source of media and the extraction method. The most accurate and robust landmark extraction approach is Convolutional Experts Constrained Local Model (CE-CLM) that depends on Convolutional Experts Network (CEN) as a local detector. This ensures a precise calculation of the facial width to height ratio FWHR from the input sources. The exact face features are fetched by Matlab for calculations and analysis. Since face tilt has a high effect on miscalculation of the FWHR, our code ensures to correct the tilt of every captured face. Finally, the FWHR is calculated as shown in Figure 3. The source image used for this example is from the public Chicago face database (Ma et al., 2015b).

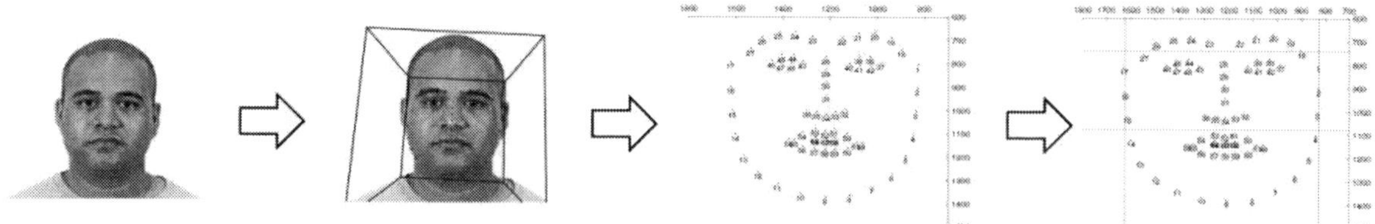

Figure 3. FWHR extraction from face landmarks

4.2 NVG-RAM

The NVG-RAM is a high efficient weightless neural network classifier that requires a onetime training only. Moreover, the classifier functions with decimal numbers instead of binary numbers as in VG-RAM. The training process is equivalent to storing the pairs of inputs and outputs in the virtually created RAM. Later, these pairs are called for classification. A high number of pairs results in improved accuracy for classification because of the higher efficiency training process (Hasan et al., 2018).

The classifier becomes ready after the training of the NVG-RAM is completed. The classification operation depends on the calculation of minimum Manhattan Distance (MD) between the entire recorded pairs and the unidentified input according to Equation (1). The class of the closest Manhattan distance is fetched as the output for classifier depending on its class number.

$$d(p,q) = \sum_{i=1}^{n} |p_i - q_i| \quad (1)$$

where n is the number of elements in each vector, and *d(p,q)* is the Manhattan distance between two vectors *q* and *p* (Hashim & Sadah, 2015). The NVG-RAM operation is explained in the next sections.

4.2.1Databases

The group of datasets consists of 82 images with minimum resolution 250x300 pixels (40 normal faces, 42 aggressive faces) which are collected from online medical sources identifying as aggressive or normal (Inc, n.d.; Malaga, 2020; Wilber, 2015). Then, each group of the database is divided into two halves. Examples from the used database are shown in Figure 4 below. The first half of each collection is used for training the NVG classifier and the other half is used for classification. The dataset images include different ages of men and women as well as images of children. However, the data is applied to classifier without taking into consideration the gender issue.

Figure 4. Aggressive man face samples from used database

4.2.2Training Phase

The training phase uses the Numeral Virtual Generalizing Random Access Memory (NVG-RAM) classifier for preparation and classification Dataset images serves as a source for both operations as explained below.

4.2.3Classification Phase

Two types of facial expressions (aggressive, nonaggressive) are required to be classified by the NVG-RAM for two different FWHR factors. There is not an exact number to quantify the relationship between the FWHR and aggressive behavior. Therefore, training the NVG-RAM with the collected face dataset helps to predict the closest form of behavior. After the training process, the NVG-RAM is tested on the training data to determine the 100% accuracy of the classification. The evaluation of the classification method is calculated in terms of specificity, sensitivity, accuracy, and testing time.

The sensitivity reflects the correctly identified proportion of the actual positives. It identifies humans with aggressive behavior positively depending on facial expression. The sensitivity can be calculated as in Equation (2) (Kuruvilla & Gunavathi, 2014).

$$Sensitivity = \frac{T_P}{T_P + F_N} \quad (2)$$

Where T_P (True Positive) is the number of samples that have aggressive behavior and identified properly, F_N (False Negative) is the number of samples that were aggressive but incorrectly identified as normal. Equation (3) measured the proportion of people who are correctly identified as normal, i.e. a condition where a set of normal humans is correctly identified as not having aggressive behavior (Panday & Godara, 2012).

$$Specificity = \frac{T_N}{T_N + F_P} \quad (3)$$

Where F_P (False Positive) is the number of samples that are normal but identified aggressive incorrectly. T_N (True Negative) is the number of samples that are normal and properly identified. In addition to sensitivity and specificity, the success rate can also be evaluated with Positive Prediction Value (PPV) which calculates how accurate a positive test result determines that the aggressive behavior is present as illustrated in Equation (4).

$$PPV = \frac{T_P}{T_P + F_P} \times 100 \quad (4)$$

Equation (5) measures the Negative Prediction Value (NPV), which calculates how well a negative test result determines that the aggressive behavior is absent (Olaniyi et al., 2015).

$$NPV = \frac{T_N}{T_N + F_N} \times 100 \quad (5)$$

The statistical evaluation of a classifier is called accuracy and it measures the precision of identification or exclusion of a condition. It is the proportion of true results (both true positive and true negative) in the population, it can obtained as in Equation (6) (Kuruvilla & Gunavathi, 2015).

$$Accuracy = \frac{(T_P + T_N)}{(T_P + T_N + F_N + F_P)} \quad (6)$$

5. Results and Analysis

The NVG-RAM performance results for the classification phase of the provided image database are shown in Table 1. These results were obtained from the second half of dataset which was selected for classification. The analyzed results showed that aggressive faces have FWHR ration mostly greater than 2.0. The maximum accuracy reached 87.8% with a sensitivity of 86.3% and classification time 0.026 seconds. The acquired number exceeds the maximum previously recorded number 85% which depended on facial expressions analysis and used the Support Vector Machine (SVM) classifier (Piątkowska & Martyna, 2012). The increased accuracy resulted from the correction of the face tilt and the usage of the NVG-RAM as classifier. None of the previous studies mentioned correcting the tilt error and its effect on the miscalculations of FWHR. Moreover, all previous studies deepened on other humans assessed or SVM classifier for the organization process.

Table 1. NVG-RAM Performance Results

Testing Data samples	TN	FN	Test time Sec	Specificity	Sensitivity	NPV%	PPV%	Accuracy%
	FP	TP						
41	17	3	0.0268	0.8947	0.863	85	90.4762	87.8049
	2	19						

6. Conclusions

The conducted research successfully confirmed that the human aggressive behavior can be predicted by relying on the facial expression as previous studies. However, this is the first reported study to show that the NVG-RAM can be used as facial landmark classifier for facial aggressive detection. Facial landmarks are extracted from aggressive and normal individuals and are used for calculating facial width to height ratio FWHR. The output results were obtained from the NVG-RAM classifier after training the classifier on a previously prepared dataset. The reported outcomes showed better accuracy from previously recorded result. The FWHR estimation accuracy is increased by correcting the tilting error produced from the incorrect angle of the human face. In the future, the proposed method will be tested on live video cameras for aggressive behavior prediction.

References

[1] Arriaga, P., & Aguiar, C. (2019). Gender differences in aggression: The role of displaying facial emotional cues in a competitive situation. *Scandinavian Journal of Psychology*, *60*(5), 421–429. https://doi.org/10.1111/sjop.12568

[2] Baltrusaitis, T., Robinson, P., & Morency, L. P. (2016). OpenFace: An open source facial behavior analysis toolkit. *2016 IEEE Winter Conference on Applications of Computer Vision, WACV 2016*. https://doi.org/10.1109/WACV.2016.7477553

[3] Buss, A. H., & Perry, M. (1992). The Aggression Questionnaire. *Journal of Personality and Social Psychology*, *63*(3), 452–459. https://doi.org/10.1037/0022-3514.63.3.452

[4] Carré, J. M., & McCormick, C. M. (2008). In your face: Facial metrics predict aggressive behaviour in the laboratory and in varsity and professional hockey players. *Proceedings of the Royal Society B: Biological Sciences*, *275*(1651), 2651–2656. https://doi.org/10.1098/rspb.2008.0873

[5] Carré, J. M., McCormick, C. M., & Mondloch, C. J. (2009). Facial structure is a reliable cue of aggressive behavior: Research report. *Psychological Science*, *20*(10), 1194–1198. https://doi.org/10.1111/j.1467-9280.2009.02423.x

[6] Darwin, C., & Darwin, F. (2009). The expression of the emotions in man and animals. In *The Expression of the Emotions in Man and Animals*. https://doi.org/10.1017/CBO9780511694110

[7] Fischer, P., Asal, K., & Krueger, J. I. (2013). *Aggression* (pp. 61–78). https://doi.org/10.1007/978-3-642-30272-5_5

[8] Goetz, S. M. M., Shattuck, K. S., Miller, R. M., Campbell, J. A., Lozoya, E., Weisfeld, G. E., & Carré, J. M. (2013). Social Status Moderates the Relationship Between Facial Structure and Aggression. *Psychological Science*, *24*(11), 2329–2334. https://doi.org/10.1177/0956797613493294

[9] Hasan, T. T., Jasim, M. H., & Hashim, I. A. (2018). FPGA Design and Hardware Implementation of Heart Disease Diagnosis System Based on NVG-RAM Classifier. *2018 Third Scientific Conference of Electrical Engineering (SCEE)*, 33–38. https://doi.org/10.1109/SCEE.2018.8684125

[10] Haselhuhn, M. P., Ormiston, M. E., & Wong, E. M. (2015). Men's facial width-to-height ratio

predicts aggression: A meta-analysis. *PLoS ONE*, *10*(4), 1–10. https://doi.org/10.1371/journal.pone.0122637

[11] Haselhuhn, M. P., & Wong, E. M. (2012). Bad to the bone: facial structure predicts unethical behaviour. *Proceedings of the Royal Society B: Biological Sciences*, *279*(1728), 571–576. https://doi.org/10.1098/rspb.2011.1193

[12] Hashim, I. A., & Sadah, J. W. A. (2015). Efficient Numeral VG-RAM Pattern Recognition Using Manhattan Distance Calculation and Minimization Algorithm. *Kasmera*, *43*(2), 111–122.

[13] Hehman, E., Leitner, J. B., Deegan, M. P., & Gaertner, S. L. (2015). Picking teams: When dominant facial structure is preferred. *Journal of Experimental Social Psychology*, *59*, 51–59. https://doi.org/10.1016/j.jesp.2015.03.007

[14] Inc, D. (n.d.). *Aggressive Man Stock Photos*. https://www.dreamstime.com/photos-images/aggressive-man.html

[15] Krenn, B., & Meier, J. (2018). Does Facial Width-to-Height Ratio Predict Aggressive Behavior in Association Football? *Evolutionary Psychology*, *16*(4), 1–8. https://doi.org/10.1177/1474704918818590

[16] Kuruvilla, J., & Gunavathi, K. (2014). Lung cancer classification using neural networks for CT images. *Computer Methods and Programs in Biomedicine*, *113*(1), 202–209. https://doi.org/10.1016/j.cmpb.2013.10.011

[17] Kuruvilla, J., & Gunavathi, K. (2015). Lung cancer classification using fuzzy logic for CT images. *International Journal of Medical Engineering and Informatics*, *7*(3), 233. https://doi.org/10.1504/IJMEI.2015.070128

[18] Lefevre, C. E., & Lewis, G. J. (2014). Perceiving Aggression from Facial Structure: Further Evidence for a Positive Association with Facial Width-to-Height Ratio and Masculinity, but not for Moderation by Self-Reported Dominance. *European Journal of Personality*, *28*(6), 530–537. https://doi.org/10.1002/per.1942

[19] Ma, D. S., Correll, J., & Wittenbrink, B. (2015a). *The Chicago face database : A free stimulus set of faces and norming data*. https://doi.org/10.3758/s13428-014-0532-5

[20] Ma, D. S., Correll, J., & Wittenbrink, B. (2015b). The Chicago face database: A free stimulus set of faces and norming data. *Behavior Research Methods*, *47*(4), 1122–1135. https://doi.org/10.3758/s13428-014-0532-5

[21] Malaga, R. (2020). *Biological Components of Human Aggression*. Hubpages.Com/Psychology & Psychiatry. https://hubpages.com/education/Sub-cortical-Neurobiological-Components-of-Human-Aggression

[22] Mehrabian, A. (1968). Communication without Words. *Psychology Today*, *2*(4), 53–56.

[23] Oladipo, O., Osamor, I. P., Osamor, V. C., Abiodun, T. N., Omoremi, A. O., Odim, M. O., & Ekpo, R. H. (2019). Face-Age Modeling: A Pattern Recognition Analysis for Age Estimation. *Proceedings - 2019 IEEE International Conference on Bioinformatics and Biomedicine, BIBM 2019*, 2421–2427. https://doi.org/10.1109/BIBM47256.2019.8983347

[24] Olaniyi, E. O., Oyedotun, O. K., & Adnan, K. (2015). Heart Diseases Diagnosis Using Neural Networks Arbitration. *International Journal of Intelligent Systems and Applications*, *7*(12), 75–82. https://doi.org/10.5815/ijisa.2015.12.08

[25] Özener, B. (2012). Facial width-to-height ratio in a Turkish population is not sexually dimorphic and is unrelated to aggressive behavior. *Evolution and Human Behavior*, *33*(3), 169–173. https://doi.org/10.1016/j.evolhumbehav.2011.08.001

[26] Panday, P., & Godara, N. (2012). *Decision Support System for Cardiovascular Heart Disease*

Diagnosis using Improved Multilayer Perceptron. *45*(8), 12–20. https://doi.org/10.5120/6799-9134

[27] Piątkowska, E., & Martyna, J. (2012). Spontaneous facial expression recognition: Automatic aggression detection. *Lecture Notes in Computer Science (Including Subseries Lecture Notes in Artificial Intelligence and Lecture Notes in Bioinformatics)*, *7208 LNAI*(PART 1), 147–158. https://doi.org/10.1007/978-3-642-28942-2_14

[28] Pinto, A., Bhasi, M., Bhalekar, D., Hegde, P., & Koolagudi, S. G. (2019). A deep learning approach to detect drowsy drivers in real time. *2019 IEEE 16th India Council International Conference, INDICON 2019 - Symposium Proceedings*, 10–13. https://doi.org/10.1109/INDICON47234.2019.9030305

[29] Sell, A., Cosmides, L., Tooby, J., Sznycer, D., Von Rueden, C., & Gurven, M. (2009). Human adaptations for the visual assessment of strength and fighting ability from the body and face. *Proceedings of the Royal Society B: Biological Sciences*, *276*(1656), 575–584. https://doi.org/10.1098/rspb.2008.1177

[30] Spivak, I., Krepych, S., Faifura, V., & Spivak, S. (2019). Methods and tools of face recognition for the marketing decision making. *2019 IEEE International Scientific-Practical Conference: Problems of Infocommunications Science and Technology, PIC S and T 2019 - Proceedings*, 212–216. https://doi.org/10.1109/PICST47496.2019.9061229

[31] Taylor, A. J. G., & Jose, M. (2014). Physical aggression and facial expression identification. *Europe's Journal of Psychology*, *10*(4), 650–659. https://doi.org/10.5964/ejop.v10i4.816

[32] Weston, E. M., Friday, A. E., & Liò, P. (2007). Biometric evidence that sexual selection has shaped the hominin face. *PLoS ONE*, *2*(8). https://doi.org/10.1371/journal.pone.0000710

[33] Wilber, R. (2015). *5 Important Strategies For Dealing With A Passive-Aggressive Ex*.

[34] Ya, W., Bao, T., Chunhui, D., & Zhu, M. (2017). Face recognition in real-world surveillance videos with deep learning method. *2017 2nd International Conference on Image, Vision and Computing, ICIVC 2017*, 239–243. https://doi.org/10.1109/ICIVC.2017.7984553

Bioactivity of CaSiO3 and Porecalin-CaSiO3 Composites Prepared by Sol-Gel Method for Dental Applications

Asawer Mohsen Hamza [1], Eman Jabber Abed Alshaibani[2,]* and Rajaa R. Abbas[3]

[1]Dept. of Materials Engineering, Faculty of Engineering, University of Kufa, Najaf, Iraq. E-mail : asawerm.jabeer@uokufa.edu.iq
[2]Dept. of Materials Engineering, Faculty of Engineering, University of Kufa, Najaf, Iraq. E-mail: iman.alshaibani@uokufa.edu.iq
[3]Faculty of Dentistry, University of Kufa, Najaf, Iraq. E-mail: rajaa_74@yahoo.com
* Corresponding author: Faculty of Engineering, University of Kufa, Iraq.
Tel: +9647810088057.

Abstract. In this study, the bioactivity of $CaSiO_3$ and Porecalin-$CaSiO_3$ composites have been investigated utilizing X-ray diffraction (XRD) and scanning electron microscopy-Energy-dispersive X-ray spectroscopy (SEM-EDS). $CaSiO_3$ and P-$CaSiO_3$ are prepared by Sol-Gel method for different weight percentages of $CaSiO_3$ (P90CS10, P80CS20 and P70CS30 wt. %). The in vitro bioactive properties have been carried out by immersed the powder and sintered samples on different compositions in simulated body fluids at 32 day to evaluate their bioactivity. Ca/P ratio have been calculated through EDS measurements. Structural characterization revealed the amorphous phase of $CaSiO_3$ when heated at 700 °C and phase transition to the crystalline phase when heated at 950 °C, where pseudowollastonite (α-$CaSiO_3$) and $Ca_2(SiO_4)$ phases have been appeared. $CaSiO_3$ showed better bioactive ability in comparison with its composites. In general, the bioactivity increases with increasing the $CaSiO_3$ time of immersion in stimulated body fluid (SBF). The closest value to the stoichiometric hydroxyapatite, $Ca_{10}(PO_4)_6(OH)_2$ has been observed with P70CS30 sintered specimen where Ca/P was 1.694.

Keywords: *bioactive ceramics, Calcium Silicate, Porcelain, Sol-gel, Hydroxyapatite, SBF.*

1. Introduction.

Bioactive ceramics may be categorized into bioactive glass and bioactive glass-ceramic [1]. Bioactive glasses are silicate-based ceramics with amorphous structure. The first system of bioactive glass was studied by Hench in 1969, it included SiO_2, Na_2O, CaO and P_2O_5 [2] and followed by a series of bioactive glasses with different ratios and additional elements such as $CaO-B_2O_3$ - P_2O_5 [3] and $Na_2O-K_2O-MgO-CaO-B_2O_3-P_2O_5-SiO_2$ [4].

Binary system of silicate bioceramic (calcium silicate, $CaSiO_3$) was published as a potential biomaterial in 1990 [5]. Calcium silicate can be obtained by reaction of calcium oxide (CaO) with silicon dioxide (SiO_2) [6]:

In situation of bioactive bioceramics, characterized by their ability to bond directly with the hard tissue by forming a hydroxyapatite layer at the interface between the bioactive ceramic and tissue [7]. Hydroxyapatite material is similar in composition with the minerals phase which involved in bone or tooth structure [8].

In dentistry, when a tooth is missing due to decay, damage, dental carries or trauma, restorative materials have been used to replace the damaged part by an artificial new one [9]. The material which is used as a restorative material should be biocompatible and have no side effects. For many years, porcelain is one of the most restorative materials which has been used for treatment [10]. Dental ceramic(porcelain) known by its low ability to form the apatite layer when it is in contact with living tissue. This is due to its high chemical stability that made it classified as a bioinert ceramic. In fact, the disability of a porcelain to react with the tissue considered as one of the requirement of dental biomaterials that will be in contact with oral tissues to resist degradation in environment with changeable acidic level and temperature. However,using of porcelain as crown, veneer and prosthetic restorative material may have some disadvantages such as the marginal gap that may occur between tooth and biomaterial, failure of restoration and secondary carries [11,12].

In recent years, many studies have been carried out to improve the bioactivity of porcelain by adding a definite amount of bioactive ceramic such as calcium silicate ceramics.It have been found that composites of porcelain-bioactive ceramic have enhanced bioactive and mechanical properties [13,14].

Bioactive dental glass-ceramics (BDGCs) are characterized by the ability to form a good bonding at the material-tissue interface [15]. In general, to use a ceramic in dental applications, it should be biocompatible, durable, good wear resistant, aesthete and easy to customize. Also, it must have the ability to simulate the natural tooth in translucency and strength [16]. Practically, the biological response of tissues to the bioactive materials and in vivo process of hydroxyapatite layer formation can be studied in vitro by immersing the biomaterial in a simulated body fluid (SBF) which has ion concentrations similar to that of blood or cellular fluids [17, 18].

The bioactive behavior of bioactive ceramics comes from the components that existence on the surface [19]. With regarding to calcium silicate ceramics, CaO and SiO_2 play a rule in apatite formation process [20].

The bioactivity of CS and formation HA results from the reaction between the released bioceramic's ions Ca2- and Si^{4+} (in the form of H_4SiO_4) by the partial dissolution of glass network and that presents in the cellular fluid (or SBF), phosphor ions PO_4^{3-} and OH^{1-}. As a result, silica gel will be formed on the surface to be the basic structure for HA formation [21, 22].

Bioceramics can be synthesized by two process; the traditional ceramic process (melt-cast method) and the chemical based process (sol-gel method). In the traditional method, the glass oxides are mixed and melted together at high temperatures up to about 1300°C [23, 24]. Simply, sol-gel rout can be defined as a chemical process to fabricate metal oxide, such as oxides of silicon (Si) and titanium (Ti). It involves formation of a solution (sol: liquid phase), gelation of the sol (gel: diphase, liquid and solid) and removing the solvent by heat treatment. The sol-gel method goes through three steps: hydrolysis,polycondensation and drying process. Chemically, for preparing silicate oxides using silicon alkoxide $Si(OR)_4$, sol-gel processing based on the transformation of Si-OR and Si-OH to siloxane compounds by condensation processes [25, 26]. Glasses and ceramics that prepared by sol-gel method appear more enhanced biocompatibility and bioactivity with higher purity and homogeneity than that fabricated by melting method [27]. The high bioactivity comes from the large surface area and the presence of hydroxyl (-OH) groups which stimulate the hydroapatite layer to nucleate when it is in contact with body fluids [27,28]. However, the study of bioactivity of melt-derived Si ceramic systems displayed that the biological response is limited by the composition, where SiO_2 must be lower than 60 mol% [29]. On the other side,many sol-gel Si ceramic systems such as tri-system (SiO_2-CaO-P_2O_5), di-system (SiO_2-CaO)and even pure silica show high bioactivity[30]. and quick development of hydroxyapatite layer. This research is centered on the decrease the marginal gap that may occur between tooth and biomaterial, failure of restoration and secondary carries by improvement the bioactivity of $CaSiO_3$ and Porecalin-$CaSiO_3$ composites prepared by sol-gel method.

2. Experimental part

The procedures of the experimental part as shown in following steps:

2.1. Material preparation

2.1.1. The $CaSiO_3$ bioactive ceramic preparation

The procedures of the preparation process of 80% SiO_2-20% CaO by sol gel method can be described by the following steps:

1. The sol was prepared by using a magnetic stirrer, to mix 25 ml Tetraethyl orthosilicate (TEOS) ($Si(OC_2H_5)_4$)(China,99.0%purity)in solution of 3.05 ml nitric acid (Mumbai, India,71% purity)(2M) and 15.8ml H_2O. Then after 1hr, amount of 6.48g of $Ca(NO_3)_2.4H_2O$ (Mumbai, India,99.0% purity) was added to the mixture with continuous mixing for 4hr.
2. The sol was left in a closed container for three days at room temperature to form a transparent gel mass.
3. The resultant gel was transferred to a furnace at 70°C aging for three days. In this step, the size of the gel is decreased and completely transformed to a solid phase.
4. The solid samples, were placed in the oven for drying for two days at 150 °C until the color of the sample become white which then grinded using pestle and mortar to form a powder.
5. To nitrate elimination and stabilization, the powder heat treated at 700 °C for 3hr in electrical resistance furnace with heating rate 13 °C/min and then left in the furnace to cool to room temperature (5°C/min). In this step the weight of the produced powder was 8.25g.
6. The final step of the preparation process, the resultant materials were grinded using pestle and mortar and then milled by planetary ball mill (SMF-Disk-Top Planetary Ball Miller) for 30 min with milling speed 2700rpm to obtain a ceramic powder with particle size > 63μm.

2.1.2 The porcelain-$CaSiO_3$ composites preparation

Porcelain-$CaSiO_3$ composites samples have been prepared with different weight ratios in the same preparation steps of $CaSiO_3$ ceramic. Porcelain (IPS InLine supplied by Ivoclar Vivadent Company, Liechtenstein) has been added to the solution in three weight concentrations (70%, 80%, and 90% porcelain), while maintaining the mixing process with magnetic stirrer during gelation process for about 2-3 hours until the gelatin was completed. The mixing process should be continued to prevent precipitation of the porcelain particles and ensure mixing all the components homogenously. After that condensation method to prepare all specimens was made according to ISO 6872[31] by mixing the ceramic powder with distilled water (molding liquid) with liquid to powder weight ratio 0.335 and then pour the mixture in a steel mold. To homogenize disrupt the mixture and remove the excessed liquid and compact the powder, a vibrator with vibration for 90 sec was used. The molded samples were dried by using a filter paper for many times. After making sure that the sample dried completely we take out the sample by hand pressure. Then, the samples were sintered in the furnace (VITA VACUMAT 6000 M) through a firing cycle as mentioned in the manufacturer's instructions of dental porcelain [32], Table (1) shows firing cycle of the samples.

Table 1: Firing cycle of the samples according to manufactures instructions [32].

Tmax (°C)	950
Heating rate °C / min	60
Stand by temperature °C	403
Drying time (min)	4
Vacuum start °C	450
Vacuum release °C	949
Holding time (min)	1

2.1.3. Simulated body fluid (SBF) solution

To study the in vitro bioactive properties of the fabricated biocomposites, SBF should be used. The inorganic ions concentrations in SBF are similar to that existing in the blood plasma as listed in Table (2). Frequently, to accelerate the formation of the apatite layer on the material surface [33], a concentrated form of simulated body fluid is used. However,the concentrated SBF contains ions with concentrations 1.5 times that of conventional SBF [33,34]. 1.5SBF works on increasing the degree of supersaturation while maintaining the ratio of Ca to P [33].

The simulated body fluid (SBF) is prepared as published by Kokubo and Takadama [34]. All materials that involved in the preparation of the SBF and 1.5SBF were weighed by the sensitive balance with accuracy (± 0.0001%). The weights of the reagents are listed in Table (3). The procedures to prepare 2000 ml of SBF can be described as following:

1-First of all, 1400 ml of deionized water have been put in a glass beaker 2000 ml on a hot plate stirrer. After heating the water to the required temperature, the heated water was transferred to a baker plastic and covered with plastic film, while continuing in mixing during the preparation process.

2-The reagents from 1st to 8th one was added consequently to the solution with making sure that each material added is completely dissolved before adding the next material.

3-The next step was adding drops of the 9th and 10th reagents (Tris and 1M-HCl) little by little alternatively until the pH of the final solution became 7.4 at temperature 36.5 ± 0.5°C.

4-The pH of the solution finally has been adjusted to 7.4 exactly at 36.5 °C. The prepared solution was transferred from the beaker to polyethylene bottles and kept at room temperature until its temperatures were about 20 °C. After cooling, the distilled water was added up to the total capacity of the beaker (2000ml).

5-The $CaSiO_3$ and composite specimens were soaked in SBF at room temperature for 4, 8, 16 and, 32 days. Before doing the test, the specimens were rinsed with ethanol and water and dried at 70 °C for (6-8) hr. To provide constant chemical composition of solution, the simulated body fluid was changed every 4 days.

The bioactivity of the ceramic powder was investigated through XRD and SEM-EDS tests by soaking 600mg of $CaSiO_3$powder that sintered at 700 °C in 1000 ml SBF and by soaking the prepared sample which was sintered at 950 °C in 100 ml 1.5 SBF. It is important to notice that the glass containers should be avoided, because apatite nucleation can be induced at the surface of a glass container [34].

Table 2: Ionic concentrations of different solutions (mol/m³) [34].

	Ion concentrations (Mm)								
	Na^+	K^+	Mg^{2+}	Ca^{2+}	Cl^-	$H(CO)^{2-}$	$H(PO_4)^{2-}$	$(SO_4)^{2-}$	pH
Blood plasma	142	5	1.5	2.5	103	27	1	0.5	7.2-7.4
SBF	142	5	1.5	2.5	147.8	4.2	1	0.5	7.4

Table 3: Reagents required for preparation of SBF and 1.5 x SBF solution [34].

Seq.	Reagents	SBF in 1 liter	SBF in 2 liter	1.5*SBF in 1 liter	1.5*SBF in 2 liter
1	NaCl	8.035 g	16.07	12.053 g	24.105
2	$NaHCO_3$	0.355 g	0.71	0.533g	1.065
3	KCl	0.225 g	0.45	0.338 g	0.675
4	K_2HPO_4 $3H_2O$	0.231 g	0.462	0.347 g	0.693
5	$MgCl_2$ $6H_2O$	0.311 g	0.622	0.467 g	0.933
6	1MHCl	39 g	78	56.5 ml	117
7	$CaCl_2$	0.292 g	0.584	0.438 g	0.876
8	Na_2SO_4	0.072 g	0.144	0.113 g	0.216
9	Tris	6.118	12.236	9.177 g	18.354
10	1 M-HCl	0-5 ml	0-10	0.7.5 ml	0-15

2.1.4. Analytical Methods for Characterization

The pure and composites prepared samples, before and/or after soaking in SBF, were characterized by many analytical methods. In order to characterize the crystalline structure of the prepared samples before and after dipping in SBF solution, XRD (6000 SHIMADZU X-Ray diffractometer, Japan). Microstructure, morphology and chemical compositions of the prepared sample were examined using scanning electron microscopy (inspect S50 SEM, Japan) equipped with an energy dispersive X-ray spectroscopy (EDX) system.

3. Results and Discussion

XRD was performed before soaking in SBF to characterize the crystalline structure of bioceramics that used in this work. Fig. (1a, b) shows the x ray diffraction pattern of $CaSiO_3$ powder heat treated at 700 °C and 950 °C. From Figure (1a), one can see that the amorphous phase is predominant in this spectrum which refers to there is no phase and it is to written as below of amorphous structure. No peaks indicate to the crystalline phase, either α or β-wollastonite $CaSiO_3$. XRD results of amorphous $CaSiO_3$ agree with the results of many publications where the researchers prepared $CaSiO_3$ bioceramics by by sol-gel rout under the same thermaltreatment conditions [35,36].

However, Fig.(1b) Fig.(1b) presented many peaks of new phase, it was the crystalline phase of the calcium silicate ceramics.

Thermal treatment of $CaSiO_3$ at 950°C lead to transform of the amorphous phase to the crystalline structure. The crystalline form pseudowollastonite(α-$CaSiO_3$) are identified by the peaks at 2θ = 25.89°, 27.47°,31.73°, 36.45°, 45.29°, 45.71°, 46.09°, 49.57°,54.2°, 56.79° that can be assigned, respectively, to the (-210), (112), (-222), (008), (-218), (-300), (031), (119), (-422) and (-424) planes of α-$CaSiO_3$ phase according to the XRD card (00-031-0300). Other small peak at 2θ = 22.13°, 45.01° that can be assigned, respectively, to the (101) and (202) planes may be belonged to the calcium silicate(Ca_2SiO_4) phase according to the card (00-052-0069). These results are in agreement with that reported by Julian, et al [37] and Chun-Cheng, et al [38].

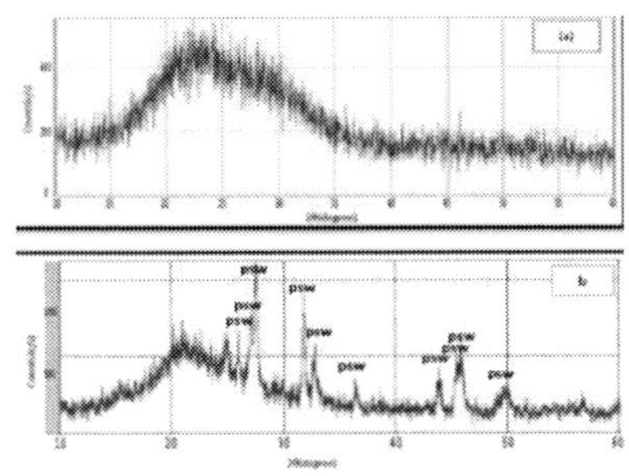

Figure 1: The X-radiation spectrum of $CaSiO_3$ powder heat treated (a) At 700 °C (b) At 950 °C.

Fig. (2) displays the XRD pattern of the dental ceramic (porcelain) that have been used in this study. Numbers of peaks have been appeared on a wide background. These peaks represent the reflections of X-radiation from the tetragonal Leucite phase ($KAlSiO_2O_6$) where their positions at 2θ =16.38°, 25.95°, 27.18°, 30.48°, 31.46°, 33.6° and 38° that can be assigned, respectively, to the (211), (004), (400), (420), (323), (314) and (404) reflections of leucite phase according to the XRD the card (00-038-1423). The same results have been obtained by the Refs. [39, 40 ,41].

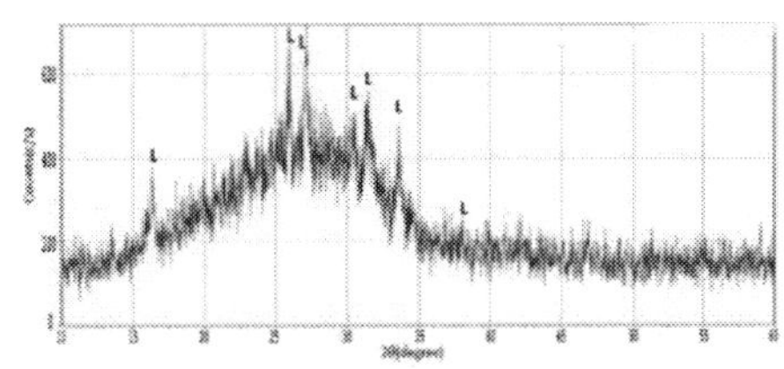

Figure 2: The XRD pattern of the dental ceramic (porcelain).

Because, in practice, porcelain needs to be treated thermally (sintered) at a temperature of 950 °C to form a solid material suitable for the work of the teeth restorative materials, so the crystalline structure of the Porcelain- Calcium Silicate composites should be investigated when treated at this temperature. Fig. (3) represents the XRD spectra of pure porcelain and the three composites that were studied in this research. Peaks at diffraction angles 16.9°, 26.44°, 30.91°, 31.77°, 34.22° and 38.51° are attributed to existence of leucite ($KAlSi_2O_6$) crystals within pure porcelain and porcelain composite, overlapping on an amorphous background.

One can observe that increasing of CS bioceramic content in the composites leads to gradually disappearing of leucite effect and appearing new additive peaks with high intensities. These new x-ray reflections at 27.7° and 46.21° 2θ may be belonging to the (-122) and (031) reflected planes of α-$CaSiO_3$ that crystallized as a result to the high temperature of sintering (950°C). Also, two major reflections at 22° and 45.56° of the calcium silicate (Ca_2SiO_4) phase according to the card (00-052-0069) have been seen in the pattern.

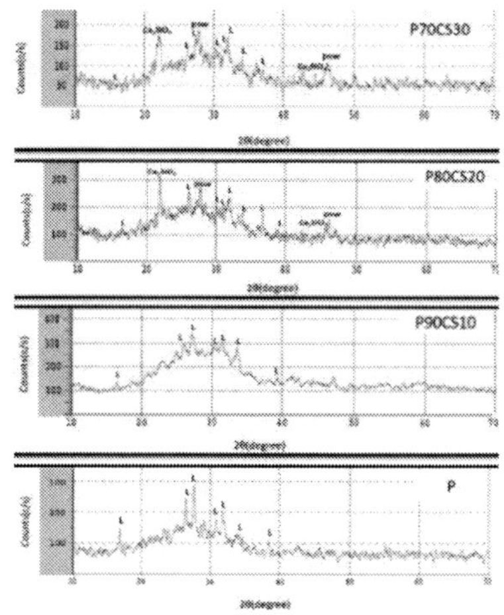

Figure 3: XRD spectra of pure porcelain and the porcelain-$CaSiO_3$ composites sintered at 950 °C.

XRD spectra of $CaSiO_3$ bioceramics in the powder form with different immersion time in SBF; 8, 16 and 32 days, have been displayed in Fig. (4). As can be seen in 8 day, major peaks reflections at 25.83°, 28°, 31.66°, 34.31°, 39.53°, 46.5°, 49.4° and 53.15° in 2θ, and in 16 day, major peaks at 25.86°, 28° , 31.96°, 34.45°, 39.64°, 46.61°, 49.53° and 53.22° in 2θ, and in 32 day, major peaks reflections at 25.86°, 28° , 31.95°, 34.47°, 39.67°, 46.62°, 49.44° and 53.21° in 2θ, that can be assigned, respectively, to the (002), (102), (211), (202), (310),(222),(320), (004) reflections of HAP according to ASTM card [JCPDS pattern NO. 09-0432] which are indicated to the "formation of calcium phosphate" crystals on the surface of the sample, overlapping on a random background

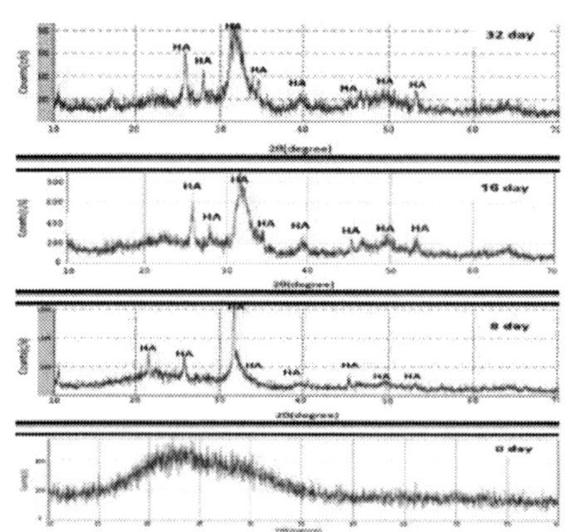

Figure 4 XRD spectra of $CaSiO_3$ bioceramics with different immersion time in SBF.

This result is in a good agreement with that published by Xanthippi et. al.in Ref. [42]. Salinas et. al. [36] and Yong Kim et. al. [43]. Hydroxyapatite; HAP (calcium phosphate phase) is formed as a result to the chemical interaction between the Ca^{+2} and P^{+4} ions that are present in SBF with the Ca^{+2} that solute from $CaSiO_3$ in the solution. However, it can be observed that increasing of immersion time lead to form more HAP crystals which indicated by increasing the reflected X-ray intensity. However, in order to investigate the bioactivity of the composite ceramic of sintered bulk sample, the composite with the largest amount of $CaSiO_3$ have been chosen to test through XRD technique. In spite of the low concentration of $CaSiO_3$ in the P70CS30 composites, but the bioactive effect was very clear. This result may be due to the effect of concentrated SBF (1.5SBF) which makes the solution saturated with the ions needed to create and accelerate the formation of apatite layer. However, no leucite peaks were noticed. On the other side, XRD measurements of P70SC30 sol-gel sintered specimen in 950 °C have been displayed in Fig. (5). X-ray profile revealed the formation of HAP crystals after soaking at 1.5SBF for 32 days. The bioactivity of P70SC30 bioceramics are characterized by observing peaks at 25.76°, 28.16°, 31.74°, 32.03°, 34.1°, 39.84° and 46.82° that attributed to the (002), (102), (211), (112), (202), (310), (222) and (320) reflections of HA according to ASTM card [JCPDS pattern NO. 09-0432]. These results agree with Ref. [44].

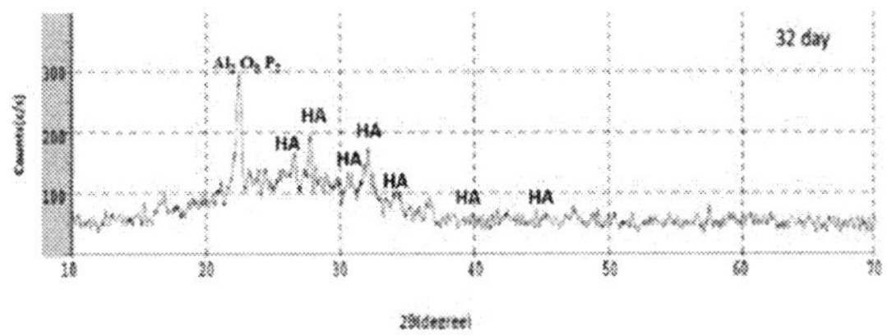

Figure 5: XRD spectra of P70CS30 composites sintered at 950 °C immersion 32 day in 1.5 x SBF.

SEM-EDS of $CaSiO_3$ before and after soaking in SBF. Figures (6,7) display the (SEM-EDS) results of $CaSiO_3$. In Fig. (6) and Fig. (7), the morphology of $CaSiO_3$ particles before and after soaking in SBF for 32 days have been showed respectively. A thick layer of bone-like apatite was formed on the surface of the particles. This result agrees with the XRD measurements, since peaks which indicated to the presence of hydroxyapatite material have been appeared clearly. On the other side and before soaking in SBF, EDS results indicates to the presence of O_2, Si and Ca where the $CaSiO_3$ are original composed as shown in Fig.(6b). In Fig.(7b) after immersion for 32 days, appearance of phosphate traces has been detected clearly. Also, Ca content increased drastically in comparison with a significant decrease in Si concentration. The total elimination of Si from the substrate indicated the precipitation of a thick hydroxyapatite layer on the surface of the grains which the Ca/P ratio by EDS is 1.409 this result is in agreement with that reported in Ref [45].

The microstructure and the morphology surface of P70CS30 bulk specimens (at 950 °C) before and after soaking in 1.5SBF was investigated with (SEM) as show in Fig. (8) and Fig. (9). After 32 days of soaking, in Fig.(9a) show an appetite aggregation have been noticed to be crystallized on the surface ceramics particles and formed a non-uniform layer of hydroxyapatite. Fig.(8b) and Fig.(9b) displayed the chemical composition of elements in P70CS30 sintered specimens before and after immersion in 1.5 SBF respectively. Ca/P ratio by EDS was 1.694 very close to the ratio of HAP bone in which stoichiometric hydroxyapatite ($Ca_{10}(PO_4)_6(OH)_2$) is (Ca/P=1.67).

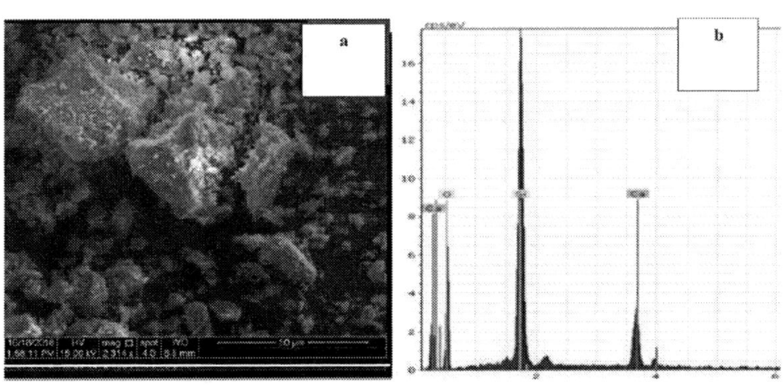

Figure 6: SEM-EDS measurements powdered $CaSiO_3$: (a) SEM and b) EDS before immersion in SBF.

IOP Conf. Series: Materials Science and Engineering **978** (2020) 012047 doi:10.1088/1757-899X/978/1/012047

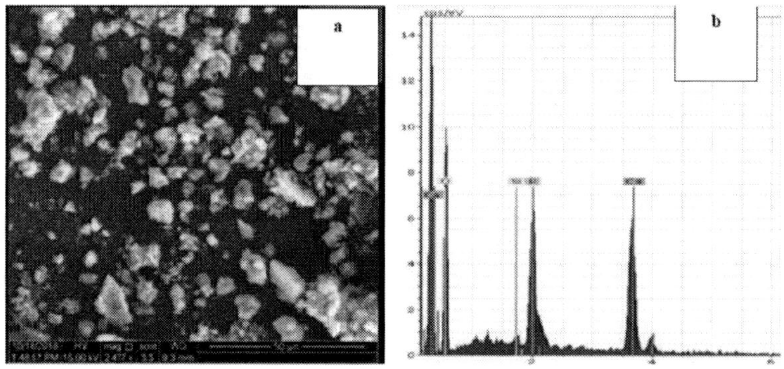

Figure 7: SEM-EDS measurements powdered $CaSiO_3$: (a,) SEM and (b) EDS after immersion in SBF for 32 day.

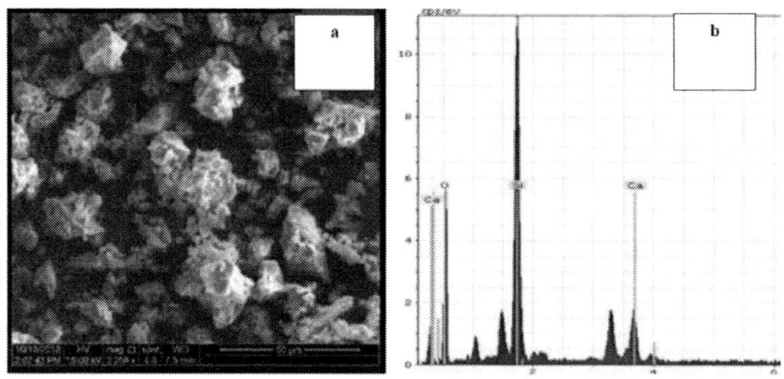

Figure 8: SEM-EDS measurements sintered specimen P70CS30: (a) SEM and (b) EDS before immersion in 1.5SBF.

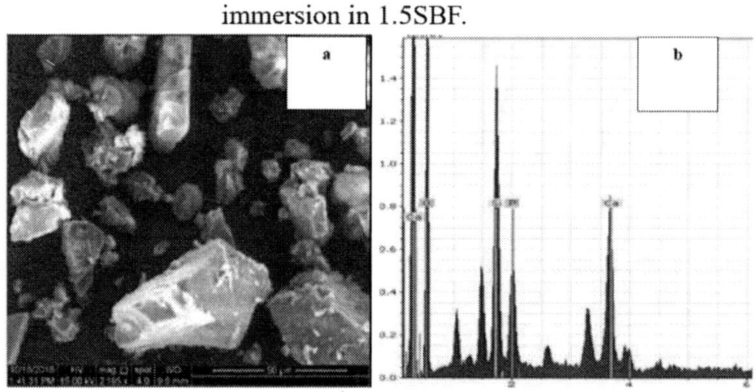

Figure 9: SEM-EDS measurements sintered specimen P70CS30: (a) SEM and (b) EDS after immersion in 1.5SBF for 32 day.

4. Conclusion

The conclusions that can be drawn from this work are summarized as following:

1. $CaSiO_3$ showed a better bioactivity in comparison with its composites and increasing bioactivity with increasing immersion time in SBF.

2. Heat treatment temperature has an effect on the crystallization phases of $CaSiO_3$ bioceramic and its composites. However,sintering at 950°C resulted in appearance of pseudowollastonite (α-$CaSiO_3$) crystalline phase and calcium silicate (Ca_2SiO_4) phase.
3. The closest value to the stoichiometric hydroxyapatite, $Ca_{10}(PO_4)_6(OH)_2$ has been observed with P70CS30 sintered specimen where Ca/P was 1.694.
4. The composite porcelain-$CaSiO_3$ can exhibit integrity bioactivity and consequently, can be potentially used in restorative dentistry.

Acknowledgment

The authors would like to express their appreciation to the University of Kufa/College of Engineering /Materials Engineering Department for its support of this work. The authors would like to thank all the technicians in the laboratories for their valuable help.

References

[1] K. avindranadh, "Bioactive glasses for technological and clinical applications," International Journal of Chemical Science, vol. 14, no. 3, pp.1339-1348, 2016.
[2] L. Hench, "Bioceramics: from concept to clinic," Journal of the American ceramic society, vol. 74, no. 7, pp. 1487-1510, 1991.
[3] Saranti, et al, "Bioactive glasses in the system CaO–B2O3–P2O5: preparation, structural study and in vitro evaluation," Journal of Non-Crystalline Solids, vol. 352, no. 5, pp. 390-398, 2006.
[4] M. Brink, et al, "Compositional dependence of bioactivity of glasses in the system Na2O-K2O-MgO-CaO-B2O3-P2O5-SiO2," Journal of Biomedical Materials Research, vol. 37, no. 1, pp. 114-121, 1997.
[5] X. Chatzistavrou, et al, "Sol-gel derived bioactive glass ceramics for dental applications" " Non-Metallic Biomaterials for Tooth Repair and Replacement" , Woodhead Publishing, PP. 194-231, 2013.
[6] N. Yamaguchi, et al, "Synthesis of CaO-SiO2 Compounds Using Materials Extracted from Industrial Wastes," Open J Inorg Non-Met Mater, vol. 5, pp. 1-10, 2014.
[7] H. Mohammadi, et al, "Bioinorganics in bioactive calcium silicate ceramics for bone tissue repair: bioactivity and biological properties," J. Ceram. Sci. Technol, vol. 5, no. 1, pp. 1-12, 2014.
[8] K. J. Anusavice, "Degradability of dental ceramics," Advances in dental research, vol. 6, no. 1, pp. 82-89, 1992.
[9] Shenoy, and S. Nina, "Dental ceramics: An update," Journal of conservative dentistry: JCD, vol.13, no. 4, pp. 195-203, 2010.
[10] E. Kontonasaki, et al, "Attachment and proliferation of human periodontal ligament fibroblasts on bioactive glass modified ceramics," Journal of oral rehabilitation, vol. 34, no. 1, pp. 57-67,2006 .
[11] O. M. Goudouri, et al, "Dental ceramics/bioactive glass composites: characterization and mechanical properties investigation," Bioceramics Development and Applications, vol. 1, 2011.
[12] O. M. Goudouri, et al, "A bioactive glass/dental porcelain system by the sol gel route: fabrication and characterization," Key Engineering Materials, vol. 396, pp 95-98, 2009.
[13] M. Manda, et al, "Material characterization and bioactivity evaluation of dental porcelain modified by bioactive glass," Ceramics International, vol. 38, no. 7, pp. 5585-5596, 2012.

[14] Z. Abbasi, et al, "Bioactive glasses in dentistry: a review," Journal of Dental Biomaterials, vol. 2, no. 1, pp. 1-9, 2015.

[15] M. Montazerian, and E. D. Zanotto, "Bioactive and inert dental glass-ceramics," Journal of Biomedical Materials Research Part A, Vol. 105, no. 2, pp. 619-639, 2017.

[16] R, Raghavan, "Ceramics in dentistry," "Sintering of Ceramics-New Emerging Techniques", IntechOpen, 2012.

[17] D. Almasi, et al, "In vitro evaluation of bioactivity of chemically deposited hydroxyapatite on polyether ether ketone," International journal of biomaterials, vol. 2015, pp. 1-5, 2015.

[18] A. Najafinezhad, et al, "A comparative study on the synthesis mechanism, bioactivity and mechanical properties of three silicate bioceramics," Materials Science and Engineering: C, vol. 72, pp. 259-267, 2016.

[19] Salinas, et al, "Bioactive ceramics: from bone grafts to tissue engineering," The Royal Society of Chemistry Advances, vol. 3, no. 28, pp. 11116-11131, 2013.

[20] H. Gheisari, and E. Karamian, "Preparation and characterization of hydroxyapatite reinforced with hardystonite as a novel bio-nanocomposite for tissue engineering," Nanomedicine Journal, vol. 2, no. 2, pp. 141-152, 2015.

[21] N. Zhang, et al, "Crystal structures of CaSiO3 polymorphs control growth and osteogenic differentiation of human mesenchymal stem cells on bioceramic surfaces," Biomaterials science, vol. 1, no. 10, pp. 1101-1110, 2013.

[22] J. Ma, et al, "In vitro degradability and bioactivity of mesoporous CaO-MgO-P2O5-SiO2 glasses synthesized by sol–gel method," Journal of sol-gel science and technology, vol. 54, no. 1, pp. 69-76, 2010.

[23] Sabaliauskiene, "Synthesis, investigation and possible applications of selected phosphates and silicates," ph.d thesis, Vilnius University, 2014.

[24] E. Marenna, "Sol-gel synthesis of functional nanocomposites oxides based on inorganic oxide", Ph.d thesis, University of Naples Federico, 2008.

[25] O. M. Goudouri, "Novel bioactive ceramics for dental applications", Aristotle University of Thessaloniki, 2012.

[26] R. Asmatulu, "Nanocoatings for corrosion protection of aerospace alloys." "Corrosion Protection and Control Using Nanomaterials". Woodhead Publishing, pp. 357-374,2012 .

[27] M. Catauro, "Advanced Glass-Ceramic Materials for Biomedical Applications", Journal of Bone Reports & Recommendations vol. 3, no. 1:2, pp. 97-108, 2017.

[28] X. V. Bui, "Bioactive ceramic powder prepared using a new sol-gel process," Processing and Application of Ceramics, vol. 11, no. 2, pp. 87-92, 2017.

[29] S. Hesaraki, et al, "Physico-chemical and in vitro biological evaluation of strontium/calcium silicophosphate glass," Journal of Materials Science: Materials in Medicine, vol. 21, no. 2, pp. 695-705, 2010.

[30] P. Sepulveda, "Characterization of melt-derived 45S5 and sol-gel–derived 58S bioactive glasses," Journal of Biomedical Materials Research: An Official Journal of the Society for Biomaterials, The Japanese Society for Biomaterials, and The Australian Society for Biomaterials and the Korean Society for Biomaterials, vol. 58, no. 6 734-740, 2001

[31] ISO,"6872: Dentistry-ceramic materials," Switzerland: International Organization for Standardization (IOS), 2008.

[32] G. Sınmazışık, and M. L. Oveçoglu, "Physical properties and microstructural characterization of dental porcelains mixed with distilled water and modeling liquid", dental materials, vol. 22, no.8, pp. 735-745, 2005.

[33] N. M. Hidzir, et al, "In vitro mineralisation of grafted ePTFE membranes carrying carboxylate groups," Bioactive materials, vol. 2, no.1, pp. 27-34, 2017 .

[34] T. Kokubo, and H. Takadama, "How useful is SBF in predicting in vivo bone bioactivity?" Biomaterials, vol. 27, no.15. pp. 2907-2915, 2006.

[35] Izquierdo-Barba, et al, "In vitro calcium phosphate layer formation on sol-gel glasses of the CaO-SiO2 system," Journal of biomedical materials research, vol. 47, no. 2, pp. 243-250, 1999.

[36] J. A. Salinas, et al, "Biomimetic apatite deposition on calcium silicate gel glasses," Journal of sol-gel science and technology, vol. 21, no. 1-2, pp. 13-25,2001 .

[37] J. R. Jones, et al, "Optimising bioactive glass scaffolds for bone tissue engineering", Biomaterials, vol. 27, no. 7, pp. 964-973, 2005.

[38] Chen, et al, "Green synthesis of calcium silicate bioceramic powders," Ceramics International, vol. 41, no. 4, pp. 5445-5453, 2014.

[39] G. Sınmazışık, and M. L. Oveçoglu, "Physical properties and microstructural characterization of dental porcelains mixed with distilled water and modeling liquid", dental materials, vol. 22, no.8, pp. 735-745, 2005.

[40] W. Holand, et al, "Radiopaque strontium fluoroapatite glass-ceramics," Frontiers in bioengineering and biotechnology, vol. 3, pp.149, 2015.

[41] J. C. M. de Souza, et al, "Effect of compaction method and multiple firing on the formation of microstructural defects in feldspar-based dental ceramics," Ceramica, vol.53, no. 327, pp. 288-294, 2007.

[42] X. Chatzistavrou, et al, "Sol–gel based fabrication and characterization of new bioactive glass–ceramic composites for dental applications," Journal of the European Ceramic Society, vol. 32, no. 12, pp. 3051-3061, 2012.

[43] Meiszterics, and K. Sinko, "Sol–gel derived calcium silicate ceramics," Colloids and Surfaces A: Physicochemical and Engineering Aspects, vol. 319, no. 1-3, pp. 143-148, 2007.

[44] Beketova, et al, "Evaluation of the micro-mechanical and bioactive properties of bioactive glass-dental porcelain composite," Journal of the mechanical behavior of biomedical materials, vol. 86, pp. 77-83, 2018 .

[45] Martinez, et al, "Bioactivity of a $CaO-SiO_2$ binary glasses system." Chemistry of Materials, vol. 12, no. 10, pp. 3080-3088, 2000.

Air Distribution Performance Inside Office Room with Combined Displacement and Personal Ventilation

Ala'a Abbas Mahdi[1,*], Qusay Rasheed Al-amir[2] and Asmaa Khudhair Yakoob[3]

[1]College of Engineering, University of Babylon, Iraq
E-mail: eng.alaa.abbas@uobabylon.edu.iq
[2]College of Engineering, University of Babylon, Iraq
E-mail: eng.qusay.rashid@uobabylon.edu.iq
[3]College of Engineering, University of Babylon, Iraq
E-mail: soma.almasoody@gmail.com
*Corresponding Author: College of Engineering, University of Babylon, Iraq.
Tel: +9647801035526.

Abstract. The main goal of the paper is to study the effect of adding a new type of ventilation device, which is personal ventilation to displacement ventilation, and know its effect on heat removal and how to reach thermal comfort for humans. Experiments were practically conducted inside a thermally insulated room with full-scale of dimensions (3 × 1.75 × 3) m. The results of the experimental work are used to validate the CFD simulations. The RNG k-ε model is utilized to simulate the cases for predicting indoor airflows. The displacement ventilation supply temperature is close to 18°C and the flow velocity of 0.25 m/s. As for the personal ventilation, the flow rates 5l/s and 10l/s were taken with a temperature close to where the air was drawn (at height 0.4 m). It was concluded that adding personal ventilation improved air quality and thermal comfort, and it was also concluded that a flow rate of 10 l/s gave better efficiency for the device to function in terms of air distribution performance index and effectiveness of temperature (ε_t) which about 71% and 1.8 respectively.

Keywords: *Displacement combined with personal ventilation; effectiveness temperature; thermal comfort; air quality, air distribution performance index*

1. Introduction

Several studies have discussed the importance of air quality in occupied space and the extent of its effect on human breathing and thermal comfort. As it has been proven that poor air quality causes

IOP Conf. Series: Materials Science and Engineering **978** (2020) 012048 doi:10.1088/1757-899X/978/1/012048

symptoms of pathological construction syndrome, such as increased headaches, reduced thinking, increased dizziness, and so on (Wargocki et al., 2000). As it has been demonstrated by and by that the nature of the air breathed in from low warm substance (low temperature and dampness) is superior to the high warm substance (Fang L. et al., 1998).

The most significant explanation behind utilizing ventilation is to give a satisfactory nearby atmosphere inside the space to be ventilated. In this manner, planners and administrators of ventilation frameworks know about the prerequisites of comfort and air quality, which are required to accomplish sound and satisfactory inner conditions, notwithstanding the warm harmony between the body of humans and the conditions encompassing it and the components that influence its temperature. Just as adequate indoor air contamination and the idea of physical movement (H. B. Awbi, 2003).

Personal ventilation (PV) is of great importance in providing clean air to the breathing area in the occupied space, and thus this increases the effectiveness of ventilation, especially when using displacement ventilation (Melikov, A.K., 2004). Providing people with cold and dry air in the respiratory area provides thermal and respiratory comfort, and therefore also increases the quality of air inside the space, as a small flow of air is used for the air provided at temperatures two degrees lower than the temperature of the space etc (Arsen K. Melikov et al., 2002). Experimentally the thermal comfort of thirty women endures to nine different of indoor conditions by supplying cold air at a level near the floor when the lower parts of their legs, ankles, and feet weren't covered. The foremost sensitive ~~a~~ part of these parts was found at the ankle. Also, showed that if designers apply a particular recommendation for thermal comfort for air-conditioning design with displacement ventilation and air distributed underfloor, they created conditions uncomfortable for ladies. An oblong displacement diffuser unit was used with dimensions of (51 cm by 20 cm) and the air was delivered to the three used diffusers from an independently isolated air-handling unit utilizing an insulated duct. The temperature drop between the entering air from the three used diffusers hasn't exceeded the value of (0.2 °C). The used diffusers were placed behind the chair and consequently, the flow of air came from behind the themes (Stefano, 2016). A displacement ventilation concept and reveal more suitable turbulence model were evaluated among four models (standard k-ε, RNG k-ε, realizable k-ε, and SST k-ω) for different case studies under Iraqi climate conditions for two case studies. Case-I at ACH equals 3.5 for Non-Isothermal ventilation office room, and Case-II at ACH equals 4.5 for Isothermal office room ventilation by using a displacement ventilation system (Aedan, 2016).

The main objectives of the present work are comparing thermal performance between two types of ventilation systems; DV only and DV in conjunction with PV to reach the acceptable system which gives suitable Indoor Air Quality (IAQ) and human comfort. Studying the optimum occupant distribution inside the attested chamber to reach the acceptable indoor air quality. Giving recommendations on the strategy of using DV in combined with PV system in a hot climate at different occupant distributions.

2. Experimental method

The experiments were carried out in a thermally insulated wooden room with glass fibers and thermal nylon within the mechanical engineering laboratories at the University of Babylon as shown in Figure 1.

It has been assumed that working conditions are stable in numerical and experimental analyses. The distribution of temperature, speed, and airflow from the displacement with personal ventilation devices was studied inside an office room and under the Iraqi climate and in different locations for the distribution of occupants where the dimensions of the room are ($3 \times 1.75 \times 3$) m, where the operation of the ventilation devices was approved on two case studies, the first state Case-I was placed the displacement device only with two situation of occupant distribution as shown in Figure 2.

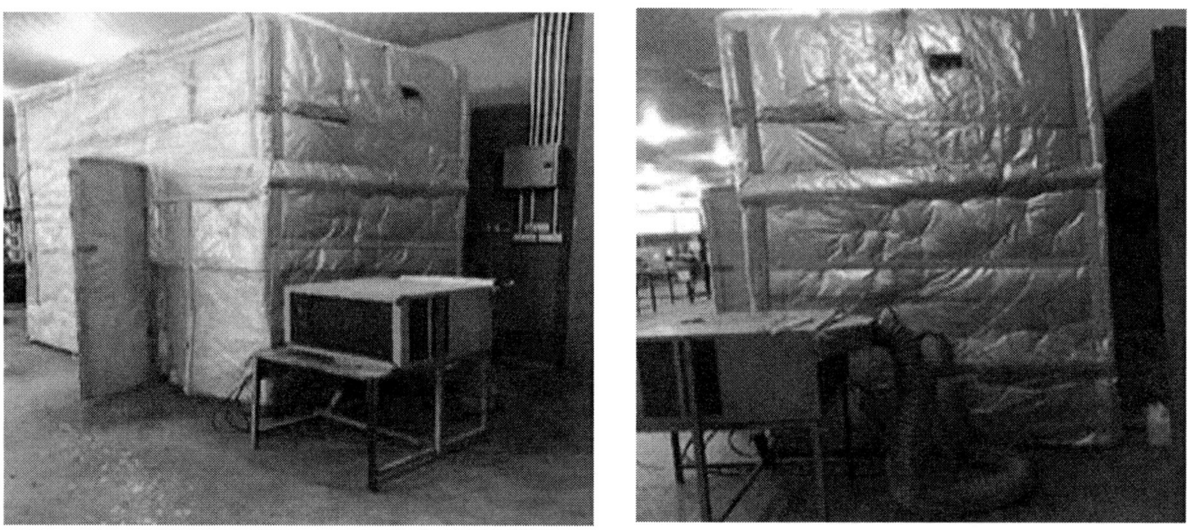

(a) (b)

Figure 1. A photograph of the test chamber room with air conditioning unit and air supply hose, (a) front view (b) side view.

The second state case-II used the personal ventilation device in addition to the displacement ventilation unit also with two situation of occupant distribution as shown in Figures 3. Many types of thermal sources were adopted, the personal computer, the room lights, and the office employee where the calculations were done according to the international ASHRAE standards.

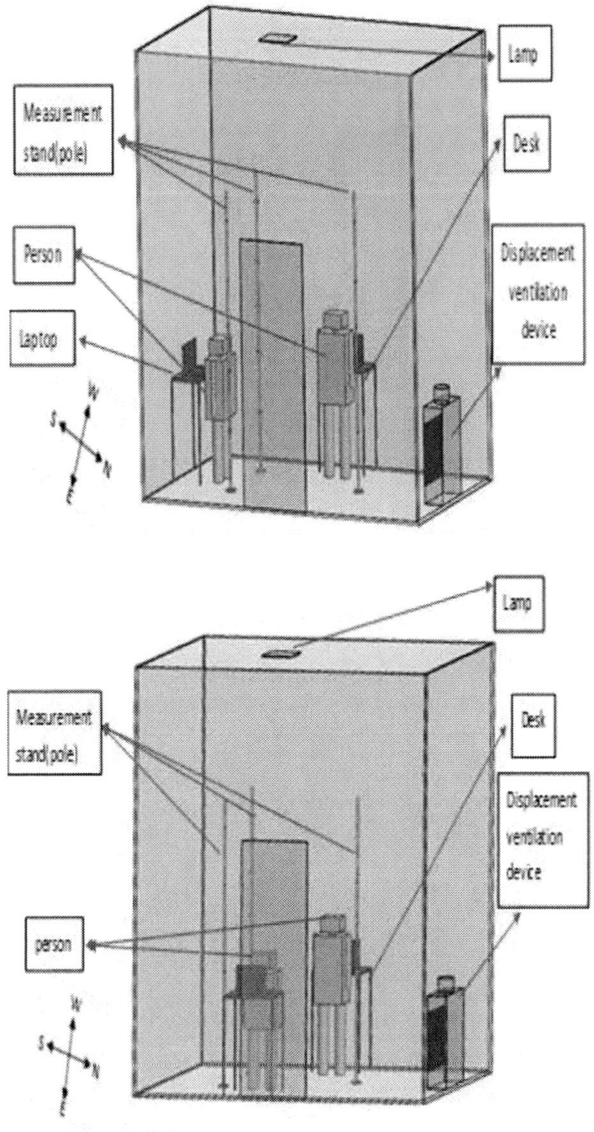

(a) (b)

Figure 2. Office room diagram for case-I of occupant distribution, (a)1st situation (b) 2nd situation

The second case includes the addition of a ductless personal ventilation device that draws air from below and at a height of 0.4m from the floor and is fixed to a table where the bottom part consists of fans at different flow rates. A flow rate of 5l/s and 10l/s is adopted as flow rates are studied to know its effect on comfort. The breathing zone for occupants from the floor level equals 1.1 m at sitting position (Iraqi cooling code, 2012), as shown in Figure 3.

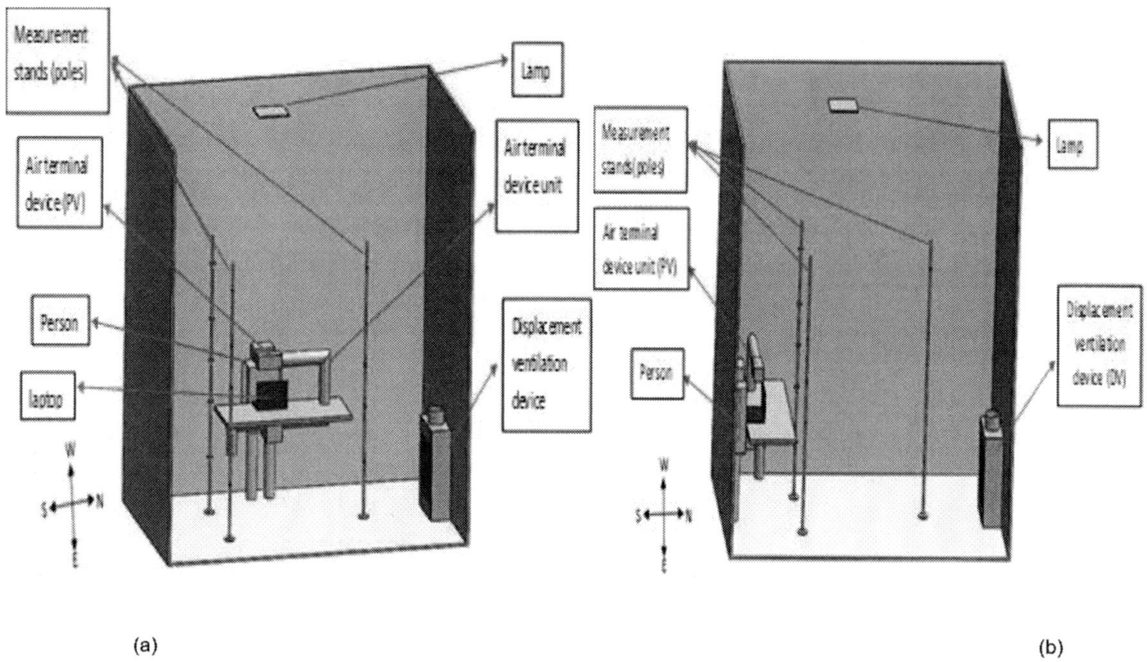

Figure 3. office room diagram for case-II of occupant distribution, (a)3rd situation (b) 4th situation

3. Mathmatical of supply system

There are some procedures adopted to calculate the design temperature of the displacement ventilation device as well as the amount of air flowing from air ventilation terminal (Seppänen, O., 2008).

a. The amount of supply air flow rate

Supply air flow rate required for cooling load is:

$$Q = \frac{0.295q_{oe}+0.132q_{l}+0.185q_{ex}}{\rho cp\Delta T_{hf}}$$

(1)

$$Q=U*A*\Delta T$$

(2)

b. Temperature of ventilation air supply

The following equation represents the design temperature of the displacement ventilation device during the experiment (A. k. Melikov et al., 2007)

$$T_s=T_{sp}-\Delta T_{hf}-[(A_f.q_t)/(0.584Q_2+1.208A_f.Q_{DV})]$$

(3)

The air change per hour (ACH) is determined by equation of Nobukazu, 2003. Its value (8.32).

$$ACH= (Q/V_{Room}) *3600 \quad (4)$$

Figure 4 represents an illustration of the displacement ventilation device in Solid Work program and represents an illustration of the number and diameter of the holes for the personal ventilation device.

c. Measurement devices

The estimating gadgets are separated into three primary frameworks:

1-Example Units: For all tests, there are three poles, and six nodes in which every node is set at various degrees of height (0.0.4, 0.8, 1.1, 1.4 & 1.8) m and knows about account parameters. The poles were arranged close to the beginning at (X =2m, Y=0m, Z=0.875m), the other poles at (X =0.7m, Y =0m, Z =0.5m) and (X =0.7m, Y=0m, Z =1.25m) for all cases as showed up in Figure 5.

2-Air Speed**:** by using the thermocouple wire anemometer sensor model YK.2005AH.

3-Temperature assessing instruments: number of K-type thermocouples used to be standard air temperature. Thermocouples measure temperature after each 600s.

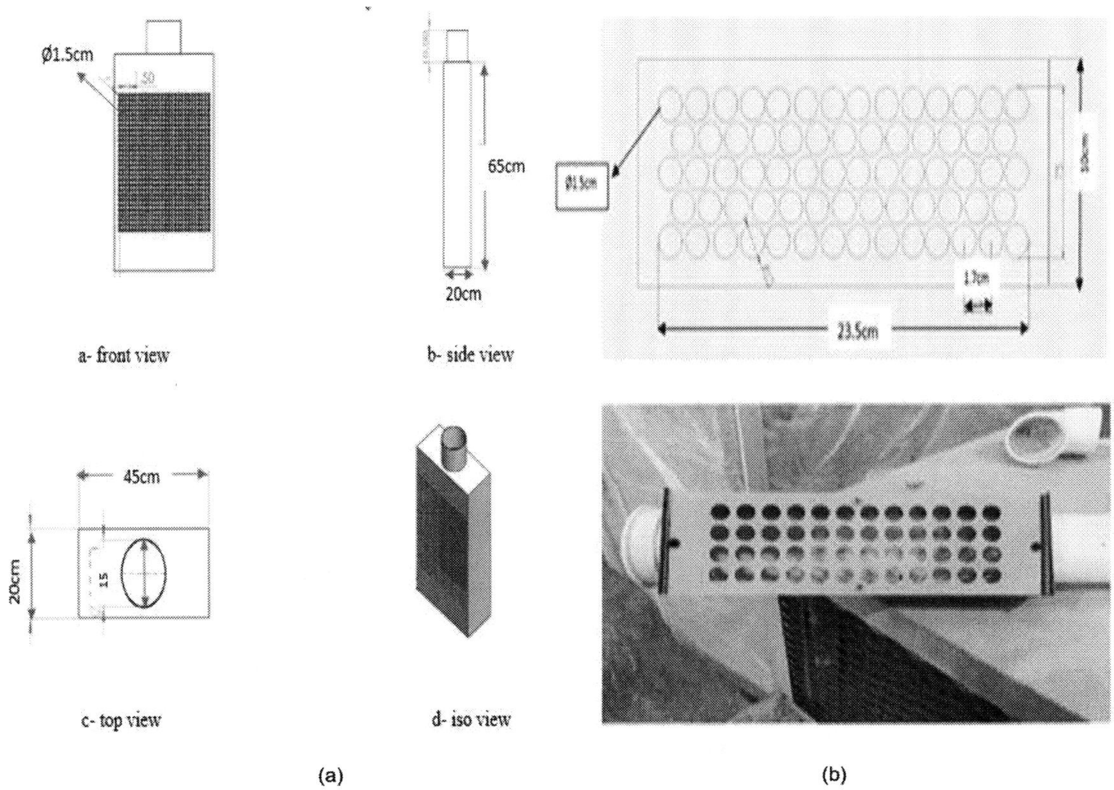

Figure 4. The supply devices ventilation, the supply air diffuser (DV) manufacturing(b) air flexibly outlet for PV diffuser

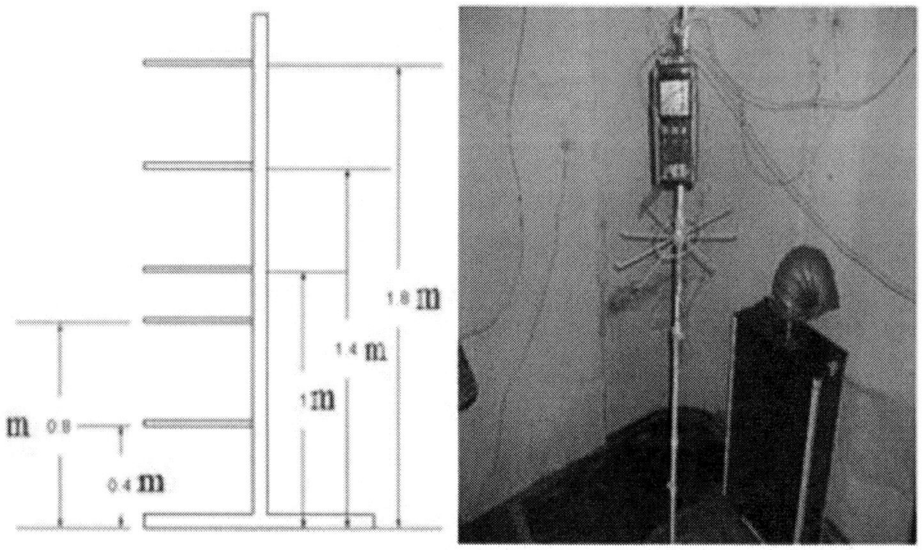

Figure 5 Poles with nodes level location and K-type thermocouples measurement.

d. Exploratory procedures

The trial was done in two phases relying upon the sort of ventilation gadget utilized, as the examination was in late March and April of 2019. The trial room is thermally protected with a high 3m from the floor. In room, temperatures were estimated an hour after turning on the cooling unit to get warm strength. The tests were rehashed multiple times in the two cases to check the exactness of the estimation.

1. Case-I

Air enters the room through a free-standing one-way supply diffuser which was called displacement ventilation device of the rectangular cross-sectional area located at the center ground of the north wall, occupants were classified in 1st and 2nd situation as shown in Figure 2.

2. Case-II (DV combined with PV)

Air enters the room through a suppling device displacement and personal ventilation. Personal ventilation draws the air from inside the room at a height of 0.4 meters from the ground level. PV system consists of fans at different flow rates 5l/s and 10l/s where the air is flowing from the bottom to the top and through (ATD) to the seating human level (1.1m). Occupants were classified in the 3rd and 4th situations as shown in Figure 3.

PV from 5l/s to 10l/s (5)

4. ANALYTICAL INVESTIGATION

a. General Governing Conservation Equations

The conservation equations which govern the continuity, momentum and energy at constant density in a Cartesian coordinate can be presented as follows.

- Preservation of mass:

Preservation of mass is given by the accompanying condition (Wargocki, P. et al., 2000)

$$\partial/\partial x(u)+\partial/\partial y(v)+\partial/\partial z(w)=0$$

(6)

Preservation of energy (Navier_Stokes equation): The resultant powers affecting a differential control volume are regularly controlled by applying energy protection law in each facilitate bearing to ask three power parts inside the x, y and z headings.

- Protection of warm energy (Fluent Inc., 2006):

$$\frac{\partial}{\partial x}(\rho UT)+\frac{\partial}{\partial y}(\rho VT)+\frac{\partial}{\partial z}(\rho WT)=\frac{\partial}{\partial x}\left(\Gamma\frac{\partial T}{\partial x}\right)+\frac{\partial}{\partial y}\left(\Gamma\frac{\partial T}{\partial y}\right)+\frac{\partial}{\partial z}\left(\Gamma\frac{\partial T}{\partial z}\right) \quad (7)$$

The terms and are the fierce warmth motions, St is a source term considering the pace of warm vitality creation.

- The centralization of species condition:

Condition of species fixation can be composed as:

$$\frac{\partial}{\partial}(\rho uc)+\frac{\partial}{\partial y}(\rho vc)+\frac{\partial}{\partial z}(\rho wc)=\frac{\partial}{\partial x}\left(\Gamma\frac{\partial c}{\partial x}\right)+\frac{\partial}{\partial y}\left(\Gamma\frac{\partial c}{\partial y}\right)+\frac{\partial}{\partial z}\left(\Gamma\frac{\partial c}{\partial z}\right)+\frac{\partial}{\partial x}\left(-\rho\overline{u'c'}\right)+\frac{\partial}{\partial y}\left(-\rho\overline{v'c'}\right)+$$

$$\frac{\partial}{\partial z}\left(-\rho\overline{w'c'}\right)+S_C \quad \text{..........} (8)$$

b. Computational Set and Numerical Scheme

This examination incorporates a confined office space for case-I DV just there are two warmth source over tables (PC) and two-man with two places and for case-II DV joined with PV there is one warmth source over a strong table and one individual additionally with two places of inhabitant, other than new numerically study case which was used displacement ventilation only with office person and one PC with two positions which named as case-III.

The effective draft temperature (EDT) was resolved in (100) centers scattered along with ten levels range between (Y=0 to Y=1.8) inside the included zone. Similarly, are performed to facilitate utilitarian results with speculative results through pole-1 at (x=2 m, y=0 m, z=0.875). The reaction to stream field issues (temperature, pressure, speed, and so on) is characterized at nodes for each plane. the number of components for case-I first and second situations were 291707, 289496, for case II third and fourth situations were 255871, 267492, and case-III third and fourth situations were 200073, 199382, individually as appeared in Figure 6. Materials arrangement incorporating material properties used in recreation like thickness (ρ), heat (Cp), and warm Conductivity (k) at values 1.189, 1005 and 0.0258, individually. The Limit conditions which includes a critical effect on CFD recreation achievement in giving a dependable outcome for the tackled issue. Limit conditions that forced in CFD reenactment executed during this work contain three sorts: speed gulf and outpouring as organized in Figure 7 divider limit included Dividers with consistent temperatures and furthermore steady warmth motions from another inside source like an individual, PC and lightweight are 44.69, 176.47 and 2000 separately.

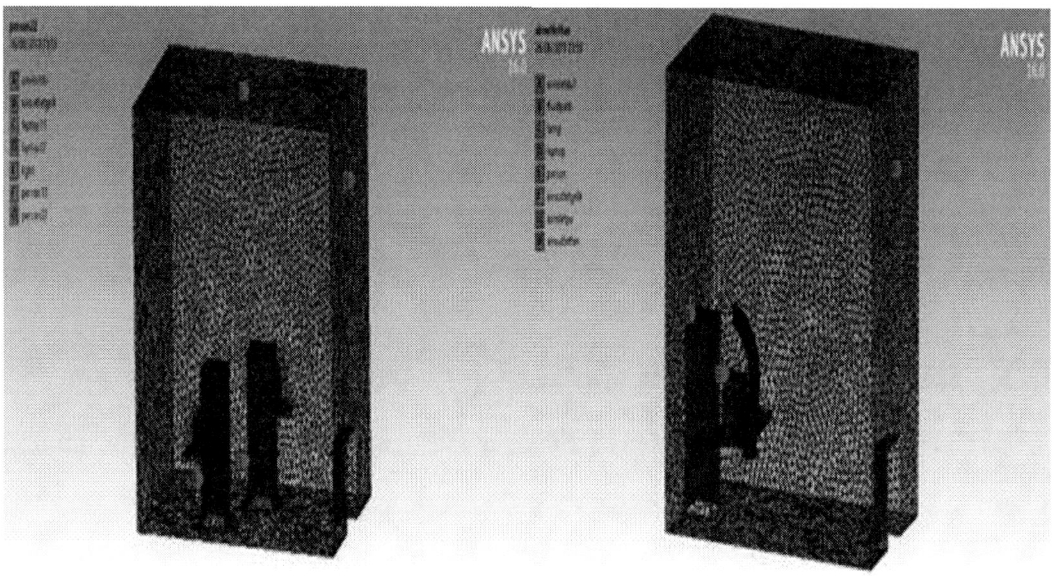

(a)

(b)

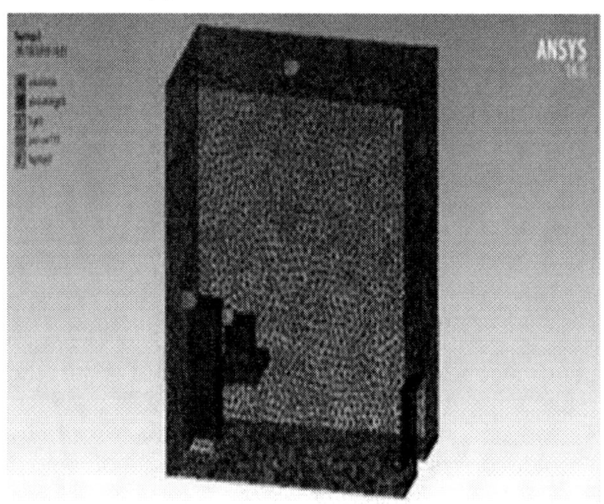

(c)

Figure 6. Fit model for exploratory space (a)First case (b)Second case (c)Third case.

When taking care of issues numerically it's difficult to encourage a specific arrangement, so acknowledged scaled mistake residuals ought to be determined for different terms like congruity, speed segments, and vitality. during this Appropriation Execution List (ADPI) was assessed to

IOP Conf. Series: Materials Science and Engineering **978** (2020) 012048 doi:10.1088/1757-899X/978/1/012048

appraisal dispersion execution of air at different cases during a ventilated space. Effective draft temperature (EDT) is determined from the condition below (M. Ahmed, 2004):

$$EDT = (Tx - \mathrm{Tav}) - 8 * (Vx - 0.15)$$

(9)

A comparison was made between the results of the CFD program from the Humans program and showed the practical results that were conducted in an isolated room during the first and second cases in the case of the presence of two displacement and personal ventilation devices or the presence of the displacement device only where the examination depends on the vertical temperature estimated in a room at six levels which are numerical investigations. General normal mistake computations on account of the resulting condition (10) (Hashimoto, 2005). the run of the mill rate mistake for case-I first and second circumstances were 6.8 % and 10.5% separately and case-II third and fourth circumstances were 7.8%, 12% individually at 5 l/s, and 13%, 13.5% individually at 10 l/s.

$$(E\%) = (1/n)\sum_{i=1}^{n} \frac{|x^i_{CFD} - x^i_{EXP}|}{x^i_{EXP}} * 100$$

(10)

5. Results and discussion

A. Experimental and Numerical Results

Results of numerical with practical were studied and a comparison was made between them for the first and second cases. It was studied theoretically and practically based on the different positions of the occupants and based on the presence of ventilation devices where the distribution of heat and speed of air was studied on three columns divided within the area designated for human comfort. The numerical prompts this examination for the tried room was received utilizing numerous computational works at different planes for all spaces. Air distribution performance index (ADPI) and reasonability temperature degree were settled. Figure 7 shows the association of air temperature at the space height through pols (1,2,3). Where the change in temperature is evident during the occupied space. This temperature thinking strategy is the addition of a good way to brilliantly air level.

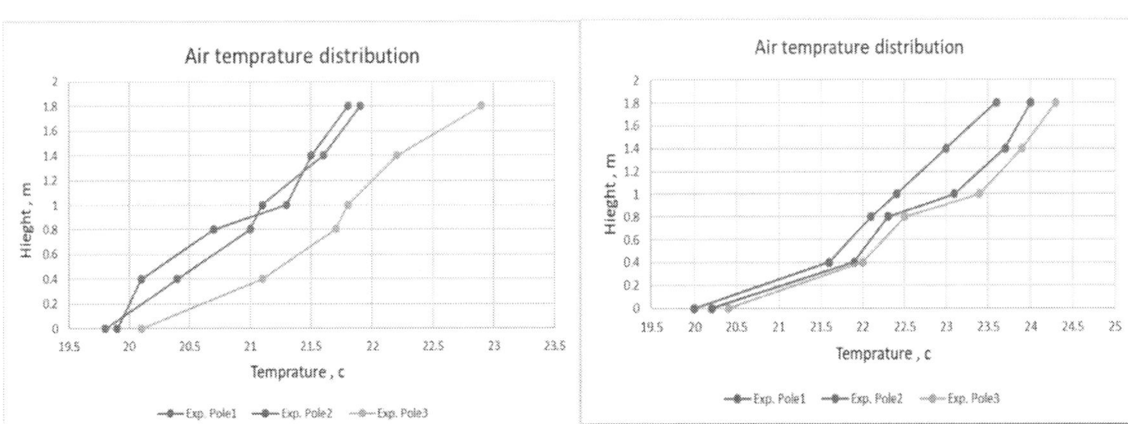

1st situation 2nd situation

(a) case-I

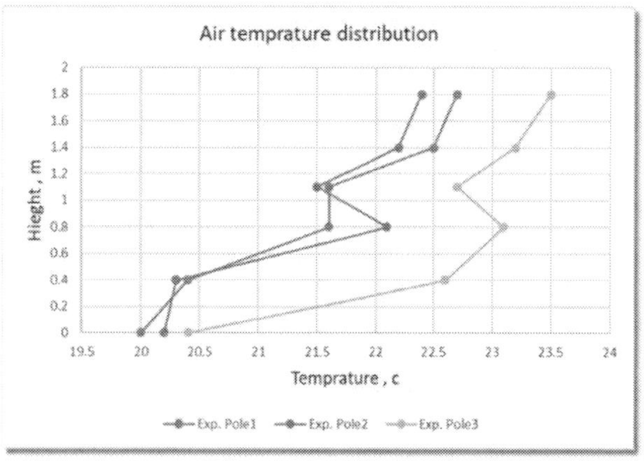

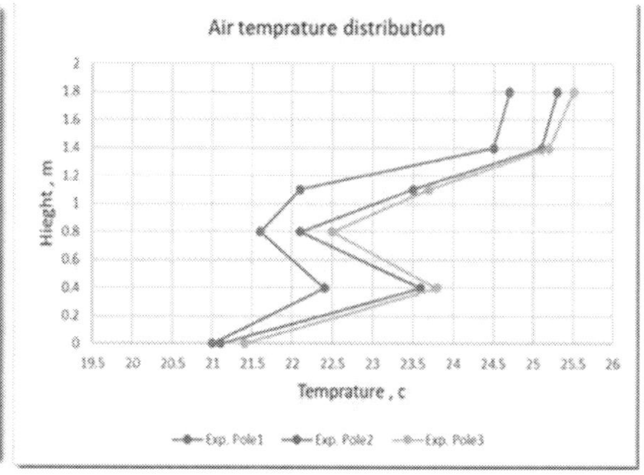

3rd situation 4th situation

(b) case-II at mass stream rate5 l/s.

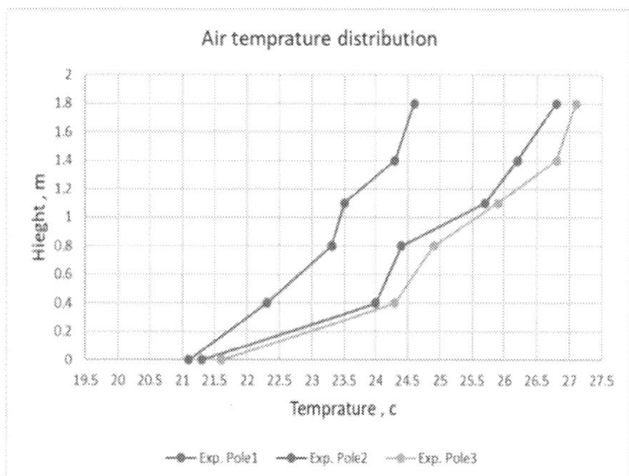

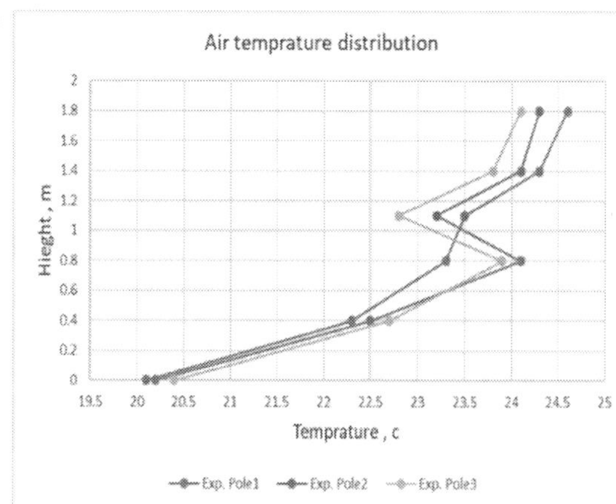

3rd situation 4th situation

(c) case-II at mass flow rate 10l/s

Figure 7. Estimated temperature dispersions for poles1,2 and 3

Table 1 and Table 2 speak to numerical consequences of Air Dispersion Execution Record (ADPI) and adequacy temperature(εt) for all cases considered. The most extreme estimation of viability and Air Circulation Execution Record (ADPI) was established at case-II for stream rate 10 l/s fourth circumstance.

Table 1. Numerical estimations for ADPI and effectiveness

Ventilation Devices	Case-I (DV only)		Case-III (DV only)	
	1st situation	2nd situation	3rd situation	4th situation
ADPI%	0.56	0.6	0.63	0.65
Effectiveness	1.3	1.32	1.33	1.34

Table 2. Numerical estimations of ADPI and effectiveness

Ventilation Devices	Case-II DV combined PV Case-II			
	At flow rate 5l/s		At flow rate 10l/s	
	3rd situation	4th situation	3rd situation	4th situation
ADPI%	0.54	0.64	0.46	0.71
Effectiveness	1.25	1.4	0.93	1.8

Figures 8 and 9 demonstrated numerical outcomes as shapes of air temperature dispersion at five designs for the testing room. Additionally, the same figures indicated that the temperature increments from 20° C near the accessibility terminal reach about 30° C close to physical bodies. It found on the off chance that case-I and case-III that the ground district close to the zone of diffuser would show low temperature esteems, as indicated by the cooling impact of the accessibility entering the air. Besides, a ceaseless augmentation of temperature gained as stature increase inside the space beside that case-III shows a little addition of temperature level close to a locale of warmth sources nearness than case-I, as for case-II exhibited that PV drawing the air at low high at 0.4 m to arrange high at 1.1 m (breathing zone) along these lines, so temperature degree at seat level is smaller than case-I and case-III. Moreover, a rising of indoor temperature is as often as possible noted near the individual, PC, lighting for all cases.

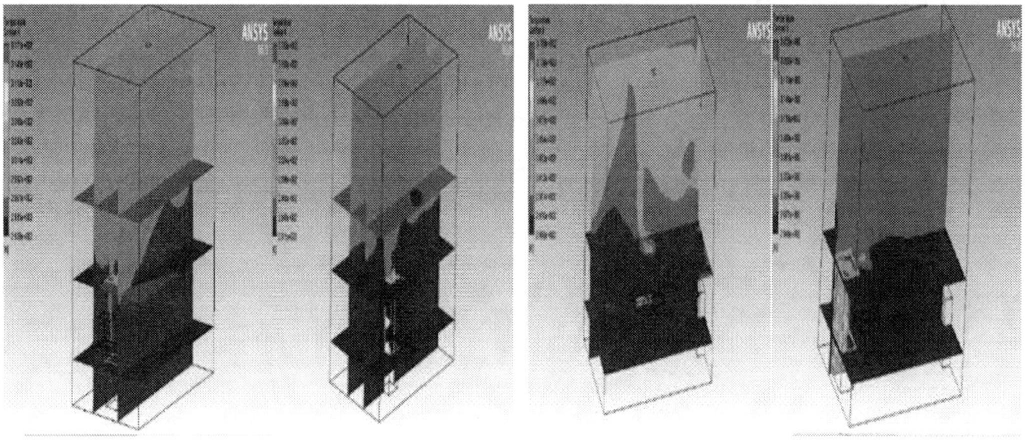

1st situation 2nd situation 3rd situation 4th situation

(a) (b)

Figure 8. Temperature forms plans (1,2,3,4&5), a-Case-I, b-Case-II

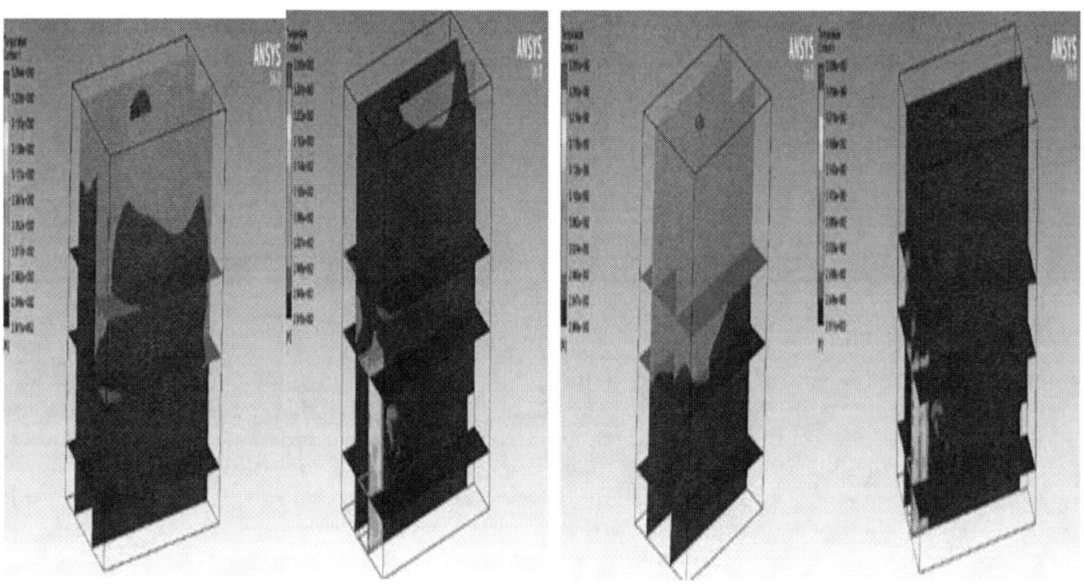

3rd situation 4th situation 3rd situation 4th situation

(a) (b)

Figure 9. Process temperature shapes plans (1,2,3,4&5) for Case-II, (a) at mass stream rate 5l/s, (b)at mass stream rate 10l/s.

For case-II (DV combined with PV), Figure 9 shows the conscious temperature apportionment which is best than two cases (case-I and case-III) gained in Figure 9. For case-I with case-III thermal manikin, PC and light effect on the environment of all room air temperature that reduces ventilation capability, for case-II suitable spread region and air course gained near the three warmth source result from the improvement of cold air from personal ventilation decrease heat transmitted by the thermal manikin, light and PC. In like manner, case-I and case-II in Figures 8 and 9 second and fourth situation showed best air dissemination over first and third condition while case-II at mass stream 10 l/s fourth situation show air movement execution document and effectively better than each and every other case and condition. Numerical results show the speed fields for the three cases are shown in Figures 10 and 11. High estimations of velocity were gotten at the lower territory where the region of air effortlessly diffuser DV exists during this zone. The air move process at this zone is progressively capable of diminishing the temperature of the hot space of warmth sources during this zone.

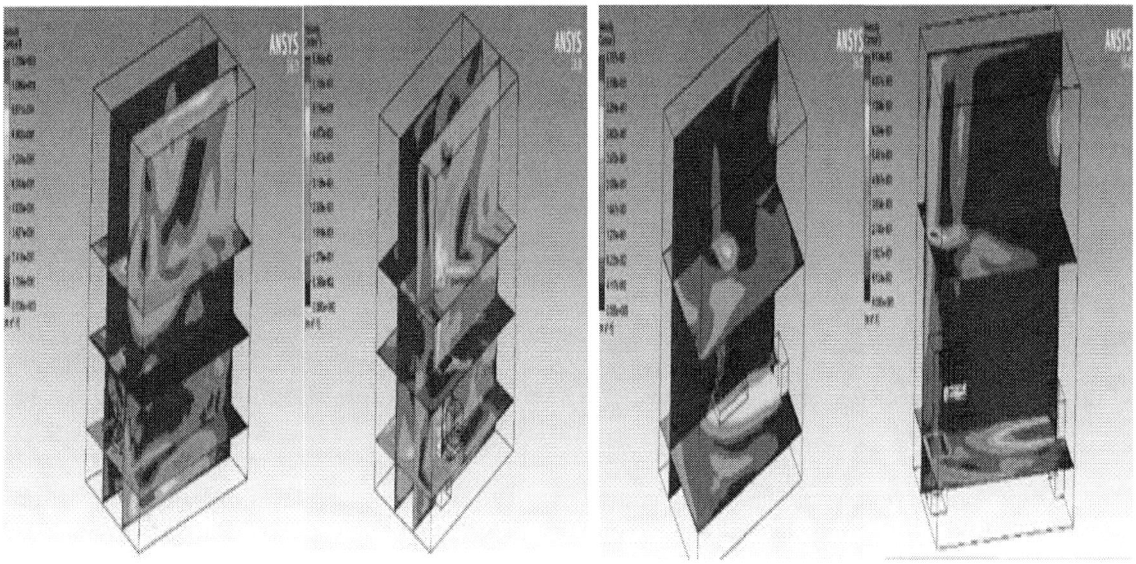

1st situation 2nd situation 3rd situation 4th situation

(a) (b)

Figure 10. velocity contours for planes 1.2,3,4&5 to (a) Case-I (b) Case-III

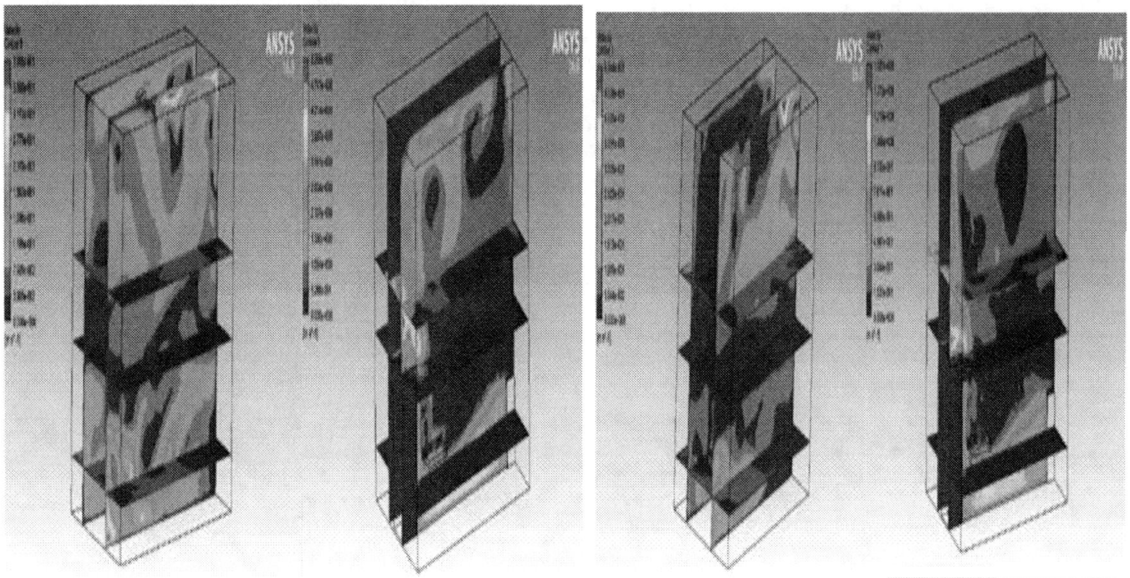

3rd situation 4th situation 3rd situation 4th situation

(a) (b)

Figure 11. Processed speed shape at planes 1.2,3,4&5 for case-II (a) at mass flow rate 5l/s (b) at mass flow rate 10l/s

The distinction between estimated temperature profile for pole-1 at (1) m from DV air diffuser uprooting ventilation with reenacted temperature profile utilizing (RNG-K-ε) disturbance model. This correlation appeared in Figure 12. the regular deviation between the trial and numerical qualities, for case-I first and second circumstances were about 6.8% and 10.5% separately. For case-II at the mass stream 5 l/s third and fourth circumstances were about 7.8%, 12%, and 13%, 13.5% for mass flow10 l/s separately.

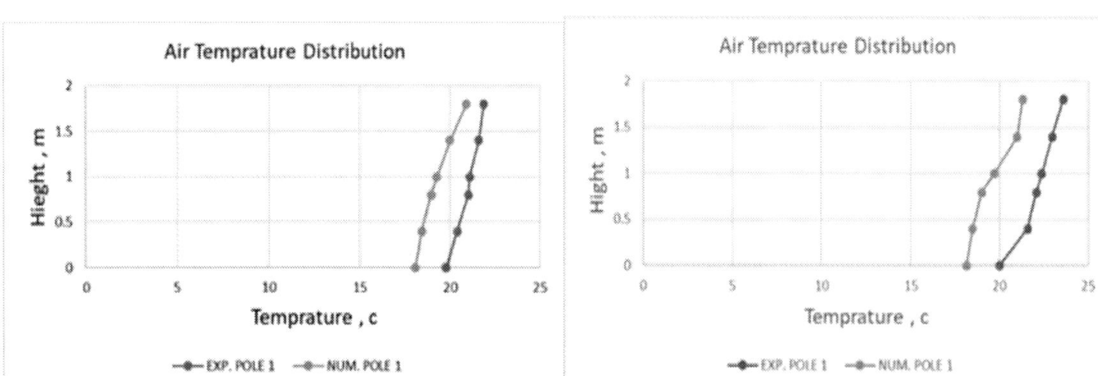

1st situation 2nd situation

(a)

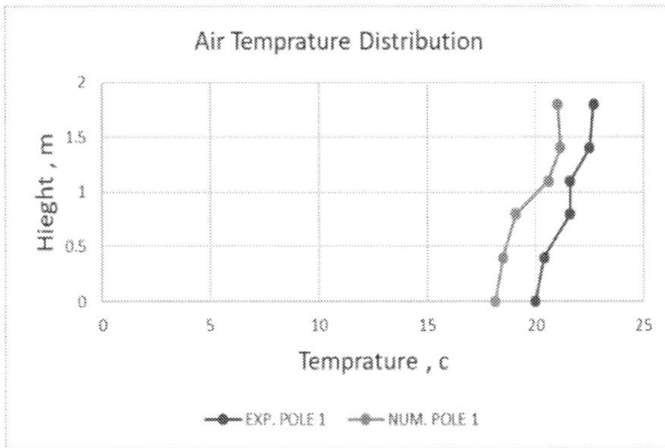

3rd situation

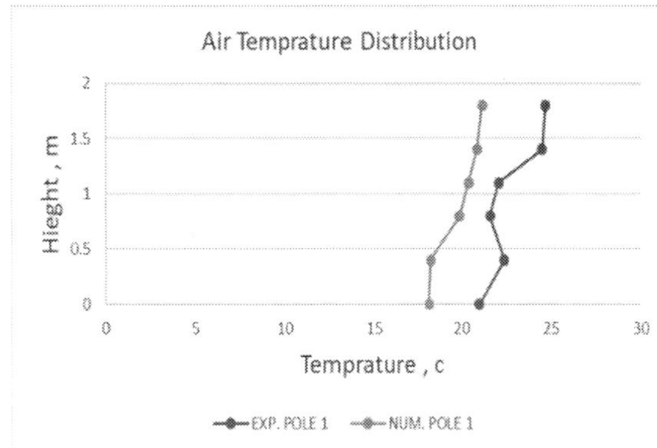

4th situation

(b)

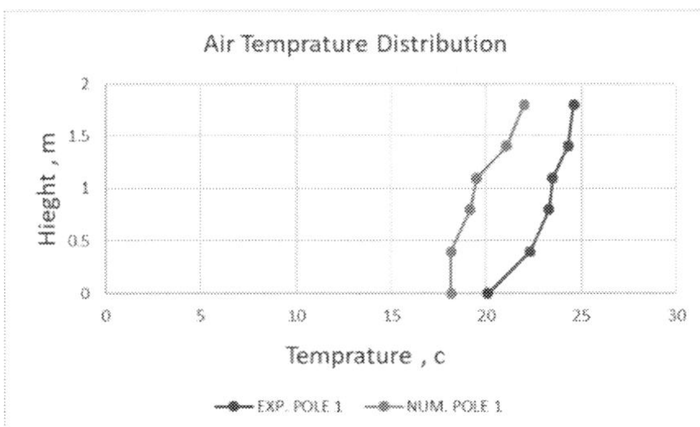

3rd situation

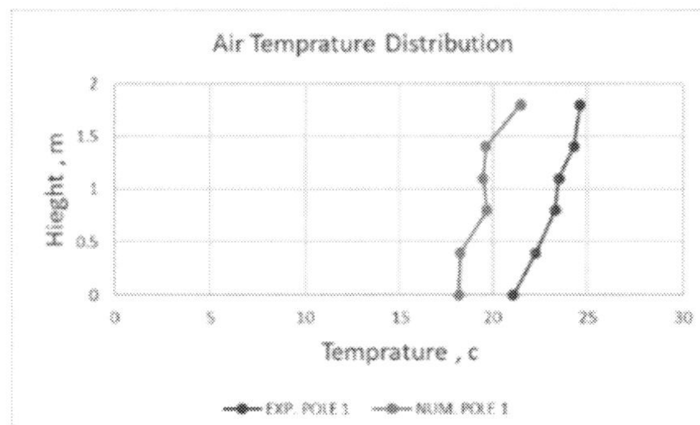

4th situation

(c)

Figure 12. Comparison between predicted and experimental results using DV and PV ventilation system. (a)Case-I(b) Case-II at mass flow rate 5l/s (c) Case-II at mass flow rate 10l/s

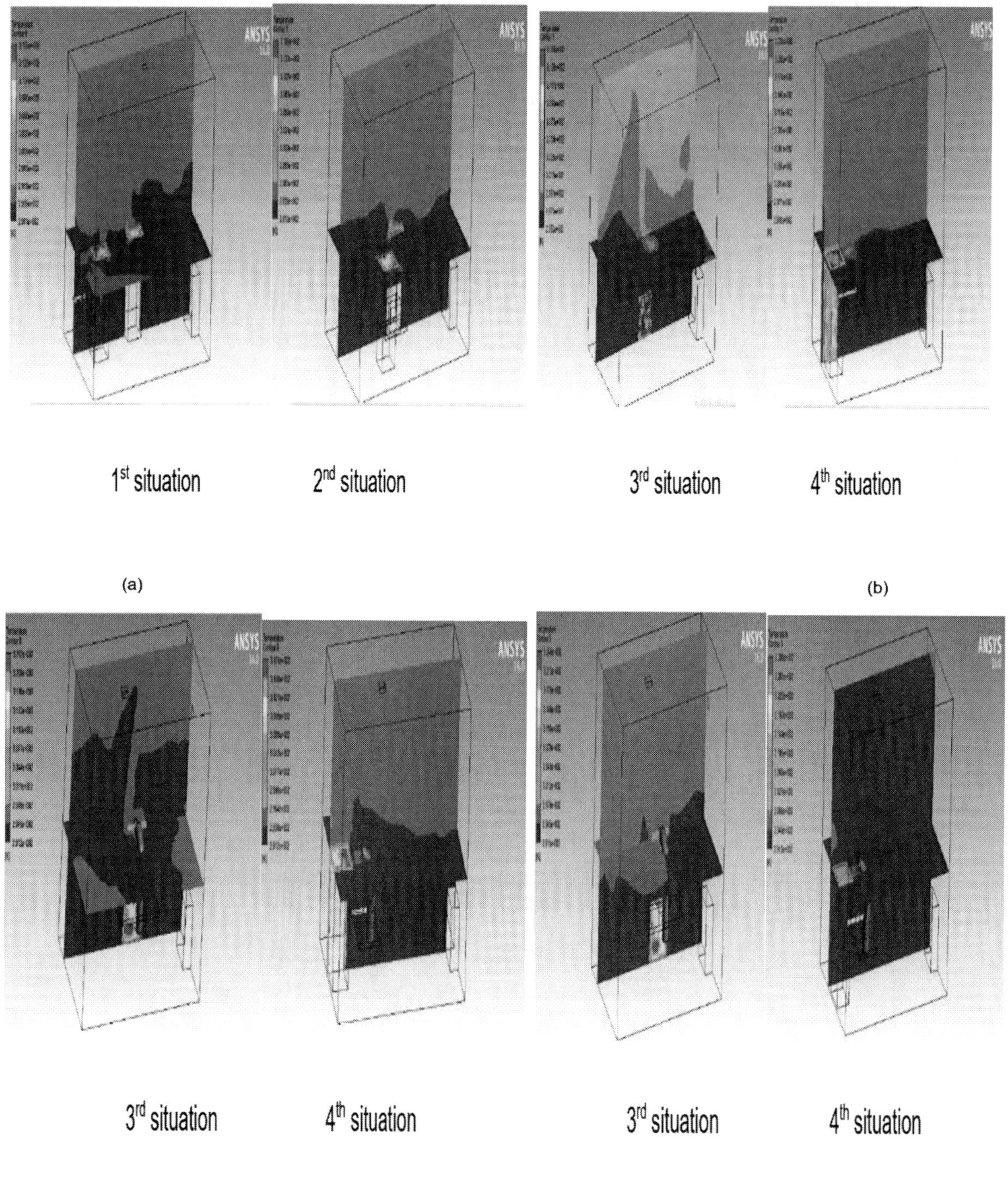

Figure 13. Temperature shape at 1.1m planes (2,5), (a)Case-I (b) Case-III (c) Case-II at mass flow rate 5l/s (d)Case-II at mass flow rate 10l/s

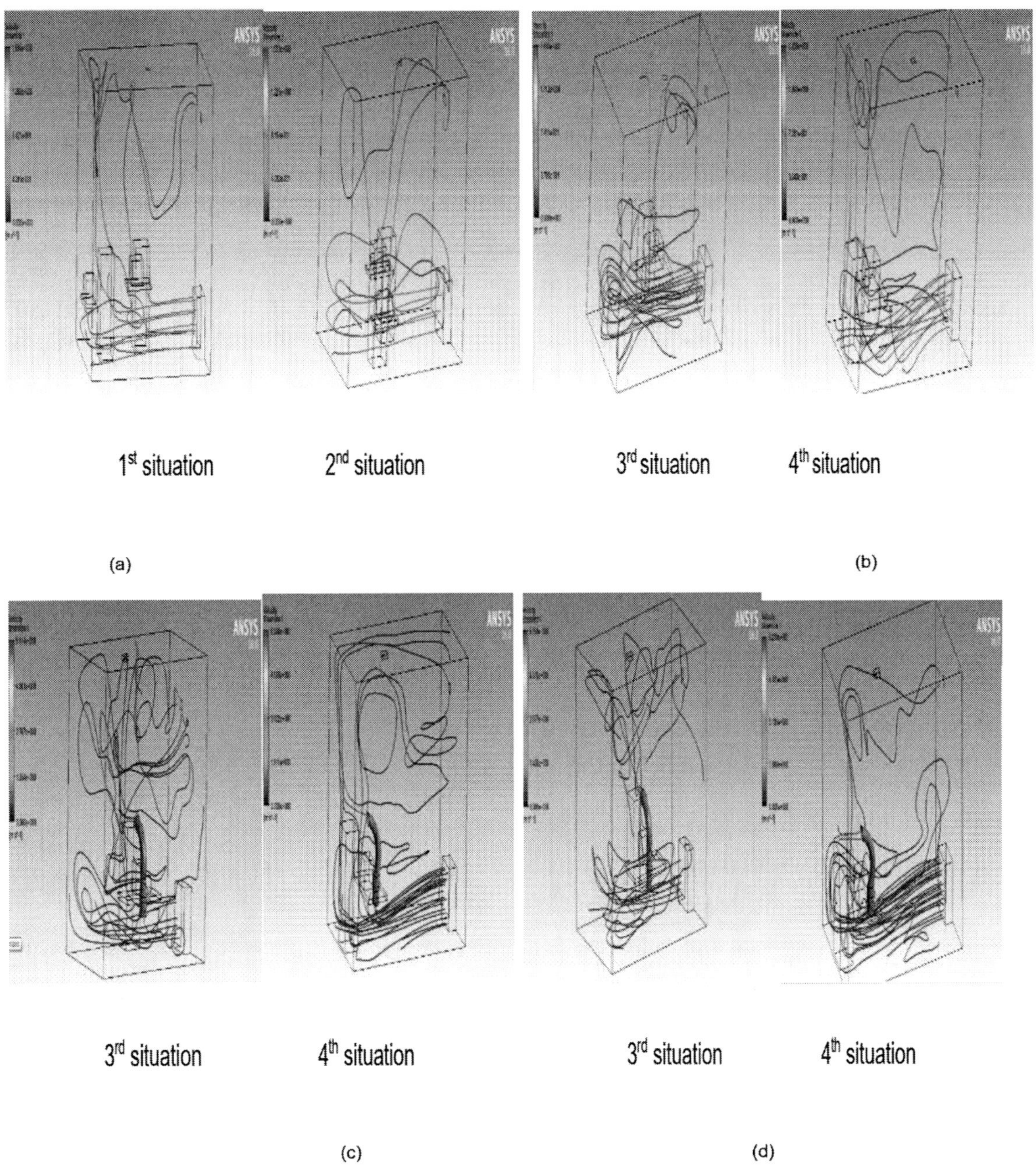

Figure 14. Streamlines flow pattern for (a)Case-I (b)Case-III (c) Case-II at mass flow rate 5l/s (d) Case-II at mass flow rate 10l/s

IOP Conf. Series: Materials Science and Engineering **978** (2020) 012048 doi:10.1088/1757-899X/978/1/012048

6. CONCLUSIONS

Experimental and numerical investigations were carried out to evaluate air distribution performance with combined displacement and personal ventilation. In this study, firstly using one-way displacement ventilation only according to two occupants' distribution situations, and secondly using combined displacement and personal ventilation system. It was concluded that the 2nd situation of occupant distribution achieved higher efficiency and good distribution than the 1st situation. Also, using the displacement ventilation combined with personal ventilation systems are more efficient than using displacement ventilation only, taking into consideration the various locations of the occupant distribution, it gives more acceptable agreement. Air mass flow 10 l/s from the personal terminal gives acceptable results for Air Distribution Performance Index (ADPI) equals to about (0.71) and effective temperature equals about (1.4). The results showed that personal ventilation (PV) system works to provide occupants with thermal and breathable comfort by increasing the ability to individually control air outlets as well as by improving indoor air quality for the purpose of use. Also will have the ability to reduce energy use by reducing the amount of air conditioning that is delivered to the interior.

Acknowledgment

The authors would like to express our appreciation to the University of Babylon / College of Engineering for providing laboratory facilities and financial support.

Nomenclature

A	Surfaces Area for Wall. m^2
Cp	Specific heats of the air at constant pressure. kJ/kg.K
$C_{1\varepsilon}$, $C_{2\varepsilon}$	Coefficient in the specific dissipation rates
D	Diameters m
dx dy dz	Control volume m
E_{ij}	The mean rate of deformation tenser
G	Gravitational acceleration m/s^2
H	Convection heats transfer coefficient $W/m^2.K$
K	Color factor correct
K_1	Conduction heats transfer coefficient for first layer of wall. W/m.K
K_n	Conduction heats transfer coefficient for final layer of wall. W/m.k
$k_{i,j,k}$	Turbulents kinetic energy at cell (i,j,k) J/s
N	Total numbers of draft temperature points measured in occupied
N_θ	Number of points of draft temperature measured in occupied zone
P	Pressure N/m^2
Pz	Zone population person
Q	Heats transfer through the wall. W
S_t	Source term for the rates of thermal energy Production J/kg
T	Temperature oC
U	Total heats transfer coefficient. $W/m^2.K$
u,v,w	Velocity components in x,y, and z-directions m/s
ui,j,k	Velocity at cell (i,j,k) m/s

x,y,z Cartesian coordinate m
V_x Local air velocity m/ s
V_{room} Volume of room m^3

Greek Symbols

ρ Air density kg/m3
ε Turbulents energy dissipation rate. J/kg.s
Pr Prandtl or Schmidt number
Σk Model constant
Σε Model constant
Ω Specific dissipation rate 1/s
Γ Diffusion coefficient (diffusivity) m^2/s
μt Turbulent viscosity $N.s/m^2$
Ω̇ Rotation velocity rad/s
ΔT_{hf} Temperatures difference from head to foot level. °C
αk, αε Coefficient in the specific dissipation rate
εt Effectiveness temperature.
τ_t Turbulent Reynolds stress

References

[1] Wargocki, P., Wyon, D.P., Sundell, J., Clausen, G., and Fanger, P.O. 2000. The effects of outdoor air supply rate in an office on perceived air quality Sick Building Syndrome (SBS) symptoms and productivity. Indoor Air, 10:222-236.

[2] Fang, L., Clausen, G., and Fanger P.O. 1998. Impact of temperature and humidity on the perception of indoor air quality during immediate and longer whole-body exposures. Indoor Air, 8:276-284.

[3] H. B. Awbi.2003. Ventilation of Buildings. Taylor and Francis group, second edition.

[4] Melikov, A.K. 2004. Personalized ventilation, Indoor Air. supplement 7, 14:157-167.

[5] Arsen K. Melikov, Radim Cermak, Milan Majer,2002. Personalized ventilation: evaluation of different air terminal devices. Energy and Buildings, 34:829–836.

[6] Stefano Schiavon.2016. Sensation of draft at uncovered ankles for women exposed to displacement ventilation and under floor air distribution systems. Building and Environment 96

[7] Ali Aedan, 2016. Experimental and numerical investigation of displacement ventilation system with assessment of four turbulence models. M.Sc. Thesis, University of Babylon, Iraq.

[8] Iraqi cooling code, 2012.

[9] Seppänen, O. 2008. Ventilation strategies for good indoor air quality and energy efficiency. International Journal of Ventilation, No.4,6:297-306.

[10] A. k. Melikov, H. Grønbæk, J. B.Nielsen, 2007. Personal Ventilation: from research to practical use. Technical University of Denmark, Building 402, DK-2800 Lyngby, Denmark.

[11] Nobukazu Kobayashi, 2003. Floor-supply displacement ventilation in a small office. Indoor and built environment, 12:281-291.

[12] Wargocki, P., Wyon, D.P., Sundell, J., Clausen, G., and Fanger, P.O. 2000. The effects of outdoor air supply rate in an office on perceived air quality. Sick Building Syndrome (SBS) symptoms and productivity, Indoor Air, 10:222-236.

[13] Fluent Inc, 2006. Fluent 6.3.26 user's guide.

[14] Abdul-Jabar M. Ahmed, 2004. Regional Airflow Characteristics in Mechanically Ventilated Rooms. M.Sc. Thesis, University of Technology.

[15] Yukihiro Hashimoto, 2005. Numerical investigation on airflow in an office room with a displacement ventilation system. Ninth International IBPSA Conference Montréal, 15-18, Canada.

Evaluating and Modeling of Modified Asphalt Binder at Elevated Temperatures

Yasameen Hadi Abed[1,*], Abdulhaq H. Abedali AL-Haddad[2]

[1]College of Engineering, University of Mustansiriayah, Baghdad, Iraq
E-mail: yasameenh.abed@gmail.com
[2]College of Engineering, University of Mustansiriayah, Baghdad, Iraq
E-mail: Abdulhaq1969@uomustansiriyah.edu.iq
* Corresponding author: College of Engineering, University of Mustansiriayah, Baghdad, Iraq. Tel: +9647802547739.

Abstract. Many regions in Iraq with high weathering temperature suffer from very hot climate during a long summer season continue to more than 5 months with temperature rise to more than 50C°.Also, increasing road traffic and axial loads causing a permanent deformation that leads to decrease pavement performance and reducing its life. Asphalt binder is considered one of the most significant keys in flexible pavement and its characteristics have a great impact on asphalt mixtures so choosing the suitable type of asphalt binder may improve its performance and control on permanent deformation failures. In this research, Daurh and Basrah asphalt binders were chosen to be characterized and enhanced their properties by using some modifiers. Three types of modifiers were selected with different percentages, styrene-butadiene-styrene (SBS), butyl rubber (BR) and BG plus anti-stripping agent. The results showed Multi Stress Creep and Recovery (MSCR) test is a suitable technique to evaluate the improvement of asphalt cement. The degrees of improvement in the percent of recovery %R and the non-recoverable compliance Jnr values and percent of reduction accumulated strain at 0.1 and 3.2 kPa after asphalt modification is higher for the SBS followed by BR then BG plus and Durah asphalt binder has better rutting resistance than Basrah asphalt cement. The behavior of modified asphalt cement modeled as viscoelastic materials and predicted the sigmoidal model as a suitable mechanical model to represent this behavior.

Keywords: *Asphalt cement; MSCR; recovery percent R; non-recoverable compliance Jnr; sigmoidal model*

1. Introduction

Rutting distress in flexible pavements appears due to the repeated traffic loading as a result of the accumulation of permanent deformation and significantly influenced by asphalt binder's rheological

IOP Conf. Series: Materials Science and Engineering **978** (2020) 012049 doi:10.1088/1757-899X/978/1/012049

properties (Cooper et al.,1985 and Sousa et al., 1991). Due to the viscoelastic behavior and thermal susceptibility of asphalt binder, permanent deformation is greatly increased at high temperatures and in the existence of heavy and slow-moving vehicles. In general, the rutting phenomenon reduces the comfortable and safety envisaged by the users of roads, therefore using asphalt cement with good rheological properties will increase the resistance of pavement to this phenomenon. To attain this aim, construction materials in flexible pavement need to be thoroughly selected by indicating the results of authoritative laboratory testing, that able to assess their irreversible strain response under repeated loading and yielding actual performance that related to ranking.

This would force the researchers to develop more advanced ways to enhance the asphalt pavement to resist the high temperatures and loads and also to enhance the evaluating methods of those materials. The main goal of these improvements is to provide asphalt pavement material with higher durability and last longer which results in cost-effective material. One of the common ways to achieve such improvements is to add some polymer modifiers to the asphalt cement to reduce rutting and cracking (D'Angelo, 2007).

When asphalt cement used as construction materials of roads, it should be tested in a series of tests to evaluate its strength and capability to resist rutting and cracking phenomena. The elastic recovery test is the main test used to evaluate the existence of elastomers in the modified asphalt cement. Increasing the percentage of elastomers improves the rutting resistance of flexible pavement. However, the state Department Of Transportation (DOT) experts were not satisfied with this test because the test needs a long time to finish (4-5 hours) and most agencies do not follow (AASHTO) standard guidelines to conduct the test (D'Angelo, 2007 and Abedali, 2017).

The traditional SHRP parameter ($G^*/\sin(\delta)$) (Anderson et al., 1994) certified in performance grading proved to be inappropriate in assessing the actual potential of asphalt cement to anti-rutting, particularly in the case of products of polymer-modified asphalt binder (Bahia et al., 2001). The limits of the SHRP approach are essentially connected to the fact that the ($G^*/\sin(\delta)$) is identified by using small-strain oscillatory loading within the linear-viscoelastic range, away from real conditions of damage. Many studies have been conducted by (Bahia et al.,(2001), Masad et al.,(2009) and Saboo and Kumar, (2016)) to get over these limitations, resulting in developing of several standard methods used to evaluate the rutting properties of asphalt binders. among these tests, the Multi-Stress Creep and Recovery Test (MSCR) had a prominent role because of its ability to simulate the discontinuous traffic loading nature and material dependence on stress (AASHTO TP 70,2013).
To replace ($|G^*|/\sin(\delta)$), the current parameter of Superpave at a high temperature of asphalt cement, a new test, named Multi Stress Creep and Recovery (MSCR), was studied extensively (D'Angelo, 2009). In this test, a series of shear creep and recovery is used. In its current form, the MSCR test composes of 1 second of creep loading followed by 9 seconds of recovery over multiple stress levels of 0.1kPa and 3.2kPa at 10 cycles for each level of stress (Wasage et al., 2011).

MSCR is considered a new test method that has been introduced to evaluate the resistance of asphalt mixture to permanent deformation. According to Saboo and Kumar (2016), the MSCR test method was introduced as a part of the new Superpave grading system (AASHTO MP 19-10) and accepted as a standard. The MSCR test output used to determine non-recoverable creep compliance (Jnr) and recovery percent (%R) for identifying the susceptibility of different types of asphalt cement to permanent deformation. The experimental results showed that there is a good correlation between the unrecoverable creep compliance (Jnr) at 3.2 kPa and the actual field performance (Kumar et al., 2017). The (Jnr) parameter determining at high values of used stresses and at non-linear material characteristics have is more accurate because the strain in asphalt binder film on aggregate surfaces is much greater than the strain that happens in asphalt mixtures (Subhy, 2017).

Eventually, (MSCR) test was created as an alternative for the presence AASHTO M 320 test at high temperature for asphalt cement. The output results obtained from the MSCR test can also be used as a replacement for the different SHRP tests. Also, to demonstrate the essential properties of the MSCR test likes an easy to use in the performance-related test. According to Golalipour, (2011), both base and modified asphalt cement were evaluated in the developed a new asphalt cement test to identify the permanent deformation property at high temperatures for asphalt cement. Equipment that used in the testing of the asphalt cement was the existing dynamic shear rheometer (DSR) which is widely accepted equipment by highway agencies that used in identifying the rheological properties of asphalt cement in specifications that related to the stress-strain response of viscoelastic materials and is very appropriate to assess asphalt types of cement (Golalipour, 2011). This study will focus on the following objectives:

1. Quantify the percent recoveries (%R) and non-recoverable compliance (Jnr) of the base and modified asphalt cement by using the Multiple Stress Creep and Recovery (MSCR) test.
2-. Examine the viscoelastic behavior of asphalt cement before and after using modifiers and the effect of modification on an elastomeric response and the recovery rates.
3. Simulate the behavior of asphalt cement as viscoelastic materials.

2. Experimental work

2.1. Materials

Two types of modified and unmodified asphalt types of cement were collected to perform the (MSCR) and frequency sweep test. To cover a wide range of viscoelastic properties of asphalt cement, ten samples of asphalt cement(base and modified) were used in this experimental work included two samples of AC 40-50 base asphalt binder from two different refineries (Al-Basrah and Al-Durah) Their physical properties are presented in Table 1, the other samples included modified asphalt cement with radial styrene–butadiene styrene (SBS) with different percentages (5and 7% by weight of asphalt cement) mixed with asphalt cement for 2 hours and 180 ˚C as a mixing temperature(Kareem, 2016), one percentage of Butyl rubber (BR) (9%)(by weight of asphalt cement) mixed at a temperature of 165 ˚C for 60 minutes (Liu and Weilong, 2003)and one modified asphalt cement with at 0.075% (by weight of asphalt cement) of a liquid anti-stripping agent (BG plus), the product from Sika IRAQ L.L.C that mixed at a temperature of 150 ˚C and for 30 minutes. The properties of BG plus and SBS are presented in Tables 2 and 3 respectively. All samples were prepared using a blending machine with 500 rpm and subjected to DSR and MSCR tests to measure the rheological properties and evaluate of non-recoverable creep compliance parameters of Jnr0.1 and Jnr3.2 correspondent to 0.1 and 3.2 kPa stress levels respectively.

Table 1. Physical and rheological properties of base asphalt cement (AC).

Property	Basrah asphalt cement	Durah asphalt cement
Penetration	41	43
Rotational viscosity (c.P)	527.1	479.01
Softening point (˚C)	52	51
Penetration Index (PI)	-1.2	-1.3
Complex modulus G*(Pa) at 40°C and 10 Hz	$5.5{*}10^5$	$5.43{*}10^5$
Phase angle (δ°)	60.79	61.13

Table 2. Properties of BG plus anti-stripping agent.

Property	Requirement
Chemical Base	Nanotechnology, Silane based
Form	Liquid
Appearance	Pale yellow
Density	1.0 Approximately @ 27° C
pH Value	Approx. 9.00
Consumption / Dosage	0.02% to 0.1%. By weight of

IOP Conf. Series: Materials Science and Engineering **978** (2020) 012049 doi:10.1088/1757-899X/978/1/012049

Table 3. Properties of SBS modifier.

Property	Unit	Requirement
Density	Kg\ m³	1240
Specific gravity	-------	0.95
Tensile Strength (σt)	MPa	min.32
Melting point	C°	170-190
Elongation	%	880
Molecular structure	-----	Radial

2.2 Laboratory Tests

2.2.1 Multiple Stress Creep and Recovery Test

RHEOTEST model RN4.3 dynamic shear rheometer (DSR) as shown in Figure 1(Abed and Al-Haddad, 2020) was used to establish a creep and recovery test concept to assess the potential of asphalt binder to rutting for both base and modified asphalt cement. Asphalt binder can be represented as a sandwiched between the DSR's parallel plates (25mm) with a 1mm gap setting, ten cycles of creep and recovery are conducted at two levels of stress as suggested in AASHIO (0.1 and 3.2 kPa). The test is conducted according to ASTM D7405–15 specifications at 64°C. The new test procedure allows two stress levels of 0.1(to describe asphalt cement's behavior within the linear viscoelastic region) and 3.2 Kpa (represent a binder's behavior within a non-linear viscoelastic region) for both base and modified asphalt cement as shown in Figure 2, to be applied for 1 second and allow recovery for analysis of Multiple Stress Creep and Recovery(MSCR) for Iraqi modified asphalt cement followed 9 seconds for 10 cycles and no time lag between cycles are applied as shown in Figure 3. The parameters that are obtained from the MSCR test are:

$$R_{0.1} = \frac{\text{sum}\ (\varepsilon_r\ (0.1,N))}{10} \quad \text{for}\quad N=1 \text{ to } 10 \qquad (1)$$

$$R_{3.2} = \frac{\text{sum}(\varepsilon_r\ (3.2,N))}{10} \quad \text{for}\quad N=11 \text{ to } 20 \qquad (2)$$

Non-recoverable creep compliance between 0.1 kPa and 3.2 kPa are determined using Equations (3)and (4) below:

$$J_{nr}\ (0.1) = \frac{\text{sum}(J_{nr}\ (0.1,N))}{10} \quad \text{for } N=1 \text{ to } 10 \qquad (3)$$

$$J_{nr}\ (3.2) = \frac{\text{sum}(J_{nr}\ (3.2,N))}{10} \quad \text{for } N=11 \text{ to } 20 \qquad (4)$$

J_{nr} : non-recoverable creep compliance and R: percent of Recovery.

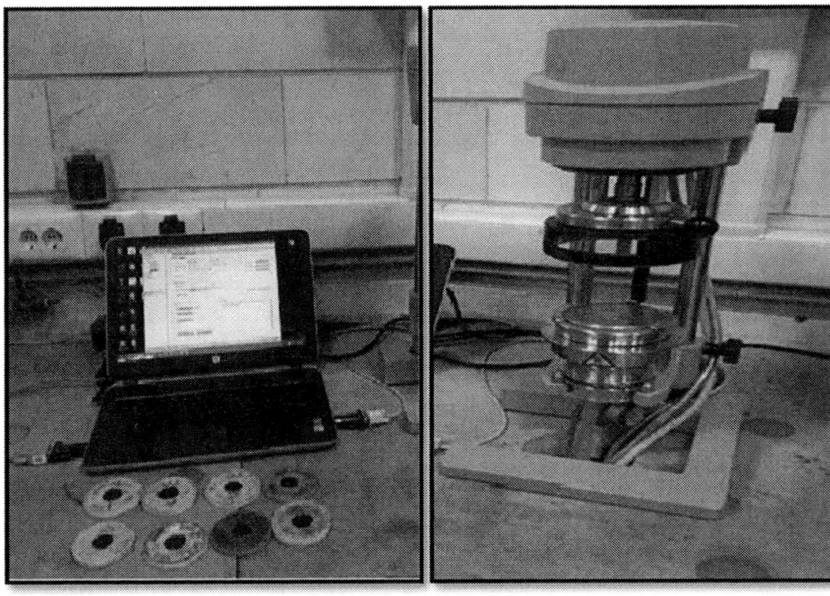

Figure 1. RHEOTEST model RN4.3 dynamic shear rheometer(DSR).

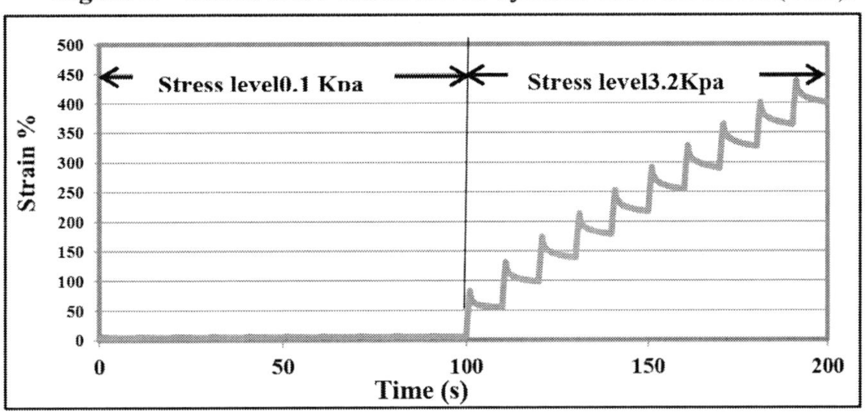

Figure 2. A typical plot of the MSCR curve for modified Durah binder with a 7 %SBS modifier.

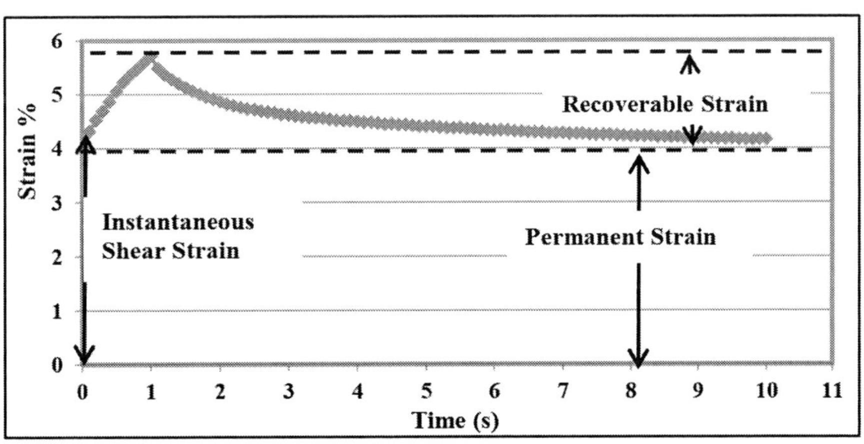

Figure 3. A typical one cycle of creep-recovery for Daurah binder modified by 7% SBS

2.2.2 Frequency Sweep Test

Rheological testing was performed on an RHEOTEST model RN4.3 dynamic shear rheometer (DSR) to evaluate the rheological behavior of modified and base asphalt binder from Durah and Basrah refineries and using parallel plate geometry based on ASTM (D7175 – 08) "Determining the Rheological Properties of Asphalt Binder Using DSR" Plates diameter was 8 with gap 2mm between them and plate with 25 mm diameter and 1 mm gap, depending on test temperatures.
The frequency sweep test is considered as a useful way of measuring the viscoelastic properties of asphalt cement in DSR. The procedure of this test composes of applying constant low-level load amplitude to avert asphalt specimen damaging overloading frequencies rang (typically from 0.1 to 10 Hz due to equipment limitations).
To assure results repeatability, all tests were performed at least twice. Viscoelastic behavior of asphalt cement was considered linear when the modulus value had not deviated more than 5% from its initial value and the results of complex modulus are used to predict a mathematical model to describe the rheological properties of binders. In this study, the rheological tests were conducted under the controlled-stress conditions at low and intermediate temperatures (4, 16 and 28) °C using plate's diameter 8 mm with 2 mm gap opening and at a higher temperature (40, 55 and 70) °C using 25 mm diameter plate to plate, and 10 rad/s of frequency and 1 mm gap opening.

3. Results and Discussion

MSCR test was carried out to study the behavior of the materials at higher stresses level in regions exceeded the linear viscoelastic limits. The output data of the MSCR test is utilized to determine the non-recoverable creep compliance (Jnr) and recovery percent (%R) for identifying the susceptibility of asphalt cement to permanent deformation or rutting. Non-recoverable creep compliance (Jnr), is calculated by dividing non-recoverable shear strain by the shear stress and used to assess the rutting potential of the asphalt cement. The results presented in Figure 3 and Figure 4 showed that the addition of the modifiers provides a high reduction in strain values compared with base asphalt binder, which is an improvement to the performance.
At both stress levels, 0.1and 3.2 KPa, it can be noticed that the reduction of the accumulated strain is more than 4.5 times for 9% BR, 13 times for 0.075% of BG Plus, 14.31 times for 5% SBS and 66.5 times for 7% SBS when compared them with base Basrah asphalt cement at 0.1 KPa stress level. While the effect of modifiers on Durah asphalt cement was better than Basrah asphalt cement, the reduction in accumulated strain is more than 6.5 times for 9% BR, 6.1 times for 0.075% of BG Plus, 58.43 times for 5% SBS and 116 times for 7% SBS when compared them with base asphalt binder.

3.1Percent of Recovery (%R)

Using the modifiers with base asphalt cement of Durah and Basrah refineries increased elastic recovery of the binders to maximum values between (21.2-36) at stress level 0.1 KPa and between (52.93 – 64.45) at stress level 3.2 KPa compared to the base binders indicating a superior polymer network in the binder as shown in Tables4 and 5 and Figures 5 and 6. The highest values of recovery percent at the 7% SBS modified asphalt binder, followed by asphalt binder modified with 5% SBS, then asphalt binder modified 9% BR and, as the last one, the base asphalt cement. In general, the unmodified asphalt binder would suffer a lower percentage of recovery than those modified with modifiers. Interestingly, the results show that at 0.1Kpa the addition of 7% SBS to the base binder for both types of asphalt binders recorded a high percentage of the recovery. However, it reduced between (33- 49) % when increasing the stress to 3.2 KPa for both types of asphalt cement. On the other hand, adding %7 of SBS to the base binder increases the recovery percent by 21.2 and 52.93 times compared to the base binder at 0.1 and 3.2 KPa respectively for Durah asphalt binder and by 36 and 64.45 times compared to the base binder for Basrah asphalt cement.

Table 4. The MSCR parameters at the two stress levels for Durah asphalt cement (AC).

Binder type	MSCR Parameters				Different percentage in Jnr
	at stress level 0.1 kPa		at stress level 3.2 kPa		
	R (%)	Jnr (1/kPa)	R(%)	Jnr (1/kPa)	
base	4.381	3.012	0.894	4.166	38.346
AC with 5%SBS	73.535	0.047	43.139	0.134	183.263
AC with 7%SBS	92.801	0.013	47.32	0.124	857.360
AC with 9%BR	29.839	0.460	27.028	0.629	36.862
AC with 0.075%BG	4.561	2.665	0.965	3.819	43.311

Table 5. The MSCR parameters at the two stress levels for the Basrah asphalt cement (AC).

Binder type	MSCR Parameters				Jnr Percent difference
	0.1 kPa Stress level		3.2 kPa Stress level		
	R (%)	Jnr (1/kPa)	R (%)	Jnr (1/kPa)	
base	2.532	4.059	0.942	5.203	28.209
AC with 5%SBS	41.799	0.254	17.198	0.475	86.557
AC with 7%SBS	91.344	0.027	60.719	0.176	555.390
AC with 9%BR	22.859	0.843	21.878	1.203	42.670
AC with 0.075%BG	8.743	3.084	0.675	2.136	44.382

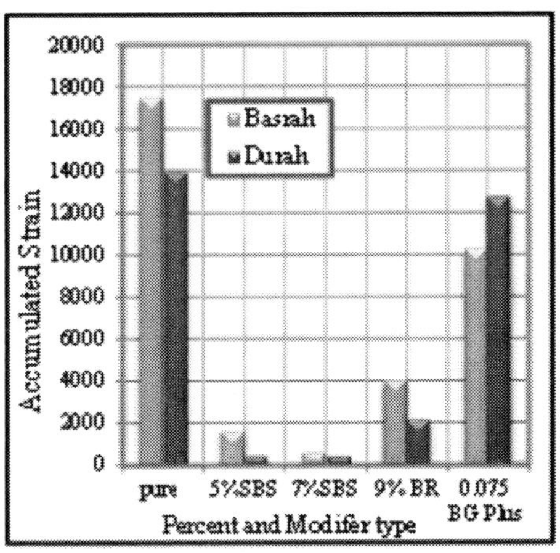

Figure 4. Accumulated strain for base and modified binders at 3.2Kpa stress level.

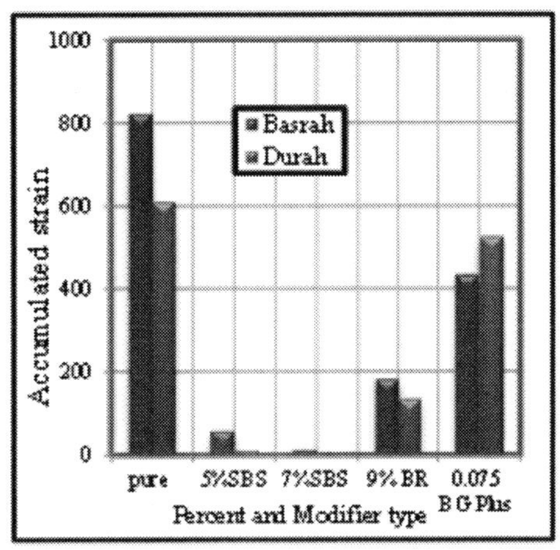

Figure 5. Accumulated strain for base and modified binders at 0.1Kpa stress level.

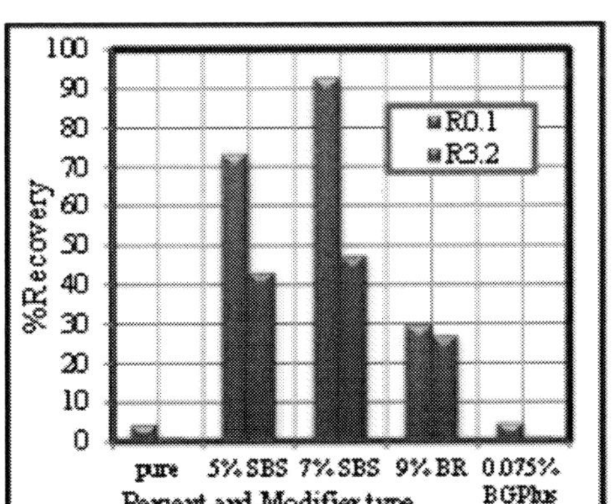

Figure 6. Variations in percent recovery at 0.1KPa and 3.2 KPa stress level for modified and pure Durah asphalt binder.

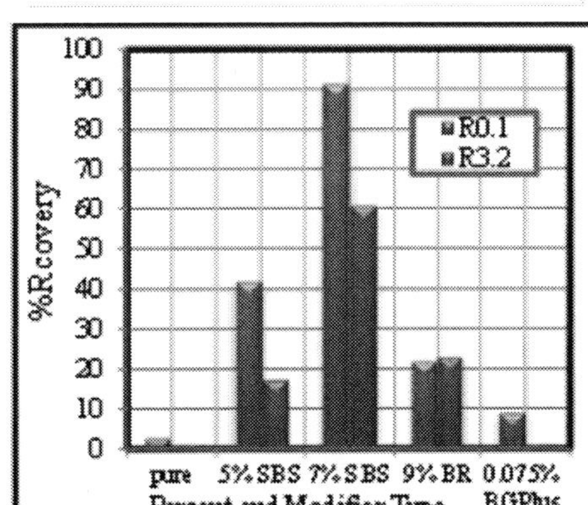

Figure 7. Variations in percent recovery at 0.1KPa and 3.2 KPa stress level for modified and pure Basrah asphalt binder.

3.2 Non-recoverable creep compliance (Jnr)

The improvement of modified asphalt cement can be estimated by the non-recoverable compliance (Jnr); which is sensitive to the applied stress and measured by 1/kPa. In this study, Jnr was calculated for the base and modified asphalt cement, and two kinds of asphalt binders, Figures 7 to 9 presents the variation of creep compliance at 0.1kPa and 3.2 kPa stress levels of asphalt binder with and without additives at 64°C. Decreased the non-recoverable compliance Jnr values of modified asphalt binder at the typical highest pavement temperature of 64°C in both stress levels, indicates that all modified asphalt cement are less susceptible to rutting than the base material because of the lower the Jnr value the stiffer the binder.

For both types of binders, Jnr was calculated for the base and modified binders as shown in Figures 7 and 8. It can be showed that increasing the stress from 0.1 to 3.2 Kpa did not make significant performance change in most of the binders especially those modified with elastomers as shown in Tables 2 and 3 for both types of asphalt cement. The Jnr results confirm the findings that the binder modification by 7% SBS resulted in the lowest value of Jnr which means that it has the best performance among the other modified binders in this study. The addition of SBS to modify asphalt cement achieved good behavior and approximately sustained that behavior with the increase of the applied stress followed by 5%SBS, 9% BR then 0,075% BG Plus when comparing them with base asphalt binder. Therefore, it could be mentioned that SBS modification is a good option for both local base binders. Depends on the results presented in Figure 10, Durah asphalt cement has better rutting resistance than Basrah cement.

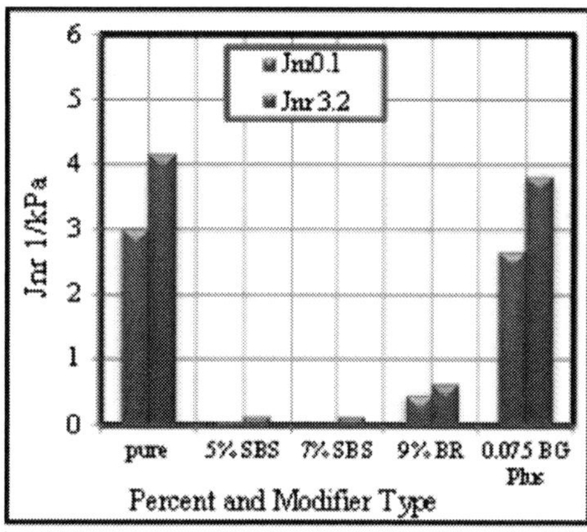

Figure 8.Variations creep compliance at 0.1kPa and 3.2 kPa stress levels of Durah asphalt binder with and without additives at 64°C.

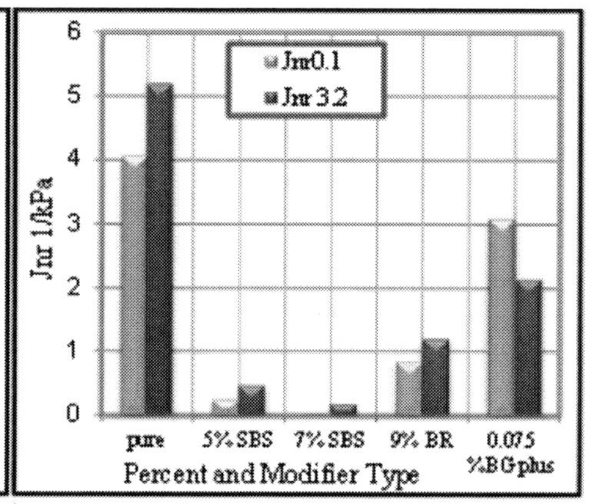

Figure 9. Variations creep compliance at 0.1kPa and 3.2 kPa stress levels of Basrah asphalt binder with and without additives at 64°C.

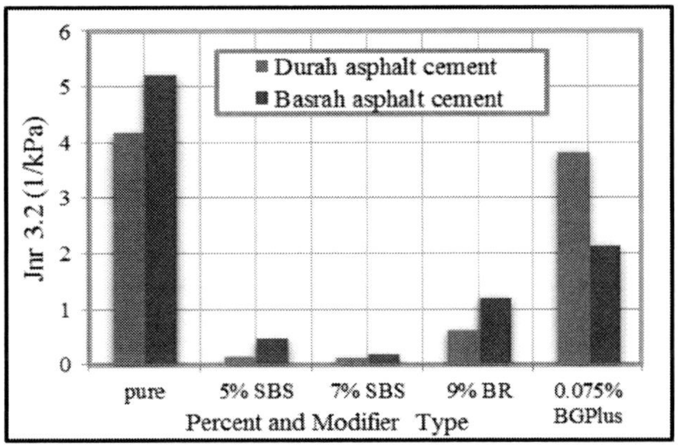

Figure 10. Variations creep compliance at 3.2 kPa stress level for both Basrah and Durah asphalt binders with and without additives at 64°C

4. Modeling of Experimental Data

This research investigates the use of one mathematical model namely the sigmoidal model, to describe the rheological properties of asphalt cement. Mathematically, the sigmoidal model can be presented in'Equation 5':

$$\log|G^*| = \delta + \frac{\alpha}{1+e^{\beta+\gamma(\log f_r)}} \quad (5)$$

Where: G* is the complex shear modulus of the asphalt cement, δ is a minimum value of G*, α is a difference between the maximum and minimum value of G*, fr is the reduced frequency (Hz) and ß,γ are S-shaped function parameters based on asphalt cement properties and characterizes the shape of sigmoidal functions-β presented the point of the turning curve halfway between the maximum and minimum value of G* and γ represent the curve slop; the standard value of γ is 1.0; if γ is more than 1.0 the curve is steeper and if less than 1.0 then the curve is shallower. In this research, the sigmoidal model is used for characterizing the master curves of complex modulus for asphalt cement. The four-parameter sigmoidal model is used for fitting the complex modulus of asphalt mixtures (Pellinen et al., 2003). However, this model has been modified to be used for asphalt cement (base and modified) data. Four fitting parameters, α, β, δ, and γ, are estimated by using the non-linear least square fitting method as shown in Tables 6 and 7. The fitting process is carried out with the assist of Graph Pad Prismver.8 Software.

Table 6. The average parameters of the sigmoidal model of different types of Durah asphalt binder.

Parameter	β	α	δ	γ
Base	3.744	21.544	-1.603	0.6216
5% SBS	2.597	11.350	1.439	1.052
7%SBS	2.447	15.271	1.624	0.8852
9%BR	3.385	8.678	1.492	0.8021
0.075%BG plus	3.739	23.915	-1.489	0.6686

Table 7. The average parameters of the sigmoidal model of different types of Basrah asphalt binder.

Parameter	β	α	δ	γ
Base	3.683	10.546	-1.552	0.7879
5% SBS	2.433	7.286	1.135	0.8929
7%SBS	3.855	8.179	1.116	0.7983
9%BR	2.814	7.865	1.540	0.8359
0.075%BG plus	3.421	10.870	-1.087	0.6225

Based on the results of the analysis process by Pad Prismver.8 Software:

1. At a given type of asphalt binder, the magnitude of δ for base and 0.075% BG plus are negative. The negative δ pointed out the value of complex shear modulus G*of asphalt cement is very small at low frequencies and high temperatures.
2. The values of δ through the analysis process are small because asphalt cement (base and modified) are acting like a viscoelastic liquid and the δ values reduce with the addition of modifiers comparing with base asphalt cement because the modifiers enhanced the elastic response at the high temperatures and/or low frequencies due to the polymer network.

3. The addition of modifiers reduces the values of β, δ, and α values and increase γ values for both types of asphalt binders comparing with base asphalt binder. This indicates that G* values are too small at low frequencies and high temperatures.

4. The γ values are observed to be steady for the base and modified asphalt binders, indicating that the value of γ did not play an essential role in affecting the slope of the master curve in the range from intermediate temperatures to high temperatures.

5. Conclusions

Several conclusions can be gain from this work and are listed below:

1. Both MSCR parameters: non-recoverable creep compliance and percentage of elastic recovery are improved as a result of adding modifiers to asphalt cement and increase with increasing modifier percent.
2. Using modifiers with base asphalt cement of Durah and Basrah refineries increased the elastic recovery of the binders to maximum values between (21.2-36) at stress level 0.1 KPa and between (52.93 – 64.45) at stress level 3.2 KPa compared to the base binders indicating a superior polymer network in the binder.

3. Decreasing the non-recoverable compliance Jnr values of Durah and Basrah modified asphalt cement at the typical highest pavement temperature of 64°C at both stress levels indicates that the modified asphalt cement is less susceptible to permanent deformation (rutting) than the base asphalt cement because of the lower the Jnr value, the stiffer the asphalt cement.
4. Regardless the asphalt type, ranking the formulations in terms of higher values of R and lower values for Jnr, i.e. higher rutting resistances, at the regular creep and recovery at the first cycle, the 7% SBS modified asphalt binder is the best, followed by asphalt binder modified with 5% SBS, the 9% BR modified asphalt binder and, then the addition of BG Plus anti-stripping agent didn't affect significantly on values of R of asphalt binders and the last one, the base AC. With increasing cycles number, the general effect observed is the reduction of %R and increase of Jnr, (i.e a lower rutting resistance), but the ranking is the same.

5. The best performance of the modified binder can be attained by mixing 7% SBS with base asphalt binder from the two sources and keep sustained with an increased level of stress to 3.2 Kpa. The modified binder was improved by 21.2 and 36 times compared to the neat Durah and Basrah asphalt respectively binders.

6. Based on the results of non-recoverable compliance Jnr, Durah asphalt cement has better rutting resistance than Basrah asphalt binder.
7. The effect of modifiers on Durah asphalt cement was better than Basrah asphalt cement. Maximum reduction in accumulated strain is more than 116 times for Durah asphalt cement modified with 7% SBS while for Basrah asphalt cement modified with 7% SBS is 66.5 times comparing with base asphalt cement.

8. The stiffness of asphalt cement increase with adding modifiers to asphalt cement and improve the properties of asphalt cement within the viscoelastic range at stress level 3.2 Kpa.
9. The addition of SBS to asphalt binder achieved good behavior and approximately sustained with increasing the applied stress. Therefore, it could be mentioned that SBS modification is a good option for both local base binders.
10. For a wide range of service temperatures between 4 °C to 70°C, using the function of the sigmoidal model to build the master curve appears to be in a good fitting with the data because it follows the physical form of measured data.

11- The addition of modifiers reduces the values of β, δ, and α values and increase γ values for both types of asphalt binders comparing with base asphalt binder. This indicates that G* values are too small at low frequencies and high temperatures.

Acknowledgments
The authors acknowledge Mustansiriayah University, College of Engineering, Highway, and Transportation Department for supporting this work. https://uomustansiriyah.edu.iq.

References

[1] Cooper K.E., Brown S.F., Pooley G.R. (1985). The design of aggregate gradings for asphalt base course, J. *Assoc. Asphalt. Pav*. 54, 324–346

[2] Sousa J.B., Craus J., Monismith C.L. (1991). Summary Report on Permanent Deformation in Asphalt Concrete, SHRP Report A/IR-91-104, Strategic Highway Research Program, Transportation Research Board, National Research Council, Washington, D.C.,

[3] D'Angelo, John A., and Raj N. Dongre, (2007). Creep and recovery for polymer modified binder with weak structure. *Public roads*, 70(5).

[4] Abedali, A.H., (2017). Evaluation of Rutting Susceptibility of Modified Asphalt Binder. *Journal of Engineering and Sustainable Development,* 21(04).

[5] Anderson D.A., Christensen D.W., Bahia H.U., Dongre R., Sharma M.G., Antle C.E. (1994). Binder Characterization and Evaluation. Volume 3: Physical Characterization, SHRP Report A-369 Strategic Highway Research Program, Transportation Research Board, National Research Council, Washington, D.C.

[6] Bahia, Hussain U., Hanson D. I., Zeng M., Zhai H., Khatri M. A., and Anderson R. M. (2001). Characterization of modified asphalt binders in Superpave mix design, NATIONAL ACADEMY PRESS WASHINGTON, D.C. NCHRP report 459.

[7] Masad E., Huang C.-W., D'Angelo J., Little D. (2009). Characterization of asphalt binder resistance to permanent deformation based on nonlinear viscoelastic analysis of multiple stress creep recovery (MSCR) test, J. *Assoc. Asphalt Pav*. 78, 535–562.

[8] Saboo N. and Kumar P. (2016). Analysis of different test methods for quantifying rutting susceptibility of asphalt binders, *J. Mater. Civil Eng*. 28 (7).

[9] AASHTO TP 70, (2013). Multiple stress creep recovery (MSCR) test of asphalt binder using dynamic shear rheometer (DSR), American Association of State Highway Transportation Officials, Washington, D.C.

[10] D'Angelo, J. A. ((2009). The relationship of the MSCR test to rutting, *Road Materials and Pavement Design*, 10(23), 61-80.

[11] Wasage, T. L. J., Stastna, J., & Zanzotto, L. (2011). Rheological analysis of multi-stress creep recovery (MSCR) test, *International Journal of Pavement Engineering*, 12(6),561-568.

[12] Kumar, R., Saboo, N., Kumar, P., & Chandra, S. (2017). Effect of warm mix additives on creep and recovery response of conventional and polymer modified asphalt binders, *Construction and Building Materials*,138, 352-362.

[13] Subhy, Ayad Tareq. (2017). Characterization and development of rubberized bitumen and asphalt mixture based on performance-related requirements, Ph.D. diss., University of Nottingham.

[14] Golalipour, A. (2011). Modification of multiple stress creep and recovery test procedure and usage in the specification, Doctoral dissertation, University of Wisconsin–Madison

[15] Yassir Nashaat A. Kareem, (. 2016). Effect of Polymer Modification on Asphalt Cement Properties, *Diyala Journal of Engineering Sciences*, 9 (03), 1-11.

[16] Yangsheng Liu and Nanjing Weilong'(2003). Reagents for swelling rubber and method of using, United States Patent.

[17] Yasameen Hadi Abed and Adabdulhaq H Abedali Al-Haddad, (2020). Temperature Susceptibility of Modified Asphalt Binders, *IOP Conf. Series, Materials Science, and Engineering 671.*

[18] ASTM D7405 (2015). Standard Test Method for Multiple Stress Creep and Recovery (MSCR) of Asphalt Binder Using a Dynamic Shear Rheometer.

[19] ASTM D7175, (2008). Determining the Rheological Properties of Asphalt Binder Using a Dynamic Shear Rheometer.

[20] Pellinen, T. K., Witczak, M. W., and Bonaquist, R. F., (2003). Asphalt Mix Master Curve Construction Using Sigmoidal Fitting Function with Non-Linear Least Squares Optimization, Geotechnical Special Publication, 83-101.

Investigating the Nonlinear Performance of Reinforced Concrete Shell Foundations

Rondik Sidqi[1,*] and Mohammad Najim Mahmood [2]

[1]College of Engineering, University of Mosul, Iraq - E-mail: rondik.sidqi@gmail.com
[2]College of Engineering, University of Mosul, Iraq
E-mail: drmohamad_najim@yahoo.com
* Corresponding author: College of Engineering, University of Mosul, Iraq.
Tel: +9647511367663.

Abstract. The present study involved a numerical investigation of two types of shell foundation having the same dimensions, these are the "Inverted" and "Upright" shell foundations. This type of foundation is considered as an alternative of the shallow foundations when the footing is subjected to a high load or used in a weak soil. The study includes a comparison between these two types of shell foundations. The results showed that the "Inverted" shell footing is superior to the "Upright" one in terms of Load carrying capacity, settlement and contact pressure. The study involved the effects of using different edge angles namely (10°, 20°, 30 and 40°). It was found that by increasing the angle the load-bearing capacity increased in case of "Inverted' shell footing, while it was decrease in the case of "Upright" shell footing. Different thickness of footing is considered, these are 160,200 and 240mm. it was found that by increasing the thickness of footing, the load-bearing capacity increased in both types of shell footing.

Keywords: *Finite element; Geotechnical effects, Shell foundation; Shell angle; Shell thickness.*

1. Introduction

The elementary necessity of foundation is to transmit loads from the superstructure to the soil/ rock below without any distress in the structure and foundation. Shell foundation has taken attention by many researchers due to the reduction in the ratio of steel reinforcement and amount of concrete as well (Salam and Mahmoud, 2017). The performance of shell foundation is better than the conventional footing through eliminating the tension zone in concrete and reducing the stresses of reinforcing steel compared with the flat footing (Mahmoud and Essam, 2017). The shell foundation is used in case of heavy loads that should be transmitted to weak soil or problematic soil (Ab Rahman, Amera Ratia, 2016). Experimental and theoretical studies (Hanna, et al.,1990) were carried out on a triangle shell footing and the results were compared with that of the flat-footing. The results showed the load-bearing capacity of triangle shell footing is higher than flat-footing. The behavior of curved shell strip footing (Thilakan, 2015) was investigated by using a finite element software OptumG2 Ver. 1.12 (2015) using a different values of friction angle of soil and compared it with flat strip footing. The results indicated that the efficiency of curved shell footing is increased by increasing the value of the angle of friction and the stresses under shell footing is less than that of a flat strip. The folded strip footing with 20° angle was numerically studied in Ref. (Salam and Mahmoud, 2017) and the results were compared with rectangular strip footing by using finite element software ADINA Ver. 9.0 (2014). The results showed that the folded strip footing is more efficient and more economical by reducing the amount of reinforcement ratio of about 26%. The hyper shell footing was numerically studied (Huat et al., 2006) by using Program

IOP Conf. Series: Materials Science and Engineering **978** (2020) 012050 doi:10.1088/1757-899X/978/1/012050

PLAXIS (2018) with various edge beam size. The results showed that increasing the embedment ratio caused improvement in the load-Settlement characteristics and the footing with a higher value of inertia has low initial settlement. In Ref. (Hassan et al., 2019), the behavior of pyramidal shell foundation rested on reinforced sandy soil were experimentally and numerically studied by using finite element software ABAQUS (2019) and its results compared with that of flat strip footing. It was observed that the pyramidal shell footings rested on reinforced or unreinforced sand have a higher ultimate load-bearing capacity than unreinforced sand of flat-footing, the aim from the present research is to carry out a comparison between the behaviors of the "Inverted" and "Upright" shell footing in terms of ultimate bearing capacity, settlement, the crack pattern, steel stress and the stress distribution in soil.

2. Finite element analysis

In the present work, the structural and geotechnical behavior of both types of shell foundation, "Upright" and "Inverted", are studied numerically using the finite element program ANSYS Ver.15 (2015) (ANSYS program is providing access to virtually any field of engineering simulation that a design process requires) as shown in Figs. 1 to 4. Each type of footing rest on (C – Ø) soil which its properties are listed in Table 1. The soil was modelled by using (SOLID45) element which can model the soil as an elastoplastic material using Drucker Prager (Drucker et al., 1952) failure criterion, this element has eight nodes with three degrees of freedom (u, v, w) at each node. The properties of concrete are listed in Table 2 which was modelled by using (SOILD65) elements. This element can model the nonlinear behavior of concrete as an elastoplastic material with failure criterion defined by William Warnke Model (Willam, 1975), this element has eight nodes with three degrees of freedom (u, v, w) same as the element (SOLID45). The properties of steel reinforcement are listed in Table 3 and modelled by (LINK180) in ANSYS as Elastic-perfectly Plastic (Owen, 1980). This element has two nodes with three degrees of freedom in each node. Standard fixities were taken as the boundary conditions which provide a partial fixity allowing vertical motion at the sides and a total fixity at the bottom of the soil. Due to the symmetry only half of the structure was considered. The dimensions of each footing are (2.4 *3.0) m using different thickness 160, 200 and 240 mm and this was increased at the section under the load to avoid local failure. The load was applied as a pressure equal to 20 MPa at the column location and the section of concrete was increased to resist the punching failure. Also, this study includes the effect of using different angle 10°, 20°, 30° and 40°. Fig. 1 and Fig.2 show the finite element mesh of inverted and upright shell footing, respectively. Figs. 3 and 4. Show the footings with the surrounding soil.

Table 1. Properties of Soil.

Parameters	C_ Ø Soil
Es (MPa)	20
Φ(deg.)	30
C (kN/m2)	25
μ(Poisson's ratio)	0.3
Density (kN/m3)	15

Table 2. Properties of Concrete.

Modulus of Elasticity (MPa)	23650
Uniaxial crushing stress (MPa)	25
Uniaxial cracking stress (MPa)	3.1
Poisson's ratio	0.2
Open shear transfer coefficient	0.3
Closed shear transfer coefficient	0.7

IOP Conf. Series: Materials Science and Engineering **978** (2020) 012050 doi:10.1088/1757-899X/978/1/012050

Table 3. Properties of Reinforcement steel.

Elastic modulus (MPa)	200,000
Yield stress (MPa)	400
Ratio of Reinforcement	480mm^2/m
Diameter of bar	12 mm

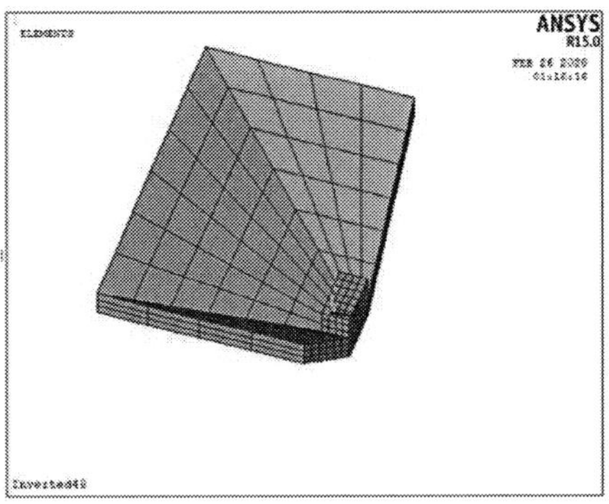

Figure 1. F.E.M. mesh of Inverted shell footing.

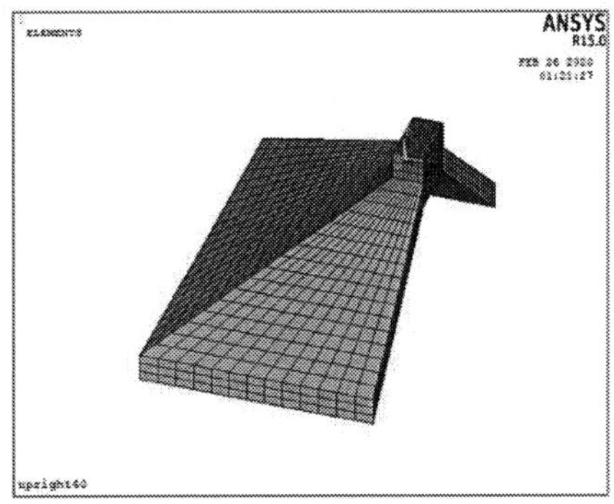

Figure 2. F.E.M. mesh of Upright shell footing.

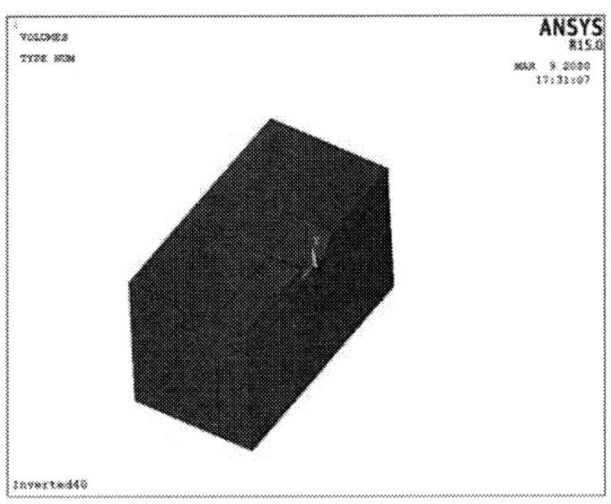

Figure 3. Soil and Inverted shell Footing model.

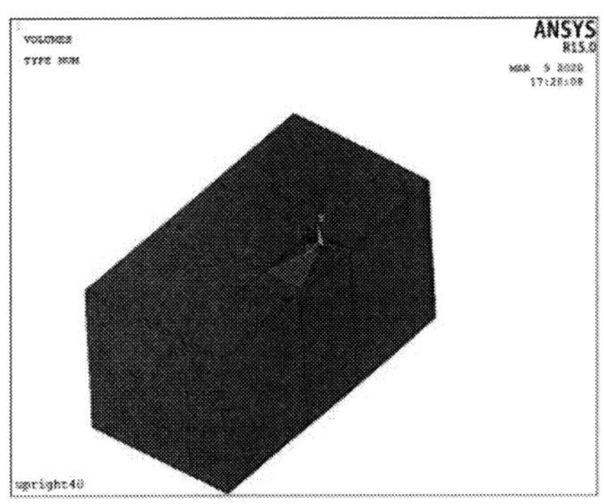

Figure 4. Soil and Upright shell Footing model.

3. Results and discussions

In the present work the analysis included the effect of increasing the thickness and angles of shell footing on the settlement, load-bearing capacity and soil contact pressure, so, a comparison between the two types of shell footing "Upright" and "Inverted" are considered in this study for all studied cases.

3.1. The Angle effects of shell footings

The predicted results, in terms of load-settlement curves, for the inverted shell show that by increasing the angle, the load-bearing capacity increased and the settlement decreased, as shown in Fig.5 for shell thickness of 160mm. the initial settlements are due to the self-weight of footing and soil, while By increasing the angle, the load-bearing capacity and settlement decreased for Upright shell footings for all thickness, Fig. 6 shows the results for 160 mm thickness.

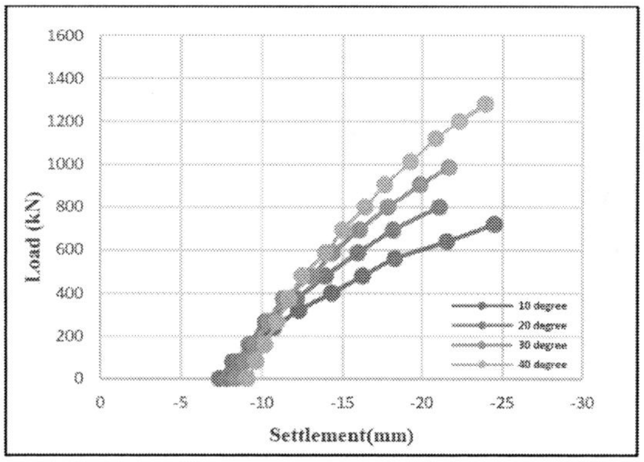

Figure 5. Load Settlement curve at the center of the "Inverted" for 160 mm thickness.

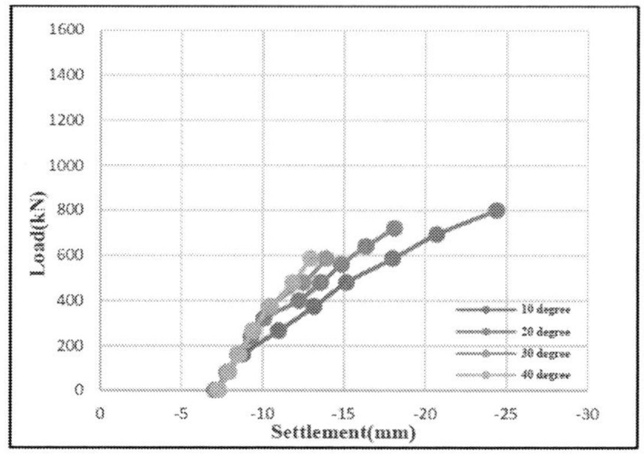

Figure 6. Load Settlement curve at the center of the "Upright" for 160 mm thickness.

From Figs. 5 and 6, it can be noticed that the response of Inverted shell footings is better than the Upright one in terms of load-bearing capacity, also, the initial settlement due to self-weight of footing and soil is increased by increasing the angle of shell footing in case of inverted shell footing, while it is almost equal for upright shell footings having different thicknesses.

The increasing of angle of both types of shell footings result in to decrease soil contact pressure, this is shown in Figs. 7 and 8 which depicted the contact pressure for "Inverted' and "Upright" shell footings for thickness 160 mm, it is also noticed from Fig.7 that the soil contact pressure of “Inverted” shell footings has a reflected point when the cracks propagated more in all parts of footing then the stress is less at the center of footing and redistributed toward the edges of footing.

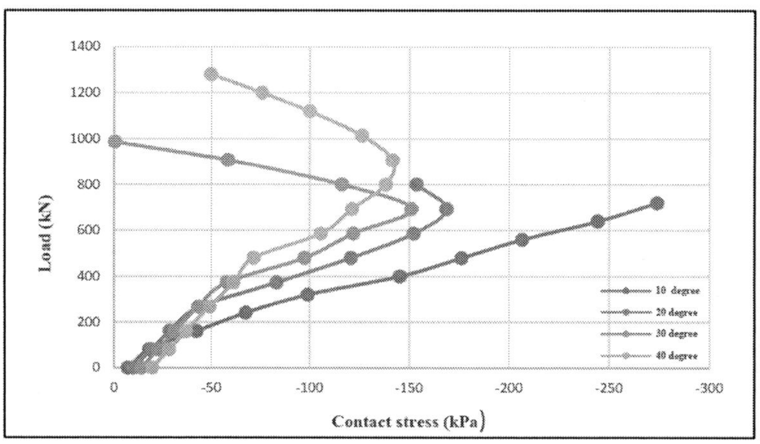

Figure 7. The relationship between the load and vertical contact pressure of soil at the center of the "Inverted" shell footing of 160 mm thickness.

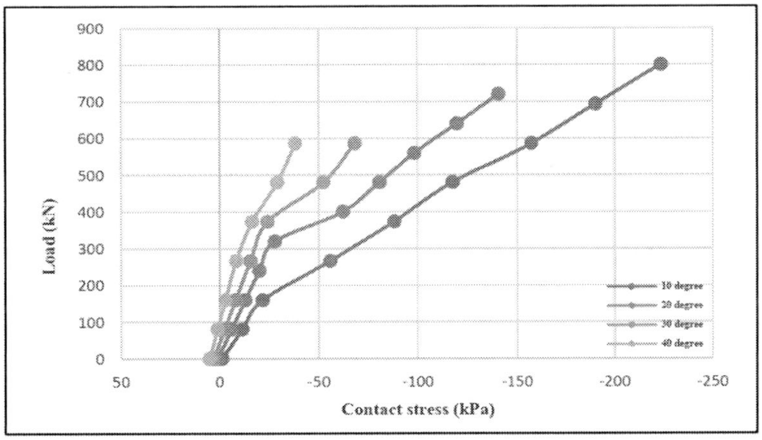

Figure 8. The relationship between the load and vertical contact pressure of soil at the center of the "Upright" shell footing of 160 mm thickness.

For shell footings having a different angle of "Inverted" and "Upright" shell footing for 160 mm thickness, it is noticed from Figs.9 and 10 that the distribution of soil stress before failure is regular under the footing in all studied cases, but, after cracking the stresses is concentrated at the center of footing due to the reduction of the footing ability to transformed the load from center to the outer edge due to the stiffness reduction of the footing after cracking. Figs.11 and 12 show the stresses of steel reinforcement at crack load and near failure load. The figures show that some places had low stress at crack load where the cracks not happened but after increased load and crack propagation, the steel stress increased and reached yielding near failure load, in addition that the stress of steel reached to the yielding at the center of "Inverted" shell footing while it reached yielding at the outer edges of "Upright" shell footing at failure load.

3.2. The thickness effect of shell footing.

With increasing the thickness, the load-bearing capacity increased for all cases under study, as shown in Figs.13 and 14 for inverted and upright shell footing respectively.

The previous results are for shell footing having 160 mm thickness for both types of shell footings which is similar to other thickness 200 mm and 240 mm, all values are enlisted in Table 4.

IOP Conf. Series: Materials Science and Engineering **978** (2020) 012050 doi:10.1088/1757-899X/978/1/012050

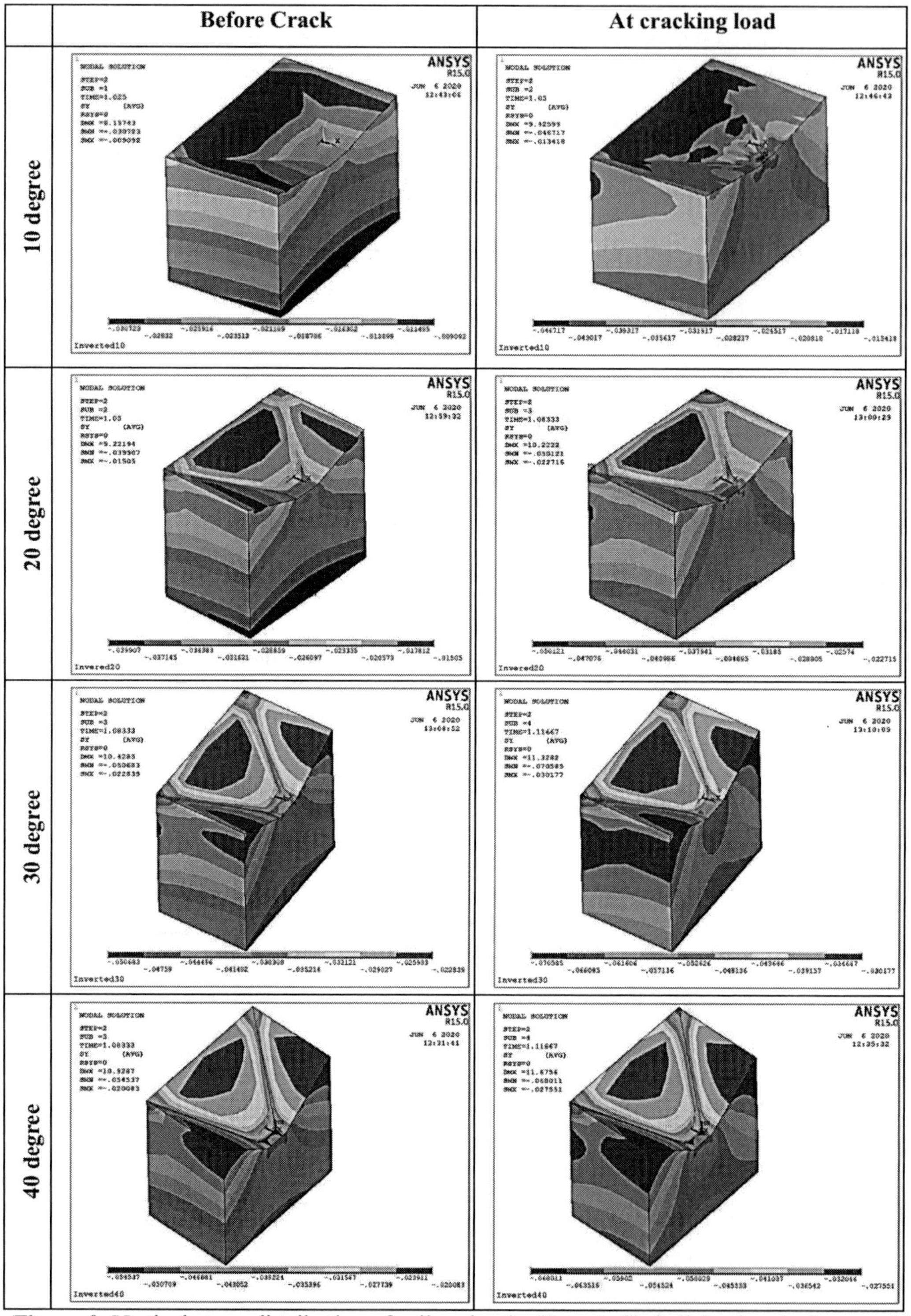

Figure 9. Vertical stress distribution of soil under the "Inverted" shell footing, before cracking and at load crack for all angle of 160 mm thickness of shell footing.

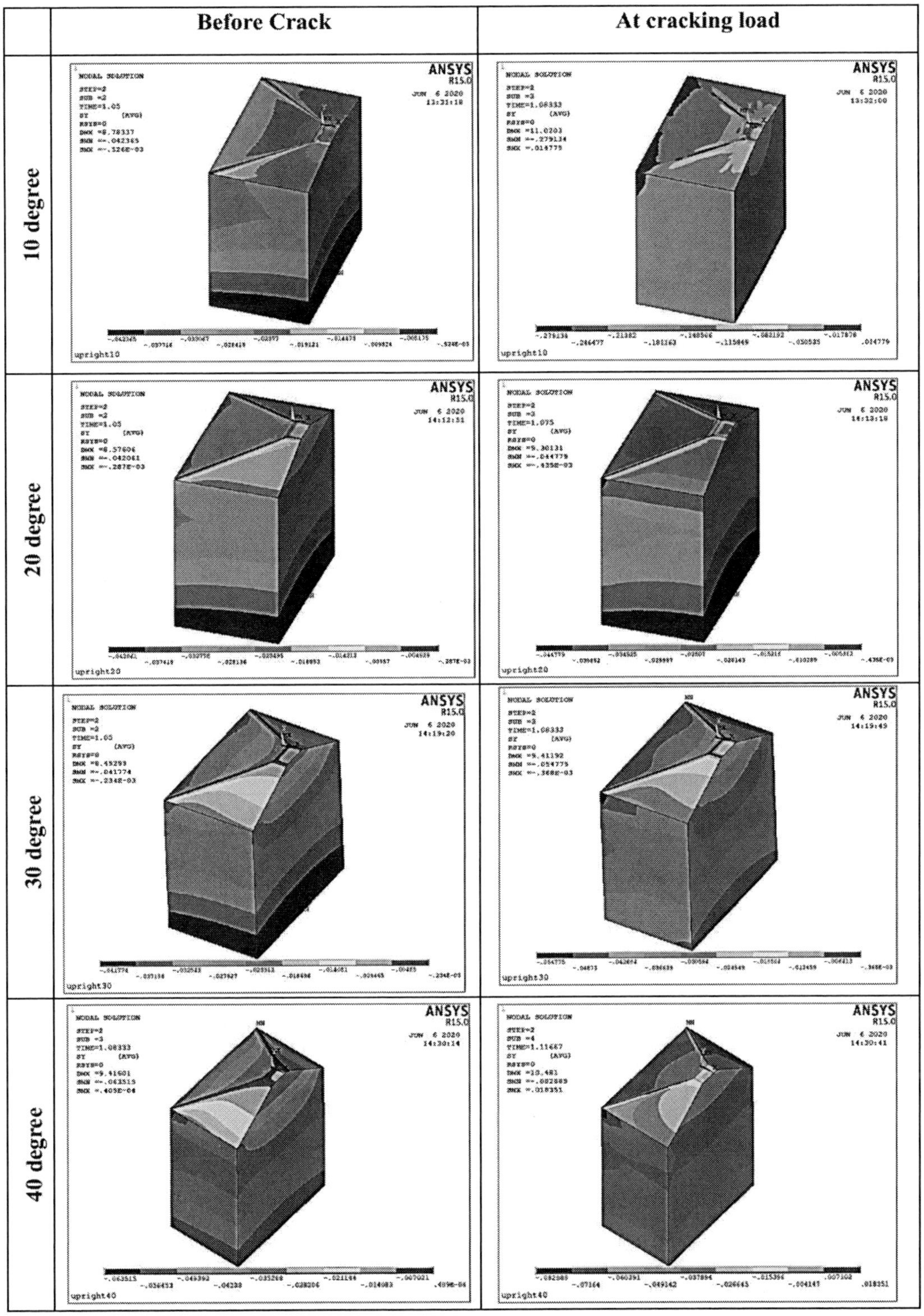

Figure 10. Vertical stress distribution of soil under the "Upright" shell footing, before crack and at cracking load for all angle of 160 mm thickness of shell footing.

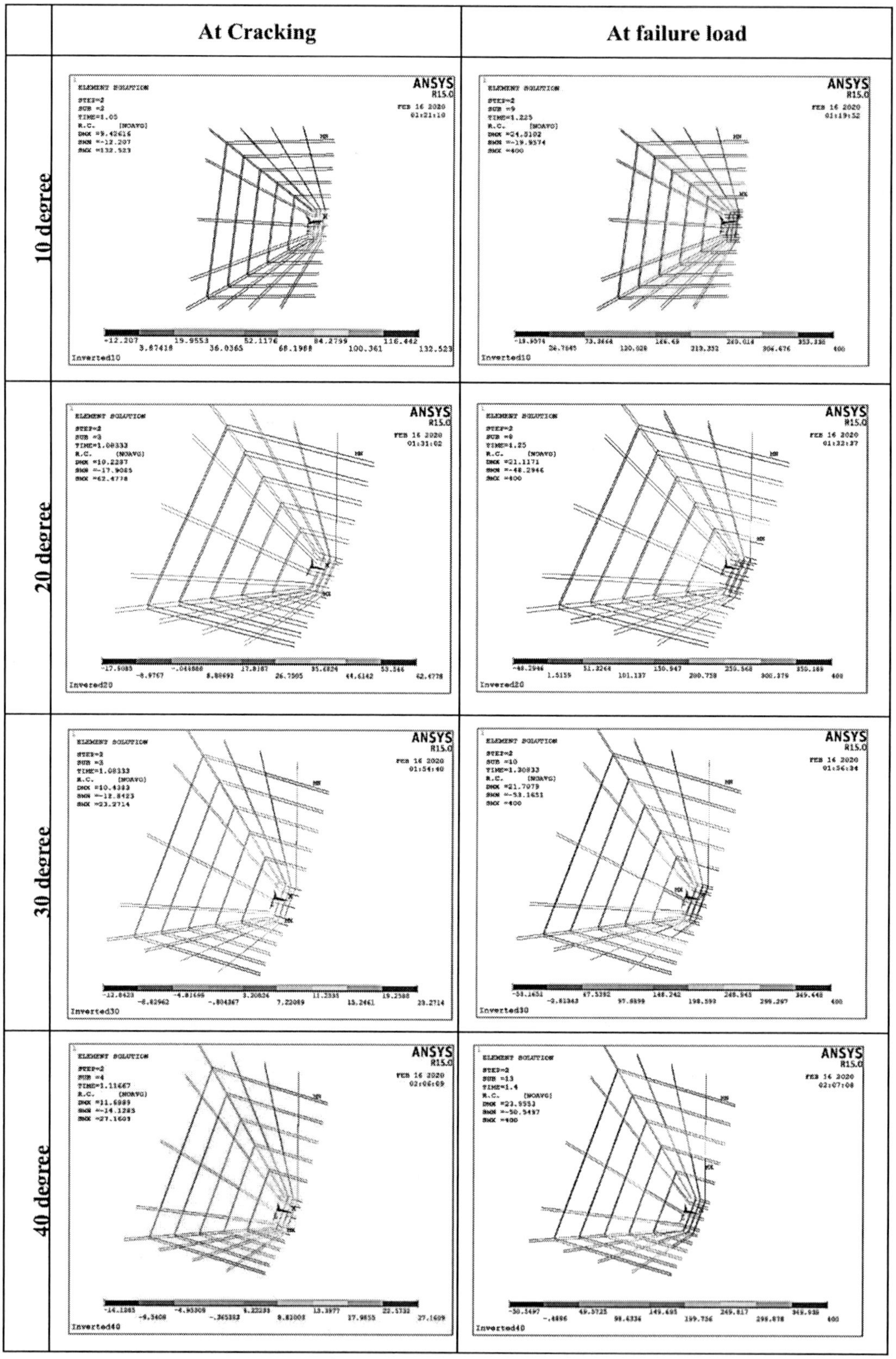

Figure 11. Stresses in reinforcement steel of "Inverted" Shell footing at crack load and at near failure load for 160 mm thickness.

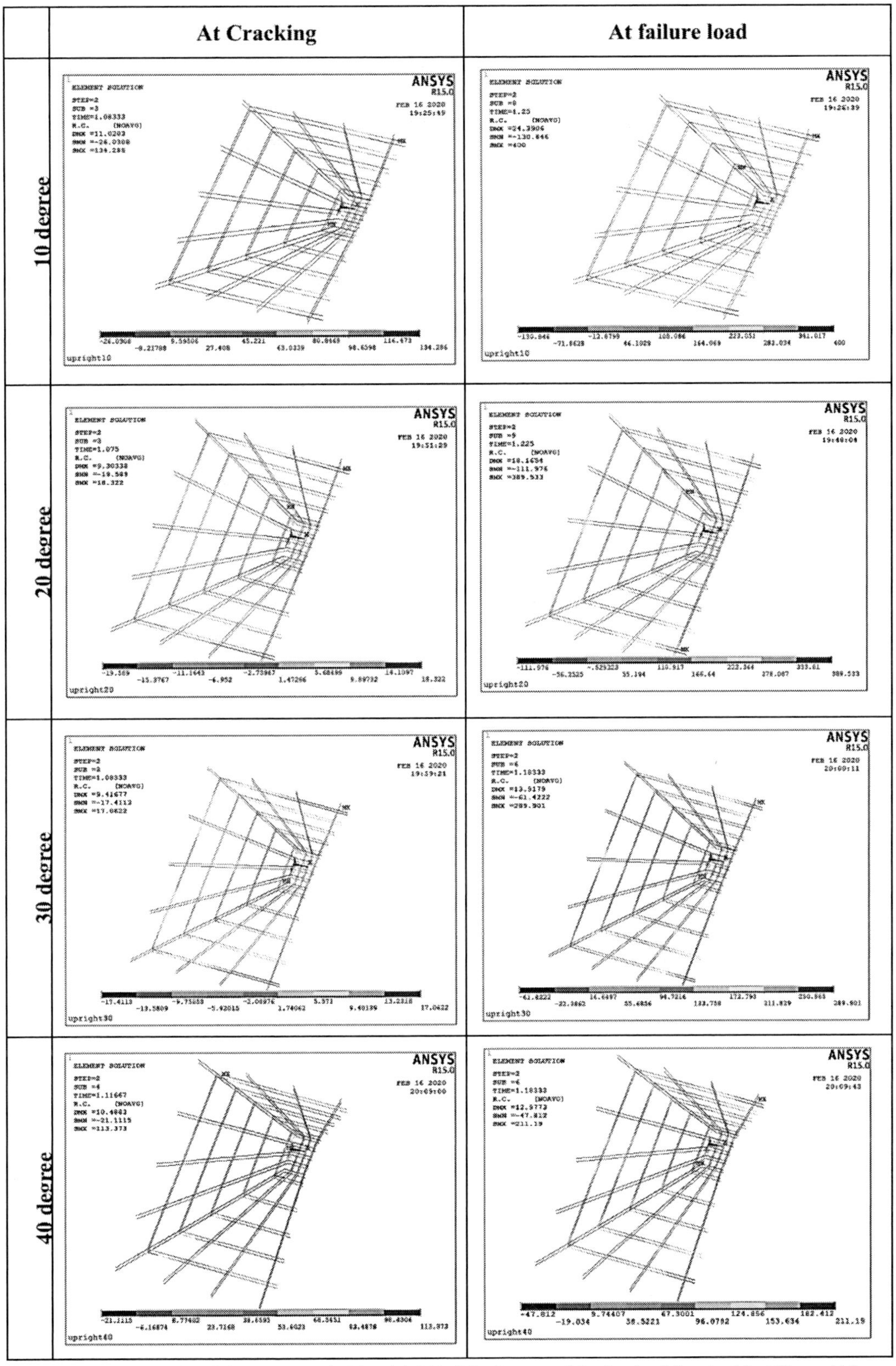

Figure 12. Stresses in reinforcement steel of "Upright" Shell footing at crack load and at failure load for 160 mm thickness.

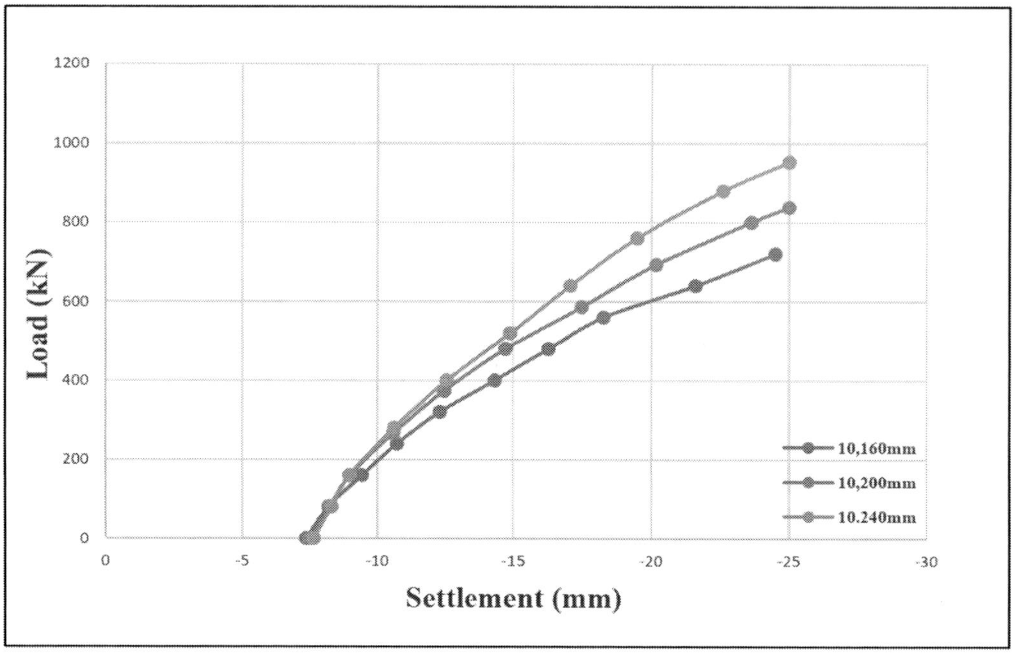

Figure 13. The load – settlement curve of 10° "Inverted" footing for all thickness.

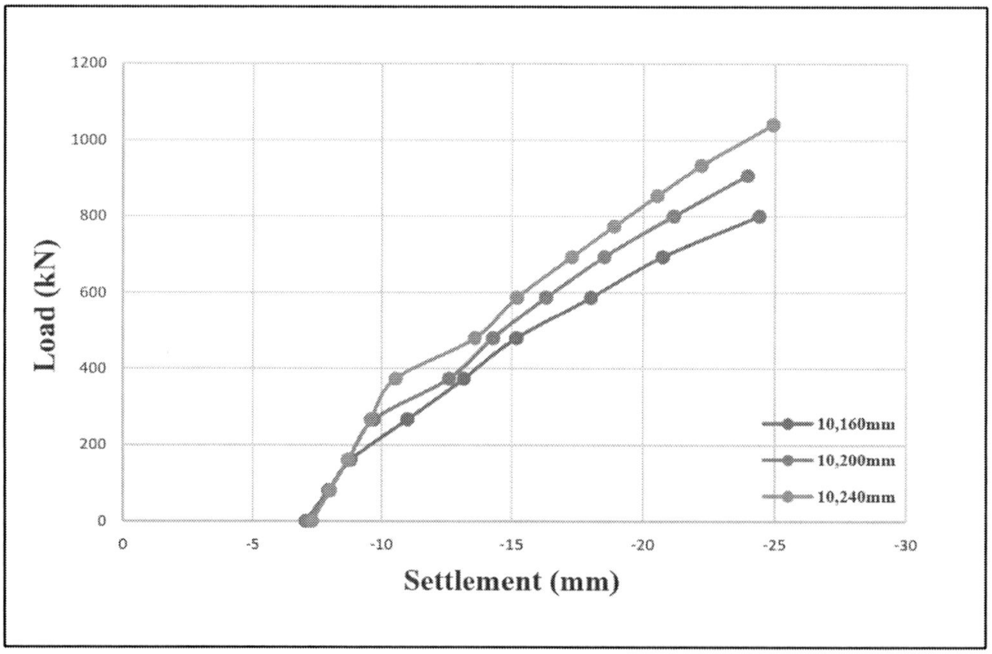

Figure 14. The load – settlement curve 0f 10° "Upright" footing for all thickness.

The value of cracking load, load failure, maximum settlement and maximum contact stress pressure are listed in Table 4 for all studied cases.

Table 4. The crack load, load failure, Max. Settlement and Max. Contact pressure for all studied cases.

Thickness of footings (mm)	Angle (Degree)	Inverted Shell Footings				Upright Shell footings			
		Maximum Settlement of footings (mm)	Failure load (kN)	Concrete Cracking load(kN)	Max. Contact pressure (kPa)*	Maximum Settlement of footings (mm)	Failure load (kN)	Concrete Cracking load(kN)	Max. Contact pressure (kPa)
160	10	24.5102	720	160	237.937	24.4073	800	266.56	223.865
	20	21.1164	640	266.656	168.599	18.1148	720	240	140.696
	30	21.7078	986.56	266.656	151.193	13.9127	586.56	373.37	68.3274
	40	23.9553	1280	373.344	141.875	12.9677	586.56	266.656	38.1891
200	10	25	838.1588	160	361.665	23.9537	906.56	373.344	211.241
	20	25	985.815	266.656	159.93	15.3563	640	373.344	100.053
	30	22.6936	1120	373.344	135.402	17.4754	880	373.344	56.3752
	40	21.5611	1226.56	373.344	137.886	13.0445	640	373.344	35.5333
240	10	25	952.8318	280	370.554	24.9203	1090	373.344	220.846
	20	25	1080.232	240	150.513	25	998.107	266.656	211.417
	30	23.1979	1200	373.344	124.667	25	1567.02	586.56	77.0939
	40	25	1433.166	480	81.149	15.4181	906.56	480	52.6685

***Before the cracks were propagated in all parts of footing and redistribution of soil stress toward the edges of footing.**

4. Conclusion

- The "Inverted" shell footing is more efficient in term of load bearing capacity and settlement than "Upright" shell footing.
- The load bearing capacity increased and settlement decreased by increasing the angle and thickness of "Inverted" shell footing.
- The load bearing capacity decreased with increasing the angle of "Upright" shell footing, but it is increased by increasing the thickness.
- The contact pressure of the soil under footing is decreased by increasing the angle and thickness in "Inverted" and "Upright" shell footings.
- The stress of steel reinforcement reached to the yielding at the center of the “Inverted” shell footing while it reached the yielding at the outer edges of “Upright” shell footing.

References

[1] Ab Rahman, A. R. (2016). Performance of physical shell foundation model under axial loading (Doctoral dissertation, Universiti Tun Hussein Onn Malaysia).

[2] ADINA a finite element program for automatic dynamic incremental nonlinear analysis, ver.9.Watertown, MA02473, USA,2014.

[3] ABAQUS Standard User's Manual, Hibbitt,Karlsson and Sorenson,Inc.vols.1_3, Version6.14,USA,2014.

[4] ANSYS Inc., Manual: Theory, Elements and Commands References, ANSYS Inc., Houston, TX,2012.

[5] Drucker, D. C., & Prager, W. (1952). Soil mechanics and plastic analysis or limit design. Quarterly of applied mathematics, 10(2), 157-165.

[6] El-kady, M. S., & Badrawi, E. F. (2017). Performance of isolated and folded footings. Journal of Computational Design and Engineering, 4(2), 150-157.

[7] Hanna, A., & El-Rahman, M. A. (1990). Ultimate bearing capacity of triangular shell strip footings on sand. Journal of geotechnical engineering, 116(12), 1851-1863.1972.

[8] Huat, B. B., & Mohammed, T. A. (2006). Finite element study using FE code (PLAXIS) on the geotechnical behavior of shell footings. Journal of Computer Science, 2(1), 104-108.

[9] Hassan, S. A., Al-Soud, M. S., & Mohammed, S. A. (2019). Behavior of Pyramidal Shell Foundations on Reinforced Sandy Soil. Geotechnical and Geological Engineering, 37(4), 2437-2452.

[10] Optum G2 computational engineering, USA,2015.

[11] Owen, D. R. (1980). Finite elements in plasticity, theory and practice.

[12] PLAXIS. Reference manual, 2D_ Version 9.0, PlAXIS. Delft,Netherland:Delft University of Technology; 2008.

[13] Salam, S. A., & El-kady, M. S. (2017). Foundations for low cost buildings. Journal of Computational Design and Engineering, 4(2), 143-149.

[14] Thilakan, S., & Naik, N. (2015). Geotechnical behaviour of shell foundations. In 50th Indian Geotechnical Conference (pp. 1-9).

[15] Willam, K. J. (1975). Constitutive model for the triaxial behaviour of concrete. Proc. Intl. Assoc. Bridge Structl. Engrs, 19, 1-30.

Manta Ray Foraging Optimization for Hyper-Parameter Selection in Convolutional Neural Network

Omar Adil Kamil[1, *], Shaymaa W. Al-Shammari[2]

[1] Department of Computer Engineering, Al-Nahrain University, Iraq
E-mail: omar.altae11@gmail.com
[2] Department of Computer Engineering, Al-Nahrain University, Iraq
E-mail: shaymaa.alshammari@gmail.com
* Corresponding author: Department of Computer Engineering, Al-Nahrain University, Iraq. Tel: +9647702582489.

Abstract. Convolutional neural networks (CNNs) have been attracting attention as one of the most common deep learning techniques used for different applications like images classifications, objects recognition, face recognition, etc. The performance and efficiency of the CNN model depend directly on their hyper-parameter, which must be selected by an expert or using one of the models that proposed and improved. This makes it very important to determine the optimal hyper-parameters. In this work, the Manta Ray Foraging Optimization (MRFO) algorithm is used to select CNN's Hyper-Parameter. We demonstrate that MRFO efficiently explores the solution space. This allowing CNN with simple architecture to achieve a good classification accuracy over Cifar_10 dataset. In the presented experiment Cifar_10 dataset used as the benchmark data sets. By optimizing CNN hyper-parameters with MRFO algorithm and comparing the obtained results with other CNN, it was approved that the accuracy was improved.

Keywords: *Convolutional neural network; manta ray foraging optimization; hyper-parameters selection;*

1. Introduction

Convolutional neural networks represent a special type of neural network. CNN is most commonly used in different visual tasks, such as classifying images, generating image descriptions, segmenting images, detecting objects in pictures, recognizing faces. CNNs' models essentially consist of different layers, which are the convolution layer, the pooling layer, and the fully-connected layer (FC layer). In a deep network stacking these layers properly can successfully solve different visual tasks (**Albawi, S., et al. 2017**). The performance of any CNN models highly depends on the hyper-parameters such as filters, the number of convolutional layers, and batch size, etc. and require experts to determine these hyper-parameters, meaning that it might be hard to find good settings for a non-expert. In this paper, we propose solving the problem of selecting the hyper-parameters by using Manta Ray Foraging

Optimization (MRFO) algorithm to obtain optimal hyper-parameters. The general hyperparameters in CNN models could be selected are the following: the number of filters, the number of convolutional layers, the stride, the size of the filters, the number of training epochs, activation function, the dropout probability, and the batch size.

This paper is organized as follows: Section 2 presents the related work, section 3 discusses the Manta Ray Foraging Optimization Algorithm, section 4 introduces the proposed method, section 5 shows the experiments and results and discusses the obtained results.

2. Related work

Recently, there have been many researches proposed a method to optimize the convolutional neural network hyper-parameter such as the research presented by (**Serizawa, T. and H. Fujita 2020**) who proposed a method for hyperparameter optimization of CNN by using linearly decreasing weights PSO (LDWPSO), which is metaheuristic algorithms and the model achieved better results than CNN without LDWPSO. Another hybrid method was presented by (**Albeahdili, H. M., et al. 2015**) called (PSO-SGD) algorithm training CNN. In (**Lorenzo, P. R., et al. 2017**) proposed a method using PSO to optimize the hyper-parameter in DNN. It shows that very simple DNN after optimizing by PSO can improve the performance of the model. Another interesting method was presented in (**Junior, F. E. F. and G. G. Yen (2019)**) the authors proposed a method to search for convolutional neural networks (CNNs) architectures using particle swarm optimization. One more novel proposed by (**R Takahashi, T Matsubara 2019**) called a weight shared multi-stage network to improve classification accuracy. (**Ismail, A., et al. 2019**) were presented a study for image classification using mini VGG Net architecture. The classification accuracy of the models increased by adding batch normalization layer to the network with learning rate decay schedulers to avoid overfitting. (**Smith, L. N. (2017)**) proposed a new method called Cyclical Learning Rates (CLR) to find best value for learning rate instead of using fixed value.

3. Manta Ray Foraging Optimization Algorithm

MRFO algorithm is a new optimization technique proposed by (**Zhao, W., et al. 2020**). The revelation of the algorithm is based on behaviors of manta rays. These creatures have three intelligent and fantastic foraging strategies.

1. Chain foraging uses this strategy by forming an orderly line and individually update the best position found can be expressed mathematically as follows:

$$x_i^d(t+1) = x_i^d(t) + r \cdot \left(x_b^d(t) - x_i^d(t)\right) + \alpha \cdot \left(x_b^d(t) - x_i^d(t)\right) \quad i = 1 \tag{1}$$

$$x_i^d(t+1) = x_i^d(t) + r \cdot \left(x_{i-1}^d(t) - x_i^d(t)\right) + \alpha \cdot \left(x_b^d(t) - x_i^d(t)\right) \quad i = 2, \dots, N \tag{2}$$

$$\alpha = 2 \cdot r \sqrt{|log(r)|} \tag{3}$$

2. Cyclone foraging: each manta swims toward the one in front of it forming spirally move towards the food, where this movement behavior may be extended for n-D space. So, the mathematical equation can be defined as:

$$x_i^d(t+1) = x_b^d(t) + r \cdot \left(x_b^d(t) - x_i^d(t)\right) + \beta \cdot \left(x_b^d(t) - x_i^d(t)\right) \quad i = 1 \tag{4}$$

$$x_i^d(t+1) = x_b^d(t) + r \cdot \left(x_{i-1}^d(t) - x_i^d(t)\right) + \beta \cdot \left(x_b^d(t) - x_i^d(t)\right) \quad i = 2, \dots, N \tag{5}$$

$$\beta = 2e^{r_1 \frac{T-t+1}{T}} \cdot \sin(2\pi r_1) \tag{6}$$

Where T represents the maximum iteration, β is the weight coefficient and r_1 a rand number in the range [0, 1].

Another mechanism focuses on exploring the search space by making each search for a new position from the current best position by assigning random position as follows:

$$x_{rand}^{d} = Lb^{d} + r \cdot (Ub^{d} - Lb^{d}) \tag{7}$$

$$x_i^d(t+1) = x_{rand}^d(t) + r \cdot \left(x_{brand}^d(t) - x_i^d(t)\right) + \beta \cdot \left(x_{rand}^d(t) - x_i^d(t)\right) \quad i = 1 \tag{8}$$

$$x_i^d(t+1) = x_{rand}^d(t) + r \cdot \left(x_{i-1}^d(t) - x_i^d(t)\right) + \beta \cdot \left(x_{brand}^d(t) - x_i^d(t)\right) \quad i = 2, \dots, N \tag{9}$$

Where L and U represent the lower and upper boundaries for search space and x_{rand}^d is the random position.

3. Somersault foraging in this strategy it swims in pivot way and update their positions always around the best position found so far. Where the mathematical model can be shown as follow:

$$x_i^d(t+1) = x_i^d(t) + S \cdot \left(r_2 \cdot x_b^d - r_3 \cdot x_i^d(t)\right), i = 1, \dots, N$$

(10) where S is the somersault factor, $r2$ and $r3$ are random numbers in [0, 1]

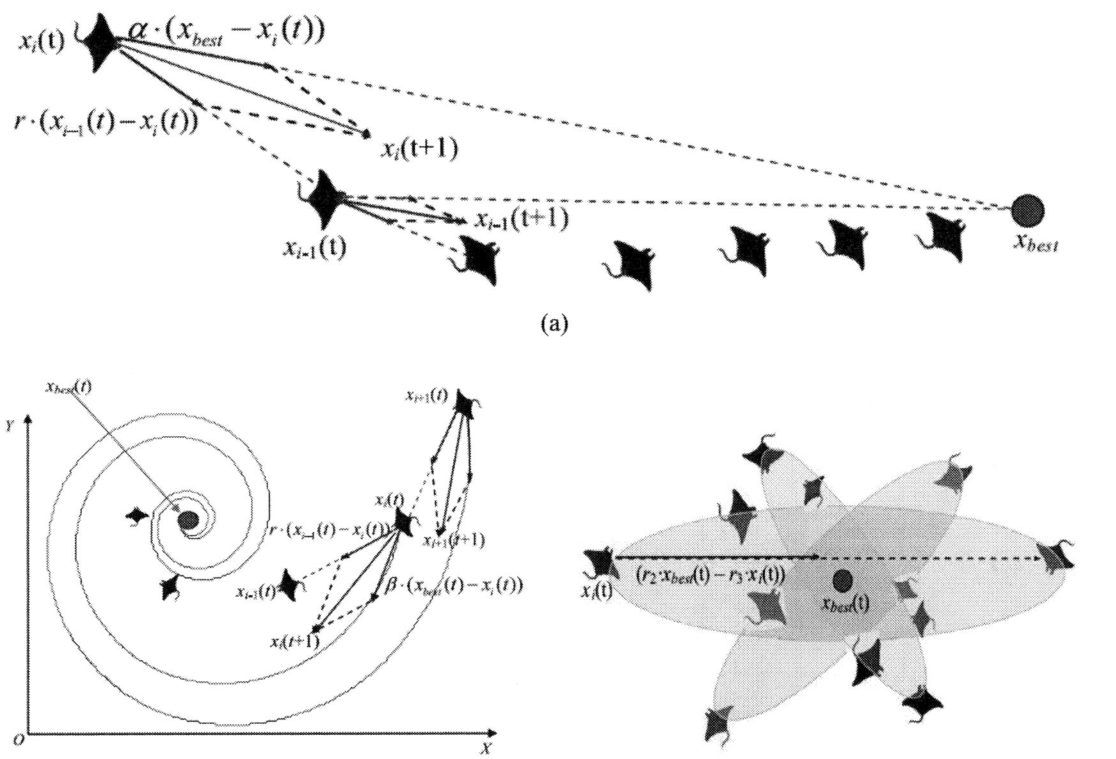

(b) (c)

Figure 1. Show the strategies of the manta (a) chain foraging (b) cyclone foraging and (c) Somersault foraging

IOP Conf. Series: Materials Science and Engineering **978** (2020) 012051 doi:10.1088/1757-899X/978/1/012051

4. Proposed Method

Generally, the CNN model consists of convolutional layer, pooling layer and at the end fully connected layer. The selection of the hyper-parameter of the CNN model is very difficult for non-expert person. In this paper, we propose a method to select the specific hyper-parameters and improve the accuracy obtained by the model. So, the goal is to find the optimum solution for available hyper-parameters. In MRFO, the lower and upper boundaries represent the range of the hyper-parameters that form the search space, the best position represent the optimum hyper-parameters and f(x) is the function value returned by CNN model to MRFO algorithm. A noteworthy proposed algorithm tailored to any CNN architecture. The proposed method starts by setting randomly lower and upper boundaries and max number of iterations. After that MRFO algorithm firstly initializes the hyper-parameters, which represents the current position. Next, it begins to update its position depending on the previous strategies to get the best position (optimum hyper-parameters) with high accuracy and less loss. where accuracy and loss calculated simply from training CNN model.

WHILE stop criterion is not satisfied do

FOR i=1 TO N DO

IF $rand$ < 0.5 THEN //*Cyclone foraging*

IF $t/T_{max} < rand$ THEN

$$x_{rand} = x_l + rand \cdot (x_u - x_l)$$

$$x_i(t+1) = \begin{cases} x_{rand} + r \cdot (x_{rand} - x_i(t)) + \beta \cdot (x_{rand} - x_i(t)) & i = 1 \\ x_{rand} + r \cdot (x_{i-1}(t) - x_i(t)) + \beta \cdot (x_{rand} - x_i(t)) & i = 2, \cdots, N \end{cases}$$

ELSE

$$x_i(t+1) = \begin{cases} x_{best} + r \cdot (x_{best} - x_i(t)) + \beta \cdot (x_{best} - x_i(t)) & i = 1 \\ x_{best} + r \cdot (x_{i-1}(t) - x_i(t)) + \beta \cdot (x_{best} - x_i(t)) & i = 2, \cdots, N \end{cases}$$

END IF.

ELSE //*Chain foraging*

$$x_i(t+1) = \begin{cases} x_i(t) + r \cdot (x_{best} - x_i(t)) + \alpha \cdot (x_{best} - x_i(t)) & i = 1 \\ x_i(t) + r \cdot (x_{i-1}(t) - x_i(t)) + \alpha \cdot (x_{best} - x_i(t)) & i = 2, \cdots, N \end{cases}$$

END IF.

Compute the fitness of each individual $f(x_i(t+1))$. IF $f(x_i(t+1)) < f(x_{best})$

THEN $x_{best} = x_i(t+1)$

//*Somersault foraging*

FOR i=1 TO N DO

$$x_i(t+1) = x_i(t) + S \cdot (r_2 \cdot x_{best} - r_3 \cdot x_i(t))$$

Compute the fitness of each individual $f(x_i(t+1))$. IF $f(x_i(t+1)) < f(x_{best})$

THEN $x_{best} = x_i(t+1)$

END FOR.

END WHILE.

Return the best solution found so far x_{best}.

Figure 2. Pseudocode of MRFO algorithm

5. Experiments and Results

In this study, an experiment is studied with cifar10 dataset. Figure (3) shows an example of the dataset.

Figure 3. Cifar10 dataset example

5.1 Experiment 1 using CIFAR-10 dataset with the proposed model

The cifar-10 dataset contains 60000 32X32 RGB images belonging to 10 classes (6000 images per class) and divided into 50000 images for training portion and 10000 images for testing. It is used as a benchmark dataset for image classification. In the experiment, optimization is performed with MRFO algorithm to select the hyper-parameters. The proposed model used is shown below:

1- Convolution layer with 32 and 3x3 filter.
2- Convolution layer with c2 and 3x3 filter. c2 range (27-37)
3- Max pooling layer with 2x2 filter.
4- Dropout layer with 0.25 to avoid overfitting
5- Convolution layer with c3 and sf x sf filter. c3= (60-72) sf= (2-9)
6- Max pooling layer with 2x2 filter.
7- Dropout layer with 0.25 to avoid overfitting
8- FC layer with d neurons. d= (192-1024)
9- Dropout layer with 0.5 to avoid overfitting
10- FC layer with ten neurons.
11- Learning rate = k
12- batch size = p rang (8-86)

The hyper-parameters configure changes every five epochs by MRFO until getting optimum configuration.

5.2 Experiment 2 Applied MRFO for Lenet5 model

In this experiment, MRFO algorithm is applied for Lenet5 model with cifar10 and Mnist dataset and the changes in the models are shown in Table 1 below:

Table 1. Shows the change in hyper-parameters values for Lenet5 model

Hyper-parameters	baseline	range	New value
Number of filters in C2	16	13-30	21
Number of neurons in FC2	84	64-84	83
Learning rate	0.001	0.0001-0.005	0.003
batch size	128	32-128	92
Activation function	tanh	Tanh, ReLU	ReLU

Figure 4 and Figure 5 shows a comparison between Lenet5 after (LenMRFO) and before (Lenet5) using the MRFO optimization method algorithm. It can be seen that there was an improvement in terms of increasing accuracy and decreasing the loss values.

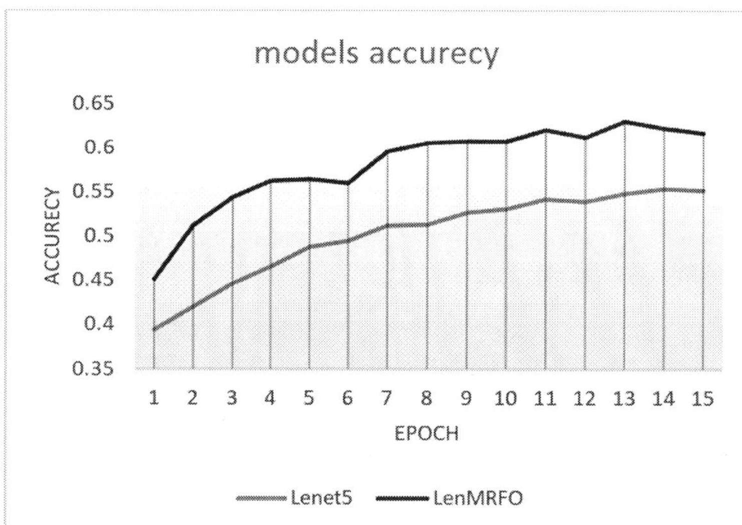

Figure 4. Shows the accuracy before and after using MRFO optimization method for Lenet5 model with cifar10 dataset

IOP Conf. Series: Materials Science and Engineering **978** (2020) 012051 doi:10.1088/1757-899X/978/1/012051

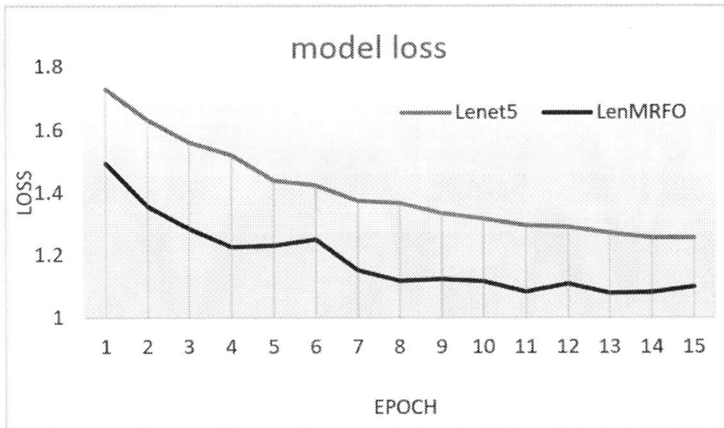

Figure 5. Shows the loss values before and after using the MRFO optimization method for Lenet5 model with cifar10 dataset

The accuracy and loss are studied for Mnist dataset, the results are shown in the figures 6 and 7:

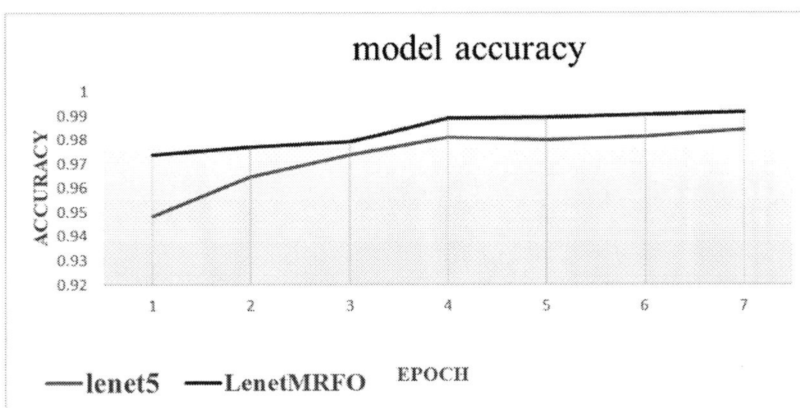

Figure 6. Shows the accuracy before and after using MRFO optimization method for Lenet5 model with Mnist dataset

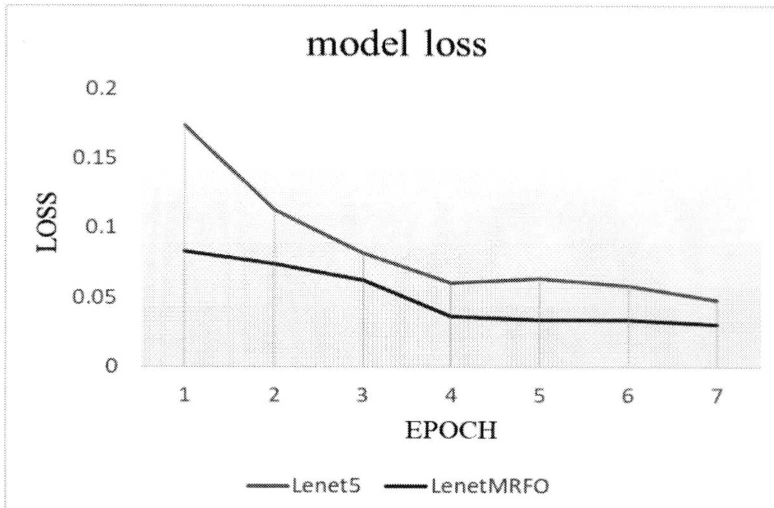

Figure 7. Shows the loss values before and after using the MRFO optimization method for Lenet5 model with Mnist dataset

5.3 Experiment 3

In the third experiment AlexNet model is optimized by the MRFO algorithm with cifar10 dataset and compared its results to the original model. Dropout layer added after both first and second fully connected layers. Table (2) shows the changes in the hyper-parameters of AlexNet model after applying the optimization algorithm with the upper and lower boundaries used for each one.

Table 2. Show the change in hyper-parameters for Lenet5 model

Hyper-parameters	baseline	range	New value
Strides in C1	4	2-11	2
Kernel size in C1	11	2-15	5
Number of filters in C2	256	128-260	189
Kernel size in C2	5	2-11	10
Number of filters in C4	384	264-396	325
Number of neurons in	4096	192-4096	847
Learning rate	0.001	0.0001-0.005	0.0002
batch size	128	32-128	60
Activation function	tanh	Tanh, ReLU	ReLU

Then optimized model is trained with cifar10 dataset for 15 epochs and from the results we can see that the optimized model has higher accuracy and fine loss comparing with same model before optimization as shown in the figures 8 and 9:

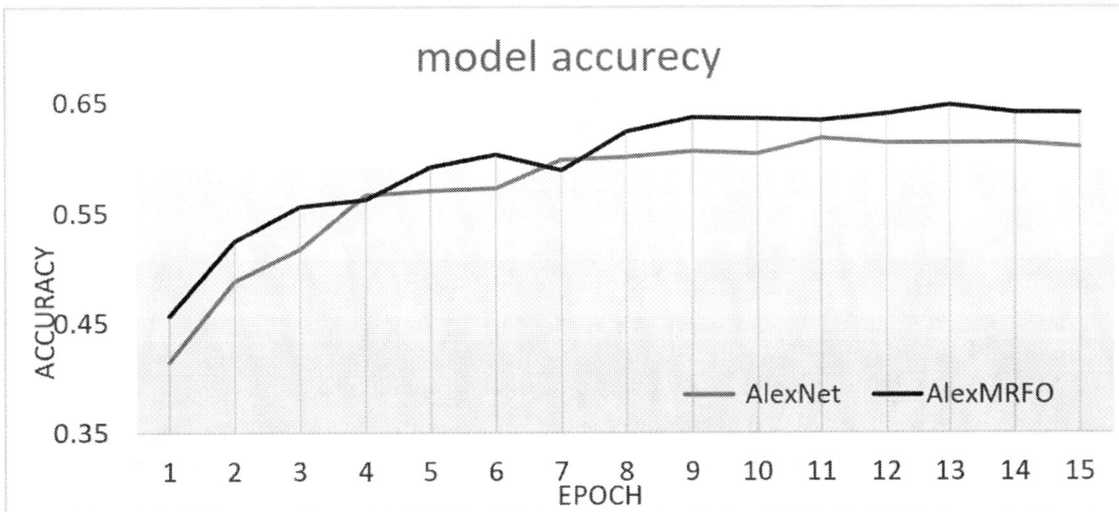

Figure 8. Shows the accuracy before and after using MRFO optimization method for AlexNet model with cifar10 dataset

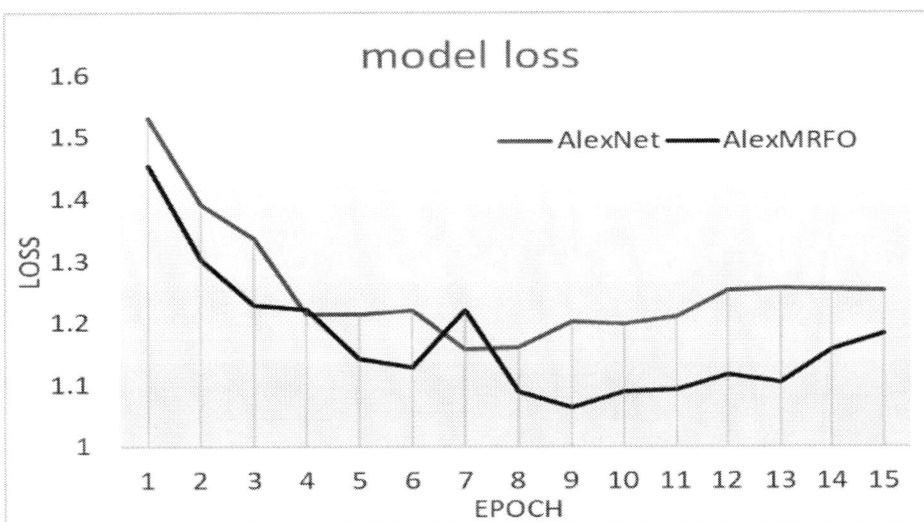

Figure 9. Shows the loss values before and after using the MRFO optimization method for AlexNet model with cifar10 dataset

6. Conclusion and Future work

This paper presented MRFO as an optimization method to select the hyper-parameters values of CNN. The proposed method has been studied and experimented on Lenet5 and AlexNet models using cifar-10 dataset. The obtained results showed that using MRFO has improved the performance in terms of increasing the accuracy and decreasing the loss values. This research can be extended in terms of exploring this method in more CNN models like VGG-16.

IOP Conf. Series: Materials Science and Engineering **978** (2020) 012051 doi:10.1088/1757-899X/978/1/012051

References

[1] Albawi, S., Mohammed, T. A., & Al-Zawi, S. (2017). Understanding of a convolutional neural network. Paper presented at the 2017 International Conference on Engineering and Technology (ICET).

[2] Albeahdili, H. M., Han, T., & Islam, N. E. (2015). Hybrid algorithm for the optimization of training convolutional neural network. International Journal of Advanced Computer Science & Applications, 1(6), 79-85.

[3] Ismail, A., Ahmad, S. A., Soh, A. C., Hassan, K., & Harith, H. H. (2019). Improving Convolutional Neural Network (CNN) architecture (miniVGGNet) with Batch Normalization and Learning Rate Decay Factor for Image Classification. International Journal of Integrated Engineering, 11(4).

[4] Junior, F. E. F., & Yen, G. G. (2019). Particle swarm optimization of deep neural networks architectures for image classification. Swarm and Evolutionary Computation, 49, 62-74.

[5] Lorenzo, P. R., Nalepa, J., Kawulok, M., Ramos, L. S., & Pastor, J. R. (2017). Particle swarm optimization for hyper-parameter selection in deep neural networks. Paper presented at the Proceedings of the genetic and evolutionary computation conference.

[6] Serizawa, T., & Fujita, H. (2020). Optimization of Convolutional Neural Network Using the Linearly Decreasing Weight Particle Swarm Optimization. arXiv preprint arXiv:2001.05670.

[7] Smith, L. N. (2017). Cyclical learning rates for training neural networks. Paper presented at the 2017 IEEE Winter Conference on Applications of Computer Vision (WACV).

[8] Takahashi, R., Matsubara, T., & Uehara, K. (2018). A novel weight-shared multi-stage CNN for scale robustness. IEEE Transactions on Circuits and Systems for Video Technology, 29(4), 1090-1101.

[9] Zhao, W., Zhang, Z., & Wang, L. (2020). Manta ray foraging optimization: An effective bio-inspired optimizer for engineering applications. Engineering Applications of Artificial Intelligence, 87, 103300.

Effect of Silica Fume and Super-Plasticizer on Mechanical Properties of Self- Compacting Concrete: A Review

Waseem Khairi Mosleh Frhaan[1*], Badorul Hisham Abu Bakar[2], Nahla Hilal[3], Abdulkader Ismail Al-Hadithi[3]

[1]School of Civil Engineering, Universiti Sains Malaysia, Main Campus, 14300 Nibong Tebal, Penang, Malaysia, Email : wkm_dams82@yahoo.com
[2]School of Civil Engineering, Universiti Sains Malaysia Main Campus, 14300 Nibong Tebal, Penang, Malaysia. Email: cebad@usm.my
[3]University of Fallujah, Department of international Relation office, Iraq. Email: nahla.naji@uofallujah.edu.iq
[4]University of Anbar, College of Engineering, Department of Dams and Water Resources Engineering, Main Campus, Ramadi, Anbar, Iraq. Email: al_hadithi2000@yahoo.com
* Corresponding author: School of Civil Engineering, Universiti Sains Malaysia, Main Campus, 14300 Nibong Tebal, Penang, Malaysia - Tel: +964 7710007876.

Abstract. Self-compacting concrete (SCC) is among the high-performance and modern concretes in the concrete industry. The inclusion of chemical and mineral admixtures and other aggregates causes better workability, flow ability, better compressive strength and high resistance to segregation. Silica fume (SF) is a mineral admixture which improves mechanical properties and reduced permeability. Secondly, the super-plasticizer (SP) is a chemical admixture and used for water-reducing, increase cohesiveness, improve passing and filling ability of SCC. However, SF affecting the physical properties of SCC if added in excess quantity in cement. Furthermore, the substitution of SP causes to deliver a negative surface charge on the concrete particles which in turns to result in electrostatic repulsion. Moreover, the higher quantity addition of both SF and SP will badly affect the fresh and hardened properties of SCC. Thus, an ample review was conducted to study the influence of SF and SP on the fresh and hardened properties of SCC. More than 50-previous research articles have been reviewed systematically to argue on the fresh ad hardened properties of SCC. Based on the reviewed literature, the conclusion has been drawn some future recommendations have been achieved related to the title of the research.

Keywords: *Self-compacting concrete; Silica Fume; Super-plasticizer; Fresh properties; Hardened properties.*

1. Introduction

SCC was firstly introduced in Japan in the 1980's decade as a means of producing a regular quality of concrete and to address any problem caused by poor labor. Moreover, the list of problems such as intricate designs, the density of reinforcing steel in the structural members, reducing the numbers of skilled labor, the rapid growth of construction industry and quality of construction necessitated introducing SCC to overcome these problems. Japanese researchers 1986 at the University of Tokyo firstly produced SCC and Ozawa et al., (1989) and his research team from the University of Tokyo presented the first model of SCC in 1989. During the 1990's decade and further than that, the knowledge and application of SCC increased. The main advantage of the application of SCC is its capability of flowing to those portions of the mold where compaction cannot be applied, especially; congested reinforcement areas and this makes it more advantageous than traditional concrete (Aslani et al. 2018). Other lists of advantages include flow and compact under its own weight (Bogas et al. 2012), completely filling out mold by preserving homogeneity (Salman, 2018), the high degree of fluidity (Abdul and Khushpreet, 2018), highly cost-effective (Alyamac et al. 2017) and productivity improvement will be offset by reduced labor and non-use of vibrators (Nehdi, M. and Rahman, 2004). Furthermore, due to no-use of vibrators; SCC is environmentally friendly as it eliminates the vibration noise. Also due to better passing and filling ability; SCC provides more design flexibility and increases productivity (Beigi et al. 2013).

Many authors and international standard organizations define SCC in their own way: according to American Concrete Institute (ACI) 237 (2007), it is a concrete type which does not need to be compacted when placed in the mold because of its flowability and spread due to its own weight and fill the mold completely even in the case of reinforcing high-density steel, the resulting concrete after hardening is homogenous, high durability, and possessed engineering properties similar to normal concrete. Whereas, the definition of SCC according to British Standard European Norm (BS EN) 206-9 (2010) is that the concrete which has self-flowability and compatibility because of its own weight, filling mold completely along with the reinforces materials whilst preserving homogeneity. Rabar et al. (2020) define that SCC is a concrete which can flow easily in different sections of mold due to its own weight and created a significant consolidation within the directed shape of the mold, it does not require any external vibration and provides defect fewer parts due to no-bleeding or segregation (Abdulkader et al. 2016; Al-Hadithia et al. 2019 and Mahmoud et al. 2019). And to achieve full compaction, SCC should have better cohesion and high flowing and passing ability and cohesive (Djelal, et al. 2004). However, the inclusion of viscosity modifying admixtures (VMA) provides resistance to segregation cohesiveness, while the addition of SP can increase the passing and flowing ability of SCC (Han and Yao, 2004).

Furthermore, the inclusion of SF, limestone (LS) and fly ash (FA) results improvement in stability and workability of SCC (Felekoglu et al. 2006 and Sahmaran et al. 2006). Though the addition of these materials in SCC will enhance durability and compressive strength, still SCC will fail in brittle mode (Deilami et al. 2019). Thus, the selection of these materials that mix and or reinforce with the design of SCC requires high accuracy because they are characterized by a high degree of fluidity and must reduce viscosity. And such can achieve by employing chemical (super-plasticizer SP) and mineral admixtures (silica fume SF) (Mahmoud, 2002). Most researchers have confirmed that the use of chemical and mineral admixtures gives better fresh and hardened of the concrete (Al-Hadithia and Hilal, 2016; Al-Hadithia et al. 2019; Mahmoud et al. 2019 and Rabar et al. 2020). SP inclusion reduced the use of water up to 30% without decreasing the workability of SCC which in turn increase productivity, flowing ability, self-leveling, self-compaction, higher performance and strength of the concrete. Furthermore, the addition of SP can improve the specific gravity, fresh and hardened properties of concrete when it added to concrete paste during mixing process. Generally, SP is classified as high-range water-reducing admixture (HRWRA) which produced self-compaction in concrete and produced high compressive strength up to 150 N/mm^2 (ASTM C494/C494M, 2004), also affects the fresh concrete positively, such as increasing slump without adding a quantity of water, reducing the proportion of water to cement, improved workability and thus easy pumping for concrete (ACI–Committee 212, 2007). Whereas, SF can affect the different concrete properties of concrete like "setting time characteristics", "water requirement for specified consistency", "improve workability including cohesiveness and bleeding", etc (Vivek and G. Dhinakaran, 2015). Additionally, the bleeding in fresh concrete can be stopped physically with the addition of SF, however, its appropriate addition may not change the unit weight of the concrete (Mohammadi et al. 2015). Nonetheless, the production of the SCC test depends on "trial and error" method, thus, finding optimal proportions of these components of the concrete mix, including mineral and chemical admixtures is important so as to reduce the experiments, cost, time and effort.

Therefore, in this research, the effort for an ample review has been made to emphasize the latest research on the assessment of using SF and SP in the production of SCC. Furthermore, in this ample review work, the latest previous research until 2019 was reviewed on the use of SF and SP for SCC production. In contrast to argued different properties of concrete, this review work was focused on the influence of addition SF and SF on the fresh properties and compressive strength of SCC. At the end of this ample review work, future recommendations were also drawn related to the title of the research.

2. Mineral Admixtures

The addition of mineral admixtures in SCC production decreases the proportion of use of cement in the concrete mix and can save the cost of cement. And their inclusion also decreases the heat resulting from the chemical reactions during the mixing process. The mineral admixtures like "SF", "FA", "blast furnace slag (BS)" and "metakaolin (MT)" improve the concrete porous structure and by that can thus increase durability. Most of the materials used are in a powder form that exceeds the thickness of the pozzolan cement, thus affecting the physical properties of the soft cement paste in a similar cement style. These admixtures are used in any type of concrete. Keeping such facts, in this research, the mineral admixture: SF was considered for arguing its impact on fresh and hardened properties of SCC.

2.1 Silica Fume (SF)

The first practice of using silica fume in Portland cement was in 1952. Afterward, SF was used in concrete in 1971 at Fisko metallurgical plant in Norway and in 1978, the standard of using SF in concrete was adopted by Norway's government. Later, in 1981, the use of SF in concrete was approved by the Canada government also and then the practice was continued in other countries (Stanislav et al. 2017). SF is usually characterized as supplementary cementitious material and exhibited pozzolanic and cementitious properties. The other common names such as condensed SF, micro SF ad volatilized SF are frequently referred for SF. SF is a by-product of "ferro-silicon alloys" or only "silicon metal" produced through the smelting method by electric arc furnaces as shown in Figure 1 (Terence, 2005).

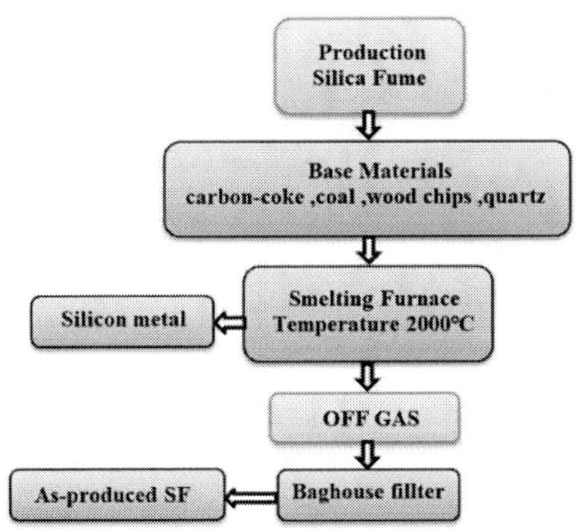

Figure 1. Silica fume production (Terence, 2005)

Among other applications of SF, it has a more prominent application in the construction industry for manufacturing better quality concrete. The primary chemical and physical properties of SF are show in Tables 1. It is commonly available in gray colored powder and somehow resembles Portland cement as shown in Figure 2 (Terence, 2005).

IOP Conf. Series: Materials Science and Engineering **978** (2020) 012052 doi:10.1088/1757-899X/978/1/012052

Figure 2. Silica fume (Terence, 2005)

Table 1. chemical and physical properties of SF (Terence, 2005)

Chemical Properties of SF	Physical Properties of SF	Value
Silicon dioxide > 85%	Particle size (typical)	< 1μm
Amorphous	Specific gravity	2.2
Trace elements depending upon type of fume	Bulk density (As-produced)	130 to 430 kg/m^3
--	Bulk density (Densified)	480 to 720 kg/m^3
---	Specific surface	15000 to 30000 kg/m^2

There is a list of important characteristics of SF which make it right choice to use and mix in concrete mixes such as the fine spherical granules of (0.1-0.2microns) which are 100 times smaller than the granules of normal Portland cement (ACI 116R-8, 1985). The fine spherical granules of SF have an average diameter of 0.1 μm as shown in Figure 3.

Figure 3. Volume of grains of (SF) (ACI 116R-8, 1985)

Therefore, they fill the gaps between the cement blocks of Portland cement, filling the sand spaces between the granules of coarse aggregate and SF granules, the spherical shape of SF particles increases the impact of slipping in cement, increasing the physical effect of SF on concrete to produce dense and cohesive concrete. SF is non-crystallized and due to that, they increase the reactivity with normal Portland cement. However, such effects are compounded because of the low specific gravity of SF which increases the size of the cement sample to reduce voids and areas of weakness, such as drainage channels and inter-transition zone (ITZ). Because of these distinct characteristics of SF, the pozzolanic reaction often begins within 24 hours, producing large amounts of secondary water calcium silicate and significantly improving the concrete microscopic structure at weak vicinity (ITZ) area. The use of SF in concrete decreases the pool of bleeding water under aggregates of iron bars, which reduces the porosity of the transition area and consequently increases the compressive and bond strengths, and corrosion resistance in reinforcing steel.

SF significantly affect the fresh properties of concrete and can resulted more cohesive and less or no bleeding (segregation) on the fresh concrete as shown in Figure 4 (Terence, 2005). The concrete with SF is less prone to segregation than concrete without SF. And such higher cohesion of SF in fresh concrete has typically resulted to increase 40-50 mm more slump than the concrete without SFin the same placement (Terence, 2005).

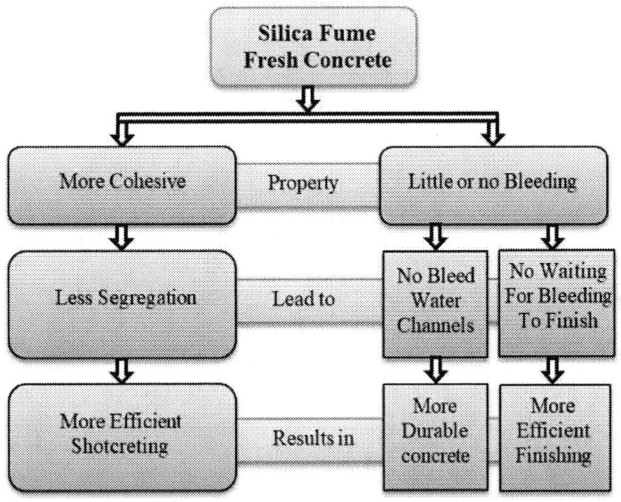

Figure 4. Effects of silica fume on fresh concrete (Terence, 2005)

While the effect of SF on hardened concrete, there are also two main positive impacts such as better mechanical properties and reduce permeability (enhance durability) as shown in Figure 5 (Terence, 2005). The enhancement in mechanical properties such as compressive and flexural strengths, and modulus of elasticity. Though the concrete with SF could improve other mechanical properties, the most important focus is of concrete compressive strength in the construction industry. Furthermore, the durability is directly corresponded with the concrete permeability. The reduction in permeability resulted to extend the time for any chemical (aggressive chemical) to proper mix to the concrete and can damaged permeability as shown in Figure 6 (Terence, 2005).

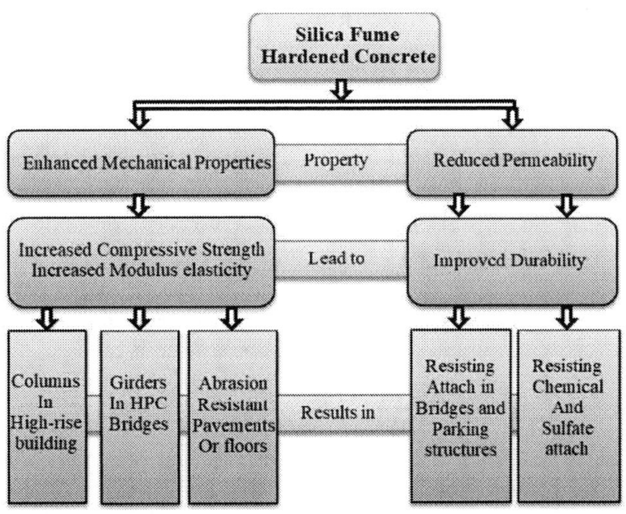

Figure 5. Effects of silica fume on hardened concrete (Terence, 2005)

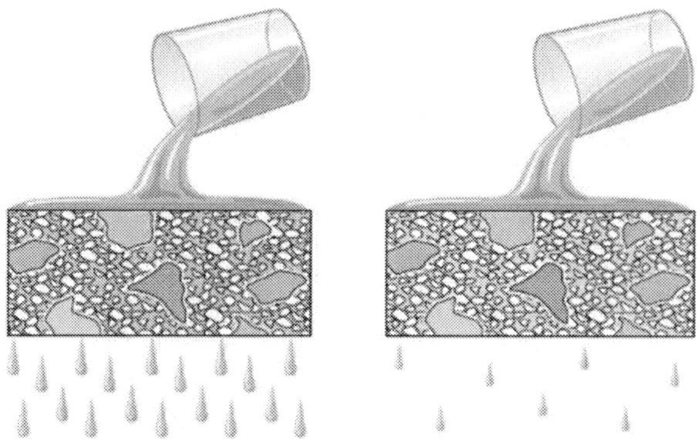

Figure 6. Concrete permeability of concrete. Higher concrete permeability (left) permits water to move readily through concrete. Adding SF and lowering the can decrease permeability to zero (Terence, 2005).

The influence of adding SF on fresh and hardened properties of concrete is explained in detail following previously published research work. However, all the previous research work was reviewed systematically.

2.2 Influence of Silica Fume on Fresh and Hardened Properties.

Mehmet Gesog˘lu et al. (2009) conducted experimental investigation of SCC properties with mineral additives. Twenty-two concrete mixes were designed, the ratio of water to powders was fixed at 0.44 and the percentage of powders in the main mix was fixed at 450 kg / m3. The main mix contains Portland cement only and the cement ratio used for all mixes was ranged from 180 to 450 kg /m3, and SF was used by 5%, 10% and 15% of cement weight of 22.5, 45 and 67.5 kg / m3 respectively with SP 4.9, 5.2 and 7.8 kg / m3 respectively. The results showed the tests slump flow time, slump flow diameter, V-funnel flow time, and L-box height ratio for mixes containing SF by 5%, 10%, and 15% were 5, 4 and 4 seconds, 67, 68 and 69.5 cm, 10, 10 and 10 seconds, and 0.732, 0.824, 0.918 respectively. Whereas for the control mix; slump flow time, slump flow diameter, V-funnel flow time, and L-box height ratio were 1 second, 67 cm, 3.2 second, and 0.706 respectively. The replacement of SF in all mixes resulted more viscous than the control mix but within the EFNARC specifications [18]. The mechanical test results for the control mix and 5%, 10% and 15% of SF mixes were 73.6 MPa and 71.2, 76.1 and 74.8 MPa respectively.

Heba, (2011) used three kinds of mixes to experimentally investigate the properties of SCC. The first mix have different FA ratios, second have different percentages of SF and third mix was combined of FA and SF. Total 9 specimens (cylinder) were created for each kind mix, three of them were cured for 28 days in water, another three were cured for 7 days in water, while the remaining three were left in air for 28 days. The mixture proportions for cement used in their research are shown in Table 2. The results of SCC fresh properties revealed that increasing the ratio of "SF" and "FA" resulted in improvement in passing and filling ability together with segregation resistance. The replacement of cement with 15% of SF results higher compressive strength than replacing cement by 30% of FA for 550 kg/m^3 and 450 kg/m^3 cement content as 12% and 10% respectively.

Table 2. Mix proportions (Heba, 2011)

Cement (Kg/m^3)	450
F-agg (Kg/m^3)	1109
C-agg (Kg/m^3)	612
Water (Kg/m^3)	189
Water-to-cement ratio	0.42

Dehwah, (2012) conducted experimental investigation on SCC and produced 15 SCC specimens. All the 15 SCC specimen were trial tested for V-flow, slump flow, L-box flow, and U-box flow and tests. Only five mixes out of these 15 mixes meet the general requirement flow criteria and accepted for further evaluation. All these five mixes have water to cement and powder ratio as: (M1): "w/cm ratio of 0.40" and "8% QDP (quarry dust powder)", (M2): "w/cm ratio of 0.38 and 8% QDP", (M3): "w/cm ratio of 0.4 and 10% QDP", (M4): "w/cm ratio of 0.40 and 8% QDP" plus "5% of SF" and (M5): "w/cm ratio of 0.40" and "30% FA". Moreover, all these five mixes have cement quantity of 400 kg/ m^3. Results revealed that M2 and M3 have the maximum compressive strength and was more than that of M1, M4, and M5. However, M5 had the lowest compressive strength. The compressive strength of "62.7", "70.1", "65", "64.4", and "52.1" MPa were obtained for M1, M2, M3, M4, and M5 specimens respectively for curing at 28-days and were "64.8", "79.0", "70.7", "68.5", and "56.1" MPa respectively for curing at 90-days curing. The flexural strength results for "M1", "M2", "M3", "M4", and "M5" were 6.6, 6.9, 6.7, 6.6, and 5.2 MPa respectively for curing at 28-days age. And split tensile strength was5.5, 6.5, 6.4, 5.8, and 4.9 MPa for M1, M2, M3, M4, and M5 respectively for curing at 28-days age.

Ali (2012) conducted experimental research on polymer-modified SCC (PMSCC). A total of 12 concrete mixes were used and cured for 90 days for evaluation and compared with traditional concrete and SCC. The SF and limestone (LS) were used as fillers and SP such as naphthalene and modified polycarboxylic were used to increase the flow properties. Polyvinyl acetate (PVC) and styrene-butadiene rubber (SBR) were used in different dosage from 0%, 5%, 10%, and 15%. The mixture design used in this research is shown in Table 3. The results revealed that PVC has a significant effect on compressive strength when curing at an early age and later increase as 27.3% and 23.3% when curing at 90-days age for PMSCC with 5.0% and 10.0% of PVC respectively as compared with traditional SCC. And the effect of mineral additives on all concrete mixes showed that the LS gave a higher compressive strength with concrete than SF at the same age. Furthermore, higher compressive strength for modified polycarboxylic for both traditional SCC and concrete was obtained when used naphthalene. The use of 10% of SBR with SP has little effect at different referred ages. The tensile strength and modulus of elasticity for PMSCC were higher than the traditional SCC at different ages. LS powder and chemical admixture based on naphthalene usually have a progressive effect on tensile strength and the use of naphthalene caused higher tensile strength than with modified polycarboxylic. While the use of LS powder (as a filler) provides a higher dynamic modulus of elasticity compared for concrete mixes with SF. Thus, it can be determined that the partial replacement of cement with fine materials reduces the segregation process in SCC and the use of mineral additives with SCC gives satisfactory results.

Table 3. Description of concrete mixes Ali (2012)

Cement	Water	C/F-agg	Filler	Polymer	Plasticizer	Characterize	Filler Type	Polymer Type	Plasticizer	W/B	Plasticizer /C%	Polymer /C%
400	180	875	-	-	2	TC	-	-	MPCE	0.45	0.5	-
400	180	855	40	-	13.6	TSCC	SF	-	MPCE	0.41	3.4	-
400	180	845	40	20	9.1	PSCC	SF	SBR	MPCE	0.43	2.28	5
400	180	825	40	40	5.8	PSCC	SF	SBR	MPCE	0.45	1.45	10
400	180	820	40	60	5.5	PSCC	SF	SBR	MPCE	0.48	1.38	15
400	180	840	40	20	10.6	PSCC	SF	PVA	MPCE	0.43	2.65	5
400	180	825	40	40	8	PSCC	SF	PVA	MPCE	0.45	2	10
400	180	855	40	-	13	TSCC	LP	-	MPCE	0.41	3.25	-
400	180	825	40	40	10	PSCC	LP	SBR	MPCE	0.45	2.5	10
400	180	875	-	-	2	TC	SF	-	NAPH	0.45	0.5	-
400	180	855	40	-	15	TSCC	SF	-	NAPH	0.41	3.75	-
400	180	820	40	40	12	PSCC	SF	SBR	NAPH	0.45	3	10

Anhad, (2012) carried out an experimental study to SCC, 7 concrete mixing mixes were designed, the control mix contains only the concrete cement without the powders. The remaining mixes contain different proportions of powders. The cement ratio used for all mixes ranges from 467 to 550 kg / m3, SF was used by 5%, 10% and 15% of the weight of cement 28, 55 and 83 respectively and 1.8% of SP was used for manufacturing SCC. The mixture design is shown in Table 4. Fresh tests showed that slump flow, V-funnel flow time, and L-box height ratio for mixes containing 5%, 10% and 15% of SF were 610, 660, 685 mm, 0.9, 0.9 and 0.9 and 11, 10 and 8 seconds respectively. The control mix slump flow, V-funnel flow time, and L-box height ratio were 585 mm, 0.9 and 11 seconds respectively. This means that the control mix and mix containing 5% of SF is not within the required limits for slump flow only, while the remaining mixes are within the required limits and for all tests. The compressive strength results with the water-to-powder ratio of 0.48 were 38.7, 43.5 and 39.3 MPa for 5%, 10%, and 15% respectively for 28-days curing age and were higher than the control mix (31.7 MPa). These results indicated that 15% of SF can be used for possible design SCC mixes.

Table 4. Mix proportions Anhad, (2012)

Mix	Cement (Kg/m^3)	MK (Kg/m^3)	SF (Kg/m^3)	C-agg (Kg/m^3)	F-agg (Kg/m^3)	Water (Kg/m^3)	W/P	SP (Kg/m^3)
Control	550	0	0	590	910	265	0.48	9.91
5%SF	522	-	28	590	910	265	0.48	9.91
10%SF	495	-	55	590	910	265	0.48	9.91
15%SF	467	-	83	590	910	265	0.48	9.91
5%MK	522	28	-	590	910	265	0.48	9.91
10%MK	495	55	-	590	910	265	0.48	9.91
15%MK	476	83	-	590	910	265	0.48	9.91

E. Güneyisi et al. (2012) conducted an experimental investigation of fresh properties of SCC mixed with SF and FA. Total nine mixtures were designed having "water to binder ratio" of "0.35" and total binder of 550 kg/m3 was used. The results revealed that V-funnel time and slump flow time decreased, and L-Box ratio was improved significantly with the addition of SF and FA together.

Kennouche et al. (2013) produced SCC by using locally available materials to demonstrate the influence of SF on the concrete rheological behavior. The SF was replaced with 15% of cement. Two types of Portland cement and SP were used. Table 5 shows the mixture design. The results showed that the diameter of the spread obtained from 650-750mm, L-box test was easy traffic and flow without obstruction and isolation due to high liquidity o account of using SP. The minimum compressive strength was 20 MPa when curing at an age of 7-days and the highest is 40.93 MPa when curing at an age of 28-days.

Table 5. Quantity (kg) of the components in each formulation (Kennouche et al. (2013)

Mode Code	Cement (Kg/m^3)	SF (Kg/m^3)	Aggregate 3/8mm	Aggregate 8/15mm	C-Sand (Kg/m^3)	F-Sand (Kg/m^3)	Water (Kg/m^3)	Super-plasticizer
F1	400	60	533.14	265.04	582.92	243.78	193.2	7.2
F2	400	60	534.27	264	582.41	241.66	193.2	8
F3	400	60	534.08	264.99	584.64	242.55	193	6
F4	400	60	534.27	264	243.69	241.46	193.1	8

Fareed, et al. (2013) conducted experimental investigation on SCGC and produced a total 4-mixture specimen with a geopolymer and SF addition. 3-mixes were prepared with different percentages of SF (5, 10 and 15%) and 1-without SF (control mix). FA was replaced with SF, the total powder content was fixed at 400 kg/m3 and the "water to geo-polymer solids (W/Gs) ratio" was maintained at "0.33" for all the mixes. 12% of water and 6% of SP by mass for binders were used to attain the desired workability characteristics of SCC. Table 6 shows the mixture design. The workability-related fresh and hardened properties of SCGC were assessed at age of "1", "7", and "28" days after curing 48-h in oven. The results of fresh properties indicated that the partial replacement of FA with SF in SCGC provides loss of workability and revealed poor results than the control mix. However, the values of fresh properties of all mixes were within EFNARC specifications (2002) except for the mix contain

IOP Conf. Series: Materials Science and Engineering **978** (2020) 012052 doi:10.1088/1757-899X/978/1/012052

15% of SF. The compressive strength values for SCGC were 53.38, 55.02, and 53.96 MPa for 5, 10 and 15% of SF respectively, all these values were higher than the compressive strength (51.43 MPa) for mix without SF (control mix) at all curing age. The compressive strength values indicated that 10% of SF in SCGC is the optimum percentage. The same trend was also observed for flexural strength; the maximum value achieved at 10% of SF. Thus the overall conclusion from their research is to replace SF with FA should be maximum up to 10% for producing better SCGC.

Table 6. Mix proportions Fareed, et al. (2013)

Mix	FA (Kg/m³)	SF (Kg/m³)	C-agg (Kg/m³)	F-agg (Kg/m³)	Sodium hydroxide		Sodium silicate (Kg/m³)	water to geo-polymer solids (W/Gs) ratio	SP/ wt %	Extra water/ wt%	Oven treatment	
					(Kg/m³)	(mol/l)					Time/h	Temperature/ °C
M1 (0 wt% SF)	400	0	950	850	57	10	143	0.33	6	12	48	70
M2 (5 wt% SF)	380	20	950	850	57	10	143	0.33	6	12	48	70
M3 (10wt% SF)	360	40	950	850	57	10	143	0.33	6	12	48	70
M4 (15wt% SF)	340	60	950	850	57	10	143	0.33	6	12	48	70

Karthikeyan et al. (2013) carried out and an experimental study on SCC with the inclusion of two mineral admixtures such as SF (2.5, 5, 7.5 and 10%) and FA (5, 10 and 15%) by weight of cementitious material with a reference mix design (without mineral additives). While reducing the "water to powder ratio" as 0.36 l / m3 and with SP at 0.013 l / m3. Total 8-SCC mixes were designed according to the required requirements (EFNARC, 2002) as shown in Table 7. The results of fresh properties indicated that all the mixes were within the specified specifications, except the slump flow for the mix containing the 10% SF. The compressive strengths were 62.48, 64.85, 64.95 and 65.60 MPa for 2.5, 5, 7.5 and 10% of SF respectively and 62.20, 63.40 and 63.90 MPa for 5, 10 and 15% of FA respectively, while the reference mix has 61.79 MPa. As based on these compressive strength values, 10% of SF mixture provides the highest values and concluded for the optimum percentage of SF in their research. However, at 10% of SF mixture, flowing, filling and passing ability decreased.

Table 7. Mix proportion Karthikeyan et al. (2013)

Water (L/m³)	Cement (Kg/m³)	C-agg (Kg/m³)	F-agg (Kg/m³)	SP (L/m³)	VMA(L/m³)
0.36	1	1.97	2.14	0.013	0.001

Salem Alsanusi, (2013) conducted experimental investigation on SCC and nine concrete mixes were produced with different percentages of SF as 3, 6, 9% and "water to powder ratio" were 0.3, 0.34 and 0.37 for each percentage of SF. The results indicated that the "slump flow", "slump flow T50 cm", "U-Box", "L-box height ratio", "V-funnel flow time" and "slump flow with J-Ring" for all mixes with water to powders 0.34 were "530, 660 and 550 mm", "2, 3 and 3 seconds", "0.76, 0.9, 0.95", "0.78, 0.87, 0.87", "11, 5 and 8 seconds" and "510, 510 and 520 mm" respectively and were optimum for the "water-to-powder ratio" of "0.34". Also, the compressive strength for mixes with the "water-to-powder ratio" of "0.34" was 36.84, 47.63 and 46.18 MPa for each percentage of SF respectively and these values of compressive strength were optimum as compared to other water to powder ratio. Moreover, on these results basis, 6% of SF was concluded as the optimum percentage.

Ramanathan et al. (2013) conducted experimental investigation on SCC and produced 9-specimen and 1-control mix. Portland cement was replaced with 30%, 40% and 50% of mineral admixture (SF), the "water/powder mass ratio" was considered as 0.35. The total powder content was added with different quantity as "400 kg/m3", "450 kg/ m3", "500 kg/m3" for iteration trials and finally, 500 kg/m3 was fixed. All concrete mixes showed slump flows and V funnel tests value obtained through experimental investigation were within European standards (2005). Concerning mechanical results; 30% of SF showed the best performance when curing at 7 and 28-days age.

Vivek et al. (2015) produced 4-SCC mixes by replacing cement with various percentages of SF from "5% to 20%". The SP (Conplast SP430) at the optimal dose equal to (1%) and the "water-to-binder ratio (w/b) ratio" of "0.8" was used for producing SCC mixes. The design mixture used in their research is shown in Table 8. The better fresh properties were achieved at replacing the cement with 15% of the SF, while the highest compressive strength was achieved at the replacement of cement by 5% of SF at 7 and 14 days curing age.

Table 8. Materials required of SCC (Vivek et al. 2015).

Materials	SF5	SF10	SF15	SF20
Cement (Kg/m^3)	4.75	4.5	4.25	4
Water (Liters)	4	4	4	4
W/P	0.8	0.8	0.8	0.8
SF (Kg/m^3)	0.25	0.5	0.75	1
C-agg (Kg/m^3)	8.2	8.2	8.2	8.2
F-agg (Kg/m^3)	10	10	10	10
HRWR %(Kg/m^3)	1	1	1	1

Mehrdad, et al. (2015) conducted experimental investigation on SCC and produced a total of 14-SCC mixes containing a high volume of LS, SF, zeolite (ZL), metakaolin (MT), and VMA. The results showed that the mixes containing MT, SF, and ZL did not show segregation of these mixes, the results L-box height ratio results were unsatisfying for mixes containing SF and ZL. The lower compressive strength of the mix containing ZL was achieved as compared to other mixes. They concluded that SF compensates for poor sand quality and the SCC made of SF can pass the tests successfully. The mixtures contain SF and ZL deteriorated the mechanical properties because of the high absorption of water as compared to reference mix.

Jhansirani et al. (2015) conducted experimental investigation on SCC and produced total 11-SCC mixes with various partial replacements of FA (10, 20 and 30%), SF (5, 10 and 15%) and combination of both FA and SF (SF5+FA30, SF10+FA20, SF15+FA10%). From the results, it was identified that SF shows more resistance against the acid, alkaline, chlorine penetration. It was also observed that by increasing mineral admixes in SCC both fresh and durability properties were increased. Results revealed that the "slump flow", "T50 (sec)", "V – Funnel T50 min (sec)", "L - Box (H2/H1)" were within the limits of the EFNARC specifications (2002). Moreover, the replacement of cement by 20% of FA and 15% SF and the combination of both SF15 + FA10% shows that the maximum durability factor. And replacement of cement by 15% of SF and 20% of FA and a combination of SF15% and FA10% shows that lower absorption.

Faseyemi, (2015) conducted experimental investigation on SCC and produced a total 6-SCC specimen. 5-SCC specimen were containing micro SF (10, 8, 6, 4 and 2%) together with FA (35, 30, 25, 15 and 10%) by weight of cement and 1-SCC mix without mineral additive (control mix). The "w/c ratio" ranged from 0.35-0.30 for SCC contain mineral additives and 0.40 for the control. The total binder ratio was 450kg/m3 and SP and rheomatix (III) were used as a viscous modifier. Table 9 shows he mixture design used in this research. Based on hardened properties in their research, they conclude that the highest compressive strength was obtained at 2% of micro SF and 10% of FA, while all other values were above the design strength expected for SCC at all curing ages.

Table 9. Mix Proportion for 50Npa SCC (Faseyemi. 2015).

Materials	Mix Proportion					
	Reference	10%M/S +30%FA	8%M/S +25%FA	6%M/S +20%FA	4%M/S +15%FA	2%M/S +10%FA
Cement (Kg/m^3)	450	324	384	400	437	475
Fly Ash (Kg/m^3)	-	162	113	108	81	54
MSF (Kg/m^3)	-	54	43.2	32.4	22	11
C-agg (Kg/m^3)	1037	1037	1037	1037	1037	1037
W/Sand (Kg/m^3)	880	880	880	880	880	880
W/C (Kg/m^3)	0.37	0.4	0.4	0.4	0.4	0.4
Water %	200	218	218	218	218	218
SP %	5	3.5	3.5	3.5	3.5	3.5
Rheomatrix %	-	0.5	0.5	0.5	0.5	0.5

E. Güneyisi et al. (2015) conducted experimental investigation on hardened properties of SCC produced cold bond FA lightweight aggregates. Total nine specimen of SCC mixed with SF and FA were produced having constant "water to binder ratio" of "0.35" and total quantity of binder quantity of 550 kg/m3 was used. The hardened properties of SCC specimens were determined for 28 and 56-days curing age. The results revealed that the addition of mineral admixture significantly improved permeability of the SCC. Moreover, compressive strength of SCC specimens mixed with lightweight aggregates was much higher as compared to control concrete (without mineral admixture)

3. Chemical Admixtures

In order to improve the fresh and hardened properties of concrete, chemical admixtures are usually added in small quantities. However, the addition of chemical admixtures is proportional to the increase in cost and the addition must not adversely affect the concrete. The most important types of chemical admixtures used with concrete are water-reducing admixtures, water-preventing admixtures for concrete, and air agent entraining additives. In general, the main function of adding chemical admixtures in the concrete is to adjust fresh properties of concrete and improve workability without increasing water content. Superplasticizers are chemical admixtures which are used for improving workability of concrete. In other sense, they can also help to decrease the water quantity in the concrete (for the same slump and workability) and thus advantage to increase the concrete strength and durability. Keeping such facts, in this research, the chemical admixture: superplasticizers was considered for arguing its impact on fresh and hardened properties of SCC.

3.1 Super-plasticizer (SP)

SP is added to the cement mix by 1% - 3% of the weight of the cement in the liquid form (Mahmoud, 2002). SP is used to increase compressive strength in the early and late ages, increase flexure strength, increase adhesion strength, reduce permeability, increase the modulus of elasticity (Mahmoud, 2002). Mainly; SP is used in different applications of concrete like high-performance-concrete, the hot-air-concrete that requires to maintain the workability for a longer period, concrete pumping, and pre-cast concrete especially in narrow sections where the amount of armament is dense as it increases the susceptibility to spreading in all parts of the mold. SP can surround the grains of cement and aggregates and works on the lubrication of the region so that the grains slide on each other and distributed in the size of the total concrete and more than all of that is to make the concrete more homogeneous as shown in Figure 7 (Koehler, E. P. and Fowler, 2007).

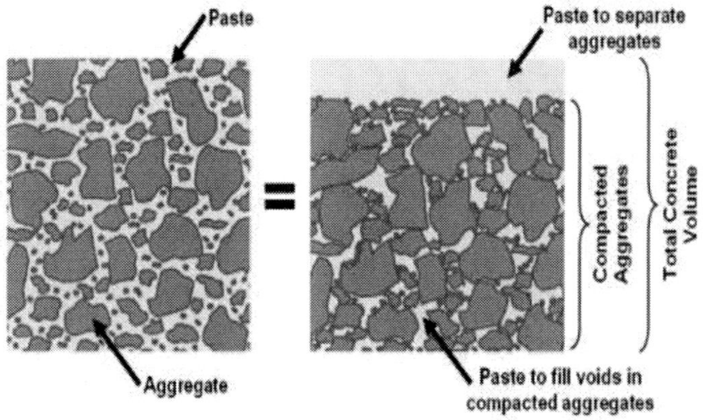

Figure 7. Embracing of SP for grains cement and aggregates (Koehler, E. P. and Fowler, 2007)

However, the application of SP gives negative surface charge on the cement particles and result in electrostatic repulsion, liberates the trapped free water and prevents flocculation and agglomeration, and as shown in Figure 8. (Wajde, 2016). For more detail knowledge, the influence of adding SP on fresh and hardened properties of concrete is explained below by considering previously published research work. However, all the previous research work was reviewed systematically.

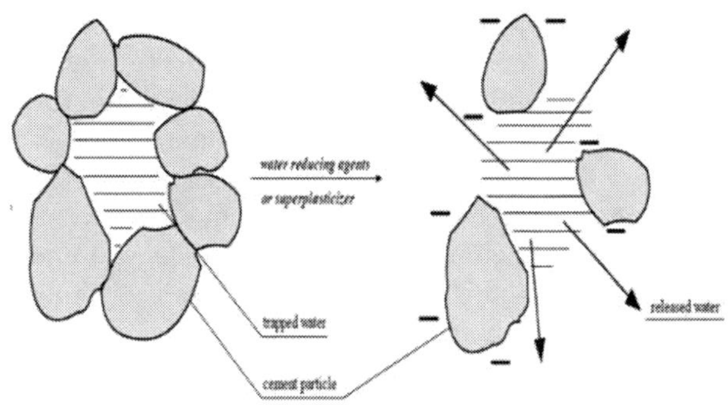

Figure 8. Flocculation of cement particles deceiving water and the dispersion effect by SP (Wajde, 2016)

3.2 Effect Superplasticizer (SP) on Fresh and Hardened Properties.

Nuruddin et al. (2011) conducted experimental research on geopolymer SCC (SCGC) and produced 8-SCGC mixtures to examine the influence of various percentages of SP on fresh and hardened properties. FA and molarity of sodium hydroxide alkaline solution were also in SCGC mixtures. Table 10 shows the mixture design used in their research. The workability and compressive strength increased with the increasing percentage of SP. All SCGC mixtures meet EFNARC specifications (2002) except SCGC mixture 1, 2 and 3. The maximum compressive strength (53.8 MPa) was obtained for SCGC mixture 5 containing 7% of SP at 28-days curing age. They concluded that a lower percentage (3, 4 and 5%) of SP had poor passing and filling ability and workability. Moreover, the mixture of 6% of SP and 12% of alkaline sodium hydroxide results in satisfactory fresh properties as per EFNARC specification (2002) and also results in the higher value of compressive strength (51.52 MPa) at 28-days curing age.

Table 10. Mix design proportion Nuruddin et al. (2011)

Mix	FA (Kg/m^3)	C-agg (Kg/m^3)	F-agg (Kg/m^3)	NaOH (Kg/m^3)	Na-Silicate (Kg/m^3)	Extra Water (Kg/m^3)	SP (Kg/m^3)	Curing	
								Time hr.	Temp °C
S1	400	950	850	57	143	48	12	48	70
S2	400	950	850	57	143	48	16	48	70
S3	400	950	850	57	143	48	20	48	70
S4	400	950	850	57	143	48	24	48	70
S5	400	950	850	57	143	48	28	48	70
S6	400	950	850	57	143	48	24	48	70
S7	400	950	850	57	143	48	24	48	70
S8	400	950	850	57	143	48	24	48	70

Dubey et al. (2012) conducted experimental research on SCC and produced 6-SCC mixes with different percentages of SP (2% to 10%) of cementitious material. Table 11 shows the mixture design used in their research. Concrete setting time was achieved as 1, 3, 7, 7 and 11 for 2, 4, 6, 8 and 10% of SP respectively. The compressive strength of the SCC was reduced for 2% of SP, on the other hand significant increment was observed at 4% of SP for higher curing age. Furthermore, a minimal increase in compressive strength was observed until 8% of SP addition at any curing age and 10% of SP; another significant increment in compressive strength was obtained at higher curing age. Thus, they concluded that increasing the curing age period causes an increment in compressive strength and 4% and 10% of SP provide higher values of compressive strength in their research.

Table 11. Mix proportions used in the trials Dubey et al. (2012)

Mix	Cement (Kg/m^3)	SF (Kg/m^3)	FA (Kg/m^3)	C-agg (Kg/m^3)	F-agg (Kg/m^3)	W/Powder
1	400	30	170	700	790	0.38

Ali Mardani et al. (2013) conducted experimental research on SCC and produced 4-SCC mixtures by adding the different molecular weight of SP to examine the SCC fresh and hardened properties. However, the SP incorporated in their research had the same polymer structure and same chain. The "w/c ratio" of "0.4" and "slump flow" of "730±10 mm" was used for producing SCC mixes. Table 12 shows the mixture design used in their research. The results showed that the addition of SP has a significant influence on the fresh properties of SCC. Moreover, the inclusion of SP resulted that the slump flow remained unchanged, V-funnel times obtained from 27-47s. Also, less compressive strength obtained at curing age of 28-days and it is due to more loss in slump loss due to the inclusion of SP.

Table 12. Mix proportions of the SCC mixes (kg/m3) Ali Mardani et al. (2013).

Mix	Cement (Kg/m^3)	L.P	Aggregate		Water (Kg/m^3)	SP (Kg/m^3)
		0-0.125 mm	0-5 mm	5-15 mm		
A	465	173	793	793	186	6.6
B	459	171	783	783	183	6.1
C	461	172	786	786	184	5.7
D	442	165	754	754	177	7.4

Dumne, (2014) carried out experimental research on SCC and produced 6-SCC mixtures with M20 mixes grade. The w/c ratio was 0.55, 10% replacement of cement with FA and different percentage of SP (0.25, 0.30%) of cement content were used for producing SCC. One normal concrete (without SP) was also produced to compare the SCC fresh and hardened properties. Table 13 shows the mixture design used in their research. Fresh properties results showed that normal concrete mix has a slump value of 120 mm, whereas for a mix containing 10% FA and 0.35% SP; the slump value achieved as 285 mm which is very high. Moreover, at 10% of FA and addition of SP; the compressive strength also accelerates and achieved very high.

Table 13. Mix proportion for various trials mixes of M20 concrete grade Dumne, (2014).

Mix	Cement (Kg/m^3)	FA (Kg/m^3)	C-agg (Kg/m^3)	F-agg (Kg/m^3)	Water (Liters)	SP (Liters)
A	348	-	1207	558	191.4	-
A1	313.2	0.11	1207	558	191.4	0.87
A2	313.2	0.11	1207	558	191.4	1.04
A3	313.2	0.11	1207	558	191.4	1.22

Salahaldein Alsadey, (2015) carried out experimental research on SCC and produced 6-SCC mixtures with M35 mixes grade. The "w/c ratio" was "0.475" and different percentage of the SP (0.6, 0.8, 1.2, 1.8 and 2.5%) was used to study the influence on fresh properties and compressive strength of SCC. However, a control mix (without SP) was produced to compare the above-said properties. Table 15 shows the mixture design used in their research. The results indicated that increasing the amount of SP leads to improve workability. However, the highest value of compressive strength (40.24 MPa) was achieved for the control mix (without SP) as compared to other SCC mixtures except for the SCC mixture having 0.8% SP. Moreover, the SCC mixture containing 0.8% of SP provides higher than the desired characteristics strength for SCC as per EFNARC specifications (2002).

Shahiron. S et al. (2017) conducted experimental investigation on SCC to investigated the mechanical and physical properties of SCC. Different percentages of MT (0%, 5%, 10%, and 15%) were replaced with cement for the design of four SCC mixes. The addition of MT helps to increase the mechanical properties, but it turns to reduce the workability of SCC. However, to counter this problem, the addition of SP is recommended.

Nataraja et al. (2018) conducted experimental investigation on SCC and produced 25-SCC mixes to get the relationship between "compressive strength" and "water cementitious ratio". The compressive strength was studied from 20 to 60 MPa using SP (poly corboxilic). They found that the water ratio of 0.37 to 0.47 was sufficient to obtain compressive strength values of 25 to 60 MPa and the results of the strength obtained were well matched with the required design strength.

Nagaraju et al. (2018) conducted experimental investigation on SCC and produced 7-SCC mixtures with M40 mix grade to obtain the required admixture dosage of ADVA960 (Poly-carboxylic-ether based) admixture and its effect on fresh properties and compressive strength of SCC at 7 and 28-days curing age. Table 14 shows the mixture design used in their research. Results obtained in their research showed that only SCC-mix-7 having water to cement ratio of 1.1 and the 0.5% dose of SP achieved all the requirements of the SCC as per EFNARC specifications (2002) as compared to other SCC mixes. SCC-mix-1 (500Kg/m3 of powder, 0.8 of w/c and 0.35% of SP) and SCC-mix-2 (500Kg/m3 of powder, 0.8 of w/c and 0.4% of SP) had unsatisfactory fresh properties results and bleeding and segregation phenomenon was observed due to high dose of SP especially in SCC-mix-2 with lower w/c ratio (0.8). SCC-mix-4 to 6 contain 0.4%, 0.45% and 0.5% of SP respectively also failed and did not meet EFNARC specifications for SCC, it was mainly because of the high dose of SP and lower w/c ratio (0.8). Thus, they concluded that the combination of the higher dose of SP (0.5%) with a higher w/c ratio (1.1) can be used to produce SCC as per EFNARC specifications (2002). Moreover, such a combination of SP (0.5%) and w/c ratio (1.1) targeted compressive strength values (at 7 and 28 days curing age) very near to EFNARC specification [35] for compressive strength for SCC.

Table 14. Mix Quantities Nagaraju et al. (2018)

Mix	Cement (Kg/m^3)	FA (Kg/m^3)	C-agg (Kg/m^3)	F-agg (Kg/m^3)	Water (Kg/m^3)	W/C (based on vol.)	Admixture
Mix1	350	150	902.066	788.76	177	1	1.75
Mix2	350	150	902.066	788.76	177	1	2.5
Mix3	350	150	950.85	823.65	163.4	0.92	2
Mix4	400	150	840.85	721	188.18	1	2.2
Mix5	400	150	855.71	748.57	207.14	1	2.47
Mix6	400	150	865.71	671.42	207.14	1	2.75
Mix7	400	150	933.571	664.28	215	1.1	2.75

Mehena, et al. (2018) conducted experimental research to examine the rheological and hardened properties of SCC. The mixture paste was designed as "70% of cement", "w/c ratio (w/c + f) of 0.35", "30% of pozzolana" and "three types of SP (poly naphthalene sulfonate (PNS), poly melamine sulfonate (PMS) and polycarboxylate (Poly)" with different percentages were used for producing SCC. The rheological results were achieved as slump flow cone was 72, 72 and 73cm and L-box (H1/H2) was higher than 90% for the SP dose: PNS= 1.75, PMS= 2.5 and Poly=1% respectively. The compressive strength was 39, 47 and 41 MPa for the SP dose: PNS= 1.75, PMS= 2.5 and Poly=1% respectively. They concluded that for rheological properties; Polytype of SP is best and for hardened properties; PMS type of SP is best.

Benaichaa et al. (2019) produced total 9-concrete including 1-normally-vibrated carried (N) concrete and 8-SCC with different percentage of SP (0, 0.3, 0.4, 0.5, 0.6, 0.7, 0.8, 0.9, 1). The design mixture used in their research is shown in Table 15. They found in their results that all SCC mixes containing SP were within the EFNARC specification (2002) limits except the N concrete. Moreover, the inclusion of SP causes to decrease the time taken by the mixes to empty the funnel and yield stress and plastic viscosity decreases. The best mechanical properties (74.4 MPa) were achieved for the mix SCC-SP1 (0.3 SP in SCC) at an age of 28-days as compared to other SCC mixtures and N concrete.

Table 15. Mix proportions of concretes (Benaichaa et al. (2019)

Mix	W/P	Cement (Kg/m^3)	Limestone filler (kg/m3)	C-agg (Kg/m^3)	F-agg (Kg/m^3)	(SP)%
N	0.37	350	170	900	890	-
SCC-SP1	0.37	350	170	900	890	0.3
SCC-SP2	0.37	350	170	900	890	0.4
SCC-SP3	0.37	350	170	900	890	0.5
SCC-SP4	0.37	350	170	900	890	0.6
SCC-SP5	0.37	350	170	900	890	0.7
SCC-SP6	0.37	350	170	900	890	0.8
SCC-SP7	0.37	350	170	900	890	0.9
SCC-SP8	0.37	350	170	900	890	1

4. Conclusion

Recently, extensive research works have been conducted to comprehend the influence of the inclusion of SF and SP on the SCC fresh and hardened properties. In this review work, the previous research works has been reviewed and presented. On the basis of the previous research revision which was argued and reported in this research work, the notable conclusions were drawn and presented as follows:

- The production of the SCC test depends on the trial and error based methods, thus, it is utmost necessary to determine the optimal proportions of the components of the concrete mix including mineral and chemical admixtures.
- Such optimal proportions of concrete admixtures reduce the experiments, cost and time for producing SCC.
- The use of SF as an alternative to cement improves the ITZ as it is considered a filler material. Because of the smaller surface area SF particles; the porosity decreases, increases density and provides more cohesion than non-SF mixes. SF imparts to increase the cement content, thus reducing the percentage of water to the cement which in turn reduced bleeding and segregation of the SCC.
- The use of SF as an alternative to cement decreases the workability and therefore increases the need to increase the amount of water.
- The use of SP with SF improves the workability of the SCC which in turn decreases in the amount of cement required for producing SCC. Such a reduction in cement quantity reduces the cost of producing SCC and hence enhance sustainability in the construction.

- Previous studies have shown the compressive strength decreases with the increase in the percentage of SP. Although the inclusion of SP has slightly influenced on the compressive strength, but early curing ages; it pronounced a better effect on the compressive strength of SCC.
- The increased percentage of SP indications to an improvement in workability, but higher percentage of SP leads to damage the concrete cohesiveness.
- The increased percentage of SP will produce the phenomenon of bleeding and Segregation.
- In general, it was confirmed by researchers that the use of SF and SP increases the hardened and fresh properties of concrete, but it should be within optimal quantities and doses.

5. Future Recommendations

The application of SCC in increasing day by day in the construction industry due to many attractive characteristics. Better fresh and improved hardened properties make SCC for use in the construction industry. Better fresh and hardened properties were achieved by incorporating mineral and chemical admixtures and different aggregates. Though the review has been taken as per the title of the research and ample information was presented in this review work, but still, there are a lot of research gaps that could be investigated further in this research area.

To focus the application of SCC in the construction industry, it is recommended to investigate the effect of the addition of SF and SP on other mechanical properties such as bond strength, impact resistance, and abrasion resistance. Besides these mechanical properties; the time-dependent mechanical properties such as creep and shrinkage of SCC are also recommended for further investigation in focusing the title of the research. Though the addition of SF and SP are mineral and chemical admixtures and increase passing and flowing ability of SCC, the other types of mineral and chemical admixtures are also encouraged to investigate their effect on fresh and hardened properties of SCC. Moreover, very few research work was observed on rheological and thixotropic behavior of SCC, thus for better understanding workability, such behaviors are recommended for investigation by adding SF and SP in SCC. The major contribution and output of conducting this review work are demonstrating that the use of SF and SP in SCC beneficial and recommending their use in SCC.

References

[1] Abdul. S.K and Khushpreet.S. (2018). An Experimental Investigation by Utilizing Plastic Waste and Alccofine in Self-Compacting Concrete. *Indian Journal of Science and Technology. 11(26).*

[2] ACI–Committee 212. *Guide for the Use of High-Range Water Reducing Admixtures (Superplasticizers) in Concrete.* ACI Manual of Concrete Practice, 2017, p. 212.4R-1-13.

[3] ACI 116R-8. *Cement and Concrete Terminology*. American Concrete Institute, Michigan (1985).

[4] Al-Hadithia. A. I and Hilal. N. H. (2016). The possibility of enhancing some properties of self-compacting concrete by adding waste plastic fibers. *Journal of Building Engineering, 8*, p. 20–28.

[5] Al-Hadithia. A. I, Ahmed T. N. and Waseem. K. M. (2019). Mechanical properties and impact behavior of PET fiber reinforced self-compacting concrete (SCC). *Composite Structures, 224,* p. 111021.

[6] Ali Mardani-Aghabaglou, Murat Tuyan, Gökhan Yılmaz, Ömer Arıöz and Kambiz Ramyar. (2013). Effect of different types of superplasticizer on fresh, properties and strength of self-consolidating concrete. *Construction and Building Materials 47*, p. 1020-1025

[7] Alyamac K.E, Ghafari. E and Ince. R. (2017). Development of eco-efficient self-compacting concrete with waste marble powder using the response surface method. *Journal of Cleaner Production, 144,* p. 192–202.

[8] American Concrete Institute-ACI 237R. 2007. *Self-Consolidating Concrete*. USA.

[9] Anhad.S.G. (2012). *Self-strength properties and sulphate resistance of self-compacting concrete incorporating silica fume and metakolin*. Thapar Univerersity Patiala-147004 2012.

[10] ASTM C494/C494M. (2004). *Standard Specification for Chemical Admixtures for Concrete*. Annual Book of ASTM Standards, Vol. 04.02, pp. 1–8.

[11] B. Felekoglu, K. Tosun, B. Baradan, A. Altun, B. Uyulgan. (2006). The effect of fly ash and limestone fillers on the viscosity and compressive strength of self-compacting repair mortars. *Cement and Concrete Research, 36 (9),* p. 1719–1726.

[12] Bogas, J. A., A. Gomes, and M. F. C. Pereira. (2012). Self-compacting lightweight concrete produced with expanded clay aggregate. *Construction and Building Materials 35,* pp. 1013–1022.

[13] BS EN 206-9 2010. *Additional rules for self-compacting concrete (SCC).* British Standard Unit

[14] C. Djelal, Y. Vanhove, A. Magnin. (2004). Tribological behavior of self-compacting concrete. *Cement and Concrete Research, 34 (5),* p. 821–828.

[15] E. Güneyisi, M.Gesoğlu, E. Booya. (2012). Fresh properties of self-compacting cold-bonded lightweight fly ash aggregate concrete with different mineral admixtures. *Materials and Structures*, 45 (12), pp. 1849-1859.

[16] E. Güneyisi, M.Gesoğlu, E. Booya, K. Mermerdas. (2015). Strength and permeability properties of self-compacting concrete with cold bonded fly ash lightweight aggregate. *Elsevier's Construction and Building Materials*, 74, pp. 17–24.

[17] F. Aslani, G. Ma, D.L.Y. Wan, G. Muselin. (2018). Development of high-performance self-compacting concrete using waste recycled concrete aggregates and rubber granules. *Journal of Cleaner Production. 182*, pp. 553–566, https://doi.org/10.1016/j. jclepro.2018.02.074.

[18] Fareed, A.M, Muhd, F.N., Yazoghli, O. and Nasir, S. (2013). Effect of silica fume on the fresh and hardened properties of fly ash-based self-compacting geo-polymer concrete. *International Journal of Minerals, Metallurgy and Materials 20*(2), p. 1.

[19] EFNARC. *Specification and Guidelines for Self-Compacting Concrete*, 2002, 1-32.

[20] G.A. Salman. (2018). Density and Ultrasonic Pulse Velocity Investigation of Self-Compacting Carbon Fiber- Reinforced Concrete. *Engineering and Technology Journal, 36*, Part A, No. 1, pp. 89-99.

[21] H. A. F. Dehwah. (2012). Mechanical properties of self-compacting concrete incorporating quarry dust powder, silica fume or fly ash. *Construction and Building Materials, 26*, p. 547-551.

[22] Heba A. Mohamed. (2011). Effect of fly ash and silica fume on compressive strength of self-compacting concrete under different curing conditions. *Ain Shams Engineering Journal, 2,* p. 79–86.

[23] J. Karthikeyan and C. Arunkumar. (2013). Effect of silica fume and fly ash on fresh and hardened state of self-compacting concrete. *Cement and Concrete Composites, 61*, p. 16718-16722.

[24] K. S. Jhansirani and A. Jagannathan. (2015). Studies on Effect of Mineral Admixtures on Durability of Self Compacting Concrete. *European Journal of Advances in Engineering and Technology, 2,* p. 87-94.

[25] Kennouche. S, Zerizer. A, Benmounah. A, Hami.B., Mahdad .M, Benouali .H and Bedjou.S. (2013). Formulation and characterization of self- compacting concrete with silica fume. *Journal of Engineering and Technology, 5*(5), p. 160-169.

[26] Koehler, E. P. and Fowler, D. W. (2007). *ICAR Mixture Proportioning Procedure for Self Consolidating Concrete*. Research Report ICAR 108-1, Aggregates Foundation for Technology, Research, and Education.

[27] L.H. Han, G.H. Yao. (2004). Experimental behavior of thin-walled hollow structural steel (HSS) columns filled with self-consolidating concrete (SCC). *Thin-Walled Structure 42 (9)*, p. 1357–1377.

[28] M. Benaichaa, A. Hafidi Alaouib, O. Jalbaudc, Y. Burtschell. (2019). Dosage effect of superplasticizer on self-compacting concrete: correlation between rheology and strength. *Journal of Materials Technology, 8*(2), p. 2063–2069.

[29] M. C. Nataraja, Anvit Gadkar and Giridhar Jogin. (2018). A simple mix proportioning method to produce SCC based on compressive strength requirement by modification to IS 10262:2009. *The India Concrete Journal, 92*, p. 15-23.

[30] M. Fadhil Nuruddin, Samuel Demie, M. Fareed Ahmed, and Nasir Shafiq. (2011). Effect of Superplasticizer and NaOH Molarity on Workability, Compressive Strength and Microstructure

Properties of Self-Compacting Geopolymer Concrete. *International Journal of Geological and Environmental Engineering, 5*(3).

[31] M.H. Beigi, J. Berenjian, O.L. Omran, A.S. Nik, I.M. Nikbin. (2013). An experimental survey on combined effects of fibers and nano-silica on the mechanical, rheological, and durability properties of self-compacting concrete. *Materials Design, 50,* 1019–1029.

[32] M. Sahmaran, H.A. Christianto, I.O. Yaman. (2006). The effect of chemical admixtures and mineral additives on the properties of self-compacting mortars. *Cement and Concrete Composites, 28 (5),* p. 432–440, https://doi.org/10.1016/j. cemconcomp.2005.12.003.

[33] Mahmoud Imam. *Concrete Types, Part Three.* Egyptian Book House, First Edition, 2002.

[34] Mahmoud. K.M, Abdulkader. I. A and Marwa H. M. (2019). Production and optimization of eco-efficient self-compacting concrete SCC with limestone and PET. *Construction and Building Materials, 197.* pp. 734–746.

[35] Mehena, Q., Amar, I., And Yannick, M. (2018). Saturation Point of Superplasticizers Determined by Rheological Tests for Self Compacting Concrete. *Civil Engineering 62*(2), p. 346–352.

[36] Mehmet Gesog˘lu, Erhan Güneyisi, Erdog˘an Özbay. "Properties of self-compacting concretes made with binary, ternary, and quaternary cementitious blends of fly ash, blast furnace slag, and silica fume" Construction and building material, Vol. 23, pp. 1847-1854. (2009).

[37] Mehrdad, M. and Mohammad, S. (2015). Effect of Inert and Pozzolanic Materials on Flow and Mechanical Properties of Self-Compacting Concrete. *Journal of Materials. Volume 2015*, Article ID 239717, 11 pages.

[38] Mohammadi Y, Mousavi S. S, Rostami F, Danesh A, Sarand N. I. (2015). The Effect of Silica Fume on the Properties of Self-Compacted Light weight Concrete. *Special Issue of Current World Environment, 10.*

[39] Nagaraju, S. Vijaya Bhaskar Reddy, T. Suresh. (2018). Influence of Chemical admixture dosage on Fresh Properties of Self-compacting concrete. *International Journal of Engineering Science Invention (IJESI), 7*, p. 69-76.

[40] Nehdi, M. and Rahman, M.A. (2004). Estimating rheological properties of cement pastes using various rheological models for different test geometry, gap and surface friction. *Cement and Concrete Research 34(11)*, p. 1993–2007.

[41] Ozawa K. Maekawa K. Kunishima M. and Okamura H. (1989). Development of high performance concrete based on the durability design of concrete structures. In Proceedings: *Second East-Asia and Pacific Conference on Structural Engineering and Construction (EASEC-2) .Vol. 1,* pp. 445-450.

[42] P. Ramanathan, I. Baskar, P. Muthupriya, and R. Venkata subramani. (2013). Performance of Self-Compacting Concrete Containing Different Mineral Admixtures. *Civil Engineering, 17*(2), p. 465-472.

[43] Rabar H. Faraj, Hunar F. Hama Ali, Aryan Far H. Sherwani, Bedar R. Hassan and Hogr Karim. (2020). Use of recycled plastic in self-compacting concrete: A comprehensive review on fresh and mechanical properties. *Journal of Building Engineering 30*, 101283.

[44] Rahul Dubey and Pardeep Kumar., "Effect of superplasticizer dosages on compressive strength of self-compacting concrete" INTERNATIONAL JOURNAL OF CIVIL AND STRUCTURAL ENGINEERING Volume 3, No 2, 2012.

[45] S. Deilami, F. Aslani, M. Elchalakani. (2018). An experimental study on the durability and strength of SCC incorporating FA, GGBS and MS. *Proceedings of the Institution of Civil Engineers-Structures and Buildings 172 (5)*, p. 327–339.

[46] S. M. Dumne. (2014). Effect of Superplasticizer on Fresh and Hardened Properties of Self-Compacting Concrete Containing Fly Ash. *American Journal of Engineering Research, 03*(03), p. 205-211.

[47] S.S. Vivek and G. Dhinakaran. (2015). Effect of Silica Fume in Flow Properties and Compressive Strength of Self Compacting Concrete. *International Journal of Chem Tech Research. 8 (1),* p. 01-05, 2015.

[48] Salahaldein Alsadey. (2015). Effect of Superplasticizer on Fresh and Hardened Properties of Concrete. *Journal of Agricultural Science and Engineering, 1*(2), p. 70-74.
[49] Salem Alsanusi. (2013). Influence of Silica Fume on the Properties of Self Compacting Concrete. *Civil and Environmental Engineering, 7*(5).
[50] Shahiron. S, Bassam. A. Tayeh, A. A Jamaludin, N. A. A. S Bahari, S. S Mohd Zuki, N. Ali and F. S Khalid. (2017). Physical and mechanical properties of self-compacting concrete containing superplasticizer and metakaolin. *Materials Science and Engineering 271*, 012004.
[51] Stanislav Maslennikov, Vladimir Dmitrienko, Irina Kokunko and Nadezhda Dmitrienko. (2017). Investigating the micro silica effect on the concrete strength. *MATEC Web of Conferences 106.*
[52] Terence. C. Holland. *Silica Fume User's Manual.* FHWA-IF-05-016.April,2005.
[53] The European Guidelines for Self-Compacting Concrete Specification, Production and Use, 2005, 1-63.
[54] V. A. Faseyemi. (2015). Investigation into the use of Microsilica and fly ash in self-compacting concrete. *International Journal of Research and Reviews in Applied Sciences, 24*(2), p.67-74.
[55] Wajde. S. S. A. (2016). Self-Compacting Concrete: Mix Proportioning, Properties and Its Flow Simulation in the V-Funnel. School of Engineering Cardiff University United Kingdom.

AUTHOR INDEX

Institute of Physics
Dirac House, Temple Back
Bristol BS1 6BE UK

ISSN: 1757-8981
ISBN 978-1-7138-4793-9

CURRAN ASSOCIATES INC.
proceedings.com